QUANTUM MECHANICS

Prentice Hall Series in Solid State Physical Electronics

N. Holonyak, Jr., Editor

CHEO *Fiber Optics and Optoelectronics, 2nd Edition*

HAUS *Waves and Fields in Optoelectronics*

HESS *Advanced Theory of Semiconductor Devices*

KROEMER *Quantum Mechanics: For Engineering, Materials Science, and Applied Physics*

NUSSBAUM *Contemporary Optics for Scientists and Engineers*

PULFREY/TARR *Introduction to Microelectronic Devices*

SHUR *Physics of Semiconductor Devices*

SOCLOF *Analog Integrated Circuits*

SOCLOF *Applications of Analog Integrated Circuits*

STREETMAN *Solid State Electronic Devices, 3rd Edition*

VERDEYEN *Laser Electronics, 2nd Edition*

WOLFE/HOLONYAK/STILLMAN *Physical Properties of Semiconductors*

QUANTUM MECHANICS

FOR ENGINEERING, MATERIALS SCIENCE, AND APPLIED PHYSICS

Herbert Kroemer
University of California, Santa Barbara

Prentice Hall, Englewood Cliffs, New Jersey 07632

Library of Congress Cataloging-in-Publication Data

Kroemer, Herbert
Quantum mechanics : for engineering, materials science, and applied physics / Herbert Kroemer.
p. cm. —
Includes bibliographical references and index.
ISBN 0-13-747098-3
1. Quantum theory. I. Title. II. Series.
QC174.12.K76 1994
530.1'2—dc20 93-34286
CIP

Publisher: Alan Apt
Project Manager: Mona Pompili
Cover Designer: Design Solutions
Copy Editor: Brian Baker
Buyer: Linda Behrens
Editorial Assistant: Shirley McGuire

A Paramount Communications Company
Englewood Cliffs, New Jersey 07632

Printed in the United States of America

10 9 8 7 6 5 4 3 2 1

ISBN 0-13-747098-3

PRENTICE-HALL, INTERNATIONAL (UK) LIMITED, *London*
PRENTICE-HALL OF AUSTRALIA PTY. LIMITED, *Sydney*
PRENTICE-HALL CANADA, INC., *Toronto*
PRENTICE-HALL HISPANOAMERICANA, S. A., *Mexico*
PRENTICE-HALL OF INDIA PRIVATE LIMITED, *New Delhi*
PRENTICE-HALL OF JAPAN, INC., *Tokyo*
SIMON & SCHUSTER ASIA PTE. LTD., *Singapore*
EDITORA PRENTICE-HALL DO BRASIL, LTDA., *Rio de Janeiro*

To the memory of
Richard Becker

PREFACE

Sometime during the 50s or 60s, quantum mechanics ceased to be an esoteric specialty needed only by Theoretical Physicists (with capital *T* and *P*) actively engaged in fundamental research; it began to become an increasingly important working tool of "everyday" physicists. These include specifically those who disguise themselves under other disciplinary designations, such as electrical engineers, materials scientists, and what have you. Those applied physicists are not often the ones who teach quantum mechanics courses, much less write nonelementary textbooks about the subject. As a result, much of quantum mechanics teaching, and textbooks beyond the elementary level, continue to be dominated by the needs and perspectives of the professional theorist, more often than not of the high-energy elementary-particle persuasion. For the most advanced courses and texts, this is the way it should be. Nor do I perceive a problem at the elementary level: We have all learned from Feynman there. It is at the intermediate level where I plead for more diversity, and this is where I would like to offer an alternative, designed specifically to meet the needs of what may by now be a majority of physicists—the general and applied physicists, working in such areas as solid state research, quantum electronics, materials science, and related fields.

NONTRADITIONAL APPROACH

The quantum mechanics needs of these groups have advanced rapidly in recent years, to a level that goes significantly beyond that of most elementary texts. At the same time, these needs have become sufficiently distinct from the *traditional* intermediate-to-advanced treatment of quantum mechanics, so that the available texts satisfy them only poorly. It is a situation similar to what happened in chemistry some time ago: The chemists' highly specific needs for the quantum mechanics of the chemical bond led them to spin off their own quantum mechanics courses (including some pretty advanced ones). I believe it is time for another spinoff, to serve another need that has grown to the point where it demands to be served on its own terms.

A text with such aims must differ in both coverage and methodology from traditional texts. And the differences go both ways: Some material is covered more extensively than it would be in a traditional text on this level, while other material is omitted.

PERTURBATION THEORY

I consider the various forms of perturbation theory to be the backbone of practical quantum mechanics on the actual problem-solving level, and I develop them on a scale that goes beyond what is common. On the other hand, I have omitted much subject matter that is often included in texts on this level. For example, there is nothing about relativistic quantum mechanics or about three-dimensional scattering theory, and there are other omissions. In these departures from traditional treatments, I was guided by what I perceive to be the needs of the general and applied physicist for whom I write, as opposed to the needs of, say, the high-energy or elementary-particle physicist or of the professional theorist of whatever persuasion. The reader who wishes eventually to fill in the omitted material will not find it necessary to go back to another intermediate-level text, but should have no difficulty continuing directly with a more advanced text. In this regard, Sakurai's *Advanced Quantum Mechanics* is probably an ideal text for everything except scattering theory.

LESS ABSTRACT

Just as important as changes in coverage are differences in methodology. I believe that the treatment in many intermediate-to-advanced texts is unnecessarily abstract, making the subject difficult for those who are not highly theoretically inclined, but who need a thorough understanding of quantum mechanics all the same. Inasmuch as I am explicitly committed to meet the needs of individuals who are not professional theorists, I have always searched for new and less abstract ways to present formal concepts. I was surprised how often it could be done. Results of these efforts are present throughout the book, sometimes leading to quite nontraditional presentations. The introduction of the magnetic vector potential into the Hamiltonian is an obvious example, and the justification of the Pauli spin Hamiltonian is another, but there are more.

Minimizing abstractness does not mean sacrificing a high level of rigor. I believe that a reasonably high level of rigor is essential, but this does not require abstractness, much less an axiomatic treatment. I usually try first to justify new concepts by physical arguments and then to tighten up on the rigor, after the meaning of the concepts is understood physically and qualitatively.

Sometimes the coverage of a topic is deliberately broken up, giving an elementary treatment early in the text and returning to the topic on a more advanced level later. The treatment of the harmonic oscillator is perhaps the most obvious example of this approach, but there are others as well.

PREREQUISITES

The text is self-contained in the sense that no prior exposure to *formal* quantum mechanics is assumed. In fact, the first seven or eight chapters could be used, with some omissions, as the basis for a self-contained *undergraduate* quantum mechanics course, leading naturally into graduate-level material as the subject matter develops. If taught as a graduate course from beginning to end, the text contains material for about two semesters or three quarters, with at least the last half on the graduate level. If taught to students with a prior undergraduate-level quantum mechanics background, some of the earlier material may be treated as an accelerated review; some of it may be dropped altogether.

The most important prerequisite is a good introductory course in modern physics, with its typical heavy share of concepts in atomic physics. Second in importance is a good background in electromagnetic theory, going beyond electrostatics and containing EM wave theory and at least some exposure to EM boundary value problems. Some familiarity with magnetic vector potentials is assumed. The text does not require any classical mechanics beyond what is contained in the traditional lower division physics course, provided that the latter is taught at the level of calculus. No Hamiltonian or Lagrangian mechanics is required, nor is knowledge of either even desired.

The mathematical prerequisite is the usual one: a good traditional undergraduate math background, including familiarity with the Fourier integral, linear algebra, linear partial differential equations, some complex variables, and basic probability concepts.

EXERCISES AND EXAMPLES

Throughout the text are sprinkled numerous exercises. They are an essential part of the course, to be considered automatic assignments to be done by every student—not by the instructor—but sufficiently straightforward that they should not be turned in for grading. Subsequent course material will often draw on these exercises. The homework problems at the end of many sections are more elaborate than the exercises. Not only are they the real test of the student's understanding of the material, but it is probably impossible to gain a true understanding of quantum mechanics without working out in detail the solution to a sizable number of problems at the level at which they are presented. Yet I have not attempted to have homework problems at the end of every section. Some topics simply lend themselves more readily to homework than others, and I see no point in homework for homework's sake—or in large numbers of theorem-proving problems. Hence, the homework coverage is very uneven, yet it reflects rather accurately the style and coverage I have found useful in class.

ON THE ORGANIZATION OF THE TEXT

Numbering

All chapters are divided into sections, and most sections are divided further into subsections. In addition, there are several appendices. All pages, equations, figures, problems, tables, etc. are numbered separately within each section or appendix. For example, the notation "(3•2–1)" refers to equation 1 in section 2 of chapter 3. Analogous notation is used for the numbering of figures and tables. When the chapter number of any item is omitted, the reference is to the current chapter.

MKS Notation

In accordance with international convention, as well as engineering practice, we use SI (= MKS) notation throughout. The reader can easily convert to the CGS notation frequently used in physics texts by making the substitutions

$$\epsilon_0 \rightarrow 1/4\pi, \qquad \mu_0 \rightarrow 4\pi/c^2,$$

$$D \rightarrow D/4\pi, \qquad B \rightarrow B/c, \qquad H \rightarrow cH/4\pi.$$

Symbols for Charges

The letter e, if denoting a charge, is always used for the *magnitude* of the elementary charge, i.e.,

$$e = +1.60219 \cdot 10^{-19} \text{ coulomb}$$

An electron carries the charge $-e$, a proton $+e$. To express an unspecified charge, we use the letter q or Q.

Frequencies

The word *frequency* is used to denote both the true frequency v, and the radian frequency $\omega = 2\pi v$. Numerical values always refer to the true frequency v.

ACKNOWLEDGMENTS

I am indebted to many individuals. The text grew out of lecture notes I had been accumulating gradually over about 20 years, and I owe much to the many students who took the course over this time—and to their endless questions. The notes had initially been intended to accompany other texts, later becoming increasingly self-sufficient. Yet those notes might still be just notes had it not been for Nick Holonyak, who pestered me year after year to put it all together. Karl Hess and Kambiz Alavi reviewed the first full draft of the manuscript and made many valuable suggestions (including that I elaborate on some casual comments I had made about hidden variables, the Bell inequality, and photon correlations). Last but not least, I am deeply indebted to the patience with which my wife Marie Louise endured this seemingly endless project.

Herbert Kroemer
University of California,
Santa Barbara

BRIEF CONTENTS

CONTENTS

Chapter 1

WAVE-PARTICLE DUALITY AND THE SCHROEDINGER EQUATION

1.1 WAVE-PARTICLE DUALITY

1.1.1 Wave-Like and Particle-Like Properties of Elementary Objects

Quantum mechanics is the backbone of modern physics; indeed, some would say that modern physics *is* quantum mechanics. The objective of quantum mechanics is to give a quantitative description of the behavior of nature on a microscopic scale. Here, *microscopic* has two separate meanings: (a) pertaining to the *size* of atomic and subatomic dimensions and (b) at the *level* of the

The central idea of quantum mechanics—what distinguishes it from classical physics—is the idea of **wave-particle duality**. By the end of the 19th century, classical physics had reduced almost all macroscopic phenomena to the interactions of two classes of **elementary objects**: *particles,* which obeyed the laws of Newtonian mechanics, and *wave fields,* which obeyed Maxwell's equations. Every elementary object was *either* a particle *or* a wave field, and the two were considered mutually exclusive.

In quantum mechanics, the concepts of waves and particles have changed their character. They are not *classes of objects* themselves; they are now two distinct *modes of behavior,* shared by *all* classes of objects. Every object can behave like a particle *and* like a wave. Thus, the question "*Is* the electron a particle or a wave?" has the answer "No; it *is* an electron." And the question "Can the electron *behave* like a particle or a wave?" has the answer "Yes; it can *behave* like either."

This is the conceptual core of wave-particle duality. We will develop the idea further as we proceed. For now, we may express it more specifically as follows:

(a) *Elementary Objects.* All objects of nature are composed of *discrete* elementary objects that cannot be further subdivided. Electrons and photons are two examples. All elementary objects exhibit two distinct classes of properties, of a wave-like and of a particle-like nature.

(b) *Particle-like Properties.* The elementary objects are *discrete* and *indivisible*: They can be *counted.* This is in itself a particle-like property and, in fact, the most central one. For example, all electrons have the same finite charge and mass (here, *rest mass*); fractional electrons with a smaller charge and mass do not exist. Photons are more subtle: They carry neither charge nor mass, but are "pure energy" with a range continuous down to zero. Nevertheless, photons of any *given* energy are again discrete indivisible objects. Consider, for example, the splitting of a monochromatic light beam into a transmitted and a reflected beam by partial reflection at a dielectric interface. The *energy* of the photons in the reflected and transmitted beams is exactly the same as in the incident beam. It is the *number* of photons in the incident beam that is split between the two beams, not the energy of the individual photons. Similar examples can be given for other elementary objects: The discrete and indivisible nature of *all* elementary objects is one of the basic postulates of modern physics.

Indivisibility does not mean that the objects cannot be annihilated or generated. Electrons and positrons can annihilate each other, generating gamma-ray photons in the process, and electron-positron pairs can in turn be generated from gamma-ray photons. Photons themselves can be both annihilated and generated. However, any elementary object, once generated, will subsequently interact with other objects as a discrete, indivisible unit and, if annihilated, will be annihilated in its entirety.

(c) ***Wave-like Properties.*** On a sufficiently microscopic scale, the propagation of all elementary objects is wave-like, similar to the way a light pulse or radar pulse propagates through space or through a dielectric medium. This means that all objects are capable of exhibiting the diffraction and interference effects that *define* wave-like behavior. Familiar examples are the diffraction and interference of light and of electrons, but again, this property is postulated to be universal.

1.1.2 Wave Functions and the Planck–Einstein–de Broglie Relations

Wave-like propagation cannot be adequately described in terms of a classical particle-like **trajectory**, that is, in terms of a sharp position $\mathbf{r}(t)$of the object as a continuous function of time. Instead, it calls for the introduction of suitable field-like quantities similar to the field-like quantities of classical electromagnetic theory, usually called **wave functions** and generally designated here by the symbol Ψ. Field-like quantities always extend over a nonzero volume of space, and they are themselves functions of *both* position and time as separate independent variables, i.e., $\Psi = \Psi(\mathbf{r}, t)$.

The simplest such wave function is a plane wave of the form

$$\Psi(\mathbf{r}, t) = A \exp[i(\mathbf{k} \cdot \mathbf{r} - \omega t)]. \tag{1•1–1}$$

Here $\mathbf{k}$ is the **wave vector** associated with the wave, and ω is the **wave frequency**. These are the two properties that characterize the state of the object from the wave point of view, just as momentum $\mathbf{p}$ and energy $\mathcal{E}$ characterize it from the particle point of view. Quantum mechanics postulates that the two characteristic *particle-like* properties of each object—energy and momentum—are simply proportional to its two characteristic *wave-like* properties—frequency and wave vector, that is,

$$\boxed{\mathcal{E} = \hbar\omega, \quad \mathbf{p} = \hbar\mathbf{k},} \tag{1•1–2a,b}$$

where $\hbar$ is a universal constant called **Planck's constant.**[1] In honor of their discoverers, we refer to these relations as the **Planck-Einstein-de Broglie** (PEdB) relations. We view them together as constituting the most basic physical postulate of quantum mechanics, believed to be rigorously valid for all elementary objects of nature. We assume that the reader is familiar with at least some of the many direct experimental tests of the PEdB relations that

[1] We follow the practice of referring to $\hbar = h/2\pi$ as Planck's constant, rather than to Planck's original constant h, which we shall not use.

have been carried out.[2] But our faith in them is based not only on these *direct* tests; it is based even more on the less direct, but overwhelming, *quantitative* agreement between experiment and the detailed quantum-mechanical theory that is based on these relations.

1.1.3 Dispersion Relations and Such

The wave-like properties ω and $\mathbf{k}$ are inter-dependent via a **dispersion relation** $\omega = \omega(\mathbf{k})$. Similarly, the particle-like properties $\mathcal{E}$ and $\mathbf{p}$ are inter-dependent via an **energy-momentum relation** $\mathcal{E} = \mathcal{E}(\mathbf{p})$. Knowing one of these relations implies the other, via the PEdB relations. For example, we know already from classical electromagnetic wave theory that the free-space dispersion relation for light is

$$\omega = 2\pi\nu = 2\pi c/\lambda = ck, \qquad k = |\mathbf{k}| = 2\pi/\lambda, \tag{1•1–3}$$

where c is the speed of light. With the help of the PEdB relations, this translates into an energy-momentum relation for photons,

$$\mathcal{E} = cp, \tag{1•1–4}$$

which implies that photons carry the momentum

$$p = \hbar k = \mathcal{E}/c = 2\pi\hbar/\lambda. \tag{1•1–5}$$

In classical E&M theory, this momentum manifests itself as a **radiation pressure** exactly equal to the pressure exerted by a stream of particles with the energy-momentum relation (1•1–4). The photon momentum plays an important role in many interactions between photons and other objects. Several examples are discussed in the problems at the end of this section.

We will be mostly concerned with objects with mass, especially electrons. We know from classical mechanics that (non-relativistic) objects with finite rest mass obey the Newtonian energy-momentum relation

$$\mathcal{E} = p^2/2M, \tag{1•1–6}$$

which translates into the dispersion relation

$$\omega(\mathbf{k}) = \frac{\hbar k^2}{2M} \tag{1•1–7}$$

We shall see in section 1.3 that the Schroedinger wave equation is a direct consequence of this dispersion relation.

The simple energy-momentum relation (1•1–6) is valid only for non-relativistic values of energy and momentum. It is shown in the theory of relativity

[2] An excellent discussion of many of these tests is found in the text by French and Taylor, listed in appendix G.

that the generalization of (1•1–6) to arbitrary values of energy and momentum may be written in the form[3]

$$(\mathcal{E} + Mc^2)^2 = (cp)^2 + (Mc^2)^2. \tag{1•1–8a}$$

For $\mathcal{E} \ll Mc^2$, this reduces to (1•1–6). For $\mathcal{E} \gg Mc^2$, all terms containing M can be neglected, and we obtain

$$\mathcal{E}^2 = (cp)^2, \tag{1•1–9}$$

which is essentially the same as the relation (1•1–4) for photons. Hence, (1•1–8) is the general energy-momentum relation for objects with arbitrary mass and energy.

In practical problems involving comparisons of wave-like and particle-like properties, one often needs the dependence of the wavelength λ on the energy $\mathcal{E}$. For photons, we find that

$$\lambda = 2\pi\hbar c/\mathcal{E} = 1.240\ \mu m/\mathcal{E}\ [\text{eV}], \tag{1•1–10}$$

and for electrons (with $M = m_e$)

$$\lambda = \frac{2\pi\hbar}{\sqrt{2m_e\mathcal{E}}} = \frac{1.226\text{nm}}{\sqrt{\mathcal{E}\ [\text{eV}]}} = 1\ \text{nm} \cdot \sqrt{\frac{1.504\ \text{eV}}{\mathcal{E}}}. \tag{1•1–11}$$

These relations explain, at least partially, why classical physics treats electromagnetic radiation as a pure wave phenomenon and electrons as particles.

At long wavelengths, above the optical wavelength range, the quantum energy is so small that for most purposes the granularity of energy is negligible. For example, a wave carrying the (macroscopically) very small power of 1 μW at a microwave frequency of 10 GHz (λ = 3cm) corresponds to a quantum flow of 1.5×10^{17} quanta per second, or—more relevantly—1.5×10^7 quanta per cycle. This is a very dense flow, and its granularity is all but unobservable. Electromagnetic energy may then be treated as if it were infinitely divisible, as is done in classical electromagnetic theory.

Even at short optical wavelengths (say, λ = 300 nm, i.e., ultraviolet light), the quantum energy is still small by macroscopic standards: 6.6×10^{-19} J $\sim$ 4.1 eV. But this is large enough, for example, to eject an electron from an atom. In such processes, in which only a single photon is involved, the quantum structure of light becomes dominant. The development of an interaction theory for such cases is one of our tasks. Classical physics fails completely: It cannot explain how an electron can pick up enough energy to be ejected (see "Problems").

---

[3] In the theory of relativity the rest mass energy Mc^2 is usually included in the energy $\mathcal{E}$, thus giving (1•1–8a) the slightly simpler form

$$\mathcal{E}^2 = (cp)^2 + (Mc^2)^2. \tag{1•1–8b}$$

With electrons, the situation is the other way around. For the kinds of energies that are of interest in most scenarios involving free electrons in a vacuum, the electron wavelengths are of atomic and subatomic dimensions. As a result, diffraction and interference effects are unobservable in most macroscopic structures. The wave properties of electrons may therefore usually be neglected in calculating the propagation of electrons through such structures. As we shall see later, the laws of quantum mechanics then revert to those of classical mechanics.

The situation is quite different inside an atom or a crystal. The electron energies are often quite low, making the electron wavelength longer, while the relevant dimensions are much smaller—in fact, often smaller than the electron wavelengths. Diffraction and interference effects then dominate the electron propagation. The development of a wave propagation theory that permits the calculation of the behavior of electrons inside such structures is one of our tasks. Newtonian mechanics fails completely.

The considerations for electrons apply even more to other elementary objects with a nonzero mass. Because all other known objects with nonzero mass have a larger mass than the electron, their wavelengths for a given energy are shorter, which makes their wave properties even harder to observe.

Exercise: Impossibility of Photon Absorption by a Free Electron. By superimposing plots of the dispersion relations for photons and electrons, show that energy and momentum cannot both be conserved during the absorption or emission of a photon by a free electron, regardless of the initial velocity of the electron. (Note: The speed of the electron cannot exceed the speed of light, c.)

1.1.4 Composite Objects

The wave-particle duality exhibited by indivisible elementary objects continues to manifest itself when several such objects are bound together into a composite object. This includes not only such obviously composite objects as atoms and molecules, but also the proton and the neutron, which were once thought to be elementary, but are now known to be composite, albeit with a huge binding energy. The behavior of any composite object becomes indistinguishable from that of an indivisible object when the kinetic energy of the composite object is insufficient to excite any of its internal degrees of freedom (such as the molecular vibrations in a molecule) or to break it up into different constituents. The composite object then acts just like an indivisible elementary object, with wavelengths given by (1•1–11), where M is now the total mass of the object. Diffraction experiments with different kinds of composite objects, including entire atoms, confirm this prediction. In fact, two such diffraction techniques are increasingly used in studying crystals and surfaces: The diffraction of thermal and sub-thermal neutrons is used to supplement X-ray diffraction, and the diffraction of thermal ^{4}He atoms is used in surface studies to supplement electron diffraction.

When the kinetic energy of the composite object exceeds the limit given above, its behavior may become more complicated than that of indivisible elementary objects. In this case, a quantitative description of the interaction of such composite objects requires the complete quantum-mechanical formalism of composite quantum systems, to be developed later (chapter 11).

1.1.5 Principle of Linear Superposition

Plane waves of the form (1•1–1), infinitely extended in both space and time, with position- and time-independent wave vectors and frequencies, represent a mathematical limit. To describe more realistic situations, we must construct more complicated wave functions, even for propagation in free space. In classical electromagnetic theory, more complicated solutions of Maxwell's equations can be constructed by linear superposition of simpler solutions, and we postulate that this **principle of linear superposition** carries over into quantum mechanics:

> Let $\Psi_1(\mathbf{r}, t)$ and $\Psi_2(\mathbf{r}, t)$ be two wave functions representing two physically allowed states of a given object. Then any linear superposition of the form
>
> $$\Psi(\mathbf{r}, t) = a_1\Psi_1(\mathbf{r}, t) + a_2\Psi_2(\mathbf{r}, t), \tag{1•1–12}$$
>
> with complex position- and time-independent coefficients a_1 and a_2, also represents a physically allowed state of that object.[4]

Taken together, the PEdB relations and the principle of linear superposition serve as the foundation of the mathematical formalism of quantum mechanics.

Repeated application of the superposition principle shows that, if the principle holds for the superposition of *two* wave functions, it also holds for the superposition of an arbitrary number of wave functions. For objects in free space, the most general wave function we can construct is formed from continuous linear superpositions of an infinite number of plane waves and is of the form

$$\Psi(\mathbf{r}, t) = \iiint A(\mathbf{k}) \exp\{i[\mathbf{k} \cdot \mathbf{r} - \omega(\mathbf{k})t]\}\, d^3k. \tag{1•1–13}$$

Here $A(\mathbf{k})$ is a suitable (time- and position-independent) amplitude function, and each plane wave in the superposition satisfies the **dispersion relation** $\omega = \omega(\mathbf{k})$ for the kind of elementary object under consideration.

Amongst the superpositions of the form (1•1–13) are pulse-like superpositions, called **wave packets**, in which the wave function has an appreciable

[4] If an external potential is present, all states must be states in the same potential.

magnitude only over a relatively small volume, falling off rapidly outside this volume (**Fig. 1•1–1**). Such pulses propagate through space similarly to the way a light pulse or radar pulse propagates. If viewed from a sufficiently large distance, the internal wave structure of the pulse becomes unobservable, and the pulse appears as a point-like object, propagating through space along a particle-like trajectory. We shall argue later that the trajectories of classical mechanics are nothing other than these wave packet trajectories of quantum mechanics for objects with a nonzero mass.

Linear superpositions of plane waves contain—by definition—components with different wave numbers and frequencies. According to the PEdB relations, this implies that the associated particle-like properties of the object described by the superpositions contain different momenta and energies. What this means is the following: Just as the *kinetics* of an object, viewed on a sufficiently microscopic scale, cannot be described in terms of a sharp position $\mathbf{r}(t)$, its *dynamics* cannot be described in terms of sharp values of energy and momentum. We will later interpret this conclusion operationally, in the sense that a measurement of any of these quantities will not lead to a unique sharp value, but to a value that will fluctuate from measurement to measurement.

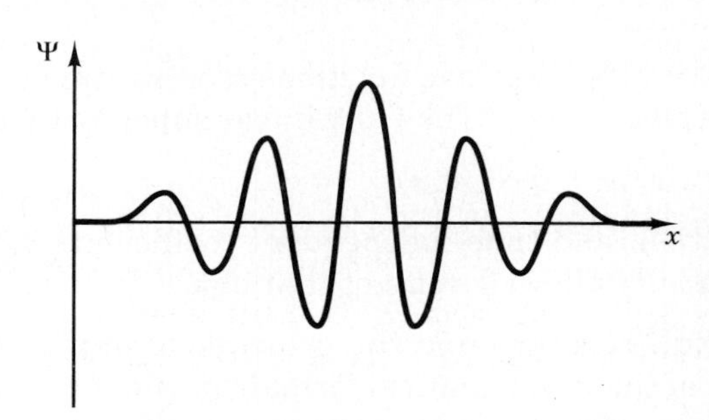

Figure 1•1–1. A simple wave packet. Because wave functions will in general be complex, the figure may be viewed as a plot of either the real or the imaginary part of the function.

1.1.6 Complex vs. Real Wave Functions

A comment is in order concerning the use of complex plane waves of the form (1•1–1), rather than real waves of the forms

$$\Psi(\mathbf{r}, t) = A \sin(\mathbf{k} \cdot \mathbf{r} - \omega t) \quad \text{and} \quad \Psi(\mathbf{r}, t) = A \cos(\mathbf{k} \cdot \mathbf{r} - \omega t), \tag{1•1–14}$$

which are used in classical electromagnetic field theory. The difference is more apparent than real: Classical E&M theory requires *two different* real fields, electric *and* magnetic fields, coupled via Maxwell's equations. Working with a single complex wave function eliminates this need for having to work with two different field-like quantities.

It would be readily possible to formulate quantum mechanics in terms of two real wave functions coupled to one another. But such a formulation would be more complicated, without any compensating benefits. In classical E&M

theory, the electric and magnetic fields have distinct meanings by themselves, but no such distinct meanings exist for the real and imaginary parts of the wave functions of quantum mechanics; hence, such a formalism is not used.

In fact, classical E&M theory itself could easily be re-formulated in terms of a single complex field

$$\mathbf{F} \equiv \sqrt{\epsilon_0/2}\mathbf{E} + i\sqrt{\mu_0/2}\mathbf{H}, \tag{1•1–15}$$

where $\mathbf{E}$ and $\mathbf{H}$ are the classical electric and magnetic fields, and ϵ_0 and μ_0 are the permittivity and permeability of space. It is left to the reader to show that Maxwell's equations in free space, in the absence of free charges, may then be written as a single complex equation of the form

$$i\frac{\partial \mathbf{F}}{\partial t} = c \cdot \text{curl } \mathbf{F}, \tag{1•1–16}$$

where $c = (\epsilon_0 \mu_0)^{-1/2}$ is the speed of light. The electromagnetic energy density u and the Poynting vector $\mathbf{S}$ assume the forms

$$u = \frac{1}{2}(\epsilon_0 \mathbf{E}^2 + \mu_0 \mathbf{H}^2) = |\mathbf{F}|^2, \tag{1•1–17}$$

and

$$\mathbf{S} = \mathbf{E} \times \mathbf{H} = -ic(\mathbf{F}^* \times \mathbf{F}) . \tag{1•1–18}$$

Note that for a purely real or purely imaginary field, there would be no energy flow, because the cross product of a vector with itself vanishes. Hence, a single real field could not describe *both* the instantaneous state of a system *and* its state of motion. We shall find a similar situation in quantum mechanics.

◆ PROBLEMS TO SECTION 1.1

#1•1-1: Photon Momentum and Classical Radiation Pressure

When light is absorbed in a conducting medium, the photon momentum is transferred to the medium, where it acts as a force on the medium,the radiation pressure. This force may be described classically as the Lorentz force of the magnetic field $\mathbf{B}$ of the wave, acting on the electrical current in the medium. The time-averaged force per unit volume is

$$\bar{\mathbf{f}} = \overline{\mathbf{j} \cdot \mathbf{B}}, \tag{1•1–19}$$

where $\mathbf{j}$ is the current density and where the overbars indicate time averaging. By comparing the magnitude $\bar{f}$ of the force with the time-averaged power absorbed per unit volume,

$$\overline{w} = \overline{\mathbf{j} \cdot \mathbf{E}}, \tag{1•1–20}$$

it is possible to relate the photon momentum to the photon energy.

For simplicity, assume that the dielectric and magnetic properties of the medium are those of free space (i.e., $\epsilon = \epsilon_0$, $\mu = \mu_0$) and that the conductivity of the absorbing medium is sufficiently small that reflection is negligible. Show that in this limit,

$$\overline{f} = \overline{w}/c, \tag{1•1–21}$$

and that this implies that the photon momentum $p = \mathcal{E}/c$, as in (1•1–5).

Note: Photons that are reflected rather than absorbed transfer twice their momentum, and under conditions where reflections are not negligible, the extraction of the photon momentum from a classical radiation force argument becomes more complicated.

#1•1-2: Compton Scattering

Suppose a photon with momentum $\mathbf{p} = \hbar\mathbf{q}$ is absorbed by an electron that is initially at rest, and simultaneously a photon with the different momentum $\mathbf{p}' = \hbar\mathbf{q}'$ is emitted. The overall effect of this two-photon process is the same as if the incident photon is scattered into a new state with different momentum (**Compton Scattering**; **Fig. 1•1–2**).

If the photon is scattered by the electron, the change in photon momentum must be taken up by the electron. This implies that the electron acquires kinetic energy, which must also be provided by the photon. If the energy change of the photon is expressed in terms of a wavelength change $\Delta\lambda$, one finds that the latter depends only on the angle θ, according to the relation

$$\Delta\lambda = 2\lambda_C \sin^2(\theta/2), \tag{1•1–22a}$$

where the quantity

$$\lambda_C = \frac{2\pi\hbar}{m_e\, c} = (2.42631 \cdot 10^{-12}\mathrm{m}), \tag{1•1–22b}$$

is called the **Compton wavelength**. Derive this relation. Use the relativistic energy-momentum relation (1•1–8) for electrons, not the simpler non-relativistic form (1•1–6), which actually leads to a more complicated (and less accurate) result than (1•1–22).

Eq. (1•1–22a) is a simple and experimentally readily testable relation. Its experimental verification has historically played a large role in the acceptance of the quantum structure of light. Compton scattering is the dominant process by which high-energy radiation (e.g., X-rays and γ-rays) gets degraded and ultimately absorbed in passing through matter.

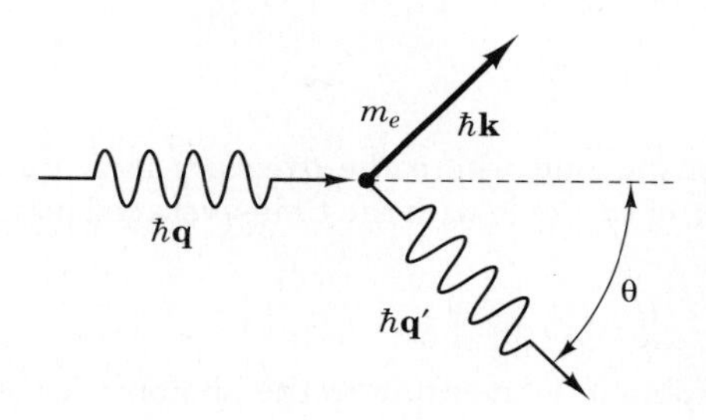

Figure 1•1–2. Compton scattering. A photon with momentum $\hbar\mathbf{q}$ (magnitude $\hbar\omega/c$) is scattered into a new photon with momentum $\hbar\mathbf{q}'$, and the electron recoils with momentum $\hbar\mathbf{k}$.

The Compton wavelength is one of several fundamental **natural units of length** occurring in quantum mechanics; we will encounter others later. Note that λ_C is much smaller than the size of an atom. This means that for "ordinary" X-rays, for which λ is of the order of atomic dimensions, the wavelength shift during Compton scattering is only a small fraction of the initial wavelength, and hence, the photon energy loss is only a small fraction of the initial energy. This changes drastically for short-wavelength radiation, such as γ-rays.

#1•1-3: Photodissociation

Composite objects, such as atoms, molecules, and atomic nuclei, can be dissociated by absorbing a photon of sufficient energy. Let $\mathcal{E}_0$ be the dissociation energy, and assume that $\mathcal{E}_0 \ll (M_1 + M_2)c^2$, where M_1 and M_2 are the masses of the two constituents of the object and c is the speed of light. Assume that the object is initially at rest. Because the incident photon carries momentum, which must be taken up by the masses M_1 and M_2, the photon energy $\mathcal{E}$ required to break up the nucleus exceeds $\mathcal{E}_0$ by the kinetic energy of the recoiling fragments.

(a) Calculate the minimum *excess* energy $(\Delta\mathcal{E})_{\min} = \mathcal{E}_{\min} - \mathcal{E}_0$ required to break up the object, under the assumptions stated.

(b) For the specific case of *minimum* photon energy, what are the magnitudes of both the longitudinal and the transverse components of the recoil momenta and recoil velocities of the two masses M_1 and M_2.

(c) Calculate numerical values for all quantities of interest for the photo-dissociation of a deuteron ($M_1 \sim M_2 \sim 1$ amu, $\mathcal{E}_0 \sim 2.2$ MeV), and compare the minimum recoil energies with thermal energies at 300K. Repeat the numerical calculations for a hydrogen atom ($\mathcal{E}_0 \sim 13.6$ eV), to which the same formal equations apply.

Note: The seemingly straightforward approach of first solving the recoil problem for *arbitrary* energy is unnecessarily complex for determining just the *minimum*-energy recoil conditions. Instead, after setting up the equations for energy and momentum conservation, *first* take differentials, *next* set $d\omega = 0$, which is the condition of minimum photon energy, and *then* solve the equations that are left.

#1•1-4: Energy Exchange between Photons and Free Electrons: Classical and Quantum Limits

Consider a classical electromagnetic wave, traveling in the z-direction and polarized in the x-direction, with an electric field strength

$$E_x(z, t) = E_0 \sin(kz - \omega t). \tag{1•1–23}$$

Assume that this wave transfers energy an electron that is initially ($t = 0$) at rest at $x = z = 0$. At sufficiently low frequencies, the energy transfer may be calculated classically from Newton's Law, yielding a certain maximal kinetic energy $\Delta\mathcal{E}_c$ that cannot be exceeded.

Determine the conditions under which $\Delta\mathcal{E}_c = \hbar\omega$. This condition may be viewed as representing the limit of the validity of the classical calculation. Express it in the form of a functional relation $\overline{S}(\lambda)$ between the time- and space-averaged power density $\overline{S}$ of the wave and its wavelength λ. Make a plot of this relation. Indicate a typical microwave field ($\lambda \approx 1$ cm) and sunlight ($\lambda \approx 600$ nm), both at the power density of the latter ($\overline{S} \sim 100$ mW/cm^2). Comment on the applicability of the classical treatment for these two cases.

Note: The comparison of $\Delta\mathcal{E}_c$ with the photon energy ignores the momentum conservation constraints on the energy exchange between a photon and a free electron, discussed in problem 1 in the context of Compton scattering. In effect, the present comparison assumes that the electron can somehow exchange the small photon momentum via weak coupling to other objects, an assumption that is often satisfied in practice.

1.2 INDETERMINACY

1.2.1 Probability Densities

Neither the particle-like discreteness and indivisibility of elementary objects nor their wave-like extension through space is by itself noteworthy. What is non-classical is their combination. All *classical* waves are by their nature infinitely divisible. The idea that something can be distributed over a nonzero volume of space and at the same time be indivisible is foreign to classical physics. In fact, it is hard to visualize, making quantum mechanics inherently a more abstract discipline than classical physics. This psychological difficulty is something we must learn to live with.

The nonclassical combination of wave-like and particle-like properties inevitably leads to the idea that the laws of physics on the level of individual elementary objects *necessarily* contain an element of indeterminacy.

To illustrate the problem, we consider a classical optical beam splitter from both a wave and a particle point of view. From a classical wave point of view, the beam splitter distributes the power W in the input beam in a certain ratio W_2:W_1 (often 1:1) over the two output beams. From a photonic particle point of view, the beam splitter distributes the incident photons over the two output beams.

It is useful to express the latter point of view in operational terms. Suppose we place into each of the output beams a **quantum counter**, that is, a detector that has the capability of registering and counting individual photons as discrete events (**Fig. 1•2–1**). Let N_1 and N_2 be the number of photons registered by the two quantum counters during a certain time interval Δt. With each photon carrying the same energy, we must evidently have

$$\frac{N_2}{N_1} \rightarrow \frac{W_2}{W_1}, \qquad (1{\bullet}2\text{–}1)$$

where the arrow represents the limit of a sufficiently large number of photons having passed through the beam splitter that the granularity of the photon energy has become negligibly small compared to the overall energy that has passed through the setup.

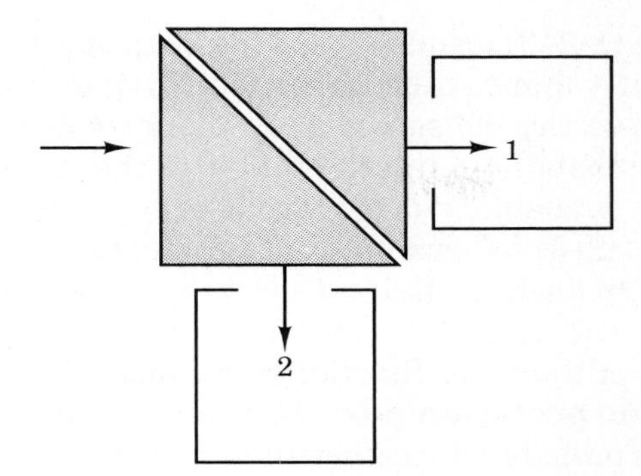

Figure 1·2–1. Optical beam splitter with quantum counters placed into the two output beams.

But with N_1 and N_2 necessarily being integers, we clearly have a compatibility problem between the wave and particle points of view in the limit of a small number of photons. Consider specifically the extreme limit that the incident beam contains only a single photon.[5] It is clear that only one of the quantum counters can respond, but which one? Any contemplation of the answer to this question inexorably leads to the conclusion that the exact outcome of any *single* capture attempt *inherently* cannot be predicted, and it must, in fact, fluctuate probabilistically if the attempt is repeated on identical waves under identical conditions. Only the statistical probabilities P_1 and $P_2 = 1 - P_1$ of capture by the two detectors can be predicted, and the classical power partition ratio is nothing other than the macroscopic manifestation of the probability ratio,

$$\frac{W_2}{W_1} = \frac{P_2}{P_1}. \tag{1·2–2}$$

Although the preceding discussion is in terms of photons, we postulate that this probabilistic behavior is universal, for all classes of indivisible objects.

The two probabilities must somehow be related to the "strength" of the wave, that is, to the value of the wave function $\Psi(\mathbf{r}, t)$. Accordingly, we postulate the following:

(a) The incident wave packet representing a single indivisible object is split into two wave packets Ψ_1 and Ψ_2, one in each output beam. The two partial waves represent the indivisible object *collectively,* in the sense that any interaction with one of the partial waves represents an interaction with the whole object.

(b) The probability ratio in (1·2–2) is given by

$$\frac{P_2}{P_1} = \frac{\int |\Psi_2|^2 \, d^3r}{\int |\Psi_1|^2 \, d^3r}. \tag{1·2–3}$$

[5] We ignore here the nontrivial problems involved in how an incident pulse containing exactly one photon might actually be generated. We will return to that question in section 1.5. See especially problem 1·5-3.

What the postulate (1•2–3) means is that the quantity $|\Psi|^2$ is interpreted as a probability per unit volume, called a **probability density**, just as the classical quantity $|\mathbf{F}|^2$ in section 1.1 served as an energy density—see (1•1–17). Given the need for a probabilistic interpretation of the wave function, this proportionality between probability density and energy density is of course inevitable, and hence, (1•2–3) follows naturally from the probabilistic postulate. The wave function Ψ itself is often referred to as a **probability amplitude**.

The interpretation of the wave function Ψ contained in (1•2–3) is called the **Born statistical interpretation postulate** for the waves associated with elementary objects. This probabilistic rather than deterministic interpretation of $|\Psi|^2$ is believed to reflect a true indeterminacy of the laws of nature on the level of interactions with individual elementary objects. Quantum mechanics is in this regard fundamentally different from classical statistics, where the probabilities simply reflect a lack of knowledge about all the relevant parameters of the problem.

The relation (1•2–3) makes a statement only about the ratio between two probabilities. When dealing with single indivisible objects, the object must be somewhere; hence, all local probabilities must add up to unity:

$$\int_{\text{all space}} dP = 1. \tag{1•2–4}$$

It is common practice to normalize the wave function in such a way that $|\Psi|^2$ represents the probability density without any additional proportionality factor:

$$\boxed{\int_{\text{all space}} |\Psi|^2 \, d^3r = 1.} \tag{1•2–5}$$

We note that infinitely extended plane waves, as in (1•1–1) are inherently not normalizable and hence cannot correspond to a single object. We shall argue later that they represent streams of statistically independent objects. However, it is possible to construct normalizable wave packets, as in Fig. 1•1–1, by linear superpositions of plane waves, as in (1•1–13). We will discuss such wave packets extensively in chapter 4.

Given a wave function thus normalized, (1•2–3) may be broken up into two individual probabilities,

$$P_{1,2} = \int |\Psi_{1,2}|^2 \, d^3r. \tag{1•2–6}$$

The latter form is easily generalized further, beyond the case where an incident single-object wave packet was split into two spatially separated parts by a beam splitter. Let $\Delta\Omega$ be an arbitrary volume, and let us assume that there is a way to intercept that part of the wave that is contained in $\Delta\Omega$ and inspect it with a quantum counter. Evidently, in this case, we must interpret

$$P = \int_{\Delta\Omega} |\Psi|^2 \, d^3r \tag{1•2–7}$$

as the probability that the object will materialize in $\Delta\Omega$.

If the volume $\Delta\Omega$ is sufficiently small that Ψ may be treated as constant throughout $\Delta\Omega$, we may omit the integral and write

$$dP = |\Psi|^2 \, d^3r \ (\ll 1). \tag{1•2–8}$$

1.2.2 Diffraction and Interference

Diffraction and interference are *the* characteristic manifestations of wave-like propagation. The statistical interpretation postulate explains the occurrence of diffraction and interference patterns in terms of the spatial distribution of the capture events of *many* individual elementary objects.

Consider a stream of elementary objects impinging on a double-slit diffraction screen, as shown in **Fig. 1•2–2**. For simplicity, assume that each object may be represented by a wave packet of a finite length, as in Fig. 1•1–1, and that the stream is sufficiently sparse that the wave packets describing different objects do not overlap, a condition that is usually satisfied in actual interference experiments.

For the objects whose associated wave passes through the screen, that wave is split into two parts, Ψ_1 and Ψ_2. Recall that we are explicitly assuming that the *wave* corresponding to an indivisible object *can* be split: The two (or more) spatially disconnected parts of the wave represent the object *collectively*.

Interference effects occur whenever two or more partial waves belonging to *the same* object overlap. According to the postulate of linear superposition, the overall wave function in the region of overlap is then simply the sum of the two partial waves,

$$\Psi(\mathbf{r}, t) = \Psi_1(\mathbf{r}, t) + \Psi_2(\mathbf{r}, t), \tag{1•2–9}$$

and the probability density for capturing an object will be

$$|\Psi(\mathbf{r}, t)|^2 = |\Psi_1(\mathbf{r}, t) + \Psi_2(\mathbf{r}, t)|^2. \tag{1•2–10}$$

Depending upon whether the two waves reinforce or weaken each other, the net probability density is enhanced or diminished relative to the presence of only a single partial wave.

Consider, then, the recording of an interference pattern by a photographic emulsion (**Fig. 1.2–3**). Such an emulsion is essentially an assembly of a very large number of very small detectors, the photographic grains. Each grain intercepts only a very small fraction of each object's wave, and hence has only a very small probability of capturing the object, even though, for a suitably thick emulsion (and for radiation that is not highly penetrating), the object will be captured *somewhere* within the emulsion.

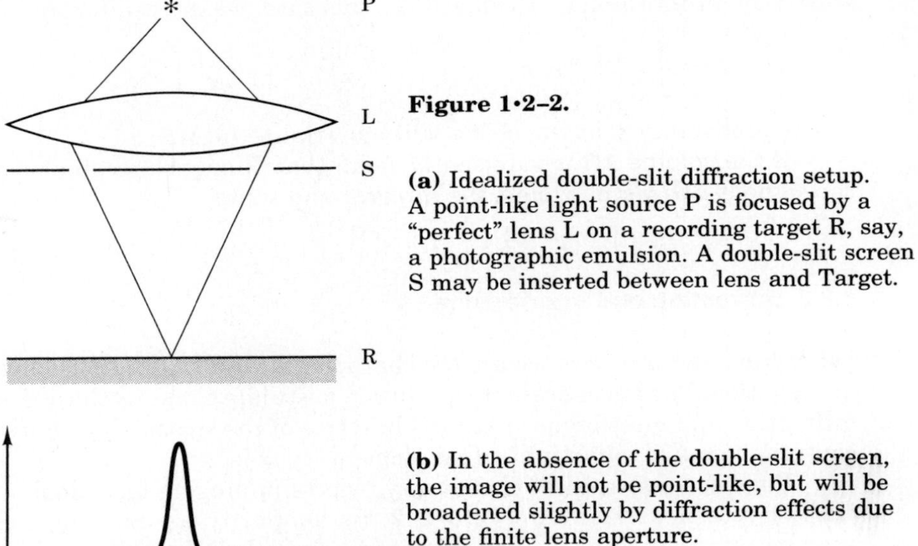

Figure 1•2–2.

(a) Idealized double-slit diffraction setup. A point-like light source P is focused by a "perfect" lens L on a recording target R, say, a photographic emulsion. A double-slit screen S may be inserted between lens and Target.

(b) In the absence of the double-slit screen, the image will not be point-like, but will be broadened slightly by diffraction effects due to the finite lens aperture.

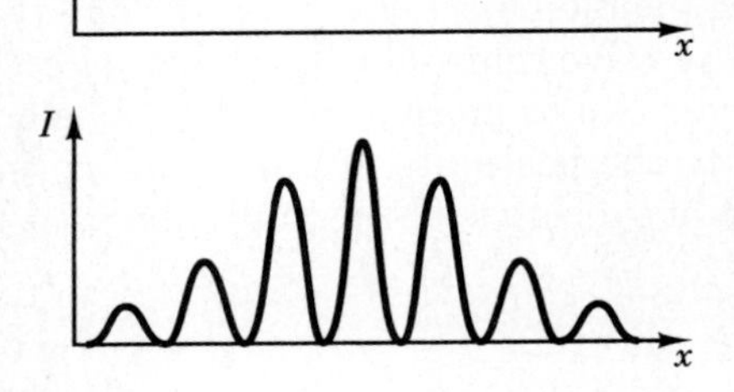

(c) The diffraction broadening increases if the double-slit screen is inserted with only one *or* the other of the two slits open, and not both simultaneously. Because of the lens, the two diffraction images from both individual slits coincide.

(d) If both slits are opened *simultaneously*, interference effects occur. No interference effects take place if both slits are opened *in succession*, with only one slit open at any one time.

However, a single captured object reveals almost nothing about the underlying probability pattern contained in the spatial variation of the probability density $|\Psi|^2$. An observable interference pattern results when there is a very large number of successive capture events, each with the same probability distribution. In those regions of space where $|\Psi|^2$ is large, a large number of random capture events will take place, while in those regions where $|\Psi|^2$ is small, the number of capture events will be small. If a sufficiently heavily exposed photographic emulsion is viewed with a resolution that does not reveal the individual grains and hence the discrete nature of the capture events, the density of exposure will appear to be a continuous function of position, following the position dependence of the probability density.

Note that the interference pattern is *not* caused by interference between *different* individual objects. Rather, *each object interferes with itself:* Every

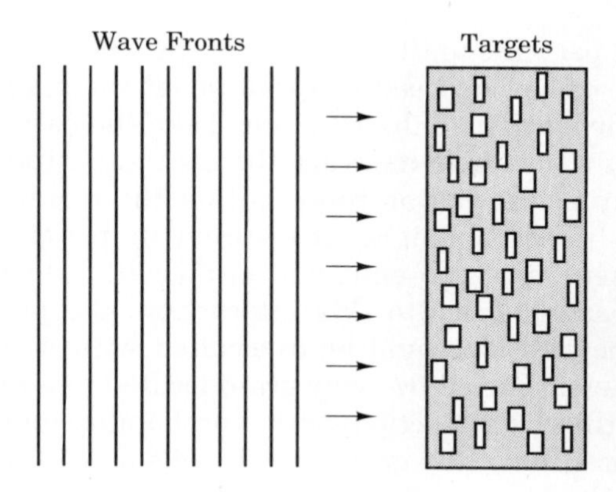

Figure 1•2–3. Interception of photons by a dense array of many individual targets, like the grains in a photographic emulsion. Any one of the grains may intercept a given photon.

partial wave of each individual object interferes with the other partial wave(s) of the same object. A large number of objects is needed only to make the pattern stand out above the statistical fluctuations; it is not part of the interference itself.

Interference effects dispense with any notions of a continuous classical trajectory during an interference experiment. If such a trajectory existed, it would have to pass through one of the two slits, and one would have to make bizarre assumptions to explain why the trajectory downstream from that slit depends on whether a second slit is also open and why opening that second slit would actually reduce the capture probability in some regions of space, namely those near the minima of the diffraction pattern.

An essential ingredient of the foregoing interpretation is the idea that an elementary object may be captured by a target that is small compared to the wave packet of the object. In fact, such a small detector volume is necessary for a high-resolution, detailed interference pattern. In effect, the object becomes localized by the interception event to within the volume of the target, which may be almost arbitrarily small. It is important to realize that this localization is a property of the interception process itself.

It was for this reason that, following (1•2–7), we spoke of the elementary object *materializing* in the intercepted volume $\Delta\Omega$, rather than of *finding* the object there. The danger with using the everyday word *finding* is that it is all too easily misinterpreted, in an everyday sense, as implying that the object was localized somewhere already before the interception event and that this event merely led to *finding* that point of localization. Unfortunately, it is precisely this everyday terminology that is commonly used throughout the quantum mechanics literature, where the quantity $|\Psi|^2$ is all but universally referred to as the probability density for *"finding"* the object. It would be futile to try to ignore this deeply entrenched usage, but it is important to warn the reader not to make the everyday association to a pre-localized object with a classical trajectory. We have here another one of those instances where an everyday word is used in quantum mechanics, but with a meaning that is changed in a subtle, but essential, way relative to its everyday meaning.

A related comment pertains to the meaning of the word *measurement.* In quantum mechanics, we will often need to speak about the results of a set of measurements, and their statistical distribution. Take the case of our photographic plate, viewed as a device to measure the spatial position of photons. Each exposed grain may be viewed as representing the result of a position measurement. But strictly speaking, it is a measurement, not of the position of a photon, but of the position of the emulsion grain with which the photon *happened* to have interacted. Prior to this interaction, the photon was not localized, and no specific position could be associated with it. It could have interacted with—and been localized by—*any* grain located in a region of space with a nonzero value of the wave function, with a probability given by (1•2–8). A measurement of the position of the grain simply tells us with which specific grain the photon "elected to interact"; it is a measurement of the position of interaction and localization, rather than of a position the photon had prior to the localization. Put differently, the "measurement" *creates* the position value, rather than determining a pre-existing value.

1.2.3 Light Beams as Sparse Streams of Photons

In the preceeding discussion, we specifically assumed that the light beam is sparse, meaning that the wave packets of the individual photons in a light beam do not overlap. It is instructive to develop a feeling for the densities that might actually occur in a typical beam of elementary objects and to express these densities in their natural units, as numbers of objects per volume λ^3, where λ is the wavelength associated with the objects.

Consider a light beam with the energy flux density S (i.e., the magnitude of the Poynting vector, averaged over the oscillation period). The average energy density is then $u = S/c$, where c is the speed of light. If the light is monochromatic, with a wavelength λ and a photon energy $\hbar\omega = 2\pi\hbar c/\lambda$, then it follows that the number of photons within λ^3 is

$$N_\lambda = (u/\hbar\omega)\lambda^3 = S\lambda^4/2\pi\hbar c^2. \tag{1•2–11}$$

As an everyday example, consider sunlight. It has about $S \sim 10^3\ \mathrm{W/m^2}$ and an average wavelength around $\lambda \sim 600$ nm. This yields $N_\lambda \sim 2 \times 10^{-6}$. Similarly low (or lower) numbers are obtained for representative electron beams.

Such numbers dispense with any notion that the elementary objects of nature are "really" particles in the classical sense, whose wave properties arise only secondarily, similar to the way sound waves arise in air. In reality, sound waves are simply density waves in a system of particles in which the number N_λ of particles per volume λ^3 is so large ($\geq 10^{20}$), that the discontinuous atomic nature of the gas becomes irrelevant. By contrast, the wave nature of elementary objects is evidently of a different kind: Every *individual* elementary object has wave-like propagation properties.

History: Taylor's 1909 Experiment In 1909, four years after Einstein's photon hypothesis, G.I. Taylor reported interference experiments with very weak light.[6] The weakest intensity required an exposure of "about three months," and the light intensity "was the same as that due to a standard candle burning at a distance slightly exceeding a mile." Taylor estimates the energy flux density for the weakest intensity as 5×10^{-13} W/cm^2, which would be 5×10^{-12} times that of sunlight, corresponding to $N_\lambda \sim 10^{-17}$! Taylor's experiment was not a double-slit diffraction experiment; it involved interference effects in the shadow of a needle. Assuming a needle length of a few times $10^4\lambda$, and an effective strip a few times λ wide around the needle contributing to the diffraction, the effective beam area contributing to the diffraction is at most $10^6\,\lambda^2$, corresponding to about 10^{-11} photons per oscillation cycle, or an average photon separation—if they could be viewed as localized objects—of $10^{11}\lambda$, that is, about 6 kilometers! Nevertheless, there was no "diminution of sharpness of the pattern" compared to the highest intensity, about 10^6 times stronger, and already corresponding to a very sparse photon stream. This is exactly what one would expect both from the linearity of Maxwell's equations and from the probabilistic interpretation of quantum mechanics.

Ironically, the experiment was apparently undertaken to refute the idea of light quanta. It was performed some 14 years before the existence of wave-particle duality was recognized, not to mention the statistical interpretation. Light quanta could then be visualized only as being localized in space and time, moving along classical trajectories. Taylor writes: "...if the intensity of light in a diffraction pattern were so greatly reduced that only a few of these indivisible units should occur on a Huygens zone [*meaning:* in one oscillation period of the effective beam cross section] at once the ordinary phenomena of diffraction would be modified."

After presenting his observations and his estimate of the energy flux density for the weakest light intensity, Taylor concludes tersely: "According to Sir J. J. Thomson this value sets an upper limit to the amount of energy contained in one of the indivisible units mentioned above." No numerical estimate for this maximum amount of energy is given. It is an oddly understated way to deal with a discrepancy by at least a factor of 10^{11}. Did Taylor not wish to dignify the quantum hypothesis with more words than necessary? Or is the reference to Sir J. J. Thomson a disclaimer of responsibility for a conclusion about which Taylor himself wasn't so sure?

Exercise: Electron Density in an Electron Beam. Electron beams in television picture tubes typically are accelerated by voltages on the order of 20 kV, with beam currents on the order of 1 mA and beam cross sections on the order of 10^{-3} cm^2. Calculate N_λ for such a beam.

[6] G. I. Taylor, Interference fringes with feeble light, *Proc. Cambridge Philosoph. Soc.*, 1909, pp. 114–115.

1.2.4 The "Collapse of the State Function"

Our statistical interpretation postulate is still incomplete. Consider once again the beam splitter scenario of Fig. 1•2–1. If one of the quantum counters captures an object even though it has intercepted only a partial wave packet, the indivisibility of elementary objects implies that the *entire* object has been captured. This naturally leads to the question: What happens to the partial wave packet in the other beam?

Once the object has been accounted for by one counter, there is no way for it to be registered once more by any additional counters. Hence, we are forced to conclude that the capture process itself changes the wave function in *both* beams in such a way that the wave function collapses to zero *outside* the volume of the responding counter. Otherwise, the statistical interpretation would imply a nonzero probability of a second capture outside that volume. Evidently, the two partial wave functions are not independent of each other, but are correlated; they describe the object collectively.

Note that we must view the collapse of the wave function as a non-local process that takes place instantaneously throughout space, even if the two partial wave functions have become separated by a large distance. At first glance, that appears to violate the principles of relativity, according to which no signal can propagate faster than the speed of light. On closer inspection, however, no such violation occurs: An observer at counter 1 has no control whatsoever over the outcome of any interception events there and, hence, no control over the correlated interception events at counter 2 . But this means that the correlation between counters 1 and 2 cannot be used for faster-than-light signaling between observers at those counters. Both observers are strictly passive onlookers of whatever it is nature decides to do. The most the observers can do is compare their observation records afterwards and confirm that the predicted correlation did indeed occur. We will return to this point repeatedly later.

The collapse of the wave function does not preclude the subsequent buildup of the wave function outside the counter volume *if* the object is subsequently released from that volume.

◆ PROBLEMS TO SECTION 1.2

#1•2-1: Double-Slit Interference

Consider the double-slit diffraction setup of **Fig. 1•2–4**. Assume that near the centerline of the recording plane (i.e., equidistant from both slits) the two partial waves Ψ_1 and Ψ_2, emerging from slits 1 and 2, respectively, may be approximated by plane waves with the same amplitude A, but different wave vectors $\mathbf{k}_1$ and $\mathbf{k}_2$:

$$\Psi_\nu(x, y, z, t) = A \cdot \exp[i(\mathbf{k}_\nu \cdot \mathrm{r} - \omega t)] \; ; \qquad \nu = 1{,}2. \qquad (1•2–12)$$

Express the components of $\mathbf{k}_1$ and $\mathbf{k}_2$ relative to the coordinate axes shown, in terms of λ, d, and D, assuming $\lambda,\, d \ll D$. Give the total wave function $\Psi(x, y, 0, t)$ in the record-

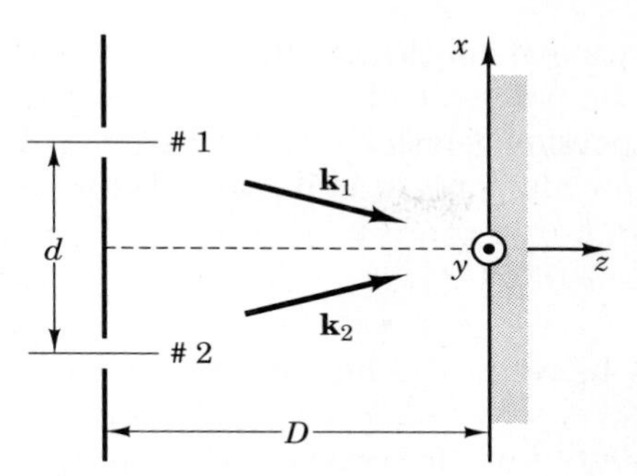

Figure 1•2–4. Double-slit interference.

ing plane $z = 0$, as well as $|\Psi|^2$. What is the spacing between the diffraction fringes, again in terms of λ, d, and D, with $D \gg d, \lambda$? Give numerical values for electrons with a wavelength $\lambda = 0.1$ nm, for a slit spacing $d = 1\ \mu m$, and, a slit-to-screen distance $D = 10$ cm. According to (1•1–11), such a wavelength corresponds to an electron energy of 150 eV.

1.3 SCHROEDINGER WAVE EQUATION

1.3.1 Wave Packets in Free Space

We return to our brief discussion in Section 1.1 of wave packets in free space, of the mathematical form (1•1–13) and illustrated in Fig. 1•1–1. As we pointed out, if such a wave packet is viewed from a sufficiently large distance, the internal wave structure of the pulse becomes unobservable, and the pulse appears as a point-like object, propagating through space along a particle-like trajectory.

In the particle-like limit, when all wave properties can be neglected, the laws of propagation for objects with mass must reduce to the laws of classical mechanics, just as for photons they reduce to the laws of geometric optics. Or, put another way, classical mechanics is the geometric-optics limit of quantum mechanics for objects with a finite mass!

We note first that the propagation velocity of a wave packet obeying the dispersion relation (1•1–7) for objects with mass is indeed the same as their Newtonian particle velocity. The propagation velocity of linear wave superpositions of the form (1•1–13) is of course the **group velocity** of the waves, obtained from the wave dispersion relation via

$$\mathbf{v} = \nabla_{\mathbf{k}}\omega \equiv \left(\frac{\partial \omega}{\partial k_x}, \frac{\partial \omega}{\partial k_y}, \frac{\partial \omega}{\partial k_z}\right). \tag{1•3–1}$$

The differential operator $\nabla_{\mathbf{k}}$, defined by the identity in (1•3–1), is called the **gradient in k-space**. The result (1•3–1) is independent of the physical nature of the waves; it is a purely mathematical result following from (1•1–13), holding for all waves for which the principle of linear superposition holds.

Insertion of the dispersion relation (1•1–7) into (1•3–1), together with the PEdB relation (1•1–2b) for momentum and wave vector, leads immediately to

$$\mathbf{v} = \frac{\hbar \mathbf{k}}{M} = \frac{\mathbf{p}}{M}, \tag{1•3–2}$$

the well-known relation between the momentum and the velocity of a Newtonian particle.[7]

For linear superpositions of the form (1•1–13), the group velocity depends on neither position nor time. This is as expected for a pulse representing an object in free space, in the absence of any forces acting on the object (or, for photons, in a uniform dielectric). But when forces are present, classical mechanics predicts an accelerated motion. This result must somehow be contained within quantum mechanics. Evidently, plane-wave superpositions of the simplest form (1•1–13), with a fixed dispersion relation, are too simple to describe such a behavior. This should come as no surprise: The plane waves (1•1–1) that make up (1•1–13) have a constant wave vector throughout space, corresponding to a constant momentum. But in the presence of forces, the momentum of an object will change as the object propagates. We evidently need a generalization of the form (1•1–13) that permits such a change. In classical electromagnetic theory, waves in a non-uniform medium must satisfy Maxwell's equations. What we need is a wave equation that plays the same role for objects with a nonzero mass that Maxwell's equations play for photons.

1.3.2 Wave Equation in Free Space

We retain, for now, the assumption that no forces act on the object, and consider a *single* plane wave in one dimension, propagating in the x-direction,

$$\Psi(x, t) = A \cdot \exp[i(kx - \omega t)], \tag{1•3–3}$$

where ω and k are related via the dispersion relation (1•1–7), which we re-write as

$$\hbar\omega = \frac{\hbar^2 k^2}{2M}. \tag{1•3–4}$$

Given a plane wave of the form (1•3–3), both the *temporal* frequency ω and the *spatial* frequency k (i.e., the wave number), may be extracted from Ψ by simple differentiation:

$$\omega\Psi = i\frac{\partial\Psi}{\partial t} \quad \text{and} \quad k\Psi = -i\frac{\partial\Psi}{\partial x}. \tag{1•3–5a,b}$$

---

[7] We shall see later that (1•3–2) must be modified in the presence of magnetic fields.

Furthermore,

$$k^2\Psi = -\frac{\partial^2\Psi}{\partial x^2}. \qquad (1{\bullet}3\text{–}5c)$$

If we multiply the dispersion relation (1•3–4) by Ψ and insert the expressions (1•3–5a,c), we obtain the differential equation

$$i\hbar\frac{\partial\Psi}{\partial t} = -\frac{\hbar^2}{2M}\frac{\partial^2\Psi}{\partial x^2}. \qquad (1{\bullet}3\text{–}6)$$

This equation no longer contains ω and k explicitly; hence, it holds for *all* plane waves that satisfy the dispersion relation. Furthermore, because the equation is linear and homogeneous in Ψ, it also holds for arbitrary linear superpositions of plane waves with *different* values of ω and k, so long as each of the plane waves in the superposition satisfies the dispersion relation (1•3–4). In effect, the two **differential operators**

$$\hat{\omega} \equiv +i\frac{\partial}{\partial t} \qquad \text{and} \qquad \hat{k} \equiv -i\frac{\partial}{\partial x} \qquad (1{\bullet}3\text{–}7a,b)$$

contained in (1•3–6) extract the values of ω and k from each plane wave of the superposition, and test whether or not that plane wave satisfies (1•3–6). In (1•3–7), the carets (^) over the operator symbols indicate that the symbols represent operators rather than numbers.

It is a simple exercise in Fourier analysis (left to the reader) to show that *any* solution of (1•3–6) can be written in the form (1•1–13), with ω and k interrelated by (1•3–4). Hence, in force-free space the wave equation (1•3–6) is mathematically equivalent to the dispersion relation (1•3–4), and it is the simplest possible form of a wave equation for objects with mass in force-free space, in one dimension.

If we generalize (1•3–6) to a plane wave of the original form (1•1–1) in three dimensions, propagating in the direction given of the wave *vector* **k**, then the relations (1•3–5b,c) must replaced by

$$\mathbf{k}\Psi = -i\nabla\Psi \qquad \text{and} \qquad k^2\Psi = -\nabla^2\Psi\,, \qquad (1{\bullet}3\text{–}8a,b)$$

and (1•3–6) takes on the form

$$i\hbar\frac{\partial\Psi}{\partial t} = -\frac{\hbar^2}{2M}\nabla^2\Psi\,. \qquad (1{\bullet}3\text{–}9)$$

Exercise: Show that plane waves of the real sine or cosine form (1•1–14) do not satisfy (1•3–9). Construct the simplest wave equation for the linear superposition of waves of the form (1•1–14) that satisfy the dispersion relation (1•3–4).

An Alternative Point of View

Once (1•3–9) has been derived, we may adopt an alternative point of view that will turn out to be very fruitful: We note that the differential equation (1•3–9) is a **local** equation, drawing only on the properties of the wave function in an infinitesimal vicinity of each space-time point $(\mathbf{r}, t)$, independently of its behavior far away or at a different time. For example, if (in a one-dimensional case) we were given the real and imaginary parts of $\Psi(x, t)$ only in a narrow interval $(x_0, x_0+\Delta x)$, $(t_0, t_0 + \Delta t)$, we would be able to determine whether or not (1•3–9) is satisfied in this interval and, hence, whether or not this "space-time slice" of $\Psi(x, t)$ could belong to a physically possible state of the object.

Suppose next that we attempt to fit the (one-dimensional) wave function $\Psi(x, t)$ over such a narrow space-time interval as well as possible to a simple plane wave of the form

$$\Psi_0(x, t) = A \cdot \exp[i(Kx - \Omega t)] , \tag{1•3–10}$$

with a *fixed* complex amplitude A, a *fixed* wavelength $\Lambda = 2\pi/K$, and a *fixed* frequency Ω (see **Fig. 1•3–1**). The question then is, which values of A, K and Ω should we choose for an optimum fit?

The fitting wave Ψ_0 obeys the two differential equations

$$K^2\Psi_0 = -\frac{\partial^2\Psi_0}{\partial x^2} \quad \text{and} \quad \Omega\Psi_0 = i\frac{\partial\Psi_0}{\partial t}. \tag{1•3–11a,b}$$

The best possible local fit is obtained if we select K and Ω such that, at the fitting point (x_0, t_0), fitting wave Ψ_0 satisfies the same differential equations as the *original* wave Ψ, with the same values of K and Ω. That is, we select K and Ω such that

$$K^2\Psi \equiv -\frac{\partial^2\Psi}{\partial x^2} \quad \text{and} \quad \Omega\Psi \equiv i\frac{\partial\Psi}{\partial t}, \tag{1•3–12a,b}$$

with the *original* wave function Ψ and its partial derivatives all *evaluated at the fitting point.* By determining the values of K and Ω from the original wave $\Psi(x, t)$ via (1•3–12a) and (1•3–12b), and then choosing an appropriate value of the complex amplitude A, we can make a plane wave of the form (1•3–10) match

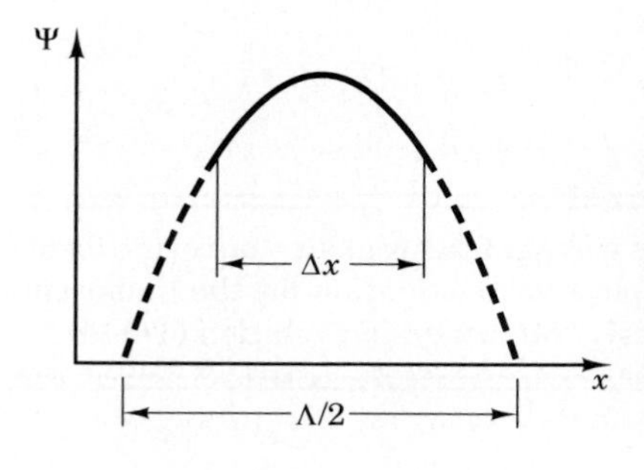

Figure 1•3–1. Matching a section of a wave function to a pure plane wave. The thin gray line is a sine wave that matches the full wave function over the interval Δx. Note that the fit may be performed on any part of the wave, not just at one of the extrema.

the original slice of $\Psi(x, t)$ in magnitude, slope, curvature, and rate of change with time.

Note that the values of K and Ω defined by (1•3–12a,b) are themselves functions of the location (x_0, t_0), in time and space, of the fitting point. We therefore refer to them as the **local and momentary** values of the wave number and the frequency at the point (x_0, t_0); they characterize the plane-wave behavior of the wave in the immediate vicinity of the spatial position x_0 and of the time t_0.

Our wave-fitting procedure is easily generalized to three dimensions, at a fitting point $(\mathbf{r}_0, t_0)$, by defining a local and momentary wave *vector* $\mathbf{K}$ such that its Cartesian components satisfy

$$K_x^2\Psi \equiv -\frac{\partial^2\Psi}{\partial x^2}, \qquad K_y^2\Psi \equiv -\frac{\partial^2\Psi}{\partial y^2}, \qquad K_z^2 \equiv -\frac{\partial^2\Psi}{\partial z^2}, \qquad \text{3–13a,b,c)}$$

with the wave function and its partial derivatives again all taken at the fitting point $(\mathbf{r}_0, t_0)$. From (1•3–13a-c),

$$K^2\Psi \equiv -\nabla^2\Psi, \qquad (1\bullet3\text{–}14)$$

in obvious analogy to (1•3–8b).

In terms of the definitions (1•3–14) and (1•3–12b), the wave equation (1•3–9) simply states that the local and momentary values of wave vector and frequency satisfy the same dispersion relation as plane waves. This concept of a local and momentary dispersion relation will shortly prove very useful in generalizing the wave equation (1•3–9) to the case where external forces are present.

1.3.4 Full Schroedinger Wave Equation

We now turn to the case where an external force $\mathbf{F}$ acts on the object. We assume that the force can be expressed as the gradient of a potential energy $V(\mathbf{r})$:

$$\mathbf{F} = -\nabla V(\mathbf{r}). \qquad (1\bullet3\text{–}15)$$

For electrons in an electrostatic potential $\Phi(\mathbf{r})$,

$$V(\mathbf{r}) = -e\Phi(\mathbf{r}) \qquad (1\bullet3\text{–}16)$$

In classical mechanics, the sum of the kinetic energy and the potential energy of a Newtonian particle is constant along its trajectory and equal to the total energy $\mathcal{E}$ of the particle:

$$\mathcal{E} = p^2(\mathbf{r})/2M + V(\mathbf{r}). \qquad (1\bullet3\text{–}17)$$

This classical result must somehow be a consequence of the underlying exact wave equation, and the formal structure of (1•3–17) must reflect the formal structure of this wave equation. The simplest hypothesis we can make is that in the presence of a potential $V(\mathbf{r})$ the *local* and *momentary* values of the frequency Ω and of the square of the wave vector, K^2, defined in (1•3–14) and (1•3–12b), satisfy the modified dispersion relation

$$\hbar\Omega = \frac{\hbar^2 K^2}{2M} + V(\mathbf{r}), \tag{1•3–18}$$

which follows from the dynamical $\mathcal{E}(\mathbf{p})$ relation (1•3–17) via the PEdB relations. Multiplying (1•3–18) by Ψ and using (1•3–12b) and (1•3–14) yields the full **Schroedinger wave equation** (SWE),

$$\boxed{i\hbar\frac{\partial\Psi}{\partial t} = -\frac{\hbar^2}{2M}\nabla^2\Psi + V(\mathbf{r})\Psi,} \tag{1•3–19}$$

as the simplest possible equation meeting our requirements.

Our argument leading to (1•3–19) is not a rigorous mathematical derivation. It is only a plausibility argument, justifying (1•3–19) as a hypothesis, the validity of which can be corroborated only experimentally, not by theoretical derivations. What form might such a verification take? To put (1•3–19) to a truly meaningful test, we must consider situations in which, as a result of forces, the kinetic energy changes by an appreciable fraction of itself over a distance on the order of the wavelength of the wave. We saw earlier that an electron with an energy of, say, 1 eV has a wavelength on the order of 10^{-9} m. This suggests that for a meaningful test of the Schroedinger wave equation we should look at the behavior of electrons in fields approaching or exceeding 10^9 V/m. This exceeds most laboratory fields, but approaches the local fields inside atoms. The simplest such test system is an electron in the simplest of all atoms, the hydrogen atom. Comparisons between the measured and the theoretically calculated spectral lines of atomic hydrogen have proven to be one of the most severe tests of quantum-mechanical theories, and they have confirmed beyond the shadow of a doubt that the Schroedinger wave equation (1•3–19) is indeed the correct wave equation for an elementary object of mass M with a non-relativistic energy $\mathcal{E}$ in a fixed external potential $V(\mathbf{r})$.

Total vs. Kinetic Energy

Our assumption that the energy $\mathcal{E}$ in the Planck-Einstein relation $\mathcal{E} = \hbar\omega$ is the *total* energy of the object, rather than just its kinetic energy, deserves some comments. The propagation of an elementary object with mass through a position-dependent potential is quite analogous to the propagation of a classical electromagnetic wave through a non-uniform dielectric. We know that in the latter case, the frequency of the wave along its path is constant and the wavelength changes. Presumably, the conservation of frequency for classical waves and the conservation of total energy for classical particles arise from a common quantum mechanical origin, calling for the identification of the wave frequency with the *total* particle energy. We do not view this as a new and separate hypothesis, but simply as a clarification of the exact meaning of the basic hypothesis contained in the Planck-Einstein postulate.

Example: Reflection by a Moving Wall As a simple example, we consider the reflection of objects with a momentum $p_1 = Mv_1 = \hbar k_1$ by an infinitely high potential wall **(Fig. 1•3–2)**, which itself moves with the uniform speed v_0, so that its position at time t is

$$x_0 = v_0 t. \tag{1•3–20}$$

The problem could be readily treated classically. However, it is instructive to treat it purely quantum-mechanically, by actually solving the Schroedinger wave equation with a suitable boundary condition at the moving wall.

Inside the infinitely high wall ($x > x_0$), the only way to satisfy the SWE is by setting $\Psi = 0$. The simplest wave function that could describe our situation to the left of the wall ($x < x_0$) consists of a superposition of two plane waves, of the form

$$\Psi(x, t) = A_1 \exp[i(k_1 x - \omega_1 t)] + A_2 \exp[i(k_2 x - \omega_2 t)], \tag{1•3–21}$$

where one of the two terms represents the incident object and the other represents the reflected object. We leave it open which one is which. Note that a positive k-value represents an object moving to the right.

If both ω-k pairs obey the dispersion relation (1•3–4), then the wave function (1•3–21) evidently satisfies the SWE in the region $x < x_0$, to the left of the wall.

We will show later that the solutions of the SWE cannot have any discontinuities. Hence, with $\Psi = 0$ inside the wall, the solution (1•3–21) must go to zero at the edge of the wall, i.e.,

$$\Psi(x_0, t) = \Psi(v_0 t, t) = 0, \tag{1•3–22}$$

for all values of t. If the wall moves, this condition cannot be met for $|k_1| = |k_2|$ and $\omega_1 = \omega_2$; hence, we see that reflection from a moving wall requires the reflected object to have a momentum and energy different from those of the incident object, just as in classical mechanics.

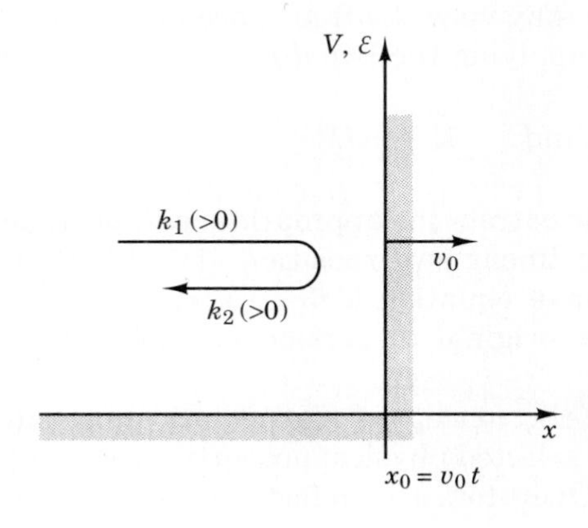

Figure 1•3–2. Reflection of a wave by a moving, infinitely high potential wall.

If we subject (1•3–21) to the boundary condition (1•3–22), we obtain the requirement

$$\Psi(x, t) = A_1 \exp[i(k_1 v_0 - \omega_1)t] + A_2 \exp[i(k_2 v_0 - \omega_2)t] = 0. \qquad (1\text{•}3\text{–}23)$$

In order for this to be satisfied for all values of t, we must have

$$A_1 + A_2 = 0 \qquad (1\text{•}3\text{–}24)$$

and

$$k_1 v_0 - \omega_1 = k_2 v_0 - \omega_2. \qquad (1\text{•}3\text{–}25)$$

If we replace the ω's via the dispersion relation (1•3–4) and drop a common factor $(k_1 - k_2)$ from the result, we may write (1•3–25) as

$$\frac{\hbar k_2}{M} + \frac{\hbar k_1}{M} = 2v_0, \qquad (1\text{•}3\text{–}26a)$$

or, in terms of group velocities,

$$v_1 + v_2 = 2v_0. \qquad (1\text{•}3\text{–}26b)$$

Evidently, one of the two velocities is larger than v_0, representing the incident object catching up with the wall, while the other velocity is smaller, representing the reflected object. Let v_1 be the incident velocity. We must then distinguish two cases:

If $v_1 > 2v_0$, then $v_2 < 0$, and the reflected object moves to the left, with a lower speed than the incident speed.

If $v_0 < v_1 < 2v_0$, the incident object still catches up with the moving wall and is reflected by it, moving to the left *relative to the wall,* but with a speed less than that of the wall itself, leading to a *net* motion to the right in the rest frame, again with a lower speed than the incident speed.

It is left to the reader to show that these results are exactly what one would have expected from a purely classical treatment.

1.3.3 Operators

As indicated earlier, we may view the free-space Schroedinger wave equation (1•3–9) as the result of applying the two *differential operators*

$$\hat{\omega} = +i\frac{\partial}{\partial t} \qquad \text{and} \qquad \hat{\mathbf{k}} = -i\nabla \qquad (1\text{•}3\text{–}27a,b)$$

to the wave function Ψ, to extract the appropriate pair of values of $\mathbf{k}$ and ω from each plane wave in the linear superposition (1•1–13) that forms the wave function $\Psi(\mathbf{r}, t)$. The wave equation (1•3–9) then ensures that this pair of parameters satisfies the original dispersion relation (1•3–4) for each plane wave.

The significance of the concept of mathematical operators acting on the wave function to extract selected physical properties goes far beyond the example we have given here. Operators are, in fact, one of the central concepts of the mathematical formalism of quantum mechanics. We call the operator $\hat{\mathbf{k}}$ in (1•3–27b) the *wave vector operator* and the operator $\hat{\omega}$ in (1•3–27a) the (angular) *frequency operator,* and we will follow the practice of placing a caret (^) above the operator symbols to signal that they are not ordinary numbers. All operators we have introduced here are *differential* operators. We will encounter other kinds of operators later.

According to the PEdB relation, the energy of an object is proportional to the frequency of the associated wave, and the momentum is proportional to the wave vector. This immediately suggests introducing the (total) **energy operator**

$$\hat{\mathcal{E}} \equiv \hbar\hat{\omega} = +i\hbar\frac{\partial}{\partial t} \tag{1•3–28}$$

and the **momentum operator**

$$\hat{\mathbf{p}} \equiv \hbar\hat{\mathbf{k}} = -i\hbar\nabla. \tag{1•3–29}$$

These two operators evidently extract the energy and the momentum vector of an object from a plane wave representing that object. In Cartesian coordinates, the vector operator $\hat{\mathbf{p}}$ has the scalar components

$$\hat{p}_x \equiv -i\hbar\frac{\partial}{\partial x}, \qquad \hat{p}_y \equiv -i\hbar\frac{\partial}{\partial y}, \qquad \hat{p}_z \equiv -i\hbar\frac{\partial}{\partial z}. \tag{1•3–30}$$

In terms of these operators, we may say that the Schroedinger wave equation (1•3–19) is obtained from the energy-momentum relation

$$\mathcal{E} = p^2/2M + V \tag{1•3–31}$$

by replacing the particle properties contained in this relation by the operators associated with them, i.e.,

$$\mathcal{E} \to \hat{\mathcal{E}} = +i\hbar\frac{\partial}{\partial t}, \qquad \mathbf{p} \to \hat{\mathbf{p}} = -i\hbar\nabla, \tag{1•3–32a,b}$$

and

$$V(\mathbf{r}) \to \hat{V} \quad [=V(\mathbf{r})], \tag{1•3–33}$$

and by applying the result of this substitution to the wave function Ψ. Note that the potential energy "operator" is simply the potential energy function itself.

It is often convenient to write the right-hand side of the Schroedinger wave equation in the form of an operator operating on the wave function Ψ, that is,

$$\boxed{i\hbar\frac{\partial\Psi}{\partial t} = \hat{H}\Psi} \tag{1•3–34}$$

where we have defined

$$\boxed{\hat{H} = -\frac{\hbar^2}{2M}\nabla^2 + V(\mathbf{r}).} \tag{1•3–35}$$

The operator $\hat{H}$ is called the **Hamilton operator**, or the **Hamiltonian**, because of its formal similarity to the Hamilton *function* in classical mechanics. It is perhaps the most important operator in quantum mechanics. As (1•3–34) indicates, its role goes beyond being simply the operator for the sum of kinetic and potential energy: It also controls the time evolution of the state Ψ of the object. This close interrelation between the total energy of a system and its time evolution is a characteristic feature of all of quantum mechanics.

◆ PROBLEMS TO SECTION 1.3

#1•3-1: Harmonic Oscillator Wave Packet

The potential energy for a harmonic oscillator may be writen

$$V(x) = \frac{1}{2}M\omega^2 x^2, \tag{1•3–36}$$

where M is the mass of the object and ω is the (angular) frequency of oscillation. Show that amongst the solutions of the Schroedinger wave equation for such a potential are solutions with a probability density of the form

$$|\Psi(x, t)|^2 \propto \exp\left[-\frac{M\omega}{\hbar}(x - a \cdot \sin \omega t)^2\right], \tag{1•3–37}$$

a Gaussian wave packet oscillating sinusoidally with frequency ω and amplitude a, resembling the classical behavior of a harmonic oscillator. Determine the actual wave function $\Psi(x, t)$, including any time- and position-dependent phase factor.

1.4 TIME-INDEPENDENT SCHROEDINGER EQUATION

1.4.1 Stationary States

We shall be particularly interested in solutions of the Schroedinger wave equation that have the simple form

$$\Psi(\mathbf{r}, t) = \psi(\mathbf{r})e^{-i\omega t}, \tag{1•4–1}$$

where $\psi(\mathbf{r})$ is a function of position only. Evidently, states represented by such wave functions have a time-independent probability density $|\psi(\mathbf{r})|^2$, and we shall see later that *all* observable properties of such states are time-independent. Hence, such states are called **stationary states**; we shall find that their understanding forms the backbone of the understanding of *all* states.

We interpret $\mathcal{E} = \hbar\omega$ as the (total) energy of the state, and we may express this by writing (1•4–1) in the more "physical" form

$$\Psi(\mathbf{r}, t) = \psi(\mathbf{r}) \cdot \exp\left(-\frac{i\mathcal{E}t}{\hbar}\right). \tag{1•4–2}$$

Inserting (1•4–2) into the Schroedinger wave equation (1•3–19), and dropping the common phase factor $\exp(-i\mathcal{E}t/\hbar)$, leads to an equation for the time-independent function $\psi(\mathbf{r})$ in (1•4–2):

$$\boxed{-\frac{\hbar^2}{2M}\nabla^2\psi + V(\mathbf{r})\psi = \mathcal{E}\psi.} \tag{1•4–3}$$

This is called the **time-independent Schroedinger equation**. We will often write it in the compact form

$$\boxed{\hat{H}\psi = \mathcal{E}\psi} \tag{1•4–4}$$

employing the Hamilton operator.

1.4.2 Superpositions of Stationary States

Given several solutions of the time-independent Schroedinger equation with different values $\mathcal{E}_n$ of the energy, any linear superposition of the form

$$\Psi(\mathbf{r}, t) = \sum_n c_n \psi_n(\mathbf{r}) \cdot \exp\left(-\frac{i\mathcal{E}_n t}{\hbar}\right) \tag{1•4–5}$$

will also be a solution of the Schroedinger wave equation, and we will see later that *all* physically admissible solutions may be written in this way. The solution of the time-independent Schroedinger equation is therefore central to *all* of quantum mechanics.

Mathematically, the superposition (1•4–5) represents a solution of the Schroedinger wave equation—a partial differential equation involving both space and time—by the familiar technique of **separation of variables**. However, the physical reasons for our interest in this form are even more important than the mathematical ones. States of the form (1•4–5) are inherently non-stationary. It is instructive to consider the simple example of a superposition state that contains just two frequencies:

$$\Psi(\mathbf{r}, t) = \psi_1(\mathbf{r}) \cdot \exp(-i\omega_1 t) + \psi_2(\mathbf{r}) \cdot \exp(-i\omega_2 t). \tag{1•4–6}$$

The probability density associated with this state,

$$\begin{aligned} |\Psi(\mathbf{r}, t)|^2 &= |\psi_1|^2 + |\psi_2|^2 \\ &+ \psi_1^*\psi_2 \cdot \exp[i(\omega_1 - \omega_2)t] + \psi_1\psi_2^* \cdot \exp[-i(\omega_1 - \omega_2)t], \end{aligned} \tag{1•4–7}$$

is evidently time-dependent. However, we note that the probability density depends only on the frequency *difference* ($\omega_1 - \omega_2$); the *absolute* frequencies ω_1 and ω_2 never show up! This remains true for arbitrary linear superpositions of the form (1•4–5), and we shall see later (chapter 8) that this is true not just for the probability density, but for *all* observable properties.

Unobservability of Absolute Frequencies

The foregoing means that the *absolute* frequencies contained in the wave functions are fundamentally unobservable experimentally; only frequency *differences* are operationally meaningful. The formalism is evidently redundant in this regard. Because $\mathcal{E} = \hbar\omega$, this behavior is simply the quantum-mechanical corollary to the arbitrariness of the zero of the energy in classical mechanics, where potential energies are not defined absolutely, but only relative to an arbitrary zero, which may be shifted at will. This reference energy cancels out in all observable properties—in both classical mechanics and quantum mechanics. It does not matter relative to which reference the potential energy is defined when we express the quantum-mechanical wave frequency in terms of the energy via the relation $\mathcal{E} = \hbar\omega$: With classical mechanics being a limiting case of quantum mechanics, we may view the classical cancellation as a consequence of the quantum-mechanical one.

Evidently, the quantum-mechanical wave functions behave in this regard differently from the classical electromagnetic fields, for which the absolute wave frequency has an observable meaning. We have reached here one of the limits of the analogy between the wave functions of quantum mechanics and the fields of electromagnetic theory. The reason for the breakdown of the analogy is that the classical fields are not true quantum-mechanical probability amplitudes, but are still purely classical quantities, even the complex field $\mathbf{F}$ we introduced in (1•1–15). We shall see later (chapters 10 and 20) that there exist true quantum-mechanical probability amplitudes for the electromagnetic fields. When the classical fields are calculated from these underlying quantum-mechanical probability amplitudes, the classical frequencies arise again, as beat frequencies between different-frequency components in the quantum-mechanical probability amplitudes, just as in the case of the probability densities for objects with mass.

Readers with an electrical engineering background may perceive a limited analogy to a radio broadcast signal, where all the information is contained in the difference between the actual signal frequency and a carrier frequency that is itself irrelevant to the information. However, the physics in the two cases is quite different: In quantum mechanics there is no carrier frequency.

Non-Stationary States as States with an Unsharp Energy

At the beginning of this section we identified the energy $\mathcal{E}$ in (1•4–2) and (1•4–3) with the energy of the object. No single sharp value of the energy can be associated with wave functions of the superposition form (1•4–5), whose expansion contains more than one energy. The states represented by such wave functions must be interpreted probabilistically as having an energy that is not sharply determined, in exactly the same way in which the *position* of an object is not sharply determined for any wave function that is spread out over space. Just as an object may materialize in space wherever the wave function is nonzero, we postulate that an object may materialize on the energy scale at any of the energy values that occur in the superposition (1•4–5), with a probability related to the magnitude of the associated expansion coefficient c_n. This interpretation is a natural—and inevitable—extension of the probabilistic interpretation of the wave function to properties other than the position of an object. We will elaborate on it in later chapters.

Note that the possibility of states with an uncertain energy implies neither that energy is not conserved, nor that it is conserved only on the average.

1.4.3 Example: Scattering at a Potential Step; Connection Rules

As a first example of the time-independent Schroedinger equation, we solve here the problem of the propagation of a mono-energetic wave through a one-dimensional potential $V(x)$ that contains an upward step of height V_0 at $x = 0$ (**Fig. 1.4–1**):

$$V(x) = \begin{cases} 0 & \text{for} \quad x < 0 \\ V_0 & \phantom{\text{for}} \quad x > 0 \end{cases}. \tag{1•4–8}$$

We assume that an object with total energy $\mathcal{E} > V_0$ is incident from the left. In classical mechanics, a particle encountering such a step would simply pass on. In quantum mechanics, we expect that the wave describing the object will be partially reflected and partially transmitted at the potential discontinuity, analogously to an electromagnetic wave at a dielectric discontinuity. The strength of the reflection cannot be obtained from the dispersion relation alone; it is probably the simplest problem requiring the actual wave equation. For simplicity, we treat the problem as one-dimensional, ignoring the y- and z-coordinates.

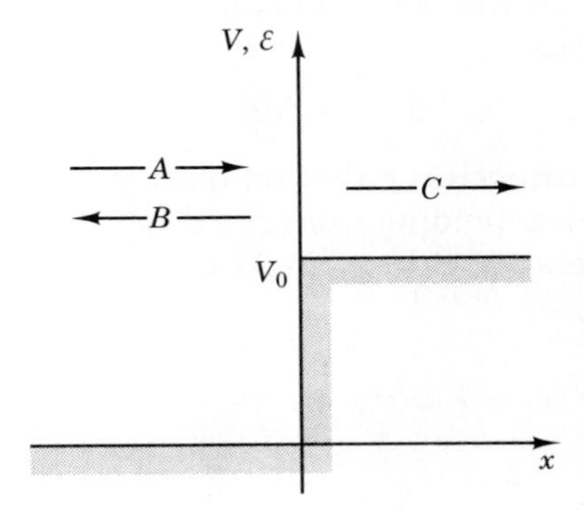

Figure 1•4–1. Reflection at an upward potential step of a wave incident from the left. There will be an incident wave (amplitude A), a reflected wave (B) and, a transmitted wave (C).

In the region $x < 0$, the general solution of the one-dimensional Schroedinger equation for a given value of $\mathcal{E}$ consists of linear superpositions of two plane waves of the form

$$\psi_-(x) = Ae^{+ikx} + Be^{-ikx} \qquad (x < 0), \tag{1•4–9}$$

with

$$k = \left[\frac{2M}{\hbar^2}\mathcal{E}\right]^{1/2}. \tag{1•4–10}$$

The term $\exp(+ikx)$ in (1•4–9) corresponds to the wave that is incident from the left, the term $\exp(-ikx)$ to the reflected wave. We would like to know the amplitude of the reflected wave relative to that of the incident wave, that is, the ratio B/A.

In the region $x > 0$, the general solution of the Schroedinger equation again consists of plane waves,

$$\psi_+(x) = Ce^{+ik'x} + De^{-ik'x} \qquad (x > 0), \tag{1•4–11}$$

where the wave number is now

$$k' = \left[\frac{2M}{\hbar^2}(\mathcal{E} - V_0)\right]^{1/2}. \tag{1•4–12}$$

The two terms in (1•4–11) again correspond to waves traveling to the right and to the left. If the object is incident from the left of the step, only a transmitted wave traveling to the right will be present for $x > 0$. Hence, we set $D = 0$:

$$\psi_+(x) = Ce^{ik'x} \qquad (x > 0). \tag{1•4–13}$$

We would like to know the strength of the transmitted wave relative to that of the incident wave, that is, the ratio C/A.

Connection Rules at a Potential Step

To determine the two ratios B/A and C/A, we must know how the wave functions $\psi_-(x)$ and $\psi_+(x)$ are connected across the potential discontinuity at $x = 0$. We claim that both the wave function $\psi(x)$ and its derivative $\psi'(x) \equiv d\psi/dx$ must be continuous:

$$\psi_-(0) = \psi_+(0), \qquad \psi_-'(0) = \psi_+'(0). \tag{1•4–14a,b}$$

To derive these **connection rules** we treat the abrupt potential step as the limit of a *graded* step extending from $x = a$ to $x = b$, as in **Fig. 1•4–2**. We re-write the time-independent Schroedinger equation (1•4–3), reduced to one dimension, in the form

$$\psi''(x) = \frac{2M}{\hbar^2}[V(x) - \mathcal{E}]\psi(x). \tag{1•4–15}$$

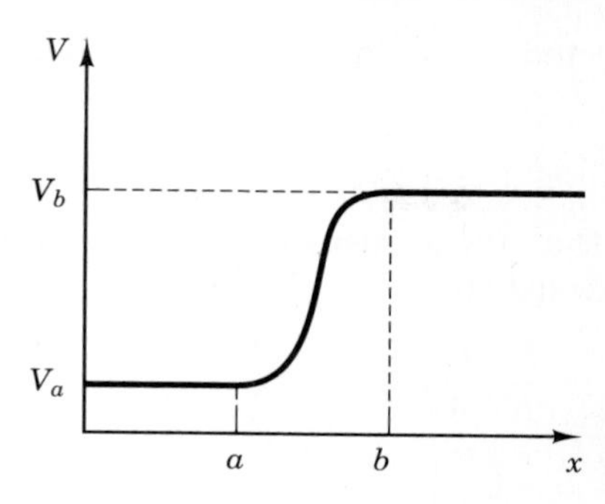

Figure 1•4–2. Abrupt step as a limiting case of a gradual step.

We integrate once, with the lower integration limit at $x = a$, the beginning of the step:

$$\psi'(x) = \psi'(a) + \frac{2M}{\hbar^2}\int_a^x [V(y) - \mathcal{E}]\psi(y)\,dy. \qquad (1•4–16a)$$

Specifically, for $x = b$,

$$\psi'(b) = \psi'(a) + \frac{2M}{\hbar^2}\int_a^b [V(y) - \mathcal{E}]\psi(y)\,dy. \qquad (1•4–16b)$$

A second integration, of (1•4–16a), from $x = a$ to $x = b$, leads to

$$\psi(b) = \psi(b) + \psi'(a)\cdot(b - a) + \frac{2M}{\hbar^2}\int_a^b dx \int_a^x [V(y) - \mathcal{E}]\psi(y)\,dy. \qquad (1•4–17)$$

We now go to the limit of an abrupt step, $b \to a$. For a potential $V(x)$ that remains finite everywhere in the integration interval, the integrals in (1•4–16b) and (1•4–17) vanish when $b \to a$, and we obtain

$$\psi'(b) = \psi'(a), \qquad \psi(b) = \psi(a), \qquad (1•4–18)$$

which is equivalent to (1•4–14). These connection rules apply not only to the solutions of the time-*independent* Schroedinger equation, but also to those of the Schroedinger wave equation.

We return to the scattering problem at our step. If we apply the matching conditions (1•4–14) to the two wave functions ψ_- and ψ_+, we obtain the two conditions for the amplitudes A, B, and C:

$$A + B = C, \qquad ik(A - B) = ik'C. \qquad (1•4–19a,b)$$

By eliminating either B or C from (1•4–19a), we obtain the ratio of the transmitted to the incident amplitude,

$$\left(\frac{C}{A}\right)_\uparrow = \frac{2k}{k + k'}, \qquad (1•4–20a)$$

and the ratio of the reflected to the incident amplitude,

$$\left(\frac{B}{A}\right)_{\uparrow} = \frac{k - k'}{k + k'}. \tag{1•4–20b}$$

The arrow (↑) indicates that these results are for an upward step. They are easily applied to a downward step by simply interchanging k and k':

$$\left(\frac{C}{A}\right)_{\downarrow} = \frac{2k'}{k' + k} = \frac{k'}{K}\left(\frac{C}{A}\right)_{\uparrow}, \tag{1•4–21a}$$

$$\left(\frac{B}{A}\right)_{\downarrow} = \frac{k' - k}{k' + k} = -\left(\frac{B}{A}\right)_{\uparrow}. \tag{1•4–21b}$$

We see that the incident wave splits into a reflected and a transmitted wave. From a classical *wave* point of view, there is nothing remarkable about this observation: It is characteristic of all wave phenomena. What makes it a remarkable prediction is wave-particle duality, with its claim that the waves represent the wave properties of *indivisible* objects. How can a wave describe a property of an indivisible object when the wave itself splits into two parts that travel in opposite directions? This calls again for a probabilistic interpretation, in terms of probabilities that the incident object is either reflected or transmitted. We will elaborate on this point in section 1.5.

1.4.4 Tunneling

Tunneling into a Step

In our discussion of scattering by an upward step, we assumed that the incident energy $\mathcal{E}$ was *above* the top of the step, $\mathcal{E} > V_0$. Assume now that the incident energy $\mathcal{E}$ is below the top, $\mathcal{E} < V_0$. In this case, the Schroedinger equation predicts that the wave function penetrates into the region with $\mathcal{E} < V_0$, which implies that there is a finite probability that the object could be "found" inside the barrier. This is in striking contrast to the behavior of a *classical* particle, which cannot penetrate into regions where it would have a negative kinetic energy. But such a penetration is precisely what one would expect from a pure wave theory. An example is the exponentially evanescent wave that penetrates into the region of lower refractive index during total internal reflection at a dielectric discontinuity. Applied to particle-like objects, this penetration is a striking new prediction of quantum mechanics called **tunneling**, which forms the basis of many important physical phenomena, from radioactive decay to charge transport in electronic devices.

When $\mathcal{E} < V_0$, the solution of the time-independent Schroedinger equation is a sum of two exponentials,

$$\psi(x) = C \exp[-\kappa x] + D \exp[+\kappa x], \tag{1•4–22}$$

where

$$\kappa = \left[\frac{2M}{\hbar^2}(V_0 - \mathcal{E}) \right]^{1/2} \tag{1•4–23}$$

If the potential step extends infinitely far to the right, as we have implicitly assumed here, then the growing exponential in (1•4–22) is unphysical. Hence, we must again set $D = 0$, and the correct solution assumes the form of a pure decaying exponential:

$$\psi(x) = C \exp[-\kappa x]. \tag{1•4–24}$$

Evidently, we may take over the entire formalism for the $\mathcal{E} > 0$ case by simply making everywhere the substitution

$$ik' \rightarrow -\kappa \qquad \text{or} \qquad k' \rightarrow i\kappa, \tag{1•4–25}$$

with κ given by (1•4–23).

Exercise: Calculate the amplitude ratio B/A for this case. Show that always, $|B/A|^2 = 1$. Explain the phase of B/A.

The distance after which the probability density has decayed by $1/e$ is

$$\delta = \frac{1}{2\kappa} = \left[\frac{\hbar^2}{8M \cdot (V_0 - \mathcal{E})} \right]^{1/2}$$

$$= 0.097 \text{ nm} \times \left[\frac{m_e}{M} \cdot \frac{1 \text{ eV}}{V_0 - \mathcal{E}} \right]^{1/2}. \tag{1•4–26}$$

For an electron with an energy 1 eV below the top of the step, δ is about 0.1 nm = 1A, a simple and useful rule to remember.

Tunneling through a Barrier

If we had assumed a finite barrier width, there would be a finite probability that an object impinging upon the barrier would emerge on the other side.

We consider a barrier of width w (**Fig. 1•4–3**), extending from $x = 0$ to $x = w$. In the region $x < 0$, we have again the solution (1•4–9), and inside the barrier $(0 < x < w)$, we have again (1•4–22), but now with a non-vanishing amplitude D. To the right of the barrier $(x > w)$, we have an outgoing plane wave, which we may write as

$$\psi(x) = Ee^{ikx} \; (x > w). \tag{1•4–27}$$

Note that the amplitude E must not be confused with the energy $\mathcal{E}$.

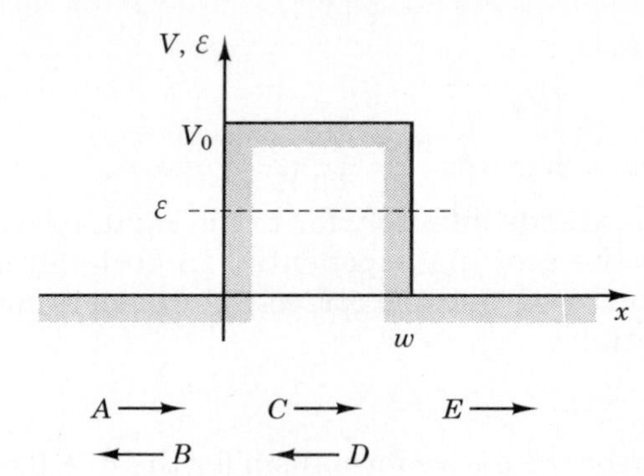

Figure 1•4–3. Tunneling through a barrier.

Continuity of the wave function and its derivative at both $x = 0$ and $x = w$ requires that the amplitudes satisfy the following four conditions:

$$A + B = C + D, \tag{1•4–28a}$$

$$ik(A - B) = -\kappa(C - D), \tag{1•4–28b}$$

$$Ce^{-\kappa w} + De^{+\kappa w} = Ee^{ikw}, \tag{1•4–29a}$$

$$-\kappa(Ce^{-\kappa w} - De^{+\kappa w}) = ikEe^{ikw}. \tag{1•4–29b}$$

We are interested here only in the ratio $|E/A|^2$, the ratio of the probability density in the transmitted wave to that in the incident wave, which we will later interpret as the probability that an object incident on the barrier will tunnel through it. By elementary elimination of the amplitudes B, C, and D from the above equations, one finds the relation

$$4ik\kappa A = [-(\kappa - ik)^2 e^{\kappa w} + (\kappa + ik)^2 e^{-\kappa w}] \cdot Ee^{ikw}. \tag{1•4–30}$$

We shall restrict ourselves further to barriers that are sufficiently thick that $\kappa w \gg 1$. In this case, the attenuation of the wave function inside the barrier is strong, and we may neglect the $e^{-\kappa w}$ term in (1•4–30), which arises from the reflection of the wave at $x = w$. When this is done, we obtain

$$\left|\frac{E}{A}\right|^2 = \frac{(4k\kappa)^2}{(k^2 + \kappa^2)^2} \cdot e^{-2\kappa w}. \tag{1•4–31}$$

It is useful to express the factor preceding the exponential in terms of $\mathcal{E}$ and V_0, using (1•4–10) and (1•4–23):

$$\left|\frac{E}{A}\right|^2 = \frac{16\mathcal{E} \cdot (V_0 - \mathcal{E})}{V_0^2} \cdot e^{-2\kappa w}. \tag{1•4–32}$$

The pre-factor varies between 0 and 4, with the maximum of 4 occurring for $\mathcal{E} = V_0/2$. The exponential factor is the important one. It shows that the tunneling probability decreases very rapidly with increasing barrier thickness, but also with increasing barrier height, which leads to a decreasing tunneling depth $\delta = 1/2\kappa$. The numerical estimate for δ given in (1•4–26) shows that tunneling is restricted to barriers of atomic and subatomic dimensions.

Exercise Generalize the above treatment to the case where the $e^{-\kappa w}$ term in (1•4–30) is not neglected.

PROBLEMS TO SECTION 1.4

#1•4-1: Connection Rules for a Delta-Function Potential

In our derivation of the connection rules (1•4–18) at a step, the potential remained everywhere finite. Another form of a singularity, often encountered in quantum mechanics, would be exhibited by a potential that becomes locally infinite, but in such a way that the integral across the singularity remains finite. The simplest such potential, often used in quantum-mechanical calculations as a computational model to illustrate the effect of infinities, is a Dirac δ-function potential of strength S,

$$V(x) = \pm S\,\delta(x)\,, \tag{1•4–33}$$

where $\delta(x)$ is the **Dirac δ-function**, defined by the requirements that $\delta(x)$ vanish except at $x = 0$, and that it be "infinitely strong" at $x = 0$, in such a way that

$$\int_{<0}^{>0} \delta(x)\,dx = 1 \tag{1•4–34}$$

over *any* interval that contains $x = 0$. (For a rigorous treatment, see appendix A.) By extension of our derivation of the connection rules at a step, show that for δ-function potentials, the wave function itself remains continuous, but with a kink at the δ-function such that

$$\boxed{\Delta\psi' - \psi'(+0) - \psi'(-0) = \pm\frac{2MS}{\hbar^2}\psi(0),} \tag{1•4–35}$$

where the sign on the right-hand side is the same as in (1•4–33).

#1•4-2: Reflection under Oblique Incidence

A mono-energetic plane wave representing particles with mass M and energy $\mathcal{E}$ *is* incident on a potential step of height V_0, just as in Fig. 1•4–1, except that now the plane wave is not incident perpendicularly to the plane of the step, but under an angle θ relative to the **x**-direction. (Make a graph.) Give closed-form expressions for the wave functions on both sides of the step and for the transmission probability T, as functions of the incident energy and of the angle of incidence θ. Make a plot of the angular dependence of the transmission probability for a fixed incident energy $\mathcal{E} = 2V_0$.

#1•4-3: Resonant Transmission across a Potential Well

Consider a potential well of depth V_0 and width w, as shown in **Fig. 1•4–4**. A wave with energy $\mathcal{E} > 0$ is incident from the left. For certain combinations of the incident energy $\mathcal{E}$ and the depth V_0 and width w of the well, there will be no reflected wave present.

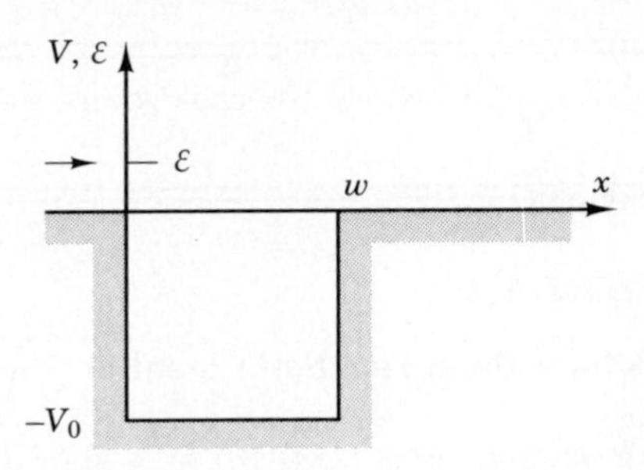

Figure 1•4–4. Transmission of a wave across a potential well.

Determine these conditions, avoiding unnecessary generality in your solution. (Do *not* first solve the problem for *arbitrary* combinations of $\mathcal{E}$, V_0, and w.) Explain the perfect transmission as a destructive interference between the two waves reflected at the downward step at $x = 0$ and at the upward step at $x = w$. What is the round-trip phase shift of the wave inside the well under resonant conditions? How can this create destructive interference between the two reflected waves?

If $\psi(x) = A \exp(ikx)$ is the incident wave for $x < 0$, what are the wave functions inside the well and for $x > w$, under resonant conditions? Assume that A is real.

1.5 PROBABILITY CURRENT DENSITIES

1.5.1 Continuity equation

If we are dealing with indestructible objects, such as electrons, we must evidently require that a wave function, once normalized, remain normalized:

$$\frac{d}{dt}\int_{\text{all space}} |\Psi|^2\, d^3r = 0. \tag{1•5–1}$$

We show here that any wave function $\Psi(\mathbf{r}, t)$ that is a solution of the Schroedinger wave equation satisfies (1•5–1): We first evaluate the quantity

$$\frac{\partial|\Psi|^2}{\partial t} = \Psi^*\frac{\partial\Psi}{\partial t} + \Psi\frac{\partial\Psi^*}{\partial t} \tag{1•5–2}$$

by replacing the time derivatives on the right-hand side via the Schroedinger wave equation (1•3–19). When this is done, the $V(\mathbf{r})\Psi$- term cancels, and we obtain

$$\frac{\partial|\Psi|^2}{\partial t} = \frac{i\hbar}{2M}(\Psi^*\nabla^2\Psi - \Psi\nabla^2\Psi^*) = \frac{i\hbar}{2M}\nabla\cdot(\Psi^*\nabla\Psi - \Psi\nabla\Psi^*). \tag{1•5–3}$$

This is of the form of a standard continuity equation

$$\frac{\partial\rho}{\partial t} = -\operatorname{div}\mathbf{j}, \tag{1•5–4}$$

where

$$\rho = |\Psi|^2 \tag{1•5–5}$$

is the probability density and

$$\boxed{\mathbf{j} = -\frac{i\hbar}{2M}(\Psi^*\nabla\Psi - \Psi\nabla\Psi^*)} \tag{1•5–6}$$

is called the **probability current density**.

With the aid of (1•5–4) and of Gauss' integral theorems, the left-hand side of (1•5–1) may be written

$$\frac{d}{dt}\int \rho d^3r = -\int \operatorname{div} \mathbf{j}\, d^3r = -\oiint \mathbf{j}\cdot d\mathbf{A}. \tag{1•5–7}$$

Here the last integral goes over the surface at infinity, enclosing the volume. But if Ψ is to be normalized, both Ψ and its derivatives must go to zero at infinity. Hence, $\mathbf{j}$ goes to zero at infinity, and the surface integral in (1•5–7) vanishes, which gives (1•5–6).

Exercise: Show that Ψ falls off sufficiently rapidly that the integral in (1•5–7) vanishes even though the surface *area* becomes infinite itself.

Continuity equations of the form (1•5–4) occur in many branches of physics, whenever there is a physical quantity Q whose total amount is conserved, be it mass, energy, electrical charge, or what have you. The density ρ is then the amount of the quantity Q per unit volume, and $\mathbf{j}$ is the flow density of Q, that is, the amount of Q per unit area and per unit time passing through a cross-section perpendicular to $\mathbf{j}$. A familiar example is the Poynting theorem of classical electromagnetic theory,

$$\frac{\partial u}{\partial t} = -\operatorname{div} \mathbf{S}\,, \tag{1•5–8}$$

where u is the electromagnetic energy density and $\mathbf{S}$ is the energy flux density (the Poynting vector), both introduced in section 1.1—see (1•1–17) and (1•1–18).

The operational interpretation of the probability *current* density $\mathbf{j}$ is similar to that of the probability density. The quantity

$$P = \int_{\Delta t}\left[\iint_{\Delta A} \mathbf{j}\cdot d\mathbf{A}\right] dt \tag{1•5–9}$$

is the probability that the object materializes inside a detector with cross-sectional area ΔA during the time Δt. The warnings issued in section 1.2 about misinterpretations of localization in space apply to localization in time as well. Suppose we intercept an elementary object with a suitable electronic detector that has a resolving time δt that is much shorter than the time it takes for the whole wave packet to pass the detector. Then the indivisibility of the objects assures us that the detector will respond, if at all, with a discrete event whose time is localized to within an interval of length δt. Again, it is important to realize that this localization in time is a property of the interception process itself and that we must not view the object as, say, a pre-localized object passing through a particular plane in space at a sharply defined time.

Scattering at a Step and Tunneling through a Barrier: The Interpretation

We are now ready to interpret our results of section 1.4 for the scattering of a wave at a potential step. By inserting the two plane waves from (1•4–9) and the plane wave from (1•4–13) into (1•5–6) we obtain the three probability current densities

$$j_I = \frac{\hbar k}{M}|A|^2, \qquad j_R = -\frac{\hbar k}{M}|B|^2, \qquad j_T = \frac{\hbar k'}{M}|C|^2. \tag{1•5–10}$$

Here j_I is the incident current density, j_R the reflected current density, and j_T the transmitted current density. Evidently,

$$R = -\frac{j_R}{j_I} = \left|\frac{B}{A}\right|^2 = \left|\frac{k - k'}{k + k'}\right|^2 \tag{1•5–11}$$

is the **reflection coefficient**, defined as the probability that an object is reflected at the step. Similarly,

$$T = \frac{j_T}{j_I} = \frac{k'}{k}\left|\frac{C}{A}\right|^2 = \frac{4kk'}{(k + k')^2} = 1 - R \tag{1•5–12}$$

is the **transmission coefficient**, defined as the probability that the object will be transmitted past the step.

Similarly, the quantity $|E/A|^2$ in (1•4–31) and (1•4–32) must be interpreted as the ratio of the transmitted to the incident probability current density and, hence, as the probability that an object incident on the barrier will tunnel through it.

1.5.2 Streams of Statistically Independent Objects

Our probabilistic interpretation of the wave function for a *single* object required that the overall wave function $\Psi(x, t)$ be normalized—or at least that it *can* be normalized. However, it is obvious that the plane waves used in our discussion of scattering at a step do not go to zero at infinity and, hence, are not normalizeable.

What this means is simply that Ψ cannot correspond to a single object. In fact, scattering experiments that might be described by such wave functions are performed not with single objects, but with streams of many objects. As we have seen, such streams tend to be sparse, and it is natural to assume that the objects are statistically independent of each other. In that case their probabilities simply add, and we should interpret the quantity $|\Psi|^2$ as the probability density, not for a *specific* single object materializing, but for *any arbitrary* (unspecified) single object out of the entire stream to materialize. Analogous considerations apply to the probability current densities.

Wave functions representing streams of statistically independent objects may be normalized by choosing the amplitude A of the incident wave in such a way that the incident probability current density $\mathbf{j}$ equals the time-averaged actual incident current density.

When many elementary objects are present, the integrals in (1•2–7) and (1•5–9) may become larger than unity. Evidently, they no longer represent *probabilities.* Instead, they must be interpreted as the *average* number of objects that materialize per interception experiment, in the limit of a large number of such experiments.

The question then often arises how the *actual* numbers of materialization events are distributed. If the events are indeed statistically independent, as assumed here (or else our entire interpretation would be inapplicable), their distribution is simply the **Poisson Distribution**, well-known from probability theory and derived in appendix B. To put the Poisson distribution in a context of interest to us, consider a uniformly but sparsely exposed ideal photographic emulsion. By ideal, we mean that only grains that have intercepted a photon become sensitized, and one photon is sufficient to sensitize a grain. We also assume that the grains are sufficiently small or the exposure sufficiently weak that the probability for any grain to intercept two or more photons can be neglected. Given these assumptions, suppose we sub-divide the emulsion into a large number of equal areas ΔA and count the number of exposed grains N within each area. This number will fluctuate from area to area. Let $\langle N \rangle$ be the average number of exposed grains per area ΔA. Then the probability that a given area ΔA contains exactly N exposed grains is

$$\boxed{P(N) = \frac{\langle N \rangle^N}{N!} \exp(-\langle N \rangle.} \tag{1•5–13}$$

The same law applies to the time distribution: If a photon counter registers, on the average, $\langle N \rangle$ photons per time interval Δt, then (1•5–13) gives the probability that the counter registers exactly N photons in a given interval.

In quantum mechanics, averages taken over many measurements, such as $\langle N \rangle$, are commonly referred to as **expectation values**.

Exercise. Suppose a light beam or electron beam is chopped into pulses of such a length that $\langle N \rangle$ is the *average* number of objects per pulse. What should be the *average* number $\langle N \rangle$ of objects contained in a pulse, to maximize the fraction containing *exactly* $N = 100$? What *is* the maximum fraction of pulses that can be generated yielding exactly $N = 100$ objects? Use Stirling's formula for the factorial of a large number,

$$N! \cong (2\pi N)^{1/2} N^N e^{-N}. \tag{1•5–14}$$

Next to its mean value $\langle N \rangle$, the most important property of a distribution is its **standard deviation** ΔN, defined by

$$(\Delta N)^2 = \langle (N - \langle N \rangle)^2 \rangle = \langle N^2 \rangle - \langle N \rangle^2. \tag{1•5–15}$$

For Poisson-distributed events, from appendix B,

$$\Delta N = \langle N \rangle^{1/2}, \ \Delta N / \langle N \rangle = \langle N \rangle^{-1/2}. \tag{1•5–16}$$

◆ PROBLEMS TO SECTION 1.5

#1•5-1: State of Cavity after Capturing One Photon

Suppose an electromagnetic cavity initially contains the amount $U = \langle N \rangle \hbar\omega$ of classical field energy (of frequency ω), where $\langle N \rangle$ is the expectation value of the number of photons in the cavity and the probability distribution $P(N)$ is a Poisson distribution. Assume that a photon-counting device is inserted into the cavity until it has registered (and in the process annihilated) its first photon and that it is immediately withdrawn thereafter, before it has any chance to annihilate another photon. Also, assume that, if no photon has materialized after a certain length of time t_0, this may be viewed as proof that the cavity was indeed empty. Eliminating those cases in which no photon has materialized, what is the expectation value $\langle N \rangle$ for the number of photons left in the cavity for the remaining cases? Evaluate numerically for $\langle N \rangle = 1$.

#1•5-2: Reflections at Semiconductor Hetero-Interfaces

In semiconductor physics it is shown that the reflection of electrons at the interface between two different semiconductors may be *modeled*, to the first order, as the reflection of *free* electron waves at an abrupt potential step, as in Fig. 1•4–1, but with the important difference that the mass used in the calculation is not the mass of a *free* electron, but an effective mass m^*, the value of which depends on the semiconductor. If the effective masses on the two sides of the interface differ from each other, the connection rules at the interface can no longer be the simple pair (1•4–14a,b): Because the effective mass also enters the current density expression (1•5–6), the connection rules (1•4–14) would imply that the probability current density across the interface is not conserved. The simplest and most widely used approximation is to retain the continuity of the effective-mass wave function, but to replace the slope condition (1•3–50) by

$$\frac{1}{m_{-}^{*}}\psi'(-0) = \frac{1}{m_{+}^{*}}\psi'(+0), \tag{1•5–17}$$

where m_{-}^{*} and m_{+}^{*} are the effective masses on the left and right side of the interface.

What are the transmission and reflection coefficients at such an interface? Under which condition will there be no reflection, even for a step of finite height? Express this condition in terms of the electron velocities on both sides of the interface.

#1•5-3: Generation of Single-Photon Pulses

Give a detailed prescription for a hypothetical setup designed to generate photon pulses, each of which contains *exactly* one photon, rather than Poisson-distributed pulses containing an *average* of one photon. As a suggestion, consider the emission of a photon from an excited atom or Compton recoil. Others?

Comment on any limitations to the precision with which the *timing* of an exact one-photon pulse can be specified.

1.6 SUBTLETIES AND REFINEMENTS

1.6.1 Wave Functions as Statistical Generating Functions

Although we have, throughout this chapter, often argued by analogy with classical electromagnetic waves, ultimately these analogies reach an end to their validity. Nothing showed this better than our recognition, in section 1.4, that the absolute frequencies of quantum-mechanical waves are not observable, in contrast to the frequencies of electromagnetic waves. The two electromagnetic fields **E** and **B** are directly measurable quantities—at least from the classical point of view. But the wave functions Ψ have no *directly* measurable physical meaning. They are quantities of a new kind that simply does not occur in classical physics. They contain within themselves the probability distributions of the measurable physical properties of the object, and each of these probability distributions can be extracted from Ψ by certain mathematical operations. But the connection is quite indirect; the simplest quantity extractable from Ψ is the probability density $|\Psi|^2$. We shall see later how to extract other so-called **observables** from the wave function. The procedure in all cases involves both the wave function and its complex conjugate, as well as an appropriate operator that extracts the observable of interest from the wave function.

Mathematical functions that have little or no meaning by themselves, but from which other quantities of interest can be extracted by suitable mathematical operations, are often called **generating functions**. Probably the most familiar examples of pure generating functions in physics are the various partition functions of statistical thermodynamics. Given the partition function *Z(T)* of a thermodynamic system, it is possible to calculate from this function the energy of the system, the various free energies, the entropy, and, by extension, all other quantities that can in turn be derived from these. But the partition function itself has no direct "natural" meaning.

A borderline case between a directly meaningful quantity and a generating function is the magnetic vector potential **A** of classical electromagnetic theory. At least from an elementary point of view, it appears to have no obvious physical meaning by itself; its principal role seems to be to serve as a convenient mathematical repository for both the electric and the magnetic field of an electromagnetic wave, both of which can be extracted from **A** by the operations

$$\mathbf{B} = \nabla \times \mathbf{A}, \qquad \mathbf{E} = -\frac{\partial A}{\partial t}. \tag{1•6–1}$$

In fact, a given set of fields may be described by an infinite set of quite different vector potentials, so long as the curl and the time derivatives of the latter are the same.

From the point of view of generating functions, we may view the wave functions of quantum mechanics as *statistical* generating functions, for *probabilities* rather than deterministic quantities.

1.6.2 Indeterminacy versus Hidden Variables

The probabilistic interpretation of the wave function makes quantum mechanics an inherently statistical theory. Presumably, this is because the laws of nature are themselves partially probabilistic rather than fully deterministic on the level of interactions of elementary objects. Quantum mechanics differs in this respect from classical statistical theories. For example, in the classical kinetic theory of gases, the velocity of any selected gas atom is treated as if it fluctuates randomly with a certain probability distribution. This is simply an approximation, made because the calculation of the exact variation would be both hopelessly difficult—and uninteresting. But nowhere does classical physics assume that the interaction between gas molecules is itself indeterministic. Quantum mechanics is the first theory in the history of modern physics that seriously considers the possibility that the laws of physics might themselves not be perfectly deterministic.

This idea represents an even larger departure from classical physics than wave-particle duality; quite possibly, it is the greatest revolution in post-Galilean science. Ever since Galileo and Newton, it has been considered the essence of physics to reduce all natural phenomena to quantitative cause-and-effect relationships. The acceptance of an indeterministic relationship appeared to many 20th-century scientists and philosophers to be a renunciation of science itself. This includes some of the founding fathers of quantum mechanics, most notably, Einstein. Their view was that quantum mechanics could not be a complete description of nature.

Unfortunately, it is not for man to decide which form the laws of nature must take. The notion that they must be strictly deterministic is itself a hypothesis. Whether or not this hypothesis is correct can be decided only empirically—if at all. Nobody said this better than Feynman, in whose celebrated *Lectures on Physics*[8] the reader can find the following delightful putdown:

> A philosopher once said, "It is necessary for the very existence of science that the same conditions always produce the same results." Well, they don't.

Still, one does not throw more than 300 years of scientific tradition overboard without the most compelling reason. Therefore, ever since the indeterministic aspects of quantum mechanics became appreciated, there have been numerous attempts to reconcile it with strict determinism, by postulating the existence of so-called **hidden variables**, which influence the observations in a strictly deterministic way, but are not under the control of the experimenter and therefore fluctuate from measurement to measurement. However, we shall see later that certain experimentally testable predictions of quantum mechanics are inherently incompatible with *any* deterministic hidden-variable theory, regardless of the nature of the hidden variables. In all those cases, the experimental results are in good agreement with the predictions of quantum mechanics, thus ruling out the existence of hidden variables. We will say more about this in chapter 20.

Denying the existence of hidden variables, orthodox quantum mechanics postulates that the wave function gives a *complete* description of each system, in the sense that the probability distributions of *all* observables can be extracted from Ψ and that properties not extractable from Ψ cannot be observed. Put differently, "The wave function is all there is."

1.6.3 On "Understanding" Quantum Mechanics: Wave-Particle Duality as a Unifying Concept

In his *Lectures on Physics,* Feynman introduced wave-particle duality as "a phenomenon which is impossible, *absolutely* impossible to explain in any classical way." We might add to Feynman's words:

> The reason it is impossible to explain in terms of anything else is because it contains within itself the explanation of everything else.

The point is this: It can never be a meaningful objective to try to "explain" quantum mechanics in terms of classical physics, and we have, in fact, carefully refrained from attempting to do so. The problem of reconciling quantum concepts with classical physics can only mean that we must understand how the

[8] See appendix G.

concepts and laws of classical physics arise as consequences of those of quantum mechanics, and *never* the other way around.

We will frequently do this. In particular, we will be very much interested in understanding how classical Newtonian mechanics arises from quantum mechanics. Qualitatively, the basic idea is exceptionally simple, and it was stated earlier: Classical mechanics is the geometric-optics limit of quantum mechanics. We will see later (chapter 8) how this works out in quantitative detail.

While not easy to visualize, wave-particle duality is actually a unifying concept that simplifies physics in many ways, although the simplification is on a more abstract level. It is only the latest in a long sequence of such unifying concepts, a sequence that began in the 17th century, when Newton subjected astronomy to the laws of physics by postulating that the motion of celestial bodies is governed by the same terrestrial laws that govern the fall of an apple from a tree. The unification process continued, accelerating during the 19th century. Early in that century, heat was recognized as a form of energy rather than as a substance, and toward the end of the century, Boltzmann made thermodynamics a branch of mechanics. Also, early in the 19th century, Oerstedt, Ampere, and Faraday showed that electricity and magnetism were only two aspects of a single phenomenon. In 1864, Maxwell made optics a branch of electromagnetism.

Around the turn of the 20th century, classical physics had been unified to the point where it had only two truly independent branches: classical electromagnetic theory and classical mechanics. These two branches attempted to explain all physical phenomena in terms of the interaction of two totally different classes of *objects*: particles and fields, the latter often occurring in the form of waves.

The first breach into this rigid "wave-particle dichotomy" was made in 1905, when Einstein unified mass and energy in his famous formula $\mathcal{E} = Mc^2$. He assigned what was until then a pure particle property, mass, to *all* forms of energy, including field energy. But this conceptual breach was neither noticed nor pursued further until 18 years later, when de Broglie postulated wave-particle duality.

Each of the historical unification steps was accompanied by an increase in abstractness, the price that had to be paid to achieve the deeper and more unified understanding.

Chapter 2

INTRODUCTION TO BOUND STATES

2.1 SOME GENERAL PRINCIPLES

2.1.1 What Is a Bound State?

In this chapter and the next, we extend our consideration to the **bound states** of selected simple systems in which an external potential confines a single object to—essentially—a finite volume, in such a way that the wave functions vanish at infinity:

$$\Psi(\mathbf{r}) \to 0 \qquad \text{as} \qquad \mathbf{r} \to \infty. \tag{2•1–1}$$

Inasmuch as we are dealing with a single object, we must request that the wave function be normalizable, and we shall usually assume that it is in fact normalized,

$$\int |\Psi|^2 \, d^3r = 1, \tag{2•1–2}$$

where the integral goes over all space.

We will be mostly interested in **stationary** bound states, defined as bound states whose wave functions are of the form (1•4–2),

$$\Psi(\mathbf{r}, t) = \psi(\mathbf{r}) \cdot \exp\left(-\frac{i\mathcal{E}t}{\hbar}\right), \tag{2•1–3}$$

where $\psi(\mathbf{r})$ satisfies the time-*independent* Schroedinger equation (1•4–3),

$$\boxed{-\frac{\hbar^2}{2M}\nabla^2\psi + V(\mathbf{r})\psi = \mathcal{E}\psi,} \tag{2•1–4a}$$

and $\mathcal{E}$ is the energy of the state.

Such states evidently have a time-independent probability density $|\Psi|^2$—hence the term **stationary states**. In fact, even in the absence of a qualifier like "non-stationary," the simple term "bound states" often refers to states that are *both* bound *and* stationary.

Most of our examples in this chapter will be one-dimensional ones, in which case the time-independent Schroedinger Equation (2•1–4a) reduces to the ordinary differential equation

$$-\frac{\hbar^2}{2M}\frac{d^2\psi}{dx^2} + V(x)\psi = \mathcal{E}\psi. \tag{2•1–4b}$$

The central point is now that the boundary condition (2•1–1) can be satisfied only for certain discrete values of the energy $\mathcal{E}$ in the Schroedinger equation (2•1–4a,b). The energy of stationary bound states can have only one of these discrete values. All other values are forbidden; there is no continuous range of allowed energies. The allowed energy values are called the **energy eigenvalues** of the particular potential. The associated wave functions are called **energy eigenfunctions**, and the states themselves are often referred to as **energy eigenstates**, a term synonymous with *stationary states*.

The occurrence of discrete eigenvalues in boundary value problems is well known from other areas of mathematical physics. For example, the electromagnetic oscillations of a cavity with perfectly conducting walls have certain specific discrete values of the oscillation frequency. Quantum mechanics extends this concept to the energies of the stationary bound states. The idea that these energies form a discrete set is one of the central *physical* consequences of quantum mechanics, quite different from the situation in classical dynamics. A large portion of quantum mechanics consists of the study of the properties of these states, which will turn out to form the backbone of the treatment of time-*dependent* states as well.

The restriction to discrete energy eigenvalues does not mean that solutions of the time-independent Schroedinger equation for other energies do not exist; they simply do not describe stationary confined objects.

2.1.2 Degeneracy and Current-Carrying Stationary States

In the special case of a one-dimensional problem, the associated Schroedinger equation (2•1–4b) is an ordinary (rather than partial) differential equation, which has only two linearly independent solutions. It is easily shown that, if one of these solutions corresponds to a bound state, the other solution cannot vanish at infinity and hence cannot represent a bound state. Thus, a one-dimensional system can have *at most* one bound state for any value of the energy. We say that the energy eigenvalues of one-dimensional problems are **non-degenerate**. Note that, with the Schroedinger equation (2•1–4b) being real, the stationary-state wave functions $\psi(x)$ may always be chosen to be real.

Exercise: Let ψ_1 and ψ_2 be two linearly independent solutions of the one-dimensional Schroedinger equation (2•1–4b). Consider the quantity

$$W \equiv \psi_1' \cdot \psi_2 - \psi_2' \cdot \psi_1, \tag{2•1–5}$$

called the **Wronskian** in the theory of second-order linear differential equations. Show that W must be a position-independent nonzero constant and that this implies the above claims that one-dimensional bound states cannot be degenerate.

In two- or three-dimensional systems, the time-independent Schroedinger equation becomes a partial differential equation, which is not subject to the restriction to only two linearly independent solutions. The energy eigenvalue problem (2•1–4a) may then have more than one (linearly independent) solution with the same energy eigenvalue. When this is the case, we say that the energy eigenvalue is **degenerate**; we shall find that degeneracies are common.

Any linear superposition of two or more linearly independent eigenfunctions belonging to the same energy eigenvalue is again an energy eigenfunction with that eigenvalue. As a result, the set of eigenfunctions belonging to a given energy eigenvalue is not unique: From a given set, it is always possible to construct a new set, by linear superposition of functions from the original set. In particular, it is always possible to construct complex superpositions that correspond to a nonzero probability *current* density

$$\mathbf{j} = -\frac{i\hbar}{2M}(\psi^* \nabla \psi - \psi \nabla \psi^*), \tag{2•1–6}$$

from (1•5–6). Stationarity does not the imply absence of currents; according to the continuity equation (1•5–4), it only means that any current flow must be divergence free,

$$\nabla \cdot \mathbf{j} = 0, \tag{2•1–7}$$

or else the probability density $\rho = |\Psi|^2$ would be non-stationary.

In *one* dimension, (2•1–7) reduces to the condition that the probability current density must be position-independent, and inasmuch as it must be zero at infinity, it must be zero everywhere. But no such restriction holds in two or three dimensions, where even a stationary bound state may exhibit closed, divergence-free current loops. In fact, such states are of great importance in many problems.

Given a current-carrying stationary bound state ψ, this state must in turn belong to a degenerate energy eigenvalue. From (2•1–6) it follows that a nonzero current density requires a wave function that is inherently complex, with a position-dependent phase. Inasmuch as the Schroedinger equation in the form (2•1–4) is a *real* equation, ψ^* must also be an energy eigenfunction along with ψ, corresponding to a reversed current density and clearly linearly independent of ψ. We will frequently encounter such complex conjugate degenerate pairs. Both the real and the imaginary parts of ψ and ψ^* must then be solutions of the Schroedinger equation by themselves, with neither part corresponding to any current.

2.1.3 Orthogonality

One of the most important properties of energy eigenfunctions is their mutual **orthogonality**: Two wave functions ψ_1 and ψ_2 are said to be **orthogonal** when they satisfy the condition

$$\boxed{\int \psi_1^* \psi_2 \, d^3r = 0,} \tag{2•1–8}$$

where the integral again goes over all space, as in (2•1–2). We claim that two energy eigenfunctions ψ_1 and ψ_2 belonging to *different* energy eigenvalues $\mathcal{E}_1$ and $\mathcal{E}_2$ are necessarily orthogonal.

To show this, we write down the Schroedinger equation for ψ_2, multiply it with the complex conjugate of ψ_1, and integrate over all space:

$$-\frac{\hbar^2}{2M}\int \psi_1^* \nabla^2 \psi_2 \, d^3r + \int \psi_1^* V(\mathbf{r}) \psi_2 \, d^3r = \mathcal{E}_2 \int \psi_1^* \psi_2 \, d^3r. \tag{2•1–9a}$$

Similarly, starting with the Schroedinger equation for ψ_1^*, multiplying with ψ_2, and integrating yields:

$$-\frac{\hbar^2}{2M}\int \psi_2 \nabla^2 \psi_1^* d^3r + \int \psi_2 V(\mathbf{r}) \psi_1^* \, d^3r = \mathcal{E}_1 \int \psi_2 \psi_1^* \, d^3r. \tag{2•1–9b}$$

The two potential energy integrals on the left-hand sides are evidently equal to each other. With the help of the boundary condition (2•1–1), it is not difficult to show that the two kinetic energy integrals are also equal to each other. Hence, if we subtract (2•1–9b) from (2•1–9a), we obtain

$$0 = (\mathcal{E}_2 - \mathcal{E}_1)\int \psi_1^* \psi_2 \, d^3r. \tag{2•1–10}$$

But for $\mathcal{E}_1 \neq \mathcal{E}_2$, this can be true only if (2•1–8) is satisfied.

The foregoing proof does not apply to different linearly independent eigenfunctions belonging to the same degenerate energy eigenvalue. Such functions need not be orthogonal to each other. However, in such a case, it is always possible to form a new set, by linear superpositions of the original functions, that *is* orthogonal. For example, let ψ_1 and ψ_2 be two linearly independent *real* functions that are *not* orthogonal. It is a simple exercise to show that the two replacement functions

$$\psi_+ = \psi_1 + \psi_2 \quad \text{and} \quad \psi_- = \psi_1 - \psi_2 \tag{2•1–11}$$

are orthogonal. Procedures for orthogonalizing arbitrarily large sets of linearly independent functions are given in the mathematical literature, as well as in some quantum mechanics texts.

If the set of orthogonal eigenfunctions has also been normalized, it is referred to as an **orthonormal** set.

2.1.4 Inversion Symmetry and Parity

Many of the potentials of interest in quantum mechanics do not change if all coordinate directions are inverted,

$$V(-\mathbf{r}) = V(\mathbf{r}). \tag{2•1–12}$$

We say that the potential exhibits **inversion symmetry** and that the coordinate origin $\mathbf{r} = 0$ acts as a **center of inversion**.

Given inversion symmetry, if $\psi(\mathbf{r})$ is an energy eigenfunction, then $\psi(-\mathbf{r})$ must also be an energy eigenfunction with the same eigenvalue. If the energy eigenstate in question is nondegenerate, then the two functions can differ from one another only by a factor ± 1. Put differently, *nondegenerate* energy eigenfunctions must be either even or odd functions about the center of inversion,

$$\psi(-\mathbf{r}) = \pm\psi(\mathbf{r}). \tag{2•1–13}$$

We say the wave functions have even or odd **parity**.

If the energy eigenvalue is degenerate, the different eigenfunctions need not automatically satisfy (2•1–13), but they can always be rearranged in such a way that they do. Let $\psi(\mathbf{r})$ be a function that does *not* obey the parity requirement (2•1–13). In this case, $\psi(\mathbf{r})$ and $\psi(-\mathbf{r})$ are linearly independent, and inspection shows that the two new functions

$$\psi_+(\mathbf{r}) = \psi(\mathbf{r}) + \psi(-\mathbf{r}) \quad \text{and} \quad \psi_-(\mathbf{r}) = \psi(\mathbf{r}) - \psi(-\mathbf{r}) \tag{2•1–14}$$

constructed from the original pair are of even and odd parity. As a rule, wave functions are chosen to obey parity symmetry unless there is a specific reason not to do so.

Note that wave functions of opposite parity are automatically orthogonal to each other.

2.2 PARTICLE IN A SQUARE WELL

2.2.1 One-Dimensional Well with Infinitely High Barriers

The simplest possible bound-state problem is a particle confined to a one-dimensional square well—a "box"—of width L with infinitely high walls, but free to move inside the well (**Fig. 2•2–1**):

We have placed the coordinate origin at the left edge of the well,[1] in which case we may write

$$V(x) = \begin{cases} 0 & \text{for } 0 < x < L, \\ \infty & \text{elsewhere.} \end{cases} \tag{2•2–1}$$

Inside the well we have essentially a free-particle problem, with wave functions of the form

$$\psi(x) = Ae^{+ikx} + Be^{-ikx}, \tag{2•2–2}$$

where

$$\mathcal{E} = \frac{\hbar^2 k^2}{2M} \tag{2•2–3}$$

is the corresponding energy.

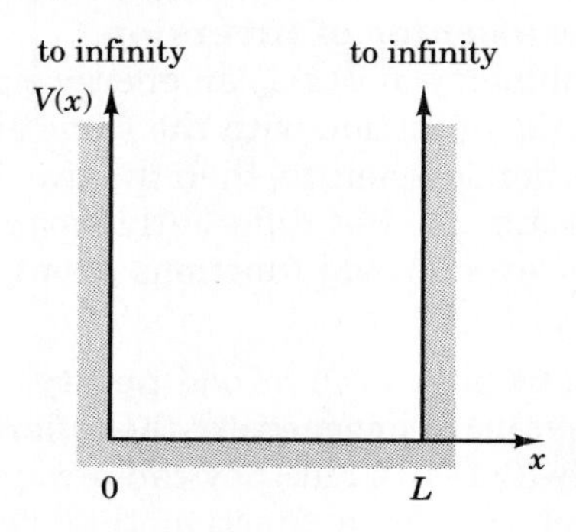

Figure 2•2–1. One-dimensional square well.

[1] The one-dimensional square well with *infinitely* high walls has a center of inversion at $x = L/2$, but is so simple that the benefits of placing the origin at the center are outweighed by the need to keep track of the even- and odd-parity wave functions separately. It is suggested as an exercise that the reader repeat our treatment with a coordinate origin placed at the center of the well.

Outside the well, the wave function must be zero, or else the term $V\psi$ in the Schroedinger equation would be infinite. Because the wave function cannot have any discontinuities, it must vanish at both edges of the well,

$$\psi(0) = \psi(L) = 0. \tag{2•2–4}$$

The boundary condition at $x = 0$ requires that in (2•2–2) we have $B = -A$, leading to eigenfunctions of the form

$$\psi(x) = C \sin kx. \qquad (C = 2iA) \tag{2•2–5}$$

To satisfy the boundary condition at $x = L$, the wave number k must satisfy

$$kL = n\pi, \qquad n = 1, 2, 3, \ldots \tag{2•2–6}$$

The normalization condition (2•1–2) will be met if the amplitude is selected as

$$C = \sqrt{2/L}. \tag{2•2–7}$$

The first three eigenfunctions are shown in **Fig. 2•2–2**.

The integers n in (2•2–6), used to identify and label the different allowed states, are called the **quantum numbers** of the problems. In our case they happen to be integers starting with $n = 1$, but this is simply the way they appeared naturally in our counting scheme. In other problems, different sets of identifiers might be preferable.

Because of (2•2–6), the energy eigenvalues of the stationary states are restricted to the discrete set

$$\mathcal{E}_n = \frac{\hbar^2\pi^2}{2ML^2} \cdot n^2. \tag{2•2–8}$$

To appreciate the transition from quantum mechanics to classical physics, it is instructive to insert numbers into (2•2–8). All energies are integer

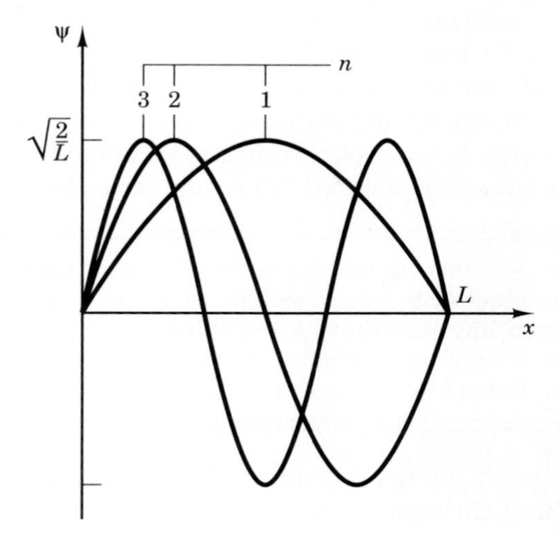

Figure 2•2–2. The three lowest energy eigenfunctions of a particle in a one-dimensional square well with infinitely high walls.

multiples of the energy of the lowest state,

$$\mathcal{E}_1 = \frac{\hbar^2 \pi^2}{2ML^2}, \tag{2•2–9}$$

which is, in effect, the **natural unit of energy** for the system. Typical atomic dimensions are on the order of a few times 10^{-10} m. If we insert, say, $L = 2 \times 10^{-10}$ m, and, for M, the mass m_e of an electron, we obtain $\mathcal{E}_1 \approx 1.5 \times 10^{-18}$ J ≈ 9.4 eV, on the order of typical atomic energies. This is a large energy, making the discreteness of atomic energy levels readily observable experimentally. But for macroscopic dimensions, say, $L = 1$ cm, we obtain $\mathcal{E}_1 \approx 3.8 \times 10^{-15}$ eV, an energy far too small to be measurable by any known means explaining why the energies of macroscopic systems *appear* to be continuously variable.

Zero-Point Energy

The lowest energy state has an energy above the minimum of the *potential* energy, in striking contrast to classical mechanics, where the lowest possible *total* energy coincides with the minimum *potential* energy, with zero *kinetic* energy. The quantum-mechanical excess energy is called the **zero-point energy**; it occurs in all quantum-mechanical bound-state problems, and it has far-reaching consequences. For example, without it, the electrons in atoms would fall into the nuclei; the zero-point energy is therefore responsible for the existence of stable atoms. We will return to this point later.

The origin of the zero-point energy is easily understood: An object confined to a finite space requires a wave function containing contributions with nonzero local wave numbers K. These make nonzero contributions to the internal kinetic energy of the state. In the case of an object confined by two infinitely high walls, the wave function must go to zero at the walls. The smallest kinetic energy is evidently obtained if we pick as a wave function a single standing wave with the longest possible wavelength such that the wave will fit into the well, $\lambda = 2L$. But this is exactly the state ψ_1, with energy $\mathcal{E}_1$. Any contributions to the wave function from standing waves of shorter wavelength would have raised the energy (quite apart from the fact that such superpositions would not obey the time-independent Schroedinger equation).

Exercise: According to (2•1–8), the eigenfunctions of all higher energy eigenstates must be orthogonal to the ground-state wave function. Show that the energy $\mathcal{E}_2$ *is the lowest possible* kinetic energy of any wave function (including any superposition of plane waves) that vanishes at the walls of the well and that, *in addition*, is orthogonal to the ground-state wave function. Extend the argument to higher energy states.

Note that the zero-point energy is *internal* kinetic energy, with no net *external* motion of the object through space. The standing wave that makes up the net wave function is a superposition of two counter-propagating traveling waves, each with the same kinetic energy, but opposite momentum. When we superimpose the two waves to meet the boundary conditions (2•2–4), the momentum contributions cancel, but the energies do not. One might be tempted to compare the ground state of our system to the bouncing back-and-forth of a confined particle in classical mechanics. However, this analogy misses the central point, that the probability density $|\Psi|^2$ of the lowest energy state is *stationary*, rather than showing any oscillation. Solutions of the time-*dependent* Schroedinger wave equation corresponding to bouncing states exist, but those states are never states of the lowest possible energy.

2.2.2 Three-Dimensional Well

We generalize the previous discussion to a particle confined to a three-dimensional cube-shaped square well,

$$V(x, y, z) = \begin{cases} 0 & \text{for } 0 < x, y, z < L, \\ \infty & \text{elsewhere.} \end{cases} \tag{2•2–10}$$

Inside the well, the energy eigenfunctions must be linear superpositions of plane waves. At the surface of the well, the eigenfunctions must again vanish. That is,

$$\psi(0, y, z) = \psi(x, 0, z) = \psi(x, y, 0) = 0, \tag{2•2–11a}$$

$$\psi(L, y, z) = \psi(x, L, z) = \psi(x, y, L) = 0, \tag{2•2–11b}$$

for *all* x, y, and z in the interval $(0, L)$. These boundary conditions sharply restrict the admissible superpositions. One finds easily that, in order to meet (2•2–11a), all wave functions must *either* be of the form

$$\psi(x, y, z) = A(\mathbf{k}) \sin(k_x x) \sin(k_y y) \sin(k_z z), \tag{2•2–12}$$

or they must be linear superpositions of such terms with different $\mathbf{k}$, such that $k^2 = k_x^2 + k_y^2 + k_z^2$ has the same value for every $\mathbf{k}$ in the superposition. The corresponding energies are

$$\mathcal{E} = \frac{\hbar^2 k^2}{2M} = \frac{\hbar^2}{2M}(k_x^2 + k_y^2 + k_z^2). \tag{2•2–13}$$

By applying the second set of boundary conditions, (2•2–11b), one finds that the $\mathbf{k}$ are further restricted to values that satisfy

$$k_x L = n_x \pi, \qquad k_y L = n_y \pi, \qquad k_z L = n_z \pi, \tag{2•2–14a,b,c}$$

where n_x, n_y, and n_z are three independent positive integers. The corresponding energy eigenvalues are

$$\mathcal{E}(n_x, n_y, n_z) = \frac{\hbar^2 \pi^2}{2ML^2} \cdot (n_x^2 + n_y^2 + n_z^2) = \mathcal{E}_1 n^2, \tag{2•2–15}$$

where $\mathcal{E}_1$ is again the one-dimensional ground-state energy, as in (2•2–9), and where

$$n^2 = n_x^2 + n_y^2 + n_z^2. \qquad (2•2–16)$$

Compared to the one-dimensional case, each state is now associated with three identifying quantum numbers rather than just one. This is typical of all three-dimensional single-particle problems; we will later add a fourth quantum number for the electron spin.

Most of the energy eigenvalues in (2•2–15) are degenerate. If the three quantum numbers differ from each other, then each of the $3! = 6$ possible permutations of the three numbers corresponds to a different (i.e., linearly independent) three-dimensional eigenfunction and, therefore, to a different state. The simplest example is

$$\begin{aligned}(n_x, n_y, n_z) &= (1, 2, 3); (1, 3, 2); (2, 1, 3);\\ &= (2, 3, 1); (3, 1, 2); (3, 2, 1).\end{aligned}$$

All six of these states have the same energy eigenvalue $\mathcal{E}(1, 2, 3) = 14\mathcal{E}_1$.

Because there are no other states with this eigenvalue, this energy level is six-fold degenerate. If two of the quantum numbers are the same, and the third is different, there will still be three different states.

Part of this six-fold degeneracy that is associated with a permutation of three different quantum numbers occurs because the three coordinate axes are completely equivalent. Such degeneracies are therefore called **symmetry degeneracies**. However, we shall see later, in chapter 16, that this would lead to at most a three-fold degeneracy and that part of the six-fold degeneracy is **accidental**, in the sense that it is due to the extreme simplicity of the potential and would not occur for a less trivial potential (such as a cubic well with walls of finite height). There are numerous additional accidental degeneracies. Take the energy level $\mathcal{E} = 27\mathcal{E}_1$. The factor 27 can be obtained in two different ways:

$$1^2 + 1^2 + 5^2 = 27 \quad \text{and} \quad 3^2 + 3^2 + 3^2 = 27.$$

Thus, there are four different states leading to this energy level: three with the quantum numbers (1, 1, 5) and one with (3, 3, 3). Those two sets are unrelated, and the degeneracy between the two sets is again accidental, due once more to the extreme simplicity of the potential in a cube-shaped well. Deforming the potential in any way will usually split the accidental degeneracy, but if the deformation retains the equivalence of the three Cartesian axes, the symmetry degeneracies will remain.

Exercise: Find all cubic box states with the energy level $\mathcal{E} = 54\mathcal{E}_0$. Investigate the nature of the degeneracies of this level.

The number of additional degeneracies increases drastically with increasing energy. The degree of degeneracy does not follow any simple pattern. However, in the limit $n^2 \gg 1$, it is possible to estimate the *average* degeneracy of the energy levels in the vicinity of a specified energy. We view n_x, n_y, and n_z as the components of a vector $\mathbf{n}$ in a space whose Cartesian coordinates are n_x, n_y, and n_z. Each state corresponds to a point with positive integer coordinates in this space, and the volume per state is 1. This is illustrated in **Fig. 2•2–3**, for two rather than three dimensions.

According to (2•2–15), the number of states below a certain energy is equal to the number of states inside a sphere of radius n in this space. The volume of such a sphere would be $4\pi n^3/3$ if both positive and negative values of the components of $\mathbf{n}$ were counted. However, negative components produce only an irrelevant sign change, rather than additional states; hence, we must count only the states in the positive octant of the sphere. With a volume of unity per state, the total number N of states in that octant in the limit $\mathcal{E} \gg \mathcal{E}_1$ is simply the volume of the octant,

$$N = \frac{1}{8} \cdot \frac{4\pi}{3} n^3 = \frac{\pi}{6}\left(\frac{\mathcal{E}}{\mathcal{E}_1}\right)^{3/2}. \qquad (2•2–17)$$

provided we neglect rounding errors at the octant boundaries.

Because all possible energy values occur in steps of $\mathcal{E}_1$, the *average* degeneracy of the energy levels near $\mathcal{E}$ is equal to the increase in N if the energy $\mathcal{E}$ is increased by $\mathcal{E}_1$:

$$\Delta N \cong \frac{dN}{d\mathcal{E}} \mathcal{E}_1 = \frac{\pi}{4}\left(\frac{\mathcal{E}}{\mathcal{E}_1}\right)^{1/2} = \frac{\pi}{4} n. \qquad (2•2–18)$$

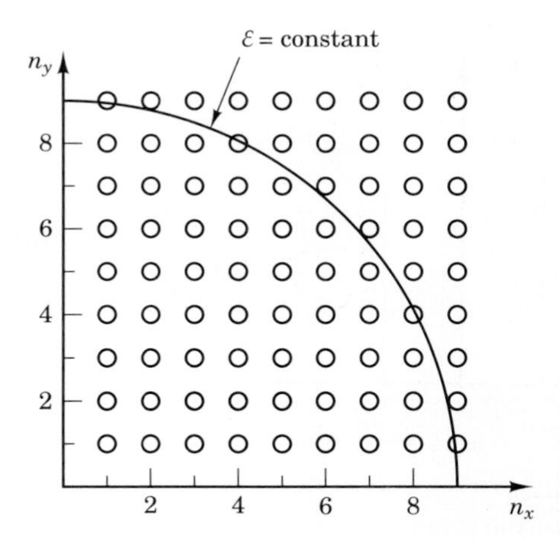

Figure 2•2–3. The states of a particle in a square well, in a space with the coordinates n_x and n_y, in two dimensions.

Evidently, the average degeneracy increases with increasing energy. For macroscopic-sized wells with their very small value of $\mathcal{E}_1$, the degeneracy at macroscopic energies (i.e., large n) may be huge.

The **density of states** $D(\mathcal{E})$ is defined as the number of states per unit energy interval in the vicinity of the energy $\mathcal{E}$, *and* per unit volume of the well:

$$D(\mathcal{E}) = \frac{1}{L^3}\frac{dN}{d\mathcal{E}} = \frac{\pi}{4L^3}\frac{\mathcal{E}^{1/2}}{\mathcal{E}_1^{3/2}} = \frac{1}{4\pi^2}\left(\frac{2M}{\hbar^2}\right)^{3/2}\mathcal{E}^{1/2}. \qquad (2{\bullet}2\text{–}19)$$

Here L^3 is the volume of the well. Note that $D(\mathcal{E})$ is a density in a two-fold sense: both per unit energy *and* per unit volume.[2] Also, note that we have obtained a volume-independent final result.

The density of states plays an important role in the quantum mechanics of gases and of free electrons in metals and semiconductors.

2.2.3 Well with Finite Barrier Height

Tunneling into the Barrier

The assumption of infinitely high walls for a square well is an oversimplification that does not occur in nature. A more realistic model of an actual potential well would be one with walls of height V_0, as shown in **Fig. 2•2–4** for a simple symmetric one-dimensional well. In fact, wells of this simple form play an important role in modern semiconductor technology. Note that we are now placing the coordinate origin in the center of the well, to utilize the inversion symmetry present in the problem.

As we pointed out in section 1.5, the finite barrier height leads to tunneling into the barrier. The appropriate solutions of the Schroedinger equation for

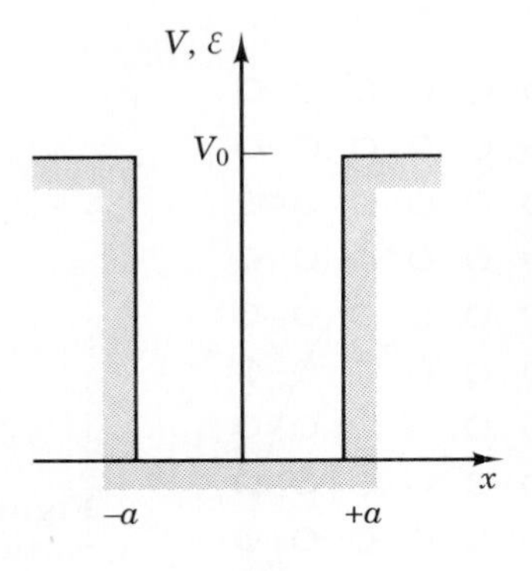

Figure 2•2–4. Symmetric potential well with finite barrier height V_0. Compared to the well with infinitely high walls, we have placed the coordinate origin in the center of the well and have expressed the overall well width as $L = 2a$.

[2] Some authors define the density of states not as a "double density," per unit energy interval *and* per unit volume, but simply per unit energy interval, obtaining a volume-proportional result.

$\mathcal{E} < V_0$ are exponentials that are evanescent to the left or to the right. In the region $x > a$, we have

$$\psi(x) = C \exp(-\kappa x), \tag{2•2–20a}$$

where, instead of (2•2–3), we have

$$V_0 - \mathcal{E} = \frac{\hbar^2 \kappa^2}{2M}. \tag{2•2–21}$$

Similarly, in the region $x < a$,

$$\psi(x) = D \exp(+\kappa x). \tag{2•2–20b}$$

Energy Eigenvalues

With the potential being symmetric about the center of the well,

$$V(-x) = V(x), \tag{2•2–22}$$

the energy eigenfunctions must be either even or odd functions about this center,

$$\psi(-x) = \pm\psi(x). \tag{2•2–23}$$

Evidently, the even- and odd-parity solutions of the Schroedinger equation *inside* the well are of the forms

$$\psi(x) = A \cos kx \qquad \text{and} \qquad \psi(x) = A \sin kx, \tag{2•2–24}$$

where k continues to be related to the energy via (2•2–3). The quantities k and κ are not independent of each other, but are inter-related via

$$k^2 + \kappa^2 = k_0^2 \equiv \frac{2MV_0}{\hbar^2}. \tag{2•2–25}$$

The wave function inside the well no longer vanishes at the wall, but must be connected smoothly to the outside wave function, with continuous magnitude and slope, as in the case of a simple step discussed in section 1.5. If we express the well width as $L = 2a$, we have, analogously to (1•4–14a,b),

$$\psi_-(\pm a) = \psi_+(\pm a), \qquad \psi_-'(\pm a) = \psi_+'(\pm a). \tag{2•2–26a,b}$$

Although those connection rules were derived in section 1.4 in the context of a wave with an energy larger than height of the barrier, there was nothing in the derivation that relied on this context, and the rules apply for arbitrary energies.

Evidently, the bound-state wave functions are of the shape shown in **Fig. 2•2–5**. We consider only the connection at $x = +a$; that at $x = -a$ follows by parity. It is left to the reader to show that the wave function matching conditions (2•2–26) lead to the requirements

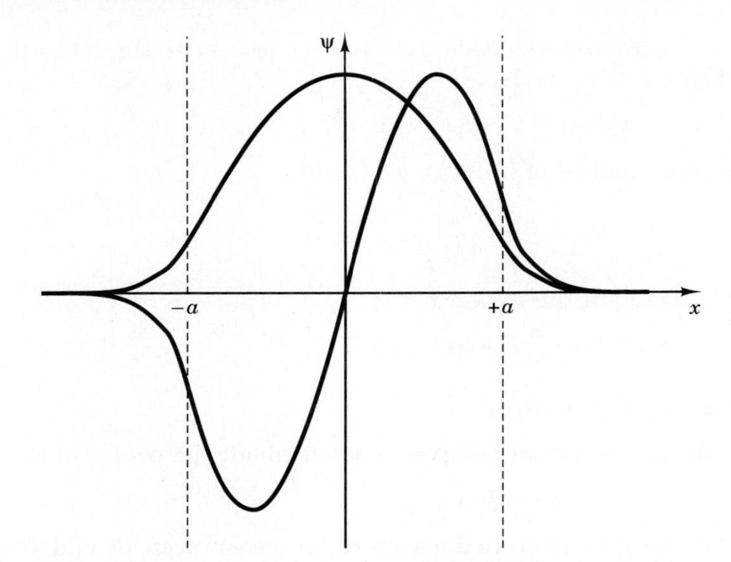

Figure 2•2–5. Eigenfunctions of the two lowest bound states in a well with finite barrier height.

$$\kappa a = \left[\sqrt{k_0^2 - k^2}\right] \cdot a = +ka \cdot \tan ka \quad \text{(even parity)} \qquad (2•2–27a)$$

and

$$\kappa a = \left[\sqrt{k_0^2 - k^2}\right] \cdot a = -ka \cdot \cot ka. \quad \text{(odd parity)} \qquad (2•2–27b)$$

With the help of elementary trigonometric identities, these conditions may be simplified to

$$\cos ka = \pm ka/k_0 a \quad \text{(even parity)} \qquad (2•2–28a)$$

and

$$\sin ka = \pm ka/k_0 a \quad \text{(odd parity).} \qquad (2•2–28b)$$

The left-hand sides of (2•2–28a,b) are basically the wave functions themselves, but viewed as functions of the phase angle $\phi = ka$ rather than of the position coordinate x for fixed k. Hence, the conditions (2•2–28) imply that energy eigenstates occur for those energies for which the wave functions, taken at the edge of the well, intersect one or the other of the two straight lines $\pm ka/k_0 a$, as illustrated in **Fig. 2•2–6**.

However, care is in order: For a bound state, the wave function at the entrance into the barrier *must* slope *toward* the horizontal axis, rather than away from it. Hence, only those intersections shown in Fig. 2•2–6 correspond

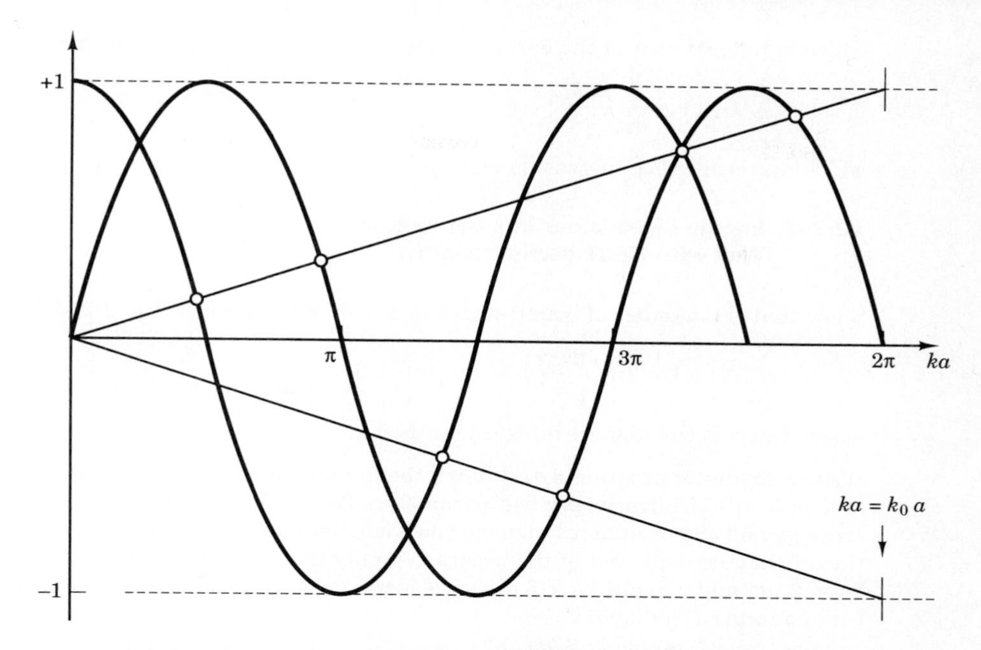

Figure 2•2–6. The conditions (2•2–28a) and (2•2–28b) for the energy eigenvalues of a particle in a potential well with finite barrier height. The plot is for a combination of barrier height and width such that $k_0a = 3\pi$, which evidently yields three even-parity and three odd-parity states.

to admissible solutions for which the sine or cosine functions slope toward the axis at the point of intersection with one of the two straight lines.

Exercise. Show that the intersections omitted from Fig. 2•2–6 do *not* in fact correspond to solutions of the *original* conditions (2•2–27a,b), but lead to a sign conflict if inserted into those relations.

The conditions (2•2–28) cannot be solved in closed form, but are easily solved numerically (see "Problems," next).

◆ PROBLEMS TO SECTION 2.2

#2•2–1: Density of States for Reduced Dimensionality

Redo the derivation of the density-of-states function $D(\mathcal{E})$ for a particle confined to a two- or one-dimensional square well. Show that, for two degrees of freedom,

$$D_2(\mathcal{E}) = \frac{M}{2\pi\hbar^2}, \qquad (2\text{•}2\text{–}29)$$

which is independent of the energy. Show that for only one degree of freedom,

$$D_1(\mathcal{E}) = \frac{1}{4\pi}\left(\frac{2M}{\hbar^2}\right)^{1/2} \mathcal{E}^{-1/2}, \tag{2•2–30}$$

which decreases with increasing energy.

#2•2–2: Energy Eigenvalues in a Symmetric Well with Finite Barrier Height

Show that the number of bound states in a well of the form of Fig. 2•2–4 is

$$N = 1 + \mathrm{Int}\left(\frac{2k_0a}{\pi}\right) = 1 + \mathrm{Int}\left(1.63 \cdot \frac{2a}{1\ \mathrm{nm}} \cdot \sqrt{\frac{M}{m_e} \cdot \frac{V_0}{1\ e\mathrm{V}}}\right), \tag{2•2–31}$$

where Int(x) is the highest integer less than x.

Write a computer program determining the numerical values of all energy eigenvalues of a well with arbitrary specified parameters $2a$ and V_0. You may use any numerical strategy and any computer language you wish, but must give a detailed verbal description of that strategy and of the program. Apply the program to an electron well with $2a = 2$ nm and a depth $V_0 = 5$ eV; give energy eigenvalues in eV, with an uncertainty $|\Delta\mathcal{E}|$ no larger than $0.001V_0$.

#2•2–3: Bound State Expulsion from an Unsymmetric Well

The potential well of **Fig. 2•2–7** has an infinitely high wall on one side, and a flat potential on the other.

If such a well is too shallow or too narrow, it will not have any bound states. Make an accurate sketch of the shape of the wave function in the limit of vanishing binding energy. (Ignore any normalization problems.) Reduce that limit to the problem of an equivalent symmetric square well with infinite barrier height, and give the necessary and sufficient conditions, in terms of V_0 and w, under which the well has at least one

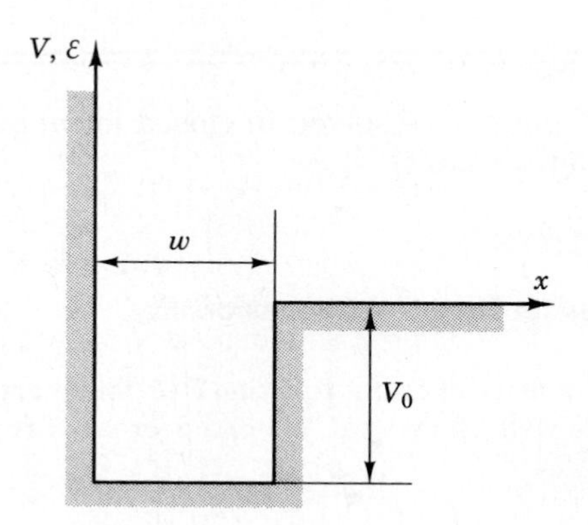

Figure 2•2–7. Unsymmetric potential well.

bound state. Give a numerical value for the minimum width w that an electron well with depth $V_0 = 10$ eV must have so that it has at least one bound state.

#2•2–4: Bound State of a Dirac Delta Function Well

A limiting case of a deep, but narrow, potential well is a δ-function well (**Fig. 2•2–8**), with a potential energy given by

$$V(x) = -S\delta(x), \tag{2•2–32}$$

where $S(>0)$ is the "strength" of the well and $\delta(x)$ is the Dirac δ-function (see appendix A).

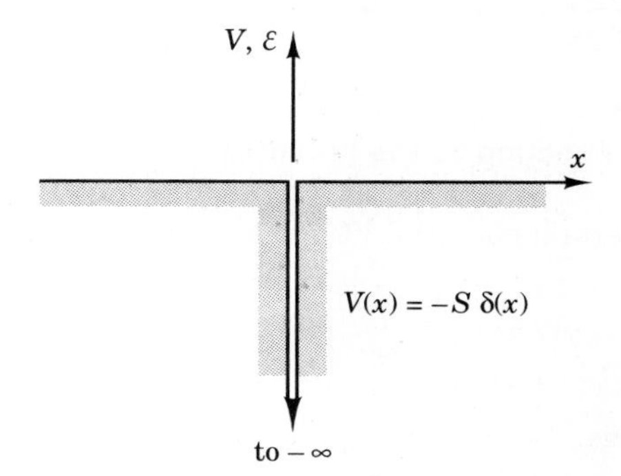

Figure 2•2–8. A Dirac δ-function well.

With the help of the wave function connection rule (1•4–35) across a δ-function potential, show that a single δ-function well always has *exactly* one bound state, regardless of the strength of the well. Give closed-form expressions for the energy of that state and for its (normalized) energy eigenfunction. Make a sketch of that function. Also, give a closed-form relation for the distance over which the wave functions falls off by e^{-1}.

The δ-function well is a useful first-order approximation for any narrow potential well that has only a single bound state. Atomic potential wells are on the order of $1\ \text{Å} = 10^{-10}$ m wide and a few eV deep. Comment on the applicability of this approximation for a well with $S = 10^{-10}\ \text{m} \times 10\ \text{eV} = 1.6 \times 10^{-28}\ \text{m} \cdot \text{J}$.

#2•2–5: Well with Added Delta Barriers

Consider a particle in a symmetric square well with barriers of finite height, as in **Fig. 2•2–9**. Suppose δ-function barriers of strength $S > 0$ are added right at the potential steps. Suppose also that the well is sufficiently shallow or narrow that, in the absence of the δ-barriers, it would have only one bound state. With increasing strength S of the δ-barriers, the bound-state energy gets pushed up by the barriers, and at a certain maximal strength S_{max}, the bound state gets pushed out of the well. Make a graph of the qualitative shape of the ground-state wave function, for two or more values of $S < S_{max}$, including the limit $S \to S_{max}$, showing the effect of the added δ-barriers, but ignoring normalization problems in that limit. Drawing on the plot, give a closed-form expression for S_{max}, as a function of a, V_0, and whatever other parameters are relevant.

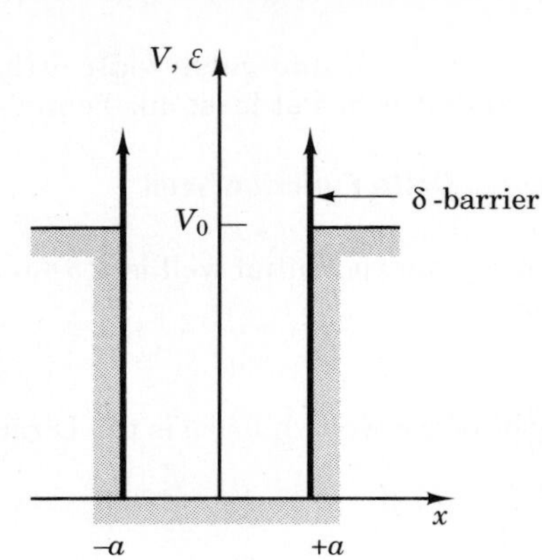

Figure 2•2–9. Symmetric potential well with finite barrier height V_0, with delta-function barriers added to the potential steps.

#2•2–6: From the Wave Function to the Potential

A particle confined by a certain potential $V(x)$ has the very peculiar ground-state wave function

$$\psi(x) = \begin{Bmatrix} A \cdot \sin(\pi x/2a) \\ 2A \cdot (1 - x/2a) \end{Bmatrix} \quad \text{for} \quad \begin{Bmatrix} 0 < x < a \\ a < x < 2a \end{Bmatrix} \tag{2•2–33}$$

and zero outside this range.

Make a plot of the wave function. Determine the potential $V(x)$ in which the particle moves, and the energy of the state relative to that potential.[3]

2.3 HARMONIC OSCILLATOR

2.3.1 Ground State

There are many physical systems that are capable of purely sinusoidal oscillations with an amplitude-independent frequency. Most electromagnetic oscillations are of this kind, as are the oscillations of some idealized mechanical systems, including the motion of an otherwise free charged particle in a magnetic field. Collectively, these systems are called **harmonic oscillators**; they form one of the most important classes of physical systems. Many additional systems can be approximated to a first order as harmonic oscillators.

The prototype of a harmonic oscillator is a Newtonian particle of mass M, bound to its equilibrium position by an elastic restoring force proportional to the displacement x of the particle,

$$F = -Kx, \tag{2•3–1}$$

[3] The text by French and Taylor, cited in Appendix G, contains an excellent selection of homework problems relating simple wave functions to simple potentials and vice versa.

where K is the "spring constant" of the binding force. In classical mechanics, such a particle undergoes oscillations at the frequency

$$\omega = (K/M)^{1/2}, \tag{2•3–2}$$

and it is useful to characterize the oscillator in terms of this frequency rather than in terms of the purely mechanical spring constant.

The potential energy corresponding to the force (2•3–1) is

$$V(x) = \frac{1}{2}Kx^2 = \frac{1}{2}M\omega^2 x^2, \tag{2•3–3}$$

a parabolic potential (**Fig. 2•3–1**). It leads to the Schroedinger equation

$$\boxed{-\frac{\hbar^2}{2M}\frac{d^2\psi}{dx^2} + \frac{1}{2}M\omega^2\psi = \mathcal{E}\psi.} \tag{2•3–4}$$

The solution of this eigenvalue problem is the task of this section.

The occurrence of a term $x^2\psi$ in (2•3–4) suggests trying a solution in the form of a Gaussian, which may always be written

$$\psi_0(x) = A_0 \exp(-x^2/2L^2), \tag{2•3–5}$$

where L is a characteristic length. For later convenience, we have split off a factor 2 from L^2 in the exponent. From (2•3–5),

$$-\frac{\hbar^2}{2M}\frac{d^2\psi_0}{dx^2} = -\frac{\hbar^2}{2ML^4}(x^2 - L^2)\psi_0. \tag{2•3–6}$$

The $x^2\psi$-term in (2•3–6) will cancel the potential energy term in (2•3–4) if L^2 is selected according to

$$L^2 = \hbar/M\omega, \text{ or } L = (\hbar/M\omega)^{1/2}. \tag{2•3–7}$$

The energy $\mathcal{E}$ follows by equating $\mathcal{E}\psi$ in (2•3–4) to the remainder in (2•3–6). We obtain

$$\mathcal{E} = \mathcal{E}_0 \equiv \hbar^2/2ML^2 = \hbar\omega/2. \tag{2•3–8}$$

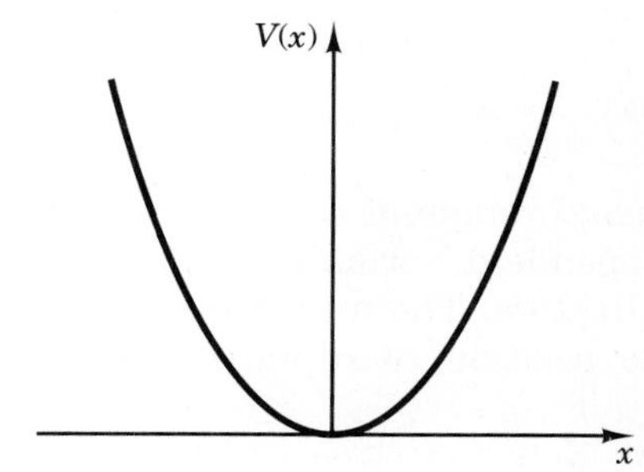

Figure 2•3–1. Harmonic oscillator potential.

Evidently, if L is chosen as in (2•3–7), the Gaussian (2•3–5) is indeed a bound-state solution of (2•3–4), with the energy $\mathcal{E} = \frac{1}{2}\hbar\omega$. We shall see shortly that the state ψ_0 thus defined is in fact the lowest energy bound state of the harmonic oscillator.

It is left to the reader to show that the wave function ψ_0 will be normalized,

$$\int_{-\infty}^{+\infty} |\psi_0|^2 dx = 1, \tag{2•3–9}$$

if we set

$$A_0{}^2 = 1/(L\sqrt{\pi}) = (M\omega/\pi\hbar)^{1/2}. \tag{2•3–10}$$

The quantity L characterizes the spatial size of the state ψ_0 of the system and may therefore be regarded as the **natural unit of length** for the harmonic oscillator. It is instructive to develop some feeling for the numerical magnitude of L. For electrons, with $M = m_e$, bound sufficiently tightly that their oscillation frequency falls into the range of visible light, say, $\omega/2\pi = 10^{15}$ Hz, we find that $L = 1.35 \times 10^{-10}$ m, roughly the size of a small atom. Oscillators with different oscillation frequencies scale according to (2•3–7).

To understand both the physical origin and the magnitude of L better, we re-write (2•3–7) in the form

$$\frac{1}{2}M\omega^2L^2 = \frac{\hbar^2}{2ML^2}. \tag{2•3–11}$$

This is an equality between two energies. The left-hand side is the *potential* energy of the oscillating object at the characteristic distance L from the center of oscillation, and the right-hand side is the *kinetic* energy of a plane wave with the wave number $k = 1/L$. Evidently, L is such that these two characteristic energies are the same.

The form (2•3–5) of the eigenfunction suggests introducing the dimensionless position variable

$$Q = x/L. \tag{2•3–12}$$

In terms of Q rather than x, the Schroedinger equation (2•3–4) may be written as

$$\left(-\frac{d^2}{dQ^2} + Q^2\right)\psi = \frac{2\mathcal{E}}{\hbar\omega}\psi. \tag{2•3–13}$$

This is the form we will use throughout the remainder of the section.

When switching independent variables from x to Q, attention must be paid to the correct normalization. The normalization in (2•3–9) and (2•3–10) was in terms of the true position coordinate x; it may be referred to as

***x*-normalization**. It is often useful to employ ***Q*-normalization** of ψ_0 instead, defined by

$$\int_{-\infty}^{+\infty} |\psi_0|^2 \, dQ = 1. \tag{2•3–14}$$

This requires that

$$A_0^2 = 1/\sqrt{\pi}, \tag{2•3–15}$$

differing from (2•3–10) by a factor L, the scale factor in (2•3–12). Both kinds of normalization have their advantages, as long as it is clear which one is being used. In cases of ambiguity, it is therefore essential to state the normalization employed.

2.3.2 Higher States: Energy Eigenvalues

Once the ground state has been determined, the higher energy eigenvalues and eigenfunctions are easily obtained by the following trick. We define the differential operator

$$\hat{a}^+ \equiv \frac{1}{\sqrt{2}}\left(-\frac{d}{dQ} + Q\right). \tag{2•3–16}$$

If this operator is applied to both sides of the Schroedinger equation (2•3–13), the result may be re-arranged to read

$$\left(-\frac{d^2}{dQ^2} + Q^2\right)(\hat{a}^+\psi) = \frac{2(\mathcal{E} + \hbar\psi)}{\hbar\omega}(\hat{a}^+\psi). \tag{2•3–17}$$

Exercise: Prove (2•3–17). Write the left-hand side of (2•3–17) as $\hat{h}(\hat{a}^+\psi)$, where

$$\hat{h} \equiv \left(-\frac{d^2}{dQ^2} + Q^2\right). \tag{2•3–18}$$

Next, show that

$$\hat{h}(\hat{a}^+\psi) - \hat{a}^+(\hat{h}\psi) = \sqrt{2}\left(-\frac{d\psi}{dQ} + Q\psi\right) = 2\hat{a}^+\psi. \tag{2•3–19}$$

Then show that this implies (2•3–17).

Evidently, if ψ is an energy eigenfunction with the energy eigenvalue $\mathcal{E}$, then the function $\hat{a}^+\psi$ is also an energy eigenfunction, but with the higher energy eigenvalue $\mathcal{E} + \hbar\omega$. The process may obviously be repeated. By starting with the already known eigenfunction ψ_0 and its energy eigenvalue $\mathcal{E}_0 = \hbar\omega/2$, we can construct an infinite set of successive eigenfunctions, which we call $\psi_n(Q)$, with the eigenvalues

$$\boxed{\mathcal{E}_n = \left(n + \frac{1}{2}\right)\hbar\psi, \qquad n = 0,1,2,3,\ldots.} \tag{2•3–20}$$

This is Planck's famous equidistant energy-level scheme with energy-level spacings of $\hbar\omega$, the scheme that gave birth to quantum mechanics.[4] We will show shortly that there are no additional energy eigenvalues.

Note that we are labeling the harmonic oscillator states with integer quantum numbers that start with $n = 0$, rather than with $n = 1$, the choice for the particle in a box. This difference in counting schemes is purely a matter of mathematical convenience, to which no physical significance should be attached.

The functions $\hat{a}^+\psi$ in are not yet normalized. It is left to the reader to show that normalization is retained if at every step we split off a factor $\sqrt{n+1}$, such that

$$\hat{a}^+\psi_n(Q) = \sqrt{n+1}\,\psi_{n+1}(Q). \tag{2•3–21}$$

If $\psi_n(Q)$ was normalized, then $\psi_{n+1}(Q)$ will also be normalized.

Because $\hat{a}^+$ is a *differential* operator, we call the relation (2•3–21) the **differential upward recursion relation** for the harmonic oscillator eigenfunctions. The operator $\hat{a}^+$ itself is often called a **raising operator**.

For later purposes, we also define a second operator

$$\hat{a}^- \equiv \frac{1}{\sqrt{2}}\left(+\frac{d}{dQ} + Q\right), \tag{2•3–22}$$

which differs from $\hat{a}^+$ only by an internal sign. It is left to the reader to show that $\hat{a}^-$ acts as a **lowering operator** with the **differential downward recursion relation** (for $n > 0$)

$$\hat{a}^-\psi_n(Q) = \sqrt{n}\,\psi_{n-1}(Q). \tag{2•3–23}$$

For $n = 0$, the right-hand side of (2•3–23) vanishes, and the condition $\hat{a}^-\psi_0 = 0$ has as its solution the ground-state Gaussian, $\exp(-Q^2/2)$, from (2•3–5). The two operators $\hat{a}^+$ and $\hat{a}^-$ form the basis for much of the mathematical theory of the harmonic oscillator.

Returning to the topic of energy levels, **Fig. 2•3–2** shows a superposition of the energy levels onto the parabolic potential.

The points x_n where $V(x) = \mathcal{E}_n$ are the points where, in classical mechanics, the particle motion would turn around; hence, they are called **classical turning points**. One finds easily that

$$x_n = \sqrt{2n+1}\,L. \tag{2•3–24}$$

[4] Planck did not yet have the zero-point energy term $\hbar\omega/2$.

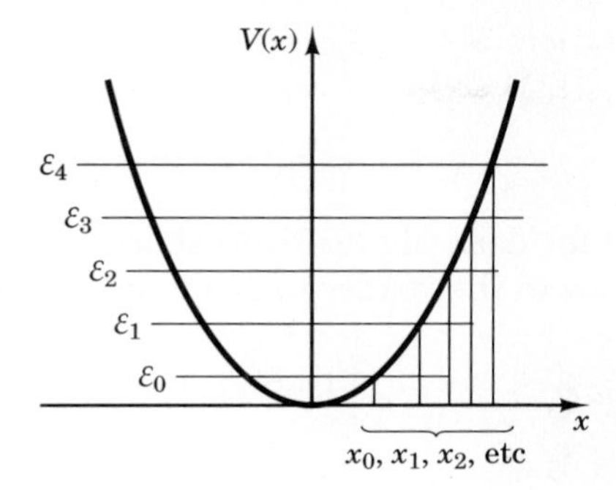

Figure 2•3–2. Bound-state energies relative to the potential energy. Also shown are the classical turning points.

The lowest-energy classical turning point is at L itself.

In classical mechanics, a particle with the energy $\mathcal{E}_n$ cannot travel beyond the classical turning points. In quantum mechanics, the wave function becomes evanescent there, representing a tunneling of the particle into the potential barrier.

The set of energy eigenstates generated by the upward recursion relations is a **complete set** in the sense that no other energy eigenstates outside this set can exist. The proof of this statement is based on our observation that the downward differential recursion relation (2•3–23) leads to a zero result only if it is applied to the wave function ψ_0. If additional eigenfunctions outside the set of ψ_n's existed, with different energy eigenvalues, repeated application of downward recursion to these would *not* lead to ψ_0, and hence not to truncation, and therefore to energy eigenfunctions with negative energy eigenvalues. But the Schroedinger equation cannot have any energy eigenvalues below the bottom of the potential energy. Hence, there cannot be any eigenfunctions outside the set of ψ_n's.

2.3.3 Higher States: Eigenfunctions

Inspection of the differential recursion relation (2•3–21) shows that the energy eigenfunctions obtained are all of the form

$$\psi_n(Q) \propto H_n(Q) \exp(-Q^2/2), \tag{2•3–25}$$

where each H_n is a certain polynomial of degree n. The lowest energy eigenfunction ψ_0 is the Gaussian of (2•3–5).

Except for a proportionality factor, the polynomials $H_n(Q)$ in (2•3–25) are the **Hermite polynomials** of the mathematical literature, named after the 19th-century French mathematician Charles Hermite. They occur in many problems in science and engineering. The Hermite polynomials may be defined in several equivalent ways; for our purposes, the most useful definition is via the upward *algebraic* recursion relation

$$H_n(Q) = 2Q \cdot H_{n-1}(Q) - 2(n-1) \cdot H_{n-2}(Q), \tag{2•3–26}$$

with the two starting polynomials

$$H_0(Q) = 1, \tag{2•3–27a}$$

$$H_1(Q) = 2Q. \tag{2•3–27b}$$

We will give instructions for deriving (2•3–26) shortly.

It is left as an exercise to the reader to show that the next four polynomials are

$$H_2(Q) = 4Q^2 - 2, \tag{2•3–27c}$$

$$H_3(Q) = 8Q^3 - 12Q, \tag{2•3–27d}$$

$$H_4(Q) = 16Q^4 - 48Q^2 + 12, \tag{2•3–27e}$$

$$H_5(Q) = 32Q^5 - 160Q^3 + 120Q, \tag{2•3–27f}$$

The polynomials are either purely even or purely odd, depending on whether n is an even or odd number. The coefficient of the highest power term is 2^n.

To normalize the wave functions, we should re-write (2•3–25) as

$$\psi_n(Q) = A_n H_n(Q) \exp(-Q^2/2), \tag{2•3–28}$$

where A_n is a normalization coefficient. The normalization coefficients for $n > 0$ can be shown to be given by the recursion relation

$$A^2{}_n = \frac{A^2{}_{n-1}}{2n} = \frac{A^2{}_{n-2}}{4n(n-1)} = \ldots = \frac{A_0^2}{2^n n!}, \tag{2•3–29}$$

where A_0 is as in (2•3–10) or (2•3–15), depending upon whether x or Q is used as the integration variable in the normalization integral. We shall not pursue the relation of the energy eigenfunctions to the Hermite polynomials here. In quantum-mechanical computations, it is usually more convenient to work directly with the energy eigenfunctions, rather than making the detour via the Hermite polynomials.

By eliminating the derivative terms from (2•3–21) and (2•3–23), one derives easily the two-step **algebraic upward recursion relation** for the normalized wave functions,

$$\boxed{\psi_n(Q) = \sqrt{\frac{2}{n}} \cdot \left[Q \cdot \psi_{n-1}(Q) - \sqrt{\frac{n-1}{2}} \cdot \psi_{n-2}(Q)\right],} \tag{2•3–30a}$$

with the two starting functions

$$\psi_0(Q) = A_0 \cdot \exp(-Q^2/2) \text{ and}$$
$$\psi_1(Q) = A_0\sqrt{2} \cdot Q \cdot \exp(-Q^2/2). \tag{2•3–30b,c}$$

Here, again, A_0 is taken from (2•3–10) or (2•3–15), depending on the desired normalization. The algebraic upward recursion relation is the most convenient way to compute numerical values of the normalized energy eigenfunctions. It also provides a convenient link to the Hermite polynomials.

Exercise: Eigenfunctions and Hermite Polynomials. Insert the form (2•3–25) for ψ_n into (2•3–30a), along with the relation (2•3–29) for the normalization coefficients, and show that this leads to the recursion relation (2•3–26) for the Hermite polynomials.

Graphs of the lowest four normalized eigenfunctions, as well as of the associated probability densities ψ_n^2, are shown in **Fig. 2•3–3.**

In **Fig. 2•3–4** we show ψ_n^2 for $n = 20$ as an example for a large value of n, indicating how the probability density distribution, averaged over the rapid *spatial* oscillations, approaches the *time-averaged* classical probability density distribution. The time-averaged classical probability density for finding the oscillating particle at the position x is inversely proportional to the magnitude of the velocity v at this point; it is easily found to be given by

$$\rho(x) = \frac{\omega}{\pi|v(x)|}, \tag{2•3–31}$$

where ω is the (radian) frequency of oscillation and where the velocity is viewed as a function of position, $v = v(x)$, rather than as a function of time.

Consider now a classical sinusoidal oscillation with amplitude a: $x = a \sin \omega t$. The associated particle velocity is

$$|v| = a\omega|\cos \omega t| = \omega(a^2 - x^2)^{1/2}. \tag{2•3–32}$$

If this is inserted into (2•3–31), we obtain:

$$\rho(x) = \frac{1}{\pi\sqrt{a^2 - x^2}}. \tag{2•3–33}$$

Note that this expression assumes x-normalization; to obtain the probability density for Q-normalization, the right-hand side of (2•3–33) must be multiplied by the scale factor L. If one averages over the rapid spatial oscillations of the probability density in Fig. 2•3–4, the result closely represents the classical probability density from (2•3–33), given by the dashed line.

2.3.4 Oscillating States

As with all stationary states, the probability density distributions for the harmonic oscillator bound states are time-independent and therefore do not fit the classical picture of a particle oscillating back and forth. To construct oscillating states, we must look at linear superpositions of bound states with *different* energies, with time-dependent coefficients chosen in such a way that the superposition satisfies the Schroedinger wave equation.

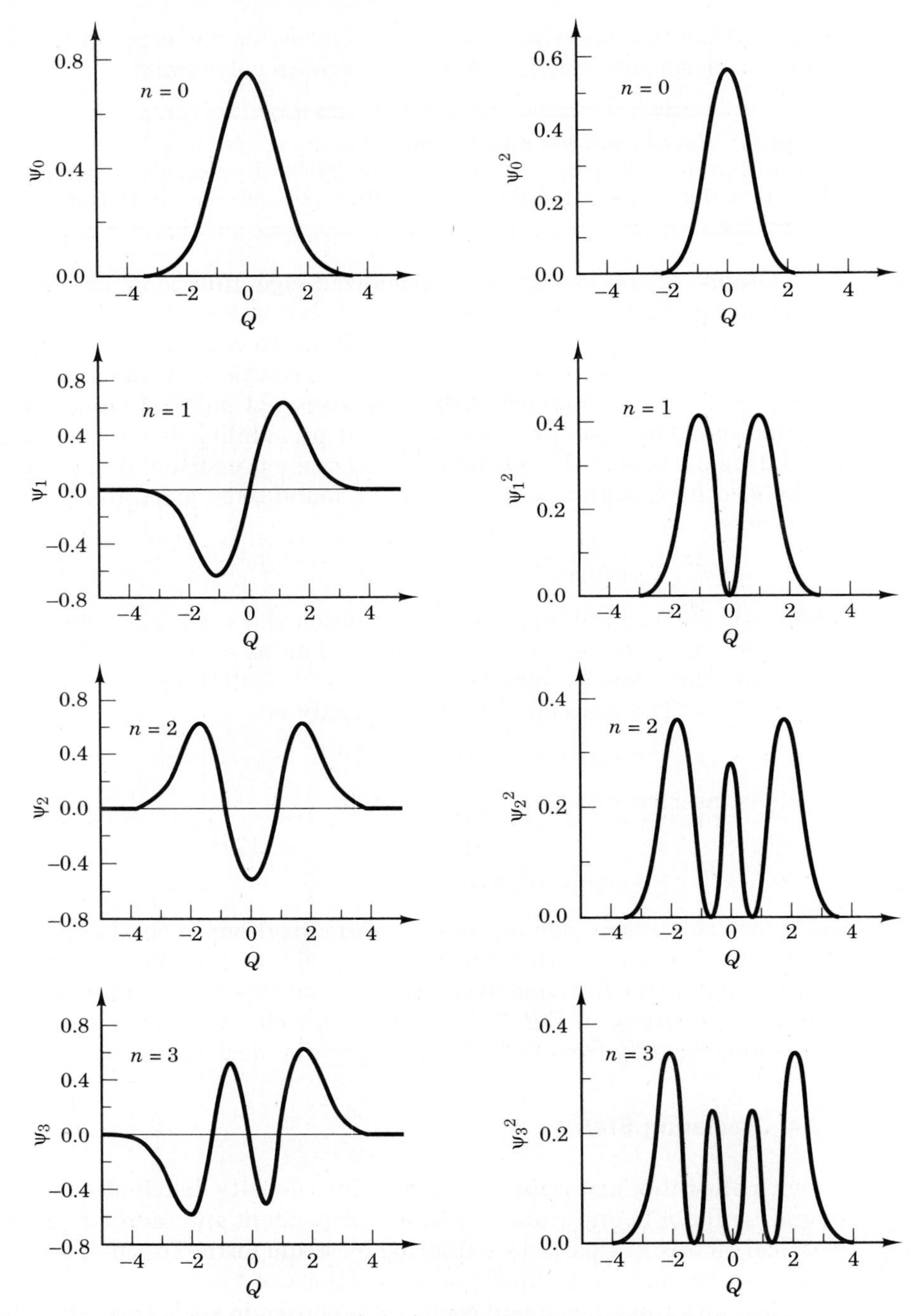

Figure 2•3–3. Wave function $\psi(Q)$ and probability density $\psi^2(Q)$ for the four lowest states ($n = 0$ through $n = 3$) of the harmonic oscillator. The wave functions are Q-normalized.

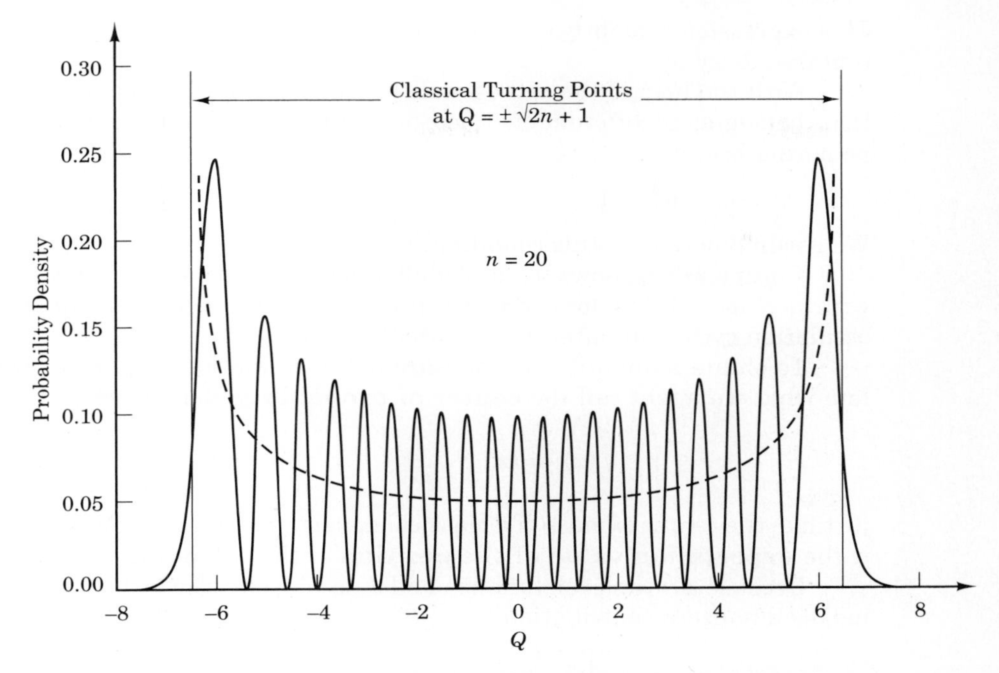

Figure 2•3–4. The Q-normalized probability density of the harmonic oscillator for $n = 20$. The dashed line represents the *classical* probability density, given by (2•3–33).

We study here the simplest possible case, a linear superposition of the states $n = 0$ and $n = 1$ for the harmonic oscillator. Consider first the two *time-dependent* wave functions

$$\Psi_0(x, t) = \psi_0(x) \exp\left(-\frac{i}{2}\omega t\right) \qquad \text{and}$$

$$\Psi_1(x, t) = \psi_1(x) \exp\left(-\frac{3i}{2}\omega t\right). \tag{2•3–34}$$

Both wave functions are of the form (2•1–3), and hence, they satisfy the Schroedinger wave equation. But in this case any linear superposition of the two also satisfies the Schroedinger wave equation—for example,

$$\Psi(x, t) \equiv a_0 \Psi_0(x, t) + a_1 \Psi_1(x, t), \tag{2•3–35}$$

where we may assume that the coefficients are chosen to be real.

The probability density distribution associated with (2•3–35) is easily found to be

$$\rho(x, t) = |\Psi(x, t)|^2 = a_0^2 \psi_0^2(x) + a_1^2 \psi_1^2(x) + 2a_0 a_1 \psi_0(x) \psi_1(x) \cdot \cos \omega t. \tag{2•3–36}$$

This expression evidently contains a term oscillating with the classical oscillation frequency ω.

With the help of the orthogonality relation (2•1–8) for energy eigenfunctions belonging to different energies, one finds easily that the wave packet will be normalized if

$$a_0^2 + a_1^2 = 1. \tag{2•3–37}$$

We assume here that this condition is satisfied.

Figure 2•3–5 shows the probability density distribution of a wave packet with $a_0 = a_2 = 1/\sqrt{2}$, for three instances of time: at $t = 0$, after a quarter of an oscillation cycle, and after half an oscillation cycle.

To obtain a quantitative measure of the oscillation amplitude, we calculate what we might call the **center of probability**, defined as

$$\langle x \rangle \equiv \int_{-\infty}^{+\infty} \rho(x) \cdot x dx, \tag{2•3–38}$$

just like the center of mass of a classical density distribution. We will later call it the **expectation value of the position** of the oscillating wave packet.

Because ψ_0 is an even function and ψ_1 is odd, only the last term in (2•3–36) makes a nonzero contribution to $\langle x \rangle$ in (2•3–38), leading to

$$\langle x \rangle = 2a_0a_1\left[\int_{-\infty}^{+\infty} \psi_0(x)\psi_1(x) \cdot x\, dx\right] \cdot \cos \omega t. \tag{2•3–39}$$

Note that this formulation assumes that the wave functions are x-normalized. From (2•3–30b,c), we have

$$x\psi_0(x) = \frac{L}{\sqrt{2}}\psi_1(x). \tag{2•3–40}$$

This turns the integral in (2•3–39) into $L/\sqrt{2}$ times the x-normalization integral for ψ_1, leading to

$$\langle x \rangle = \sqrt{2}a_0a_1\, L \cos \omega t. \tag{2•3–41}$$

The largest value of the product a_0a_1 under the constraint (2•3–37) is 1/2. If we choose this value, we obtain, finally,

$$\langle x \rangle = \frac{L}{\sqrt{2}} \cos \omega t. \tag{2•3–42}$$

Our treatment is readily generalized to linear superpositions of the general form

$$\Psi(x, t) = \sum_{n=0}^{\infty} a_n\psi_n(x) \cdot \exp\left[-i\left(n + \frac{1}{2}\right)\omega t\right], \tag{2•3–43}$$

containing an arbitrary number of energy eigenfunctions ψ_n, with coefficients subject only to the constraint that the overall superposition state be normal-

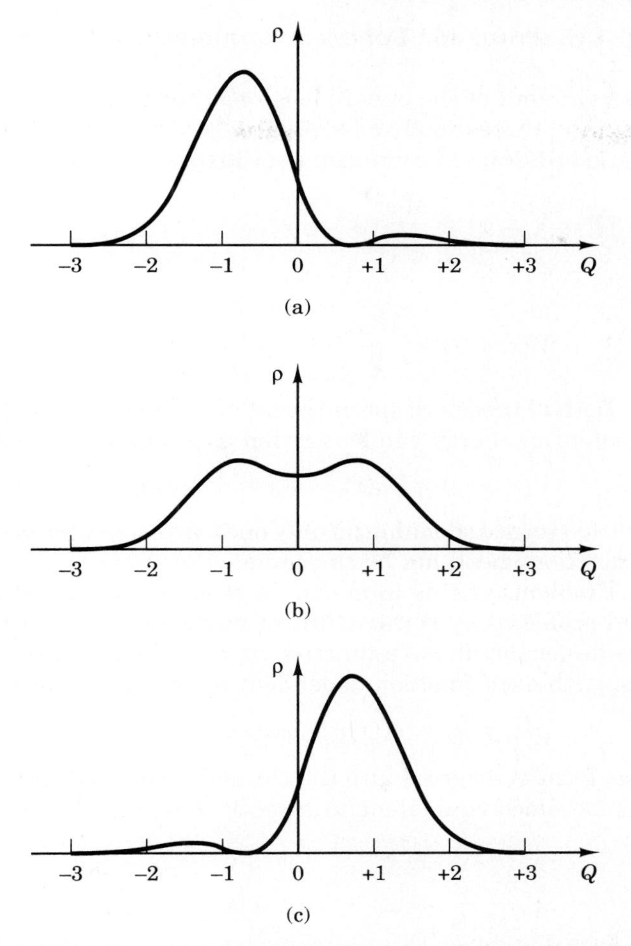

Figure 2·3–5. **(a)** Probability density for oscillating wave packet at $t = 0$. **(b)** The same, for $\omega t = \pi/2$, after one-quarter of an oscillation cycle. The wave packet is perfectly symmetric at this instant. **(c)** The same, for $\omega t = \pi$, after one-half of an oscillation cycle. The probability distribution is reversed relative to that at $t = 0$.

ized. In this way, an infinite variety of oscillating states can be constructed, with a far greater richness of oscillation behavior than in classical mechanics, where two numbers, an amplitude and a phase, fully specify the oscillator. By utilizing this freedom, it is possible to adapt the oscillating state to essentially arbitrary initial conditions for the wave function at $t = 0$. The problems at the end of the section give an example, and we will return to this topic later, where we will look at the quantum-mechanical oscillating states coming closest to the classical harmonic oscillator.

2.3.5 Cylindrical and Spherical Harmonic Oscillators

Our treatment of the one-dimensional harmonic oscillator is readily extended to two and three dimensions, especially to the rotationally invariant **cylindrical** and **spherical harmonic oscillators**, with the potential energy functions

$$V(x, y) = \frac{M\omega^2}{2}(x^2 + y^2) \tag{2•3–44a}$$

and

$$V(x, y, z) = \frac{M\omega^2}{2}(x^2 + y^2 + z^2). \tag{2•3–44b}$$

Both of these are special cases of a general category of problems in which the potential energy can be written as a sum of three terms,

$$V(x, y, z) = V_x(x) + V_y(y) + V_z(z), \tag{2•3–45}$$

with each term depending on only one Cartesian coordinate. The various potential functions need not be the same.

Problems of this kind can be reduced to a set of equivalent one-dimensional problems by **separation of variables**. In that method, one writes the three-dimensional wave function as a product of three one-dimensional functions, with each function dependent on only one coordinate:

$$\psi(x, y, z) = \psi_x(x)\psi_y(y)\psi_z(z). \tag{2•3–46}$$

If this form is inserted into the three-dimensional Schroedinger equation, the latter becomes equivalent to a *set of three* one-dimensional equations of the form

$$-\frac{\hbar^2}{2M}\frac{d^2}{dx^2}\psi_x(x) + V_x(x)\psi_x(x) = \mathcal{E}_x\psi_x(x), \tag{2•3–47}$$

with analogous equations in y and z. The total energy eigenvalue for the full three-dimensional problem is

$$\mathcal{E} = \mathcal{E}_x + \mathcal{E}_y + \mathcal{E}_z. \tag{2•3–48}$$

The reader is asked to work out the details.

Applied to our harmonic oscillator problem, the total energy eigenvalues of the cylindrical oscillator are easily found to be given by

$$\mathcal{E} = \hbar\omega \cdot (n + 1), \qquad \text{where} \qquad n \equiv n_x + n_y, \tag{2•3–49a}$$

and those of the spherical one by

$$\mathcal{E} = \hbar\omega \cdot (n + 3/2), \qquad \text{where} \qquad n \equiv n_x + n_y + n_z. \tag{2•3–49b}$$

With the energies depending only on the sum of the one-dimensional quantum numbers, all energy levels except the lowest one are degenerate. In the cylindrical case, (2•3–49a), a given value of n can evidently be obtained by

$n + 1$ different combinations of n_x and n_y, corresponding to an $(n + 1)$-fold degeneracy

$$g_2(n) = n + 1. \tag{2•3–50}$$

For the spherical harmonic oscillator, the degeneracy is even higher; it is left as an exercise for the reader to show that

$$g_3(n) = \frac{1}{2} \cdot (n + 1)(n + 2). \tag{2•3–51}$$

Normalization

If the overall three-dimensional wave function $\psi(x, y, z)$ is to be normalized, the individual one-dimensional wave functions must be normalized in a compatible way. It is usually simplest to normalize each function by itself to unity;

$$\int_{-\infty}^{+\infty} |\psi_i(x_i)|^2 \, dx_i = 1, \qquad \text{where} \qquad x_i = x, y, z. \tag{2•3–52}$$

Alternative Coordinate Systems

The cylindrical and spherical harmonic oscillators are unique in the sense that they are the only systems that can be solved by separation of variables in Cartesian coordinates and also have rotationally invariant potential energy functions. Such rotationally invariant systems can also be solved by separation of variables in cylindrical or spherical coordinate systems. We will discuss such systems in detail later.

◆ PROBLEMS TO SECTION 2.3

#2•3–1: Computation of Eigenfunctions

Write a computer program that uses the upward algebraic recursion relation (2•3–30) to compute the Q-normalized harmonic oscillator energy eigenfunctions $\psi_n(Q)$ for arbitrary specified values n and Q. Use the program to compute and plot both the wave function and the probability density for the equal-amplitude linear superposition

$$\psi(Q) = \frac{1}{\sqrt{2}}[\psi_4(Q) + \psi_5(Q)], \tag{2•3–53}$$

over the range $-5 \leq Q \leq +5$.

#2.3–2: Pulsating Harmonic Oscillator Wave Packet

Construct a time-dependent linear superposition of two harmonic oscillator states similar to the superposition (2•3–35), but with the following different properties:

(1) The probability density distribution is at all times symmetric about $x = 0$.

(2) The value of the probability density at $x = 0$ oscillates between zero and a maximum value.

Construct the *simplest* superposition you can find that satisfies these two specifications. Give a complete discussion of the properties of this superposition, on the level of the discussion of the superposition (2•3–35). Make a series of three or more plots that show the probability density distribution at several representative times during a full oscillation cycle.

2.4 POTENTIAL AND KINETIC ENERGY CONTRIBUTIONS TO BOUND-STATE ENERGIES

2.4.1 Expectation Values of Potential and Kinetic Energy

It is often important to understand the relative contributions of the potential and kinetic energy terms in the Schroedinger equation to the total energy. To obtain such a breakdown, we multiply the Schroedinger equation by the complex conjugate ψ^* of the (normalized) energy eigenfunction and integrate over all space. The result may be rearranged to read

$$\mathcal{E} = \langle \mathcal{E}_{\text{pot}} \rangle + \langle \mathcal{E}_{\text{kin}} \rangle, \tag{2•4–1}$$

where we have defined two new quantities:

$$\boxed{\langle \mathcal{E}_{\text{pot}} \rangle = \int \psi^* V(\mathbf{r}) \psi \, d^3r,} \tag{2•4–2}$$

called the **expectation value of the potential energy**, and

$$\boxed{\langle \mathcal{E}_{\text{kin}} \rangle = -\frac{\hbar^2}{2M} \int \psi^* \nabla^2 \psi \, d^3r,} \tag{2•4–3}$$

called the **expectation value of the kinetic energy**. Evidently, this division accomplishes the desired objective.

The two quantities have a simple meaning. Recall that we interpret $\psi^* \psi \, d^3r$ as the probability that the object will materialize in the volume element d^3r at the position $\mathbf{r}$ during a suitable localization experiment. Hence, the integral in (2•4–2) is simply the weighted average of the potential energy, taken over all possible positions of localization, with the local probability density $\psi^* \psi$ as the weighting factor. The integrand itself, $\psi^* V \psi$, serves as a probabilistic *potential energy density*.

Similarly, the integral in (2•4–3) may be interpreted as the weighted average of the *kinetic* energy. To see this more clearly, recall the definition of a position-dependent *local* wave number $K(\mathbf{r})$ in (1•3–14), in our original "derivation" of the Schroedinger wave equation. With this definition,

$$-\frac{\hbar^2}{2M} \psi^* \nabla^2 \psi = \frac{\hbar^2 K(\mathbf{r})^2}{2M} \psi^* \psi. \tag{2•4–4}$$

The term $\hbar^2 K(\mathbf{r})^2/2M$ may be viewed as the local kinetic energy at $\mathbf{r}$, similar to the way $V(\mathbf{r})$ is the local potential energy. If this is done, the integral in (2•4–3) becomes the weighted average of the kinetic energy, again with the local probability density $\psi^*\psi$ as the weighting factor. The expression (2•4–4) serves as a probabilistic *kinetic energy density*.

The reason for the name *expectation value* is the following. Suppose that, during a suitable localization experiment, the object materializes in a volume element d^3r around the position $\mathbf{r}$. Suppose further that we were *somehow* able to measure the potential energy $V(\mathbf{r})$ of the now-localized object, taken at that position of localization. Because of the spread of the wave function over space before localization, different experiments would lead to different places of localization and, hence, to different values of the measured potential energy. These measurement values will have a certain probability distribution, with the average $\langle\mathcal{E}_{\text{pot}}\rangle$. This average is the value we might "expect" in an as-yet unperformed localization experiment; hence the terminology *expectation value*. We will see later (chapter 7) that $\langle\mathcal{E}_{\text{kin}}\rangle$ may similarly be interpreted as the average value of the kinetic energy, taken over many measurements.

Exercise. **(a)** By treating a kink in a real wave function as a limiting case of a smooth wave function, show that the kink makes a contribution

$$\Delta\langle\mathcal{E}_{\text{kin}}\rangle = -\frac{\hbar^2}{2M}\psi\Delta(\psi') \tag{2•4–5}$$

to the expectation value of the kinetic energy, where ψ is the wave function at the kink and $\Delta(\psi')$ is the difference between the slopes of the wave function on the two sides of the kink.

(b) Show that a *discontinuity* in the wave function would make an infinite contribution to the expectation value of the kinetic energy. This is an alternative proof that the wave function cannot have discontinuities in its magnitude.

It is instructive to apply the expectation value formalism to the ground-state wave function of the harmonic oscillator, (2•3–5). One finds readily that the two expectation values are given by

$$\langle\mathcal{E}_{\text{pot}}\rangle = \frac{1}{4}M\omega^2L^2 \tag{2•4–6a}$$

and

$$\langle\mathcal{E}_{\text{kin}}\rangle = \frac{\hbar^2}{4ML^2}. \tag{2•4–6b}$$

If we now insert the value characteristic length L from (2•3–7), we obtain

$$\langle\mathcal{E}_{\text{kin}}\rangle = \langle\mathcal{E}_{\text{pot}}\rangle = \frac{1}{4}\hbar\omega. \tag{2•4–7}$$

This is the quantum-mechanical generalization—at least for the ground state—of a result familiar from classical physics, namely, that the *time-averaged* kinetic and potential energies of the harmonic oscillator are equal to each other.

2.4.2 Variational Principle

Although we have introduced the concept of expectation values for the potential and kinetic energy in the context of energy eigenstates, these expectation values may be calculated for essentially arbitrary (normalizable) wave functions, regardless of whether the latter are solutions of the Schroedinger equation. For example, the two expressions (2•4–6a,b) for the expectation values of a Gaussian wave function hold for any Gaussian of the form (2•3–5), independently of whether or not the characteristic length L is chosen according to (2•3–7). If both expectation values are viewed as functions of L for arbitrary L, we see from (2•4–6a,b) that they depend on L in opposite ways. The potential energy increases proportionally to L^2 with increasing L. This is as we would expect, because the potential itself increases quadratically with distance, and the more the wave function is spread out, the larger will be the potential energy contribution. Conversely, the kinetic energy *decreases* proportionally to $1/L^2$ with increasing L. This, too, is as expected, because the kinetic energy increases quadratically with wave number, and the more the wave function is spread out, the larger will be the wavelengths contained in the wave function, and the smaller will be the wave numbers.

The sum of the two expectation values,

$$\langle \mathcal{E} \rangle = \langle \mathcal{E}_{\text{kin}} \rangle + \langle \mathcal{E}_{\text{pot}} \rangle = \frac{\hbar^2}{4ML^2} + \frac{1}{4} M\omega^2 L^2, \tag{2•4–8}$$

is called the expectation value of the total energy. It evidently has a minimum for some value of L. One finds easily that this minimum occurs at the point where the two energies are equal, that is, when

$$\frac{1}{4} M\omega^2 L^2 = \frac{\hbar^2}{4ML^2}, \tag{2•4–9}$$

which is essentially the same as (2•3–11). Evidently, the actual energy eigenfunction ψ_0 is indeed the Gaussian with the lowest possible expectation value of the total energy.

The foregoing argument was based on the knowledge that the actual wave function is a Gaussian. What if we had not known that? We will show later that the sum of the two expectation values for *arbitrary* wave functions can never fall below the lowest energy eigenvalue of the system, and it can be equal to the lowest energy eigenvalue only if the wave function ψ is in fact the lowest energy eigenfunction. This is the **variational theorem** of quantum mechanics. It is useful to state this theorem already now, before we have acquired the full formalism required for a rigorous proof of it, because it will help us understand several aspects of bound states.

The variational theorem provides an extraordinarily powerful tool for estimating the ground-state energy of a system without knowing the exact ground-state wave function. Even for an arbitrary wave packet that is in some fashion concentrated around the minimum of the potential energy, we can always define a characteristic length L that somehow measures the overall size of the wave packet, and we would expect that the expectation value of the potential energy would scale with L^2, with a magnitude roughly on the order of $M\omega^2L^2/2$, but probably differing by some "fudge factor" on the order of unity. Similarly, we would expect that the expectation value of the kinetic energy would scale with $1/L^2$, with a magnitude roughly on the order of $\hbar^2/2ML^2$, differing by a "fudge factor" of its own. Unless these two "fudge factors" differ greatly from each other, the overall energy minimum will occur in the vicinity of the value defined by the condition (2•3–11).

Evidently, such energy scaling and minimization arguments are a very powerful tool to predict the general size of quantum systems. Crude qualitative estimates like the one given here can be very useful even in the absence of quantitative results, and we will employ them repeatedly later.

2.5 GENERAL ONE-DIMENSIONAL POTENTIAL WELLS

2.5.1 Bound States in Monotonic One-Dimensional Wells

We show now that many of the qualitative aspects of the bound states of a particle in a one-dimensional box or a harmonic oscillator remain the same for a wide class of one-dimensional potentials. The only assumption we will make about the potential $V(x)$ is that it has a *single* minimum somewhere, increasing *monotonically* away from the location of the minimum. We allow steps in the potential.

Without loss of generality, we may place the zero of both the x-axis and the potential energy at the minimum of potential energy, as in **Fig. 2•5-1**. Furthermore, we define

$$V_L = V(-\infty), \qquad V_R = V(+\infty). \tag{2•5–1a,b}$$

One or both of these quantities may be infinite.

A potential satisfying the above requirements is not the most general form we might discuss, and many of our conclusions remain valid for more general potentials. But we wish to keep our assumptions simple, rather than getting involved in mathematical subtleties.

Under our assumptions, the Schroedinger equation, which we write in the form

$$\psi''(x) = \frac{2M}{\hbar^2}[V(x) - \mathcal{E}]\psi(x), \tag{2•5–2}$$

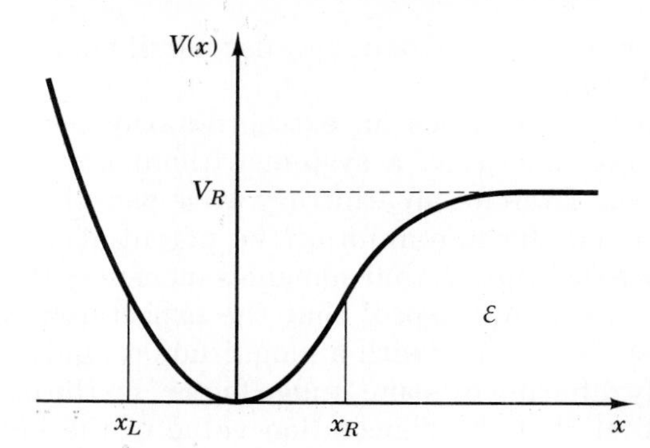

Figure 2•5–1. A general one-dimensional potential.

has two well-behaved linearly independent solutions for every $\mathcal{E}$. If we specify, at an arbitrary point x_0, both ψ and its first derivative ψ', these initial conditions determine one particular solution. In **Fig. 2•5–2**, we assume that values of ψ and ψ' have been specified at the point $x = x_R$, one of the two points where $[V(x) - \mathcal{E}]$ changes its sign. As stated earlier, such points are called **classical turning points**, because in classical dynamics, a particle would turn around when it reached one of them. Quantum-mechanically, there always exists a finite penetration probability into the classically inaccessible ranges, with an exponential-like decay of the wave function in the classically forbidden region.

We investigate the behavior of the wave function for various values of the derivative $\psi'(x_R)$ at $x = x_R$. Without loss of generality, we may assume that $\psi(x_R)$ is positive. According to (2•5–2), for $x > x_R$, the second derivative of ψ has the same sign as ψ itself. This means that ψ will always curve away from the x-axis, as shown in Fig. 2•5–2 for five different values of the slope $\psi'(x_R)$.

For a positive or zero value of $\psi'(x_R)$, the wave function grows monotonically with increasing x (curves 1 and 2). For a sufficiently small negative value of the slope, the wave function decreases initially, but because of its positive curvature, it reaches zero slope before reaching the x-axis (curve 3) and then increases monotonically beyond its minimum. On the other hand, if $\psi'(x_R)$ is sufficiently negative, ψ reaches the x-axis while it still has a negative slope,

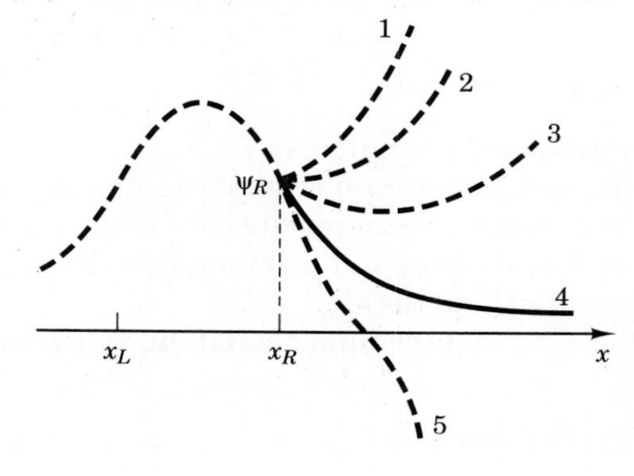

Figure 2•5–2. For a given value of ψ at $x = x_R$, only one value of ψ' leads to a bound state.

after which it turns away from the x-axis toward increasing negative values (curve 5). Somewhere between curves 3 and 5, there must exist a case where the slope $\psi'(x_R)$ has a negative value of just the right magnitude that $\psi(x)$ approaches the x-axis asymptotically (curve 4). Because the different functions with a common value of ψ_R cannot cross each other to the right of $x = x_R$, there will be exactly one such function for that energy. We call it $\psi_R(x)$.

Of the five different functions shown, only function 4 can correspond to a bound state. The exact value ψ_R' of the slope at $x = x_R$ depends, of course, on the arbitrary value we assigned to $\psi(x_R)$. What matters is the *logarithmic* derivative ψ_R'/ψ_R taken at $x = x_R$, i.e.,

$$\sigma_R \equiv \psi_R'(x_R)/\psi_R(x_R), \tag{2•5–3}$$

which is independent of the scaling of ψ_R. For every value of the energy $\mathcal{E}$, there will be exactly one value of σ_R that leads to a wave function that vanishes for $x \to +\infty$.

By similar reasoning, we conclude that there will also be, for every value of the energy $\mathcal{E}$, exactly one value σ_L of the logarithmic derivative evaluated at the *left* turning point, $x = x_L$, such that the wave function vanishes for $x \to -\infty$. We call the associated wave function $\psi_L(x)$.

For an *arbitrary* value of $\mathcal{E}$, ψ_L will in general not vanish for $x \to +\infty$, nor will ψ_R vanish for $x \to -\infty$. Hence, neither function can represent an energy eigenstate. Because the two functions are clearly linearly independent, and because there cannot be more than two linearly independent solutions to the one-dimensional Schroedinger equation (2•5–2), an arbitrary $\mathcal{E}$ value will in general not correspond to a bound state.

However, for certain discrete energy values, $\psi_L(x)$ and $\psi_R(x)$ will differ at most by an irrelevant constant factor, which might as well be assumed to be unity. The common wave function $\psi = \psi_L = \psi_R$ vanishes then for both $x \to -\infty$ and $x \to +\infty$. It therefore represents a bound state, and the corresponding energy $\mathcal{E}$ is the energy eigenvalue of that bound state.

For a well of nonzero width, with both V_L and V_R infinite, there must exist an infinite number of bound states with non-degenerate discrete energy eigenvalues. To show this, we consider the qualitative shape of the wave function $\psi_L(x)$ in the range $x > x_L$ for various energies. Without loss of generality, we may assume that both $\psi_L(x_L)$ and $\psi_L'(x_L)$ are positive. The evolution of the qualitative shape of $\psi_L(x)$ with increasing energy must then be of the kind shown in **Fig. 2•5–3**, where increasing labels represent increasing energies.

We note first that in the classical range $x_L < x < x_R$, we have $\mathcal{E} - V(x) > 0$, as a result of which the solutions of (2•5–1) have a second derivative ψ'' with a sign opposite to that of ψ itself. This means that ψ always curves toward the x-axis. If the energy is large enough, the curvature will ultimately lead to a sign change, first of ψ' and later also of ψ, and for sufficiently large values of $\mathcal{E}$, it will lead to an oscillating wave function. We follow this evolution in detail.

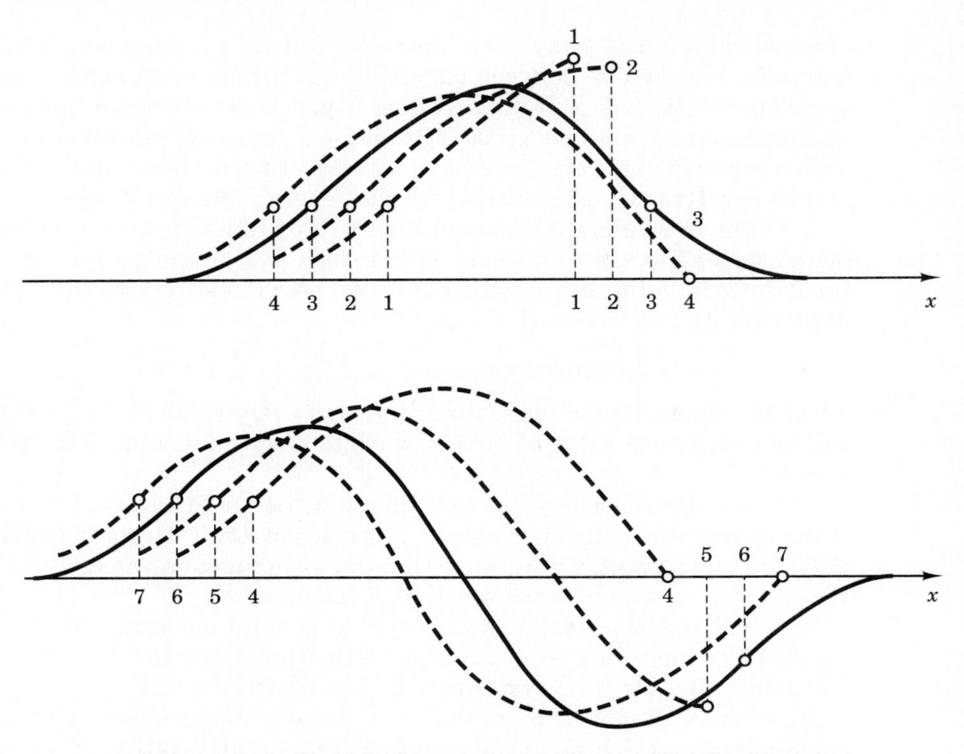

Figure 2•5–3. Change of the wave functions with increasing energy.

For energies just above $\mathcal{E} = 0$ the classical range is narrow and $\mathcal{E} - V(x)$ is small, that is, the curvature of the wave function is small. As a result, up to some energy $\mathcal{E} = \mathcal{E}_1'$, both $\psi_L'(x)$ and $\psi_L(x)$ remain positive throughout the entire classical range (curves 1 and 2). There is, therefore, no bound state for $\mathcal{E} \leq \mathcal{E}_1'$.

As the energy is increased, the curvature (for a given magnitude of ψ) increases throughout the classical range, and the classical range itself becomes wider. In consequence, $\psi_L'(x_R)$ becomes increasingly negative until, for some energy $\mathcal{E}_1''$, the wave function itself becomes zero at $x = x_R$ (curve 4). At some energy $\mathcal{E}_1$ in the range $\mathcal{E}_1' \leq \mathcal{E} \leq \mathcal{E}_1''$, the logarithmic derivative of ψ_L, taken at $x = x_R$, must coincide with the logarithmic derivative σ_R for ψ_R at $x = x_R$:

$$\psi_L'(x_R)/\psi_L(x_R) = \sigma_R. \tag{2•5–4}$$

At that energy, $\psi_R(x)$ and $\psi_L(x)$ can differ only by an irrelevant proportionality factor. Evidently, we then have a bound state. The qualitative shape of the corresponding eigenfunction is that of curve 3 in Fig. 2•5–3.

If the energy is raised beyond $\mathcal{E}_1''$, *both* $\psi_L(x_R)$ and $\psi_L'(x_R)$ become negative, and no new bound state is possible until the energy has been raised beyond the energy $\mathcal{E}_2'$ at which again $\psi_L'(x_R) = 0$. At some energy $\mathcal{E}_2$ between $\mathcal{E}_2'$ and the energy $\mathcal{E}_2''$ for which again $\psi_L(x_R) = 0$, there must lie a second bound state, with a wave function such as curve 6 in Fig. 2•5–3.

Obviously, this process can be continued ad infinitum. As the energy is raised and the potential well between the classical turning points gets deeper and wider, the number of oscillations increases. At each value of $\mathcal{E}$, there is only one value of σ_R for which $\psi(\infty) \to 0$, and because a specified value of σ_R determines the wave function to within an irrelevant constant, there can be at most one bound-state wave function for any given energy, a conclusion we had already reached in section 1 by a purely mathematical argument.

Finally, it is evident from Fig. 2•5–3 that the lowest bound-state eigenfunction has only one extremum and no nulls, and the numbers of both the extrema and the nulls increase by one for every higher state.

2.5.2 Generalizations

Most of our considerations remain valid if we drop the assumption of infinitely high potential walls—including specifically the conclusion about the numbers of nulls in the wave function. Only the conclusion that an infinite number of bound states must exist can no longer be drawn. The energy of the bound states is limited by V_L or V_R, whichever is lower. This, in turn, restricts the number of oscillations that a wave function can have over a given distance, and, consequently, in a potential well of finite width, this limits the number of bound states that can occur. If the potential approaches its limiting value only asymptotically, the classical range approaches an infinite width as the potential approaches its asymptotic value. If the latter approach is sufficiently slow, even a potential well of finite depth can exhibit an infinite number of bound states. The energies of these states are then crowded near the top of the potential well. We will encounter a case of this type later: The bound states of the hydrogen atom are those of a one-dimensional equivalent potential well that approaches its asymptotic value for $x \to +\infty$ as $1/x$.

If a potential well is sufficiently shallow *and* $V_L \neq V_R$, the well need not have any bound states at all. However, if $V_L = V_R$, it can be shown that even an arbitrary shallow potential well must have *at least* one bound state, if the potential approaches its limiting value monotonically.

Exercise. Prove this assertion. *Hint*: Look at the logarithmic derivatives ψ_L'/ψ_L and ψ_R'/ψ_R at the classical turning points as the energy approaches the top of the well.

2.6 OSCILLATIONS IN COUPLED WELLS

2.6.1 Coupled Wells

A number of problems in quantum mechanics involve potentials with two or more minima. The case of greatest interest is that of a simple symmetrical double-well, as shown in **Fig. 2•6–1**, with a central barrier sufficiently high that the energy of the lowest energy eigenstates is actually below the top of the barrier, implying a negative kinetic energy and evanescent wave functions inside the barrier. Such a well occurs, for example, in the ammonia molecule, in which the potential energy of the nitrogen atom relative to the plane of the three hydrogen atoms of the molecule (**Fig 2•6–2**) has a double minimum.

The *stationary* states of such symmetric coupled wells alternate between even- and odd-parity states. By a time-dependent linear superposition of adjacent even- and odd-parity states, similar to the superposition of two harmonic oscillator states in section 2.3, it is possible to construct *oscillating* states in which the object tunnels back and forth through the barrier between the two wells. In the cases of greatest interest, the barrier between the two wells is nearly impenetrable, making the wells only weakly coupled and leading to oscillations at relatively low frequencies, sometimes in the microwave range or lower. For example, the oscillation frequency of the nitrogen atom in the ammonia molecule is 23.87 GHz; it is the operating frequency of the ammonia maser. The purpose of the present section is to understand this kind of oscillation, in terms of the properties of the underlying even- and odd-parity *stationary* states.

Consider the qualitative sketch of the lowest even- and odd-parity eigenfunctions in **Fig 2•6–3(a)**. The even-parity function ψ_+, although not going to zero at the center of inversion, has a minimum there. If the barrier is sufficiently strong, the magnitude of ψ_+ at its minimum will be very small, almost vanishing. (For clarity, Fig 2•6–3(a) shows a less extreme situation.) In that case, our discussion of the qualitative behavior of wave functions in the preceding section suggests that a very small increase in energy will be

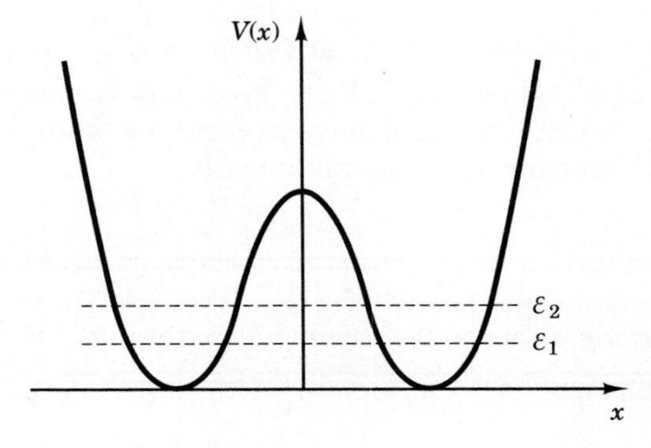

Figure 2•6–1. Double well. The energies $\mathcal{E}_1$ *and* $\mathcal{E}_2$ represent the energies of the two lowest states.

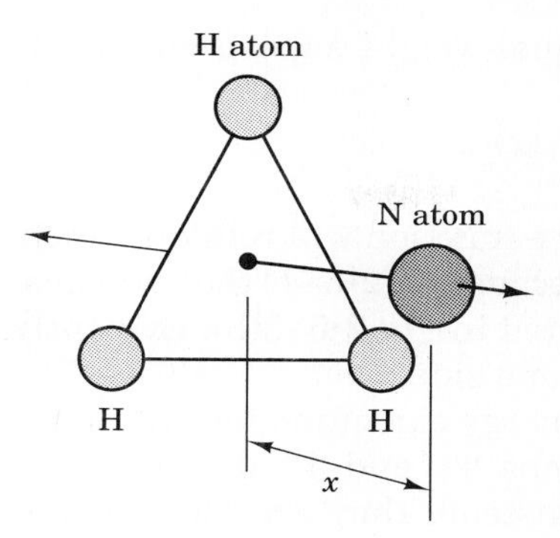

Figure 2•6–2. Ammonia molecule.

sufficient to "bend over" the even-parity function to the point that it actually goes through zero at the center of inversion, thus turning it into the odd-parity wave function ψ_-.

But if the energy difference between the two states is small, then the two functions will have a very similar shape outside the central barrier, except for a sign change on one of the two sides. If the signs of the two wave functions are chosen as in Fig 2•6–3(a), then, to a first order, the wave functions outside the barrier will satisfy

$$\psi_-(x) \approx \begin{cases} +\psi_+(x) & \text{for} \quad x < 0, \\ -\psi_+(x) & \text{for} \quad x > 0. \end{cases} \tag{2•6–1}$$

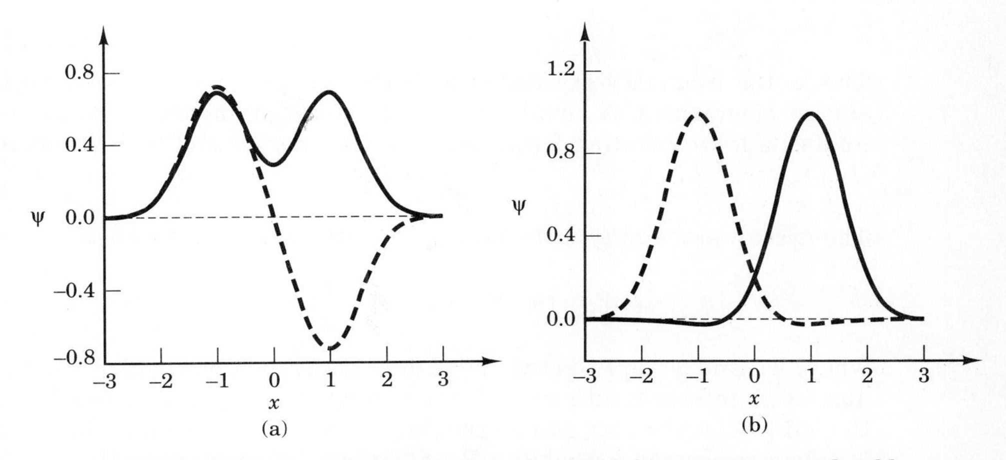

Figure 2•6–3. (a) Qualitative shape of lowest even- and odd-parity wave functions for a symmetric double-well, with its two halves separated by a barrier sufficiently high to cause the wave functions to be evanescent inside the barrier. **(b)** Equal-weight linear superpositions of the two energy eigenfunctions of (a), showing the shift of most of the wave function into one of the two wells.

Consider now the two equal-weight superpositions of ψ_- and ψ_+ of the form

$$\Psi_\pm(x) = \frac{1}{\sqrt{2}}[\psi_+(x) \pm \psi_-(x)]. \tag{2•6–2}$$

The two terms in this wave function will reinforce each other on one of the two sides of the barrier, but because of (2•6–1) they will nearly cancel each other on the other side, as illustrated in Fig 2•6–3(b). Evidently, the entire probability distribution is shifted to one side.

However, with the energy eigenfunctions ψ_- and ψ_+ belonging to different energies, the superpositions Ψ_- and Ψ_+ do not satisfy the *time-independent* Schroedinger equation. Instead, they are "snapshots" of the time-dependent solution

$$\Psi(x, t) = \frac{1}{\sqrt{2}}[\psi_+(x)e^{-i\omega_+ t} + \psi_-(x)e^{-i\omega_- t}] \tag{2•6–3}$$

of the Schroedinger wave equation, where $\omega_+ = \mathcal{E}_+/\hbar$ and $\omega_- = \mathcal{E}_-/\hbar$ are the frequencies associated with the energies of the two *stationary* states. The two terms in (2•6–3) have a different time dependence, leading to a time-dependent probability density distribution

$$|\Psi|^2 = \frac{1}{2}[\psi_+^2 + \psi_-^2 + 2\psi_+\psi_- \cos(\Delta\omega)t], \tag{2•6–4}$$

which contains a term that oscillates with the frequency

$$\Delta\omega = \frac{\Delta\mathcal{E}}{\hbar} = \frac{1}{\hbar}(\mathcal{E}_- - \mathcal{E}_+). \tag{2•6–5}$$

This is the frequency associated with the energy *difference* between the two energy eigenstates. A small energy difference, predicted for weak coupling, implies a low oscillation frequency. The time period T of the oscillations is

$$T = 2\pi/\Delta\omega = 2\pi\hbar/|\Delta\mathcal{E}|. \tag{2•6–6}$$

The overall probability distribution evidently oscillates between

$$\frac{1}{2}[\psi_+ + \psi_-]^2 = |\Psi_+|^2 \quad \text{and} \quad \frac{1}{2}[\psi_+ - \psi_-]^2 = |\Psi_-|^2, \tag{2•6–7}$$

where Ψ_+ and Ψ_- are the two "snapshot" superpositions of (2•6–2). For $t = 0$, and for all integer multiples of T, the two eigenfunctions very nearly cancel on the left side, while they add on the right side. This means that for $t = 0$, there is only a very small probability P_L of "finding" the particle in the left well and a correspondingly large probability $P_R = 1 - P_L$ of "finding" it in the right well. For $t = T/2$, and for all half-integer multiples of T, the situation is exactly inverted. The two probabilities P_L and P_R thus oscillate between a minimum value P_{min} and a maximum value $P_{max} = 1 - P_{min}$ (**Fig 2•6–4**). The stronger the barrier, the closer will P_{min} be to zero and P_{max} to unity.

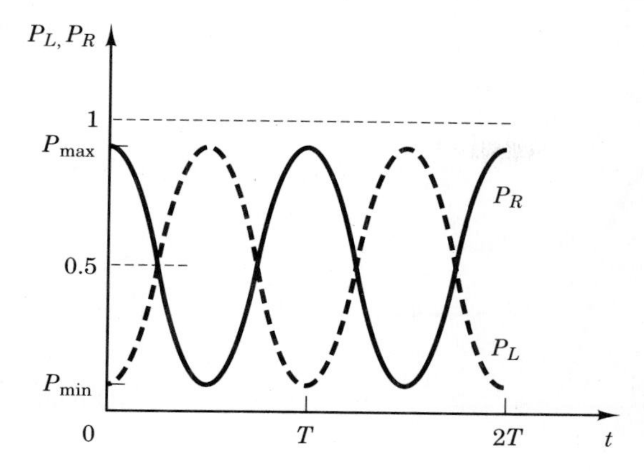

Figure 2•6–4. Probability oscillations between two coupled wells.

The oscillation frequency of the nitrogen atom in the ammonia molecule, 23.87 GHz, corresponds to an energy level separation $\Delta\mathcal{E}$ between even- and odd-parity states of 9.87×10^{-5} eV. Note that the usual chemical picture of an ammonia molecule, with four atoms fixed at the four corners of a tetrahedron, does not correspond to a stationary state with a sharp energy; it is only a momentary snapshot of an oscillatory situation, with an energy that is uncertain by $\pm\Delta\mathcal{E}/2$. For a stationary state with a sharp energy, the nitrogen atom is de-localized. It is then at all times completely undetermined on which side of the hydrogen plane the nitrogen atom would materialize in a suitably designed experiment; both sides are equally probable.

An ammonia maser is a device that first generates an ammonia molecular beam with all atoms in the higher odd-parity state, with its sharp energy eigenvalue. Some of the molecules then make a transition to the even-parity state, with its sharp lower energy eigenvalue. The energy difference $\Delta\mathcal{E}$ appears as a quantum of electromagnetic radiation with a frequency $\Delta\mathcal{E}/\hbar$ equal to the bouncing frequency of the nitrogen atom in a nonstationary state. Neither the initial nor the final state of each ammonia molecule is such an oscillating state. Both are pure stationary states with a de-localized nitrogen atom.

2.6.2 Quasi-Bound States

Consider the non-monotonic one-dimensional potential shown in **Fig 2•6–5**. Such potentials occur in many physical problems—for example, in the theory of radioactive decay and in the field emission of electrons from metals.

For energies $\mathcal{E} < V_\infty$ all our considerations of section 2.3 for monotonic potentials apply, but in the energy range

$$V_\infty < \mathcal{E} < V_B, \qquad (2\text{•}6\text{–}8)$$

no truly bound states exist. However, if the cross-hatched barrier is sufficiently wide and high, the problem can, to the first order, be treated as if the potential

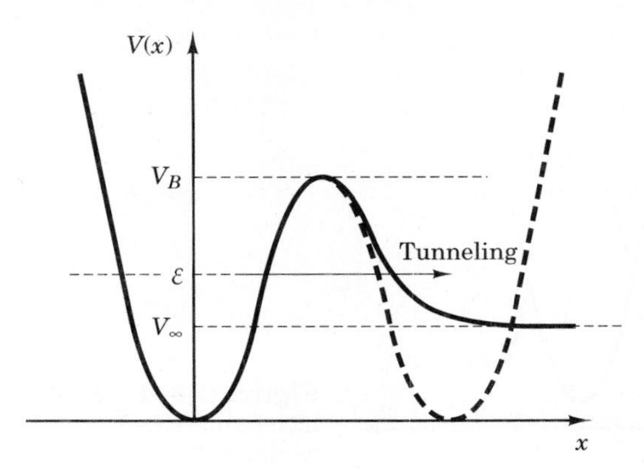

Figure 2•6–5. Quasi-bound states.

leveled out at the top of the barrier, rather than decreasing again to $V = V_\infty$. If this first-order treatment leads to bound states in the range (2•6–8), these states correspond to **quasi-bound** states of the actual problem, that is, to states for which the wave function inside the potential well decays slowly with time, due to tunneling through the potential barrier. The rate of decay can be roughly estimated by treating the problem as a symmetrical double-well problem, as indicated by the gray potential line in Fig. 2•6–5. Each quasi-bound state is then converted into two closely spaced double-well states of opposite parity. The energy difference between the two states defines a double-well oscillation frequency, and the length $T/2$ of half an oscillation period is a useful approximate measure for the time it takes for the particle to escape from the quasi-bound state.

◆ PROBLEMS TO SECTION 2.6

#2•6–1: Oscillations in a Double Square Well

Study the oscillations a one-dimensional square well of width $L = 2a$, with infinitely high barriers, divided in the middle by a Dirac δ-function barrier of strength $S > 0$. Show that the odd-parity states are not affected by the barrier. Show also that the wave functions of the even-parity states have a kink at the barrier, but are free-particle wave functions in the space between the barrier and the walls, with a wave number k that is a solution of the transcendental equation

$$-\tan ka = c \cdot ka, \tag{2•6–9}$$

where the parameter c, a measure of the strength of the coupling between the two wells, is defined as

$$c = \frac{\hbar^2}{MSa}. \tag{2•6–10}$$

Make a plot of the lowest even- and odd-parity eigenfunctions and of their superposition Ψ_-, assuming the specific value $\pi/k = 1.1a$. Also, make a plot of the probability density for $t = 0$, $t = T/4$, and $t = T/2$.

The condition (2•6–9) may be visualized graphically in terms of the intersections of $-\tan ka$ with a straight line. From such a graph, show that the energies of the even-parity states get pushed up close to those of the next higher odd-parity states if the coupling between the wells is decreased.

In the limit of weak coupling, $c \ll 1$, an *approximate* analytic solution of (2•6–9) can be given. Show that in this limit, the energy of the lowest even-parity state may be approximated as

$$\mathcal{E}_1 = \frac{\hbar^2 k^2}{2M} = \frac{\hbar^2 (ka)^2}{2Ma^2} \approx \frac{\hbar^2 \pi^2}{2Ma^2}(1 - c)^2 \approx \mathcal{E}_2 - \Delta\mathcal{E}, \qquad (2\text{•}6\text{–}11)$$

where $\mathcal{E}_2$ is the energy of the lowest odd-parity state and

$$\Delta\mathcal{E} \approx 2c\,\mathcal{E}_2 \ll \mathcal{E}_2 \qquad (2\text{•}6\text{–}12)$$

is the approximate energy splitting between the two states.

Chapter 3

ROTATIONALLY INVARIANT POTENTIALS: HYDROGEN ATOM AND BEYOND

3.1 SPHERICALLY SYMMETRIC POTENTIALS

3.1.1 Separation of Variables in Spherical Polar Coordinates

Most of the problems discussed in the preceding chapter involved one-dimensional potentials; we treated only a few two- or three-dimensional potentials that could be immediately reduced to a set of one-dimensional problems by separation of variables in Cartesian coordinates.

The most important class of three-dimensional potentials are those with **spherical symmetry**, where the potential energy depends only on the *magnitude* of the distance from a point, i.e., $V = V(r)$. One such potential is the Coulomb potential between, say, an electron and a proton. In general, the mathematical analysis of spherically symmetric potentials calls for spherical polar coordinates (r,θ,ϕ). Transformed to such coordinates, the Laplace operator ∇^2 becomes

$$\nabla^2 = \frac{\partial^2}{\partial r^2} + \frac{2}{r}\frac{\partial}{\partial r} - \frac{1}{r^2}\hat{\Lambda}, \tag{3•1–1}$$

where $\hat{\Lambda}$ is itself a differential operator that operates on the two angular variables,

$$\hat{\Lambda} = -\left[\frac{1}{\sin\theta}\frac{\partial}{\partial\theta}\left(\sin\theta\frac{\partial}{\partial\theta}\right) + \frac{1}{\sin_2\theta}\frac{\partial^2}{\partial\phi^2}\right]. \tag{3•1–2}$$

Note that ∇^2 can no longer be written as a sum of three terms, each of which depends on only one coordinate. Instead, we have three layers of operators within operators: The innermost operator is $\partial^2/\partial\phi^2$; contained within the operator $\hat{\Lambda}$, which is in turn contained within ∇^2.

In these kinds of cases—and spherical symmetry is not the only one[1]—separation of variables can still be achieved. In the specific case of spherical polar coordinates, we look for functions of the form

$$\psi(r,\theta,\phi) = \frac{1}{\sqrt{2\pi}}R(r)f(\theta)e^{im\phi}. \tag{3•1–3}$$

Here m must be a positive or negative integer (or zero), or else the wave function would not go over into itself under rotation by 2π about the polar axis. We have split off a factor $(2\pi)^{-1/2}$ for later convenience in normalizing.

The factor $e^{im\phi}$ is evidently an eigenfunction of the operator $\partial^2/\partial\phi^2$, with the eigenvalue $-m^2$. Hence, for the subset of functions with a given value of m, we may replace the operator $\partial^2/\partial\phi^2$ in (3•1–2) by its eigenvalue $-m^2$,thereby reducing $\hat{\Lambda}$ to a set of one-dimensional operators that operate on the polar angle θ only. Evidently, the function $f(\theta)$ in (3•1–3) must be an eigenfunction of that reduced operator; that is,

$$-\left[\frac{1}{\sin\theta}\frac{\partial}{\partial\theta}\left(\sin\theta\frac{\partial}{\partial\theta}\right) - \frac{m^2}{\sin^2\theta}\right]f(\theta) = \Lambda f(\theta) \tag{3•1–4}$$

where we have written Λ (without the operator caret ^) for the as-yet unknown eigenvalue.

Once we have determined the eigenvalues of $\hat{\Lambda}$, we may further replace $\hat{\Lambda}$ in (3•1–1) by one of its eigenvalues and request that the remaining function in (3•1–3), $R(r)$, be an eigenfunction of the resulting ordinary differential equation in r,

$$-\frac{\hbar^2}{2M}\left(\frac{d^2}{dr^2} + \frac{2d}{rdr} - \frac{\Lambda}{r^2}\right)R(r) + V(r)R(r) = \mathcal{E}(\Lambda)R(r). \tag{3•1–5}$$

[1] For a complete discussion of the conditions under which separation of variables can be achieved, see P. M. Morse and H. Feshbach, *Methods of Theoretical Physics* (New York: McGraw-Hill, 1953).

The last expression evidently constitutes a reduction of the original three-dimensional Schroedinger equation to a *set* of one-dimensional equations, with *a different equation for each of the different eigenvalues* Λ.

Evidently, everything hinges on the eigenvalues Λ. The eigenvalue problem (3•1–4) for Λ does not depend on the shape of the potential $V(r)$; it occurs in many problems of spherical symmetry throughout mathematics, physics, and engineering. The problem and its solution have been studied in every conceivable detail, and the interested reader is referred to the mathematical literature.[2] The only property of interest to us at this point is that the eigenvalues Λ of (3•1–4) are all of the form

$$\Lambda = l(l+1)\,, \tag{3•1–6}$$

where l must be a non-negative integer that satisfies

$$l \geq |m|. \tag{3•1–7}$$

Hence, for a given value of m, there is an infinite number of values of l and, therefore, of the eigenvalue Λ in (3•1–4) and (3•1–5). We shall not derive the relations (3•1–6) and (3•1–7) here; their proof will emerge later in a more general context.

Comment: Multiple Symmetries and Degeneracy. We pointed out in chapter 2 that for the cylindrical and the spherical harmonic oscillator, the separation of variables can be achieved in more than one coordinate system. Inspection shows that, as a rule, the different coordinate systems lead to different eigenfunctions. Clearly, the physical properties of the system cannot depend on the coordinate system used to describe that physics. What happens in all such cases is that those energy levels that permit a multiple description are degenerate and that each of the energy eigenfunctions in one coordinate system can be written as a linear superposition of the energy eigenfunctions obtained in the other coordinate system, with all energy eigenfunctions involved belonging to the same energy eigenvalue. We will return to this point later.

3.1.2 Centrifugal Potential and Angular Momentum

The radial equation (3•1–5) can be brought into a simpler and more useful form by the substitution

$$rR(r) = \chi(r), \qquad \text{or} \qquad R(r) = \chi(r)/r. \tag{3•1–8}$$

Executing the differentiations in (3•1–5) and inserting (3•1–6) leads to

---

[2] See again the book by Morse and Feshbach, as well as many texts on quantum mechanics, such as those by Schiff and by Merzbacher, quoted in appendix G.

$$-\frac{\hbar^2}{2M}\frac{d^2\chi}{dr^2} + \left[\frac{\hbar^2\, l(l+1)}{2Mr^2} + V(r)\right] = \mathcal{E}\chi, \tag{3•1–9}$$

which has the form of a true one-dimensional Schroedinger equation for the motion of an object in a one-dimensional **effective potential**

$$V_{\text{eff}}(r) = \frac{\hbar^2\, l(l+1)}{2Mr^2} + V(r) \text{ (for r > 0).} \tag{3•1–10a}$$

Because $R(r)$ cannot go to infinity as $r \to 0$, Eq. (3•1–8) implies that $\chi(r)$ must satisfy the boundary condition

$$\chi(0) = 0. \tag{3•1–11}$$

This is equivalent to requesting that the one-dimensional effective potential $V_{\text{eff}}(r)$ be terminated at $r = 0$ by an infinitely high wall, even for $l = 0$:

$$V_{\text{eff}}(r) = \infty \qquad \text{for} \qquad r \leq 0. \tag{3•1–10b}$$

The overall effective potentials for the four lowest values of l are shown in **Fig. 3•1–1**, for the case of a Coulomb potential. Note that different values of l in (3•1–10a) correspond to different effective potentials. Only for $l = 0$ is $V_{\text{eff}}(r) = V(r)$.

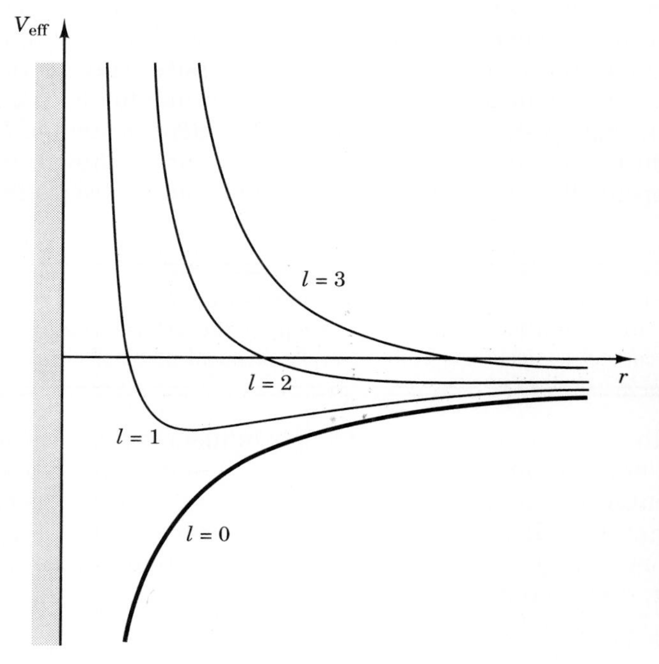

Figure 3•1–1. Effective one-dimensional potential for four different values of the quantum number l. The effective potential for $l = 0$ is equal to the real potential for $r > 0$. At $r = 0$ the effective potential goes to $+\infty$ for all values of l. The true potential shown here is the Coulomb potential, as it occurs in the hydrogen atom.

To understand the physical meaning of the additional term in V_{eff}, we define a new quantity L via

$$L^2 \equiv \hbar^2\Lambda = \hbar^2 l(l+1) \tag{3•1–12}$$

and rewrite (3•1–10a) as

$$V_{\text{eff}}(r) = \frac{L^2}{2Mr^2} + V(r) \qquad \text{for} \qquad r > 0. \tag{3•1–13}$$

If the L-term were a classical potential energy, it would correspond to a force

$$F_L = -dV_{\text{eff}}/dr = +L^2/Mr^3, \tag{3•1–14}$$

directed away from the coordinate origin. Now we know that in classical mechanics, for a particle moving along a circular trajectory with radius r in a potential $V(r)$, the speed of motion v is such that the inward-directed centripetal force exerted by the potential is balanced by the outward-directed inertial centrifugal force:

$$F_c = +Mv^2/r. \tag{3•1–15}$$

If we equate (3•1–14) and (3•1–15), we obtain the result

$$L = Mvr, \tag{3•1–16}$$

which is nothing other than the classical **angular momentum**! Thus, we see that the quantity L, which is clearly related to the orbital motion of the object, may be viewed as a quantum-mechanical generalization of the classical angular momentum and the extra term in V_{eff} as a **centrifugal potential**. Moreover, we see that this angular momentum is quantized according to (3•1–12), similar to the way energy eigenvalues are quantized. This identification of L with the angular momentum, and its quantization, is a very far-reaching concept, to which we will repeatedly return. For now, we simply note these observations and proceed.

Exercise: The relation (3•1–14) implies that the angular momentum $L = Mvr$ is to be held constant during the differentiation, rather than keeping the orbital velocity v constant. Why is that the correct choice even from a purely classical point of view?

The magnitude of the angular momentum strongly influences the way the one-dimensional equivalent wave function χ goes to zero as $r \to 0$. Suppose that $L \neq 0$ and that the potential $V(r)$ either remains finite as $r \to 0$, or that it goes to infinity less rapidly than $1/r^2$. In this case both the $V(r)$-term and the energy term in (3•1–9) may be neglected relative to the centrifugal potential term as $r \to 0$, and in that limit, (3•1–9) reduces to

$$-\frac{d^2\chi}{dr^2} + \frac{l(l+1)}{r^2}\chi = 0. \tag{3•1–17}$$

We would expect $\chi(r)$ to be an analytic function of r, in which case it must have a power series expansion. The lowest power term is not necessarily a constant, but must be of the order r^n, where $n \geq 0$ is an integer:

$$\chi(r) = c_n r^n + c_{n+1} r^{n+1} + c_{n+2} r^{n+2} + \ldots \tag{3•1–18}$$

If we insert this into (3•1–17), and compare the lowest-power terms, we find that

$$n = l + 1. \tag{3•1–19}$$

Note that the resulting function always satisfies the boundary condition (3•1–11), even for $l = 0$, in which case our argument leading to (3•1–17) is not applicable.

In fact, we could have used the power series expansion (3•1–18) to argue that the quantity Λ in (3•1–5) *had to* be of the form (3•1–6): For a sufficiently well-behaved potential, like the spherical harmonic oscillator potential, we *know* that the overall three-dimensional eigenfunctions are all analytic functions. It is not difficult to show that in that case $\chi(r)$ must also be an analytic function, with a power series expansion of the form (3•1–18). This would have restricted the possible values of Λ to the form $\Lambda = n(n-1)$. Because of (3•1–19), this is equivalent to (3•1–6). But with the eigenvalue problem (3•1–4) being independent of the potential $V(r)$, the set of eigenvalues must also be independent.

Inasmuch as it is Λ rather than m that shows up in the radial Schroedinger equation, the condition (3•1–7) on what values of l are allowed for a given m is better re-expressed as a condition on what values of m are allowed for a given l:

$$-l \leq m \leq +l\,. \tag{3•1–20}$$

There is a different solution of (3•1–4) for every value of m allowed by (3•1–20). Because the parameter m does not appear in (3•1–5), the energy eigenvalues of spherically symmetric potentials are necessarily at least $(2l+1)$-fold degenerate. Sometimes the degree of degeneracy is even higher.

We close this subsection with a remark on terminology. The states corresponding to the various values of l are often designated, not by the numerical values of l, but by the letters *s, p, d, f, g,* etc., the correspondence being as follows:

$l =$	0	1	2	3	4	5
letter:	*s*	*p*	*d*	*f*	*g*	*h*

This terminology dates back to late-19th century spectroscopy. The first four letters of the series originally stood for *sharp, principal, diffuse,* and *fundamental,* characterizing certain spectral line series. The rest of the sequence is alphabetic. The energy levels involved in the series were later found to involve the foregoing values of l. Although the original motivation for this terminology has long since disappeared, it continues to be widely employed.

3.1.3 Directional Dependence: Spherical Harmonics

It is common practice to lump the two angle-dependent factors in (3•1–3) together by defining a new set of functions

$$Y_l^m(\theta,\phi) = \frac{1}{\sqrt{2\pi}} f(\theta) e^{im\phi}, \qquad (3\text{•}1\text{–}21)$$

called the **spherical harmonics**. These functions play a role in two-dimensional Fourier analysis on the surface of a sphere that is similar to that of sin ϕ and cos ϕ in one-dimensional Fourier analysis. With the definition (3•1–21), (3•1–3) becomes

$$\psi(r,\theta,\phi) = R(r) \cdot Y_l^m(\theta,\phi). \qquad (3\text{•}1\text{–}22)$$

The spherical harmonics are usually normalized to unity, according to

$$\int_0^{2\pi} d\phi \int_0^{\pi} |Y_l^m(\theta, \phi)|^2 \sin\theta d\theta = \int_0^{\pi} |f(\theta)|^2 \sin\theta d\theta = 1. \qquad (3\text{•}1\text{–}23)$$

We list here—without derivation—the three spherical harmonics for $l = 0$ and $l = 1$:

$l = 0$:

$$Y_0^0 = \frac{1}{2}\sqrt{\frac{1}{\pi}}; \qquad (3\text{•}1\text{–}24)$$

$l = 1$:

$$Y_1^0 = +\frac{1}{2}\sqrt{\frac{3}{\pi}} \cos\theta; \qquad (3\text{•}1\text{–}25)$$

$$Y_1^{\pm 1} = \mp\frac{1}{2}\sqrt{\frac{3}{2\pi}} e^{\pm i\phi} \sin\theta. \qquad (3\text{•}1\text{–}26)$$

A more extensive table is given in **Appendix C**.

Exercise: Cartesian Form of Spherical Harmonics for l=1: Replace cos θ in (3•1–25) with the equivalent expression z/r, to obtain the Cartesian form of the function Y_0^0,

$$P_z \equiv Y_1^0 = +\frac{1}{2}\sqrt{\frac{3}{\pi}} \cdot \frac{z}{r}. \qquad (3\text{•}1\text{–}27)$$

Now split up $Y_1^{\pm 1}$ into real and imaginary parts according to

$$Y_1^{\pm 1} = \frac{1}{\sqrt{2}}(\pm P_x - iP_y). \qquad (3\text{•}1\text{–}28)$$

Show that P_x and P_y differ from P_z only by a permutation of the Cartesian coordinate axes, that is, by 90° rotations in space. Evidently, the three Cartesian functions P_x, P_y, and P_z form a triplet of symmetry-related functions.

Except for Y_0^0, the spherical harmonics are highly anisotropic, which means that, for a given radial dependence, the electron probability distribution is concentrated in the vicinity of certain preferred directions and actually vanishes along certain discrete other directions. This anisotropy plays an important role in many quantum-mechanical problems, especially problems of chemical bonding in molecules and crystals. **Fig. 3•1–2** illustrates the anisotropy for Y_0^0, Y_1^0, and Y_2^0.

The figure presents a perspective view of polar plots of $|Y_l^m(\theta,\phi)|^2$ as functions of the angles θ and ϕ. Note that the shapes shown are *not* the shapes of the overall wave functions themselves in three-dimensional space: The radial distance from the center of each plot is not the radial distance r in three-dimensional (r,θ,ϕ) coordinate space, but simply represents the relative magnitude of the probability density for any fixed r, as a function of direction. The anisotropy increases with increasing quantum number l.

Exercise: Generate plots analogous to Fig. 3•1–2 for the directional dependence of the probability distributions for $Y_1^{\mp 1}$ P_x and P_y.

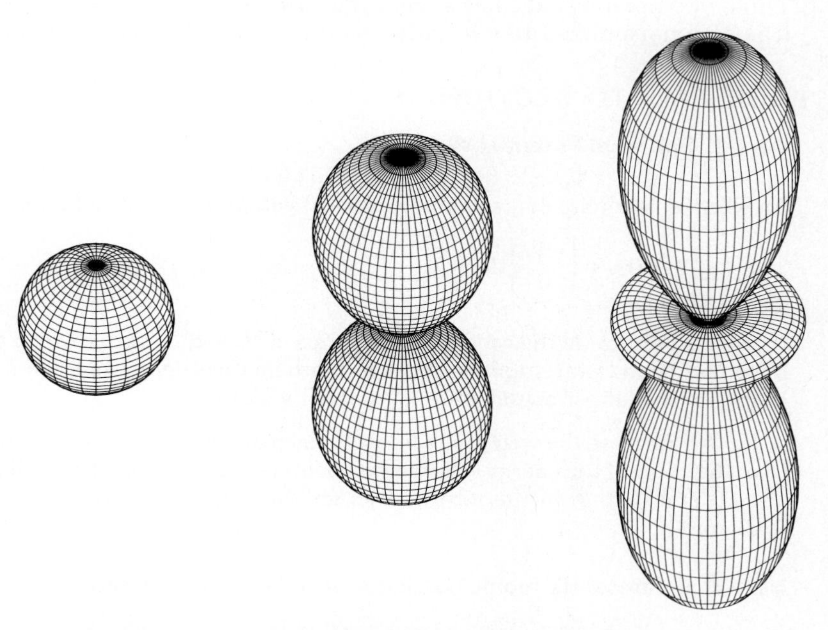

Figure 3•1–2. Perspective view of the directional dependence of the probability density for Y_0^0, Y_1^0, and Y_2^0, the three lowest-l spherical harmonics with $m = 0$.

3.1.4 Radial Normalization

If the wave functions are of the separated form (3•1–3), the normalization condition

$$\int_{\text{all space}} |\psi|^2\, d^3 r = 1, \tag{3•1–29}$$

written in spherical polar coordinates, factors into a product of two normalization integrals:

$$\int_0^\infty |R|^2 r^2\, dr \cdot \int_0^{2\pi} d\phi \int_0^\pi Y_l^m(\theta,\phi) \sin\theta\, d\theta = 1. \tag{3•1–30}$$

With the normalization condition (3•1–23) for the spherical harmonics, this reduces to

$$\int_0^\infty |R|^2 r^2 dr = 1. \tag{3•1–31}$$

If we again substitute $rR(r) = \chi(r)$—see (3•1–8)—we find that the effective one-dimensional wave function $\chi(r)$ must satisfy

$$\int_0^\infty |\chi(r)|^2\, dr = 1. \tag{3•1–32}$$

Thus, $\chi(r)$ not only satisfies a one-dimensional effective Schroedinger equation; it is also normalized like a truly one-dimensional wave function.

◆ PROBLEMS TO SECTION 3.1

#3•1-1: Spherical Potential Well

An electron is confined in a simple spherical potential well of radius R (make a graph):

$$V(r) = \begin{Bmatrix} -V_0 \\ 0 \end{Bmatrix} \text{ for } \begin{cases} r < R\,, \\ r > R\,. \end{cases} \tag{3•1–33}$$

(a) Determine the conditions, in terms of V_0 and R, under which this well has at least one bound state, and set up the mathematical formalism for determining the energy of the lowest bound state when such a state exists.

(b) Suppose the well is one Rydberg deep and has a radius twice the Bohr radius. Give the value of the energy of the lowest bound state, expressed in Rydbergs. (*Note:* See (3•2–7) and (3•2–9) in the next section for the definition of the Bohr radius and the Rydberg.)

#3•1-2: Spherical Harmonic Oscillator in Different Coordinates

We showed in Section 2.3 that the spherical harmonic oscillator energy eigenvalues, obtained by separation of variables in Cartesian coordinates, are of the form (2•3–49b),

$$\mathcal{E} = \hbar\omega(n + 3/2), \tag{3•1–34}$$

where

$$n = n_x + n_y + n_z \tag{3•1–35}$$

is an integer. The same result must also be obtainable by separation of variables in spherical polar coordinates.

Consider specifically the energy level $\mathcal{E} = \frac{7}{2}\hbar\omega$.

(a) Give the degeneracy of this level, and write down all triplets (n_x, n_y, n_z) that belong to it.

(b) Without attempting to determine the energy eigenfunctions in spherical polar coordinates, determine, from the degeneracy alone, which values of the angular momentum quantum number l **must** be present amongst the states with the specified energy. (Would this procedure work for higher energy states, without any additional inputs?)

(c) Consider next the two Cartesian states

$$(n_x, n_y, n_z) = (1, 1 ,0) \tag{3•1–36}$$

and

$$(n_x, n_y, n_z) = (0, 0, 2). \tag{3•1–37}$$

Convert their wave functions from Cartesian coordinates to spherical polar coordinates. By drawing on the table of spherical harmonics in appendix C, show that each state can be written as a linear superposition of two polar-coordinate states, Give the l- and m-values of all polar-coordinate states involved, and show that the radial portions of the wave functions thus found satisfy the radial Schroedinger equation in the form (3•1–5) with the appropriate value of Λ.

(d) Amongst the polar-coordinate states found under (c) should be one with $l = 2$, $m = 0$. Write this state as a linear superposition of Cartesian states.

3.2 HYDROGEN ATOM

3.2.1 Energy Eigenvalues

Ground State

Historically, the problem of the energy levels of an electron in the Coulomb potential inside the hydrogen atom,

$$V(r) = -\frac{e^2}{4\pi\epsilon_0 r}, \tag{3•2–1}$$

has been one of the "test beds" of quantum mechanics. The reasons for this are three-fold: (*i*) The problem is sufficiently simple that it can be solved rigorously in closed form; (*ii*) it is the only such case that also occurs in nature in almost its exact form,[3] not just as a rough approximation; and (*iii*) its properties can

[3] There exist a few very small corrections; we will discuss them at the end of the section.

be measured to a fantastic accuracy. Consequently, theories stood or fell with their ability to describe the observed energy-level spectrum of the hydrogen atom.

For simplicity, we ignore initially the motion of the proton and treat the proton as being fixed at $\mathbf{r} = 0$. Our task then is the solution of the quasi-one-dimensional radial Schroedinger equation (3•1–9) for the electron alone, with $M = m_e$ and with the potential (3•2–1),

$$-\frac{\hbar^2}{2m_e}\frac{d^2\chi}{dr^2} + \left[\frac{\hbar^2 l(l+1)}{2m_e r^2} - \frac{e^2}{4\pi\epsilon_0 r}\right]\chi = \mathcal{E}\chi. \tag{3•2–2}$$

Keeping the proton fixed at $\mathbf{r} = 0$ is equivalent to assigning an infinite mass to it; we will correct for the finite mass of the proton later.

Before actually solving the eigenvalue problem (3•2–2), it is instructive to discuss, purely qualitatively, the general kind of solution we might expect, at least for the ground state, and to estimate the order of magnitudes for both the spatial dimensions and the energy of the ground state. Qualitatively, the ground-state wave function must have the general shape shown in **Fig. 3·2–1**. We know from section 3.1—see (3•1–18) and (3•1–19)—that the wave function for $l = 0$ starts out linearly at $r = 0$; however, it must decay in an exponential-like fashion at large distances. Being the ground-state wave function, it cannot

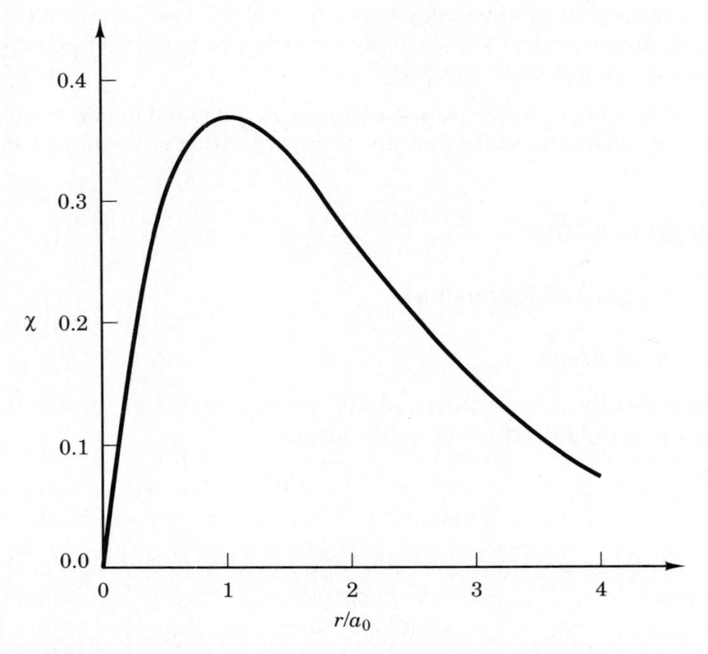

Figure 3·2–1. Qualitative shape of the ground state radial χ-function for the hydrogen atom.

have any additional nulls. The simplest trial function with that general behavior would be a function of the form

$$\chi(r) = Cr \cdot \exp(-r/a_0), \tag{3•2–3a}$$

where a_0 is an as yet unknown characteristic radius.

Normalization according to (3•1–32) requires that

$$C^2 a_0^3 = 4 \,, \tag{3•2–3b}$$

and one finds easily that the expectation values of potential and kinetic energy for the equivalent one-dimensional problem are given by

$$\langle \mathcal{E}_{\text{pot}} \rangle = -\frac{e^2}{4\pi\epsilon_0 a_0} \,. \tag{3•2–4}$$

and

$$\langle \mathcal{E}_{\text{kin}} \rangle = \frac{\hbar^2}{2m_e a_0^2} \tag{3•2–5}$$

As a curiosity, we note that $\langle \mathcal{E}_{\text{kin}} \rangle$ is the kinetic energy of a plane wave with the wave number $k = 1/a_0$, or a wavelength $\lambda = 2\pi a_0$, exactly the length of a classical circular orbit with radius a_0. This was de Broglie's point of departure for inventing wave mechanics.

As in the case of the harmonic oscillator, $\langle \mathcal{E}_{\text{pot}} \rangle$ increases with increasing size, and $\langle \mathcal{E}_{\text{kin}} \rangle$ decreases, although the detailed dependence of $\langle \mathcal{E}_{\text{pot}} \rangle$ on size is different than for the harmonic oscillator. For the specific a_0-dependences given in (3•2–5) and (3•2–4), the sum of the two expectation values has a minimum at that value of a_0 for which

$$\langle \mathcal{E}_{\text{pot}} \rangle = -2\,\langle \mathcal{E}_{\text{kin}} \rangle, \qquad \text{or} \qquad \frac{e^2}{4\pi\epsilon_0 a_0} = \frac{\hbar^2}{2m_e a_0^2}, \tag{3•2–6}$$

corresponding to

$$\boxed{a_0 = 2\frac{\hbar^2}{2m_e} \cdot \frac{4\pi\epsilon_0}{e^2} = \frac{4\pi\epsilon_0 \hbar^2}{e^2 m_e} (= 5.29177 \cdot \times 10^{-11}\text{m})\,.} \tag{3•2–7}$$

From the variational principle, we would expect the radius of the hydrogen atom to be of this order of magnitude, and we use this figure as the natural unit of length for the hydrogen atom. In fact, a_0 is nothing other than the **Bohr radius,** the radius of the innermost orbit in the 1912 Bohr atomic model of the hydrogen atom, which played a large role in the early history of quantum mechanics, predating the Schroedinger equation by more than 10 years.

Because of (3•2–6), the minimum expectation value of the total energy is

$$\langle \mathcal{E}_{\text{tot}} \rangle = \langle \mathcal{E}_{\text{pot}} \rangle + \langle \mathcal{E}_{\text{kin}} \rangle = -\langle \mathcal{E}_{\text{kin}} \rangle = -R_\infty \,, \tag{3•2–8}$$

where we have defined the quantity

$$R_\infty = \frac{\hbar^2}{2m_e a_0^2} = \frac{1}{2}\frac{e^2}{4\pi\epsilon_0 a_0} = \frac{m^e}{2}\left(\frac{e^2}{4\pi\epsilon_0\hbar}\right)^2 (= 13.6057\text{eV}), \qquad (3{\bullet}2\text{–}9)$$

called the **Rydberg energy**, also occurring already in Bohr's theory of the hydrogen atom. The subscript '∞' refers to the fact that our treatment is equivalent to a nucleus with infinite mass.

Invoking the variational principle once more, we would expect the energy of the ground state to be close to the value (3•2–8), possibly somewhat lower (more negative). However, our guess (3•2–3a) for the trial wave function was a lucky one: With the choice (3•2–7) for a_0, the function (3•2–3a) is an *exact* solution of (3•2–2) for $l = 0$, and the expectation value for the total energy is the true ground-state energy eigenvalue. The proof of this claim is left as an exercise to the reader.

Exercise. Show that the minima in the effective potential for $l > 0$ occur at

$$r_l = a_0 \cdot l(l+1)\,, \qquad (3{\bullet}2\text{–}10)$$

with a minimal energy

$$V_{\text{eff}}(r_l) = -R_\infty / l(l+1)\,. \qquad (3{\bullet}2\text{–}11)$$

Higher Energies

To obtain the full set of energy eigenvalues, we must solve (3•2–2) more systematically. To simplify the notation, we note first that the combinations of fundamental constants occurring in (3•2–2) may be expressed in terms of a_0 and R_∞, as

$$\frac{\hbar^2}{2m_e} = R_\infty a_0^2\,, \quad \frac{e^2}{4\pi\epsilon_0} = 2R_\infty a_0\,. \qquad (3{\bullet}2\text{–}12\text{a,b})$$

With the help of these, we may rewrite (3•2–2) as

$$a_0^2\frac{d^2\chi}{dr^2} - \left[a_0^2\frac{l(l+1)}{r^2}\frac{2a_0}{r} - \frac{\mathcal{E}}{R_\infty}\right]\chi = 0. \qquad (3{\bullet}2\text{–}13)$$

A bound state wave function must vanish at infinity. In the limit $r \to \infty$, the $1/r^2$-term and the $1/r$-term in (3•2–13) may be neglected next to the term $\mathcal{E}/R_\infty$. The resulting truncated equation has the simple asymptotic solution

$$\chi(r) \propto \exp(-r/na_0), \qquad (3{\bullet}2\text{–}14)$$

where we have defined

$$n \equiv \sqrt{R_\infty/|\mathcal{E}|}\,. \qquad (3{\bullet}2\text{–}15)$$

We will see later that n must be an integer—hence the notation n—but it should at this point still be treated as an unknown quantity.

The asymptotic behavior (3•2–14) suggests introducing the dimensionless variable

$$s = \frac{2r}{na_0}, \tag{3•2–16}$$

which leads to

$$\chi(s) \propto \exp(-s/2) . \tag{3•2–17}$$

We have split off a factor 2 for later convenience in normalizing $\chi(s)$. Written in terms of s rather than r, the differential equation (3•2–13) becomes

$$\frac{d^2\chi}{ds^2} - \left[\frac{l(l+1)}{s^2} - \frac{n}{s} + \frac{1}{4}\right]\chi = 0 . \tag{3•2–18}$$

It now turns out that (3•2–18) has bound-state solutions only if n is an integer. Furthermore, these solutions can be written as the product of a polynomial of order n and the asymptotic exponential (3•2–17), similar to the way the energy eigenfunctions of the harmonic oscillator could be written as the product of a polynomial and an asymptotic Gaussian.

To derive these results, we note first, from (3•1–18) and (3•1–19), that in the limit $s \to 0$, $\chi(s)$ must vanish according to a power law of the form

$$\chi(s) \propto s^{l+1} . \tag{3•2–19}$$

This suggests splitting off such a power term from the polynomial and writing the solutions of (3•2–18) in the form of a triple product

$$\chi(s) = A \cdot L(s) \cdot s^{l+1} \cdot \exp(-s/2) , \tag{3•2–20}$$

where $L(s)$ is an unknown function that remains to be determined, and A is a normalization constant, split off from $L(s)$ for later convenience. By inserting (3•2–20) into (3•2–18), one obtains the equation for $L(s)$:

$$s\frac{d^2L}{ds^2} - [s - 2(l+1)]\frac{dL}{ds} + [n - (l+1)]L = 0 . \tag{3•2–21}$$

We solve (3•2–21) by a power series of the form

$$L(s) = \sum_{\nu=0}^{\infty} c_\nu s^\nu , \text{ with } c_0 \neq 0 . \tag{3•2–22}$$

If we insert (3•2–22) into (3•2–21), and collect all terms of the same power, we obtain

$$\sum_\nu \{[\nu(\nu-1) + 2\nu(l+1)]c_\nu - [(\nu-1) + (l+1) - n]c_{\nu-1}\}s^\nu = 0 . \tag{3•2–23}$$

Because this must be true for all values of s, each power of s must satisfy (3•2–23) separately, leading to the upward recursion relation

$$c_\nu = \frac{\nu + l - n}{\nu(\nu + 2l + 1)} c_{\nu-1}, \qquad \nu = 1, 2, 3, \ldots, \tag{3•2–24}$$

which permits us to calculate each coefficient c_ν from the preceding one, starting with $c_0 \neq 0$.

Exercise: Derive (3•2–21), (3•2–23), and (3•2–24).

This recursion relation evidently truncates *if and only if* the number n in (3•2–24) is an integer that satisfies

$$n > l. \tag{3•2–25}$$

In this case, the right-hand side of (3•2–24) vanishes for $\nu = n - l$, and $L(s)$ will be a polynomial of order

$$q = n - (l + 1)\,. \tag{3•2–26}$$

Inasmuch as the asymptotic behavior of a product of an exponential and a polynomial is governed by the exponential, those energies for which the number n in (3•2–15) is an integer are indeed bound state energies, as claimed. The energy of the states follows readily by solving (3•2–15) for $\mathcal{E}$, leading to

$$\boxed{\mathcal{E}_n = -\frac{R_\infty}{n^2}.} \tag{3•2–27}$$

This is Bohr's key 1912 result. Because of its central role, the quantum number n is often called the *main* or *principal* quantum number.

It remains to be shown that there are no other bound states. For those energies for which the series (3•2–23) does *not* truncate, the ratio of successive expansion coefficients in the limit of very large values of ν satisfies

$$\frac{c_\nu}{c_{\nu-1}} \to \frac{1}{\nu}, \tag{3•2–28}$$

which is the same behavior as for the exponential function e^s. But the behavior of a (converging) power series expansion for large arguments is governed by its high-order coefficients. Hence, (3•2–28) implies an asymptotic behavior of $L(s)$ like e^s. In terms of $\chi(s)$, this would imply an exponential *increase* like

$$\chi(s) \to e^{+s/2}\,, \tag{3•2–29}$$

clearly not representing a bound state.

Allowed Quantum Numbers, and the l-Degeneracy

We turn now to the constraint imposed by (3•2–25). Because the lower limit on the angular momentum quantum number is $l = 0$, the number n can assume all positive integer values. However, for any value $l > 0$, only integer values $n > l$ can occur.

Except for the lowest energy state ($l = 0$; $n = 1$), states with a given energy $\mathcal{E}_n$ occur for more than one l-value, implying that those energy levels are degenerate. This makes it useful to turn (3•2–25) around and reinterpret it as a constraint on l for a given n, rather than as a constraint on n for a given l: Evidently, there are n different values of the angular momentum quantum number l for which the $\mathcal{E}_n$ energy can (and will) occur:

$$l = 0, 1, \ldots, n - 1. \tag{3•2–30}$$

This implies an n-fold ***l*-degeneracy** of each energy level.

Exercise: l-Degeneracy and Effective Potential Make a semiquantitative drawing, similar to Fig. 3•1–1, of the effective potentials V_{eff} inside the hydrogen atom for $l = 0$, 1, and 2. Show negative energies only, with an expanded scale. Show the various energy eigenstates by horizontal lines superimposed on the effective potentials, drawn between the appropriate classical turning points and using appropriate graphic means to distinguish different states belonging to the same energy.

Because each state with a given l also exhibits a $(2l + 1)$-fold m-degeneracy, each energy level $\mathcal{E}_n$ in (3•2–27) is in fact n^2-fold degenerate:

$$\sum_{l=0}^{n-1} (2l + 1) = n^2. \tag{3•2–31}$$

We will see later that the electron spin causes another factor of two in the degeneracy.

3.2.2 Corrections

Mass Corrections and Hydrogen-Like Spectra

Given the energy levels expression (3•2–27), the wavelengths of the spectral lines of the hydrogen atom follow from the relation

$$\frac{2\pi\hbar c}{\lambda_{nm}} = \hbar\omega_{nm} = \mathcal{E}_n - \mathcal{E}_m. \tag{3•2–32}$$

The experimentally observed wavelengths are in excellent overall agreement with this prediction, the largest discrepancy being that all observed wavelengths are slightly larger, by about 0.0547%. This is simply a consequence of our having treated the proton as infinitely heavy, with no motion of its own. It is not difficult to correct for the proton's motion. In classical mechanics, the correction is achieved by simply replacing the true mass m_e of the electron by a *reduced mass* m_r, defined via

$$\frac{1}{m_r} = \frac{1}{m_e} = \frac{1}{m_p}. \tag{3•2–33}$$

We will show later (chapter 11) that this correction carries over into quantum mechanics unchanged. The net effect is simply to change the value of the Rydberg constant from R_∞ to

$$R_H = \frac{m_p}{m_p + m_e} R_\infty = 0.99456\, R_\infty. \tag{3•2–34}$$

Making this change leads to almost perfect agreement with observations. Moreover, the mass correction formalism also predicts slight spectral differences between hydrogen and deuterium atoms: The nucleus in deuteron has roughly twice the mass of the proton, leading to a mass correction roughly one-half that for ordinary hydrogen. These differences are easily observed, and the predicted differences have been confirmed experimentally with very high accuracy.

A variety of hydrogen-like structures containing unstable particles have also been studied, such as the positronium atom, in which the heavy proton has been replaced by a light positron, and the muonium atom, in which the electron has been replaced by a much heavier muon. These systems all have hydrogen-like spectra, with suitably modified Rydberg constants.

A slightly different class of such spectra occurs in atoms from which all electrons except one have been stripped, leading to the problem of the motion of an electron in the Coulomb potential of a charge Ze rather than e, where Z is the atomic number of the atom. Inspection of the Coulomb potential (3•2–1) shows that the effect of replacing the nuclear charge by Ze, while keeping a single electron, is the same as replacing ϵ_0 by ϵ_0/Z. Inasmuch as the Rydberg energy (3•2–9) is proportional to ϵ_0^{-2}, all energy levels scale with Z^2, converting (3•2–27) into

$$\mathcal{E}_n = -\frac{Z^2 R_Z}{n^2}, \tag{3•2–35}$$

where we have also replaced R_∞ by R_Z, the slightly larger Rydberg constant appropriate for the mass of the heavier nucleus. Note that, except for this small difference between the Rydberg constants, the energy levels for n-values divisible by Z coincide with energy levels of the hydrogen atom, and hence, the ion spectra contain some lines very close to those of hydrogen. The only (small) difference is that the finite-mass correction is smaller for the heavier ions, shifting the corresponding spectral lines somewhat toward shorter wavelengths. The predicted line shifts agree perfectly with observations. The case of singly-ionized helium ($Z = 2$) is the simplest. The solar atmosphere contains a large quantity of helium, and the hydrogen-like spectral lines of He^+ had been known to occur copiously in the spectrum of sunlight, long before the reasons for this phenomenon were understood.[4] It was one of the triumphs of the "old quantum theory" of Bohr and Sommerfeld to explain this observation.

Finally, a rich variety of *approximately* hydrogen-like structures occurs in solid-state physics. Ionized donors and acceptors in semiconductors often behave approximately hydrogen-like, differing from the hydrogen atom in two ways: (*i*) The permittivity of space, ϵ_0, occurring in Coulomb's law, and hence in the Rydberg energy, must be replaced by the permittivity of the semiconductor; and (*ii*) the mass m_e of the free electron must be replaced by an effective mass m^* of either an electron or a hole. Because the dielectric constants of semiconductors tend to be large, and the effective masses small, the binding energies are small, usually between 5 meV and 50 meV, thus permitting the purely thermal ionization of the donors and acceptors that is an important feature of a good semiconductor. **Excitons** in semiconductors are structures in which a hole takes on the role of the proton. For more information on hydrogen-like structures in solids—especially on the limitations of the simple hydrogenic model—the interested reader is referred to texts on solid-state physics.

Fine Structure

Perhaps the most characteristic feature of the energy levels of the hydrogen atom is their n-fold l-degeneracy, a consequence of the Coulomb potential. On closer inspection, this degeneracy is often split, at least partially, into levels that are separated from each other by very small amounts, on the order of at most a few parts per million relative to the separation between levels with different values of the quantum number n. The principal origin of this **fine structure** is the so-called **spin-orbit interaction** between the electron and the proton: Just as the electron moves relative to the proton, the proton moves relative to the electron. But this implies a magnetic field seen by the electron, and the small potential energy of the intrinsic magnetic moment that is associated with the electron spin causes a small amount of splitting between some of the l-degenerate states.

Strictly speaking, the spin-orbit interaction is a relativistic effect, and a rigorous theory of the fine structure must also take into account the relativistic increase in the mass of the electron with increasing speed. This requires a treatment based on the relativistic generalization of the Schroedinger equation, the so-called **Dirac equation**.[5] Such a treatment lies outside the scope of

---

[4] Hence the name for helium: *Helios* is Greek for *sun*.

[5] It is one of the great ironies in the history of physics that a correct expression for the fine structure splittings in the hydrogen spectrum was obtained already by the precursor of quantum mechanics, the Bohr-Sommerfeld atomic model, corrected for the relativistic mass variation of the electron, even though that theory subsequently turned out to be a blind alley.

this text, and the interested reader is referred to more advanced texts.[6] An approximate treatment of the hydrogen fine structure, ignoring the mass variation, is given in chapter 21.

Beyond Fine Structure

Even the Dirac theory does not yet give a full account of all energy-level splittings. Certain energy levels that remain degenerate in the Dirac theory split up slightly, a phenomenon first observed experimentally in 1947 by W. Lamb, earning him the 1955 Nobel prize in physics. In his honor, the phenomenon is called the **Lamb shift**. It is sufficiently small that it can best be measured by observing the *direct* microwave-frequency transitions between the no-longer degenerate levels, rather than by the splitting of shorter wavelength lines.

The Lamb shift is due to so-called **quantum-electrodynamic corrections**. Quantum electrodynamics is an extension of relativistic quantum mechanics that includes the radiation field as part of the electron quantum system itself. In both the Schroedinger theory and the Dirac theory, the energy levels of the atom are calculated as if the atom resided in a field-free vacuum, ignoring any coupling to external radiation fields. But the very fact that electromagnetic transitions *can* take place means that there is such a coupling. The external electromagnetic radiation field is really a quantum system in its own right, which can interact with the electron. A “field-free” vacuum must be viewed as a quantum system that happens to be in its ground state, rather than being ignored. When the radiation fields and their coupling to the atom are included in the quantum mechanics, the interaction of the electron with the radiation field system changes the electron energy levels themselves very slightly, even when the radiation field is in its ground state. These changes are the origin of the Lamb shift. Precision measurements of such effects agree very well with quantum-electrodynamic theory.

Although we will give a very elementary account of field quantization in chapter 20, that treatment will stay far below the level of formal quantum electrodynamics, and the interested reader must again be referred to more advanced texts.[7]

Finally, there are corrections due to the fact that the proton is not simply a mathematical point charge without any further electrical properties, but has a spin and therefore a magnetic moment, two properties we will discuss later. This leads to an additional **hyperfine splitting** of the energy levels, depending upon whether the spins of the proton and the electron are parallel or

---

[6] See, for example, the texts by Merzbacher, Schiff, or Sakurai listed in appendix G.

[7] See, for example, Sakurai, loc. cit. A delightful, very readable elementary account is found in E. G. Harris, *A Pedestrian Approach to Quantum Field Theory* (New York: Wiley, 1972).

antiparallel to each other. The splitting is measurable only for states for which the electron wave function has a finite magnitude at the location of the proton, that is, for states with $l = 0$. In those states, the splitting is on the order of a few times 10^{-6} eV. These splittings are too small to be measured with high accuracy from the *optical* spectral lines, all of which involve transition between levels with different n. But, at least for the hydrogen *ground* state, the splitting is very easily measured *directly,* from the radiation associated with a flip in the relative orientation of the electron and proton spins while the electron remains in the state with $n = 0$. The accompanying radiation is copiously emitted (and absorbed) by the hydrogen in stars and in interstellar space; it is the "21-cm line" that forms the backbone of radio astronomy. The transition frequency, about 1.42 GHz, is very easily and accurately measured by electronic means. This is also the operating frequency of the hydrogen maser, which can be measured with almost unbelievable accuracy; the accepted value[8] is

$$\nu_H = 1{,}420{,}405{,}751.766 \pm 0.001\ Hz\ , \tag{3•2–36}$$

to an accuracy of 1 part in 10^{13}!

It is readily possible to include the spin of the proton in the quantum mechanics of the hydrogen atom and to calculate the splitting frequencies to at least the accuracy with which the fundamental constants themselves are known (about seven significant figures). The observed splitting frequencies are again in excellent agreement with those predictions.

The picture that emerges from all this is one of a complete quantitative understanding of the hydrogen atom within the limits of existing measurement techniques. This perfect agreement represents an extremely severe test of the validity of quantum mechanics itself, at least for the quantum mechanics of electrons and of light. It is a test the theory has so far passed with flying colors. In fact, the faith that most of today's physicists have in the *quantitative* validity of quantum mechanics is no longer based merely on such gross phenomena as the photoelectric effect and electron diffraction; it is largely based on the accuracy with which the theory has been able to describe every single aspect of the hydrogen spectrum. There is, of course, no absolute assurance that more refined measurements might not show up discrepancies, just as was the case eventually with Newtonian classical mechanics.

3.2.3 Radial Eigenfunctions

The polynomials $L(s)$ defined by the recursion relation (3•2–24) are what is known in the mathematical literature as **associated Laguerre polynomials**. Their properties have been studied in every conceivable detail, and the reader

---

[8] E. R. Cohen and B. N. Taylor, *Revs. Mod. Phys., 59* (4), 1121–1148, October 1987.

interested in going beyond the treatment presented here is referred to the literature.[9] A number of different normalizations are in use. The most convenient choice is obtained by setting the highest-order coefficient to equal ± 1, according to

$$c_q = (-1)^q \,. \tag{3•2–37}$$

If the upward recursion relation (3•2–24) is inverted and written as a *downward* recursion relation, one finds that, with the choice (3•2–37) for the highest-order coefficient, the remaining coefficients can be written explicitly as

$$c_{q-\mu} = \frac{(-1)^{q-\mu}}{\mu!} \cdot \frac{q!}{(q-\mu)!} \cdot \frac{1}{(n-l-1)!} \,. \tag{3•2–38}$$

Exercise: Derive (3•2–38), and show that all coefficients are integers.

The normalization coefficients A in (3•2–20) corresponding to the choice (3•2–37) can be shown to be given by

$$A^2 = \frac{1}{n^2} \cdot \frac{1}{(n+l)!} \cdot \frac{1}{(n-l-1)!} \,, \tag{3•2–39}$$

assuming r-normalization. For s-normalization, a factor $r/s = na_0/2$ must be added. The derivation of (3•2–39) is tedious and will not be attempted here; the interested reader is referred to the literature quoted earlier.

We give here the functions χ_{nl} for $n = 0$ and $n = 1$:

$n = 1$ $(s = 2r/a_0)$:

$$l = 0: \quad \chi_{10}(r) = \frac{1}{\sqrt{a_o}} s\, e^{-s/2} ; \tag{3•2–40}$$

$n = 2$ $(s = r/a_0)$:

$$l = 0: \quad \chi_{20}(r) = \frac{1}{2\sqrt{2a_0}} s(2-s)\, e^{-s/2} ; \tag{3•2–41}$$

$$l = 1: \quad \chi_{21}(r) = \frac{1}{2\sqrt{6a_0}} s^2\, e^{-s/2} . \tag{3•2–42}$$

The functions shown are r-normalized, even though the right-hand sides are expressed in terms of s, to keep the notation more compact. A more extensive table is given in **appendix D**.

[9] See, for example, the book by Morse and Feshbach, cited earlier, as well as many quantum mechanics texts, such as those by Schiff and by Merzbacher, quoted in appendix G. There does not exist agreement on either terminology or normalization. Our usage is that of Merzbacher.

Exercise: Confirm that the polynomials $L(s)$ contained in these functions satisfy the recursion relation (3•2–24) and that the χ-functions are indeed r-normalized.

If needed, the full three-dimensional wave functions may be obtained by first multiplying by r, to make the transition from $\chi(r)$ back to $R(r)$, and then multiplying by the appropriate spherical harmonic contained in the original product wave function (3•1–22). We give here only the result for the ground state, for which $f(\theta) = 1/\sqrt{2}$:

$$\psi(r) = \frac{1}{\sqrt{\pi}} \frac{1}{a_0^{3/2}} \exp\left(-\frac{r}{a_0}\right). \tag{3•2–43}$$

The transformations of the other functions are left to the reader.

Note that for any *given* value of l, the principal quantum number n is chosen in such a way that the lowest energy state for that value of l is labeled with $n = l + 1$, in order to give all states with the same energy the same principal quantum number. It might have been mathematically more consistent to use the number q to label the states, reflecting the order of the states within a set of the same l, starting with $q = 0$. This terminology is indeed used by a few authors, but we do not follow it here, preferring a labeling that reflects the energy levels.

We show in **Fig. 3•2–2** the radial probability densities $|\chi|^2$ for the six lowest energy states, with $n = 1$ through 3. Note that $|\chi|^2 dr$ is the probability of "finding" the electron in a spherical shell of radius r and thickness dr, already integrated over the angular variables. A number of features of these density distributions are readily apparent from the figure. Not counting the null at $s = 0$, which is due to the s^{l+1} term in $\chi(s)$, each distribution exhibits $q = n - l - 1$ nulls and the same number of maxima. The outermost maximum is always the most pronounced one, dominating the distribution. For a given value of l, this outer maximum shifts rapidly outward with increasing value of n, roughly as $n^2 a_0$. The dominant outer maxima for different values of l, but belonging to the same n, are very nearly of the same height, and they are clustered together in radius, moving inward slightly with increasing azimuthal quantum number l.

The associated Laguerre polynomials depend on the two quantum numbers l and n. From a quantum-mechanical point of view, it would have been preferable to label the polynomials with the two subscripts l and n, or possibly l and q, leading to a terminology such as L_{ln} or L_{lq}. However, the established practice, which pre-dates quantum mechanics,[10] is to label them with a single subscript $q = n - (l + 1)$ and a superscript $p = 2l + 1$:

$$L_q^p = L_{n-l-1}^{2l+1}. \tag{3•2–44}$$

[10] The practice is not universal: Some authors denote as $L^p{}_{q+p}$ what we call $L^p{}_q$.

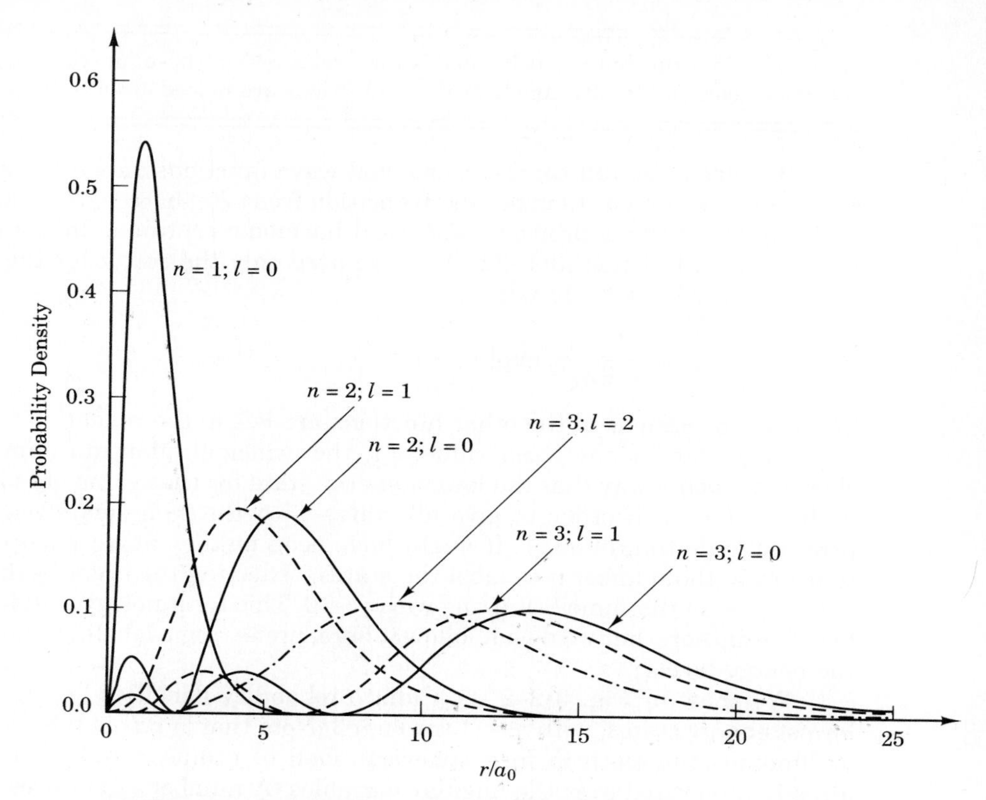

Figure 3•2–2. Probability density distributions for the lowest energy hydrogen eigenstates.

3.3 SINGLE-AXIS ROTATIONAL INVARIANCE

Although less important than problems of full spherical symmetry, one often encounters problems in which the potential energy is rotationally invariant about only a single axis. Separation of variables can still be achieved in cylindrical polar coordinates (ρ,ϕ, z), with the z-axis taken as axis of rotation. The potential energy then assumes the form

$$V(\rho,\phi,z) = V_\rho(\rho) + V_z(z)\,, \tag{3•3–1}$$

with no ϕ-dependent term. In many problems of this kind, there is not even a z-dependence, effectively reducing the problem to one of two dimensions.

In the absence of any z-dependence, the Laplace operator ∇^2 reduces to

$$\nabla^2 = \frac{\partial^2}{\partial\rho^2} + \frac{1}{\rho}\frac{\partial}{\partial\rho} + \frac{1}{\rho^2}\frac{\partial^2}{\partial\phi^2}\,. \tag{3•3–2}$$

Separation of variables is then achieved by setting

$$\psi(\rho,\phi) = R(\rho)e^{im\phi}, \tag{3•3–3}$$

where m must again be an integer, or else the wave function would not go over into itself under rotation by 2π. It is left to the reader to show that the overall wave function $\psi(\mathbf{r})$ will be satisfied if $R(\rho)$ is normalized according to

$$2\pi\int_0^\infty |\mathrm{R}(\rho)|^2\,\rho d\rho = 1. \tag{3•3–4}$$

Equation (3•3–3) is of the form (3•1–3), with a special choice for $\psi_\phi(\phi)$. With (3•3–2) and (3•3–3) the Schroedinger equation reduces to

$$-\frac{\hbar^2}{2M}\left(\frac{d^2}{d\rho^2} + \frac{1}{\rho}\frac{d}{d\rho} - \frac{m^2}{\rho^2}\right)R + V_\rho(\rho)R = \mathcal{E}_\rho\,(m)R\ . \tag{3•3–5}$$

Note that on the right-hand side we have written $\mathcal{E}_\rho(m)$ rather than simply $\mathcal{E}_\rho$, to stress that the radial Schroedinger equation, although one-dimensional, contains the quantity m that governs the angular dependence of the overall wave function. Hence, we have not a *single* radial Schroedinger equation, but an infinite *set* of such equations, one for each value of $|m|$, and with a different set of energy eigenvalues for each $|m|$.

If the z-motion cannot be ignored, we must write, instead of (3•3–3),

$$\psi(\rho\phi,z) = Z(z)R(\rho)e^{im\phi}. \tag{3•3–6}$$

Equation (3•3–5) remains unchanged, but the additional equation

$$-\frac{\hbar^2}{2M}\frac{d^2Z}{dz^2} + V_zZ = \mathcal{E}_zZ \tag{3•3–7}$$

must be satisfied. The total energy is then

$$\mathcal{E} = \mathcal{E}_\rho(m) + \mathcal{E}_z\ . \tag{3•3–8}$$

Chapter 4

WAVE PACKETS AND UNCERTAINTY RELATIONS

4.1 WAVE PACKETS AND THEIR REPRESENTATIONS

In this chapter, we return to wave packets in free space, of the form (1•1–13). For simplicity, we restrict ourselves to one dimension, and for later convenience, we split off a factor $(2\pi)^{-1/2}$ from the amplitude function $a(k)$ and write

$$\Psi(x, t) = \frac{1}{\sqrt{2\pi}} \int_{-\infty}^{+\infty} a(k) \exp[i(kx - \omega t)]\, dk. \tag{4•1–1}$$

We are interested in objects with mass; hence, ω and k related via the freespace dispersion relation (1•1–7),

$$\omega = \frac{\hbar k^2}{2M}. \tag{4•1–2}$$

We lump the factor $\exp(-i\omega t)$ together with $a(k)$ by defining a time-dependent amplitude function

$$A(k, t) = a(k)e^{-i\omega t}, \tag{4•1–3}$$

and we rewrite (4•1–1) as

$$\Psi(x, t) = \frac{1}{\sqrt{2\pi}} \int_{-\infty}^{+\infty} A(k, t) e^{ikx}\, dk. \tag{4•1–4}$$

The functions $\Psi(x, t)$ and $A(k, t)$ form a *spatial* Fourier transform pair; the time t is an external parameter in both. From Fourier transform theory (appendix E), the inverse transform of (4•1–4) is

$$A(k, t) = \frac{1}{\sqrt{2\pi}} \int_{-\infty}^{+\infty} \Psi(x, t) e^{-ikx}\, dx, \tag{4•1–5}$$

and the two function $A(k, t)$ and $\Psi(x, t)$ have the same normalization:

$$\int_{-\infty}^{+\infty} |A(k, t)|^2\, dk = \int_{-\infty}^{+\infty} |\Psi(x, t)|^2\, dx. \tag{4•1–6}$$

Our reason for splitting off the factor $(2\pi)^{-1/2}$ in (4•1–1) was to achieve the symmetry between (4•1–4) and (4•1–5), as well as the equal normalizations.

The functions $\Psi(x, t)$ and $A(k, t)$ form completely equivalent descriptions of the state of the object: From either of these functions, the other can be obtained. We say they form two different **representations** of the same state of the object, and we call the functions describing that state the **state functions** of the object in that particular representation. The different representations are named by the physical meaning of the independent variable occurring in the state function (besides the time t). The variable x designates the position of the object; hence, the ordinary wave function $\Psi(x, t)$ is the state function in the **position representation**. Similarly, $A(k, t)$ is the state function in the **wave number representation**. We will encounter additional representations later.

What is the physical meaning of the amplitude function $A(k, t)$? We interpreted $|\Psi(x, t)|^2\, dx$ as the probability that the object will materialize in the interval $(x, x + dx)$ during any position measurement made at time t. Thus, the right-hand side of (4•1–6) is a sum over probabilities, summed over all possible positions of the object. Now, if $A(k, t)$ is simply an alternative but equivalent representation of the state of the particle, the left-hand side of (4•1–6) must also be a sum over probabilities, summed over all possible values of k of the object described by either $\Psi(x, t)$ or $A(k, t)$. The wave number k is a wave property; the associated particle property is the momentum $p = \hbar k$. Thus, we postulate that

$$\rho_k(k)\, dk = |A(k, t)|^2\, dk \tag{4•1–7}$$

is the probability that the object described by the amplitude function $A(k, t)$ will materialize with a momentum in the interval $(\hbar k, \hbar k + \hbar dk)$ during any momentum measurement made at time t. The quantity $\rho_k(k)$ is the **probability density of the wave number *k***. We view this postulate as a simple generalization of the Born interpretation postulate for $|\Psi|^2$, rather than as an additional new postulate.

It is often convenient to replace the wave variable k in $A(k, t)$ by the particle-like momentum variable $p = \hbar k$ and to re-scale $A(k, t)$ by defining a wave function in the **momentum representation**,

$$\Phi(p, t) \equiv \frac{1}{\sqrt{\hbar}} A(k, t) = \frac{1}{\sqrt{2\pi\hbar}} \int_{-\infty}^{+\infty} \Psi(x, t) e^{-ikx}\, dx; \quad p = \hbar k. \tag{4•1–8}$$

Because

$$dp = \hbar dk, \tag{4•1–9}$$

we have

$$\rho_p(p)\, dp = |\Phi(p, t)|^2\, dp = |A(k, t)|^2\, dk. \tag{4•1–10}$$

Evidently, $\rho_p(p) = |\Phi(p, t)|^2$ is the probability density of the momentum of the object.

The conventional Schroedinger wave equation governs the time evolution of the state in the position representation. It is a partial differential equation in both x and t. The equivalent equations in the wave number and momentum representation, *in free space,* are

$$i\hbar \frac{\partial A(k, t)}{\partial t} = \frac{\hbar^2 k^2}{2M} A(k, t) \tag{4•1–11a}$$

and

$$i\hbar \frac{\partial \Phi(p, t)}{\partial t} = \frac{p^2}{2M} \Phi(p, t). \tag{4•1–11b}$$

The proof is left to the reader.

We have written the time derivatives in (4•1–11a,b) as partial derivatives, $\partial/\partial t$, because $A(k, t)$ and $\Phi(p, t)$ each depend on two independent variables. However, Eqs. (4•1–11) are really *ordinary* rather than *partial* differential equations, containing only time derivatives. They are algebraic in the other variable. Unfortunately, this simplicity is destroyed when a position-dependent potential $V(x)$ is present, and we shall therefore rarely use the wave number or momentum representation. Nevertheless, it is conceptually important to realize that the position representation is only one of several equivalent representations.

The treatment in this section is easily extended to three dimensions. Instead of the factors $(2\pi)^{1/2}$ and $\hbar^{1/2}$ in the relation between $\Psi(x, t)$ and $A(k, t)$ and between $A(k, t)$ and $\Phi(p, t)$ there are now factors $(2\pi)^{3/2}$ and $\hbar^{3/2}$ between $\Psi(\mathbf{r}, t)$ and $A(\mathbf{k}, t)$ and between $A(k, t)$ and $\Phi(\mathbf{p}, t)$. And, of course, all integrals become triple integrals. Everything else remains unchanged.

◆ PROBLEM TO SECTION 4.1

#4•1-1: Two Exponential Wave Packets

Consider the two wave packets shown in **Fig. 4•1–1**, given by the two wave functions distributions of the position and the momentum of the particle represented by the wave packet, defined according to

$$(\Delta x)^2 = \langle (x - \langle x \rangle)^2 \rangle = \langle x^2 \rangle - \langle x \rangle^2, \tag{4•3–2a}$$

$$(\Delta p)^2 = \langle (p - \langle p \rangle)^2 \rangle = \langle p^2 \rangle - \langle p \rangle^2. \tag{4•3–2b}$$

The relation (4•3–1) is called the **momentum-position uncertainty relation**. It states a fundamental limitation on the kinds of states of a free object that can occur: Nature does not have states for which both the position and the momentum of an object have simultaneously sharp values! In other words, states for which the position has a sharp value correspond to an unsharp momentum and vice versa. The uncertainty relation places a quantitative limit on what degrees of sharpness of the two variables are compatible with each other.

It is important to give a precise *operational* meaning to this statement. Let $\Psi_0(\mathbf{r}, t)$ be the wave function representing the state of an object. Suppose that a measurement of the position of the object in this state is made with as high a precision as possible. A certain value x is found. Let the object then be returned to the state Ψ_0—or another object of the same kind be placed in the state Ψ_0—and let the position measurement be repeated. In general, a different value of x will result, as a consequence of the indeterminacy of nature. We assume that the precision of the position measurement itself is high enough that the scatter in measurement values truly represents the properties of the state Ψ_0, rather than the inaccuracies of the measuring equipment. If the sequence state preparation/position measurement is repeated a large number of times, a distribution of position values is obtained, with a certain average $\langle x \rangle$ and a certain standard deviation Δx about this average, defined by (4•3–2a).

If, *instead* of the position, the momentum of the object had been measured many times, again with a high precision, a distribution of momentum values would have been found, characterized by an average $\langle p \rangle$ and standard deviation Δp defined by (4•3–2b).

4.3.2 Limits on *Successive* Complementary Measurements

The uncertainty relation (4•3–1), as derived above, is primarily a relation between the standard deviations of two kinds of *alternative* complementary measurements in which one measures *either* the momentum *or* the position of the object in the original state Ψ_0, but not both. Whichever property is measured first, the object was implicitly assumed to be returned to its original state before the second measurement is made. The relation puts a lower limit on the data scatter obtainable in such alternative measurements, even if both kinds of measurements are performed with a precision much finer than the widths in the data distributions themselves. This limit is *not* a limit on the precision with which the individual measurements themselves can be performed.

We now turn to the question of *successive* complementary measurements, in which the second measurement is performed not on the original state Ψ_0, but on the different state Ψ_1 that is present *after* the first measurement. We shall see that a very precise first measurement will itself introduce an uncertainty in the results of the second, complementary measurement.

To be specific, consider a wave packet Ψ_0 with a very small initial uncertainty Δp_0 in its momentum and a correspondingly large uncertainty Δx_0 in its position. Suppose that a very precise position measurement is performed that determines the position of the object to within about $\pm\delta x_1$, where $\delta x_1 \ll \Delta x_0$. Because of the large position uncertainty Δx_0 of the original state, it is not possible to predict exactly where the object will materialize. If many position measurements on the same initial state Ψ_0 were performed, the positions found would scatter with a standard deviation $\Delta x_0 \gg \delta x_1$. But—and this is the crucial point—once the object has materialized at, say, $x = x_1$, its position is now known with an uncertainty on the order of the precision δx_1 of the measurement: this is what is *meant* by the precision of a measurement.

In the language of quantum states, the first position measurement creates a new state $\Psi_1(x, t)$. Any subsequent high-precision position measurement performed on the new state will again yield a probability distribution whose mean value $\langle x \rangle_1$ is that value of x that was actually fixed during the first measurement: $\langle x \rangle_1 = x_1$. The uncertainty Δx_1 of x in the new state is the precision of the first measurement of x: $\Delta x_1 = \delta x_1$.

Suppose now that a momentum measurement is performed on the new state Ψ_1. Then the data scatter Δp_1 in its results is *not* governed by the position uncertainty Δx_0 of the *original* state, but by that of the *new* state Ψ_1. Hence, we have

$$\Delta p_1 \Delta x_1 \geq \hbar/2. \tag{4•3–3}$$

If $\Delta x_1 \ll \Delta x_0$, this means that $\Delta p_1 \gg \Delta p_0$.

Thus, even if the first position measurement is arranged in such a way that it does not change the *average* momentum value, it will change the scatter about this average.

It is instructive to illustrate these somewhat abstract arguments by considering how the wave functions of the object might look both before and after the first position measurement. Suppose that the original wave function is a simple (real) Gaussian with $\Delta x_0 = 0.5$, as shown in **Fig. 4•3–1**.

A crude way to measure the position of the object would be to trap the object in one or another of a row of narrow boxes. Unless the width of the boxes is large compared to Δx_0, it will be quite uncertain in which box the object will materialize. But once it has materialized in one box, the object now has a wave function corresponding to the confinement in that box, and the wave function *outside* this box must be set to zero.

In the figure, we have assumed that the boxes themselves have the width $w = 0.5$ ($= \Delta x_0$) and that the object has materialized in the box spanning the

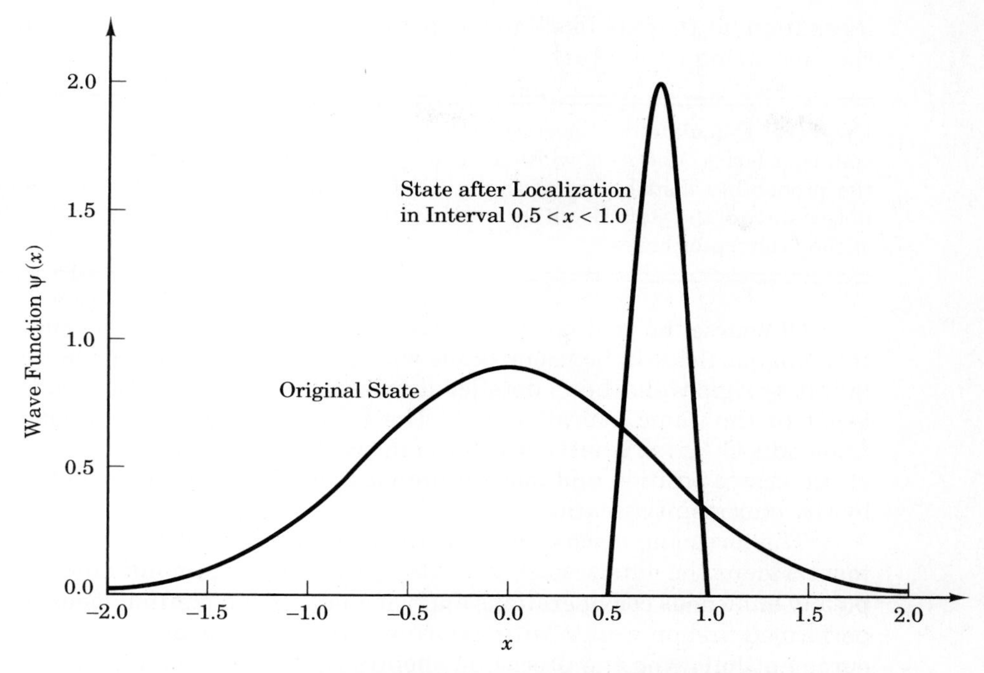

Figure 4•3–1. Narrowing of a wave packet by performing a position measurement consisting of trapping the object in a narrow box.

interval $0.5 < x < 1.0$ and now has a wave function corresponding to one of the bound states of the box.[1] This new wave function clearly has a narrower spread than the old one. Note that, although we have assumed that the wave function for the captured object is that of the ground state of the box, this choice was again made only for simplicity: The object may materialize in *any* of the states of the box, and that state need not even be a *stationary* state. In general, it will be a time-dependent linear superposition of such stationary states, corresponding to an object oscillating within the box, in a manner similar to the oscillations inside a harmonic oscillator well, discussed in section 2.3. The exact details depend on the exact details of the localization process assumed. But none of these details alter the bottom line of the present argument: The uncertainty found in any subsequent momentum measurement will be subject to the uncertainty relation (4•3–1), where Δx is now not the initial Δx_0, but the RMS position uncertainty of the object in the box, *after* having been localized by the position measurement.

Note also that the state of the object after the first measurement need not have the same energy as before. This simply reflects the fact that in any

[1] Note that we have re-normalized the wave function in the box to unity.

measurement there is likely to be an energy exchange between the object and the measuring apparatus.

Exercise: Calculate the uncertainty product $\Delta x \, \Delta p$ for the capture of an initial Gaussian by a series of boxes of width w, as in Fig. 4•3–1. Interpret Δx as the RMS width of the probability distribution of the initial Gaussian and Δp as the RMS momentum uncertainty of the state resulting from capturing the object into the ground state of one of the interceptor boxes.

Of course, no real position measurement works in such a crude way by trapping particles in fictitious boxes with infinitely thin and at the same time infinitely rigid walls, but a detailed discussion of more realistic measurements leads to the same overall result. The book by Heisenberg, quoted in the "appendix G," gives what is still one of the best discussions of how various kinds of successive position and momentum measurements are invariably governed by the uncertainty relation.

The foregoing discussion concerned the way a precise position measurement widens the data scatter of a subsequent momentum measurement. Completely analogous considerations apply if a precise momentum measurement is performed first on a state with a narrow position uncertainty. A position measurement following the precise momentum measurement will then exhibit a wide data scatter Δx_1 related to the momentum measurement precision Δp_1 via (4•3–3).

4.3.3 Momentum Uncertainty and Kinetic Energy

We note that the momentum uncertainty Δp, as defined in (4•3–2b), is related to the expectation value of the kinetic energy, as defined in chapter 2. The kinetic energy of an object of mass M is of course related to the square of the momentum p via

$$\mathcal{E}_{\text{kin}} = \frac{p^2}{2M}. \tag{4•3–4}$$

Taking the average over many experimental values of the square of the momentum is therefore equivalent to taking the average over many experimental values of the kinetic energy. Hence, the first term on the far right-hand side of (4•3–2b) may be replaced by

$$\langle p^2 \rangle = 2M \langle \mathcal{E}_{\text{kin}} \rangle. \tag{4•3–5}$$

We will often be interested in the uncertainty relation for the stationary bound states in an external potential. For such states, the expectation value $\langle p \rangle$ of the momentum itself is of course zero, and (4•3–2b) becomes

$$(\Delta p)^2 = 2M \langle \mathcal{E}_{\text{kin}} \rangle. \tag{4•3–6}$$

Evidently, a stationary bound state with a large expectation value of the kinetic energy has a large momentum uncertainty, and vice versa.

4.3.4 The Energy-Time Uncertainty Relation

The relation $\Delta k \Delta x \geq 1/2$ that formed the basis of the momentum-position uncertainty relation arose as a direct consequence of the general wave property that any wave phenomenon that is concentrated in space must contain a broad spectrum of spatial frequencies. Conversely, any wave phenomenon with a narrow spectrum of spatial frequencies must be extended in space.

Completely analogous reasoning applies to the relation between the spreads in time and in temporal frequency: Any wave phenomenon that is concentrated in time must contain a broad spectrum of temporal frequencies, and any wave phenomenon with a narrow spectrum of temporal frequencies must be extended in time. The mathematical argument is analogous to that in section 4.2. Rather than considering $\Psi(x, t)$ as a function of x for fixed t, we may consider $\Psi(x, t)$ as a function of t for fixed x and calculate its temporal Fourier transform $B(\omega)$. The entire mathematical procedure is exactly the same as if we had simply exchanged variables in section 4.2:

$$x \leftrightarrow -t, \qquad k \leftrightarrow \omega. \tag{4•3–7}$$

For a Gaussian pulse, this immediately leads to the relation

$$\Delta\omega \, \Delta t = 1/2 \tag{4•3–8}$$

between the RMS width $\Delta\omega$ of the temporal frequency spectrum of a wave pulse and the RMS width Δt of its spread in time. For a non-Gaussian pulse,

$$\Delta\omega \, \Delta t \geq 1/2. \tag{4•3–9}$$

The relation (4•3–9) is a statement about wave-like properties. It becomes a statement about particle-like properties by setting $\mathcal{E} = \hbar\omega$:

$$\boxed{\Delta\mathcal{E} \, \Delta t \geq \hbar/2.} \tag{4•3–10}$$

This is the **energy-time uncertainty relation**. As we have derived it here, it is a relation between the uncertainty in the energy of an object and the uncertainty in its time of arrival at a target. We will see later, when we discuss the formalism of the time evolution of arbitrary quantum systems, that the significance of (4•3–10) goes far beyond this particular case: It is a general relation between the uncertainty in the energy of a system and the uncertainty in the exact time at which a specified event will occur.

The most common form in which the energy-time uncertainty relation appears is the following. Suppose a quantum system makes a transition from a state with the energy $\mathcal{E}_1$ to a state with the energy $\mathcal{E}_2$, emitting the energy difference as a photon with energy $\hbar\omega = \mathcal{E}_1 - \mathcal{E}_2$. The more sharply the two energies and the energy of the photon are defined, the more uncertain it will be when the photon will materialize. But a large uncertainty in this (necessarily positive) time means, of course, that the average time to emission will be long. This leads to a reciprocal relation between what is called the *lifetime* τ of the upper state and the spectral width $\Delta\omega$ or the energetic width $\Delta\mathcal{E} = \hbar\Delta\omega$ of the emitted light.

◆ PROBLEMS TO SECTION 4.3

#4•3-1: Uncertainty Product for the Hydrogen Atom

Calculate the uncertainty product $\Delta x \cdot \Delta p_x$ involving the x-components of position and momentum for the ground state of the hydrogen atom.

#4•3-2: Non-Gaussian Wave Packets

Calculate the position and momentum uncertainties and the uncertainty products for the two exponential wave packets of Problem 1 of section 4.1. To obtain the momentum uncertainties, proceed in two different ways and show that they lead to the same result: (a) Transform the wave function to the momentum representation, and determine Δp from that representation, via (4•3–2b). (b) Working entirely within the position representation, determine Δp by drawing on the relation (4•3–5) between $\langle p^2 \rangle$ and the expectation value for the kinetic energy.

4.4 DYNAMICS OF A WAVE PACKET IN THE MOMENTUM REPRESENTATION

4.4.1 Schroedinger Wave Equation in the Momentum Representation

The statement that the function $\Phi(p, t)$ is an alternative representation of the state of the quantum system, equivalent to the representation in terms of the wave function $\Psi(x, t)$, suggests that there should exist a wave equation for this function that is equivalent to the Schroedinger equation for $\Psi(x, t)$. We gave such an equation in (4•1–11b) for the case of propagation through free space; we now generalize that result to propagation in the presence of forces. We restrict ourselves to the one-dimensional case, generalizing to three dimensions at the end.

We begin with the one-dimensional Schroedinger wave equation

$$i\hbar\frac{\partial\Psi}{\partial t} = -\frac{\hbar^2}{2M}\frac{\partial^2\Psi}{\partial x^2} + V(\mathbf{r})\Psi. \tag{4•4–1}$$

We multiply the entire equation by the factor

$$\frac{1}{\sqrt{2\pi\hbar}}e^{-ikx}$$

and integrate over x, from $-\infty$ to $+\infty$. On the left-hand side, this leads to

$$\frac{i\hbar}{\sqrt{2\pi\hbar}}\frac{\partial}{\partial t}\left[\int_{-\infty}^{+\infty}\Psi e^{-ikx}\,dx\right] = i\hbar\cdot\frac{\partial\Phi}{\partial t}. \tag{4•4–2}$$

On the right-hand side, the first term of (4•4–1) takes the form

$$-\frac{\hbar^2}{2M\sqrt{2\pi\hbar}}\int_{-\infty}^{+\infty} e^{-ikx}\frac{\partial^2\Psi}{\partial x^2}\,dx = \frac{p^2}{2M}\Phi - \frac{\hbar^2 e^{-ikx}}{2M\sqrt{2\pi\hbar}}\left[\frac{\partial\Psi}{\partial t} + ik\Psi\right]\Bigg|_{-\infty}^{+\infty}, \tag{4•4–3}$$

where we have integrated by parts twice and used (4•2–5) and (4•2–8). If we restrict ourselves to normalizable states, both the Schroedinger wave function and its derivative must vanish at infinity. In this case, only the first term on the right-hand side of (4•4–3) is left.

The last term on the right-hand side of (4•4–1) leads to

$$I_V = -\frac{1}{\sqrt{2\pi\hbar}}\int_{-\infty}^{+\infty} V(x)\Psi(x)e^{-ikx}\,dx. \tag{4•4–4}$$

The integral cannot be evaluated in closed form for an arbitrary $V(x)$, but is readily evaluated in certain special cases. The simplest non-trivial possibility is

$$V(x) = -Fx, \tag{4•4–5}$$

representing an object under the influence of a uniform force F. In that case, (4•4–4) becomes

$$I_V = -\frac{iF}{\sqrt{2\pi\hbar}}\int_{-\infty}^{+\infty}\Psi(x)\frac{\partial}{\partial k}e^{-ikx}\,dx = -i\hbar F\frac{\partial\Phi}{\partial p}. \tag{4•4–6}$$

Inserting everything leads to the differential equation

$$i\hbar\frac{\partial\Phi}{\partial t} = \frac{p^2}{2M}\Phi - i\hbar F\frac{\partial\Phi}{\partial p}, \tag{4•4–7}$$

a simple first-order equation for Φ. We will discuss the solutions of this equation in the next sub-section.

Our result is easily extended to potential energies of the form

$$V(x) = a_n x^n, \tag{4•4–8}$$

where n is a positive integer. An elementary calculation shows that in this case,

$$I_V = a_n(i\hbar)^n\,\frac{\partial^n\Phi}{\partial p^n}. \tag{4•4–9}$$

This relation may also be written symbolically as

$$I_V = V\left(i\hbar\frac{\partial}{\partial p}\right)\Phi(p, t), \tag{4•4–10}$$

implying the replacement of the position *coordinate* x in $V(x)$ by the differential *operator*

$$\hat{x} = i\hbar\frac{\partial}{\partial p}. \tag{4•4–11}$$

The form (4•4–10) is the most general way in which the integral (4•4–4) can be expressed in closed form; it remains valid, not only if $V(x)$ is given in the form (4•4–8), but also for a linear superposition of such terms, that is, whenever $V(x)$ is an analytic function of x. For such analytic potentials, we thus finally obtain a general wave equation for the momentum representation of the state function $\Phi(p, t)$:

$$i\hbar\frac{\partial\Phi}{\partial t} = \frac{p^2}{2M}\Phi + V\left(i\hbar\frac{\partial}{\partial p}\right)\Phi. \tag{4•4–12}$$

Had we performed our calculation in three dimensions, we would have obtained

$$i\hbar\frac{\partial\Phi}{\partial t} = \frac{p^2}{2M}\Phi + V(i\hbar\nabla_{\mathbf{p}})\Phi, \tag{4•4–13}$$

where $\Phi = \Phi(\mathbf{p}, t)$ and where, instead of the operator substitution (4•4–11), we have used

$$V(\mathbf{r}) \rightarrow V(\hat{\mathbf{r}}), \tag{4•4–14}$$

with

$$\hat{\mathbf{r}} = i\hbar\nabla_{\mathbf{p}} \equiv i\hbar\left(\frac{\partial}{\partial p_x}, \frac{\partial}{\partial p_y}, \frac{\partial}{\partial p_z}\right). \tag{4•4–15}$$

Equations (4•4–12) and (4•4–13) are the momentum representation analogs of the Schroedinger wave equation. For any potential that is more complicated than a second-order polynomial, (4•4–12) and (4•4–13) are differential equations of higher than second order, which usually makes them more difficult equations than the Schroedinger wave equation to which they are equivalent. As a result, the momentum representation is much less frequently employed for the solution of quantum-mechanical problems than is the position representation.

However, our reason for presenting the momentum representation here was not its utility: The momentum representation is one of the simplest examples to illustrate the important idea that Schroedinger's position representation is only a special case of a more general formalism—even if it is the most important case. A full understanding of the mathematical formalism of quantum theory is not possible, unless the idea of multiple representations is appreciated and until the Schroedinger wave function has been de-throned from its (merely apparent) central role, into a more egalitarian one in which *all* representations are equal—although "some may be more equal than others."

4.4.2 The Motion of a Wave Packet in a Uniform Field of Force

Consider now the integration of the Schroedinger wave equation (4•4–7) in the momentum representation for a uniform field of force. The equation is a linear partial differential equation of the first order in both p and t. We may write it in the form

$$i\hbar\left(\frac{\partial}{\partial t} + F\frac{\partial}{\partial p}\right)\Phi = \frac{p^2}{2M}\Phi. \tag{4•4–16}$$

The differential operator on the left leads to zero if applied to any arbitrary differentiable function that depends on p and t only through the combination

$$p' = p - Ft. \tag{4•4–17}$$

If $\Theta(p')$ is such a function, then

$$\left(\frac{\partial}{\partial t} + F\frac{\partial}{\partial p}\right)\Theta(p') = \frac{d\Theta}{dp'}(-F + F) = 0. \tag{4•4–18}$$

This means that we can solve (4•4–16) by splitting up $\Phi(p, t)$ into a product of the form

$$\Phi(p, t) = f(p)\Theta(p - Ft). \tag{4•4–19}$$

If this is inserted into (4•4–16), the function $\Theta(p - Ft)$ drops out, and we are left with the very simple ordinary differential equation

$$i\hbar F\frac{df}{dp} = \frac{p^2}{2M}f, \tag{4•4–20}$$

which is readily integrated to

$$f(p) = \exp\left(-\frac{ip^3}{6\hbar MF}\right). \tag{4•4–21}$$

Thus, we obtain, as solution of (4•4–16),

$$\Phi(p, t) = \exp\left(-\frac{ip^3}{6\hbar MF}\right)\Theta(p - Ft), \tag{4•4–22}$$

where $\Theta(p - Ft)$ is *any* differentiable function of its argument.

The function $\Theta(p')$ is determined by the initial or boundary conditions. For example, we may specify that for $t = 0$, the wave packet should be given by the Gaussian form

$$\Phi(p, 0) = \Phi_0 \exp\left[-\left(\frac{p}{2\Delta p}\right)^2\right], \tag{4•4–23}$$

which is the same as (4•2–1) for $k_0 = 0$, except for the change from k to p as independent variable. By setting $t = 0$ in (4•4–22), and then inserting (4•4–23) on the left-hand side, we find that

$$\Theta(p') = \Phi_0 \exp\left[-\left(\frac{p'}{2\Delta p}\right)^2 + \frac{ip'^3}{6\hbar MF}\right], \tag{4•4–24}$$

and that $\Phi(p, t)$ itself becomes

$$\Phi(p') = \Phi_0 \exp\left[-\left(\frac{p'}{2\Delta p}\right)^2 + \frac{i(p'^3 - p^3)}{6\hbar MF}\right]. \tag{4•4–25}$$

The momentum wave functions (4•4–22) have a very simple interpretation, which is immediately seen by considering $|\Phi|^2$ rather than Φ itself. The phase factor in (4•4–22) then drops out, and we retain

$$|\Phi(p, t)|^2 = |\Theta(p - Ft)|^2. \tag{4•4–26}$$

This corresponds to a probability distribution for the momentum that simply shifts with time, with a uniform rate

$$\frac{\partial p}{\partial t} = F. \tag{4•4–27}$$

But this is of course Newton's second law! Thus, we see that at least in the case of a *uniform* force, Newton's second law of classical mechanics emerges as the law according to which the entire quantum-mechanical momentum distribution of a state shifts with time. We shall generalize this result to arbitrary forces in chapter 8.

It is possible to return to the position representation by the Fourier transformation (4•1–4). To do so is the task of problem 1 at the end of the section.

In addition to illustrating the momentum representation, our example shows that problems that in classical mechanics are amongst the simplest of all problems may in quantum mechanics require a rather formidable mathematical apparatus.

◆ PROBLEM TO SECTION 4.4

#4•4-1: Gaussian Wave Packet under the Influence of a Uniform Force, in the Position Representation

Perform a Fourier transform of the kind (4•1–4) on the wave function (4•4–25) to obtain the wave function in the position representation. Compare the final result with that in Section 4.2 for the time evolution of a Gaussian wave packet without an external force.

Chapter 5

SCATTERING BY SIMPLE BARRIERS

5.1 SCATTERING STATES AND THEIR NORMALIZATION

In this chapter, we consider solutions of the Schroedinger equations that correspond to particles coming from infinity and being scattered by a non-constant potential in their path. Because the particles have an infinite amount of space available, the single-particle energy eigenfunctions for the problem are not normalizable and hence do not correspond to physically admissible single-particle states. Nevertheless, they remain mathematically and physically meaningful for the following reasons:

(a) Mathematically, it is always possible to construct normalizable, and hence physically admissible, time-dependent wave packets from the energy eigenfunctions by linear superposition. The situation is a direct generalization of that in chapter 4, where we constructed normalizable, and hence physically admissible, wave packets from the un-normalizable plane waves. The energy eigenfunctions extending to infinity are the generalizations of the earlier plane waves.

(b) Physically, it is inherent in the nature of scattering problems that one deals with a very large number of particles with a nonzero density. As a rule, the interaction between the particles is negligible, so that each particle behaves as if it were present alone. Furthermore, the single-particle state functions for all particles are essentially the same. Under these circumstances, it is convenient to re-normalize ψ in such a way that $|\psi|^2$ represents the *overall* density of particles,

$$\rho = |\psi|^2. \tag{5•1–1}$$

This means that $|\psi|^2 \, d^3r$ is the average value $\langle dN \rangle$ of the number of particles that materialize in the volume d^3r, taken over many different measurements of this number, all on the same state ψ. With this normalization, the quantity

$$\mathbf{j} = -\frac{i\hbar}{2M}(\psi^* \nabla \psi - \psi \nabla \psi^*) \tag{5•1–2}$$

is the *overall* particle current density. As a rule, in scattering problems, it is the incident *current* density that is known and specified, rather than the spatial density. The wave function is then most conveniently used with whatever normalization leads to the desired incident current density.

5.2 MATRIX FORMALISM FOR SCATTERING BY ONE-DIMENSIONAL BARRIERS

5.2.1 Scattering Matrix and Propagation Matrix

In this section, we extend the treatment of scattering at a step, treated briefly in section 1.4, to scattering by one-dimensional barriers of finite thickness with a more complicated potential profile, such as the schematic barrier shown in **Fig. 5•2–1.** Such one-dimensional barriers are more than simple textbook exercises; they actually play an important role in modern semiconductor electronics.

We consider a potential barrier of unspecified shape, separating two regions of constant, but not necessarily equal, potential energy, as in the figure. We define two reference planes $x = x_{\mathrm{I}}$ and $x = x_{\mathrm{II}}$ somewhere within the constant-potential regions on either side of the barrier, each in a range where the kinetic energy is positive,

$$\mathcal{E} > V(x_{\mathrm{I}}), V(x_{\mathrm{II}}) \tag{5•2–1}$$

In those two constant-potential regions—which may be arbitrarily narrow—the wave functions are superpositions of plane waves, which may always be written in the form

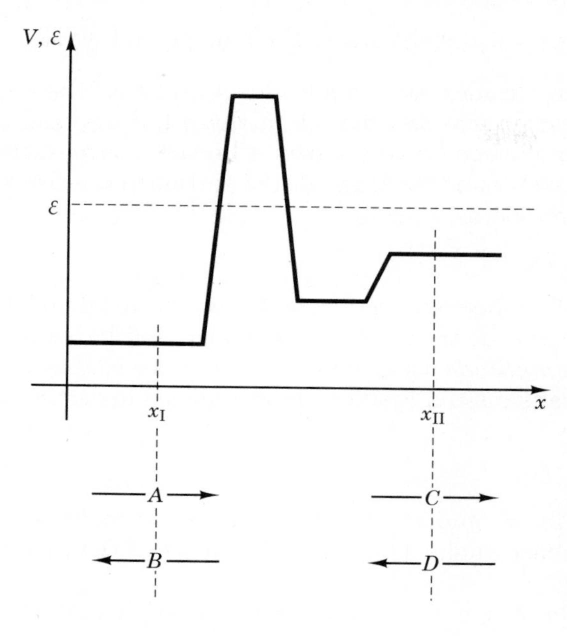

Figure 5•2–1. Scattering at an arbitrary barrier can be described by defining four complex wave amplitudes in two reference planes. We define the amplitudes in the two reference planes at x_{I} and x_{II} and assume that the total energy $\mathcal{E}$ exceeds the potential energy at least in the two reference planes.

$$\psi_{\mathrm{I}}(x) = \frac{1}{\sqrt{K_{\mathrm{I}}}}[A \cdot e^{iK_{\mathrm{I}}(x-x_{\mathrm{I}})} + B \cdot e^{-iK_{\mathrm{I}}(x-x_{\mathrm{I}})}], \tag{5•2–2a}$$

$$\psi_{\mathrm{II}}(x) = \frac{1}{\sqrt{K_{\mathrm{II}}}}[C \cdot e^{iK_{\mathrm{II}}(x-x_{\mathrm{II}})} + D \cdot e^{-iK_{\mathrm{II}}(x-x_{\mathrm{II}})}]. \tag{5•2–2b}$$

Here K_{I} and K_{II} are the *local* wave numbers at the two reference planes.[1] Compared to our treatment of the scattering at a simple step in section 1.4, we have split off amplitude factors of the form $1/\sqrt{K}$ from the wave amplitudes, along with phase factors that ensure that the complex amplitude coefficients contain whatever phases the waves have directly at their reference planes.

The reason for re-normalizing the amplitudes is the following. We are principally interested in the *current* densities of the four waves, and by splitting off the factors $1/\sqrt{K}$, we make these factors cancel out in the current density expressions, leading to the following simple expressions for the four partial probability current densities:

$$j_{\mathrm{I}}^{+} = +\frac{\hbar}{M} \cdot |A|^2, \quad j_{\mathrm{II}}^{+} = +\frac{\hbar}{M} \cdot |C|^2, \tag{5•2–3a,b}$$

$$j_{\mathrm{I}}^{-} = -\frac{\hbar}{M} \cdot |B|^2, \quad j_{\mathrm{II}}^{-} = -\frac{\hbar}{M} \cdot |D|^2. \tag{5•2–3c,d}$$

[1] Throughout this chapter, we use uppercase K to designate *local* wave numbers, reserving the use of lowercase k for the wave numbers of extended plane waves and for the so-called Bloch wave numbers for periodic potentials (defined in section 5.4).

These represent the current contributions of the four partial waves.

Of the four wave amplitudes, the amplitudes A and D of the two waves traveling *toward* the barrier may be viewed as the *physically* independent variables. The B-wave may then be considered a linear superposition of the reflected portion of the A-wave and the transmitted portion of the D-wave. This dependence may be expressed by writing

$$B = \rho_+ A + \tau_- D, \tag{5•2–4a}$$

where ρ_+ is an *amplitude* reflection coefficient for waves incident in the $+x$ direction and τ_- is an *amplitude* transmission coefficient for waves incident in the $-x$ direction. These *amplitude* coefficients should not be confused with the particle reflection and transmission coefficients introduced in section 1.5—see (1•5–11,12). Similarly,

$$C = \tau_+ A + \rho_- D, \tag{5•2–4b}$$

with obvious meanings for τ_+ and ρ_-. The four scattering coefficients are in general complex; their phase angles depend on the choice of the two reference planes.

Equations (5•2–4) may be lumped together into a single matrix equation

$$\begin{pmatrix} B \\ C \end{pmatrix} = \begin{pmatrix} \rho_+ & \tau_- \\ \tau_- & \rho_- \end{pmatrix} \begin{pmatrix} A \\ D \end{pmatrix} = \hat{S} \begin{pmatrix} A \\ D \end{pmatrix}, \tag{5•2–5}$$

where

$$\boxed{\hat{S} = \begin{pmatrix} \rho_+ & \tau_- \\ \tau_- & \rho_- \end{pmatrix}} \tag{5•2–6}$$

is the **scattering matrix**, often called the **S-matrix**. We have placed a caret (ˆ) on top of S to indicate that $\hat{S}$ is not a numerical factor, but an operator, operating on the column matrix consisting of the amplitudes A and D.

The study of the S-matrix is a central problem in three-dimensional scattering theory. In simple *one-dimensional* problems such as ours, it is more useful to express the wave amplitudes on *one* side of the barrier in terms of the wave amplitudes on the *other* side: The central problem in one-dimensional scattering theory is the overall scattering by a *sequence* of several barriers. The wave amplitudes on the right-hand side of one barrier then also serve as the wave amplitudes on the left-hand side of the next barrier. By using a formalism that expresses the amplitudes on *one* side as functions of those on the *other*, we can work through the sequence of barriers one by one. It will turn out to be useful to work "backwards," by expressing the amplitudes in the left-hand reference plane I of Fig. 5•2–1 as linear functions of the amplitudes in the right-hand reference plane II, writing

$$A = P_{11} C + P_{12} D, \tag{5•2–7a}$$

$$B = P_{21}C + P_{22}D. \tag{5•2–7b}$$

These two equations may again be lumped together in the form of a matrix equation,

$$\begin{pmatrix} A \\ B \end{pmatrix} = \begin{pmatrix} P_{11} & P_{12} \\ P_{21} & P_{22} \end{pmatrix} \begin{pmatrix} C \\ D \end{pmatrix} = \hat{P}\begin{pmatrix} C \\ D \end{pmatrix}, \tag{5•2–8}$$

where

$$\boxed{\hat{P} = \begin{pmatrix} P_{11} & P_{12} \\ P_{21} & P_{22} \end{pmatrix}} \tag{5•2-9}$$

is what we call the **propagation matrix** or, briefly, the **P-matrix**[2] of the barrier.

To appreciate this formalism, consider the combined scattering by two separate potential barriers with the propagation matrices $\hat{P}_1$ and $\hat{P}_2$, separated by a length L of field-free space (**Fig. 5•2–2**).

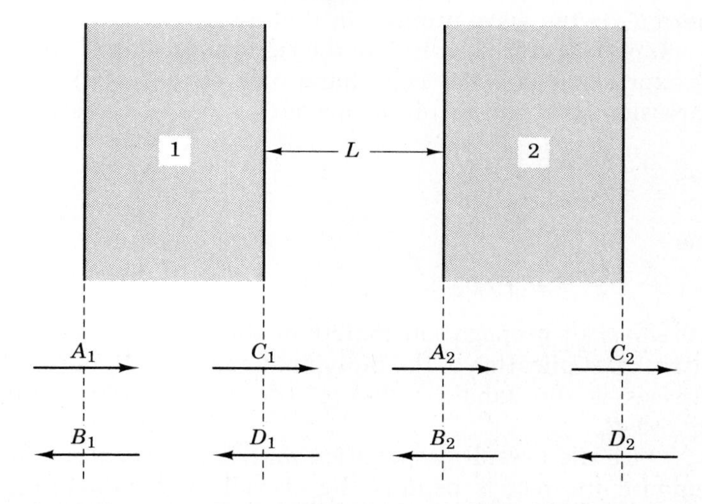

Figure 5•2–2. Combined scattering effect of a composite barrier that consists of two individual barriers separated by a distance L.

The vertical lines shown in the figure are the reference planes of the two barriers, with respect to which the wave amplitudes and the scattering coefficients are defined. The distance L is the distance between the two innermost reference planes. The problem is characterized by eight complex wave amplitudes at the four reference planes, A_1 through D_2.

[2] Various authors employ either $\hat{P}$ or its inverse $(P)^{-1}$ under various names, such as *transfer matrix.*

Our goal is to express the two leftmost wave amplitudes A_1 and B_1 as functions of the two rightmost ones, C_2 and D_2. For the amplitude relation across the two barriers, we have

$$\begin{pmatrix} A_1 \\ B_1 \end{pmatrix} = \hat{P}_1 \begin{pmatrix} C_1 \\ D_1 \end{pmatrix} \text{ and } \begin{pmatrix} A_2 \\ B_2 \end{pmatrix} = \hat{P}_2 \begin{pmatrix} C_2 \\ D_2 \end{pmatrix}. \qquad (5{\bullet}2\text{–}10\text{a,b})$$

The relations between the wave amplitudes at the two ends of the field-free space section can be written as

$$\begin{pmatrix} C_1 \\ D_1 \end{pmatrix} = \hat{P}_\Delta \begin{pmatrix} A_2 \\ B_2 \end{pmatrix}, \qquad (5{\bullet}2\text{–}11)$$

where $\hat{P}_\Delta$ is the **free-space propagation matrix**, which accounts for the phase shifts of the two waves in going from the entrance plane of barrier 2 back to the exit plane of barrier 1. It is left to the reader to show that

$$\hat{P}_\Delta = \begin{pmatrix} e^{-iK_\Delta L} & 0 \\ 0 & e^{+iK_\Delta L} \end{pmatrix}, \qquad (5{\bullet}2\text{–}12)$$

where K_Δ is the wave number in that space.

If we insert (5•2–10b) into the right-hand side of (5•2–11), and the resulting expression into the right-hand side of (5•2–10a), we obtain the desired expression for A_1 and B_1 in terms of C_2 and D_2, namely

$$\begin{pmatrix} A_1 \\ B_1 \end{pmatrix} = \hat{P} \begin{pmatrix} C_2 \\ D_2 \end{pmatrix}, \qquad (5{\bullet}2\text{–}13)$$

where

$$\hat{P} = \hat{P}_1 \hat{P}_\Delta \hat{P}_2 \qquad (5{\bullet}2\text{–}14)$$

is the overall propagation matrix of the combination, obtained by ordinary matrix multiplication of the individual matrices. Note that the sequence of the matrices is the same as that of the corresponding scattering sections in Fig. 5•2–2.

Once the overall propagation matrix has been obtained, the quantity of dominant interest is usually the overall **transmission probability** of the composite barrier, defined as the ratio of the transmitted current density to the incident current density, in the absence of any current incident from the outgoing side. Evidently,

$$T \equiv j_{\text{II}}^+ / j_{\text{I}}^+ = |C/A|^2 \quad \text{while} \quad D = 0. \qquad (5{\bullet}2\text{–}15)$$

Setting $D = 0$ in (5•2–7a) yields

$$\boxed{T = \frac{1}{|P_{11}|^2}.} \qquad (5{\bullet}2\text{–}16)$$

5.2.2 Relations between Matrix Coefficients

The matrix manipulations are greatly simplified by the fact that the four coefficients of each propagation matrix are not independent of each other. We show here that, if the kinetic energy is positive at both reference planes, as specified in (5•2–1), the coefficients must obey the relations

$$P_{22} = P_{11}^* \quad \text{and} \quad P_{21} = P_{12}^* \tag{5•2–17a,b}$$

and

$$\det \hat{P} \equiv P_{11}P_{22} - P_{12}P_{21} = 1. \tag{5•2–18}$$

The first two relations, (5•2–17a,b), are a consequence of **time reversal invariance**: If $\psi(x)$ is a solution of the time-independent Schroedinger equation, then its complex conjugate $\psi^*(x)$ must also be a solution. But in this second solution, the wave propagation directions are reversed: Under the condition (5•2–1), all four waves are propagating waves, and the B^*-wave and the C^*-wave have become the incident waves, while the A^*- and D^*-waves have become the outgoing scattered waves. This means that (5•2–8) must remain valid, with unchanged matrix coefficients, if we make the substitutions

$$A \to B^*, B \to A^*, C \to D^*, D \to C^*. \tag{5•2–19}$$

Written out, (5•2–8) then becomes

$$\begin{pmatrix} B^* \\ A^* \end{pmatrix} = \begin{pmatrix} P_{11} & P_{12} \\ P_{21} & P_{22} \end{pmatrix} \begin{pmatrix} D^* \\ C^* \end{pmatrix}. \tag{5•2–20}$$

By taking the complex conjugate and re-arranging rows and columns, this is converted to

$$\begin{pmatrix} A \\ B \end{pmatrix} = \begin{pmatrix} P_{22}^* & P_{21}^* \\ P_{12}^* & P_{11}^* \end{pmatrix} \begin{pmatrix} C \\ D \end{pmatrix}. \tag{5•2–21}$$

The matrix on the right-hand side must evidently be the same as the original matrix $\hat{P}$. Comparison of the matrix coefficients then leads to (5•2–17a,b).

A second requirement is **current conservation**: The amplitudes of the incident, reflected, and transmitted waves must be such that the net probability current density in reference plane I equals that in reference plane II. Because of the current relations (5•2–3), this implies that

$$A^*A - B^*B = C^*C - D^*D. \tag{5•2–22}$$

This condition must be satisfied for any combination of incoming or outgoing waves, including specifically the case of no incoming wave from the right, that is, $D = 0$. From (5•2–7a,b), we obtain, for $D = 0$,

$$A^*A - B^*B = (P_{11}^*P_{11} - P_{21}^*P_{21})C^*C. \tag{5•2–23}$$

Comparing this with (5•2–22) for $D = 0$ yields the condition

$$P_{11}^{*}P_{11} - P_{21}^{*}P_{21} = 1, \tag{5•2-24}$$

which, because of (5•2–17a,b), is equivalent to the determinantal relation (5•2–18).

The conditions (5•2–17) and (5•2–18) are powerful conditions in the analysis of scattering properties. In applying them, it is important, however, to recall that they were derived under the assumption that the kinetic energy at *both* reference planes is positive. In the analysis of scattering by barriers, it is often necessary to break up the overall propagation matrix of a barrier into products of several factor matrices representing successive portions of the overall barrier, some of which may involve reference planes where the kinetic energy is negative. In those cases, the relations (5•2–17) and (5•2–18) will not, in general, be satisfied for the factor matrices. We will discuss such cases later.

5.3 SCATTERING BY A SQUARE BARRIER AND A SQUARE WELL

5.3.1 The Propagation Matrix

To illustrate the propagation matrix formalism, we apply it here to the scattering properties of a square barrier of height V_0 and width L (**Fig. 5•3–1**) and a square well of depth $-V_0$. In both cases, the scattering can be represented as scattering by two steps separated by a free-space section. In the case of a barrier, we assume initially that the energy of the incident particle exceeds the barrier height, $\mathcal{E} > V_0$; we will drop this assumption later.

In section 1.4, we calculated the amplitude ratios B/A and C/A for scattering at a simple upward potential step (see 1•4–20a,b) for the specific case of no returning wave ($D = 0$). However, we wrote the individual plane waves in the simple form $\psi \sim \exp(\pm ikx)$, without splitting off the re-normalizing factors of the form $\sqrt{K}$, as in (5•2–2a,b). In terms of the "old" amplitudes of section 1.4, our "new" amplitudes are

$$A = \sqrt{k}A_{\text{old}}, \quad B = \sqrt{k}B_{\text{old}}, \quad \text{and} \quad C = \sqrt{k'}C_{\text{old}}. \tag{5•3–1}$$

With the help of these conversions, we obtain, from (1•4–20a,b),

$$P_{11} = \frac{A}{C} = \sqrt{\frac{k}{k'}}\left(\frac{A}{C}\right)_{\text{old}} = \frac{k + k'}{2\sqrt{kk'}} \tag{5•3–2a}$$

and

$$P_{21} = \frac{B}{C} = \sqrt{\frac{k}{k'}}\left(\frac{B}{C}\right)_{\text{old}} = \frac{k - k'}{2\sqrt{kk'}}. \tag{5•3–2b}$$

Here, k is the wave number on the downside of the step, and k' is the wave number on the upside. The remaining two matrix coefficients follow from (5•3–2a,b) via (5•2–17a,b). If we write K and K' instead of k and k', we obtain

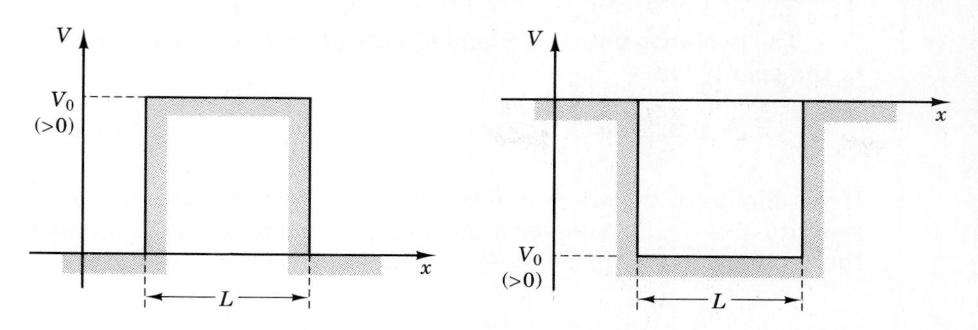

Figure 5•3–1. Square barrier ($V_0 > 0$) (left) and square well ($V_0 < 0$) (right).

the overall propagation matrix for the *upward* step:

$$\hat{P}_{\uparrow} = \frac{1}{2\sqrt{KK'}} \begin{pmatrix} K + K' & K - K' \\ K - K' & K + K' \end{pmatrix}. \tag{5•3–3a}$$

The propagation matrix for a *downward* step follows by simply interchanging K and K':

$$\hat{P}_{\downarrow} = \frac{1}{2\sqrt{KK'}} \begin{pmatrix} K' + K & K' - K \\ K' - K & K' + K \end{pmatrix}. \tag{5•3–3b}$$

If we insert (5•3–3) and the free-space matrix (5•2–12) into the matrix product (5•2–14) for the overall propagation matrix, and replace k_Δ by K', we obtain

$$\hat{P} = \frac{1}{4KK'} \begin{pmatrix} K + K' & K - K' \\ K - K' & K + K' \end{pmatrix} \begin{pmatrix} e^{-iK'L} & 0 \\ 0 & e^{+iK'L} \end{pmatrix} \begin{pmatrix} K' + K & K' - K \\ K' - K & K' + K \end{pmatrix}. \tag{5•3–4}$$

Execution of the matrix multiplications—left as an exercise to the reader—leads to

$$P_{11} = P_{22}^{*} = \cos K'L - i\frac{K^2 + K'^2}{2KK'} \sin K'L, \tag{5•3–5}$$

$$P_{12} = P_{21}^{*} = +i\frac{K^2 - K'^2}{2KK'} \sin K'L. \tag{5•3–6}$$

5.3.2 Transmission Resonances

The transmission probability for $\mathcal{E} > V_0$ follows from (5•3–5) by insertion into (5•2–16). One finds, after some manipulation, that

$$T = \frac{1}{|P_{11}|^2} = \left[1 + \left(\frac{K^2 - K'^2}{2KK'}\right)^2 \sin^2 K'L\right]^{-1}. \tag{5•3–7}$$

The two wave numbers K and K' outside and inside the barrier are related to the energy via

$$\mathcal{E} = \frac{\hbar^2 K^2}{2M} = \frac{\hbar^2 K'^2}{2M} + V_0. \tag{5•3–8}$$

If we eliminate the wave number K *outside* the barrier via (5•3–8), we can re-write (5•3–7) in a simpler form, as a function of the wave number K' inside the barrier only,

$$T = \left[1 + \frac{1}{4}\frac{\beta^2}{\beta + (K'L)^2}\frac{\sin^2 K'L}{(K'L)^2}\right]^{-1}, \tag{5•3–9}$$

where we have defined a dimensionless **barrier parameter**

$$\beta = \frac{2M}{\hbar^2} V_0 L^2, \tag{5•3–10}$$

to characterize the "strength" of the barrier or the well.

Viewed as a function of $K'L$, the transmission probability oscillates (**Fig. 5•3–2**) between full transmission, where

$$T_{\max} = 1, \tag{5•3–11}$$

and a minimum transmission curve shown as a dotted line in the figure, and given by

$$T_{\min} = \left[1 + \frac{1}{4}\frac{\beta^2}{\beta + (K'L)^2}\frac{1}{(K'L)^2}\right]^{-1} = 1 - \frac{\beta^2}{[2(K'L)^2 + \beta]^2}. \tag{5•3–12}$$

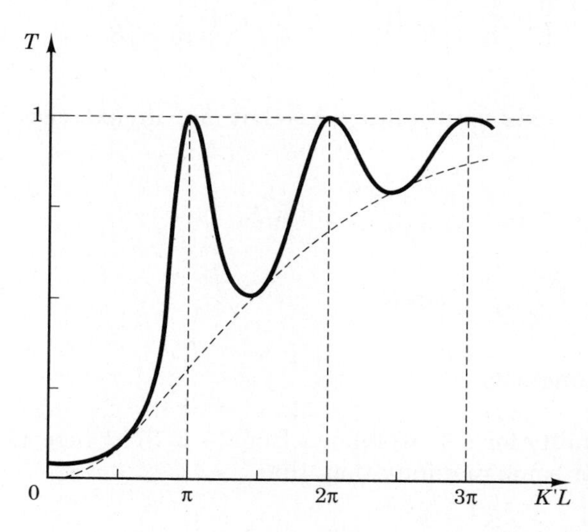

Figure 5•3–2. Transmission resonances for a rectangular barrier, for $\beta = 100$.

Written as a function of the energy $\mathcal{E}$ rather than of $K'L$, this minimum curve is simply

$$T_{\min} = 1 - \frac{V_0^2}{(2\mathcal{E} - V_0)^2}. \tag{5•3–13}$$

Full transmission ($T = 1$) is reached for those energies for which

$$K'L = n\pi, \quad \text{with} \quad n = 1, 2, 3, \ldots . \tag{5•3–14a}$$

The minimum-transmission curve $T_{\min}$ is reached for

$$K'L = n\pi, \quad \text{with} \quad n = \frac{1}{2}, \frac{3}{2}, \frac{5}{2}, \ldots . \tag{5•3–14b}$$

The occurrence of the transmission resonances is the result of an interference between the wave reflected by the upward step at $x = 0$ and that reflected by the downward step at $x = L$. We saw in section 1.4 that the amplitude reflection coefficients for the two steps are equal in magnitude, but opposite in sign. This means that the two reflected waves are equally strong, but they have a relative phase shift of π, measured in their respective planes of reflection. To this is added the additional phase shift $2K'L$ that the wave reflected at $x = L$ undergoes on its round trip from $x = 0$ to $x = L$ and back. Under the resonance condition (5•3–14a), this round-trip phase shift is a multiple of 2π. The total phase shift between the two reflected waves is therefore an odd multiple of π, and because the two reflected waves are equally strong, this leads to complete mutual annihilation of the two reflected waves. By contrast, under the antiresonance condition (5•3–14b), the two reflected waves enhance each other.

Exercise: Compare the antiresonance transmission coefficient $T_{\min}$ of (5•3–12) with the value for a simple step for the same pair K, K'.

The details of the transmission resonances depend on the magnitude of the barrier parameter β defined in (5•3–10). For electrons scattered by a barrier 1 eV high and 1 nm wide, we have $\beta = 26.25$. The curve in Fig. 5•3–2 is for $\beta = 100$. The larger the value of β, the more rapid will be the oscillations and the sharper will be the resonances.

Although we stated our calculations for the case of a potential *barrier* (V_0, $\beta > 0$), they are equally valid for a square potential *well* (V_0, $\beta < 0$). The case of a well represents a simple one-dimensional model of the scattering of a particle wave by the potential well of an atom or an atomic nucleus. Note that for scattering by a well, the condition $\mathcal{E} > 0$ outside the well implies the condition

$$(K'L)^2 > -\beta > 0 \tag{5•3–15}$$

on the wave number inside the well. Very sharp transmission resonances occur when $-\beta \gg 1$ and when the leftmost inequality in (5•3–15) is satisfied only by a small excess,[3] making the quantity $\beta + (K'L)^2$ in the denominator inside the expression (5•3–9) a small number.

5.3.3 Tunneling through the Barrier for $\mathcal{E} < V_0$

Our treatment of wave propagation across a square barrier for $\mathcal{E} > V_0$ is readily extended to the case of tunneling *through* that barrier when the incident energy is less than the height of the barrier ($\mathcal{E} < V_0$). We discussed this case already briefly in section 1.4, using a brute-force fitting of five wave functions at the two steps. It is instructive to consider the problem again, this time in the framework of the propagation matrix formalism.

We may continue to treat each of the two steps as a separate barrier with a propagation matrix of its own, just as for $\mathcal{E} > V_0$, but now one of the two reference planes for each barrier is in a region of negative kinetic energy, in contrast to the assumptions we made at the beginning of the chapter. Also, the drift space separating the two barriers has a negative kinetic energy throughout.

When $\mathcal{E} < V_0$, the solution of the Schroedinger equation at the reference planes may still be written in the form (5•2–2a,b), but now the wave number K' becomes imaginary,

$$K' \to +i\kappa, \tag{5•3–16}$$

where

$$\kappa = \left[\frac{2M}{\hbar^2}(V_0 - \mathcal{E})\right]^{1/2}. \tag{5•3–17}$$

As we pointed out in section 1.4, all we have to do to adapt our earlier wave function for $\mathcal{E} > V_0$ to the case $\mathcal{E} < V_0$ is to make the substitution (5•3–16) everywhere. We recall that this converts any wave propagating to the right into a wave evanescent to the right, i.e.,

$$\exp(+iK'x) \to \exp(-\kappa x), \tag{5•3–18}$$

automatically yielding the correct boundary conditions at infinity, if the region of negative kinetic energy extends that far: The requirement that there be no wave coming in from $+\infty$ is transformed into the requirement that the wave vanish at infinity.

However, with re-normalized plane wave amplitudes as in (5•2–2a,b), making the substitution (5•3–16) introduces a nuisance factor $\sqrt{i}$ into the

[3] The text by Merzbacher, quoted in appendix G, contains an excellent discussion of scattering by a square well, including the topic of resonances in the transmission delay.

denominators. To avoid cluttering up the subsequent math, we absorb this factor into the complex amplitudes A through D and write the substituted waves in the form

$$\psi_{\mathrm{I}}(x) = \frac{1}{\sqrt{\kappa_{\mathrm{I}}}}[A \cdot e^{-\kappa_{\mathrm{I}}(x-x_{\mathrm{I}})} + B \cdot e^{+\kappa_{\mathrm{I}}(x-x_{\mathrm{I}})}], \tag{5•3–19a}$$

$$\psi_{\mathrm{II}}(x) = \frac{1}{\sqrt{\kappa_{\mathrm{II}}}}[C \cdot e^{-\kappa_{\mathrm{II}}(x-x_{\mathrm{II}})} + D \cdot e^{+\kappa_{\mathrm{II}}(x-x_{\mathrm{II}})}]. \tag{5•3–19b}$$

Consider now the propagation matrix at the upward step on the left-hand side of the barrier shown in figures 1•4–3 or 5•3–1, assuming that $0 < \mathcal{E} < V_0$. If we had not absorbed a factor $\sqrt{i}$ into the wave amplitudes C and D, we could obtain the propagation matrix by simply making the substitution (5•3–16) in the expression (5•3–3a) for the matrix for $\mathcal{E} > V_0$. If we absorb the factor $\sqrt{i}$ in the denominator of the matrix coefficients into the right-hand amplitudes, the resulting matrix becomes

$$\hat{P}_{\uparrow} = \frac{1}{2\sqrt{K\kappa}}\begin{pmatrix} K + i\kappa & K - i\kappa \\ K - i\kappa & K + i\kappa \end{pmatrix}. \tag{5•3–20a}$$

The propagation matrix for the downward step at the end of the barrier requires more care. Now, a factor $\sqrt{i}$ has been absorbed into the amplitudes A and B at the *left* reference plane. This calls for an *additional* factor $\sqrt{i}$ in the denominator of the matrix coefficients, leading to

$$\hat{P}_{\downarrow} = \frac{1}{2i\sqrt{K\kappa}}\begin{pmatrix} i\kappa + K & i\kappa - K \\ i\kappa - K & i\kappa + K \end{pmatrix}. \tag{5•3–20b}$$

Finally, the free-space propagation matrix (5•2–12) evidently becomes

$$\hat{P}_{\Delta} = \begin{pmatrix} e^{+\kappa L} & 0 \\ 0 & e^{-\kappa L} \end{pmatrix}. \tag{5•3–21}$$

When the three propagation matrices are multiplied together, the $\sqrt{i}$ corrections to $\hat{P}_{\uparrow}$ and $\hat{P}_{\downarrow}$ cancel out, and the final result is exactly the same as if we had made the substitution (5•3–16) everywhere in (5•3–4). We may therefore obtain the transmission probability T by simply making the substitution (5•3–16) in the final result (5•3–9) of the calculation for $\mathcal{E} > V_0$:

$$\begin{aligned} T &= \left[1 + \frac{1}{4}\frac{\beta^2}{\beta - (\kappa L)^2} \cdot \frac{\sinh^2 \kappa L}{(\kappa L)^2}\right]^{-1} \\ &= \left[1 + \frac{V_0^2}{4\mathcal{E} \cdot (V_0 - \mathcal{E})} \cdot \sinh^2 \kappa L\right]^{-1}. \end{aligned} \tag{5•3–22}$$

Note the replacement of the trigonometric sine by the hyperbolic sine, from whose properties it is evident that the tunneling probability exhibits no resonances and that it decreases rapidly as the particle energy drops below the top of the barrier.

The result (5•3–22) becomes particularly simple when the barrier becomes relatively opaque, when $\kappa L \gg 1$. In that case,

$$T \cong \frac{16\mathcal{E} \cdot (V_0 - \mathcal{E})}{V_0^2} \cdot e^{-2\kappa L}, \tag{5•3–23}$$

the same as (1•4–32), except that there we called the barrier width w rather than L.

5.3.4 The δ-Function Limit

The propagation matrix becomes particularly simple in the limit of a δ-function well or a δ-function barrier. We go to the limits

$$L \to 0,\ V_0 \to \pm\infty,\ LV_0 = S = \text{const.} \tag{5•3–24}$$

With these,

$$K',\ \kappa \to \infty; \quad K'L,\ \kappa L \to 0, \tag{5•3–25a,b}$$

$$K'^2L,\ \kappa^2 L \to 2M|S|/\hbar^2 = 2\kappa_0, \tag{5•3–26}$$

where we have introduced the convenient abbreviation

$$\kappa_0 = MS/\hbar^2. \tag{5•3–27}$$

If the limits (5•3–25) through (5•3–26) are inserted into (5•3–5) through (5•3–7), we obtain, for the δ-function well,

$$P_{11} = 1 - i\kappa_0/K, \qquad P_{12} = -i\kappa_0/K, \tag{5•3–28a,b}$$

and for the δ-function barrier,

$$P_{11} = 1 + i\kappa_0/K, \qquad P_{12} = i\kappa_0/K. \tag{5•3–29a,b}$$

We shall use these limits in the next section.

5.3.5 Relations between Propagation Matrix Coefficients for Negative Kinetic Energies

Inspection of the propagation matrices (5•3–20a,b) and (5•3–21) for $\mathcal{E} < V_0$ shows that the coefficients of these matrices do not obey the relations (5•2–17a,b) and (5•2–18) derived earlier for the case where $\mathcal{E} > V_0$ in both reference planes. The reason for this discrepancy is that both the time reversal argument and the current continuity argument in section 5.2 take on different forms when the waves are evanescent rather than propagating at one or both of the two reference planes. Because of the general usefulness of the relations between the propagation matrix coefficients, we derive here the analogs of the relations (5•2–17a,b) and (5•2–18) for the other possible energetic relations.

We must consider three cases:

(a) "Upward barrier," with $V(x_{\text{I}}) < \mathcal{E} < V(x_{\text{II}})$.
(b) "Downward barrier," with $V(x_{\text{I}}) > \mathcal{E} > V(x_{\text{II}})$, the inverse of case (a).
(c) $\mathcal{E} < V$ at both reference planes.

We treat only case (a) in detail; the other two cases are left to the reader, and we only state the results at the end.

Time Reversal

When taking the complex conjugate of a wave function, right-evanescent and left-evanescent waves are not interchanged, in contrast to right- and left-propagating waves. In case (a), the wave functions in the two reference planes will be of the forms (5•2–2a) to the left and (5•3–19b) to the right. The time-reversed wave function is then no longer obtained by the substitutions (5•2–19), but by

$$A \to B^*, B \to A^*, C \to C^*, D \to D^*. \tag{5•3–30}$$

In terms of the propagation matrix coefficients, this implies that

$$P_{22} = P_{12}^* \quad \text{and} \quad P_{21} = P_{11}^*, \tag{5•3–31a,b}$$

instead of (5•2–17a,b).

Current Continuity

For wave functions with imaginary wave numbers, the probability current density is no longer given by the forms (5•2–3). Instead,

$$j_{\text{I}} = -\frac{i\hbar}{M}(A^*B - B^*A) \quad \text{or} \quad j_{\text{II}} = -\frac{i\hbar}{M}(C^*D - D^*C), \tag{5•3–32}$$

where the second relation is applicable to case (a). Current continuity then demands that

$$A^*A - B^*B = -i \cdot (C^*D - CD^*). \tag{5•3–33}$$

We insert, on the left-hand side, the basic relations (5•2–7) that express A and B in terms of C and D, and we request that the resulting expression hold for all combinations of C and D. This means that both the C^*C terms and the D^*D terms on the left-hand side must vanish and that the C^*D terms on the left-hand side must add up to $-i$:

$$P_{11}^*P_{11} - P_{21}^*P_{21} = P_{12}^*P_{12} - P_{22}^*P_{22} = 0, \tag{5•3–34}$$

$$P_{11}^*P_{12} - P_{21}^*P_{22} = -i. \tag{5•3–35}$$

If we now use the earlier result (5•3–31a,b), we see that (5•3–34) is indeed satisfied and that (5•3–35) may be written as

$$\det \hat{P} = +i. \tag{5•3–36}$$

It is left to the reader to show that the following relationships hold in the remaining two cases:

CASE (b):

$$P_{22} = P_{21}^* \quad \text{and} \quad P_{12} = P_{11}^*, \tag{5•3–37a,b}$$

$$\det \hat{P} = -i. \tag{5•3–38}$$

CASE (c):

All coefficients are real, and

$$\det \hat{P} = 1. \tag{5•3–39}$$

Inspection of the individual propagation matrices for tunneling through a barrier shows that these matrices do indeed satisfy the conditions derived here.

◆ PROBLEMS TO SECTION 5.3

#5•3-1: Zero-Reflection Conditions at a Double-Step Barrier

Devise a quantum-mechanical "antireflection coating"; that is, determine the proper width L and height V_1 of an intermediate potential step to suppress the reflection of a wave of a specific incident energy at a barrier of height $V_2 < \mathcal{E}$ (**Fig. 5•3–3**).

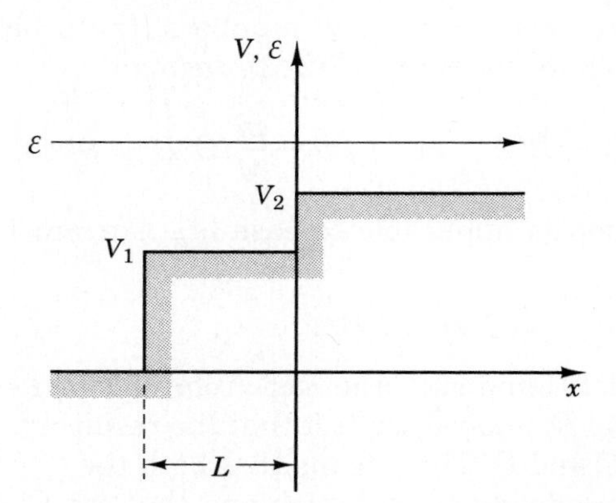

Figure 5•3–3. Zero-reflection barrier.

#5•3-2: Pair of δ-Functions

Calculate the propagation matrix for two δ-function barriers of equal strength S, separated by a distance L. Plot and discuss the transmission probability T as a function of particle energy up to 10 eV, (assuming electrons), for the specific numerical values $S = 10^{-7}$ eV · cm, $L = 2 \times 10^{-8}$ cm.

#5•3–3: Sharpness of Transmission Resonances for Deep Wells

Give a detailed discussion of the sharpness of the transmission resonances for scattering by a deep well, assuming that

$$-V_0 \gg 4\mathcal{E}. \tag{5•3–40}$$

Specifically, derive and discuss a suitable approximation for the energetic width $\Delta\mathcal{E}$ of the resonance as a function of V_0 and L under the condition (5•3–40), where $\Delta\mathcal{E}$ is defined as the width of that energy interval over which $T \geq 1/2$, the so-called full width at half-magnitude, commonly referred to by the abbreviation FWHM.

#5•3–4: Density Resonances

Investigate the probability density present in the range of a barrier, in the energetic vicinity of the transmission resonances. Express the probability density $\rho(x)$ relative to the probability density in the incident beam, $|A|^2/K$, by setting $A = \sqrt{K}$. Plot the result graphically for the lowest two transmission resonances of a barrier with $\beta = 100$.

5.4 ENERGY BANDS IN PERIODIC POTENTIALS

An important type of wave propagation is that of electrons through the periodic potential in a crystal. We apply here the propagation matrix formalism to study this problem, for the idealized case of a one-dimensional "crystal." While strictly one-dimensional crystals do not occur in nature, a solution of this simplified problem exhibits many important features of real, three-dimensional crystals. In particular, it shows the existence of allowed and forbidden energy bands.

We consider the case of a one-dimensional periodic potential $V(x)$, with period a (**Fig. 5•4–1**). We treat each cell as a potential barrier that is characterized by the propagation matrix

$$\hat{P} = \begin{pmatrix} P_{11} & P_{12} \\ P_{21} & P_{22} \end{pmatrix}. \tag{5•4–1}$$

As reference planes with respect to which the various wave amplitudes and matrix coefficients are defined, we use the boundaries of the cell. We single out one particular cell and denote the complex wave amplitudes at its left boundary with A and B and those at its right boundary with C and D, as in (5•2–8):

$$\begin{pmatrix} A \\ B \end{pmatrix} = \begin{pmatrix} P_{11} & P_{12} \\ P_{21} & P_{22} \end{pmatrix} \begin{pmatrix} C \\ D \end{pmatrix} = \hat{P} \begin{pmatrix} C \\ D \end{pmatrix}. \tag{5•4–2}$$

However, we are not dealing, with a *single* cell, but with a periodic arrangement of an infinite number of cells. Because of the translational periodicity of the potential, all bulk-like physical properties of the crystal must be periodic, too. In particular, the probability density and the probability current density

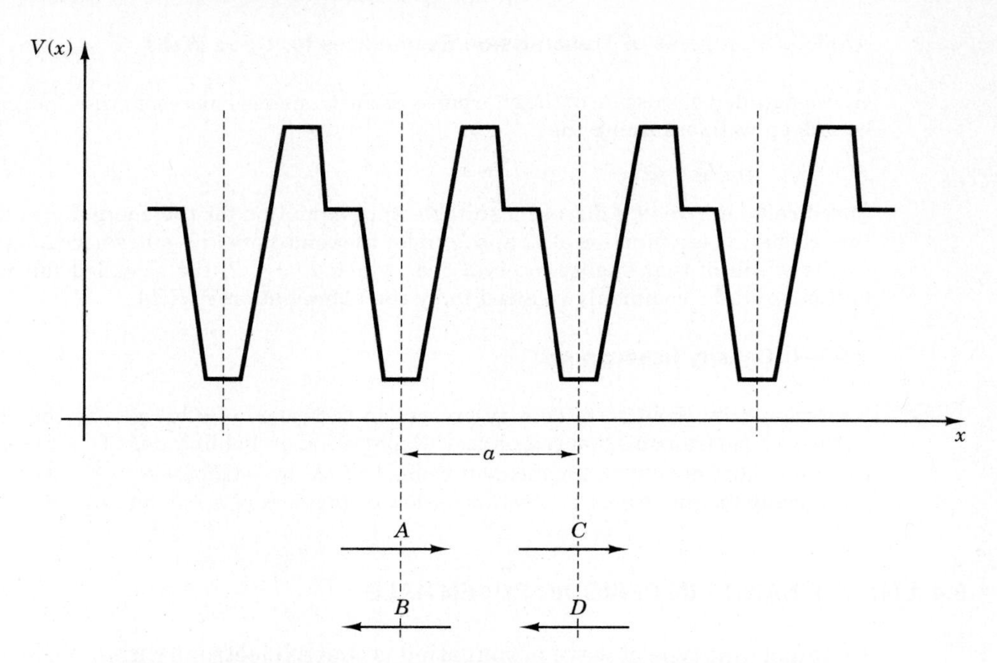

Figure 5•4–1. One-dimensional periodic potential. Each cell is represented by a potential barrier with a propagation matrix $\hat{P}$; the cell boundaries serve as reference planes.

must be periodic:

$$\psi^*(x)\ \psi(x) = \text{periodic with period } a, \tag{5•4–3a}$$

$$\operatorname{Im}\left(\psi^* \frac{d\psi}{dx}\right) = \text{periodic with period } a. \tag{5•4–3b}$$

From (5•4–3a), it follows that $\psi(x)$ must be of the form

$$\psi(x) = u(x) \exp[i\alpha(x)], \tag{5•4–4}$$

where $u(x)$ is periodic and $\alpha(x)$ is a real function that can always be chosen in such a way that it is *not* periodic, because any periodic part could be lumped into $u(x)$. When (5•4–4) is inserted into (5•4–3b), we conclude that

$$\operatorname{Im}\left(u^2 + i|u|^2 \frac{d\alpha}{dx}\right) = \text{periodic with period } a. \tag{5•4–5}$$

Because $u(x)$ is periodic, the first term in the parentheses is certainly periodic, and (5•4–5) reduces to

$$\frac{d\alpha}{dx} = \text{periodic with period } a. \tag{5•4–6}$$

But the only way a non-periodic function $\alpha(x)$ can have a "periodic" derivative is if the derivative is simply a constant, which we call k. This means that

$$\alpha(x) = kx. \tag{5•4–7}$$

We have neglected any integration constant C, because a term e^{iC} can always be incorporated into $u(x)$. Thus, we conclude that $\psi(x)$ must be of the form

$$\boxed{\psi(x) = u(x) \cdot e^{ikx}.} \tag{5•4–8}$$

From this, it follows that

$$\boxed{\psi(x + a) = \psi(x) \cdot e^{ika}.} \tag{5•4–9}$$

Conversely, any function $\psi(x)$ that satisfies (5•4–9) for all x can always be written in the form (5•4–8). The two equivalent forms (5•4–8) and (5•4–9) constitute the one-dimensional form of **Bloch's theorem**, the most fundamental theorem of the electron dynamics in crystals. The quantity k is called the **Bloch wave number**. We shall extend Bloch's theorem to three dimensions in chapter 17.

Applied to the wave amplitudes A through D, Bloch's theorem implies that

$$C = Ae^{ika}, \ D = Be^{ika}, \tag{5•4–10}$$

or

$$\begin{pmatrix} A \\ B \end{pmatrix} = e^{-ika} \begin{pmatrix} C \\ D \end{pmatrix}. \tag{5•4–11}$$

If this is inserted into (5•4–2), we obtain the condition

$$\hat{P} \begin{pmatrix} C \\ D \end{pmatrix} = e^{-ika} \begin{pmatrix} C \\ D \end{pmatrix}, \tag{5•4–12}$$

which may be re-written in the form

$$\begin{pmatrix} P_{11} - e^{-ika} & P_{12} \\ P_{21} & P_{22} - e^{-ika} \end{pmatrix} \begin{pmatrix} C \\ D \end{pmatrix} = 0. \tag{5•4–13}$$

This is a set of two linear homogeneous equations for the two unknowns C and D. Such a set has a non-trivial solution only if its determinant vanishes:

$$\begin{vmatrix} P_{11} - e^{-ika} & P_{12} \\ P_{21} & P_{22} - e^{-ika} \end{vmatrix} = (P_{11} - e^{-ika})(P_{22} - e^{-ika}) - P_{12}P_{21} = 0. \tag{5•4–14}$$

With the help of the relations (5•2–17) and (5•2–18) between the propagation matrix coefficients, this may be re-arranged to read

$$e^{-2ika} - 2Re(P_{11})e^{-ika} = -1. \tag{5•4–15}$$

If we take the real and imaginary parts of this and simplify the results using elementary trigonometric identities, we obtain the two relations

$$2 \cos ka \cdot [\cos ka - \mathrm{Re}(P_{11})] = 0 \tag{5•4–16a}$$

and

$$2 \sin ka \cdot [\cos ka - \mathrm{Re}(P_{11})] = 0, \tag{5•4–16b}$$

both of which must be satisfied. Evidently, this requires that

$$\cos ka = \mathrm{Re}(P_{11}). \tag{5•4–17}$$

However, this condition has a solution with a real value of k only if

$$\boxed{|\mathrm{Re}(P_{11})| \le 1.} \tag{5•4–18}$$

If one now plots $\mathrm{Re}(P_{11})$ as a function of energy, one invariably finds that there are energy bands of finite width, alternating between **allowed bands**, where (5•4–18) is satisfied, and **forbidden bands**, where it is not.

In every energy range where (5•4–18) is satisfied, there exist two solutions of the Schroedinger equation of the required Bloch wave form (5•4–8), with two real values of k differing from each other only in sign.

In energy ranges where (5•4–18) is not satisfied, k is imaginary. The solutions of the Schroedinger equation then do not satisfy Bloch's theorem, and therefore, they cannot correspond to legitimate bulk states actually occurring inside a crystal.[4]

We illustrate this formation of energy bands here for the simplest of all periodic potentials, a periodic array of positive δ-functions with the spacing a (**Fig. 5•4–2**), introduced by Kronig and Penney in the early days of quantum mechanics to demonstrate some of the quantum-mechanical consequences of a periodic potential.

We showed in (5•3–29b) that the coefficient P_{11} of a δ-function is of the form

$$P_{11} = 1 + i\kappa_0/K, \tag{5•4–19}$$

where κ_0 is given by (5•3–27). However, the coefficient P_{11} calculated in section 5.3 corresponded to a set of reference planes that coincided with each other and with the δ-function itself. For our purposes, we must use reference planes

[4] Under certain circumstances, such solutions may correspond to so-called **surface states** or to other effects associated with an imperfect crystal periodicity.

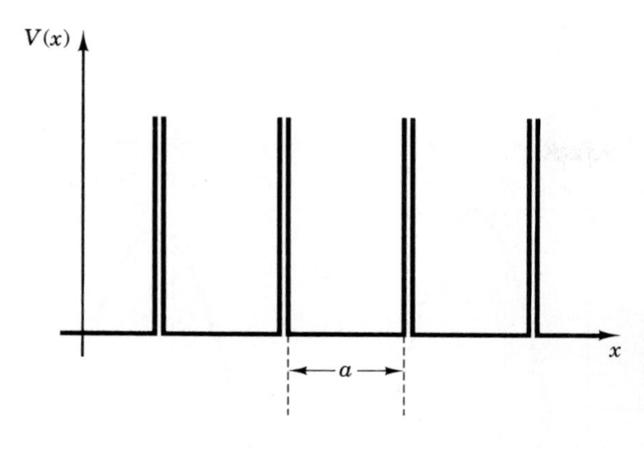

Figure 5•4–2. The Kronig-Penney potential: A periodic potential built up from a periodic array of δ-function barriers.

separated by the distance a. This adds a phase factor e^{-iKa} to the coefficient P_{11}, and we obtain

$$P_{11} = (1 + i\kappa_0/K)e^{-iKa}. \tag{5•4–20}$$

The real part of this may be written

$$\mathrm{Re}(P_{11}) = \cos Ka + P \cdot \frac{\sin Ka}{Ka}, \tag{5•4–21}$$

where we have introduced the parameter

$$P = \kappa_0 a \tag{5•4–22}$$

as a dimensionless measure of the strength of the periodic potential.

With (5•4–21), the condition (5•4–18) may be written

$$-1 \leq \cos Ka + P \cdot \frac{\sin Ka}{Ka} \leq +1. \tag{5•4–23}$$

This condition is illustrated in **Fig. 5•4–3,** where we have plotted the right-hand side of (5•4–21) as a function of Ka for $P = 5$. Only the positive-K branch is shown; the negative-K branch is its mirror image.

It is evident that the curve alternates between Ka-regions where (5•4–23) is satisfied and regions where it is not. One readily sees that the upper edges of the allowed bands (and the lower edges of the forbidden ones) occur at the values $Ka = n\pi$, where n is an integer. At those points, the cosine terms in (5•4–21) and (5•4–23) equal ± 1, with zero slope, while the sine term equals zero, with a sign of the slope equal to the sign of the cosine term. Thus, for Ka-values just below the multiples of π, the condition (5•4–23) is satisfied, while for values just above them, it is not. It is also readily seen that the widths of the forbidden bands decrease with increasing Ka, due to the decrease in the sine term, which is the term that is responsible for the formation of the forbidden bands.

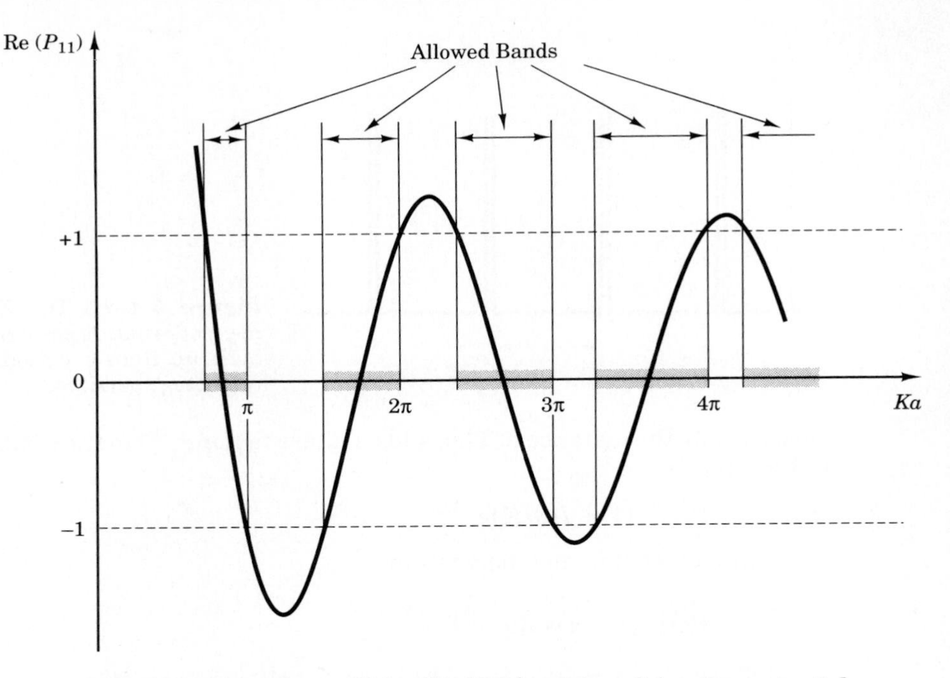

Figure 5•4–3. The condition $|\text{Re}(P_{11})| \leq 1$ for a δ-function potential with a dimensionless barrier strength parameter $P = 5$.

5.5 BOUND STATES AS A SCATTERING PROBLEM

5.5.1 The Propagation Matrix for a Bound State

Although we developed the propagation matrix formalism for the purpose of treating the scattering of waves coming in from infinity, the formalism remains applicable—and surprisingly useful—for the determination of the energies of *bound* states of one-dimensional potential wells. From the propagation matrix point of view, the wave functions of bound states are simply waves whose wave numbers happen to be purely imaginary at any pair of reference planes placed sufficiently far toward $\pm\infty$ and for which two of the amplitudes in those reference planes vanish, namely, A and D. The evanescent plane "wave" to the left of the leftmost reference plane then has a representation of the form

$$\begin{pmatrix} 0 \\ B \end{pmatrix}, \tag{5•5–1a}$$

whereas the evanescent plane "wave" to the right of the rightmost reference plane has a representation of the form

$$\begin{pmatrix} C \\ 0 \end{pmatrix}. \tag{5•5–1b}$$

In order for such a pair of waves to be connected by a propagation matrix $\hat{P}$,

$$\begin{pmatrix} 0 \\ B \end{pmatrix} = \hat{P}\begin{pmatrix} C \\ 0 \end{pmatrix}, \tag{5•5–2}$$

this propagation matrix must have vanishing diagonal matrix coefficients, especially

$$\boxed{P_{11} = 0.} \tag{5•5–3}$$

Inasmuch as P_{11} is a function of the energy, (5•5–3) is an equation for the energy, the roots of which are the energies of the bound states of that well whose propagation matrix is $\hat{P}$.

Note that this approach does not require an actual determination of the energy eigenfunctions, which is often a more difficult task than assembling the propagation matrix numerically. The method lends itself well to complicated potentials. A convenient approach consists of calculating and multiplying together the individual propagation matrices for a given energy and then repeating this process for suitably spaced energy values in the range of interest. A simple plot of P_{11} reveals the roots, which may then be refined by more precise numerical calculations.

By comparing the condition (5•5–3) for a bound state with the condition (5•4–18) for an allowed energy band in a potential composed of a periodic array of quantum wells, we see that the allowed bands may be viewed as the result of broadening each bound state of a single well into a band.

Rather than plotting P_{11}, it is often more instructive and more convenient to plot the quantity $T = 1/|P_{11}|^2$. As we saw in (5•2–16), this quantity plays the role of a transmission probability for propagating states, and a plot of T for energies corresponding to propagating states is usually of interest anyway. If such a plot is extended to bound-state energies, the bound states show up dramatically as poles in the plot. Note, however, that the quantity T can no longer be interpreted as a transmission probability when the incident and transmitted "waves" are evanescent rather than propagating waves. Having T go to infinity simply means that there is a right-evanescent wave present on the right-hand side, without any right-evanescent wave on the left-hand side.

5.5.2 Example: Bound States of a Square Well of Finite Depth

We apply the preceding formalism here to the bound states of a single well of depth $|V_0|$, as shown in Fig. 5•3–1b. The propagation matrix coefficient P_{11} for the case of a particle going over a barrier, with $0 < V_0 < \mathcal{E}$, was given in (5•3–5). The only difference for the case of bound states inside a well is that the wave number K *outside* the well is now imaginary. From (5•3–8), which remains valid for negative kinetic energies inside the barrier,

$$K = i\sqrt{2M|V_0|/\hbar^2 - K'^2} = i\kappa, \tag{5•5–4}$$

or, by re-introducing the barrier parameter β from (5•3–10),

$$KL = i\sqrt{|\beta| - (K'L)^2} = i\kappa L. \tag{5•5–5}$$

Substituting $K \rightarrow +i\kappa$ in (5•3–5) yields

$$P_{11} = \cos K'L + \frac{|\beta| - 2(K'L)^2}{2K'L\sqrt{|\beta| - (K'L)^2}} \sin K'L. \tag{5•5–6}$$

Exercise: Construct the overall propagation matrix as a product of the matrices in (5•3–20b), (5•2–12), and (5•3–20a), in that order, with $K_\Delta = K'$. Show that the resulting product is the same as that obtained by making the substitution $K \rightarrow +i\kappa$ in the original product in (5•3–4).

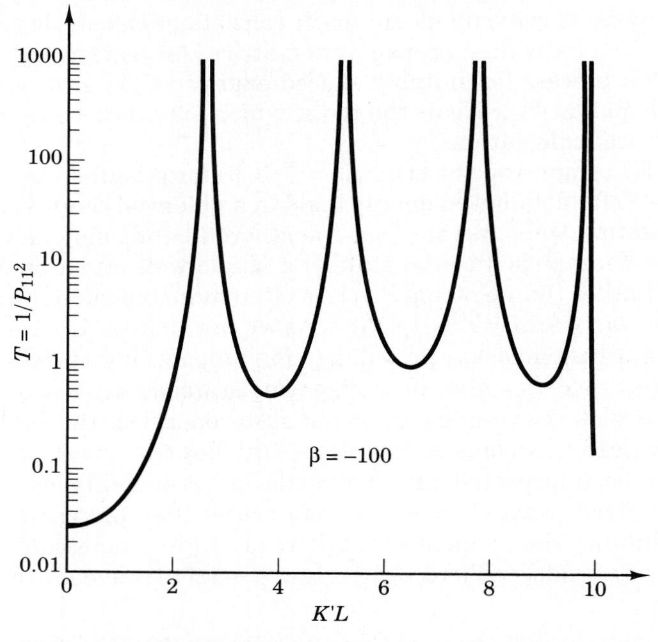

Figure 5•5–1. Semilogarithmic plot of $T = 1/|P_{11}|^2$, as a function of the wave phase $K'L$ inside the well, for a square well with the barrier parameter $\beta = -100$, up to $K'L = \sqrt{100} = 10$, corresponding to energies up to the top of the well. The plot clearly shows four poles, representing bound states; an accurate numerical calculation yields the values $K'L = 2.613$, 5.191, 7.675, and 9.813.

An example of a typical plot of T vs. $K'L$ is shown in **Fig. 5.5–1.**

PROBLEM TO SECTION 5.5

#5•5-1: Transmission Properties of a Composite Barrier

Consider a composite barrier of the general type shown in **Fig. 5•5–2**, containing N discontinuities separating by $N + 1$ flat-potential sections.

The propagation matrix of such a structure has the form

$$\hat{P} = \hat{D}^{(1)} \cdot \hat{F}^{(1)} \cdot \hat{D}^{(2)} \cdot \ldots \cdot \hat{F}^{(N-1)} \cdot \hat{D}^{(N)}, \tag{5•5–7}$$

where the $\hat{D}$'s are the propagation matrices associated with the discontinuities and the $\hat{F}$'s are associated with the flat-potential sections.

(a) Write a computer program that determines the current transmission probability T as a function of the incident energy $\mathcal{E}$, for a potential that contains an arbitrary number of discontinuities with arbitrary lengths L_n and arbitrary potential energies V_n.

(b) Execute the program for the following four-step potential

n:	0	1	2	3	4
L_n[nm]:	0	1	2	3	0
V_n[eV]:	0	+1	−2	0	−1

Calculate and plot T for this potential, in the incident energy range $0 \leq \mathcal{E} \leq 2$ eV, in steps no coarser than 0.1 eV. *Note:* The choice of the length L_N of the exit side of the potential is irrelevant, but specifying a zero length is a convenient way to signal the end of the composite barrier to the program, without having to specify the number of potential steps beforehand.

(c) Re-run the program in the energy range appropriate for bound states of the structure. Plot T vs. $\mathcal{E}$ in that range, and determine the energy of any bound states that might be present, to better than ±0.05 eV. Also, explore the existence of quasi-bound states showing up as resonances in T, with $T < \infty$ in the energy range $0 > \mathcal{E} > V_4$.

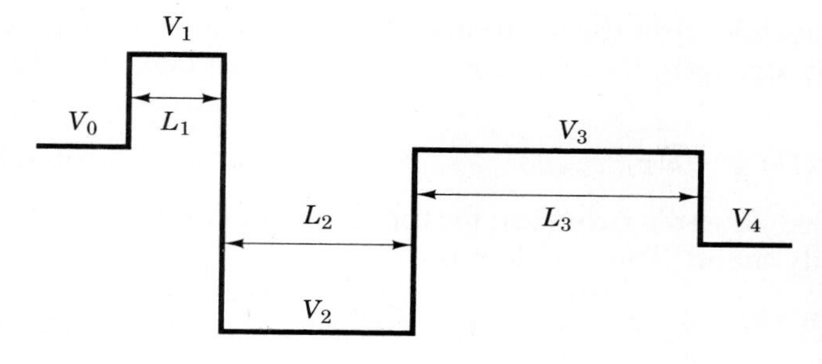

Figure 5•5–2. Barrier containing several flat-potential regions of varying lengths and heights.

5.6 THREE-DIMENSIONAL SCATTERING PROBLEMS: THE BORN APPROXIMATION

5.6.1 The Schroedinger Equation as an Integral Equation

The propagation matrix method developed in sections 5.2 through 5.5, while ideally suited for the treatment of one-dimensional scattering problems, is unsuited for scattering in three dimensions. With a few exceptions—for example, the Coulomb potential—such problems require approximation methods. We present here, in outline, the most important of these methods, the **Born approximation**.

To derive this approximation, we first re-write the Schroedinger equation as

$$\nabla^2\psi(\mathbf{r}) + \frac{2M\mathcal{E}}{\hbar^2}\psi(\mathbf{r}) = -\int s(\mathbf{R})\delta(\mathbf{R}-\mathbf{r})\,d^3R, \tag{5•6–1}$$

where

$$s(\mathbf{R}) = -\frac{2M}{\hbar^2}V(\mathbf{R})\psi(\mathbf{R}), \tag{5•6–2}$$

and $\delta(\mathbf{R}-\mathbf{r})$ is a three-dimensional Dirac δ-function located at $\mathbf{R}=\mathbf{r}$.

In the absence of the integral on the right-hand side, (5•6–1) would simply be the Schroedinger equation for plane waves whose wave vector $\mathbf{k}$ satisfies

$$k^2 = 2M\mathcal{E}/\hbar^2, \tag{5•6–3}$$

and we could simplify (5•6–1) further by writing it as

$$\nabla^2\psi_0(\mathbf{r}) + k^2\psi_0(\mathbf{r}) = 0. \tag{5•6–4}$$

The integral term in (5•6–1) acts as a perturbation, in the form of a sum over an infinite number of δ-function scattering sources, each with a strength $s \cdot d^3\mathbf{R}$. The source strength depends both on the local potential $V(\mathbf{R})$ and on the *actual* wave function $\psi(\mathbf{r})$ at each point. It is therefore an unknown quantity itself.

As a preliminary, consider first the solution of a simpler problem, a single δ-function source of unit strength, located at $\mathbf{r}=0$. That is, we look for solutions of the equation

$$\nabla^2 G(\mathbf{r}) + k^2 G(\mathbf{r}) = -\delta(\mathbf{r}). \tag{5•6–5}$$

The function $G(\mathbf{r})$ is called **Green's function** for the scattering problem. Solutions of (5•6–5) are easily found. The simplest is

$$G(\mathbf{r}) = \frac{\exp(i\mathbf{k}\cdot\mathbf{r})}{4\pi r}. \tag{5•6–6}$$

Exercise: Show that (5•6–6) satisfies (5•6–5). For $r > 0$, this is easily done by insertion. For $r \to 0$, (5•6–5) implies that

$$\lim_{\Omega \to 0} \int_{\Omega} \nabla^2 G d^3 r = -1, \tag{5•6–7}$$

where the integral goes over any volume Ω that includes $\mathbf{r} = 0$. By Gauss' integral theorem, the integral can be converted to a surface integral:

$$\int_{\Omega} \nabla^2 G d^3 r = \oint_{S} (\nabla G) \cdot d\mathbf{S}. \tag{5•6–8}$$

By choosing as the surface a sphere of vanishing radius, show that (5•6–5) is satisfied.

Consider next the function

$$\psi(\mathbf{r}) = -\int G(\mathbf{r} - \mathbf{R}) \cdot s(\mathbf{R})\, d^3 R, \tag{5•6–9}$$

where $G(\mathbf{r})$ satisfies (5•6–5). By inserting $\psi(\mathbf{r})$ on the left-hand side of (5•6–1) and using (5•6–5), one confirms easily that $\psi(\mathbf{r})$ is a solution of (5•6–1). It is not the most general solution, however: We may add to it any arbitrary solution $\psi_0(\mathbf{r})$ of the *homogeneous* equation (5•6–4), such as

$$\psi_0(\mathbf{r}) = A \cdot \exp(i\mathbf{k} \cdot \mathbf{r}). \tag{5•6–10}$$

If we add ψ_0, and insert $s(\mathbf{R})$ from (5•6–2), we obtain the integral equation

$$\boxed{\psi(\mathbf{r}) = \psi_0(\mathbf{r}) - \int G(\mathbf{r} - \mathbf{R}) V(\mathbf{R}) \psi(\mathbf{R})\, d^3 R.} \tag{5•6–11}$$

Note that this equation is not based on choosing, for ψ_0 and G, the specific examples given; all that is required is that ψ_0 and G satisfy (5•6–4) and (5•6–5).

The integral equation (5•6–11) is mathematically equivalent to the time-independent Schroedinger equation, but it is much better adapted to the treatment of scattering problems. This is because the two functions ψ_0 and G may be chosen in such a way that the ψ_0-term represents an incident wave and the integral the scattered wave. To achieve this separation, we choose ψ_0 to be a plane wave, as in (5•6–10), and G to be a wave going outward from each scattering point $\mathbf{r} = \mathbf{R}$, as in (5•6–6).

Note that (5•6–11) is a re-formulation of the scattering problem, not its solution: The unknown wave function is still contained inside the integral. This simply reflects the physical fact that the scattered intensity depends on the magnitude of the actual wave function at each scattering point, rather than on the magnitude of the incident wave alone. Multiple scattering effects are therefore automatically included in (5•6–11).

5.6.2 The Born Approximation

So far, (5•6–11) is exact. Suppose now that the scattering potential is so weak that the scattered-wave contribution of the integral in (5•6–11) is everywhere small compared to the incident plane wave. The contribution of the integral may then be approximated by replacing the *actual* wave function ψ with the *incident* wave ψ_0:

$$\psi(\mathbf{r}) \cong \psi_1(\mathbf{r}) \equiv \psi_0(\mathbf{r}) - \int G(\mathbf{r} - \mathbf{R})V(\mathbf{R})\psi_0(\mathbf{R})\, d^3R. \qquad (5\bullet6\text{–}12)$$

This is the **first-order Born approximation**. Physically, the approximation of replacing ψ by ψ_0 inside the scattering integral implies neglecting all multiple-scattering effects.

It is sometimes possible to improve the approximation by re-inserting the first-order wave function ψ_1 into the integral, leading to a second-order wave function ψ_2. The process can be repeated ad infinitum. Because only straightforward integrations over known functions are involved, this kind of iterative improvement of the wave function is well adapted to computer calculations.

Unfortunately, higher-order Born approximations are no panacea. Mathematically, the iteration represents a disguised power series expansion of the wave function, in terms of the powers of the overall strength of the perturbing potential. Each order adds a new power to the series. Like other power series expansions, the Born approximation may diverge. Indeed, it is usually only semi-convergent: The higher-order corrections tend to stop decreasing past a certain optimum order and then oscillate with increasing amplitude, leading to an alternating divergent series. For weak scattering potentials without singularities, this ultimate divergence is usually of no practical consequence, because the overshoot oscillations do not set in until long after one has cut off the series. They are, however, a problem with scattering potentials that are very strong or that contain singularities, such as the Coulomb potential $V(\mathbf{r}) \propto 1/r$. In the case of the Coulomb potential, a very peculiar phenomenon occurs: The first-order Born approximation happens to coincide with the known *exact* solution of the scattering problem, but the second-order integral diverges. Truncating the expansion after the first order then leads to the correct answer. Unfortunately, one cannot count on such lucky accidents. Alternative methods have been developed for such cases, but their discussion lies outside the scope of this text.

Chapter 6

WKB APPROXIMATION

6.1 WKB WAVE FUNCTIONS

6.1.1 Plane Waves with Variable Wavelength and Amplitude

Except for the Born approximation in the preceding chapter, we have so far always been concerned with *exact* solutions of the Schroedinger equation. We now turn to another approximation method, the so-called **WKB approximation**, which is an excellent approximation for slowly varying one-dimensional potentials and hence bridges the gap between classical and quantum mechanics. It is named after Wentzel, Kramers, and Brillouin, who, around 1926, were the first to employ this approximation in quantum mechanics, even though as a basic mathematical technique, it is much older.[1]

We take as our point of departure the remark in section 1.3 that in the presence of a force, both the *local* wave number K and the probability density

[1] For complete references, and an in-depth treatment, see N. Fröman and P. O. Fröman, "JWKB Approximation" Amsterdam: North-Holland, 1965. Another excellent discussion, centered around numerous examples, is found in Flügge, cited in appendix G.

ρ of an object wave *must* be functions of position. Consider, therefore, a time-independent wave function of a form similar to the terms in (5•2–2a), i.e.,

$$\psi_{\pm}(x) = \frac{A}{\sqrt{K(x)}} \cdot \exp\left[\pm i \int^{x} K(y)\, dy\right], \tag{6•1–1}$$

with a wave number $K(x)$ that depends on position according to

$$K(x) = \sqrt{\frac{2M}{\hbar^2}[\mathcal{E} - V(x)]} \qquad \text{if} \qquad \mathcal{E} > V. \tag{6•1–2}$$

In the case of a constant potential, this is of course an exact solution of the Schroedinger equation, with the amplitude A being a constant.

We might expect that in the case of a sufficiently slowly varying potential, a wave function of the form (6•1–1), with $K(x)$ given by (6•1–2), might remain at least a good approximation. We note first that the position-denominator in (6•1–1) leads to a probability density inversely proportional to K and, hence, to the local velocity $v = \hbar K/M$, as is needed in order to have a divergence-free probability current density, an essential requirement for any approximation.

Exercise: Show that

$$\frac{dj_{\pm}}{dx} = \frac{d}{dx}\left(\psi_{\pm}^{*}\frac{d\psi_{\pm}}{dx} - \psi_{\pm}\frac{d\psi_{\pm}^{*}}{dx}\right) = 0, \tag{6•1–3}$$

which means that current conservation is obeyed exactly. Show further that this result depends, not on the choice (6•1–2) for $K(x)$, but only on $K(x)$ being real and nonzero.

The most general wave function involving terms of the form (6•1–1) is a linear superposition of the two terms with different signs in the exponent, such as

$$\psi_{WKB}(x) = \frac{1}{\sqrt{K(x)}}\left\{A \cdot \exp\left[+i \int^{x} K(y)\, dy\right] + B \cdot \exp\left[-i \int^{x} K(y)\, dy\right]\right\}, \tag{6•1–4}$$

where both A and B are constants. The (unstated) lower integration limits in (6•1–4) depend on the choice of the phase of the two amplitudes A and B. In effect, those limits establish reference planes for the waves.

It is left to the reader to show that the probability current density associated with (6•1–4)—for positive K—is

$$j = \frac{\hbar}{M} \cdot (|A|^2 - |B|^2). \tag{6•1–5}$$

6.1.2 Validity Conditions

In order to determine how good an approximation the WKB wave function is, we look at its second derivative. The reader may confirm that

$$\psi'' = -\left[K^2 + \frac{1}{2}\left(\frac{K''}{K}\right) - \frac{3}{4}\left(\frac{K'}{K}\right)^2\right] \cdot \psi \tag{6•1–6}$$

for both ψ_+ and ψ_-. The WKB approximation is obtained if we specifically select $K(x)$ according to (6•1–2), assuming that

$$\mathcal{E} > V(x), \tag{6•1–7}$$

to make sure that $K(x)$ is real. The case $\mathcal{E} < V(x)$ will be discussed later.

If we choose the form (6•1–2) for $K(x)$, the $K^2\psi$-term in (6•1–6) represents the Schroedinger equation by itself. The other terms are extra terms; their magnitude is a measure of the degree of deviation of the approximation (6•1–1) from an exact solution. In order for (6•1–1) to be a good approximation, these extra terms must remain small compared to the $K^2\psi$-term, i.e.,

$$\frac{1}{4}\left|2\left(\frac{K''}{K}\right) - 3\left(\frac{K'}{K}\right)^2\right| \ll |K|^2. \tag{6•1–8}$$

To understand the meaning of this condition better, we convert it to the form

$$\left|\frac{V''}{\mathcal{E} - V} + \frac{5}{4}\left(\frac{V'}{\mathcal{E} - V}\right)^2\right| \cdot |\Lambda|^2 \ll 16\pi^2, \tag{6•1–9}$$

where $\Lambda = 2\pi/K$ is the *local* de Broglie wavelength. Both (6•1–8) and (6•1–9) are complicated conditions. While they might in principle be satisfied by a mutual cancellation of the terms on the left-hand sides, this is of little importance in practice. We therefore impose the stronger condition that both of the terms on the left-hand side of (6•1–8) be *separately* small. This leads to two conditions that may be written

$$\Delta V_1 \equiv |V' \cdot \Lambda| \ll \frac{8\pi}{\sqrt{5}}|\mathcal{E} - V| \approx 11.2|\mathcal{E} - V| \tag{6•1–10}$$

and

$$\Delta V_2 \equiv \tfrac{1}{2}|V'' \cdot \Lambda^2| \ll 8\pi^2|\mathcal{E} - V| \approx 79.0|\mathcal{E} - V|. \tag{6•1–11}$$

We refer to these as the **first and second WKB conditions.** Both have a simple meaning, which we can bring out as follows. If the potential V varies slowly enough with position, the quantity ΔV_1 represents the change in potential along one wavelength (**Fig. 6•1–1(a)**); similarly, ΔV_2 represents that change in potential that would build up due to the *curvature* of the potential along one wavelength, starting with $V' = 0$ (**Fig. 6•1–1(b)**). The relations (6•1–9) and (6•1–10) then state that the changes in potential, *both* due to a finite

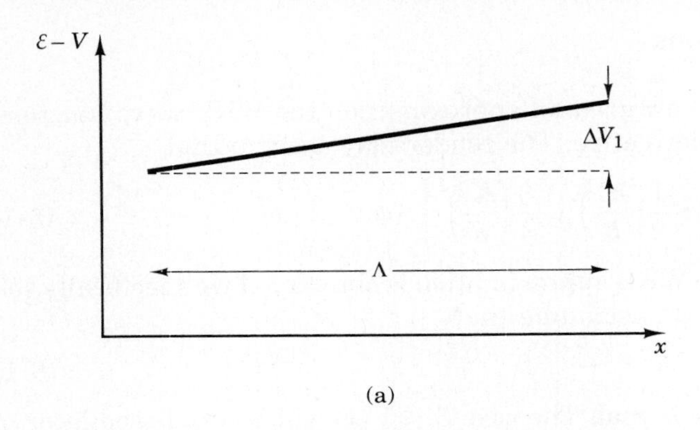

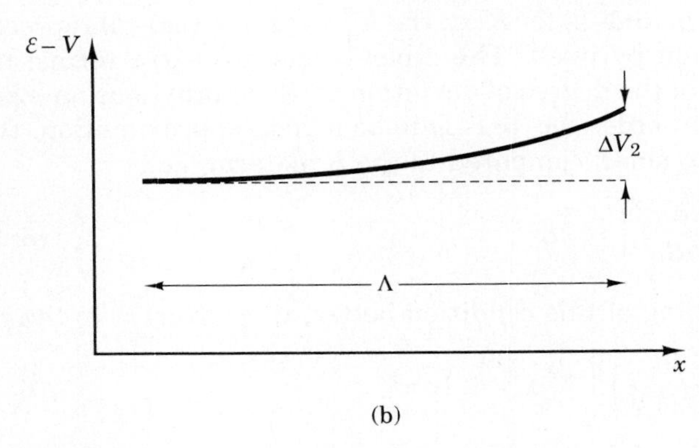

Figure 6•1–1. (a) The first WKB condition is that the potential change along a wavelength is sufficiently small compared to the minimal kinetic energy along the path of that wavelength. **(b)** The second WKB condition demands that the curvature of the potential be small.

slope and due to a finite curvature, taken per wavelength, must stay small compared to the stated large multiples of the kinetic energy of the particle.

Both conditions are easily met, and whenever they are satisfied, the WKB wave function (6•1–4) *with constant amplitudes A and B* is a good approximate solution of the Schroedinger equation.

If the correction terms in (6•1–6) cannot be neglected, the true wave function can still be written in the form (6•1–4), but then the amplitudes A and B no longer can be treated as constants, but become themselves position dependent. It is left to the reader to show that in this case the continuity equation (6•1–3) requires that

$$|A(x)|^2 - |B(x)|^2 = \text{constant}. \tag{6•1–12}$$

Position-dependent amplitudes mean that scattering is taking place between the forward- and backward-propagating waves; *the WKB approximation itself is a zero-scattering approximation.*

It is in principle possible to extend the WKB approximation to higher orders, but they are hardly ever used. A much more practical approach to the calculation of any reflected waves is simply inserting the first-order WKB wave function into the Born approximation.

6.1.3 Exponentially Growing and Decaying Approximations for Negative Kinetic Energy; the Connection Problem

The preceding discussion assumed that the total energy $\mathcal{E}$ of the particle exceeds the potential energy, i.e., $\mathcal{E} > V(x)$. However, the relation (6•1–6) holds regardless of this assumption, as do the validity conditions (6•1–8) through (6•1–11) that follow from (6•1–6). This means that wave functions of the form (6•1–1) with imaginary K remain good approximations even for $\mathcal{E} < V(x)$, under the same conditions (6•1–8) through (6•1–11). In fact, this applicability to both positive and negative kinetic energies forms the basis for one of the most important applications of the WKB approximation: tunneling through barriers.

The only change we make in the negative-energy case is again a replacement of K by $i\kappa$, where

$$\kappa(x) = \sqrt{\frac{2M}{\hbar^2}[V(x) - \mathcal{E}]} \text{ if } \mathcal{E} < V. \tag{6•1–13}$$

With this change, we obtain, instead of (6•1–4),

$$\psi_{WKB}(x) = \frac{1}{\sqrt{\kappa(x)}}\left\{C \cdot \exp\left[-\int^x \kappa(y)\,dy\right] + D \cdot \exp\left[+\int^x \kappa(y)\,dy\right]\right\}. \tag{6•1–14}$$

The conditions (6•1–9) through (6•1–11) still hold, except that the wavelength Λ is replaced by

$$\Lambda' \rightarrow 2\pi/\kappa. \tag{6•1–15}$$

What *does* assume a significantly different form is the expression for the probability current density. It is again left to the reader to show that, instead of (6•1–5), we now have

$$j = \frac{i\hbar}{M} \cdot (CD^* - C^*D), \tag{6•1–16}$$

Note that a current can flow only if both a right-evanescent (C) and a left-evanescent (D) wave are present and the ratio of the two amplitudes is a complex number.

In bound-state problems, the range for which $V(x) > \mathcal{E}$ extends to infinity. In those cases, one of the two coefficients in (6•1–14) must be zero. For $x \rightarrow +\infty$, we must have $D = 0$, or else the wave function would become infinite there. Similarly, for $x \rightarrow -\infty$, we must have $C = 0$, for the same reason. In either case, there is, of course, no current.

In order for the Schroedinger equation to have any physically meaningful solutions at all, we must have $V(x) < \mathcal{E}$ over at least *some* range of x, and this means that there must be at least one *classical turning point* $x = a$ where $V(x) - \mathcal{E}$ changes its sign. In realistic problems, $V(x) - \mathcal{E}$ will go through zero as a continuous function, rather than as a step. In the immediate vicinity of the classical turning point, neither the first nor the second WKB condition can be satisfied, and both the oscillating form (6•1–4) and the evanescent form (6•1–14) of the WKB approximation diverge, signaling the collapse of the approximation.

Suppose, however, that sufficiently far from $x = a$, the WKB approximation is applicable on both sides of a, on one side in the form (6•1–4) and on the other in the form (6•1–14), with neither form remaining a valid approximation if it is extended right up to a. This raises the question: what is the relationship between the coefficients A and B on one side to C and D on the other? This is the *connection problem* of the WKB approximation, to be treated in the sections that follow.

6.2 EXAMPLE: HARMONIC OSCILLATOR

6.2.1 Phase Connection Rule

It is useful to discuss the WKB wave functions and their connection rules for a case for which we know already the exact wave functions: the energy eigenstates of the harmonic oscillator.

We write the potential energy in the form

$$V(x) = \tfrac{1}{2}M\omega^2 x^2, \tag{6•2–1}$$

and we recall that the energy eigenvalues are then given by

$$\mathcal{E}_n = (n + \tfrac{1}{2})\hbar\omega, \tag{6•2–2}$$

where the n are the nonnegative integers. We will also need the classical turning points from (2•3–24), i.e.,

$$x_n = \sqrt{2n + 1} \cdot L, \tag{6•2–3}$$

where

$$L = \sqrt{\hbar/M\omega} \tag{6•2–4}$$

is the natural unit of length for the oscillator (see (2•3–7)).

Because for a one-dimensional stationary bound state there can be no net current flow, the WKB wave function (6•1–4) must be a pure standing wave. In this case, the two propagating waves in that function may be lumped together into a real cosine wave, which may be written in the form

$$\psi(x) = \frac{2A}{\sqrt{K}} \cdot \cos\left[\int_{-x_n}^{x} K\, dy - \alpha\right], \tag{6•2–5}$$

with a real amplitude A and a suitable phase angle α. In (6•2–5), we have selected the phase angle α in such a way that the left classical turning point serves as one of the integration limits and, hence, as a reference plane for the phase of the wave, even though at that point (6•2–5) is no longer a good approximation. Without loss of generality, we may restrict α to the range

$$-\pi/2 < \alpha < +\pi/2. \tag{6•2–6}$$

For $\mathcal{E} = \mathcal{E}_n$, the local WKB wave number K in Eq. (6•1–2) in the classically allowed range $|x| < x_n$ is given by

$$K(x) = \sqrt{\frac{2M}{\hbar^2}(\mathcal{E}_n - \tfrac{1}{2}M\omega^2 x^2)} = \frac{1}{L^2}\sqrt{x_n^2 - x^2}. \tag{6•2–7}$$

For later use, we note that

$$K(0) = \frac{x_n}{L^2} = \frac{\sqrt{2n+1}}{L}. \tag{6•2–8}$$

We insert (6•2–7) into (6•2–5), evaluate the result at $x = 0$, and compare it with the known exact wave function at that point. To be specific, we assume that n is an even number.[2] In this case, the exact wave function is symmetric about the point $x = 0$, and it has $n/2$ nulls on each side of the plane $x = 0$. In order for the WKB wave function to be itself an even function of x, with the same number of nulls, it is necessary that the argument of the cosine function in Eq. (6•2–5), evaluated at $x = 0$, be an integer multiple of π, with the multiplier $n/2$:

$$\int_{-x_n}^{0} K\,dx - \alpha = \frac{1}{L^2}\int_{-x_n}^{0} \sqrt{x_n^2 - x^2}\,dx - \alpha = \frac{n}{2}\pi. \tag{6•2–9}$$

The integral is simply the area of a quarter-circle of radius x_n:

$$\begin{aligned}\int_{-x_n}^{0} \sqrt{x_n^2 - x^2}\,dx &= \int_{0}^{+x_n} \sqrt{x_n^2 - x^2}\,dx = \frac{\pi}{4}\cdot x_n^2 \\ &= \frac{\pi}{4}\cdot(2n+1)\cdot L^2.\end{aligned} \tag{6•2–10}$$

Here, in the last equality, we have used (6•2–3). Insertion into (6•2–9) yields

$$\boxed{\alpha = \frac{\pi}{4},} \tag{6•2–11}$$

independent of the quantum number n harmonic oscillator. This is the **WKB phase connection rule.** Inserted into (6•2–5), (6•2–11) gives

$$\psi(x) = \frac{2A}{\sqrt{K}}\cdot\cos\left[\int_{-x_n}^{x} K\,dy - \frac{\pi}{4}\right]. \tag{6•2–12a}$$

[2] The reader is invited to carry through the argument for odd values of n. While the details are different, the final results are the same.

Although we have derived this result here for the harmonic oscillator, we shall see that it holds generally, provided that the potential energy varies sufficiently smoothly across the classical turning point, that it may be approximated by a harmonic oscillator parabola until well into the ranges on both sides where the WKB wave functions are good approximations.

Switch of Reference Plane

In the preceding treatment, we arbitrarily used the left-hand classical turning point as reference plane. We could just as well have used the right-hand classical turning point. It is left to the reader to show that in this case, the WKB wave function between the two turning points may be written as

$$\psi(x) = \frac{2A'}{\sqrt{K}} \cdot \cos\left[\int_x^{+x_n} K\,dy - \frac{\pi}{4}\right]. \tag{6•2–12b}$$

where $A' = (-1)^n A$. Note that in this formulation, the variable x appears as the lower rather than the upper integration limit, and that the integral remains positive. The phase shift remains $\pi/4$.

6.2.2 The WKB Amplitudes and Their Connection Rule[3]

To determine the WKB amplitudes A and C, and especially their ratio C/A, we match the WKB wave functions to the true harmonic oscillator wave functions at $x = 0$ and for large values of x.

From the second term in the recursion relations (2•3–30) for the harmonic oscillator wave functions, we find, for even order n, that

$$\psi_n(0) = A_0 \frac{(-1)^{n/2}\sqrt{n!}}{2^{n/2}\,(n/2)!}, \tag{6•2–13}$$

where A_0 is the normalization coefficient of the ground-state wave function, from (2•3–10) or (2•3–15), depending on the normalization employed. On the other hand, the WKB wave function (6•2–5), taken at $x = 0$, is

$$\psi_{WKB}(0) = \frac{2A}{\sqrt{K(0)}} \cdot \cos(n\pi/2) = A \cdot \frac{2(-1)^{n/2}\sqrt{L}}{(2n+1)^{1/4}}, \tag{6•2–14}$$

where, in the second equality, we have drawn on (6•2–8) and (6•2–9). Equating the WKB value to the exact value yields the WKB amplitude A inside the classically allowed range:

$$A = \frac{A_0}{\sqrt{L}} \cdot \frac{(2n+1)^{1/4}}{2 \cdot 2^{n/2}} \cdot \frac{\sqrt{n!}}{(n/2)!}. \tag{6•2–15}$$

[3] The derivation of the amplitude connection rule, contained in this sub-section, is fairly tedious. The reader may safely skip the details and move directly to the final result, Eq. (6•2–26).

In the opposite limit, $Q \to \infty$, the exact wave function is dominated by the highest power term in the Hermite polynomial part of the wave function. The asymptotic limit is obtained easily from the leading term in the recursion relation (2•3–30), namely,

$$\psi_n(Q) \to A_0 \frac{2^{n/2}}{\sqrt{n!}} \cdot Q^n \exp(-Q^2/2). \tag{6•2–16}$$

where

$$Q = x/L. \tag{6•2–17}$$

In the classically forbidden ranges, the WKB wave function must be purely evanescent. In the right-hand range, $x > +x_n$, we may write it as

$$\psi_{WKB}(x) = \frac{C}{\sqrt{\kappa}} \cdot \exp\left[-\int_{x_n}^{x} \kappa dx'\right]. \tag{6•2–18}$$

The integral in the exponent has the value

$$\int_{x_n}^{x} \kappa dx' = -\frac{x_n^2}{2L^2} \cdot \ln\left[\frac{x + (x^2 - x_n^2)^{1/2}}{x_n}\right] + \frac{x \cdot (x^2 - x_n^2)^{1/2}}{2L^2}. \tag{6•2–19}$$

In the limit of large x, this goes over into

$$\begin{aligned} -\frac{x_n^2}{2L^2} \cdot \ln\left[\frac{2x}{x_n}\right] + \frac{x^2}{2L^2} - \frac{x_n^2}{4L^2} + \cdots = \\ -m \cdot \ln\left[Q\sqrt{\frac{2}{m}}\right] + \frac{Q^2}{2} - \frac{m}{2} + \cdots, \end{aligned} \tag{6•2–20}$$

where, in the second equality, we have substituted $x = QL$ and

$$\frac{x_n^2}{2L^2} = m \equiv n + \tfrac{1}{2}. \tag{6•2–21}$$

All omitted terms decrease with increasing x and Q.

If we insert (6•2–22) and (6•2–20) into (6•2–18) and simplify the pre-factor according to

$$\kappa = \frac{1}{L^2}\sqrt{x^2 - x_n^2} \xrightarrow[x \to \infty]{} \frac{Q}{L}, \tag{6•2–22}$$

we obtain the asymptotic relation

$$\psi_{WKB}(Q) \to C\sqrt{L} \cdot \left(\frac{2e}{m}\right)^{m/2} Q^n \exp\left(-\frac{Q^2}{2}\right). \tag{6•2–23}$$

Equating (6•2–16) with (6•2–23) yields

$$C = \frac{A_0}{\sqrt{L}} \frac{1}{2^{1/4}} \frac{1}{\sqrt{n!}} \cdot \left(\frac{m}{e}\right)^{m/2}. \tag{6•2–24}$$

We are principally interested in the ratio C/A of the two amplitudes:

$$\frac{C}{A} = \frac{2^{1/2}}{e^{1/4}} \cdot \left(\frac{2n+1}{e}\right)^{n/2} \cdot \frac{(n/2)!}{n!}. \tag{6•2–25}$$

This rather elaborate-looking expression is simpler than it appears: For $n = 2$, 4, 6, one obtains the numerical values 1.013, 1.006, 1.004, rapidly converging toward the (exact) limit 1.0, i.e.,

$$\boxed{\lim_{n\to\infty} \frac{C}{A} = 1.} \tag{6•2–26}$$

This is the WKB amplitude connection rule for the harmonic oscillator. Like the phase connection rule, it holds more generally, as we shall see presently.

6.3 GENERAL CONNECTION RULES ACROSS A CLASSICAL TURNING POINT

6.3.1 The Problem

We now show that the phase and amplitude connection rules (6•2–11) and (6•2–26) for the harmonic oscillator remain applicable to the connection across the classical turning points of many other "well-behaved" potentials.

Consider a particle of energy $\mathcal{E}$, moving in a potential $V(x)$, with a classical turning point at $x = c$ (**Fig. 6•3–1**). Suppose that

$$V(x) > \mathcal{E} \text{ for all } x > c, \tag{6•3–1}$$

and that the second derivative of the potential at the classical turning point is positive.

If the potential varies sufficiently smoothly with position, there will be a wide interval $(x_{\mathrm{I}}, x_{\mathrm{II}})$ straddling the classical turning point, inside which the

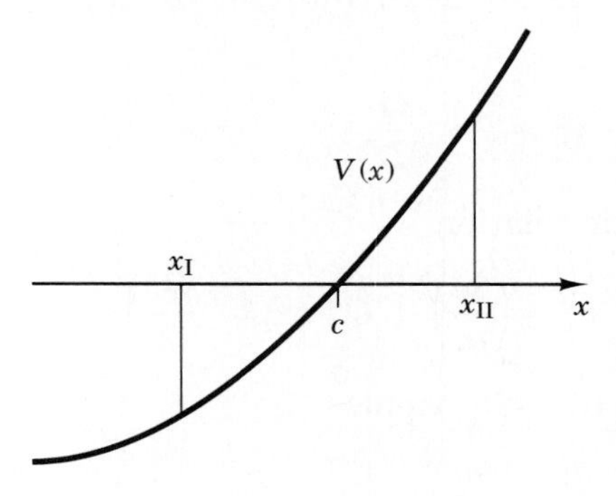

Figure 6•3–1. Connection across a classical turning point. The WKB wave function, forming a valid approximation *outside* the interval $(x_{\mathrm{I}}, x_{\mathrm{II}})$ is connected across this interval using a piece of the solution of an exactly solvable potential, such as the harmonic oscillator potential.

potential may be approximated by a section of a harmonic oscillator parabola of the form

$$V_0(x) = V_{00} + \tfrac{1}{2}M\omega^2(x - x_0)^2. \tag{6•3–2a}$$

We also assume that the energy $\mathcal{E}$ happens to coincide with one of the harmonic oscillator eigenvalues:

$$\mathcal{E} = V_{00} + \hbar\omega \cdot \left(n + \frac{1}{2}\right). \tag{6•3–2b}$$

Exercise: Given specific values of $\mathcal{E}$, a, $V'(a)$, and $V''(a)$, determine the associated fitting parameters V_{00}, ω, x_0, and n.

Suppose now that the fitting interval $(x_{\text{I}}, x_{\text{II}})$ is sufficiently wide and the potential sufficiently smooth that the WKB approximation is a good approximation at both ends of the interval *and beyond.*

Under the assumption (6•3–1), there can again be no net current flow, and the WKB wave function in the classically allowed range must once more be a pure standing wave of the form (6•2–5), with $K = K(x)$ given by (6•1–2). Most important, the phase angle α must then again be given by (6•2–11), i.e., $\alpha = -\pi/4$. This is readily seen by the following argument. If the WKB approximation is indeed applicable in the range $x > x_{\text{II}}$, then the WKB wave function must be a pure right-evanescent wave of the form (6•2–18). Because the WKB approximation is a zero-scattering approximation, a purely right-evanescent WKB wave function must connect to a purely right-evanescent harmonic oscillator wave function. But the latter is just the function (6•2–5) with $\alpha = -\pi/4$.

Similarly, one also confirms that under the stated assumptions, the amplitude connection rule also carries over. If the curvature of the potential is small, we are in the high-n limit (6•2–26), and we have

$$C = A. \tag{6•3–3}$$

6.3.2 Example: An Electron in a Uniform Electric Field

We illustrate the use of the WKB *phase* connection rule by determining the energy levels of an electron that is driven against an infinitely high abrupt potential wall by an electric field E (**Fig. 6•3–2**).

The potential energy is

$$V(x) = \begin{Bmatrix} \infty \\ eEx \end{Bmatrix} \text{ for } \begin{Bmatrix} x < 0 \\ x > 0 \end{Bmatrix}. \tag{6•3–4}$$

The quantum-mechanical problem corresponding to this potential can be solved rigorously in closed form, in terms of the so-called **Airy function**, a

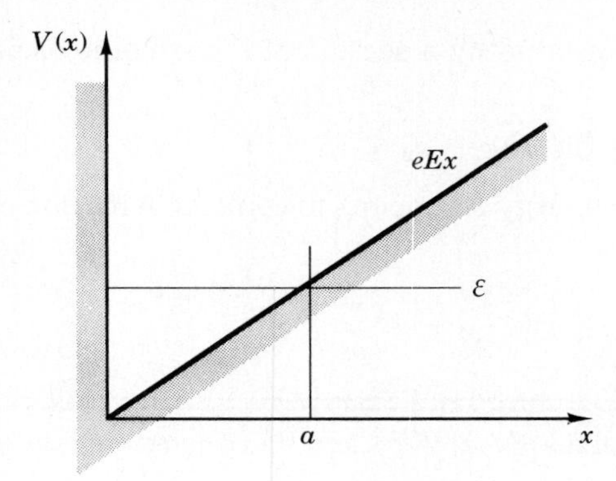

Figure 6•3–2. Potential energy for an electron in a uniform electric field at a potential wall.

linear superposition of Bessel functions of the order $\pm 1/3$. The exact values of the energy eigenvalues are

$$\mathcal{E}_n = A_n \mathcal{E}_0, \qquad n = 1, 2, 3, \ldots \tag{6•3–5}$$

where A_n is the nth root of the Airy function[4] and $\mathcal{E}_0$ is the quantity

$$\mathcal{E}_0 = \left[\frac{\hbar^2 (eE)^2}{2m_e}\right]^{1/3}, \tag{6•3–6}$$

which we may view as the **natural unit of energy** of the problem, defined in terms of the coefficients $\hbar^2/2m_e$ and eE occurring in the Schroedinger equation.

Along with the energy $\mathcal{E}_0$, a natural unit L of length may be associated with the problem, defined by the requirement

$$\mathcal{E}_0 = eEL = \frac{\hbar^2}{2m_e L^2}, \tag{6•3–7}$$

which leads to

$$L = \left[\frac{\hbar^2}{2m_e eE}\right]^{1/3}. \tag{6•3–8}$$

Numerically, for an electric field of, say, 10^6 V/cm,

$$\mathcal{E}_0 = 72.5 \text{ meV}; \qquad L = 0.725 \text{ nm}. \tag{6•3–9a,b}$$

We will use here the WKB approximation to obtain an approximate value for the various energy levels, to be compared with the exact value.

Because of the infinitely high potential wall at $x = 0$, the wave function must vanish there. This means that the argument of the cosine function in

---

[4] See, for example, problem 40 in the book by Flügge, cited earlier. For the Airy function itself and its roots, see M. Abramowitz and A. Segun, *Handbook of Mathematical Functions* (New York: Dover, 1965), chapter 10.

(6•2–12b), evaluated at $x = 0$, must be an odd multiple of $\pi/2$:

$$\int_0^a K(x)\,dx - \frac{\pi}{4} = (2n - 1) \cdot \frac{\pi}{2}. \tag{6•3–10}$$

Here we have written a instead of x_n for the as-yet unknown classical turning point,

$$a = \frac{\mathcal{E}}{eE} = L \cdot \frac{\mathcal{E}}{\mathcal{E}_0}, \tag{6•3–11}$$

and the different values of n belong to the different energy eigenvalues, starting with $n = 1$.

With the help of (6•3–7), the WKB wave number may be written

$$K(x) = \sqrt{\frac{2m_e}{\hbar^2}(\mathcal{E} - eEx)} = \frac{1}{L^{3/2}}\sqrt{a - x}, \tag{6•3–12}$$

and the integral in (6•3–10) is easily found to be

$$\int_0^a K(x)\,dx = \frac{1}{L^{3/2}}\int_0^a \sqrt{a - x}\,dx = \frac{2}{3}\left(\frac{a}{L}\right)^{3/2} = \frac{2}{3}\left(\frac{\mathcal{E}}{\mathcal{E}_0}\right)^{3/2}. \tag{6•3–13}$$

If this is inserted into (6•3–10), we obtain an expression in the form (6•3–5) for the approximate energy levels, with

$$A_n = \left[\frac{3\pi}{8}(4n - 1)\right]^{2/3}. \tag{6•3–14}$$

The three lowest values are

$$A_1 = 2.321, \qquad A_2 = 4.082, \qquad A_3 = 5.517, \tag{6•3–15a,b,c}$$

remarkably close to the first three roots of the Airy function,

$$A_1 = 2.338, \qquad A_2 = 4.088, \qquad A_3 = 5.521. \tag{6•3–16a,b,c}$$

Even the lowest level differs from the exact value by less than 1%.

6.3.3 Amplitude Connection Rules

We combine the phase connection rule (6•2–11) with the amplitude connection rule (6•3–3) to write the connection between the standing wave (6•2–5) and the right-evanescent wave (6•2–18) in the form

$$\boxed{\frac{2}{\sqrt{K}} \cdot \cos\left[\int_x^c K\,dx' - \frac{\pi}{4}\right] \leftarrow \frac{1}{\sqrt{\kappa}} \cdot \exp\left[-\int_c^x \kappa dx'\right],} \tag{6•3–17a}$$

where c indicates the classical turning point.

In the form (6•3–17a), the connection rule applies to the case where the evanescent wave decays into a barrier to the *right* of the classical turning point. In the case of an evanescent wave decaying into a barrier to the *left* of the classical turning point, we have

$$\boxed{\frac{2}{\sqrt{K}} \cdot \cos\left[\int_x^c K dx' - \frac{\pi}{4}\right] \leftarrow \frac{1}{\sqrt{\kappa}} \cdot \exp\left[-\int_c^x \kappa dx'\right].} \tag{6•3–17b}$$

Note that we have indicated the connection between the two wave functions with unidirectional arrows. What is meant by this is the following. If we start from a purely decaying wave function on the classically *forbidden* side of the turning point, the wave function on the *allowed* side will, to a good approximation, be given by the cosine-type wave function with the indicated amplitude and phase. However, if a wave function on the allowed side of the barrier is of the cosine type, as in (6•3–17a) or (6•3–17b), the continuation of this wave function as a purely evanescent wave deep into a barrier is *not necessarily* a good approximation! It would be valid only if the WKB approximation and the connection rules were exact, for arbitrary barriers. But they are only approximations, which means that an *exact* continuation of the cosine-type wave function into the interior of the barrier would very likely contain a small contribution from the exponentially growing wave function. Near the classical turning point this contribution is likely to be negligible, but unless it is *exactly* zero, it will ultimately dominate if we continue the wave function sufficiently deep into the barrier. Hence, we write the unidirectional arrow.

6.4 TUNNELING

6.4.1 The WKB Wave Function inside a Barrier

As long as we are dealing with infinitely thick classically forbidden bariers, the connection rules (6•3–17a) and (6•3–17b) are all we ever need. The situation changes, however, when we consider tunneling *through* a barrier of finite width, as in **Fig. 6•4–1**. Inside a tunneling barrier, both left-evanescent and right-evanescent waves will in general be present simultaneously, and the overall wave function will be a superposition of the form (6•1–14). Furthermore, a current will in general be flowing through the barrier, related to the two amplitudes C and D in (6•1–14) via the relation (6•1–16). Evidently, C and D cannot both be real, and the wave function becomes complex, as is appropriate for wave functions representing a current-carrying state.

The case of main interest is that of a purely outgoing wave on the exit side of the barrier, which we may write in the form

$$\psi(x) = \frac{F}{\sqrt{K(x)}} \cdot \exp\left\{i\left[\int_b^x K(y)\,dy - \frac{\pi}{4}\right]\right\}, \tag{6•4–1}$$

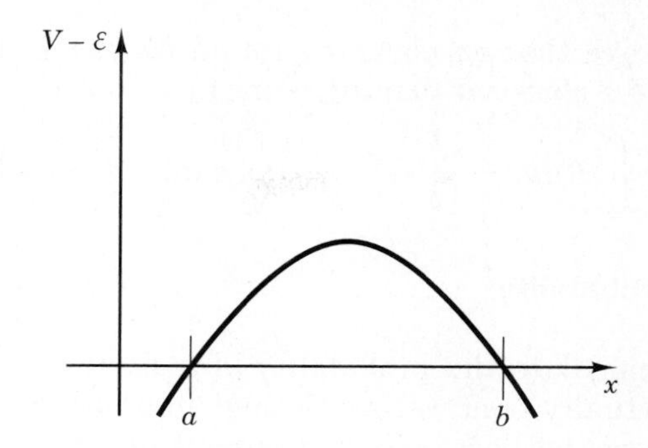

Figure 6•4–1. Smooth barrier potential.

Without loss in generality, we may assume that all phases are chosen such that F is real, in which case the probability current density to the right of the classical turning point is simply

$$j = \frac{\hbar}{M} \cdot F^2. \tag{6•4–2}$$

This current density must be equal to the current density (6•1–16) inside the barrier, which leads to the condition

$$i(CD^* - C^*D) = F^2. \tag{6•4–3}$$

With the wave function on each side of the classical turning point being complex, we need connection rules for both the real and the imaginary parts. The relation (6•3–17b) is evidently the connection rule for the real part. This implies that

$$D = F/2 \tag{6•4–4}$$

and that C is imaginary. Insertion of $D = F/2$ into (6•4–3) yields

$$C = -iF. \tag{6•4–5}$$

Hence, we obtain the overall complex connection rule across the classical turning point at $x = b$, the exit of a barrier:

$$\frac{1}{\sqrt{\kappa(x)}}\left[\frac{1}{2}\exp\left(+\int_x^b \kappa dy\right) - i \cdot \exp\left(-\int_x^b \kappa dy\right)\right] \leftrightarrow \frac{1}{\sqrt{K(x)}} \cdot \exp\left\{i\left[\int_b^x K dy - \frac{\pi}{4}\right]\right\}. \tag{6•4–6}$$

The imaginary part alone may be written

$$\frac{1}{\sqrt{\kappa(x)}} \exp\left(-\int_x^b \kappa dy\right) \leftrightarrow -\frac{1}{\sqrt{K(x)}} \cdot \sin\left[\int_b^x K dy - \frac{\pi}{4}\right]. \tag{6•4–7}$$

Note the absence of the factor 2, compared to (6•3–17b).

While (6•3–17b) is the connection rule between a standing wave and an evanescent wave that decays going into a barrier to the left, (6•4–7) is the connection rule for a wave that *grows* going into a barrier to the left. If the barrier is to the right of a classical turning point at $x = a$, we obtain

$$\frac{1}{\sqrt{K(x)}} \cdot \sin\left[\int_x^a K dy - \frac{\pi}{4}\right] \leftrightarrow -\frac{1}{\sqrt{\kappa(x)}} \exp\left(+\int_a^x \kappa dy\right). \tag{6•4–8}$$

6.4.2 The Tunneling Probability

We are now ready to calculate the probability of a particle incident on the barrier from the left actually penetrating through the barrier. We work our way from the exit side of the barrier to the entry side. We assume that an outgoing wave of the form (6•4–1) is present to the right of the barrier, with an as-yet unknown real amplitude F. As we saw, this outgoing wave is connected to the complex superposition of left- and right-evanescent waves (6•4–6). We re-write this superposition by shifting the integration limits, which serve as reference planes, to the left classical turning point, $x = a$. This leads to the form

$$\psi = \frac{F}{\sqrt{\kappa(x)}}\left[\frac{\gamma}{2} \exp\left(+\int_a^x \kappa dy\right) - \frac{i}{\gamma} \cdot \exp\left(-\int_a^x \kappa dy\right)\right], \tag{6•4–9}$$

where

$$\gamma \equiv \exp\left(-\int_a^b \kappa dy\right) \tag{6•4–10}$$

is the attenuation factor of the amplitude of each of the evanescent waves inside the barrier, in the direction of evanescence.

The case of principal interest to us is that of a relatively opaque barrier, characterized by the condition

$$\gamma \ll 1. \tag{6•4–11}$$

In this limit, the right-evanescent term in (6•4–9) will dominate near the left turning point, and we may, to the first order, neglect the left-evanescent term. But according to (6•3–17a), this right-evanescent term is connected across the left classical turning point to a standing wave of the form (6•2–5), with an amplitude

$$A = -\frac{i}{\gamma} \cdot F. \tag{6•4–12}$$

If we write the cosine function in (6•2–5) as a superposition of an incident and a reflected plane wave, we see that the incident plane wave has the amplitude A specified by (6•4–12), corresponding to an incident probability current density

$$j^+ = \frac{\hbar}{M} \cdot |A|^2 = \frac{\hbar}{M} \cdot \frac{|F|^2}{\gamma^2}. \tag{6•4–13}$$

But the ratio of the transmitted current density (6•4–2) to the incident current density is, of course, the tunneling probability T:

$$T \equiv \frac{j_T}{j_I^+} = \frac{|F|^2}{|A|^2} = \gamma^2 = \exp\left(-2\int_a^b \kappa dy\right). \qquad (6\text{•}4\text{–}14)$$

In our treatment, we assumed specifically that the potential changed smoothly through both classical turning points, so that the WKB connection rules had to be used at both ends of the tunnel. In several cases of practical interest this is not the case. It then becomes necessary to treat the discontinuous end(s) of the tunnel by the propagation matrix method developed in chapter 5. Problem 6.4–1, next, is an example.

PROBLEM TO SECTION 6.4

#6•4-1: Tunneling through a Barrier with an Applied Voltage

In solid-state physics, one encounters the problem of the tunneling of electrons through a thin oxide layer between two metals, as a function of the voltage ΔV applied between the metals. A simple model of this problem is the tunneling through the trapezoidal barrier shown in **Fig. 6.4–2**.

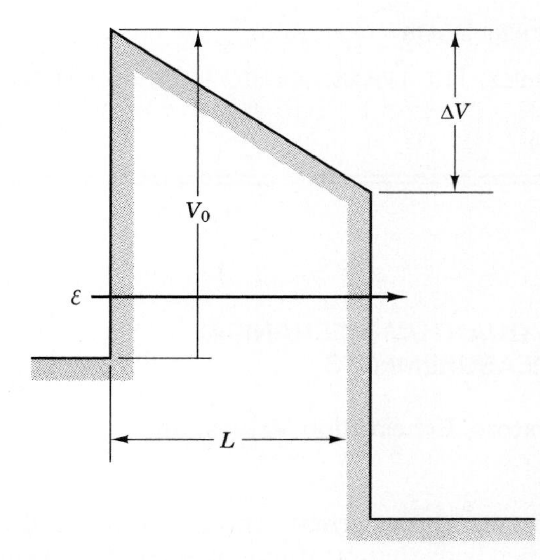

Figure 6•4–2. Trapezoidal barrier model for the tunneling of electrons through an oxide barrier between two metals.

Determine the tunneling probability as a function of ΔV, provided that ΔV remains less than $V_0 - \mathcal{E}$. For simplification, assume that L and V_0 are large enough that the total tunneling probability remains very small compared to unity.

Chapter 7

EXPECTATION VALUES AND OPERATORS

7.1 EXPECTATION VALUES AS QUANTUM-MECHANICAL AVERAGES OVER MANY MEASUREMENTS

7.1.1 Background: Operators, Expectation Values, and Representations

In this chapter, we return to three formal concepts introduced earlier and integrate them into a unified body that is intimately related to the probabilistic interpretation of quantum mechanics: (*i*) the concept of mathematical *operators* operating on wave functions, (*ii*) the concept of *expectation values,* and (*iii*) the concept of multiple equivalent *representations* for a given state of the quantum-mechanical object.

First, we review briefly the way these concepts were introduced earlier.

Operators

We introduced *operators* in section 1.3, where we defined several *differential* operators, as *extraction devices* for the purpose of extracting the properties of the individual plane wave components contained in a wave function. Among the operators, we examined specifically the *momentum operator* defined in (1•3–29)

$$\hat{\mathbf{p}} \equiv \hbar\hat{\mathbf{k}} = -i\hbar\nabla, \tag{7•1–1}$$

and its Cartesian components.

Expectation Values

We subsequently introduced *expectation values* in the context of wishing to split the energy of a bound state into a potential energy contribution (2•4–2)

$$\langle \mathcal{E}_{\text{pot}} \rangle = \int \Psi^* V(\mathbf{r}) \Psi \, d^3 r \tag{7•1–2}$$

and a kinetic energy contribution (2•4–3)

$$\langle \mathcal{E}_{\text{kin}} \rangle = -\frac{\hbar^2}{2M} \int \Psi^* \nabla^2 \Psi \, d^3 r. \tag{7•1–3}$$

Drawing on the operator concept, we may also write the latter expression as

$$\langle \mathcal{E}_{\text{kin}} \rangle = \int \Psi^*(\mathbf{r}) \hat{\mathcal{E}}_{\text{kin}} \, \Psi(\mathbf{r}) \, d^3 r, \tag{7•1–4}$$

where $\hat{\mathcal{E}}_{\text{kin}}$ is a differential operator, *defined* as

$$\hat{\mathcal{E}}_{\text{kin}} \equiv -\frac{\hbar^2 \nabla^2}{2M} = \frac{\hat{p}^2}{2M}. \tag{7•1–5}$$

Note that the algebraic relation between this operator and the momentum operator is exactly the same as the relation between the physical quantities themselves, i.e.,

$$\mathcal{E}_{\text{kin}} = \frac{p^2}{2M}. \tag{7•1–6}$$

Representations

Finally, we introduced the concept of multiple equivalent *representations* in section 4.1, where we showed that a given state of a quantum-mechanical object (in one dimension) may be described either by a wave function $\Psi(x, t)$ in the position representation or by an alternative wave function $\Phi(p, t)$ in the momentum representation. The two descriptions are completely equivalent, in the sense that each of the two wave functions may be obtained from the other via the Fourier transform pair (4•1–4) and (4•1–5), which we write here in the slightly different but equivalent form

$$\Psi(x, t) = \frac{1}{\sqrt{2\pi\hbar}} \int_{-\infty}^{+\infty} \Phi(p, t)\, e^{+ipx/\hbar}\, dp, \tag{7•1–7}$$

$$\Phi(p, t) = \frac{1}{\sqrt{2\pi\hbar}} \int_{-\infty}^{+\infty} \Psi(x, t)\, e^{-ipx/\hbar}\, dx. \tag{7•1–8}$$

In going from (4•1–4) and (4•1–5) to (7•1–7) and (7•1–8), we substituted $p = \hbar k$ and $\Phi \equiv A/\sqrt{\hbar}$, as in (4•1–8). With the symmetrized pre-factors $(2\pi\hbar)^{-1/2}$ in (7•1–7) and (7•1–8), both Ψ and Φ have the same normalization,

$$\int \Phi^*\Phi\, dp = \int \Psi^*\Psi\, dx = 1, \tag{7•1–9}$$

as was shown in (4•1–6).

The formalism is readily generalized to three dimensions. A normalizable wave function $\Psi(\mathbf{r}, t)$ can always be expressed by what is basically a three-dimensional Fourier transform,

$$\Psi(\mathbf{r}, t) = \frac{1}{(2\pi\hbar)^{3/2}} \int \Phi(\mathbf{p}, t)\, e^{i\mathbf{p}\cdot\mathbf{r}/\hbar}\, d^3p, \tag{7•1–10}$$

with the inverse transform

$$\Phi(\mathbf{p}, t) = \frac{1}{(2\pi\hbar)^{3/2}} \int \Psi(\mathbf{r}, t)\, e^{-i\mathbf{p}\cdot\mathbf{r}/\hbar}\, d^3r. \tag{7•1–11}$$

Note that the prefactors $(2\pi\hbar)^{-1/2}$ in (7•1–7) and (7•1–8) have been replaced by $(2\pi\hbar)^{-3/2}$, to ensure that both Ψ and Φ have the same normalization in *three* dimensions, analogous to (7•1–9) in one dimension:

$$\int \Psi^*\Psi\, d^3r = \int \Phi^*\Phi\, d^3p. \tag{7•1–12}$$

7.1.2 Expectation Values as Statistical Averages

The Statistical Interpretation as the Conceptual Basis

The glue that binds the above three formal concepts together, and at the same time connects them to the underlying physics, is the statistical interpretation of quantum mechanics. It implies that measurements on identical states need not lead to identical results, but only to identical probability distributions. In (1•2–8), we postulated that

$$dP_{\mathbf{r}} = |\Psi(\mathbf{r}, t)|^2\, d^3r \tag{7•1–13}$$

is the probability that the object will materialize in a small volume d^3r centered at position $\mathbf{r}$ during a suitable position measurement. In section 4.1, we extended the statistical interpretation to momentum measurements (see (4•1–10)). Generalized to three dimensions, (4•1–10) amounts to the postulate that

$$dP_{\mathbf{p}} = |\Phi(\mathbf{p}, t)|^2 \, d^3p \tag{7•1–14}$$

is the probability that the object will materialize in a small momentum interval d^3p centered at a momentum value $\mathbf{p}$ during a suitable momentum measurement. All measurement probabilities, summed over all possible outcomes of the measurement, must always add up to unity; hence the equality of the normalization integrals in (7•1–12).

We will later introduce additional representations, with probability distributions of their own.

Given a probability distribution for any measurable physical quantity A, the simplest and most important property of that distribution is the average value of A, designated as $\langle A \rangle$ and understood as an average taken over many individual measurements under identical conditions on the same initial state. Inasmuch as the state is changed by a measurement, this interpretation requires that the state be somehow restored to the state of interest before every measurement. In quantum mechanics, such averages are called **expectation values**. When we introduced that terminology earlier, it was in the somewhat different context of wishing to split the energy of a bound state into a potential energy contribution and a kinetic energy contribution. In the present section, we show that the expressions (7•1–2) and (7•1–3) are in fact in agreement with the definition of expectation values as probabilistic averages taken over many measurements.

Position and Potential Energy

Probably the simplest example is the expectation value of the position $\mathbf{r}$ of an object. From the statistical interpretation implied by (7•1–13), it follows that

$$\langle \mathbf{r} \rangle = \int \mathbf{r} \Psi^* \Psi \, d^3r = \int \Psi^* \mathbf{r} \Psi \, d^3r, \tag{7•1–15}$$

where, in the second form, we have pulled the factor $\mathbf{r}$ between Ψ^* and Ψ, *as if* it were an operator $\hat{\mathbf{r}} \equiv \mathbf{r}$ operating on the wave function Ψ to its right, the "operation" being ordinary multiplication with the vector $\mathbf{r}$, rather than a differential operation as in the differential operators $\hat{\mathbf{p}}$ and $\hat{\mathcal{E}}_{\text{kin}}$ encountered earlier. This operator point of view is perfectly legitimate, and it will turn out to be very useful. Such purely multiplicative operators are called **algebraic operators**.

Consider next the potential energy. Suppose an object has materialized in the small volume element d^3r centered at the position $\mathbf{r}$ during a suitable localization experiment. If we wish to associate a potential energy with the object, the only sensible choice is to choose $V(\mathbf{r})$, taken at the point of localization. With that *definition*, the expectation value of the potential energy is clearly given by the expression (7•1–12).

The procedure can evidently be generalized to any physical variable A that is an explicit function of the position, ($A = A(\mathbf{r})$), and that does not depend on the state of motion of the object, that is, on the momentum. The expectation value of A, *defined* as the average over many measurements, is evidently given by

$$\langle A \rangle = \int \Psi^* A(\mathbf{r}) \Psi \, d^3 r. \tag{7•1–16}$$

Momentum and Kinetic Energy

Suppose now that we were interested in the expectation value $\langle \mathbf{p} \rangle$ of the momentum. From the probability postulate (7•1–14) for the wave function in the position representation, together with the rules of probability calculus, this average must be given by

$$\langle \mathbf{p} \rangle = \int \mathbf{p} |\Phi|^2 \, d^3 p = \int \Phi^* \mathbf{p} \Phi \, d^3 p. \tag{7•1–17}$$

In the second form, we have again pulled the factor $\mathbf{p}$ between Φ^* and Φ, *as if* it were an algebraic operator $\hat{\mathbf{p}} \equiv \mathbf{p}$ operating on the wave function Φ to its right.

In order to calculate the expectation of the momentum, it is *not* necessary first to express the state function in the momentum representation: We show here that the expression (7•1–17) can be transformed into the position-representation equivalent,

$$\langle \mathbf{p} \rangle = \int \Psi^* \hat{\mathbf{p}} \Psi \, d^3 r, \tag{7•1–18}$$

where the operator $\hat{\mathbf{p}}$ is the momentum operator from (7•1–1).

For simplicity, we carry out the argument in one dimension only. We take the complex conjugate of (7•1–8) to obtain Φ^*, insert it into the one-dimensional equivalent of (7•1–7), and interchange the order of integration:

$$\langle p \rangle = \int_{-\infty}^{+\infty} \Psi^*(x, t) \left[\frac{1}{\sqrt{2\pi\hbar}} \int_{-\infty}^{+\infty} \Phi(p, t)(\hbar k) e^{+ipx/\hbar} \, dp \right] dx. \tag{7•1–19}$$

The square brackets may be re-written as

$$\begin{aligned} [\ldots] &= \frac{1}{\sqrt{2\pi\hbar}} \int_{-\infty}^{+\infty} \Phi(p, t) \left(-i\hbar \frac{d}{dx} e^{+ipx/\hbar} \right) dp \\ &= -i\hbar \frac{\partial}{\partial x} \left[\frac{1}{\sqrt{2\pi\hbar}} \int_{-\infty}^{+\infty} \Phi(p, t) e^{+ipx/\hbar} \, dp \right] \end{aligned}$$

$$= -i\hbar\frac{\partial}{\partial x}\Psi(x, t), \tag{7•1–20}$$

where, in the last equality, we have once more used (7•1–7).

If (7•1–20) is inserted into (7•1–19), the result may be written

$$\langle p \rangle = \int \Psi^* \hat{p} \Psi \, dx, \tag{7•1–21}$$

where

$$\boxed{\hat{p} = -i\hbar\frac{\partial}{\partial x}} \tag{7•1–22}$$

is the one-dimensional momentum operator, already introduced in chapter 1 (see (1•3–30)). The result (7•1–21) is evidently the one-dimensional form of (7•1–18). The generalization to three dimensions is elementary and is left to the reader.

We finally turn to the expectation value of the kinetic energy. The latter being a function of the momentum, we first express $\mathcal{E}_{\text{kin}}$ in terms of the momentum $\mathbf{p}$, according to the algebraic energy-momentum relation (7•1–6), and then express $\langle \mathcal{E}_{\text{kin}} \rangle$ in terms of the statistical interpretation postulate (7•1–14) for the momentum. This leads to the expression

$$\langle \mathcal{E}_{\text{kin}} \rangle = \int \frac{p^2}{2M} |\Phi|^2 \, d^3p = \int \Phi^* \frac{p^2}{2M} \Phi \, d^3p. \tag{7•1–23}$$

Again, the integral over the momentum may be transformed into an integral over the position, by a procedure that is a straightforward generalization of the earlier procedure for re-expressing the expectation value of the momentum in the position representation. The execution, left to the reader, shows that the right-hand sides of (7•1–3) and (7•1–23) have the same value, so that

$$\int \Phi^* \frac{p^2}{2M} \Phi \, d^3p = -\frac{\hbar^2}{2M} \int \Psi^* \nabla^2 \Psi \, d^3r. \tag{7•1–24}$$

This justifies our earlier terminology of calling the quantity defined by (7•1–3) the *expectation value* of the kinetic energy.

Comment: What Constitutes a Momentum Measurement?

Our treatment of quantum-mechanical momentum has said nothing about how a momentum measurement might actually be performed. The introduction of the momentum representation in chapter 4 and the probabilistic interpretation of the momentum wave function contained in (7•1–14) ultimately rest on the postulate of wave-particle duality—specifically, on the Einstein-De Broglie relation $\mathbf{p} = \hbar\mathbf{k}$ between momentum and wave vector. Hence, a measurement of a component of the wave vector made by passing the object through a suitable grating spectrometer constitutes a primary measurement of the corresponding momentum component. We will see later that the expectation value of the momentum thus defined obeys exactly the same laws as the classical momentum in Newtonian dynamics, such as Newton's second law and the law of conservation of momentum. But this means that any momentum measurement performed by classical means, such as measurements of recoil in collisions, as in the Compton effect, must lead to the same expectation values as a spectrometer-based wave vector measurement.

For each value of the momentum, the kinetic energy is fully specified; hence, any momentum measurement is implicitly also a measurement of the kinetic energy and of any other quantity that is a pure function of the momentum, without any other dependences.

7.1.3 Representations of Operators

We compare the two expressions (7•1–17) and (7•1–18) for the expectation value of the momentum. Both are special cases of the general form

$$\langle \mathbf{p} \rangle = \int \chi^*(\mathbf{s})\hat{\mathbf{p}}\chi(\mathbf{s})\, d^3s. \tag{7•1–25}$$

In (7•1–17), $\mathbf{s}$ is the momentum vector, $\chi(\mathbf{s})$ is the wave function $\Phi(\mathbf{p}, t)$ in the momentum representation, and $\hat{\mathbf{p}}$ stands for the ordinary factor $\mathbf{p}$. On the other hand, in (7•1–18), $\mathbf{s}$ is the position vector, $\chi(\mathbf{s})$ is the wave function $\Psi(\mathbf{r}, t)$ in the position representation, and $\hat{\mathbf{p}}$ stands for the differential operator $-i\hbar\nabla$.

Because (7•1–17) and (7•1–18) accomplish the same end, we view the two forms of $\hat{\mathbf{p}}$ as simply two different mathematical *representations* of the momentum operator itself, just as we view $\Psi(\mathbf{r}, t)$ and $\Phi(\mathbf{p}, t)$ as two different mathematical representations of the wave functions for the same physical state. Furthermore, we adopt the convention of using the same symbol $\hat{\mathbf{p}}$ for both representations, with the understanding that the representation to be used for the operator will always be the same as the representation of the function on which the operator operates. That is, we *define*

$$\hat{\mathbf{p}}\Psi(\mathbf{r}, t) \equiv -i\hbar\nabla\Psi(\mathbf{r}, t) \quad \text{(position representation)} \tag{7•1–26a}$$

and

$$\hat{\mathbf{p}}\Phi(\mathbf{p}, t) \equiv \mathbf{p}\Phi(\mathbf{p}, t) \quad \text{(momentum representation)}, \tag{7•1–26b}$$

for any state represented by either Ψ or Φ. In cases in which the representation of $\hat{\mathbf{p}}$ matters, but is not obvious from the context, we will state it explicitly.

Similar to the momentum, the two forms (7•1–3) and (7•1–23) for the expectation value of the kinetic energy may be viewed as special cases of the general form

$$\langle \mathcal{E}_{\text{kin}} \rangle = \int \chi^*(\mathbf{s})\hat{\mathcal{E}}_{\text{kin}}\chi(\mathbf{s})\, d^3s, \tag{7•1–27}$$

where the kinetic energy operator assumes the form (7•1–5) in the momentum representation and the form (7•1–6) in the position representation.

7.1.4 The Position Operator in the Momentum Representation

Just as we may express the expectation value for the momentum in the position representation, we may also express the expectation value for the position in the momentum representation. The "natural" representation for the position is, of course, the position representation. Using operator notation, we re-write (7•1–15) in the form

$$\langle \mathbf{r} \rangle = \int \Psi^* \hat{\mathbf{r}} \Psi \, d^3r, \tag{7•1–28}$$

where $\hat{\mathbf{r}} = \mathbf{r}$ evidently plays the role of a position operator in the position representation.

The position integral in (7•1–28) is readily transformed into a momentum integral of the form

$$\langle \mathbf{r} \rangle = \int \Phi^* \hat{\mathbf{r}} \Phi \, d^3p. \tag{7•1–29}$$

The procedure is the mirror image of the procedure for converting the expression for the expectation value of the momentum from (7•1–19) to (7•1–22). All we have to do is make everywhere the interchanges

$$\mathbf{r} \leftrightarrow \mathbf{p}, \qquad \Psi(\mathbf{r}, t) \leftrightarrow \Phi(\mathbf{p}, t), \qquad \text{and } i \leftrightarrow -i. \tag{7•1–30}$$

This leads directly to the momentum representation of the position operator,

$$\hat{\mathbf{r}} = +i\hbar \nabla_{\mathbf{p}}, \tag{7•1–31}$$

where the symbol $\nabla_{\mathbf{p}}$ refers to the gradient in p-space, defined as

$$\nabla_{\mathbf{p}} \equiv \left(\frac{\partial}{\partial p_x}, \frac{\partial}{\partial p_y}, \frac{\partial}{\partial p_z} \right). \tag{7•1–32}$$

Our results may be summarized in the following table:

Representation	Operator	
	$\hat{\mathbf{r}}$	$\hat{\mathbf{p}}$
$\mathbf{r}$	$\mathbf{r}$	$-i\hbar\nabla$
$\mathbf{p}$	$+i\hbar\nabla_{\mathbf{p}}$	$\mathbf{p}$

Note the different signs in the two differential operators.

Our purpose in presenting the transformation of the position operator to the momentum representation was primarily a conceptual rather than a practical one. As we pointed out already in section 4.4, the momentum representations of the operators for the potential energy tend to be complicated differential operators of higher than second order for all but the simplest potentials $V(\mathbf{r})$. This usually leads to a far more complicated formalism than the position representation, which is in fact much more often employed for the solution of quantum-mechanical problems than is the momentum representation. But that does not diminish the conceptual importance of recognizing the existence of multiple equivalent representations.

7.1.5 Generalizations

The operator formalism developed here for the expectation values of selected physical properties of an object is easily generalized to other physical properties and to other physical systems, and we postulate that the formalism applies to all quantitatively measurable physical properties of all physical systems. Such properties are called **observables**. Using this term, we state the following **general operator postulate**:

> **(a)** Associated with every observable A is a **linear operator** $\hat{A}$ that operates on the state functions χ. The mathematical form of the operator depends on the representation in which the operator and the state functions are expressed.
>
> **(b)** The expectation values of the observables are obtained by summing or integrating the product $\chi^*(\hat{A}\chi)$ over all values of all independent variables on which the state functions depend.

In regard to part (a), an operator $\hat{A}$ is linear if it satisfies the condition

$$\hat{A}(a_1\chi_1 + a_2\chi_2) = a_1\,\hat{A}\chi_1 + a_2\,\hat{A}\chi_2, \qquad (7\text{•}1\text{–}33)$$

where χ_1 and χ_2 are any two state functions for which the action of the operator is defined and a_1 and a_2 are arbitrary numerical factors. This linearity requirement is simply an extension of the principle of linear superposition postulated for the state functions themselves.

The general operator postulate makes no statements about what other conditions the operators must satisfy, nor about how to find the operator that is associated with a given observable. We will see in section 7.3 that all operators that represent an observable must be what is called **Hermitian operators.** The answers to the second question will emerge case by case, often in the form of answers to the inverse question: Given a Hermitian operator, to which observable does it correspond? In the course of studying the quantum-mechanical properties of various physical systems, we will be led naturally to introduce numerous mathematical operators. The mathematical properties of many of these operators and their expectation values will turn out to be exactly those that one would associate with certain observables. This will lead us to interpret the operators as the operators associated with those observables. In this way, the observables and their properties emerge as natural consequences of the quantum-mechanical formalism for the various physical systems.

7.1.6 Uncertainties

Although the expectation value is undoubtedly the most important property of any probability distribution, it gives no information about the *width* of that distribution, much less about more subtle details. The simplest measure of the width of the distribution of a quantity A is the **variance** $(\Delta A)^2$ of its distribution, which is an average itself, namely, the average of the quantity $(A - \langle A\rangle)^2$:

$$(\Delta A)^2 = \langle (A - \langle A\rangle)^2\rangle = \langle A^2\rangle - \langle A\rangle^2. \tag{7•1–34}$$

The square root of this quantity ΔA, called the **standard deviation** in general statistics, plays the role of the **uncertainty** of the quantity A in quantum mechanics. We discussed this point in chapter 4 for the specific case of the uncertainties of momentum and position; Eq. (7•1–34) generalizes this interpretation to arbitrary measurements.

We note that the variance may itself be calculated via the mathematical expectation value formalism; the first term on the far right of (7•1–34) is the expectation value of A^2, and the second term is the square of the expectation value of A.

Exercise: Show that the uncertainty of the momentum is related to the expectation value of the kinetic energy via

$$(\Delta p)^2 = 2M \cdot \langle \mathcal{E}_{\text{kin}}\rangle - \langle p\rangle^2. \tag{7•1–35}$$

For stationary bound states, the expectation value of the momentum must be zero, and $(\Delta p)^2$ becomes simply proportional to $\langle \mathcal{E}_{\text{kin}}\rangle$. Evidently, a large momentum uncertainty—as would be present for a high degree of spatial localization—implies a large kinetic energy.

More elaborate characterizations of the probability distributions could be obtained by drawing on more complicated averages, such as the ***n*th moment** of a distribution, $\langle A^n\rangle$, or the ***n*th centered moment**, $\langle (A - \langle A\rangle)^n\rangle$. However, such higher moments are rarely employed in quantum mechanics.

7.2 DIRAC NOTATION

7.2.1 Inner Products

The expectation value integrals we encountered in the preceding section, as well as the normalization and orthogonality integrals encountered earlier, are all special cases of the general form

$$\langle f|g\rangle \equiv \int f^*(\mathbf{s})g(\mathbf{s})\, d^3s. \tag{7•2–1a}$$

Here $f(\mathbf{s})$ and $g(\mathbf{s})$ stand either for state functions themselves or for the results of operators operating on state functions. The independent vector variable **s**—representing three independent scalar components—stands for either **r** or **p**, or for whatever other independent variable we might wish to introduce, and the integration goes over the full three-dimensional vector space.

In the case of one-dimensional problems, (7•2–1a) reduces to

$$\langle f|g\rangle \equiv \int f^*(s)g(s)\, ds. \tag{7•2–1b}$$

Integrals of the general form (7•2–1) occur so often on quantum-mechanical calculations, that it is desirable to simplify the notation by using the compact bracket symbol $\langle f|g\rangle$ for them, called the **inner product** of f and g. This notation is called **Dirac notation**, and the symbol $\langle f|g\rangle$ is also called a **Dirac bracket**.

In the quantum mechanics of several interacting particles, to be introduced later, the state function depends on more than three independent scalar variables. In anticipation of this need, we generalize the definitions (7•2–1) to arbitrary dimensionality n by writing

$$\langle f|g\rangle \equiv \int f^*(\mathbf{s})g(\mathbf{s})\, d^ns, \tag{7•2–2}$$

where **s** is now a vector in an n-dimensional space and the integration goes over that full vector space.

The inner products can be manipulated algebraically without recourse to their definitions as integrals, by making use of the following properties, which follow directly from the general definition (7•2–2):

$$\langle g|f\rangle = (\langle f|g\rangle)^*; \tag{7•2–3}$$

$$\langle (f_1 + f_2)|g\rangle = \langle f_1|g\rangle + \langle f_2|g\rangle; \tag{7•2–4a}$$

$$\langle f|(g_1 + g_2)\rangle = \langle f|g_1\rangle + \langle f|g_2\rangle. \tag{7•2–4b}$$

Furthermore, if c is a complex constant,

$$\langle f|cg\rangle = \langle c^*f|g\rangle = c\langle f|g\rangle. \tag{7•2–5}$$

If f, g, and c are real, these properties are just the same as those of the scalar product $\mathbf{f} \cdot \mathbf{g}$ of two vectors $\mathbf{f}$ and $\mathbf{g}$. With complex vectors, $\langle f|g\rangle$ is analogous to the scalar product $\mathbf{f}^* \cdot \mathbf{g}$:

$$\langle f|g\rangle \mathrel{\hat{=}} \mathbf{f}^* \cdot \mathbf{g}. \tag{7•2–6}$$

In fact, both the name *inner product* and the bracket notation have their origin in this analogy: In the mathematical literature, the scalar product is often called the *inner* product and is written in various bracket notations, such as $(\mathbf{f},\mathbf{g})$, $\langle \mathbf{f},\mathbf{g}\rangle$, and other variations on the same theme, including $\langle f|g\rangle$.

If two vectors are orthogonal, their scalar product vanishes. Hence, two state functions whose inner product vanishes are called **orthogonal**, an important concept introduced already in section 2.1.

We saw in the preceding section that both the normalization integrals and the expectation values are representation-independent. For example, with $\Psi = \Psi(\mathbf{r})$ and $\Phi = \Phi(\mathbf{p})$, the normalization conditions for the two wave functions become, in the new notation,

$$\langle\Psi|\Psi\rangle = \langle\Phi|\Phi\rangle \;(=1). \tag{7•2–7}$$

Similarly, for the expectation values of x and p,

$$\langle\Psi|\hat{x}\Psi\rangle = \langle\Phi|\hat{x}\Phi\rangle, \tag{7•2–8a}$$

$$\langle\Psi|\hat{p}\Psi\rangle = \langle\Phi|\hat{p}\Phi\rangle, \text{ etc.,} \tag{7•2–8b}$$

where $\hat{x}\Psi$ stands for the function obtained by operating with the operator $\hat{x}$ in the position representation on the position-representation wave function Ψ, etc. This representation-independence is analogous to the coordinate-system-independence of the scalar product of two vectors: Even though the components of the two vectors depend on the coordinate system in which the vector is expressed, their scalar product does not.

7.2.2 Bra and Ket Vectors

The analogy to the scalar product of two vectors suggests going one step further. The symbol $\mathbf{f}$ for a vector denotes an entity that has a geometrical meaning independently of any coordinate system used to describe it in terms of components. Similarly, a quantum state has a physical meaning that is independent of the representation used to describe it in terms of a state function. This analogy suggests introducing vector-like symbols that describe the states themselves, independently of any representation.

Following Dirac, we view the inner product $\langle f|g\rangle$ as an ordered algebraic product of two kinds of **state vectors**, written $\langle f|$ and $|g\rangle$ and called **bra** and **ket vectors**, after the first and last three letters of the word *Bracket.* This whimsical terminology has proven very useful, and is widely used because of its compactness and because of the great flexibility it offers in the choice of identifiers for the state vectors, to be enclosed between the brackets

$\langle \ldots |$ and $| \ldots \rangle$. A common choice is to use the same symbols Φ and Ψ that are employed as names for the state functions. This is convenient, but one should always keep in mind that the state *functions* are of necessity expressed in a specific representation, while the bra and ket symbols refer to the *states themselves,* independently of any representation. It is not necessary to use the state function names as identifiers. When the various state vectors occurring in a calculation differ only by what in a more traditional notation would be subscripts to a common state function name, one frequently lists simply the subscripts. For example, the energy eigenstates of the hydrogen atom, which are characterized by the three quantum numbers n, l, and m, might be denoted simply by $|n, l, m\rangle$, rather than in terms of a wave-function $\Psi_{n,l,m}(\mathbf{r})$. The designators need not be integers or even numbers: In a suitable context, the notation $|+, \uparrow\rangle$ might be perfectly clear, without any further explanation.

Except in cases where we actually need an explicit wave function, the Dirac notation is usually simpler and more convenient, and we will therefore use it preferentially. However, there will be many exceptions where we will not hesitate to fall back on the conventional wave function notation, rather than taking a purist approach. The reader should learn to handle quantum-mechanical calculations regardless of what notation is used, and what we lose in elegance by mixing notation we gain in flexibility.

The reader might perhaps wonder why we introduce different names and symbols for the bra and ket vectors, rather than using the same symbol and a more symmetric notation, say, $\langle f\rangle \cdot \langle g\rangle$. This dual notation is necessary: The integral (7•2–2) defining the inner product contains f^*, not f, and consequently, $\langle f|g\rangle$ and $\langle g|f\rangle$ are not the same (see (7•2–3)). Nor will it do to write, say, $\langle f^*\rangle \cdot \langle g\rangle$: There often exist two *different* physical states whose state functions are f and f^*, just consider e^{+ikx} and e^{-ikx}. The dual notation automatically keeps track of both the state itself, and its order in the defining integral. As we shall see later, it has numerous other advantages as well.

The state of a system may be described by either its bra or its ket vector. The all-but-universal practice is to use the ket vector. However, this is purely a convention, just as it is a convention to write the time dependence of plane waves in the form $\exp(-i\omega t)$ rather than $\exp(+i\omega t)$, while the convention in electromagnetic theory is just the opposite.

Dirac calls the bra vector $\langle f|$ the **dual** of the ket vector $|f\rangle$.

7.2.3 Dirac Brackets Containing Operators

As stated at the beginning of this section, the functions f and g inside Dirac brackets may themselves be the result of operators operating on a state function. Expectation value expressions are invariably of such a form, and we shall frequently encounter Dirac brackets of the more general form $\langle\Phi|\hat{A}\Psi\rangle$, obtained by setting $f = \Phi$ and $g = \hat{A}\Psi$. Note that in such a case, (7•2–3) takes on the form[1]

$$\langle \hat{A}\Psi | \Phi \rangle = (\langle \Phi | \hat{A}\Psi \rangle)^*. \tag{7•2–9}$$

From a state vector point of view, the ket vector $|\hat{A}\Psi\rangle$ contained in $\langle \Phi | \hat{A}\Psi \rangle$ denotes a state $|\hat{A}\Psi\rangle$, independently of whatever representation may be used to represent it. It will frequently be useful to view this state as having been generated from the state $|\Psi\rangle$ by the action of the operator $\hat{A}$. To express this view, we permit the notation $\hat{A}|\Psi\rangle$, defined by

$$\hat{A}|\Psi\rangle \equiv |\hat{A}\Psi\rangle. \tag{7•2–10}$$

Furthermore, we permit the notation $\hat{A}|\Psi\rangle$ for the ket $|\hat{A}\Psi\rangle$ even inside a Dirac bracket, leading to the double-slash notation $\langle \Phi | \hat{A} | \Psi \rangle$, defined by

$$\langle \Phi | \hat{A} | \Psi \rangle \equiv \langle \Phi | \hat{A}\Psi \rangle. \tag{7•2–11}$$

We shall use whichever notation serves our needs better in a specific case.

In the notation defined by (7•2–10), the Schroedinger wave equation and the time-independent Schroedinger equation assume the forms

$$\boxed{i\hbar \frac{\partial}{\partial t} |\Psi\rangle = \hat{H}|\Psi\rangle} \tag{7•2–12}$$

and

$$\boxed{\hat{H}|\Psi\rangle = \mathcal{E}|\Psi\rangle.} \tag{7•2–13}$$

We shall use these forms frequently.

7.3 COMMUTATORS

7.3.1 Non-Commuting Operators

When two operators are applied in succession, the combination forms an operator in its own right, called the **product** of the two individual operators. Let $\hat{A}$ and $\hat{B}$ be two arbitrary operators. We *define*

$$\hat{A}\hat{B}\Psi \equiv \hat{A}(\hat{B}\Psi), \qquad \hat{B}\hat{A}\Psi \equiv \hat{B}(\hat{A}\Psi). \tag{7•3–1a,b}$$

---

[1] From here on, the symbols Ψ and Φ no longer denote wave functions of the *same* state in the position and momentum representation of the same state, but simply any *two* states in a common representation.

That is, the operator listed to the right in the product is the operator applied first; the operator listed to the left is then applied to the result of the first operation. The order of the two factors is important, because the result of the combined operation often depends on this order, similar to the way the product of two matrices may depend on their order. For example, for the pair ($\hat{x}$, $\hat{p}_x$), we have, in the position representation,

$$\hat{x}\hat{p}_x\Psi = -i\hbar\hat{x}\frac{\partial\Psi}{\partial x}, \tag{7•3–2a}$$

$$\hat{p}_x\hat{x}\Psi = -i\hbar\hat{x}\frac{\partial}{\partial x}(x\Psi) = -i\hbar\hat{x}\frac{\partial\Psi}{\partial x} - i\hbar\Psi = (\hat{x}\hat{p}_x - i\hbar)\Psi. \tag{7•3–2b}$$

Evidently,

$$(\hat{x}\hat{p}_x - \hat{p}_x\hat{x})\Psi = i\hbar\Psi. \tag{7•3–3}$$

Analogous relations hold for the pairs $\hat{y}, \hat{p}_y$ and $\hat{z}, \hat{p}_z$. Because these results are true for *any* admissible Ψ, we may omit Ψ in (7•3–3) and view that relation as an identity between the operators themselves:

$$\boxed{\hat{x}\hat{p}_x - \hat{p}_x\hat{x} = i\hbar.} \tag{7•3–4a}$$

Similarly,

$$\boxed{\hat{y}\hat{p}_y - \hat{p}_y\hat{y} = i\hbar, \; \hat{z}\hat{p}_z - \hat{p}_z\hat{z} = i\hbar.} \tag{7•3–4b,c}$$

Note, however, that "mixed pairs," such as $\hat{x}$, $\hat{p}_y$, commute:

$$\hat{x}\hat{p}_y - \hat{p}_y\hat{x} = \hat{y}\hat{p}_y - \hat{p}_y\hat{y} = \hat{z}\hat{p}_x - \hat{p}_x\hat{z} = 0, \tag{7•3–5a}$$

$$\hat{y}\hat{p}_x - \hat{p}_x\hat{y} = \hat{z}\hat{p}_y - \hat{p}_y\hat{z} = \hat{x}\hat{p}_y - \hat{p}_y\hat{x} = 0. \tag{7•3–5b}$$

The commutation relations (7•3–4a) through (7•3–5b) are amongst the central relations of mathematical quantum mechanics.

Exercise: Confirm (7•3–5a,b), and show that (7•3–4a) through (7•3–5b) remain valid in the momentum representation.

The combination $\hat{A}\hat{B} - \hat{B}\hat{A}$ of two operators is called the **commutator** of the two operators; it is an operator in its own right, usually denoted by the symbol $[\hat{A}, \hat{B}]$:

$$\boxed{[\hat{A}, \hat{B}] \equiv \hat{A}\hat{B} - \hat{B}\hat{A}.} \tag{7•3–6}$$

Note that

$$[\hat{B}, \hat{A}] = -[\hat{A}, \hat{B}]. \tag{7•3–7}$$

Using this notation, Eqs.(7•3–4a) through (7•3–5b) may be lumped together by writing

$$\boxed{[\hat{x}_u, \hat{p}_v] = i\hbar\delta_{uv}, \quad (u, v = x, y, z).} \tag{7•3–8}$$

Exercise: Let $f(\mathbf{r})$ be a differentiable function of the Cartesian components of $\mathbf{r}$. By replacing x with $f(\mathbf{r})$ in (7•3–2a,b), show that

$$[f, \hat{p}_x] = +i\hbar\left(\frac{\partial f}{\partial x}\right); \quad [f, \hat{p}_y] = +i\hbar\left(\frac{\partial f}{\partial y}\right); \quad [f, \hat{p}_z] = +i\hbar\left(\frac{\partial f}{\partial z}\right). \tag{7•3–9a,b,c}$$

Note that the differentiations do not include any wave function following the parentheses.

Similarly, let $g(\mathbf{p})$ be a differentiable function of the Cartesian momentum components. By working in the momentum representation and drawing on (7•1–31), show that the following analog to (7•3–9) holds:

$$[g, \hat{x}] = -i\hbar\left(\frac{\partial g}{\partial p_x}\right); \quad [g, \hat{y}] = -i\hbar\left(\frac{\partial g}{\partial p_y}\right);$$
$$[g, \hat{z}] = -i\hbar\left(\frac{\partial g}{\partial p_z}\right). \tag{7•3–10a,b,c}$$

An example of a more complicated set of commutation relations is provided by the operators $\hat{a}^+$, $\hat{h}$, and $\hat{a}^-$ introduced in section 2.3 in conjunction with the recursion relations for the harmonic oscillator eigenfunctions (see (2•3–16), (2•3–18), and (2•3–22)). According to (2•3–19), the operators $\hat{a}^+$ and $\hat{h}$ obey the relation

$$\hat{h}\hat{a}^+\psi - \hat{a}^+\hat{h}\psi = 2\hat{a}^+\psi, \tag{7•3–11}$$

where ψ may be any wave function, not just a harmonic oscillator eigenfunction. But in this case, we may write (7•3–11) without ψ, as a commutation relation

$$[\hat{h}, \hat{a}^+] = +2\hat{a}^+. \tag{7•3–12a}$$

Similarly, one finds that

$$[\hat{h}, \hat{a}^-] = -2\hat{a}^-. \tag{7•3–12b}$$

The operators $\hat{a}^+$ and $\hat{a}^-$ also do not commute; from their definitions (2•3–16) and (2•3–22), one easily derives the commutation relation

$$[\hat{a}^+, \hat{a}^-] = -1. \tag{7•3–13}$$

7.3.2 Commutators Involving Products of Operators

We will frequently need commutators in which one of the operators is itself a product of two operators, as in

$$[\hat{A}\hat{B}, \hat{C}] = \hat{A}\hat{B}\hat{C} - \hat{C}\hat{A}\hat{B} \tag{7•3–14a}$$

or

$$[\hat{A}, \hat{B}\hat{C}] = \hat{A}\hat{B}\hat{C} - \hat{B}\hat{C}\hat{A}. \tag{7•3–14b}$$

By adding and subtracting either $\hat{A}\hat{C}\hat{B}$ or $\hat{B}\hat{A}\hat{C}$ on the right-hand sides of (7•3–14a,b), we easily obtain the important identities

$$[\hat{A}\hat{B}, \hat{C}] = \hat{A}[\hat{B}, \hat{C}] + [\hat{A}, \hat{C}]\hat{B} \tag{7•3–15a}$$

and

$$[\hat{A}, \hat{B}\hat{C}] = [\hat{A}, \hat{B}]\hat{C} + \hat{B}[\hat{A}, \hat{C}]. \tag{7•3–15b}$$

With the help of these two relations, it is possible to reduce the commutators of complicated operator products to simpler expressions. The most frequently needed case is when the two operators in the operator product are the same:

$$[\hat{A}^2, \hat{B}] = \hat{A}[\hat{A}, \hat{B}] + [\hat{A}, \hat{B}]\hat{A}, \tag{7•3–16a}$$

$$[\hat{A}, \hat{B}^2] = [\hat{A}, \hat{B}]\hat{B} + \hat{B}[\hat{A}, \hat{B}]. \tag{7•3–16b}$$

For example, if we apply these rules to the operator pair $\hat{A} = \hat{x}$, $\hat{B} = \hat{p}$, we obtain

$$[\hat{x}^2, \hat{p}_x] = 2i\hbar\hat{x} \quad \text{and} \quad [\, \hat{x}, \hat{p}_x^2 \,] = 2i\hbar\hat{p}_x. \tag{7•3–17a,b}$$

This result is readily generalized to commutators involving higher powers of $\hat{x}$ or $\hat{p}$. The reader is asked to show, by repeated application of (7•3–15), that

$$[\hat{x}^n, \hat{p}] = i\hbar \cdot n\hat{x}^{n-1} = i\hbar \cdot \frac{d}{d\hat{x}}\, \hat{x}^n \tag{7•3–18a}$$

and

$$[\hat{x}, \hat{p}^n] = i\hbar \cdot n\hat{p}^{n-1} = i\hbar \cdot \frac{d}{d\hat{p}}\, \hat{p}^n. \tag{7•3–18b}$$

In the rightmost equalities, the derivatives are purely formal derivatives, applying the formal rules of differentiation to the quantities $\hat{x}^n$ and $\hat{p}^n$, *as if* $\hat{x}$ and $\hat{p}$ were numbers rather than operators.

Exercise: Show that

$$[\hat{x}^2, \hat{p}^2] = 2i\hbar \cdot (\hat{x}\hat{p} + \hat{p}\hat{x}). \qquad (7{\cdot}3\text{–}19)$$

7.4 HERMITIAN OPERATORS

7.4.1 Definition and Basic Properties

The result of measuring any real property of any physical system must be a real number. The complex quantities sometimes introduced, such as the complex phasors of the form $Ee^{i\omega t}$ for electric fields, are only mathematical shorthand for packing *two* measurable real quantities, such as the amplitude and phase of a wave, into one complex quantity. Hence, if $\hat{A}$ is to be an operator corresponding to a *single* real physical quantity, its expectation value $\langle A \rangle$ must also be real; that is,

$$(\langle A \rangle)^* = \langle A \rangle. \qquad (7{\cdot}4\text{–}1)$$

In Dirac notation, (7•4–1) takes the form

$$\langle \Psi | \hat{A} \Psi \rangle = \langle \hat{A} \Psi | \Psi \rangle, \qquad (7{\cdot}4\text{–}2)$$

where we have used (7•2–3). Furthermore, if the integrals implied in (7•4–2) exist at all, then (7•4–2) must be valid for *any* state function Ψ that corresponds to a physically possible state of the system. Therefore, (7•4–2) must be valid for *any* normalizable state function for which the integrals exist.

The normalizability of the state function does not guarantee that the expectation value integral is also finite. One can easily construct state functions that are normalizable, but vanish sufficiently slowly for $x \to +\infty$ that the expectation value of the position becomes infinite, or that contain singularities causing the expectation value of the momentum or of the kinetic energy to become infinite. In realistically formulated quantum-mechanical problems, such "pathological" functions do not occur; they are of interest principally in the context of establishing the boundaries of a rigorous mathematical formalism. State functions that are normalizable *and* for which the expectation value $\langle A \rangle$ in (7•4–2) exists are said to lie in the **domain** of the operator $\hat{A}$. Unless specifically mentioned otherwise, we will always assume not only that all state functions are normalizable, but also that they lie within the domain of the operators with which they occur. We denote such state functions as **admissible**.

Any operator that satisfies (7•4–2) for *any* admissible Ψ is called a **Hermitian operator**, after the 19th-century French mathematician for whom Hermite polynomials are named.

From the definition (7•4–2), it follows that any Hermitian operator must also satisfy

$$\boxed{\langle\Phi|\hat{A}\Psi\rangle = \langle\hat{A}\Phi|\Psi\rangle,} \tag{7•4–3}$$

for any *two* admissible state functions Ψ and Φ in the same representation, regardless of what they are. To show this, we apply the condition (7•4–2) to the function

$$\Psi' = \Psi + \lambda\Phi, \tag{7•4–4}$$

where λ is an arbitrary complex number. If $\hat{A}$ is to be Hermitian, (7•4–2) must apply also to Ψ'. We insert Ψ' and execute the multiplications:

$$\begin{aligned}\langle\Psi'|\hat{A}\Psi'\rangle &= \langle\Psi+\lambda\Phi|\hat{A}|\Psi+\lambda\Phi\rangle \\ &= \langle\Psi|\hat{A}\Psi\rangle + \lambda\langle\Psi|\hat{A}\Phi\rangle + \lambda^*\langle\Phi|\hat{A}\Psi\rangle + \lambda\lambda^*\langle\Phi|\hat{A}\Phi\rangle.\end{aligned} \tag{7•4–5}$$

By the definition of a Hermitian operator, both the first and the last term on the right-hand side are real. In this case, the sum of the second plus the third term on the right must also be real, regardless of the value of λ, including particularly $\lambda = 1$ and $\lambda = i$. But the overall expression can be real for *both* $\lambda = 1$ and $\lambda = i$ only if (7•4–3) is satisfied.

Expressed in words, the condition (7•4–3) states that, in an inner product containing a Hermitian operator, it does not matter on which of the two functions the operator operates.

Warning: Recall that inside the integrals defining a Dirac bracket, moving an operator inside the bracket from the right to the left side of the bracket implies complex conjugation of *all* terms appearing on the left side of the bracket, including the operator itself.

The product of two Hermitian operators, $\hat{A}\hat{B}$, is itself Hermitian *if and only if* the two operators commute. To see this, we replace the operator $\hat{A}$ by the product $\hat{A}\hat{B}$ on the left-hand side of (7•4–3) and move the operators one by one to the other side in the inner product, utilizing the hermiticity of both $\hat{A}$ and $\hat{B}$:

$$\langle\Phi|\hat{A}\hat{B}\Psi\rangle = \langle\Phi|\hat{A}(\hat{B}\Psi)\rangle = \langle\hat{A}\Phi|\hat{B}\Psi\rangle = \langle\hat{B}(\hat{A}\Phi)|\Psi\rangle. \tag{7•4–6}$$

This is equal to $\langle\hat{A}\hat{B}\Phi|\Psi\rangle$, for arbitrary admissible Φ and Ψ, *if and only if* $\hat{B}\hat{A} = \hat{A}\hat{B}$.

Exercise: Show that the commutator of two non-commuting Hermitian operators $\hat{A}$ and $\hat{B}$ is not itself a Hermitian operator, but that, instead of (7•4–3), we have

$$\langle\Phi|[\hat{A}, \hat{B}]\Psi\rangle = -\langle[\hat{A}, \hat{B}]\Phi|\Psi\rangle = (\langle\Psi|[\hat{A}, \hat{B}]\Phi\rangle)^*, \tag{7•4–7}$$

or, written in terms of $\hat{C} = [\hat{A}, \hat{B}]$,

$$\langle \Phi | \hat{C} \Psi \rangle = -\langle \hat{C} \Phi | \Psi \rangle = -(\langle \Psi | \hat{C} \Phi \rangle)^*. \tag{7•4–8}$$

Operators with this property are called **anti-Hermitian.**

7.4.2 Examples: Position and Momentum; Potential and Kinetic Energy

The various operators introduced in section 7.1 are all Hermitian. This is obvious for any purely *algebraic* operator, such as $\mathbf{r}$ and its components, as well as $V(\mathbf{r})$, etc., provided that we restrict the set of state functions in (7•4–2) and (7•4–3) to those for which the implied integrals converge.

The hermiticity of the differential operators $\hat{\mathbf{p}}$ and $\hat{p}^2$ follows—under appropriate domain restrictions—from the representation-independence of the expectation values. In the momentum representations, these operators are numerical operators, which are automatically Hermitian. But if $\Psi(\mathbf{r})$ and $\Phi(\mathbf{p})$ are different representations of the same state $|\Phi\rangle$, a real expectation value for *one* representation implies the same value for the other—provided that the state $|\Phi\rangle$ has indeed both representations. The latter qualifier restricts us essentially to functions that have a Fourier transform, which is the case for all normalizable functions (see appendix E) under extremely wide ranges of other conditions.

It is instructive to investigate the hermiticity of the position representation of $\hat{\mathbf{p}}$ and its components directly, without recourse to their momentum representation. For simplicity, we assume that $\Psi(\mathbf{r})$ is differentiable and restrict ourselves to the x-component of $\hat{\mathbf{p}}$, $\hat{p}_x = -i\hbar\partial/\partial x$. If we execute the integrations implied by (7•4–3) in Cartesian coordinates, we merely need to consider the integration over x in the volume integral $\int \Phi^* \hat{p}_x \Psi \, d^3r$. Integration by parts leads to

$$\int_{-\infty}^{+\infty} \Phi^* \left(-i\hbar \frac{\partial \Psi}{\partial x} \right) dx = +i\hbar \int_{-\infty}^{+\infty} \left(\frac{\partial \Phi^*}{\partial x} \Psi \right) dx - \left[i\hbar \Phi^* \Psi \right]_{-\infty}^{+\infty}. \tag{7•4–9}$$

If Φ and Ψ are normalizable functions, they must both vanish for $x \to \pm\infty$. Hence, the last term in (7•4–9) must vanish, and

$$\int_{-\infty}^{+\infty} \Phi^* \left(-i\hbar \frac{\partial \Psi}{\partial x} \right) dx = \int_{-\infty}^{+\infty} \left(-i\hbar \frac{\partial \Phi}{\partial x} \right)^* \Psi \, dx, \tag{7•4–10}$$

which is of the required form (7•4–3). Hence, $\hat{p}_x$ is Hermitian.

By symmetry, the same is of course true for $\hat{p}_y$ and $\hat{p}_z$ and, therefore, for the entire vector operator $\hat{\mathbf{p}} = -i\hbar\nabla$. Note that this result depends critically on the vanishing of the state functions at infinity. For wave functions for which this would not be the case, the operator $-i\hbar\nabla$ would not be hermitian. Such wave functions do not belong to the hermiticity domain of $\hat{\mathbf{p}}$ in the position representation. Note that this domain does not even include the simple plane waves. Far from being an idiosyncracy, this is precisely what one should expect: When a uniform density distribution extends from $-\infty$ to $+\infty$, it is impossible to associate any specific value with the center of that distribution.

With the Cartesian components of $\hat{\mathbf{p}}$ being Hermitian, the hermiticity of $\hat{p}_x^2$, $\hat{p}_y^2$, and $\hat{p}_z^2$ follows from the fact that the square of any Hermitian operator is clearly Hermitian. Hence, $\hat{\mathcal{E}}_{\text{kin}}$ is Hermitian. But with the kinetic and potential energy operators being Hermitian, the sum of the two, the **Hamilton operator** $\hat{H}$, is also Hermitian:

$$\hat{H} = \frac{\hat{p}^2}{2M} + V(\mathbf{r}) = -\frac{\hbar^2}{2M}\nabla^2 + V(\mathbf{r}). \tag{7•4–11}$$

In (7•4–11), the last form is in the position representation.

7.4.3 Sharp Expectation Values: Eigenfunctions and Eigenvalues of Hermitian Operators

We inquire into the conditions under which the expectation value $\langle A \rangle$ of an observable 'A' for a state $|\Psi\rangle$ could be a *sharp* expectation value, corresponding to a zero-width probability distribution with the values of 'A'. For such a state, the variance of 'A' must be zero:

$$(\Delta A)^2 = \langle \Psi | (\hat{A} - A)^2 \Psi \rangle = 0. \tag{7•4–12}$$

Here we have simply written A for $\langle A \rangle$. If $\hat{A}$ is to correspond to an observable, it must be Hermitian. In that case, $\hat{A} - A$ is also Hermitian, and we can transform (7•4–12) into

$$\langle (\hat{A} - A)\Psi | (\hat{A} - A)\Psi \rangle = 0. \tag{7•4–13}$$

The left-hand side is simply the squared norm of the function $\Phi = (\hat{A} - A)\Psi$. But the norm of any function can vanish only if the function itself vanishes.[2] Hence, Ψ must satisfy

$$\boxed{\hat{A}\Psi = A\Psi.} \tag{7•4–14}$$

Any function Ψ that satisfies an equation of the form (7•4–14), where $\hat{A}$ is an operator and A is a number, is called an **eigenfunction**[3] of the operator $\hat{A}$, and the number A is called an **eigenvalue** of the operator $\hat{A}$, corresponding to the eigenfunction Ψ. The state $|\Psi\rangle$ represented by Ψ is called an **eigenstate**.

[2] This is not strictly true: The function may have nonzero values at discrete points, which make a zero contribution to the integral. Such pathological functions do not occur in "real" problems, and we shall ignore them here.

[3] *Eigen* is a German term, meaning "own" or "self," used here in the sense of "characteristic."

Expressed in this terminology, we conclude that:

If a state $|\Psi\rangle$ has a sharp value A of the observable 'A', it is an eigenstate of the operator $\hat{A}$ corresponding to that observable, with the eigenvalue A.

The inverse also holds:

If $\hat{A}$ is an operator corresponding to an observable 'A', and if Ψ is an eigenfunction of the operator $\hat{A}$, with the eigenvalue A, then the state $|\Psi\rangle$ is a state in which the observable 'A' has the sharp value A.

The proof of the latter follows by insertion of (7•4–14) into (7•4–12). Taken together, the two statements form the **Eigenvalue theorem**, one of the central formal theorems of quantum mechanics.

Any eigenvalue of an operator is a special case of an expectation value. The expectation values of Hermitian operators are necessarily real. We therefore conclude that:

The eigenvalues of Hermitian operators are real.

Consider next two different eigenfunctions of $\hat{A}$, with two different eigenvalues:

$$\hat{A}\Psi_1 = A_1\Psi_1, \qquad \hat{A}\Psi_2 = A_2\Psi_2. \tag{7•4–15a,b}$$

We take the following inner products:

$$\langle\hat{A}\Psi_1|\Psi_2\rangle = A_1\langle\Psi_1|\Psi_2\rangle, \qquad \langle\Psi_1|\hat{A}\Psi_2\rangle = A_2\langle\Psi_1|\Psi_2\rangle. \tag{7•4–16a,b}$$

Because $\hat{A}$ is Hermitian, the two left-hand sides are equal to each other. The same must then be true for the two right-hand sides. However, because $A_1 \neq A_2$, this can be true only if both are zero. This implies the

Orthogonality Theorem: If Ψ_1 and Ψ_2 are two eigenfunctions belonging to two *different* eigenvalues A_1 and A_2 of a Hermitian operator $\hat{A}$, then Ψ_1 and Ψ_2 are orthogonal,

$$\langle\Psi_1|\Psi_2\rangle = 0. \tag{7•4–17}$$

The converse of the orthogonality theorem need not be true: If Ψ_1 and Ψ_2 are both eigenfunctions of $\hat{A}$, and $\langle\Psi_1|\Psi_2\rangle = 0$, this does *not* imply that Ψ_1 and Ψ_2 belong to different eigenvalues.

We pointed out the orthogonality of the eigenfunctions belonging to different *energy* eigenvalues in section 2.3. We now see that this property holds for any Hermitian operator.

7.4.4 Complementarity of Non-Commuting Observables

In our discussion of the uncertainty relations in chapter 4, we saw that there are *complementary* physical properties, defined as properties that cannot simultaneously have sharp probability distributions, regardless of the state of the object. We show here that complementarity of two observables A and B is associated with a non-commuting of the operators $\hat{A}$ and $\hat{B}$ corresponding to A and B.

Two observables A and B can have simultaneously sharp values only if the state $|\Psi\rangle$ of the system is an eigenstate of *both* operators $\hat{A}$ and $\hat{B}$:

$$\hat{A}\Psi = A\Psi, \qquad \hat{B}\Psi = B\Psi. \tag{7•4–18a,b}$$

If we now apply the operator $\hat{B}$ to both sides of (7•4–18a) and the operator $\hat{A}$ to both sides of (7•4–18b), we may write the results as

$$\hat{B}(\hat{A}\Psi) = A\hat{B}\Psi = AB\Psi, \tag{7•4–19a}$$

$$\hat{A}(\hat{B}\Psi) = B\hat{A}\Psi = BA\Psi. \tag{7•4–19b}$$

Because the eigenvalues A and B on the right-hand sides are pure numbers, the two right-hand sides are equal, and we conclude that

$$\hat{B}\hat{A}\Psi = \hat{A}\hat{B}\Psi, \quad \text{or} \quad [\hat{A}, \hat{B}]\Psi = 0, \tag{7•4–20}$$

is a *necessary* condition for $|\Psi\rangle$ to be a state in which the system has a sharp value of both observables A and B. This condition will be satisfied for all states $|\Psi\rangle$ if $\hat{A}$ and $\hat{B}$ commute.

If $\hat{A}$ and $\hat{B}$ do *not* commute, the situation is simplest if the commutator of the two operators is a constant, as in the case of the position and momentum operator pair $\hat{x}$, $\hat{p}_x$. In such cases, (7•4–20) has no solution other than $\Psi = 0$. Hence, there are no states at all for which both observables have simultaneously sharp values.

In more complicated situations, the commutator of two operators may be a non-trivial operator in its own right, with nonzero eigenfunctions. We will encounter such a case in the next section.

◆ PROBLEM TO SECTION 7.4

#7•4-1: The Radial Component of $\hat{\mathbf{p}} = -i\hbar\nabla$ in Spherical Polar Coordinates

We have shown that $\hat{p}_x$, $\hat{p}_y$, and $\hat{p}_z$ and, hence, $\hat{\mathbf{p}} = -i\hbar\nabla$, are all Hermitian operators. Surprisingly, this is not the case for the components of $\hat{\mathbf{p}}$ in spherical polar coordinates.

(a) Show that the radial component of the operator $-i\hbar\nabla$,

$$(-i\hbar\nabla)_r = -i\hbar\frac{\partial}{\partial r}, \tag{7•4–21}$$

is not Hermitian and hence cannot represent whatever one might mean by the "momentum in the radial direction."

This simply reflects the fact that the concept of a "momentum in the radial direction" is itself ill-defined. The radial direction is not a *fixed* direction, but depends on where we are relative to the coordinate origin. Purely formal difficulties such as these are not uncommon in non-Cartesian coordinate systems.

(b) Show that the slightly different operator

$$\hat{p}_r = -i\hbar \frac{1}{r} \frac{\partial}{\partial r} r \tag{7•4–22}$$

is Hermitian, and give arguments suggesting that it has the properties one might expect for an operator representing the momentum in the radial direction. Consider, for example, the radial outflow of current for a spherically symmetric state. *Note:* The factor r following the derivative in (7•4–22) means that the wave function $\psi(\mathbf{r})$ must first be multiplied by r before differentiation, and the differentiation must be executed on the product $r\psi(\mathbf{r})$.

7.5 ANGULAR MOMENTUM: A FIRST LOOK

7.5.1 Angular Momentum Operators

The specific operators we have considered until now are too simple to display all the important aspects of the operator formalism we are developing. To develop a better feel for this formalism, it is useful to introduce a set of operators that will later play an important role and that exhibits a very rich set of properties, ideally suited to demonstrate the operator formalism.

We consider the vector operator

$$\hat{\mathbf{L}} = \hat{\mathbf{r}} \times \hat{\mathbf{p}} \tag{7•5–1}$$

and its Cartesian components

$$\hat{L}_x = \hat{y}\hat{p}_z - \hat{z}\hat{p}_y, \tag{7•5–2a}$$

$$\hat{L}_y = \hat{z}\hat{p}_x - \hat{x}\hat{p}_z, \tag{7•5–2b}$$

$$\hat{L}_z = \hat{x}\hat{p}_y - \hat{y}\hat{p}_x. \tag{7•5–2c}$$

The definitions (7•5–2) contain products of Hermitian operators. Note that the two factors in each product refer to Cartesian position and momentum components that are perpendicular to each other. These operators commute (see (7•3–5a,b)). The product of two commuting Hermitian operators is itself Hermitian. Hence, $\mathbf{L}$ and its components $\hat{L}_x$, $\hat{L}_y$, and $\hat{L}_z$ are Hermitian.

Being Hermitian, these operators qualify for representing observables. The nature of the observables is easily seen by simply omitting the operator carets in (7•5–1) and recalling that in classical mechanics the quantity $\mathbf{r} \times \mathbf{p}$ is the classical **angular momentum** of a particle at position $\mathbf{r}$ and with linear momentum $\mathbf{p}$, taken relative to the coordinate origin $\mathbf{r} = 0$ (**Fig. 7•5–1**). We may therefore view the operator $\hat{\mathbf{L}}$ as a quantum-mechanical *generalization* of the angular momentum. Note that this is not a "derivation" of the angular momentum operator from the classical form $\mathbf{r} \times \mathbf{p}$; rather, it is a true generalized *definition* that goes beyond the classical form, but contains the classical angular momentum as a limiting case. In particular, we shall see in the next

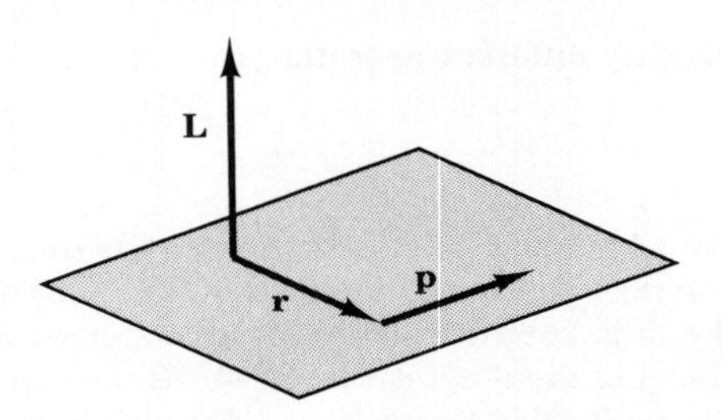

Figure 7•5–1. Angular momentum.

section that the classical law of conservation of angular momentum is a consequence of the properties of these operators.

We will later need these operators, especially $\hat{L}_z$, in spherical polar coordinates (r, θ, ϕ), related to Cartesian coordinates via

$$x = r \sin \theta \cos \phi, \qquad (7{\cdot}5\text{–}3a)$$

$$y = r \sin \theta \sin \phi, \qquad (7{\cdot}5\text{–}3b)$$

$$z = r \cos \theta. \qquad (7{\cdot}5\text{–}3c)$$

The result of this transformation is

$$\hat{L}_x = i\hbar\left[+\sin \phi \frac{\partial}{\partial \theta} + \cot \theta \cos \phi \frac{\partial}{\partial \phi}\right], \qquad (7{\cdot}5\text{–}4a)$$

$$\hat{L}_y = i\hbar\left[-\cos \phi \frac{\partial}{\partial \theta} + \cot \theta \sin \phi \frac{\partial}{\partial \phi}\right], \qquad (7{\cdot}5\text{–}4b)$$

$$\hat{L}_z = -i\hbar \frac{\partial}{\partial \phi}. \qquad (7{\cdot}5\text{–}4c)$$

For a derivation of these three relations, see the problem at the end of the section.

We have introduced here the angular momentum operators without reference to our earlier introduction of the concept of angular momentum in chapter 3 in the context of the motion of an object in a spherically symmetric potential. It turns out that the new operators are intimately related to the operator $\hat{\Lambda}$ defined in (3•1–2), the angular part of the Laplace operator. The connection between these operators is via the operator for the square of the magnitude of $\hat{\mathbf{L}}$,

$$\hat{L}^2 = \hat{L}_x^2 + \hat{L}_y^2 + \hat{L}_z^2. \qquad (7{\cdot}5\text{–}5)$$

By expressing the definitions (7•5–2) of the three angular component operators in spherical polar coordinates, one finds that $\hat{L}^2$ differs from $\hat{\Lambda}$ only by the constant factor $\hbar^2$:

$$\hat{L}^2 = \hbar^2\hat{\Lambda} = -\hbar^2\left[\frac{1}{\sin \theta}\frac{\partial}{\partial \theta}\left(\sin \theta \frac{\partial}{\partial \theta}\right) + \frac{1}{\sin^2 \theta}\frac{\partial^2}{\partial \phi^2}\right]. \qquad (7{\cdot}5\text{–}6)$$

Exercise: Drawing on the relations (7•5–4), confirm (7•5–6).

7.5.2 Quantization of Angular Momentum

We claimed in chapter 3 that the operator $\hat{\Lambda}$ has the eigenvalues

$$\Lambda = l(l+1), \tag{7•5–7a}$$

where l is any non-negative integer. Because of (7•5–6), this implies that $\hat{L}^2$ has the eigenvalues

$$L^2 = \hbar^2\Lambda = \hbar^2 l(l+1). \tag{7•5–7b}$$

Evidently, a state $|\Psi\rangle$ can have a *sharp* value of the magnitude of the angular momentum only if that value is a member of the discrete set (7•5–7b). That is, the magnitude of the angular momentum is **quantized**, in contrast to Newtonian mechanics, where angular momentum is continuously variable, even when sharp. This quantization of angular momentum is a central property of atomic physics, essential to everything from the physics of atomic spectra to understanding the periodic table of the elements and, with it, the laws of chemistry. The quantum number l is called the **orbital angular momentum quantum number**.

This quantization carries over to the components of $\hat{\mathbf{L}}$. Consider, for example, the z-component. In order for a state $|\Psi\rangle$ to have a sharp expectation value of $\hat{L}_z$, the wave function Ψ must be an eigenfunction of $\hat{L}_z$. In spherical polar coordinates, using (7•5–4c), this implies that

$$-i\hbar\frac{\partial\Psi}{\partial\phi} = L_z\Psi, \tag{7•5–8}$$

where we have written L_z—without a caret—for the eigenvalue. It is convenient to split off a factor $\hbar$ from L_z and to write

$$\boxed{L_z = m\hbar,} \tag{7•5–9}$$

in which case (7•5–8) becomes

$$-i\frac{\partial\Psi}{\partial\phi} = m\Psi. \tag{7•5–10}$$

This has the general solution

$$\Psi(r,\phi,\theta) = f(r,\theta)\exp(im\phi), \tag{7•5–11}$$

where $f(r,\theta)$ is an arbitrary function of the radial distance and the angle θ.

But Ψ must go over into itself under rotation about the polar axis by 2π. This implies that $\exp(2\pi mi) = 1$, which is satisfied only for the discrete values

$$\boxed{m = 0, \pm 1, \pm 2, \ldots .} \tag{7•5–12}$$

Evidently, the angular momentum *components* are also quantized. The quantum number m, called the **azimuthal quantum number**, is of course the same as in chapter 3. Because it plays a central role in the splitting of degenerate atomic energy levels under the influence of a magnetic field, it is also called the **magnetic quantum number**.

7.5.3 Commutation Relations

There is nothing special about the z-direction. The three Cartesian components of $\hat{\mathbf{L}}$ differ only by cyclic permutation of the coordinates and, hence, must have the same set of eigenvalues—even though the derivation of those eigenvalues for the components other than $\hat{L}_z$ might be much more tedious.

However, a state with nonzero angular momentum cannot simultaneously be an eigenstate of more than one of the component operators, because these operators do not commute: From the commutation relations (7•3–4) for the components of $\hat{\mathbf{r}}$ with those of $\hat{\mathbf{p}}$, one finds easily the set of commutation relations

$$\boxed{[\hat{L}_x, \hat{L}_y] = i\hbar\hat{L}_z \quad ; \quad [\hat{L}_y, \hat{L}_z] = i\hbar\hat{L}_x \quad ; \quad [\hat{L}_z, \hat{L}_x] = i\hbar\hat{L}_y.} \tag{7•5–13a,b,c}$$

The proof is left to the reader.

Exercise: Let Ψ be a state for which two of the angular momentum components simultaneously have sharp values. From the simultaneity condition (7•4–1) and the commutation relations (7•5–13), show that in this case the third component must also be sharp, with the value zero. Show that this is possible only if *all three* components have the sharp value zero, implying a state of no angular momentum at all.

Even though the components of $\hat{\mathbf{L}}$ do not commute with one another, they all commute with $\hat{L}^2$. From (7•5–13) and from the relations (7•3–12) for the commutators involving squares of operators, one finds readily that the various commutators cancel, leading to

$$[\hat{L}^2, \hat{L}_z] = [\hat{L}_x^2, \hat{L}_z] + [\hat{L}_y^2, \hat{L}_z] + [\hat{L}_z^2, \hat{L}_z] = 0. \tag{7•5–14a}$$

Cyclic permutation of the Cartesian indices shows that the following relations equivalent to (7•5–14a) hold for the other components as well:

$$[\hat{L}^2, \hat{L}_x] = [\hat{L}^2, \hat{L}_y] = 0. \tag{7•5–14b,c}$$

Taken together, the sets (7•5–13) and (7•5–14) of commutation relations imply that it must be possible to select the angular momentum eigenstates as simultaneous eigenstates of $\hat{L}^2$ and *one* of the three component operators, inevitably chosen to be $\hat{L}_z$. In Dirac notation,

$$\hat{L}^2 \,|l,m\rangle = \hbar^2 l(l+1)\,|l,m\rangle, \tag{7•5–15a}$$

$$\hat{L}_z \,|l,m\rangle = \hbar m\,|l,m\rangle. \tag{7•5–15b}$$

Inasmuch as the magnitude of a vector cannot be less than the magnitude of its z-component, we must evidently have $m^2 \leq l(l+1)$, which implies that

$$-l \leq m \leq +l, \tag{7•5–16}$$

a result already claimed in chapter 3, but not justified there.

7.5.4 Geometric Representation

The relation between the values of the total angular momentum and its z-component is easily represented graphically. To this end, we define

$$L_\rho^2 = L_x^2 + L_y^2 = L^2 - L_z^2. \tag{7•5–17}$$

Here, L_ρ is the component of the angular momentum perpendicular to the z-direction. The *direction* of this component within the (x,y)-plane is completely unsharp, with all in-plane directions equally probable. But because both L_z^2 and L^2 have sharp values, the *magnitude* of L_ρ also has a sharp value. If the vector **L** is plotted in the (L_z, L_ρ)-plane, its endpoint will lie on a circle of radius $\hbar\sqrt{l(l+1)}$, as shown in **Fig. 7•5–2** for the special case $l = 2$. The direction of

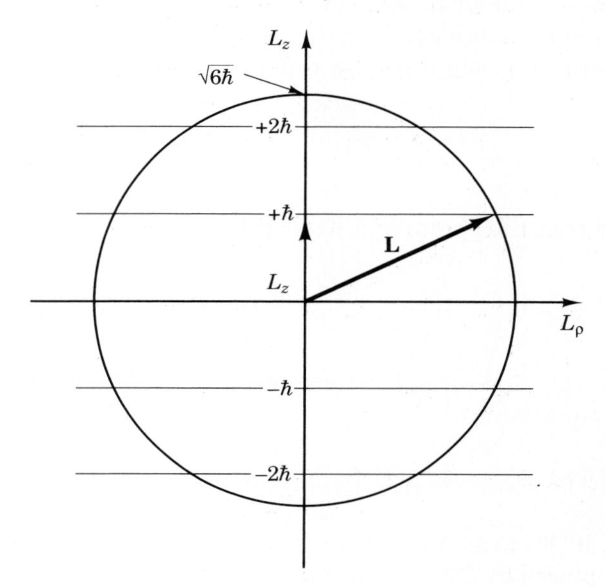

Figure 7•5–2. Geometrical relation between the vector of the angular momentum and its projection on the z-axis, shown for $l = 2$, $m = 1$.

L will be such that its z-component is an integer multiple of $\hbar$. It is clear from this construction that the ρ-component of **L** can never be less than $\hbar\sqrt{l}$.

◆ PROBLEM TO SECTION 7.5

7•5-1: Angular Momentum Operators in Spherical Polar Coordinates

The partial derivatives $\partial/\partial\theta$ occurring in (7•5–4) may be expressed in terms of Cartesian derivatives according to

$$\frac{\partial}{\partial\theta} = \left(\frac{\partial x}{\partial\theta}\right)\frac{\partial}{\partial x} + \left(\frac{\partial y}{\partial\theta}\right)\frac{\partial}{\partial y} + \left(\frac{\partial z}{\partial\theta}\right)\frac{\partial}{\partial z}, \tag{7•5–18}$$

with an analogous expression for $\partial/\partial\phi$. By drawing on (7•5–3) and on the definitions of the Cartesian momentum operators, show that the expressions (7•5–4) are the polar coordinate equivalents of the Cartesian definitions (7•5–2).

7.6 EXPECTATION VALUE DYNAMICS: THE TRANSITION TO CLASSICAL DYNAMICS

7.6.1 The Time Derivative of an Expectation Value

When a system is not in one of its stationary states, its expectation values are frequently time-dependent. A simple example would be the expectation value $\langle x \rangle$ of the position of the particle in the non-stationary superposition of two harmonic oscillator states, studied in section 2.3, or in the Gaussian wave packet of chapter 6. A general expression for the time derivative of the expectation value of *any* operator is obtained by differentiating $\langle A \rangle = \langle \Psi | \hat{A} | \Psi \rangle$; that is,

$$\frac{d}{dt}\langle A \rangle = \left\langle \frac{\partial \Psi}{\partial t} | \hat{A} | \Psi \right\rangle + \left\langle \Psi | \hat{A} | \frac{\partial \Psi}{\partial t} \right\rangle, \tag{7•6–1}$$

where we have assumed that the operator $\hat{A}$ itself is not explicitly time dependent,

$$\frac{\partial \hat{A}}{\partial t} = 0. \tag{7•6–2}$$

We transform (7•6–1) by inserting the time derivatives of Ψ and Ψ^* from the Schroedinger wave equation:

$$\frac{d}{dt}\langle A \rangle = \frac{i}{\hbar}\langle \hat{H}\Psi | \hat{A} | \Psi \rangle - \langle \Psi | \hat{A} | \hat{H}\Psi \rangle. \tag{7•6–3}$$

Note that this vanishes if $|\Psi\rangle$ is an eigenstate of the Hamiltonian, because in that case, $\hat{H}\Psi$ may be replaced by $\mathcal{E}\Psi$, and the two terms in (7•6–3) cancel. This is of course as expected, because the eigenstates of the Hamiltonian are what we called *stationary* states, with a time-independent probability density. Evidently, the stationarity applies to *all* observables.

Because $\hat{H}$ is Hermitian, the first term on the right-hand side of (7•6–3) may be re-arranged by moving $\hat{H}$ to the right-hand side of the Dirac bracket, leading to

$$\frac{d}{dt}\langle A\rangle = \frac{i}{\hbar}[\langle\Psi|\hat{H}\hat{A}|\Psi\rangle - \langle\Psi|\hat{A}\hat{H}|\Psi\rangle] = \frac{i}{\hbar}\langle\Psi|[\hat{H},\hat{A}]|\Psi\rangle. \tag{7•6–4}$$

We see, then, that the time derivative of the expectation value of any operator depends on the commutator of this operator with the Hamilton operator. Any operator that commutes with $\hat{H}$ will have time-independent expectation values. Only if the two operators do not commute *can* the expectation values depend on time, although they do not have to do so if the expectation value of $[\hat{H}, \hat{A}]$ in (7•6–4) vanishes anyway.

Note that the derivation of (7•6–4) depends, not on any specific form of the Hamiltonian, but only on its hermiticity. This point is important, because the specific form (7•4–11) of the Hamiltonian used so far applies only to the simple case of a single object with mass moving in a fixed external potential. It does not cover forces that cannot be described by a potential, such as magnetic forces, or the interaction with electromagnetic waves. Nor does it cover forces that are not fixed in space, but that originate from the interaction with other mobile elementary objects that must themselves be described by the laws of quantum mechanics. However, in all those cases, the system continues to be described by a single state vector $|\Psi\rangle$ whose time evolution satisfies an equation of the form (7•2–12),

$$i\hbar\frac{\partial}{\partial t}|\Psi\rangle = \hat{H}|\Psi\rangle, \tag{7•6–5}$$

where $\hat{H}$ is a Hermitian operator. Hence (7•6–5) holds completely generally, and it is in fact an extraordinarily powerful relation.

7.6.2 Conservation Laws

Conservation of Probability

The simplest of all operators must be $\hat{A} = 1$; its expectation value is simply the normalization integral. Obviously, $\hat{A} = 1$ commutes with $\hat{H}$; hence, normalization is conserved. In physical terms, this means that the probability that the object will materialize *somewhere* remains 100%: Objects described by a Hermitian time evolution operator $\hat{H}$ are neither created nor destroyed. We derived this result earlier for the specific form (7•4–11) of the Hamiltonian. We now see that it holds more generally, for all systems whose time evolution is governed by a law of the form (7•6–5), with a Hamiltonian that is Hermitian. In fact, hermiticity is both sufficient and necessary for conservation of matter:

Exercise: Suppose that all that is known about a quantum system is the following: (a) Its time evolution is governed by an equation of the form (7•6–5), where $\hat{H}$ is an unknown linear operator; (b) the objects described by Ψ can be neither created nor destroyed, implying that

$$\frac{d}{dt}\langle\Psi \mid \Psi\rangle = 0. \tag{7•6–6}$$

Show that under these assumptions, $\hat{H}$ must be Hermitian.

Conservation of Energy

One operator that always commutes with $\hat{H}$ is $\hat{H}$ itself. Thus, the expectation value of $\hat{H}$, that is, the expectation value $\mathcal{E}$ of the total energy, must be constant. This is the classical *law of conservation of energy*. Note that the law holds even when the energy does not have a sharp value.

Conservation of Angular Momentum

Suppose the potential energy $V(\mathbf{r})$ is spherically symmetric about the point $\mathbf{r} = 0$; that is, it depends only on the *magnitude* of the distance from the coordinate origin, but not on the direction,

$$V(\mathbf{r}) = V(r). \tag{7•6–7}$$

In this case, the components of the angular momentum operator $\hat{\mathbf{L}}$ are easily shown to commute with the potential energy; the proof is left to the reader. Inasmuch as we know already from (7•5–14) that the angular momentum components commute with the Laplace operator, it follows that they commute with the entire Hamiltonian:

$$[\hat{H}, \hat{L}_x] = [\hat{H}, \hat{L}_y] = [\hat{H}, \hat{L}_z] = 0. \tag{7•6–8}$$

Because of (7•6–5), this implies that both the magnitude and the direction of the expectation value of the angular momentum are time-independent in a spherically symmetric potential. This is the quantum-mechanical form of the classical *law of conservation of angular momentum.* We note that in quantum mechanics it arises as a direct consequence of spherical symmetry. Symmetry considerations play a central role in quantum mechanics, and spherical symmetry is one of the most important of all symmetrics; it is the symmetry of all free atoms. As a result, the angular momentum plays an important role in quantum mechanics, even more so than in classical mechanics. We will return later to the problems of symmetry in general and of spherical symmetry and of angular momentum in particular.

7.6.3 Velocity-Momentum Relation

Probably the simplest example of an operator that does *not* commute with $\hat{H}$ is the position operator $\hat{\mathbf{r}}$. Application of (7•6–4) to $\hat{\mathbf{r}}$ leads to an expression for the particle velocity,

$$\langle \mathbf{v} \rangle = \frac{d}{dt}\langle \mathbf{r} \rangle = \frac{i}{\hbar}\langle [\hat{H}, \hat{\mathbf{r}}] \rangle. \tag{7•6–9}$$

We consider only the x-component,

$$\langle v_x \rangle = \frac{i}{\hbar}\langle [\hat{H}, \hat{x}] \rangle, \tag{7•6–10}$$

and write the Hamiltonian in the form

$$\hat{H} = \frac{1}{2M}(\hat{p}_x^2 + \hat{p}_y^2 + \hat{p}_z^2) + V(x, y, z). \tag{7•6–11}$$

The components of $\mathbf{r}$ clearly commute with $V(x, y, z)$. Also, $\hat{x}$ commutes with $\hat{p}_y$ and $\hat{p}_z$. Hence,

$$[\hat{H}, \hat{x}] = \frac{1}{2M}[\hat{p}_x^2, \hat{x}] = -\frac{i\hbar}{M}\hat{p}_x, \tag{7•6–12}$$

where the second equality follows from (7•3–17b). Therefore,

$$\langle v_x \rangle = \frac{1}{M}\langle \hat{p}_x \rangle. \tag{7•6–13}$$

Equivalent expressions hold for $\langle v_y \rangle$ and $\langle v_z \rangle$. Lumped together into the vector form, we have the familiar (and expected) relation between velocity and momentum,

$$\langle \mathbf{v} \rangle = \frac{1}{M}\langle \mathbf{p} \rangle. \tag{7•6–14}$$

Note that this relation *does* depend on the specific form (7•6–11) of the Hamiltonian. We will see in chapter 8 that it ceases to be valid in the presence of magnetic fields.

7.6.4 Newton's Second Law

We are finally ready to show that the backbone of classical mechanics, **Newton's second law**, is indeed a limiting case of quantum mechanics. From (7•6–5), we obtain

$$\frac{d}{dt}\langle \hat{p}_x \rangle = \frac{i}{\hbar}\langle [\hat{H}, \hat{p}_x] \rangle = \frac{i}{\hbar}\langle [V, \hat{p}_x] \rangle, \tag{7•6–15}$$

where, in the second equality, we have inserted (7•6–11) for $\hat{H}$ and where we have utilized the fact that the $\hat{p}^2$-terms in $\hat{H}$ all commute with $\hat{p}_x$. From (7•3–9a), we next obtain

$$[V, \hat{p}_x] = i\hbar \frac{\partial V}{\partial x}. \tag{7•6–16}$$

If we insert this into (7•6–15), we obtain

$$\frac{d}{dt}\langle p_x \rangle = -\left\langle \frac{\partial V}{\partial x} \right\rangle \tag{7•6–17}$$

and equivalent expressions for the other momentum components. These may be lumped together into vector form, resulting in

$$\boxed{\frac{d}{dt}\langle \mathbf{p} \rangle = -\langle \nabla V \rangle.} \tag{7•6–18}$$

The right-hand side is clearly the expectation value of the Newtonian force acting on the wave packet, and (7•6–18) therefore is the quantum-mechanical generalization of Newton's second law. In the form (7•6–18) it is often referred to as **Ehrenfest's theorem**.

Suppose we apply (7•6–18) to a compact wave packet. From our earlier discussion, we know that

$$\langle \mathbf{p} \rangle = M\langle \mathbf{v} \rangle = M \frac{d}{dt}\langle \mathbf{r} \rangle. \tag{7•6–19}$$

Hence, the left-hand side of (7•6–18) is simply M times the acceleration of the center of mass of the wave packet, just as if the wave packet were a true Newtonian particle located at $\langle \mathbf{r} \rangle$. The essential difference from Newtonian mechanics occurs on the right-hand side, where the Newtonian force has been replaced by its expectation value, which is not necessarily equal to the *local* force at the center of mass of the packet. Conventional Newtonian mechanics emerges if the force varies sufficiently slowly with position that it may be treated as a constant over the wave packet. The expectation value brackets in (7•6–18) may then be dropped. If, however, the force varies rapidly with position, deviations from classical mechanics must be expected. In extreme cases, such as scattering at an abrupt potential step, the wave packet will break up and cease to be a compact object. Even in such cases, Ehrenfest's theorem remains formally valid; it simply ceases to tell the whole story.

7.6.5 Generalization: Hamilton-Jacobi Equations

The formalism developed in the last two sub-sections, which led us to the velocity-momentum relation (7•6–14) and to Newton's law in the form (7•6–18), was based on the assumption that the Hamiltonian is of the simplest possible form (7•6–11), appropriate for a single particle with constant mass M moving in an external potential $V(\mathbf{r})$. However, this form is far too restrictive for our future needs: It does not include the effects of magnetic fields, it does not apply to systems of more than one particle, and it does not apply to non-mechanical systems, such as electromagnetic systems, which are subject to the laws of quantum mechanics, just as mechanical systems are. Relativistic effects, although not of interest in this text, are another class of phenomena not describable by the Hamiltonian, (7•6–11).

Our formalism is easily generalized to a much broader class of Hamiltonians, which covers all of these cases. We postulate that all closed quantum systems can be described by a state vector $|\Psi\rangle$ that obeys a time evolution equation of the form (7•6–5), i.e.,

$$i\hbar\frac{\partial}{\partial t}|\Psi\rangle = \hat{H}|\Psi\rangle, \tag{7•6–20}$$

where $\hat{H}$ is a Hermitian operator. As we saw earlier, its hermiticity assures us that $\hat{H}$ itself corresponds to a scalar observable with the dimension of an energy, the expectation value of which is conserved.

This operator will depend on a number of variables, which may themselves be operators, similar to the way the Hamiltonian of a particle in an external potential, written in Cartesian coordinates, depends on three momentum operators and three position "operators." We assume that for each degree of freedom of the system, there is one *pair* of operators, designated here as $(\hat{q}_u, \hat{p}_u)$, with properties similar to those of the position-momentum operator pair for each of the three Cartesian degrees of freedom for a single particle in an external potential. In particular, we assume that the operators have been chosen in such a way that they obey the same kind of commutation relation, of the form

$$[\hat{q}_u, \hat{p}_u] = i\hbar \quad \text{for all } u. \tag{7•6–21}$$

Operator pairs satisfying these conditions are called **canonically conjugate pairs**.

We assume further that the operators have been chosen in such a way that all operators belonging to *different* degrees of freedom commute,

$$[\hat{q}_u, \hat{p}_v] = 0 \quad \text{for} \quad u \neq v, \tag{7•6–22a}$$

and, of course,

$$[\hat{q}_u, \hat{q}_v] = [\hat{p}_u, \hat{p}_v] = 0 \quad \text{for} \quad u \neq v. \tag{7•6–22b,c}$$

The time dependence of the expectation value of each of the independent operators is then given by (7•6–4), written here in the two forms

$$\frac{d}{dt}\langle q_u\rangle = \frac{i}{\hbar}\langle[\hat{H}, q_u]\rangle \quad \text{and} \quad \frac{d}{dt}\langle p_u\rangle = \frac{i}{\hbar}\langle[\hat{H}, \hat{p}_u]\rangle. \tag{7•6–23a,b}$$

To evaluate the commutators on the right-hand sides, we assume finally that the Hamiltonian is an *analytic* function of the various independent operators $(\hat{q}_u, \hat{p}_u)$. In that case, we may replace the commutators by partial derivatives, yielding generalizations of (7•3–9) and (7•3–10), namely,

$$[\hat{H}, \hat{q}_u] = -i\hbar\left(\frac{\partial \hat{H}}{\partial \hat{p}_u}\right)_{\ldots} \quad \text{and} \quad [\hat{H}, \hat{p}_u] = +i\hbar\left(\frac{\partial \hat{H}}{\partial \hat{q}_u}\right)_{\ldots}. \tag{7•6–24a,b}$$

Here the derivatives again are formal derivatives, and the dots following the parentheses symbolize that *all* other variables are kept constant during the partial differentiation. We will omit such dots in the future.

Exercise: Give a rigorous justification of the generalizations of (7•3–9) and (7•3–10) to the case of a Hamiltonian that is an analytic function of canonically conjugate operator pairs $(\hat{q}_u, \hat{p}_u)$ that satisfy the commutation relations (7•6–21) and (7•6–22).

If we insert (7•6–24) into (7•6–23), we obtain

$$\boxed{\frac{d}{dt}\langle q_u\rangle = +\left\langle\left(\frac{\partial \hat{H}}{\partial \hat{p}_u}\right)\right\rangle \quad \text{and} \quad \frac{d}{dt}\langle p_u\rangle = -\left\langle\left(\frac{\partial \hat{H}}{\partial \hat{q}_u}\right)\right\rangle.} \tag{7•6–25a,b}$$

Were it not for the operator carets and the Dirac brackets, indicating that we are dealing with expectation values, these would be just the **Hamilton-Jacobi equations** of classical dynamics,

$$\frac{d}{dt}q_u = +\left(\frac{\partial H}{\partial p_u}\right) \quad \text{and} \quad \frac{d}{dt}p_u = -\left(\frac{\partial H}{\partial q_u}\right). \tag{7•6–26a,b}$$

These equations represent probably the most general formulation of classical dynamics, applicable not only to classical *mechanical* systems, but—as we shall see—to *electromagnetic* systems as well.

The classical Hamilton-Jacobi equations may be used to guess the Hamiltonian for complicated systems whose classical-limit behavior is known, by expressing this behavior in terms of its Hamilton-Jacobi equations, and then replacing the classical variables q_u and p_u in the classical Hamilton function H by canonically conjugate operator pairs. Probably the most stunning success of this procedure was the quantization of the electromagnetic field, a topic we will discuss briefly in chapter 10 and in somewhat more detail later.

Unfortunately, the procedure is full of pitfalls. For one, the commutation relations (7•6–22) do not fully specify the operators. We saw in problem 7•4–1 that, in spherical polar coordinates, the operator $-i\hbar\partial/\partial r$, which is the simplest operator having the correct commutation relation with the radial coordinate r, is not even Hermitian and hence cannot represent the radial momentum. Next, suppose the product qp occurs in the classical Hamilton function, which is practically guaranteed to happen in non-Cartesian coordinates and which—as we shall soon see—happens even in Cartesian coordinates in the presence of a magnetic field. Should this product be replaced by $\hat{q}\hat{p}$ or by $\hat{p}\hat{q}$, which are not the same if the two operators do not commute? The answer is: No! *Neither* combination is Hermitian, and, hence, cannot be the correct operator corresponding to a classical observable qp. In such cases, the simple symmetrized sum substitution

$$qp = pq \to \tfrac{1}{2}(\hat{q}\hat{p} + \hat{p}\hat{q}) \tag{7•6–27}$$

often leads to the correct Hamiltonian—at least if q and p are simple Cartesian vector components. But in many cases, especially in non-Cartesian coordinate systems, such naive substitutions often lead to incorrect results, and more careful considerations become necessary. The founding fathers of quantum mechanics worried about this point a great deal and built up a large body of theory, in which they were trying—with some success—to axicmatize the procedure, under the name *transformation theory.*

We shall not pursue that approach here. As stated before, classical physics is a limiting case of quantum mechanics, not the other way around, and that puts a logical constraint on the extent to which the laws of quantum mechanics *can* be obtained by extrapolation from their classical limit. Logically, any Hamiltonian obtained by such extrapolation, by whatever procedure and no matter how successful, can be no more than a plausible hypothesis, subject to experimental corroboration—and possibly refutation. The extrapolation procedures themselves are eminently useful as procedures for *formulating* such hypotheses, but they can never *prove* them.

Example: The Effective-Mass Hamiltonian in Semiconductor Physics. The *effective-mass theorem* of semiconductor physics, also known as the *Wannier theorem,*[4] states that under certain conditions, the dynamics of an electron near the bottom of the conduction band of a semiconductor with a position-varying band edge, may be described by an effective-mass Hamiltonian of the form

$$\hat{H}_{\text{eff}} = \frac{1}{2m_{\text{eff}}}\hat{p}^2 + V_{\text{eff}}(\mathbf{r}), \tag{7•6–28}$$

where m_{eff} is the effective mass of the electrons (see chapter 17) and V_{eff} is an effective potential that describes the variation of the band edge of the conduction band with position. When the host semiconductor itself changes with position, as in so-called semiconductor heterostructures, the effective mass becomes position-dependent and must be viewed as an operator that does not commute with $\hat{p}^2$. With the ordering given in (7•6–28), the kinetic energy term in that equation is not Hermitian and hence cannot be correct.

---

[4] For a good modern derivation and discussion, see, for example, J. M. Ziman, *Principles of the Theory of Solids,* Cambridge, 1965, and later editions.

The simplest and most widely used modification of the Hamiltonian (7•6–28) is that generated by the ad hoc substitution

$$\frac{1}{m_{\text{eff}}}\hat{p}^2 \to \hat{p}\frac{1}{m_{\text{eff}}}\hat{p}, \tag{7•6–29}$$

which is manifestly Hermitian. It leads to a simpler mathematical formalism than the use of a symmetrized sum similar to (7•6–27), namely,

$$\frac{1}{m_{\text{eff}}}\hat{p}^2 \overset{?}{\to} \frac{1}{2}\left(\frac{1}{m_{\text{eff}}}\hat{p}^2 + \hat{p}^2\frac{1}{m_{\text{eff}}}\right), \tag{7•6–30}$$

which is in fact hardly ever used. We call (7•6–29) an ad hoc substitution, because it lacks a rigorous foundation and should be viewed as no more than a first-order approximation whose principal virtue is its simplicity. More rigorous treatments of the problem lead to more complicated effective-mass Hamiltonians. To pursue these matters further lies outside the scope of a general text on quantum mechanics.

◆ PROBLEM TO SECTION 7.6

7•6-1: Precessional Motion of a Rotating Charged Sphere in a Magnetic Field

If the dynamics of a solid sphere could be treated by quantum mechanics, the sphere's *rotational* motion would presumably be described by the Hamiltonian

$$\hat{H} = \frac{1}{2I}\hat{L}^2 = \frac{1}{2I}(\hat{L}_x^2 + \hat{L}_y^2 + \hat{L}_z^2), \tag{7•6–31}$$

where I is the moment of inertia of the sphere and the $\hat{L}$'s are angular momentum operators. If the sphere carries a charge Q, there will also be a magnetic moment $\hat{\mathbf{M}}$ present, proportional to both Q and $\hat{\mathbf{L}}$,

$$\hat{\mathbf{M}} = gQ\,\hat{\mathbf{L}}, \tag{7•6–32}$$

where g is a geometrical proportionality factor, the value of which depends on the charge distribution. In a magnetic field $\mathbf{B} = \mathbf{B}_z$, this magnetic moment contributes an additional potential energy

$$\hat{V} = -\mathbf{B}\cdot\hat{\mathbf{M}} = -gQ\mathbf{B}\cdot\hat{\mathbf{L}} = -gQB_z\hat{L}_z \tag{7•6–33}$$

to the Hamiltonian (7•6–31).

Suppose that, at time $t = 0$, the expectation value of the angular momentum vector is $\langle\mathbf{L}\rangle = \mathbf{L}_0$, and that $\mathbf{L}_0$ is *not* parallel to the magnetic field. Analyze the time evolution of $\langle\mathbf{L}\rangle$, and make a suitable graph displaying the dynamics of $\langle\mathbf{L}\rangle$. Ignore the xyz-motion through space.

#7•6-2: Virial Theorem

(a) Let $\hat{H}$ be a Hamiltonian of the standard form (7•6–11). Show that

$$[\hat{\mathbf{r}}\cdot\hat{\mathbf{p}}, \hat{H}] = i\hbar\cdot(2\hat{\mathcal{E}}_{\text{kin}} - \hat{\mathbf{r}}\cdot\nabla V). \tag{7•6–34}$$

Next, let $|\Psi\rangle$ be an energy eigenstate of $\hat{H}$ with the energy eigenvalue $\mathcal{E}$. Show that for such a state,

$$\langle[\hat{\mathbf{r}} \cdot \hat{\mathbf{p}}, \hat{H}]\rangle \equiv \langle\Psi|[\hat{\mathbf{r}} \cdot \hat{\mathbf{p}}, \hat{H}]|\Psi\rangle = 0. \tag{7•6–35}$$

Combining these two results, show that for any energy eigenstate of $\hat{H}$, the following relation holds:

$$2\langle\mathcal{E}_{\text{kin}}\rangle = \langle\hat{\mathbf{r}} \cdot \nabla V\rangle. \tag{7•6–36}$$

This relation is called the **virial theorem**, after an analogous relation in classical mechanics.

(b) Apply (7•6–36) to the harmonic oscillator and to the hydrogen atom. Show that for the harmonic oscillator energy eigenstates with the energy $\mathcal{E}$, we have

$$\langle\mathcal{E}_{\text{pot}}\rangle = \langle\mathcal{E}_{\text{kin}}\rangle = \tfrac{1}{2}\mathcal{E}, \tag{7•6–37}$$

regardless of the dimensionality of the oscillator. Show that for the hydrogen atom,

$$\langle\mathcal{E}_{\text{pot}}\rangle = -2\langle\mathcal{E}_{\text{kin}}\rangle = 2\mathcal{E} \quad (<0). \tag{7•6–38}$$

Both of these results were obtained already in chapters 2 and 3 for the ground states of the two systems. The present derivation shows their generality.

(c) Generalize the preceding results to power-law potentials with arbitrary exponents, both one-dimensional and spherically symmetric ones.

Chapter 8

ELECTRONS IN MAGNETIC FIELDS

8.1 THE VECTOR POTENTIAL HAMILTONIAN

8.1.1 From the Magnetic Lorentz Force to the Hamiltonian

In section 7.6, we applied the time evolution law (7•6–4) for expectation values to the simple Hamiltonian of the form (7•4–11) used so far, to show that this Hamiltonian does indeed reproduce the correct laws in the classical limit. We now turn the problem around, to use the *known* classical limit to extend the Hamiltonian itself, to cover a class of phenomena that cannot be described by the Hamiltonian (7•4–11), namely, the effects of magnetic fields on a moving charge, especially a moving electron.

According to classical mechanics, an electron moving in a magnetic field **B** will see the **magnetic Lorentz force**

$$\mathbf{F} = -e(\mathbf{v} \times \mathbf{B}). \tag{8•1–1}$$

This force cannot be written as the gradient of a potential energy,

$$\mathbf{F} = -\nabla V(\mathbf{r}), \tag{8•1–2}$$

hence, it cannot be described by a Hamiltonian containing only a potential energy term.

In terms of Newton's law, the effect of the Lorentz force may be expressed by writing

$$m_e \frac{d\mathbf{v}}{dt} = -e(\mathbf{v} \times \mathbf{B}). \tag{8•1–3}$$

Somehow, this behavior must be the result of an appropriate underlying quantum-mechanical formalism. We postulate that (8•1–3) is really a law for quantum-mechanical expectation values, of the form

$$m_e \left\langle \frac{d\mathbf{v}}{dt} \right\rangle = -\frac{e}{2}\langle(\hat{\mathbf{v}} \times \hat{\mathbf{B}}) - (\hat{\mathbf{B}} \times \hat{\mathbf{v}})\rangle, \tag{8•1–4}$$

where $\hat{\mathbf{v}}$ and $\hat{\mathbf{B}}$ must now be viewed as operators. In going from (8•1–3) to (8•1–4), we have made the substitution

$$\mathbf{v} \times \mathbf{B} \rightarrow \tfrac{1}{2}[(\hat{\mathbf{v}} \times \hat{\mathbf{B}}) - (\hat{\mathbf{B}} \times \hat{\mathbf{v}})], \tag{8•1–5}$$

in which the classical cross product $\mathbf{v} \times \mathbf{B}$ shows up in the two operator forms $\hat{\mathbf{v}} \times \hat{\mathbf{B}}$ and $-\hat{\mathbf{B}} \times \hat{\mathbf{v}}$. This is basically the natural generalization of the scalar product substitution (7•6–27) to vector cross products. Classically, the two terms in (8•1–5) are equal to each other, but as operators they are not: If the magnetic field $\mathbf{B}$ is position-dependent, the operators $\hat{\mathbf{v}}$ and $\hat{\mathbf{B}}$ will not commute. But in that case, the operator product $\hat{\mathbf{v}} \times \hat{\mathbf{B}}$ will not be a Hermitian operator, and hence, its expectation value need not be real. The symmetrized operator in (8•1–4) is the simplest Hermitian operator that has the vector cross product as its classical limit.

If we insert, on the left-hand side of (8•1–4), the expression (7•6-4) for the time derivative of an expectation value, we obtain

$$\langle[\hat{H}, \hat{\mathbf{v}}]\rangle = \frac{ie\hbar}{2m_e}\langle(\hat{\mathbf{v}} \times \hat{\mathbf{B}}) - (\hat{\mathbf{B}} \times \hat{\mathbf{v}})\rangle. \tag{8•1–6}$$

If we write the Hamiltonian in terms of the velocity operator rather than the momentum operator,

$$\hat{H} = \tfrac{1}{2}m_e\hat{\mathbf{v}}^2, \tag{8•1–7}$$

then (8•1–6) may be further re-written as

$$\langle[\hat{\mathbf{v}}^2, \hat{\mathbf{v}}]\rangle = \frac{ie\hbar}{m_e^2}\langle(\hat{\mathbf{v}} \times \hat{\mathbf{B}}) - (\hat{\mathbf{B}} \times \hat{\mathbf{v}})\rangle. \tag{8•1–8}$$

Evidently, in the presence of a magnetic field, the different components of the velocity operator no longer commute with one another, just as the angular momentum components do not commute. Our task is to determine the set of commutators.

The relation (8•1–8) is a vector relation, which stands for three separate relations for the different components of the vectors involved on both sides. In Cartesian coordinates, for the x-component of (8•1–8), we have

$$\langle[(\hat{v}_y^2 + \hat{v}_z^2), \hat{v}_x]\rangle = \frac{ie\hbar}{m_e^2}\langle(\hat{v}_y\hat{B}_z + \hat{B}_z\hat{v}_y) - (\hat{v}_z\hat{B}_y + \hat{B}_y\hat{v}_z)\rangle, \tag{8•1–9}$$

with cyclically equivalent relations for the y- and z-components. On the left-hand side of (8•1–9), we have dropped a term $[\hat{v}_x^2, \hat{v}x] = 0$. Using the rules (7•3–16) for commutators involving squares of operators, we may re-write the left-hand side of (8•1–9) further as

$$-(\hat{v}_y[\hat{v}_x, \hat{v}_y] + [\hat{v}_x, \hat{v}_y]\hat{v}_y) + (\hat{v}_z[\hat{v}_z, \hat{v}_x] + [\hat{v}_z, \hat{v}_x]\hat{v}_z). \tag{8•1–10}$$

A comparison of (8•1–10) with the right-hand side of (8•1–9), and of their cyclic equivalents, shows that the $\hat{\mathbf{v}}$-component commutators must be proportional to the components of $\hat{\mathbf{B}}$. It then follows that, written in cyclic order,

$$\boxed{[\hat{v}_y, \hat{v}_z] = -\frac{ie\hbar}{m_e^2}B_x, \quad [\hat{v}_z, \hat{v}_x] = -\frac{ie\hbar}{m_e^2}B_y, \quad [\hat{v}_x, \hat{v}_y] = -\frac{ie\hbar}{m_e^2}B_z.} \tag{8•1–11a–c}$$

Evidently, in the presence of magnetic fields, the velocity operator can no longer have the simple form

$$\hat{\mathbf{v}} = \frac{\hat{\mathbf{p}}}{m_e} = -\frac{i\hbar}{m_e}\nabla. \tag{8•1–12}$$

The simplest way to modify (8•1–12) to meet the conditions (8•1–11) is by adding to the right-hand side of (8•1–12) a suitable term that does not commute with the gradient operator, resulting in a relation of the form

$$\hat{\mathbf{v}} = \frac{1}{m_e}[\hat{\mathbf{p}} + e\mathbf{A}(\mathbf{r})] = \frac{1}{m_e}[-i\hbar\nabla + e\mathbf{A}(\mathbf{r})]. \tag{8•1–13}$$

Here, $\mathbf{A}(\mathbf{r})$ is an as-yet unspecified **vector field,** that is, a vector whose components depend on the position $\mathbf{r}$ in space. To determine $\mathbf{A}(\mathbf{r})$, we insert (8•1–13) into (8•1–11), which leads, after some manipulation, to the condition

$$[\hat{v}_x, \hat{v}_y] = -\frac{ie\hbar}{m_e^2}\left(\frac{\partial A_y}{\partial x} - \frac{\partial A_x}{\partial y}\right) = -\frac{ie\hbar}{m_e^2}(\nabla \times \mathbf{A})_z, \tag{8•1–14}$$

with cyclically analogous relations for the other two commutators. If we now compare (8•1–14) and its cyclic equivalents with (8•1–11), we see immediately that (8•1–9) will be satisfied if we choose $\mathbf{A}$ such that

$$\boxed{\mathbf{B} = \nabla \times \mathbf{A};} \tag{8•1–15}$$

that is, the vector $\mathbf{A}$ is simply the **vector potential** of classical electromagnetic field theory!

If we insert (8•1–13) into (8•1–7) and add a conventional potential energy back into $\hat{H}$, we obtain the final Hamiltonian

$$\boxed{\hat{H} = \frac{1}{2m_e}[\hat{\mathbf{p}} + e\mathbf{A}(\mathbf{r})]^2 + V(\mathbf{r}) = \frac{1}{2m_e}[-i\hbar\nabla + e\mathbf{A}(\mathbf{r})]^2 + V(\mathbf{r}).} \qquad (8{\bullet}1\text{–}16)$$

This is the simplest possible Hamiltonian capable of describing the effect of magnetic forces in such a way that the classical magnetic Lorentz force emerges in the expectation value limit.

Our argument leading to (8•1–16) is a plausibility argument, not a proof. We are dealing here with a true extension of the Schroedinger wave equation into a new range of phenomena. Whether or not (8•1–16) is indeed the correct Hamiltonian can never be decided by purely mathematical arguments, but only by extensive *empirical* corroboration, just as was the case with the original Schroedinger wave equation. Experience has shown that (8•1–16) does in fact describe magnetic phenomena correctly.

8.1.2 Kinetic, Potential, and Total Momentum

Individuals familiar with classical mechanics on only the most elementary level are often under the mistaken impression that $\mathbf{p} = M\mathbf{v}$ is the *definition* of momentum, rather than a physical law in its own right that relates the *dynamical* quantity $\mathbf{p}$ to the *kinematic* quantity $\mathbf{v}$. The appearance of a vector potential term in the relation between momentum and velocity shows that it is necessary to distinguish between different kinds of momenta, just as it is necessary to distinguish between different kinds of energy.

In the presence of a vector potential, the quantity $m_e\mathbf{v}$ is no longer the momentum per se, but is called the **kinetic momentum**. The quantity $-e\mathbf{A}$ is called the **potential momentum**, and the sum of the two, $\mathbf{p} = m_e\mathbf{v} - e\mathbf{A}$, is called the **total momentum**. The kinetic energy depends on the kinetic momentum, rather than on the total momentum. Although the need for this distinction arose here in a purely quantum-mechanical context, it has long been known in classical mechanics, and it will be familiar to readers with a knowledge of classical mechanics on a sufficiently advanced level.

Exercise: Show that the application of (7•6–5) to $\hat{\mathbf{r}}$, using the Hamiltonian (8•1–16), restores the expression (8•1–13) for the particle velocity,

$$\langle\mathbf{v}\rangle = \frac{d}{dt}\langle\mathbf{r}\rangle = \frac{i}{\hbar}\langle[\hat{H}, \hat{\mathbf{r}}]\rangle = \frac{1}{m_e}\langle\mathbf{p} + e\mathbf{A}\rangle. \qquad (8{\bullet}1\text{–}17)$$

This shows that the changes (8•1–13) and (8•1–16) in the velocity-momentum relation and in the Hamiltonian are consistent with each other.

Note that our treatment has assumed electrons, which are negatively charged. For positively charged objects, the term $e\mathbf{A}$ must be replaced everywhere by $-e\mathbf{A}$. Also, for objects other than electrons, the mass m_e in our formalism must be replaced by the appropriate mass.

The occurrence of a vector potential term in the relations (8•1–13) and (8•1–17) between momentum and velocity has numerous consequences. One such consequence is that the probability *current* density is no longer given by the form (1•5–6),

$$\mathbf{j} = -\frac{i\hbar}{2m_e}(\Psi^*\nabla\Psi - \Psi\nabla\Psi^*). \tag{8•1–18}$$

This expression accounts only for the $\mathbf{p}/m_e$ part of the velocity; to it must be added the $e\mathbf{A}/m_e$ term, leading to

$$\mathbf{j} = \frac{1}{2m_e}[\Psi^* \cdot (-i\hbar\nabla + e\mathbf{A})\Psi - \Psi \cdot (+i\hbar\nabla + e\mathbf{A})\Psi^*], \tag{8•1–19}$$

which is readily converted to the commonly used form

$$\boxed{\mathbf{j} = -\frac{i\hbar}{2m_e}(\Psi^*\nabla\Psi - \Psi\nabla\Psi^*) + \frac{e\mathbf{A}}{m_e}\Psi^*\Psi.} \tag{8•1–20}$$

The additional term in (8•1–20) is the origin of many of the magnetic properties of matter.

Exercise: By applying the procedure of section 1.5 to the new Hamiltonian (8•1–16), show that the current density expression (8•1–20) satisfies the continuity equation (1•5–4),

$$\frac{\partial\rho}{\partial t} = -\text{div}\,\mathbf{j}, \tag{8•1–21}$$

where the probability density ρ continues to be given by

$$\rho = \Psi^*\Psi. \tag{8•1–22}$$

8.2 EXAMPLE: FREE ELECTRON IN A UNIFORM MAGNETIC FIELD

8.2.1 Landau Levels

As an example of the vector potential formalism, we study here the simplest possible case, a free electron in a uniform magnetic field. We assume the field to be in the z-direction.

Such a uniform field is invariant under translation through space and under rotation about its own direction, and our first task is the selection of a vector potential that retains as much as possible of this invariance. We have considerable freedom of choice in selecting the vector potential because specifying the curl of a function does not fully specify the function, and there is an infinite number of different vector potentials belonging to the same magnetic field. One possible vector potential is

$$\mathbf{A} = \begin{Bmatrix} 0 \\ Bx \\ 0 \end{Bmatrix}. \tag{8•2–1a}$$

An alternative form,

$$\mathbf{A} = \begin{Bmatrix} -By \\ 0 \\ 0 \end{Bmatrix}, \tag{8•2–1b}$$

interchanges the x- and y-directions. Vector potentials of the form (8•2–1a) or (8•2–1b) are said to be in the **Landau gauge.**

A third possibility is the **circular gauge** of **Fig. 8•2–1**,

$$\mathbf{A} = \frac{1}{2}\begin{Bmatrix} -By \\ +Bx \\ 0 \end{Bmatrix} = \frac{1}{2}\mathbf{B} \times \mathbf{r}. \tag{8•2–2}$$

We will address ourselves in section 8.3 to the problems introduced by this ambiguity in the choice of the vector potential. We will show there that, although the wave functions obtained with different vector potentials may differ drastically, all physically observable properties are what is called **gauge invariant.** For now, we simply proceed, using the gauge that leads to the sim-

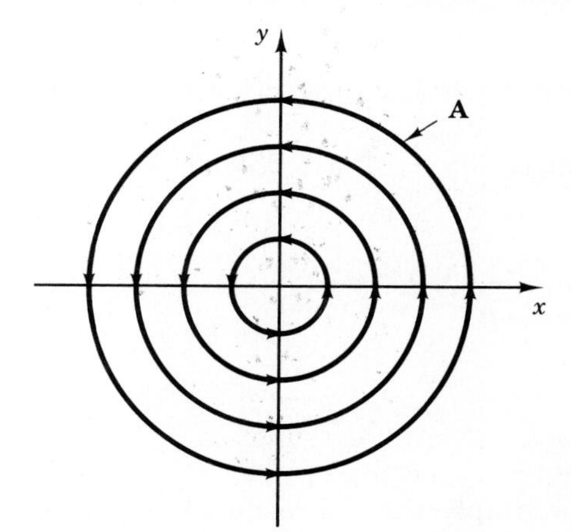

Figure 8•2–1. The vector potential (8•2–2) for a uniform magnetic field.

plest mathematical formalism, the Landau gauge of (8•2–1a). The Hamiltonian then becomes

$$\hat{H} = -\frac{\hbar^2}{2m_e}\nabla^2 - \frac{ie\hbar B}{m_e} \cdot x\frac{\partial}{\partial y} + \frac{e^2B^2}{2m_e} \cdot x^2. \tag{8•2–3}$$

It will be useful to rewrite this by introducing the **Larmor frequency**, or **cyclotron resonance frequency**,

$$\boxed{\omega_L = eB/m_e.} \tag{8•2–4}$$

This is the angular frequency with which an electron orbits in a magnetic field in classical mechanics. With this notation, (8•2–3) assumes the form

$$\hat{H} = -\frac{\hbar^2}{2m_e}\nabla^2 - i\hbar\omega_L \cdot x\frac{\partial}{\partial y} + \frac{1}{2}m_e\omega_L^2 x^2. \tag{8•2–5}$$

This Hamiltonian does not contain either y or z explicitly, but only through the momentum operators $\hat{p}_y = -i\hbar\partial/\partial y$ and $\hat{p}_z = -i\hbar\partial/\partial z$. Because $\hat{H}$ commutes with these operators, the eigenfunctions of $\hat{H}$ can—and should—be chosen in such a way that they are also eigenfunctions of $\hat{p}_y$ and $\hat{p}_z$. This calls for functions that are plane waves in y and z, of the form

$$\psi(x, y, z) = \exp[i(k_y y + k_z z)]\phi(x). \tag{8•2–6}$$

If this form is inserted into the energy eigenvalue problem

$$\hat{H}\psi = \mathcal{E}\psi, \tag{8•2–7}$$

we obtain the following equation for $\phi(x)$:

$$-\frac{\hbar^2}{2m_e}\phi'' + \left[\frac{\hbar^2k_y^2}{2m_e} + \hbar\omega_L k_y x + \frac{1}{2}m_e\omega_L^2 \cdot x^2\right]\phi + \frac{\hbar^2k_z^2}{2m_e}\phi = \mathcal{E}\phi. \tag{8•2–8}$$

By completing the square inside the brackets, this may be transformed into

$$-\frac{\hbar^2}{2m_e}\phi'' + \frac{1}{2}m_e\omega_L^2 \cdot (x - a)^2\phi = \mathcal{E}'\phi, \tag{8•2–9}$$

where

$$a = -\frac{\hbar k_y}{m_e\omega_L} = -\frac{\hbar k_y}{eB}, \tag{8•2–10}$$

and

$$\mathcal{E}' = \mathcal{E} - \frac{\hbar^2k_z^2}{2m_e}. \tag{8•2–11}$$

Equation (8•2–9) is simply the Schroedinger equation for a harmonic oscillator with the oscillation frequency ω_L, centered at $x = a$. The *effective* energy eigenvalues $\mathcal{E}'$ are, of course, the harmonic oscillator values $(n + 1/2)\hbar\omega_L$, where n is any non-negative integer. The true energy eigenvalues $\mathcal{E}$ follow from this and (8•2–11):

$$\mathcal{E} = \mathcal{E}_n(k_z) = \hbar\omega_L \cdot \left(n + \frac{1}{2}\right) + \frac{\hbar^2 k_z^2}{2m_e}. \tag{8•2–12}$$

We see that the energy consists of two parts. The first term in (8•1–34) corresponds to the energy of the motion of the electron perpendicular to the magnetic field. This energy is now quantized, just like the motion in a harmonic oscillator with an oscillation frequency equal to the classical cyclotron resonance frequency. The second term corresponds to the energy of the electron motion parallel to the magnetic field. It is continuously variable. The overall energy eigenvalue spectrum can be represented as in **Fig. 8•2–2**. The continuous energy bands for each fixed value of n are called **Landau bands**.

Note that the parameter k_y from (8•2–6) does not show up in the energy expression (8•2–12). This means that the energy eigenvalues of (8•2–12) are still infinitely degenerate, reflecting the infinite number of different possible values of k_y. The quantity k_y does not correspond to a motion in the y-direction; rather, it determines the x-component a of the position about which the cyclotron motion takes place, via (8•2–10). The infinite k_y-degeneracy of the energy eigenvalues reflects the translational invariance of the physics of the problem perpendicular to the direction of the magnetic field. A single solution of the form (8•2–6) describes a motion in which the center of the x-component of the electron motion has a sharp value, but the center of the y-component of the motion is left totally indeterminate. By linear superposition of many degenerate solutions with different k_y-values, one can construct solutions for which both the x- and the y-component of the electron motion take place about specified centers.

We close our discussion of the free electron in a magnetic field by considering the probability current distribution of the energy eigenfunctions (8•2–6).

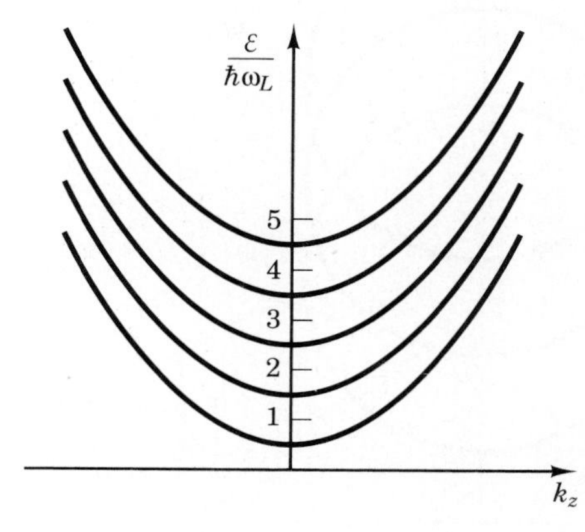

Figure 8•2–2. Landau bands of energy for an electron in a uniform magnetic field.

By inserting **A** and ψ into (8•1–18), one finds readily that

$$j_x = 0, \tag{8•2–13a}$$

$$j_y = \frac{1}{m_e}(\hbar k_y + eA_y)\psi^*\psi = \omega_L \cdot (x - a) \cdot \psi^*\psi, \tag{8•2–13b}$$

$$j_z = \frac{\hbar k_y}{m_e} \cdot \psi^*\psi. \tag{8•2–13c}$$

In going from the first form in (8•2–13b) to the second form, we have drawn on (8•2–1a), (8•2–4), and (8•2–10).

The result for j_y is the interesting one. The current density is antisymmetric about the plane $x = a$, the center plane of the x-component of the cyclotron oscillation. The current component j_y is exactly the same as for a counterclockwise classical cyclotron orbit with its center at $x = a$. The x-component of the current vanishes because our solution, in effect, averages over all possible y-components of the center of the classical cyclotron orbit, which cancels the current contribution to j_x from opposite parts of the orbit. This is illustrated in **Fig. 8•2–3.**

8.2.2 Crossed Electric and Magnetic Fields

In section 8.2.1, we considered the behavior of an electron in a uniform magnetic field. We now add a uniform electric field **E** at a right angle to the magnetic field, in the x-direction. If we continue to work with the vector potential (8•2–1a), we obtain the Hamiltonian

$$\hat{H} = -\frac{\hbar^2}{2m_e}\nabla^2 - i\hbar\omega_L \cdot x\frac{\partial}{\partial y} + [eEx + \tfrac{1}{2}m_e\omega_L^2 x^2], \tag{8•2–14}$$

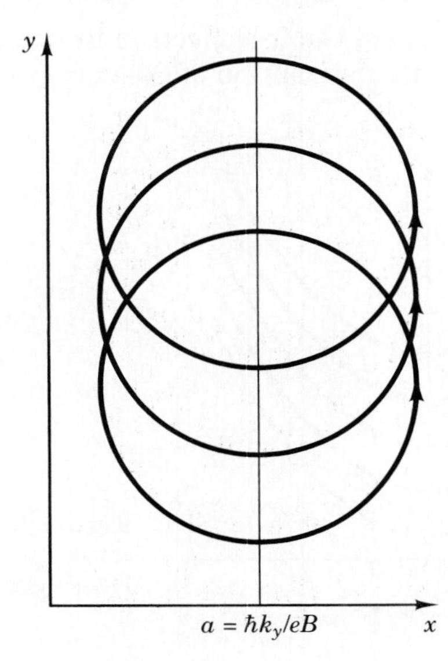

Figure 8•2–3. The eigenfunctions (8•2–6) for a given value of k_y represent an average over all possible y-positions of a classical cyclotron orbit.

which differs from (8•2–5) by the eEx-term.

Because the Hamiltonian continues to be translationally invariant in both the y-direction and the z-direction, it continues to call for eigenfunctions of the form (8•2–6). With these, the Schroedinger equation (8•2–7) becomes

$$-\frac{\hbar}{2m_e}\phi'' + [eEx + \tfrac{1}{2}m_e\omega_L^2 \cdot (x - a)^2]\phi = \mathcal{E}'\phi, \tag{8•2–15}$$

instead of (8•2–9). The content of the square brackets may be rearranged to read

$$[\ldots] = V'(x) = \tfrac{1}{2}m_e\omega_L^2(x - b)^2 + \tfrac{1}{2}m_e\omega_L^2(a^2 - b^2), \tag{8•2–16}$$

where

$$b = a - \frac{eE}{m_e\omega_L^2}. \tag{8•2–17}$$

This is evidently still the potential of a harmonic oscillator with an oscillation frequency ω_L, similar to the potential in (8•2–9), but displaced from $x = 0$ by the distance b rather than a.

An important consequence of this displacement is that there is now a net current in the y-direction, perpendicular to *both* the electric and the magnetic field. To see this, note first that the final eigenfunctions $\psi(\mathbf{r})$ of (8•2–14) are now of the form

$$\psi_n(\mathbf{r}) \propto \exp[i(k_y y + k_z z)]\,\phi_n(x - b), \tag{8•2–18}$$

where $\phi_n(x - b)$ is the nth harmonic oscillator eigenfunction, displaced by b. The z-motion is a simple (and uninteresting) uniform plane-wave motion; we ignore it here. Because the eigenfunctions (8•2–18) represent one-dimensional bound states in the x-direction, they cannot carry a current in that direction.[1]

We are interested here principally in the y-component of the motion. From (8•1–20), (8•2–1a), (8•2–17), and (8•2–18), we find readily that

$$\begin{aligned} j_y &= -\frac{i\hbar}{2m_e}\left(\psi^*\frac{\partial\psi}{\partial y} - \psi\frac{\partial\psi^*}{\partial y}\right) + \frac{e\mathbf{A}}{m_e}\psi^*\psi \\ &= \left[\frac{\hbar k_y}{m_e} + \frac{eB}{m_e}x\right]\psi^*\psi = \left[-\frac{E}{B} + \omega_L(x - b)\right]\psi^*\psi. \end{aligned} \tag{8•2–19}$$

Because the probability density for the eigenfunctions is symmetrical about $x = b$, it is clear that the antisymmetric term $\omega_L(x - b)$ does not contribute a *net* current, but corresponds to a rotary motion with the angular frequency ω_L, essentially the same as in the case without an electric field. The linear motion is contained in the first term, which shows that its speed is given by

$$\langle v_y \rangle = -E/B. \tag{8•2–20}$$

[1] It would be readily possible to construct wave packets that *oscillate* in the x-direction, by linear superposition of eigenfunctions with different quantum numbers n.

This combination of a rotary motion with the Larmor frequency and a linear motion is the quantum-mechanical manifestation of the well-known classical result that in crossed fields a charged particle undergoes a cycloidal motion—that is, a *uniform circular* (= harmonic oscillator) motion superimposed on a *uniform linear* motion perpendicular to both fields (**Fig. 8•2–4**).

Classically, the linear motion follows directly from the requirement that the time-averaged Lorentz force $-e\langle \mathbf{v} \rangle \times \mathbf{B}$ cancel the electric force $-e\mathbf{E}$. In quantum mechanics, the uniform linear motion arises from the $\mathbf{A}\,\psi^*\psi$-term in the current density expression (8•1–20). In fact, one of our motivations in presenting the crossed-field case was to show the working of this term.

The mutual orthogonality of electric field, magnetic field, and current plays a central role in the theory of the quantum Hall effect. However, this topic goes beyond the scope of the present text, and the interested reader is referred to appropriate texts on solid-state physics.

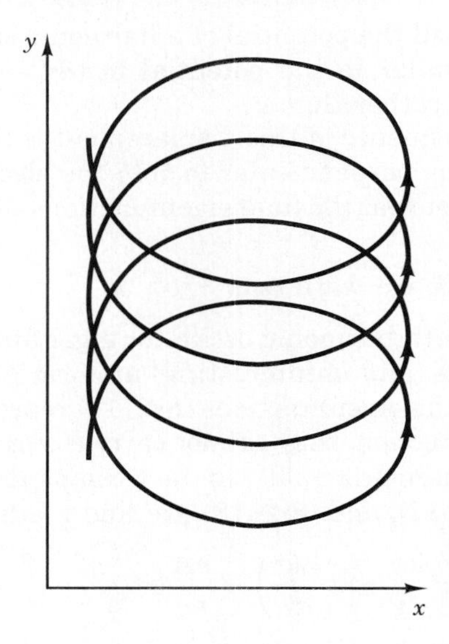

Figure 8•2–4. Classical cycloidal motion of an electron in crossed electric and magnetic fields.

8.3 GAUGE TRANSFORMATIONS, AHARONOV-BOHM EFFECT, AND ELECTRONS IN SUPERCONDUCTORS

8.3.1 Gauge Transformations and Gauge Invariance

In section 8.2, we pointed out that the relation

$$\mathbf{B} = \operatorname{curl} \mathbf{A} \tag{8•3–1}$$

between the magnetic field $\mathbf{B}$ and the vector potential $\mathbf{A}$ does not fully specify the vector potential $\mathbf{A}$ for a given field $\mathbf{B}$: The vector potential may be changed by an arbitrary additive vector *function*, so long as the curl of this function vanishes. But any curl-free function can be written as the gradient of a scalar function $\chi(\mathbf{r})$ and vice versa. Hence, if we substitute

$$\mathbf{A} = \mathbf{A}' - \nabla\chi \tag{8•3–2}$$

into (8•3–1), the magnetic field $\mathbf{B}$ remains unchanged. Such a transformation is called a **gauge transformation,** and the function χ a **gauge function**.

This arbitrariness of the vector potential goes beyond the arbitrariness of the zero of the scalar potential. The latter may be changed by an arbitrary additive *constant:*

$$V = V' - V_0. \tag{8•3–3}$$

This constant cancels out of the time-*independent* Schroedinger equation altogether if both the potential V and the energy $\mathcal{E}$ are referred to the same origin. In the time-*dependent* Schroedinger wave equation, the constant simply leads to a position-independent phase factor in the wave function, according to

$$\Psi = \Psi' \exp\left(\frac{i}{\hbar} V_0 t\right). \tag{8•3–4}$$

The time-dependent phase factor reflects the change in energy relative to the changed zero of the energy, but all other physical properties remain unchanged. It is basically a change in bookkeeping, rather than in the physics of the problem.

In the present section, we study the effects of a gauge transformation on the Schroedinger wave equation

$$i\hbar\frac{\partial\Psi}{\partial t} = -\frac{\hbar^2}{2m_e}\left(\nabla + \frac{ie}{\hbar}\mathbf{A}\right)^2\Psi + V\Psi \tag{8•3–5a}$$

and its solutions.

We claim that the transformation (8•3–2) of the vector potential is accompanied by a transformation of the wave function similar to (8•3–4), namely,

$$\Psi = \Psi' \exp\left(\frac{ie}{\hbar}\chi\right). \tag{8•3–6}$$

From (8•3–2) and (8•3–6), one finds easily that

$$\begin{aligned}\left(\nabla + \frac{ie}{\hbar}\mathbf{A}\right)\Psi &= \left(\nabla + \frac{ie}{\hbar}\mathbf{A}' - \frac{ie}{\hbar}\nabla\chi\right)\left[\exp\left(\frac{ie}{\hbar}\chi\right)\Psi'\right] \\ &= \exp\left(\frac{ie}{\hbar}\chi\right)\cdot\left(\nabla + \frac{ie}{\hbar}\mathbf{A}'\right)\Psi',\end{aligned} \tag{8•3–7}$$

and similarly,

$$\left(\nabla + \frac{ie}{\hbar}\mathbf{A}\right)^2\Psi = \exp\left(\frac{ie}{\hbar}\chi\right)\cdot\left(\nabla + \frac{ie}{\hbar}\mathbf{A}'\right)^2\Psi'. \tag{8•3–8}$$

If the gauge function χ is time-independent, insertion of (8•3–6) and (8•3–8) into the Schroedinger wave equation (8•3–5a) yields

$$i\hbar\frac{\partial\Psi'}{\partial t} = -\frac{\hbar^2}{2m_e}\left(\nabla + \frac{ie}{\hbar}\mathbf{A}'\right)^2\Psi' + V\Psi', \tag{8•3–5b}$$

which is of the same form as the original Schroedinger wave equation (8•3–5a), but with the new vector potential $\mathbf{A}'$ and with a different wave function as its solution.

Inasmuch as the physics of the problem has not been changed by the gauge transformation, the two different wave functions are again simply two different descriptions of the same physical state, similar to the way the position and momentum representations of a given state are simply different descriptions of that state. We say that the Schroedinger equation is **gauge invariant**.

As an illustration of this unchanged physics, consider the probability density and the probability current density associated with a given state. The transformation (8•3–6) evidently leaves the probability density unchanged. The invariance of the probability *current* density is most easily seen from the form (8•1–19) of the expression for the probability current density in the presence of a vector potential. Inserting (8•3–6) and (8•3–7) into that expression shows that the probability current density remains the same and that it may be expressed as

$$\mathbf{j} = -\frac{i\hbar}{2m_e}(\Psi'^*\nabla\Psi' - \Psi'\nabla\Psi'^*) + \frac{e\mathbf{A}'}{m_e}\Psi'^*\Psi', \tag{8•3–9}$$

which is the same as (8•1–20), except for the replacement of Ψ by Ψ' and $\mathbf{A}$ by $\mathbf{A}'$.

Taken together, the probability density and the probability current density fully specify the state and its time evolution, and their invariance illustrates the invariance of the state itself.

The appearance of a new position-dependent phase factor in the wave function once again illustrates a point we made in chapter 1, that the wave functions have no direct physical meaning by themselves, and that they are only generating functions from which physically observable quantities can be extracted by suitable mathematical operations. A gauge transformation changes the generating function, but it also changes the mathematical operators used to extract the observables, in such a way that the extracted values remain invariant.

8.3.2 Time-Dependent Gauge Transforms

The treatment in subsection 8.3.1 assumed a time-independent gauge function χ. This is too restrictive: We will later encounter the need to work with time-dependent vector potentials, which implies a need to consider time-dependent gauge functions χ. In such cases, insertion of the transform (8•3–6) and the relation (8•3–8) into the left-hand side of the Schroedinger wave equation (8•3–5a) yields an extra $\partial\chi/\partial t$-term:

$$i\hbar\frac{\partial\Psi}{\partial t} = i\hbar\frac{\partial}{\partial t}\left[\Psi' \exp\left(\frac{ie}{\hbar}\chi\right)\right] = \exp\left(\frac{ie}{\hbar}\chi\right)\left[i\hbar\frac{\partial\Psi'}{\partial t} - e\left(\frac{\partial\chi}{\partial t}\right)\Psi'\right]. \tag{8•3–10}$$

This term has a simple physical meaning. We recall from electromagnetic theory that, in the presence of a time-dependent vector potential, the electric field is given by

$$\mathbf{E} = -\nabla\Phi - \frac{\partial\mathbf{A}}{\partial t}, \tag{8•3–11}$$

where Φ is the electrostatic potential. Substituting $\mathbf{A} = \mathbf{A}' - \nabla\chi$ leads to

$$\mathbf{E} = -\nabla\Phi - \frac{\partial\mathbf{A}'}{\partial t} + \nabla\left(\frac{\partial\chi}{\partial t}\right). \tag{8•3–12}$$

The new vector potential $\mathbf{A}'$ evidently corresponds to a different electric field. If we wish the electric field to remain unchanged by the gauge transformation, we must supplement the gauge transformation (8•3–2) of the *vector* potential with a transformation of the *scalar* electrostatic potential Φ, such that the $\partial\chi/\partial t$ term cancels:

$$\Phi' = \Phi - \frac{\partial\chi}{\partial t}. \tag{8•3–13a}$$

In terms of the scalar potential energy function $V(\mathbf{r})$,

$$V' = V + e\frac{\partial\chi}{\partial t}. \tag{8•3–13b}$$

Evidently, the extra $\partial\chi/\partial t$ term in (8•3–10) is equal to the term in (8•3–13b), and we may eliminate it by substituting (8•3–13b) into the Schroedinger wave equation (8•3–5b), yielding finally

$$i\hbar\frac{\partial\Psi'}{\partial t} = -\frac{\hbar^2}{2m_e}\left(\nabla + \frac{ie}{\hbar}\mathbf{A}'\right)^2\Psi' + V'\Psi', \tag{8•3–5c}$$

which differs from (8•3–5b) only by the substitution of V' for V.

8.3.3 Example: Double-Slit Diffraction Revisited

Gauge invariance implies that the phase changes introduced into the wave function by a gauge transformation cannot lead to changes in any diffraction pattern. It is instructive to obtain this result here by alternative considerations, involving a simple comparison of the phase shifts introduced along different paths by a gauge transformation.

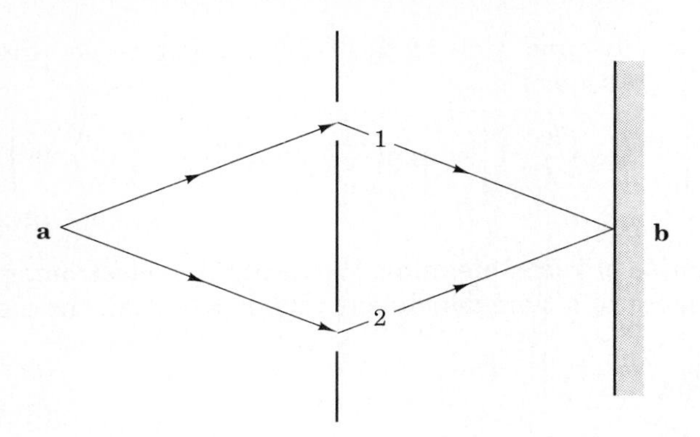

Figure 8•3–1. Double-slit diffraction and interference setup.

We consider a simple double-slit diffraction setup, as in **Fig. 8•3–1**, and assume that there is no magnetic field present *anywhere,* a case for which the natural choice of the vector potential is $\mathbf{A} = 0$. There will then be a certain diffraction pattern on the screen to the right. Suppose next that we transform to a nonzero, but still curl-free, vector potential

$$\mathbf{A}' = \nabla\chi. \tag{8•3–14}$$

According to (8•3–6), this transformation introduces a local phase change

$$\varphi(\mathbf{r}) = -\frac{e}{\hbar}\chi(\mathbf{r}) \tag{8•3–15}$$

into the transformed wave function Ψ', relative to the original wave function Ψ. Because this phase shift is not constant, but depends on the position $\mathbf{r}$, there will in general be a change in the phase *difference* between any two points $\mathbf{a}$ and $\mathbf{b}$, such as the point of origin of an electron stream and a point of observation on the diffraction screen,

$$\Delta\varphi_{\mathbf{ba}} \equiv \varphi(\mathbf{b}) - \varphi(\mathbf{a}) = -\frac{e}{\hbar}[\chi(\mathbf{b}) - \chi(\mathbf{a})]. \tag{8•3–16}$$

We write this change as a path integral,

$$\Delta\varphi_{\mathbf{ba}} = -\frac{e}{\hbar}\int_{\mathbf{a}}^{\mathbf{b}} (\nabla\chi)\cdot d\mathbf{s} = -\int_{\mathbf{a}}^{\mathbf{b}} \mathrm{A}'\cdot d\mathbf{s}, \tag{8•3–17}$$

where in the second equality we have drawn on (8•3–14). A nonzero value of $\Delta\varphi_{\mathbf{ba}}$ means that the total number of wavelengths contained in the path **ab** has been changed by the gauge transformation; $\Delta\varphi_{\mathbf{ba}}$ is simply 2π times the change in the number of wavelengths along the path. However, what matters is not the phase change along path 1 or path 2 alone, but the *difference*—if any—between the two paths:

$$\Delta\varphi = \Delta\varphi_1 - \Delta\varphi_2 = -\frac{e}{\hbar}\left[\int_1 \mathbf{A}' \cdot d\mathbf{s} - \int_2 \mathbf{A}' \cdot d\mathbf{s}\right]$$
$$= \frac{e}{\hbar}\oint \mathbf{A}' \cdot d\mathbf{s}. \tag{8•3–18}$$

The (counterclockwise) loop integral is easily evaluated by using Stokes's integral theorem and relation (8•3–1), leading to

$$\oint \mathbf{A}' \cdot d\mathbf{s} = \int_{\text{area}} (\nabla \times \mathbf{A}') \cdot d\mathbf{a} = \int_{\text{area}} \mathbf{B} \cdot d\mathbf{a} = \Phi_m, \tag{8•3–19}$$

where the last two integrals are over the area enclosed by the two paths and Φ_m is the magnetic flux—if any—enclosed by the paths. Under our assumption of no magnetic field anywhere, the enclosed flux is of course zero, and we obtain a zero phase difference between the two interference paths, thus recovering our earlier result.

8.3.4 Aharonov-Bohm Effect

Our discussion of the double-slit diffraction experiment implies that there *should* be observable interference effects if the two interfering paths enclose a region containing a nonzero magnetic flux, even if the magnetic field along the electron paths themselves is zero. This remarkable prediction—which has been confirmed experimentally—is called the **Aharonov-Bohm effect.**[2]

To be specific, consider a longitudinally magnetized iron wire, as shown in **Fig. 8•3–2**. Such a wire may contain an *internal* magnetic flux with no *external* magnetic field. According to (8•3–19), an external vector potential *must* then be present in the field-free space outside the wire.

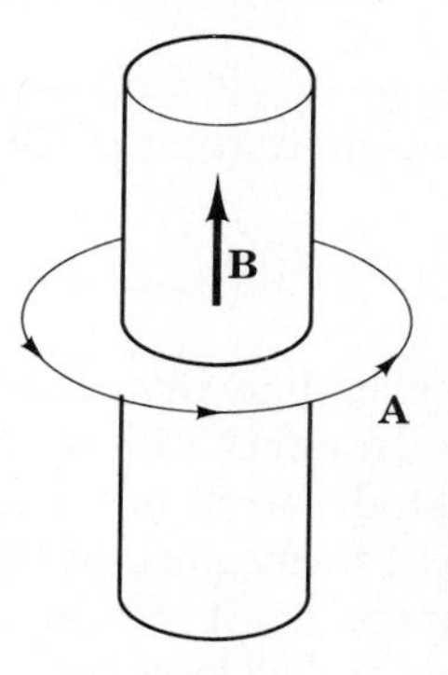

Figure 8•3–2. A longitudinally magnetized iron wire must have a vector potential **A** present on the outside, even if the magnetic field **B** is zero outside the wire.

[2] Y. Aharonov and D. Bohm, *Phys. Rev.* II **115,** 485 (1959).

Suppose next that such a wire is placed into the wave shadow between the two slits in a double-slit diffraction screen (**Fig. 8•3–3**), so that the wave function is negligible inside the wire itself. If necessary, the waves could be kept out of the wire by applying a repulsive potential to the wire.

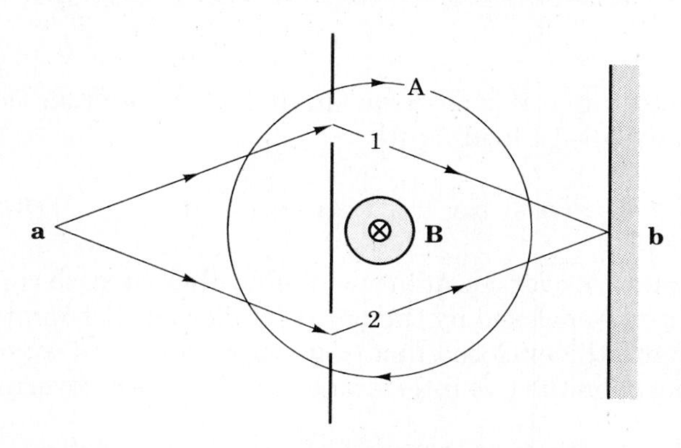

Figure 8•3–3. Double-slit diffraction and interference around a magnetic flux tube, generated by a magnetized iron wire.

If the magnetic field inside the wire points into the plane of the drawing, there will then be a clockwise circular vector potential present, as shown.

The vector potential outside the wire is of course curl-free, and may therefore still be written as the gradient of a scalar function. We may still treat this external vector-potential *as if* it had been generated from the zero vector-potential case by a gauge transformation. However, compared to the true zero-field case without any enclosed flux, the gauge function $\chi(\mathbf{r})$ is no longer single-valued. Suppose we start at some point outside the wire, and follow $\chi(\mathbf{r})$ as we go around the wire, and return to the point of origin. The gauge function $\chi(\mathbf{r})$ will then not return to its original value, but will be incremented by

$$\Delta\chi = \oint \mathbf{A} \cdot d\mathbf{s} = \Phi_m. \tag{8•3–20}$$

Associated with this change in χ is a phase difference between the two interfering waves, given by the combination of (8•3–18) and (8•3–19),

$$\Delta\varphi = \frac{e}{\hbar}\Phi_m. \tag{8•3–21}$$

Our earlier proof of gauge invariance is not applicable to such multi-valued gauge functions. In effect, the wave functions along the two different paths are associated with different branches of the gauge function, a situation outside the mathematical framework of the proof.

In classical mechanics, what matters is only the magnetic field along the actual particle trajectory, and as long as the particle stays outside the magnetic field itself, a multi-valued curl-free vector potential has no observable consequences. In quantum mechanics, this is no longer true: The propagation of particles is now wave-like rather than trajectory-like. But waves have a phase, and this phase is changed by the presence of a vector potential, even a curl-free one. The phase along the upper path in Fig. 8•3–3 is retarded by the vector potential, while that along the lower path is advanced. Eq. (8•3–21) is the resulting overall phase difference.

The Aharonov-Bohm effect is remarkably large: We will have a relative phase shift of π, corresponding to a change from constructive to destructive interference and vice versa, for a flux of magnitude

$$\Phi_0 = \frac{\pi\hbar}{e} = 2.0678 \times 10^{-15}\ \mathrm{T} \cdot \mathrm{m}^2. \tag{8•3–22}$$

This is a very small flux: The Earth's magnetic field is on the order of 40 μT; hence, the flux enclosed by an area of about 200 $\mu\mathrm{m}^2$ is sufficient to cause an interchange of fringes of constructive and destructive interference in any interference pattern.

8.3.5 Flux Quantization in a Superconducting Loop

A phenomenon closely related to the Aharonov-Bohm effect is the quantization of magnetic flux threading through a superconducting current loop. Such a loop can carry a current indefinitely, and the flux generated by the current will not decay. Now, it is one of the fundamental properties of superconductors that they will expel magnetic fields from inside the superconducting body, except for a thin surface layer, typically a few times 10^{-5} cm thick. For superconducting loops made from superconductors much thicker than this penetration depth, any closed path deep inside the superconductor will be entirely in field-free space, but will nevertheless enclose a finite flux. Hence, there will be a curl-free but multi-valued vector potential along such a path. In order for the wave function of the electrons inside the superconductor to be single-valued, the flux must be such that the vector-potential-induced phase shift of the wave function around the loop must be an integer multiple of 2π, similar to the way the wave function in a rotationally invariant system had to return to itself under rotation by 2π. As a result, the flux is quantized.

From our earlier considerations, we might expect that the flux quantum is twice the quantity Φ_0 introduced in (8•3–22) for a phase shift of π for a one-electron wave function. It turns out, however, that superconductivity involves not single electrons, but pairs of electrons, called **Cooper pairs**, whose charge is $2e$ rather than e. With the charge of the charge carriers occurring in the denominator of Φ_0, the flux quantum for a 2π phase shift for electron pairs is the same flux quantum Φ_0 as that for a phase shift of π for single electrons. The experimental data confirm this prediction, and the magnitude of the superconducting flux quantum is in fact one of the most direct proofs that supercurrents are carried by electron pairs.

Note that flux quantization is a property, not of the magnetic field itself, but of the electrons that carry the current supporting that field. In fact, it is nothing other than the quantization of the angular momentum of the electrons flowing around the flux quantum.

8.3.6 The London Equation of the Theory of Superconductivity

In 1935, some 20 years before the physical mechanism of superconductivity was understood, Fritz London showed that both the existence of zero-resistance macroscopic currents and the expulsion of magnetic fields (the Meissner effect) could be understood if one assumed that, *for some reason,* the electrons participating in the superconducting current flow contributed a local electrical current density proportional, not to the electric field **E** (Ohm's law), but to the local vector potential **A**, with a proportionality factor equal to that in the vector-potential term in the probability density expression (8•1–20):

$$\mathbf{j}(\mathbf{r}) = -\frac{e^2 n(\mathbf{r})\mathbf{A}(\mathbf{r})}{m_e}. \tag{8•3–23}$$

Here $n(\mathbf{r})$ is the local concentration of electrons participating in the superconducting current flow (the number of electrons per unit volume), and the gauge of the vector potential must be chosen to meet appropriate boundary conditions for the current density at the surface of the superconductor. The latter condition simply means that in the absence of any current flowing into or out of the superconductor, the vector potential does not have any component perpendicular to the surface of the superconductor. London postulates explicitly that any electric fields are to be included via a time-dependent vector potential, rather than as the gradient of an electrostatic potential, and that (8•3–23) remains valid for time-dependent currents and vector potentials.

Exercise: The London equation implies a change of current with time when an electric field is present. Show that the rate of change is exactly the same as if the electrons were free electrons accelerated according to Newton's law.

The idea that the supercurrent density is given by the vector potential in the probability current density expression (8•1–20) implies a cancellation of the gradient terms in that expression. In the absence of a vector potential, the Schroedinger equation is a real equation, and if $\psi(\mathbf{r})$ is an electron state, its complex conjugate $\psi^*(\mathbf{r})$ is also a state, with the same energy, but opposite current density. In thermal equilibrium, both states have equal occupation probability, and then the contributions of the gradient terms to the current do indeed cancel. However, in a normal conductor—which behaves essentially like a gas of free electrons—this cancellation ceases. Instead, the wave functions tend to change in such a way that the gradient terms cancel the vector potential term.

London postulates that it is the very essence of the superconducting state that this mutual cancellation of the gradient terms themselves persists, even in the presence of a magnetic field:

> ". . . presumably [as] a result of electronic cooperation . . . the electrons in the superconducting state differ from free electrons. . . . Superconductivity would result if the eigenfunctions of a fraction of the electrons were not disturbed at all when the system is brought into a magnetic field ($H < H_c$). It would be sufficient if the eigenfunction were to stay essentially as it was without a magnetic field, as if, so to speak, it were rigid."[3]

It has become clear since the work of London that the electronic "cooperation" leading to the rigid wave function postulated by London does in fact occur, via the formation of electron pairs, known as Cooper pairs, the essential ingredient of the Bardeen-Cooper-Schrieffer (BCS) theory of superconductivity. Suppose that there exists a coupling mechanism that causes all electrons to pair up, in such a way that in the absence of a vector potential, either both states $\psi(\mathbf{r})$ and $\psi^*(\mathbf{r})$ are occupied, or neither. Suppose further that, once this pairing is present, it somehow prevents the paired wave functions from changing when a magnetic field is added. Under this set of assumptions, the contributions from the gradient terms in the current density expression (8•1–20) cancel, but the contributions from the vector potential terms add, leading to London's equation (8•3–23).

A detailed discussion of the pair formation mechanism of the BCS theory would go far beyond the scope of this text, and the interested reader must be referred to advanced texts on solid-state physics or to texts on superconductivity itself.

◆ PROBLEMS TO SECTION 8.3

#8•3-1: Gauge Transformation of Lowest Energy State of Electron in a Uniform Magnetic Field

Solve the Schroedinger equation for the lowest energy state of an electron in a uniform magnetic field **B** in the z-direction, using the circular gauge (8•2–2) and ignoring the motion along the magnetic field. Transform the resulting wave function to the Landau gauge (8•2–1a). The transformed wave function can be expressed as a linear superposition of wave functions of the form (8•2–6), originally obtained in the Landau gauge, with different values of k_y. Derive a closed-form mathematical expression for the expansion amplitudes $a(k_y)$.

---

[3] Fritz London, *Superfluids*. Volume 1, *Macroscopic Theory of Superconductivity,* 2nd ed. New York: Dover Publications, 1961, p. 150.

Note: Watch for correct normalization, especially of the wave functions originally obtained in the Landau gauge, and of the expansion coefficients $a(k_y)$. Express the final result for the $a(k_y)$ in as simple a form as possible.

#8•3-2: Vector-Potential Formulation for an Electric Field

An electric field need not be described as the gradient of an electrostatic potential,

$$\mathbf{E} = -\nabla\Phi, \tag{8•3–24}$$

corresponding (in the case of electrons) to a potential energy $V = -e\Phi$. An alternative description is in terms of a time-dependent vector potential:

$$\mathbf{E} = -\frac{\partial \mathbf{A}}{\partial t}. \tag{8•3–25}$$

Use this formulation to study the time evolution of a Gaussian wave packet of the form (3•2–12b), under the influence of a uniform electric field. Carry out the treatment in both the position and the momentum representation.

Note that for a time-dependent vector potential, the frequency ω in the dispersion relation becomes itself time dependent. Show that in this case, plane waves must be written in the form

$$\Psi(\mathbf{r}, t) \propto \exp[i(\mathbf{k} \cdot \mathbf{r} - \int \omega\, dt)]. \tag{8•3–26}$$

Compare the result in the position representation with that for a Gaussian wave packet without an external force, in section 3.2. Compare the result in the momentum representation with that in section 3.4. Make sure that you transform back to $\mathbf{A} = 0$ before the final comparisons.

#8•3-3: Meissner Effect

In addition to being related via the London equation, the current density and the magnetic field in a superconductor are also related via one of Maxwell's equations,

$$\mathbf{j} = \text{curl}\, \mathbf{H}. \tag{8•3–27}$$

Let the plane $z = 0$ be the surface of a semi-infinite superconductor extending into the positive-z half-space. Show that the dual coupling between current and magnetic field implies that the current density and the magnetic field cannot be uniform, but must fall off exponentially from the surface of the superconductor into its bulk. Give an expression for the penetration depth, defined as the $1/e$-distance of the falloff. Give a numerical estimate for a typical electron concentration, $n = 10^{22}\ \text{cm}^{-3}$.

Chapter 9

BEYOND HERMITIAN OPERATORS

9.1 HERMITIAN CONJUGATE OPERATOR PAIRS

In chapter 7, we defined Hermitian operators by the requirement that their expectation values always be real, for any physically admissible state function. But not all operators of interest in quantum mechanics are Hermitian. For example, we saw in (7•4–6) that the product of two non-commuting Hermitian operators forms an operator in its own right, which is not Hermitian, and we will later encounter many other non-Hermitian operators.

Whenever an operator $\hat{a}$ is not Hermitian, and may hence have complex expectation values for at least some states, we can usually define a second operator $\hat{a}^\dagger$, called the **Hermitian conjugate** of $\hat{a}$, which satisfies the requirement that its expectation values are the complex conjugates of those of $\hat{a}$:

$$\langle\Psi|\hat{a}^\dagger|\Psi\rangle = (\langle\Psi|\hat{a}|\Psi\rangle)^*, \tag{9•1–1}$$

for all states $|\Psi\rangle$ in the domain of both operators. Because of the relation (7•2–3) for the complex conjugates of Dirac brackets, (9•1–1) may also be written in the two alternative equivalent forms,

$$\langle\Psi|\hat{a}^{\dagger}|\Psi\rangle = \langle\hat{a}\Psi|\Psi\rangle \quad \text{and} \quad \langle\Psi|\hat{a}|\Psi\rangle = \langle\hat{a}^{\dagger}\Psi|\Psi\rangle, \tag{9•1–2}$$

which we may view as *primary* definitions of Hermitian conjugate operator pairs, similar to the way (7•4–2) was our primary definition of Hermitian operators. As in section 7.4, the two relations (9•1–2) are special cases of a more general definition, analogous to (7•4–3), namely,

$$\boxed{\langle\Phi|\hat{a}^{\dagger}\Psi\rangle = \langle\hat{a}\Phi|\Psi\rangle \quad \text{or} \quad \langle\Phi|\hat{a}\Psi\rangle = \langle\hat{a}^{\dagger}\Phi|\Psi\rangle,} \tag{9•1–3a,b}$$

where Φ and Ψ are any *two* admissible state functions. In other words: In any inner product, an operator operating on one of the two state functions may be replaced by its Hermitian conjugate operating on the other state function. The proof is left to the reader. We shall use here the definition (9•1–3) rather than (9•1–2).

Evidently, Hermitian conjugation is a reciprocal relationship: If $\hat{a}^{\dagger}$ is the Hermitian conjugate of $\hat{a}$, then $\hat{a}$ is the Hermitian conjugate of $\hat{a}^{\dagger}$:

$$(\hat{a}^{\dagger})^{\dagger} = \hat{a}. \tag{9•1–4}$$

Any Hermitian operator $\hat{A}$ is its own Hermitian conjugate: $\hat{A}^{\dagger} = \hat{A}$.

Hermitian conjugate operator pairs are also referred to as **adjoint pairs**, or sometimes **Hermitian adjoint pairs**, and Hermitian operators are sometimes called **self–adjoint operators**. The Hermitian conjugate of an operator $\hat{a}$ is often designated by attaching a dagger superscript (†) to the original operator, as we have done here. Other common designations for Hermitian conjugate pairs are $\hat{a}^{+}$ and $\hat{a}^{-}$, or $\hat{a}_{+}$ and $\hat{a}_{-}$.

We shall frequently encounter Hermitian conjugate operator pairs. Perhaps the simplest case is that of the two products of two non-commuting operators. One confirms easily that

$$\hat{a} = \hat{A}\hat{B}, \qquad a^{\dagger} = \hat{B}\hat{A} \tag{9•1–5a,b}$$

form a Hermitian conjugate pair.

Exercise: Show that, for non-Hermitian $\hat{a}$ and $\hat{b}$,

$$(\hat{a}\hat{b})^{\dagger} = \hat{b}^{\dagger}\hat{a}^{\dagger}, \tag{9•1–6}$$

in generalization of (9•1–5).

Probably the most common occurrence of Hermitian conjugate pairs in quantum-mechanical calculations involves linear superpositions of two Hermitian operators $\hat{A}$ and $\hat{B}$, in the form

$$\hat{a} = \hat{A} + i\hat{B}, \qquad \hat{a}^{\dagger} = \hat{A} - i\hat{B}. \tag{9•1–7a,b}$$

They evidently satisfy the definition (9•1–1) of a Hermitian conjugate pair.

As a good example of such a pair, consider the differential recursion relations (2•3–21) and (2•3–23) between adjacent energy eigenfunctions of the harmonic oscillator, introduced in our discussion of the harmonic oscillator in chapter 2. Using Dirac notation, these relations may be written as

$$\hat{a}^{+}|n\rangle = C_n^{+}|n + 1\rangle \quad \text{and} \quad \hat{a}^{-}|n\rangle = C_n^{-}|n + 1\rangle, \tag{9•1–8a,b}$$

where, from (2•3–16) and (2•3–22),

$$\hat{a}^{+} \equiv \frac{1}{\sqrt{2}}\left(Q - \frac{\partial}{\partial Q}\right) \quad \text{and} \quad \hat{a}^{-} \equiv \frac{1}{\sqrt{2}}\left(Q + \frac{\partial}{\partial Q}\right), \tag{9•1–9a,b}$$

are what we called the **raising** and **lowering operators** for the harmonic oscillator, often collectively referred to as **stepping operators**. The C's are normalization factors not of interest to us at this point.

We may bring (9•1–9) into the form (9•1–7) by defining

$$\hat{P} \equiv -i\frac{\partial}{\partial Q}, \tag{9•1–10}$$

which is basically a dimensionless form of the momentum operator, differing from the conventional momentum operator only by a scale factor. This operator is clearly Hermitian, and with its help, (9•1–9a,b) may be re-written as

$$\hat{a}^{+} \equiv \frac{1}{\sqrt{2}}(\hat{Q} - i\hat{P}) \quad \text{and} \quad \hat{a}^{-} \equiv \frac{1}{\sqrt{2}}(\hat{Q} + i\hat{P}), \tag{9•1–11a,b}$$

which are of the form (9•1–7a,b), showing that $\hat{a}^{+}$ and $\hat{a}^{-}$ do indeed form a Hermitian conjugate pair.

We saw earlier that Hermitian operators represent *real* observable quantities. Evidently, Hermitian conjugate pairs of operators may represent complex conjugate pairs of quantities whose real and imaginary parts are observables:

$$a = A + iB, \qquad a^{*} = A - iB. \tag{9•1–12}$$

Note, however, that the *operator* $\hat{a}^{\dagger}$ need not be the complex conjugate of the *operator* $\hat{a}$, unless both $\hat{A}$ and $\hat{B}$ happen to be real in whatever representation is chosen. The operator pair in (9•1–9a,b) is a good example of a pair of Hermitian conjugate operators that are not complex conjugates of each other; hence the designation $\hat{a}^{\dagger}$ rather than $\hat{a}^{*}$.

Hermitian Square

Given a Hermitian conjugate pair of operators, an important operator that can be constructed from that pair is their product

$$\hat{N} = \hat{a}^{\dagger}\hat{a}. \tag{9•1–13}$$

The similarity of (9•1–7) to (9•1–12) suggests that this operator corresponds, in some sense, to the absolute square $|a|^2 = a^*a$ of the complex observable superposition $a = A + iB$ of two observables A and B. We therefore call $\hat N$ the **Hermitian square** of the operators $\hat a$ and $\hat a^\dagger$.

Note that $\hat N$ is itself Hermitian:

$$\langle\Phi|\hat N\Psi\rangle = \langle\Phi|\hat a^\dagger\hat a\Psi\rangle = \langle\hat a\Phi|\hat a\Psi\rangle = \langle\hat a^\dagger\hat a\Phi|\Psi\rangle = \langle\hat N\Phi|\Psi\rangle. \qquad (9\bullet1\text{–}14)$$

Hence, all expectation values (and any eigenvalues of $\hat N$) must be real. The important point is now that they cannot be negative. To see this, we set $\Phi = \Psi$ in (9•1–14) to obtain

$$\boxed{\langle N\rangle \equiv \langle\Psi|\hat a^\dagger\hat a\Psi\rangle = \langle\hat a\Psi|\hat a\Psi\rangle \geq 0,} \qquad (9\bullet1\text{–}15)$$

where the inequality at the end follows because the preceding form is simply the norm of $\Phi = \hat a\Psi$, which cannot be negative, and which can be zero only if

$$\hat a\Psi = 0. \qquad (9\bullet1\text{–}16)$$

It is in the sense of (9•1–15) that $\hat N = \hat a^\dagger\hat a$ is equivalent to the absolute square of an ordinary complex variable.

The property (9•1–15) of a Hermitian square is a powerful tool in the study of the eigenvalues of many operators. As an example, consider again the operators $\hat a^+$ and $\hat a^-$ for the harmonic oscillator, as written in (9•1–11a,b). We find easily that

$$\hat a + \hat a^- = \frac{1}{2}(\hat Q - i\hat P)(\hat Q + i\hat P) = \frac{1}{2}\{\hat Q^2 + \hat P^2 + i[\hat Q, \hat P]\}. \qquad (9\bullet1\text{–}17)$$

From the definition (9•1–10) of the operators $\hat Q$ and $\hat P$, the commutator contained in (9•1–17) has the value

$$[Q, P] = i, \qquad (9\bullet1\text{–}18)$$

which differs from the commutation relation (7•3–4) for the ordinary (dimensioned) momentum and position operators only by the absence of the factor $\hbar$. When (9•1–18) is inserted into (9•1–17), we obtain

$$\hat a^+\hat a^- = \frac{1}{2}[\hat Q^2 + \hat P^2 - 1]. \qquad (9\bullet1\text{–}19)$$

Inasmuch as the left-hand side of this equality is a Hermitian square, it cannot have negative expectation values. The same must then be true for the right-hand side, which means that the combination $\hat P^2 + \hat Q^2$ cannot have an expectation value less than $+1$. This lower limit is nothing other than the zero-point energy of the harmonic oscillator: A look at the Schroedinger equation (2•3–13) for the harmonic oscillator in the Q-representation shows that the second-derivative term is simply $\hat{P}^2$, and the Hamiltonian of the harmonic oscillator may be written

$$\hat{H} = \frac{\hbar\omega}{2}(\hat{Q}^2 + \hat{P}^2) = \hbar\omega\left(\hat{a}^+\hat{a}^- + \frac{1}{2}\right). \tag{9•1–20}$$

Clearly, there cannot be any energy expectation values below $\hbar\omega/2$. Furthermore, this lower limit itself is allowed: The ground-state eigenfunction does indeed satisfy the condition (9•1–16) under which the Hermitian square $\hat{a}^\dagger\hat{a}$ may have a zero eigenvalue.

None of these results are new to us, but the ease with which they were obtained here from the operator algebra alone, without having to determine the eigenfunctions themselves, demonstrates the power of such operator manipulations. We will draw extensively on such manipulations in the future.

◆ PROBLEM TO SECTION 9.1

#9•1-1: General Stepping Operators

Let $\hat{A}$ be a Hermitian operator, and assume that a Hermitian conjugate operator pair $\hat{S}^+$ and $\hat{S}^-$ can be found, such that the commutator of $\hat{A}$ with each of the operators $\hat{S}^+$ and $\hat{S}^-$ is simply proportional to the latter operator itself, with a real proportionality factor $\pm\alpha$:

$$\boxed{[\hat{A}, \hat{S}^+] = +\alpha\hat{S}^+, \qquad [\hat{A}, \hat{S}^-] = -\alpha\hat{S}^-.} \tag{9•1–21a,b}$$

Let now $|A\rangle$ be an eigenstate of the operator $\hat{A}$, with the eigenvalue A. Show that in this case, $\hat{S}^+|A\rangle$ must either be zero or be another eigenstate of $\hat{A}$, with the eigenvalue $A + \alpha$. Similarly, $S^-|A\rangle$ must either be zero or be another eigenstate of $\hat{A}$, with the eigenvalue $A - \alpha$. The different cases may be combined by writing

$$\hat{A}[\hat{S}^\pm|A\rangle] = C^\pm|A \pm \alpha\rangle, \tag{9•1–22}$$

where $C^\pm$ may be zero.

Show that the stepping operators $\hat{a}^+$ and $\hat{a}^-$ for the harmonic oscillator are of this kind, with the commutation relations

$$[\hat{H}, \hat{a}^+] = +\hbar\omega\hat{a}^+, \qquad [\hat{H}, \hat{a}^-] = -\hbar\omega\hat{a}^-. \tag{9•1–23a,b}$$

9.2 GENERAL UNCERTAINTY RELATIONS

9.2.1 Uncertainty Products and Commutators

In chapter 4, we showed that Gaussian wave packets obey the position-momentum uncertainty relation (4•3–1),

$$\Delta p\,\Delta x \geq \hbar/2, \tag{9•2–1}$$

and we claimed—without proof—that this relation holds generally. In the language developed subsequently in chapter 7, position and momentum cannot simultaneously have sharp values, because the position and momentum operators do not commute. This property—called *complementarity*—is shared by many other pairs of observables, and we must expect that some sort of uncertainty relation holds for all non-commuting pairs of observables. Presumably, these more general uncertainty relations somehow involve the commutator of the corresponding operators. This is indeed the case, and the formalism of Hermitian conjugate operator pairs, especially the properties of Hermitian squares of operators, now permit us to express that dependence quantitatively.

We prove here the following theorem:

> Let ΔA and ΔB be the standard deviations of the two observables A and B from their expectation values:
>
> $$(\Delta A)^2 = \langle (A - \langle A \rangle)^2 \rangle, \quad (\Delta B)^2 = \langle (B - \langle B \rangle)^2 \rangle. \tag{9•2–2a,b}$$
>
> Then the following inequality holds:
>
> $$\boxed{(\Delta A)(\Delta B) \geq \tfrac{1}{2} |\langle [\hat{A}, \hat{B}] \rangle |.} \tag{9•2–3}$$
>
> Here $\langle [\hat{A}, \hat{B}] \rangle$ stands for the (imaginary) expectation value of the (anti-Hermitian) operator $[\hat{A}, \hat{B}]$.

The proof of this theorem relies on constructing, from the operators $\hat{A}$ and $\hat{B}$, a certain pair of Hermitian conjugate operators in such a way that the expectation value of the Hermitian square of these operators—which cannot be negative—involves the uncertainty product $(\Delta A)(\Delta B)$.

As a preliminary, consider first the Hermitian operators

$$\hat{A}' = \hat{A} - \alpha, \qquad \hat{B}' = \hat{B} - \beta, \tag{9•2–4}$$

where α and β are two real constants, the values of which we leave open for now, but which will *eventually* be set equal to the expectation values of $\hat{A}$ and $\hat{B}$. By elementary manipulation, we find that

$$[\hat{A}', \hat{B}'] = [\hat{A}, \hat{B}]. \tag{9•2–5}$$

From $\hat{A}$ and $\hat{B}$, we now construct the Hermitian conjugate operator pair

$$\hat{a} = \lambda \hat{A}' + i\hat{B}/\lambda, \qquad \hat{a}^\dagger = \lambda \hat{A}' - i\hat{B}/\lambda, \tag{9•2–6a,b}$$

where λ is a real parameter for which a specific value will be selected later. The product $\hat{a}^\dagger \hat{a}$ is a Hermitian operator whose expectation values cannot be negative:

$$\langle \Psi | \hat{a}^\dagger \hat{a} | \Psi \rangle = \lambda^2 \langle \hat{A}'^2 \rangle + \langle \hat{B}'^2 \rangle / \lambda^2 + i \langle [\hat{A}, \hat{B}] \rangle \geq 0. \tag{9•2–7}$$

The equals sign can hold only if the state function Ψ satisfies

$$\hat{a}\Psi = (\lambda \hat{A}' + i\hat{B}'/\lambda)\Psi = 0. \tag{9•2–8}$$

All terms in (9•2–7) are real, including the commutator term. (The commutator itself is imaginary.) Without loss in generality, we may assume that the commutator term is negative (or zero); if not, it can always be made negative by interchanging $\hat{A}$ and $\hat{B}$ in (9•2–6a,b). With this ordering, (9•2–7) can be written

$$\lambda^2\langle \hat{A}'^2\rangle + \langle \hat{B}'^2\rangle/\lambda^2 \geq |\langle[\hat{A}, \hat{B}]\rangle|. \tag{9•2–9}$$

This inequality holds regardless of the values of α, β, and λ. In particular, it holds if we set

$$\alpha = \langle A\rangle, \qquad \beta = \langle B\rangle, \tag{9•2–10a,b}$$

so that the expectation values of $\hat{A}'^2$ and $\hat{B}'^2$ turn into the variances of A and B:

$$\langle \hat{A}'^2\rangle = (\Delta A)^2, \qquad \langle \hat{B}'^2\rangle = (\Delta B)^2. \tag{9•2–11a,b}$$

With this choice, (9•2–9) becomes

$$\lambda^2(\Delta A)^2 + (\Delta B)^2/\lambda^2 \geq |\langle[\hat{A}, \hat{B}]\rangle|. \tag{9•2–12}$$

The left-hand side, viewed as a function of λ^2, has a minimum at

$$\lambda^2 = \Delta B/\Delta A, \tag{9•2–13}$$

and the inequality must remain valid at that minimum. Inserting (9•2–13) into (9•2–12) leads to (9•2–3), completing the proof.

With the choice (9•2–12), the minimum-uncertainty condition (9•2–8) assumes the form

$$[(\Delta B)\hat{A}' + i(\Delta A)\hat{B}']\Psi = 0. \tag{9•2–14}$$

Example: Position-Momentum Uncertainty Product. If we apply (9•2–3) to the position-momentum pair

$$\hat{A} = \hat{x}, \qquad \hat{B} = \hat{p}, \qquad [\hat{x}, \hat{p}] = i\hbar, \tag{9•2–15a,b,c}$$

we obtain

$$(\Delta x)(\Delta p) \geq \hbar/2, \tag{9•2–16}$$

regardless of the shape of the wave function, as claimed in chapter 4.

With (9•2–15), the minimum-uncertainty condition (9•2–14) can be written

$$\frac{\partial\Psi}{\partial x} + \left(\frac{x - x_0}{2a^2} - ik_0\right)\Psi = 0, \tag{9•2–17}$$

where we have defined

$$\hbar k_0 = \langle p\rangle, \qquad x_0 = \langle x\rangle, \tag{9•2–18a,b}$$

and

$$2a^2 = \Delta x/\Delta k. \tag{9•2–19}$$

The general solution of (9•2–17) is a Gaussian of the form

$$\Psi(x) = c_0 \cdot \exp\left[-\left(\frac{x - x_0}{2a}\right)^2 + ik_0 x\right], \tag{9•2–20}$$

with a phase linear in x, essentially the same as the Gaussian wave packet in (4•2–14), except for the displacement x_0, and except for writing a instead of Δx for the RMS width of the packet. Setting $a = \Delta x$ in (9•2–19) yields $\Delta k = 1/(2\Delta x)$, thus recovering the result (4•2–9) derived earlier for Gaussian wave packets of the form (9•2–10). Hence, a Gaussian wave packet with a linear phase is a minimum-uncertainty wave packet, as we claimed in chapter 4.

The restriction to a phase *linear* in x is important. We saw in chapter 4 that a Gaussian wave packet will spread with time and that its uncertainty product will increase beyond the minimum value. This increase can be traced to the development of phase terms in the wave function that are quadratic in x, which are readily seen in (4•2–19).

We conclude our discussion with an important warning. The derivation of (9•2–3) made a tacit assumption about the state function Ψ for which the standard deviations ΔA and ΔB are calculated: Not only must Ψ belong to the domain of the operators $\hat{A}$ and $\hat{B}$, but $\hat{A}\Psi$ must belong to the domain of $\hat{B}$ and $\hat{B}\Psi$ to that of $\hat{A}$. If this condition is not satisfied, (9•2–3) need not be correct. The best known example of such a failure of (9•2–3) occurs for the uncertainty product $\Delta\phi\Delta p_\phi$ for the angular position and angular momentum in polar coordinates. The operator corresponding to the ϕ-component of the momentum is

$$\hat{p}_\phi = -i\hbar\, \partial/\partial\phi. \tag{9•2–21}$$

Its commutator with $\hat{\phi}$ clearly has the same value as $[\hat{x}, \hat{p}_x]$:

$$[\hat{\phi}, \hat{p}_\phi] = i\hbar. \tag{9•2–22}$$

Consider now the simple wave function

$$\Phi(\phi) = Ce^{i\phi}. \tag{9•2–23}$$

It is an eigenfunction of $\hat{p}_\phi$ and hence must have zero dispersion: $\Delta p_\phi = 0$. But $\Delta\phi$ is inherently finite, implying that $\Delta\phi\Delta p_\phi = 0$ for the function (9•2–23), in contradiction to (9•2–3). What happens here is that $\Delta\phi$ is finite only if we restrict the range of ϕ to one cycle, say, from 0 to 2π. But in this case, the product $\phi\Psi(\phi)$ does not belong to the hermiticity domain of $\hat{p}_\phi$; that is, the expectation value of $\hat{p}_\phi$ for $\phi\Psi(\phi)$,

$$\int_{-\infty}^{+\infty} [\phi\Psi(\phi)]^* \hat{p}_\phi [\phi\Psi(\phi)]\, d\phi, \tag{9•2–24}$$

contains an imaginary part. The reader is asked to work out the details.

9.2.2 Energy-Time Uncertainty Relation

In many physical problems, one encounters the following question: At which instant of time t will a certain event take place? If this question is to have an operational meaning, the "event" must be expressible in terms of the numerical value(s) of one or more observables. We may therefore rephrase the question: At which instant of time t will the numerical value of a certain observable A pass through a specified value A_0? But if the observable is time-dependent, then the state $|\Psi\rangle$ of the system cannot be an eigenstate of the operator $\hat{A}$ corresponding to A (unless $\hat{A}$ is itself explicitly time-dependent), and hence the observable A cannot have a sharp value, but only a probability distribution. Consequently, the time at which $A = A_0$ will itself be uncertain. Although we may associate a time t_0 with the instant at which the *expectation value* of A passes through A, the actual values will fluctuate both above and below A_0 for some time Δt before and after t_0, as illustrated in **Fig. 9•2–1**.

The most natural quantitative measure of the uncertainty in the time at which A passes through A_0 is the quantity

$$\Delta t = \frac{\Delta A}{|d\langle A\rangle/dt|}, \tag{9•2–25}$$

which is the time required for the expectation value of A to change by the uncertainty in A. We show here that the uncertainty thus defined satisfies the energy-time uncertainty relation

$$\boxed{\Delta\mathcal{E}\Delta t \geq \hbar/2,} \tag{9•2–26}$$

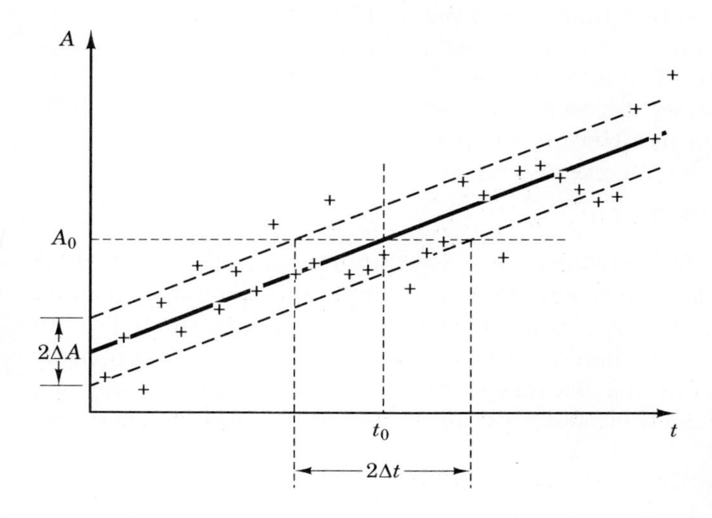

Figure 9•2–1. Meaning of energy-time uncertainty relation.

where $\Delta\mathcal{E}$ is the energy uncertainty of the system, defined quantitatively as the standard deviation of $\mathcal{E}$.

To obtain (9•2–26), we draw on the time evolution law (7•6–4) for $d\langle A\rangle/dt$, which may be written

$$\left|\frac{d\langle A\rangle}{dt}\right| = \frac{1}{\hbar}|\langle[\hat{H}, \hat{A}]\rangle|. \tag{9•2–27}$$

We next draw on the result (9•2–3) according to which the uncertainty product $(\Delta A)(\Delta B)$ of the non-commuting operators $\hat{A}$ and $\hat{B}$ is related to their commutator. This result remains valid when one of the two operators is the Hamilton operator. If we replace $\hat{B}$ by $\hat{H}$ and ΔB by $\Delta\mathcal{E}$, and insert the result into (9•2–27), we obtain

$$\left|\frac{d\langle A\rangle}{dt}\right| \leq \frac{2}{\hbar}(\Delta A)(\Delta\mathcal{E}). \tag{9•2–28}$$

Insertion of (9•2–28) into (9•2–25) yields (9•2–26). Note that the quantity A cancels out: The final result holds for all observables.

9.3 LEFT-HANDED OPERATION OF NON-HERMITIAN OPERATORS

We have so far treated all operators occurring in Dirac brackets as operating on the ket vector to their right, although we pointed out in section 7.4 that Hermitian operators may be viewed as operating in either direction.

This preference for viewing operators placed between a bra and a ket as operating to their right arises from the convention of denoting the state of a system by its ket vector rather than its bra vector. However, the formal manipulations of Dirac brackets is often simplified by being able to switch between bra and ket descriptions. The introduction of the concept of Hermitian conjugate operator pairs makes it possible to define what is meant by the **left-handed operation** of a non-Hermitian operator.

To this end, consider the expression

$$\langle\Psi|\hat{a}\,\Phi\rangle = \langle\hat{a}^{\dagger}\Psi|\Phi\rangle, \tag{9•3–1}$$

which differs from (9•1–3b) only by an irrelevant interchange of Φ and Ψ, done for later convenience. Suppose now that we view the operator $\hat{a}$ on the left-hand side of (9•3–1) not as operating on the ket $|\Phi\rangle$ to its right, but on the bra $\langle\Psi|$ to its left. The equality (9•3–1) then constitutes a definition of what is meant by such a "left-handed" operation. Because (9•3–1) must be true for any ket $|\Phi\rangle$, we may omit the ket altogether, and re-write (9•3–1) as a relation between the bras, namely,

$$\boxed{\langle\Psi|\hat{a} \equiv \langle\hat{a}^{\dagger}\Psi|.} \tag{9•3–2a}$$

Similarly,

$$\boxed{\langle\Psi|\hat{a}^\dagger \equiv \langle\hat{a}\Psi|.} \tag{9•3–2b}$$

Suppose next that two states $|\chi\rangle$ and $|\Psi\rangle$ are related to one another by the operator relation

$$|\chi\rangle = \hat{a}|\Psi\rangle. \tag{9•3–3}$$

The following question then often arises: What is the corresponding relation between the bra vectors $\langle\chi|$ and $\langle\Psi|$ that are the duals to $|\chi\rangle$ and $|\Psi\rangle$? To answer it, we multiply both sides of (9•3–3) by an arbitrary (but admissible) bra vector $\langle\Phi|$ and take the complex conjugate:

$$\langle\Phi|\chi\rangle^* = \langle\Phi|\hat{a}\Psi\rangle^*. \tag{9•3–4a}$$

Because of the basic interchange properties of Dirac brackets and Hermitian conjugate operator pairs—see (7•2–3) and (9•1–3)—this is equivalent to

$$\langle\chi|\Phi\rangle = \langle\hat{a}\Psi|\Phi\rangle = \langle\Psi|\hat{a}^\dagger|\Phi\rangle. \tag{9•3–4b}$$

Because this must be true for any admissible $|\Phi\rangle$, we may again omit $|\Phi\rangle$ altogether and write (9•3–4b) as a relation between bras, yielding the equivalence rule

$$\boxed{|\chi\rangle = \hat{a}|\Psi\rangle \Leftrightarrow \langle\chi| = \langle\Psi|\hat{a}^\dagger.} \tag{9•3–5a}$$

By interchanging $\hat{a}$ and $\hat{a}^\dagger$, we obtain, similarly,

$$\boxed{|\chi\rangle = \hat{a}^\dagger|\Psi\rangle \Leftrightarrow \langle\chi| = \langle\Psi|\hat{a}.} \tag{9•3–5b}$$

The use of such left-handed relations can often simplify the evaluation of Dirac brackets. We draw once again on the harmonic oscillator for an example. The two recursion relations (9•1–8a,b) imply that

$$\langle n|\hat{a}^- = (C_n^+)^*\langle n+1| \quad \text{and} \quad \langle n|\hat{a}^+ = (C_n^-)^*\langle n-1|. \tag{9•3–6a,b}$$

Suppose we wish to determine the values of the normalization factors C_n^+ and C_n^- without relying on the tedious way they were obtained in chapter 2. If we take the inner product of the bra (9•3–6b) with the ket (9•1–8b), we obtain the relation

$$\langle n|\hat{a}^+\hat{a}^-|n\rangle = |C_n^-|^2\langle n-1|n-1\rangle = |C_n^-|^2, \tag{9•3–7}$$

where, in the second equality, we have drawn on the normalization of the energy eigenstates. But the leftmost expression is directly related to the known energy eigenvalues of the harmonic oscillator: From (9•1–20), it follows that

$$\hat{a}^+\hat{a}^- = \hat{H}/\hbar\omega - \tfrac{1}{2}, \tag{9•3–8),}$$

and with the energy eigenvalues $\mathcal{E}_n = \hbar\omega(n + 1/2)$, we see immediately that

$$\langle n|\hat{a}^+\hat{a}^-|n\rangle = n. \tag{9•3–9}$$

The normalization coefficients C_n^+ and C_n^- may always be chosen to be real and non-negative. Insertion of (9•3–9) into (9•3–7) then yields

$$C_n^- = \sqrt{n}, \tag{9•3–10a}$$

in agreement with the result (2•3–21) from chapter 2. It is left to the reader to show that, similarly,

$$C_n^+ = \sqrt{n+1}, \tag{9•3–10b}$$

in agreement with (2•3–19).

The simplicity of this procedure should speak for itself. We shall have repeated opportunity to apply similar procedures in later chapters.

◆ PROBLEM TO SECTION 9.3

#9•3-1: Uncertainty Product for the Hydrogen Atom

Calculate the uncertainty product $\Delta x \cdot \Delta p_x$ involving the x-components of position and momentum, for the ground state of the hydrogen atom.

Chapter 10

HARMONIC OSCILLATOR: FULL OPERATOR TREATMENT

10.1 STEPPING OPERATORS

10.1.1 Review

In chapter 2, we gave an introductory treatment of the harmonic oscillator that was as elementary as possible. In the present chapter, we resume our discussion of this important system, but now drawing on the full operator formalism developed in chapters 7 through 9. Our objective in doing so goes beyond an interest in the results per se: The treatment also serves to demonstrate the power of the operator formalism. We shall see that it is not necessary to draw on the Gauss-Hermite wave functions at all: The entire treatment may be formulated in terms of the algebraic properties of the stepping operators $\hat{a}^+$ and $\hat{a}^-$ originally defined in chapter 2—see (2•3–16) and (2•3–22)—and discussed again briefly in chapter 9.

The basic stepping properties of these operators were contained in the recursion relations (2•3–21) and (2•3–23) connecting adjacent energy eigenfunctions of the harmonic oscillator. In chapter 9, we expressed these relations in the form (9•4–6a,b) and (9•4–10a,b),

$$\hat{a}^+|n\rangle = \sqrt{n+1}\,|n+1\rangle \quad \text{and} \quad \hat{a}^-|n\rangle = \sqrt{n}\,|n-1\rangle, \tag{10•1–1a,b}$$

using Dirac notation. We will also need the Hermitian conjugates of these two relations:

$$\langle n|\hat{a}^- = \sqrt{n+1}\,\langle n+1| \quad \text{and} \quad \langle n|\hat{a}^+ = \sqrt{n}\,\langle n-1|. \tag{10•1–2a,b}$$

Furthermore, we will draw extensively on the commutation relation of these operators. From the definition (9•1–11a,b) of the stepping operators in terms of the dimensionless position and momentum operators $\hat{Q}$ and $\hat{P}$, we obtain

$$\hat{a}^+ \equiv \frac{1}{\sqrt{2}}(\hat{Q} - i\hat{P}) \quad \text{and} \quad \hat{a}^- \equiv \frac{1}{\sqrt{2}}(\hat{Q} + i\hat{P}), \tag{10•1–3a,b}$$

and from the commutation relation (9•1–18) for $\hat{Q}$ and $\hat{P}$, we find easily that

$$[\hat{a}^+, \hat{a}^-] = -1. \tag{10•1–4}$$

Finally, we will need the Hamiltonian of the harmonic oscillator, expressed in terms of $\hat{a}^+$ and $\hat{a}^-$, as in (9•1–20):

$$\hat{H} = \hbar\omega(\hat{a}^+\hat{a}^- + \tfrac{1}{2}). \tag{10•1–5}$$

10.1.2 Simple Matrix Elements

In quantum-mechanical calculations, one frequently needs to evaluate Dirac brackets of the form $A_{mn} \equiv \langle m|\hat{A}|n\rangle$, where $\hat{A}$ is an operator (not necessarily Hermitian) and where $|m\rangle$ and $|n\rangle$ are any *two* state vectors, representing either the same state or two different states. The set of all A_{mn} for a given operator $\hat{A}$ may be written as a matrix; hence, these quantities are commonly called **matrix elements**. Much of the formalism in the remaining chapters of this text will work extensively with such matrix elements.

In the case of the harmonic oscillator, many matrix elements of interest are easily evaluated by expressing the operator $\hat{A}$ in terms of the stepping operators $\hat{a}^+$ and $\hat{a}^-$ and then using the stepping relations (10•1–1) and (10•1–2). The case $\hat{A} = \hat{Q}$, of interest to us anyway, illustrates the method.

From the definitions (10•1–3a,b) of $\hat{a}^+$ and $\hat{a}^-$, we have

$$\hat{Q} = \frac{1}{\sqrt{2}}(\hat{a}^+ + \hat{a}^-). \tag{10•1–6}$$

Hence, with the help of (10•1–1a,b), we obtain

$$\langle m|\hat{Q}|n\rangle = \frac{1}{\sqrt{2}}(\langle m|\hat{a}^+|n\rangle + \langle m|\hat{a}^-|n\rangle)$$

$$= \frac{1}{\sqrt{2}}(\sqrt{n+1}\cdot\langle m|n+1\rangle + \sqrt{n}\cdot\langle m|n-1\rangle). \tag{10•1–7}$$

Because of the orthogonality of states belonging to different energy eigenvalues,

$$\langle m|n\rangle = \delta_{m,n}, \tag{10•1–8}$$

we obtain

$$\langle m|\hat{Q}|n\rangle = \begin{Bmatrix} \sqrt{m/2} & \text{if } m = n+1, \\ \sqrt{n/2} & \text{if } n = m+1, \\ 0 & \text{otherwise.} \end{Bmatrix} \tag{10•1–9}$$

Evidently, only adjacent states have a non-vanishing matrix element of $\hat{Q}$. If the states *are* adjacent, the matrix element is proportional to the square root of whichever is the *larger* of the two quantum numbers m and n.

By an analogous procedure, we readily find the following relations:

$$\hat{P} = \frac{1}{\sqrt{2}}(\hat{a}^+ - \hat{a}^-); \tag{10•1–10}$$

$$\langle m|\hat{P}|n\rangle = \begin{Bmatrix} +i\sqrt{m/2} & \text{if } m = n+1, \\ -i\sqrt{n/2} & \text{if } n = m+1, \\ 0 & \text{otherwise.} \end{Bmatrix} \tag{10•1–11}$$

This treatment of the matrix elements of $\hat{P}$ and $\hat{Q}$ serves as a model for more complicated cases. Consider, for example, the operator $\hat{Q}^2$. From (10•1–6), we obtain

$$\hat{Q}^2 = \tfrac{1}{2}[(\hat{a}^+)^2 + \hat{a}^+\hat{a}^- + \hat{a}^-\hat{a}^+ + (\hat{a}^-)^2]. \tag{10•1–12}$$

The operator products $\hat{a}^+\hat{a}^-$ and $\hat{a}^-\hat{a}^+$ are not the same. Rearranging the operator sequence in $\hat{a}^-\hat{a}^+$ via the commutation relation (10•1–4) yields

$$\hat{a}^-\hat{a}^+ = \hat{a}^+\hat{a}^- + 1, \tag{10•1–13}$$

and we obtain

$$\hat{Q}^2 = \tfrac{1}{2}[(\hat{a}^+)^2 + (2\hat{a}^+\hat{a}^- + 1) + (\hat{a}^-)^2]. \tag{10•1–14}$$

From this, with the help of the stepping relations (10•1–1a,b) and of orthonormality, we find easily that

$$\langle m|\hat{Q}^2|n\rangle = \begin{Bmatrix} \tfrac{1}{2}\sqrt{m(m-1)} & \text{if} \quad m = n+2, \\ \tfrac{1}{2}(2n+1) & \text{if} \quad m = n, \\ \tfrac{1}{2}\sqrt{n(n-1)} & \text{if} \quad n = m+2, \\ 0 & \text{otherwise.} \end{Bmatrix} \tag{10•1–15}$$

Exercise: Show that, similarly to (10•1–15),

$$\langle m|\hat{P}^2|n\rangle = \begin{Bmatrix} -\tfrac{1}{2}\sqrt{m(m-1)} & \text{if} \quad m = n+2, \\ \tfrac{1}{2}(2n+1) & \text{if} \quad m = n, \\ -\tfrac{1}{2}\sqrt{n(n-1)} & \text{if} \quad n = m+2, \\ 0 & \text{otherwise.} \end{Bmatrix} \tag{10•1–16}$$

Show that the uncertainty product for the energy eigenstates of the harmonic oscillator is

$$\Delta x \cdot \Delta p = (n + \tfrac{1}{2}) \cdot \hbar. \tag{10•1–17}$$

None of the results obtained here depended for their derivation on a knowledge of the energy eigenfunctions, which were indeed nowhere used. The stepping operator treatment is in this respect quite different from the more conventional technique of solving the Schroedinger equation in the position representation, used in chapter 2. Our treatment in the present chapter did not even depend on any specific representation for the operators $\hat{Q}$ and $\hat{P}$. The only relation between $\hat{P}$ and $\hat{Q}$ actually used was their representation-independent commutation relation (9•1–18).

◆ PROBLEMS TO SECTION 10.1

#10•1-1: Stepping Operators in the Spherical Harmonic Oscillator

We saw in section 2.3 that the energy eigenstates of the spherical harmonic oscillator can be obtained by separation of variables in Cartesian coordinates. In Dirac notation, the resulting states may be designated as $|n_x, n_y, n_z\rangle$. The stepping operator formalism may be extended to this three-dimensional system, but it requires *three* pairs of raising and lowering operators, $\hat{a}_x^{\pm}$, $\hat{a}_y^{\pm}$, and $\hat{a}_z^{\pm}$, one pair for each of the three orthogonal oscillation components.

(a) Express both the Hamiltonian $\hat{H}$ and the angular momentum operator $\hat{L}_z$ in terms of the Cartesian stepping operators. Show that $\hat{H}$ and $\hat{L}_z$ commute.

(b) Identify specifically those six states $|n_x, n_y, n_z\rangle$ for which

$$n_x + n_y + n_z = 2. \tag{10•1–18}$$

(c) For each of the states found under (b), determine the result of operating with $\hat{L}_z$ on that state.

(d) Using the results found under (c), construct, by linear superposition of appropriate Cartesian states, six eigenstates $|\Psi\rangle$ of the angular momentum operator $\hat{L}_z$, satisfying

$$\hat{L}_z|\Psi\rangle = m\hbar|\Psi\rangle. \tag{10•1–19}$$

You should find two states each with $m = 0, \pm 1$, and ± 2. Normalize the states.

#10•1-2: *Normal Ordering* for Products of Stepping Operators

The evaluation of the matrix elements of products of stepping operators containing both raising and lowering operators is facilitated by using the commutation relation (10•1–4) to re-arrange the operator ordering in such a way—called **normal ordering**—that all raising operators have been moved to the left and all lowering operators to the right. For example,

$$\begin{aligned}(\hat{a}^+\hat{a}^-)^2 &\equiv \hat{a}^+\hat{a}^-\hat{a}^+\hat{a}^- = \hat{a}^+(\hat{a}^-\hat{a}^+)\hat{a}^- = \hat{a}^+(\hat{a}^+\hat{a}^- + 1)\hat{a}^- \\ &= (\hat{a}^+)^2(\hat{a}^-)^2 + \hat{a}^+\hat{a}^-.\end{aligned} \tag{10•1–20a}$$

(a) Show that, similarly,

$$(\hat{a}^+\hat{a}^-)^3 = \hat{a}^+(\hat{a}^+\hat{a}^- + 1)^2\hat{a}^- = (\hat{a}^+)^3(\hat{a}^-)^3 + 3(\hat{a}^+)^2(\hat{a}^-)^2 + \hat{a}^+\hat{a}^- \tag{10•1–20b}$$

and

$$(\hat{a}^+\hat{a}^-)^4 = (\hat{a}^+)^4(\hat{a}^-)^4 + 6(\hat{a}^+)^3(\hat{a}^-)^3 + 7(\hat{a}^+)^2(\hat{a}^-)^2 + 4\hat{a}^+\hat{a}^-. \tag{10•1–20c}$$

(b) Given normal ordering, matrix elements of the form

$$\langle m|(\hat{a}^+)^M(\hat{a}^-)^N|n\rangle \tag{10•1–21}$$

are easily evaluated by viewing all operators $\hat{a}^-$ as operating to their right, acting as conventional lowering operators, and all operators $\hat{a}^+$ as operating to their left, according to (10•1–2b), in which case they, too, act as lowering operators, rather than as raising operators. Show that:

If $m - M = n - N \geq 0$, then

$$\begin{aligned}&\langle m|(\hat{a}^+)^M(\hat{a}^-)^N|n\rangle \\ &= \underbrace{\sqrt{m(m-1)\cdots[m-(M-1)]}}_{M\text{ factors}}\cdot\underbrace{\sqrt{n(n-1)\cdots[n-(N-1)]}}_{N\text{ factors}};\end{aligned} \tag{10•1–22a}$$

otherwise

$$\langle m|(\hat{a}^+)^M(\hat{a}^-)^N|n\rangle = 0. \tag{10•1–22b}$$

10.2 OSCILLATING STATES

10.2.1 Expectation Values of Superposition States

We saw in chapter 1—see (1•4–5)—that it is possible to construct time-dependent states that satisfy the Schroedinger wave equation by the linear superposition of the stationary energy eigenstates with suitable time-dependent coefficients. Using Dirac notation, we may re-write (1•4–5) as

$$|\Psi\rangle = \sum_n c_n |n\rangle \exp(-i\omega_n t), \tag{10•2–1}$$

where we have further compacted the notation by defining frequencies ω_n via

$$\hbar\omega_n = \mathcal{E}_n. \tag{10•2–2}$$

The $|n\rangle$ are the different eigenstates of the hamiltonian $\hat{H}$ of the system, and the $\mathcal{E}_n$ are the corresponding energy eigenvalues:

$$\hat{H}|n\rangle = \mathcal{E}_n|n\rangle. \tag{10•2–3}$$

The time-dependent phase factors in (10•2–1) ensure that $|\Psi\rangle$ satisfies the Schroedinger wave equation. The sum in (10•2–1) goes over all energy eigenstates.

The coefficients c_n are arbitrary *time-independent* expansion coefficients, subject to only two constraints:

(a) The state $|\Psi\rangle$ must be normalizable, and we request that it is in fact normalized:

$$\langle\Psi|\Psi\rangle = \sum_m \sum_n c_m^* c_n \langle m|n\rangle \exp(i\omega_{mn}t) = \sum_n |c_n|^2 = 1. \tag{10•2–4}$$

In the first equality we have defined the difference frequencies

$$\omega_{mn} \equiv \omega_m - \omega_n, \tag{10•2–5}$$

and in the second equality we have utilized the orthonormality relation (10•1–8).

(b) The expectation value of the energy must be finite. It is related to the coefficients c_n via

$$\langle\mathcal{E}\rangle = \langle\Psi|\hat{H}|\Psi\rangle = \sum_m \sum_n c_m^* c_n \langle m|\hat{H}|n\rangle \cdot \exp(i\omega_{mn}t). \tag{10•2–6}$$

But

$$\langle m|\hat{H}|n\rangle = \mathcal{E}_n\langle m|n\rangle = \mathcal{E}_n\delta_{mn}, \tag{10•2–7}$$

and hence,

$$\langle\mathcal{E}\rangle = \sum_n |c_n|^2\, \mathcal{E}_n. \tag{10•2–8}$$

Because the energies $\mathcal{E}_n$ increase with increasing n, the requirement that $\langle \mathcal{E} \rangle$ remain finite will be a slightly more stringent constraint on the c_n than the requirement that $|\Psi\rangle$ be normalizable.

So far, the formalism is completely general, not specific to the harmonic oscillator. In section 2.3, we had briefly investigated a simple example of such a superposition for the harmonic oscillator, containing just two eigenstates. We now turn to the more general case.

Evidently, by different choices of the c_n, a vast diversity of time-dependent states can be constructed, much richer than in the classical limit. In many problems we will be interested, not in the full wave function, but only in selected expectation values, such as the expectation values $\langle x \rangle$ and $\langle p \rangle$ of position and momentum, and in the variances

$$(\Delta x)^2 = \langle x^2 \rangle - \langle x \rangle^2 \tag{10•2–9a}$$

and

$$(\Delta p)^2 = \langle p^2 \rangle - \langle p \rangle^2. \tag{10•2–9b}$$

The evaluation of the variances calls for the additional expectation values $\langle x^2 \rangle$ and $\langle p^2 \rangle$.

Ehrenfest's theorem of chapter 8 assures us that $\langle x \rangle$ must satisfy the differential equation

$$\frac{d}{dt}\langle p \rangle = M\frac{d^2}{dt^2}\langle x \rangle = \langle F \rangle = -K\langle x \rangle. \tag{10•2–10}$$

This is a differential equation for $\langle x \rangle$, the most general *real* solution of which may be written

$$\langle x \rangle = \tfrac{1}{2}(x_0 e^{i\omega t} + x_0^* e^{-i\omega t}), \tag{10•2–11}$$

where x_0 is an arbitrary complex amplitude and $\omega = (K/M)^{1/2}$, as in (2•3–2). The relations (10•2–10) and (10•2–11) show exactly the same behavior as the position coordinate of a harmonic oscillator in classical mechanics. In particular, none of the higher difference frequencies $\omega_{mn} = (m - n)\omega$ show up. This is a special property of the harmonic oscillator, not shared by most other quantum systems. It arises in Ehrenfest's theorem because in the harmonic oscillator the force on the object varies linearly with position. This means that the expectation value of the force—which occurs in Ehrenfest's theorem—is equal to the *local* force taken at the expectation value of the position.

It is instructive to see how this behavior arises from the energy-level spectrum and the matrix elements. To this end, we write down the complete expression for the expectation value of the position, in the same form as the expectation value (10•2–6) for the energy. It will be convenient to switch to the *dimensionless* position $Q = x/L$, introduced in chapter 2, with $L = (M\omega/\hbar)^{1/2}$ being the natural unit of length for the oscillator, as defined in (2•3–7). In terms of Q,

$$\langle Q \rangle = \langle \Psi | \hat{Q} | \Psi \rangle = \sum_m \sum_n c_m^* c_n \langle m | \hat{Q} | n \rangle \exp(i\omega_{mn} t). \tag{10•2–12}$$

We saw in (10•1–9) that only the matrix elements between adjacent states are non-zero. In this case, the frequencies remaining in (10•2–12) will all be $\pm\omega$, so that

$$\hbar\omega_{mn} = \mathcal{E}_{n\pm1} - \mathcal{E}_n = [(n \pm 1) - n]\hbar\omega = \pm\hbar\omega, \tag{10•2–13}$$

and (10•2–12) reduces to

$$\langle Q \rangle = \sum_n c_{n+1}^* c_n \sqrt{n+1}\, \exp(i\omega t) + \text{cc}, \tag{10•2–14}$$

where "cc" stands for *complex conjugate*. The relation (10•2–14) is evidently of the form (10•2–11), with

$$x_0 = 2L \sum_n c_{n+1}^* c_n \sqrt{n+1}. \tag{10•2–15}$$

Exercise: Suppose we did not already know the energy eigenvalues and the position matrix elements of the harmonic oscillator. Which conclusions about these quantities could be drawn from the Ehrenfest theorem result (10•2–11) alone? Which properties of these quantities remain open?

The result (10•2–15) shows that the relation between the oscillation amplitude x_0 of the expectation value and the expansion coefficients is a complicated one, depending strongly on the *correlation* between adjacent coefficients. For example, the oscillation amplitude will be zero if there are no adjacent pairs of expansion coefficients, as when only even-order or only odd-order coefficients are present. However, a zero amplitude x_0 of the expectation value does not mean that the wave function does not oscillate; it simply means that the oscillations are purely internal "breathing" oscillations, relative to a stationary probability center. In fact, such breathing oscillations, with frequency 2ω, superimposed on the overall oscillation with frequency ω, were clearly visible in our earlier simple example of Fig. 2•3–5, where we showed the probability density oscillations for a simple equal-weight superposition of the two lowest states. The reader is asked to consider, at least qualitatively, what happens for superpositions involving only even-order or only odd-order states.

Clearly, considerations of the expectation value of the position alone do not reveal the whole story. At the very least, we must also look at the variances. If internal density oscillations are present, the variances will in general—but not always[1]—contain an oscillating component revealing the presence of those oscillations.

[1] If all expansion coefficients are separated by gaps two or more states wide, not even the variances will oscillate, even though there will be oscillations in the higher-order moments of the probability distribution.

To calculate more complicated expectation values, we again express all operators in terms of the stepping operators $\hat{a}^+$ and $\hat{a}^-$, as in section 10.1. For example,

$$\langle Q^2 \rangle = \langle \Psi | \hat{Q}^2 | \Psi \rangle = \tfrac{1}{2} \langle \Psi | [(\hat{a}^+)^2 + (2\hat{a}^+\hat{a}^- + 1) + (\hat{a}^-)^2] | \Psi \rangle, \tag{10•2–16}$$

where we have used the form (10•1–14) for the operator $\hat{Q}^2$.

10.2.2 Quasi-Classical Oscillating States

In complicated situations such as the one just discussed, it is often useful to study first those cases that offer the simplest mathematics; they often also have special physical significance. A little reflection shows that, given normal ordering of the stepping operators, all expectation value calculations involve operations of the form $\hat{a}^- | \Psi \rangle$ and/or its Hermitian conjugate $\langle \Psi | \hat{a}^+$. Evidently, the evaluations of such expressions would become vastly simplified *if* the result of applying the lowering operator to $|\Psi\rangle$ were simply proportional to $|\Psi\rangle$, with some suitable complex proportionality constant C:

$$\boxed{\hat{a}^- | \Psi \rangle = C | \Psi.} \tag{10•2–17a}$$

We shall show later that the coefficients c_n in (10•2–1) can always be chosen such that this relation is satisfied. Mathematically, we request that $|\Psi\rangle$ be an eigenstate of the (non-Hermitian) lowering operator $\hat{a}^-$.

Given the form (10•2–17a), each of the operator products in (10•2–16) can be evaluated very easily by viewing each operator $\hat{a}^-$ as operating to its right, using (10•2–17a), and each operator $\hat{a}^+$ as operating to its left, using the Hermitian conjugate of (10•2–17a),

$$\langle \Psi | \hat{a}^+ = C^* \langle \Psi |. \tag{10•2–17b}$$

By working each operator product from the outside inwards, it is easily seen that the entire expectation value may be evaluated by simply replacing $\hat{a}^-$ with C and $\hat{a}^+$ with C^*, leading to simple results, such as

$$\langle Q^2 \rangle = \tfrac{1}{2}[C^{*2} + (2C^*C + 1) + C^2] = \tfrac{1}{2}[(C^* + C)^2 + 1]. \tag{10•2–18}$$

Similarly,

$$\langle Q \rangle = \frac{1}{\sqrt{2}} \langle \Psi \rangle | C^* + C | \Psi \rangle = \frac{1}{\sqrt{2}} (C^* + C). \tag{10•2–19}$$

From (10•2–19) and (10•2–18), we obtain the variance of Q,

$$(\Delta Q)^2 \equiv \langle Q^2 \rangle - \langle Q \rangle^2 = 1/2, \tag{10•2–20a}$$

which is independent of C. In terms of the true position $x = LQ$, (10•2–20a) implies that

$$(\Delta x)^2 = L^2/2, \tag{10•2–21a}$$

which is the same as for the ground state of the harmonic oscillator.

It is left to the reader to show that, similarly, for the dimensionless momentum,

$$(\Delta P)^2 \equiv \langle P^2 \rangle - \langle P \rangle^2 = 1/2, \tag{10•2–20b}$$

which implies that, for the true momentum $p = \hbar P/L$,

$$(\Delta p)^2 = \hbar^2/2L^2. \tag{10•2–21b}$$

Combining (10•2–20a) and (10•2–20b), we obtain the uncertainty product

$$\boxed{(\Delta x)^2(\Delta p)^2 = \hbar^2/4.} \tag{10•2–22}$$

But this is the minimum possible uncertainty product!

We recall from chapter 9 that a wave packet with a minimum uncertainty product is *necessarily* a Gaussian wave packet, just like the harmonic oscillator ground state. Thus, we see that the state we have constructed is one with a Gaussian probability density distribution equal to that of the ground state, but oscillating *rigidly*, without any *internal* breathing oscillations.

The amplitude of the overall oscillation is related to the magnitude of the coefficient C in (10•2–17). By comparing (10•2–19) with (10•2–11), we see that the relation between the coefficient C in (10•2–17a) and the amplitude x_0 is

$$C = \frac{1}{\sqrt{2}} \frac{x_0}{L} e^{i\omega t}. \tag{10•2–23}$$

The time-dependent phase factor in C is the same as the factor preceding the sum in (10•2–18). It arises because the lowering operator $\hat{a}^-$ in (10•2–17a) replaces each energy eigenstate in the superposition (10•2–1) by a state whose energy is lower by the same amount $\hbar\omega$. But this calls for an adjustment of the time factor for every state in (10•2–1) by a common factor $\exp(i\omega t)$ which may then be pulled out of the sum.

Evidently, the states $|\Psi\rangle$ specified by (10•2–17) behave as close to a classical harmonic oscillator as quantum mechanics permits. We therefore call them **quasi-classical** oscillating states.

Energy

The expectation value of the energy for the state $|\Psi\rangle$ follows readily from the Hamiltonian (10•1–5):

$$\langle \mathcal{E} \rangle = \hbar\omega \cdot (\langle \Psi | \hat{a}^+ \hat{a}^- | \Psi \rangle + \tfrac{1}{2}). \tag{10•2–24}$$

The expectation value inside the parentheses on the right-hand side has a very simple meaning: It is the expectation value $\langle n \rangle$ of the *number* of energy quanta $\hbar\omega$ contained in the energy of the oscillator, above the zero-point energy. It is easily evaluated by taking the inner product of (10•2–17a) and (10•2–17b):

$$\langle n \rangle \equiv \langle \Psi | \hat{a}^+ \hat{a}^- | \Psi \rangle = |C|^2 = \frac{|x_0|^2}{2L^2} = \frac{M\omega}{2\hbar} \cdot |x_0|^2. \qquad (10•2–25)$$

In the last two equalities, we have first used (10•2–23) and then replaced L^2 by its definition, $L^2 = \hbar/M\omega$, from (2•3–7). If we insert (10•2–25) into (10•2–24), we obtain

$$\langle \mathcal{E} \rangle = \frac{1}{2} M\omega^2 |x_0|^2 + \frac{1}{2}\hbar\omega. \qquad (10•2–26)$$

The first term on the right-hand side is exactly the total energy of a classical harmonic oscillator with mass M and frequency ω, oscillating with the amplitude $|x_0|$. Hence, except for the zero-point energy, the energy-vs.-amplitude relation for the quasi-classical states is exactly the relation we would have for a classical oscillation.

Expansion Coefficients

We finally determine the expansion coefficients implied by the condition (10•2–17a). It will be useful first to re-express the complex factor C in terms of the expectation value $\langle n \rangle$ of the number of energy quanta $\hbar\omega$. From (10•2–25), and from the fact that C must be time dependent, as specified by (10•2–23), we obtain

$$C = \sqrt{\langle n \rangle}\, e^{i(\omega t + \alpha)}, \qquad (10•2–27)$$

where α is an arbitrary constant that evidently controls the phase of the oscillation. If we replace C in (10•2–18) via (10•2–27) and compare the result with (10•2–1), we see that the condition (10•2–17a) is met if each expansion coefficient is related to the preceding one according to the recursion relation

$$c_{n+1}\sqrt{n+1} = e^{i\alpha} c_n \sqrt{\langle n \rangle}. \qquad (10•2–28)$$

The lowest coefficient, c_0, remains unspecified; it must be chosen such that the normalization condition (10•2–4) for $|\Psi\rangle$ is satisfied.

Toward that end, we transform (10•2–28) into

$$|c_{n+1}|^2 (n+1) = |c_n|^2 \langle n \rangle. \qquad (10•2–29)$$

By repeated application of this recursion relation, we find that

$$|c_n|^2 = \frac{\langle n \rangle^n}{n!} |c_0|^2. \qquad (10•2–30)$$

If this is inserted into the normalization condition (10•2–4), we obtain

$$\sum_{n=0}^{\infty} \frac{\langle n \rangle^n}{n!} |c_0|^2 = |c_0|^2 \cdot \exp(\langle n \rangle) = 1. \tag{10•2–31}$$

This specifies c_0, except for an irrelevant phase factor, and leads to

$$|c_n|^2 = \frac{\langle n \rangle^n}{n!} \cdot \exp(-\langle n \rangle). \tag{10•2–32}$$

If the wave has a finite phase $\alpha \neq 0$, the expansion coefficients themselves have a phase factor. From (10•2–28), it then follows that

$$c_n = |c_n| e^{in\alpha}. \tag{10•2–33}$$

We will return to these expansion coefficients in section 10.4.

◆ PROBLEMS TO SECTION 10.2

#10•2-1: Kinetic and Potential Energy Expectation Values via Stepping Operators

Write the operators for kinetic and potential energy in terms of the stepping operators $\hat{a}^+$ and $\hat{a}^-$. Give closed-form expressions for the expectation values of kinetic and potential energy, for the following states:

(i) The energy eigenstates $|n\rangle$.
(ii) The quasi-classical state $|\Psi\rangle$ defined by (10•2–17).

Express the two energies as fractions of the total energy of the states or of the expectation value of the total energy (whichever is applicable), and as functions of time. Compare the results with those in classical mechanics.

#10•2-2: An Odd-*n*-Only Oscillating State

A certain oscillating state $|\Psi\rangle$ consists of a superposition of the three lowest odd-n energy eigenstates, such that, at $t = 0$,

$$|\Psi\rangle = c_1|1\rangle + c_3|3\rangle + c_5|5\rangle. \tag{10•2–34}$$

(a) What condition must the expansion coefficients satisfy so that the width Δx of the oscillating wave packet, in the quantitative sense of (10•2–9a), does not contain an oscillating term, but is time-independent? Impose the additional requirement that the probability of finding the system in state $|3\rangle$ is 1/3. Give numerical values for the three c's resulting from those two requirements, assuming that all expansion coefficients may be chosen to be real.

(b) Give expressions for Δx and $\langle n \rangle$ for the wave packet.

(c) Generate a series of computer plots showing the spatial probability density distribution $\rho(x)$ at several instants (at least four) during one oscillation period, exhibiting the internal oscillations of the probability distribution. You may plot $\rho(x)$ for $x \geq 0$ only. Why is that sufficient?

(d) Discuss the relevance of all-odd superpositions such as (10•2–34) to the problem of oscillations in a "half-oscillator," that is, a harmonic oscillator with an infinitely high wall at $x = 0$, blocking one side (**Fig. 10•2–1**).

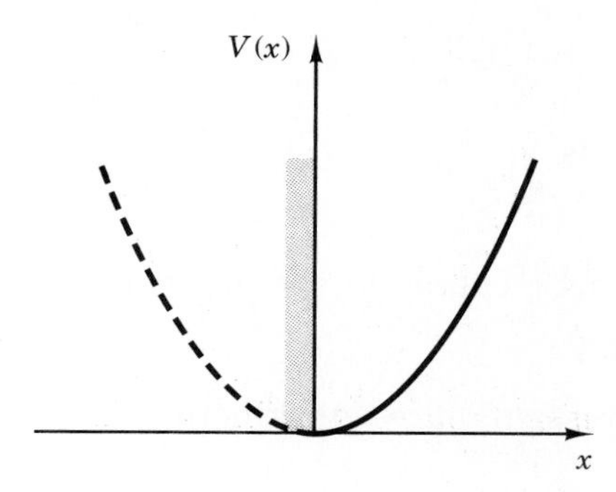

Figure 10•2–1. "Half-oscillator" potential.

#10•2-3: Difference between Kinetic and Potential Energy

In classical dynamics, one is often interested in the time evolution of the difference between the kinetic and potential energy, called the **Lagrangian**, as opposed to their sum, and this interest carries over into some quantum-mechanical problems. The appropriate operator is

$$\hat{L} = \hat{\mathcal{E}}_{\text{kin}} - \hat{V} = \frac{\hat{p}^2}{2M} - \hat{V}(x). \tag{10•2–35}$$

(a) Derive an expression for $\hat{L}$ for the harmonic oscillator, in terms of the raising and lowering operators $\hat{a}^+$ and $\hat{a}^-$.

(b) Use the result found in (a) to derive an expression for the expectation value

$$\langle \hat{L} \rangle \equiv \langle \Psi | \hat{L} | \Psi \rangle \tag{10•2–36}$$

as a function of time, for the quasi-classical oscillating state.

(c) Give a detailed verbal discussion of the result under (b), including specifically a comparison with classical physics.

10.3 ELECTROMAGNETIC HARMONIC OSCILLATORS: THE IDEAL *LC* CIRCUIT

Our treatment of the harmonic oscillator is not restricted to the mechanical case of a particle-like object bound to a point by an elastic restoring force; it can be applied to any physical system whose classical motions are purely harmonic oscillations. This includes, in particular, electromagnetic oscillations. In fact, the energy quanta $\mathcal{E} = \hbar\omega$ of the electromagnetic field—the photons—are simply the energy increments between adjacent stationary states of an electromagnetic harmonic oscillator with frequency ω.

We consider here the simplest electromagnetic harmonic oscillator, an ideal LC circuit (**Fig. 10•3–1**). The equations of motion of this circuit are

$$I = +C\frac{dV}{dt}, \quad V = -L\frac{dI}{dt}, \tag{10•3–1,2}$$

The oscillation frequency is

$$\omega = 1/\sqrt{LC}, \tag{10•3–3}$$

and the total electromagnetic energy in the circuit is

$$U = \tfrac{1}{2}LI^2 + \tfrac{1}{2}CV^2. \tag{10•3–4}$$

The first term in this expression represents kinetic energy, the second potential energy.

These equations are readily transformed into exactly the same form as the equations for the mechanical harmonic oscillator by defining

$$q \equiv CV, \qquad p \equiv IL, \qquad M \equiv L, \quad \text{and} \quad H \equiv U. \tag{10•3–5a–d}$$

In terms of these quantities, the energy, now called H, becomes

$$H = \frac{p^2}{2M} + \frac{q^2}{2C} = \frac{1}{2M}p^2 + \frac{1}{2}M\omega^2 q^2, \tag{10•3–6}$$

which has exactly the same form as the Hamiltonian of a mechanical harmonic oscillator with the position coordinate q, except that q, p, and M have different meanings and different dimensions. The equations of motion (10•3–1) and (10•3–2) also assume the same form as in the mechanical case, and they may in fact be written in universal Hamilton-Jacobi form. In the new variables, they turn into

$$\frac{dq}{dt} = +\frac{p}{M} = +\left(\frac{\partial H}{\partial p}\right)_q, \tag{10•3–7}$$

and

$$\frac{dp}{dt} = -M\omega^2 q = -\left(\frac{\partial H}{\partial q}\right)_p. \tag{10•3–8}$$

Just as in the mechanical case, these equations must somehow be the classical limit of the underlying quantum laws. The identical formal structure

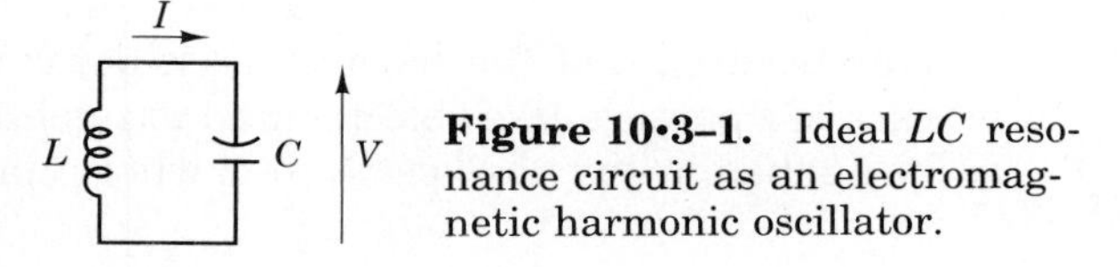

Figure 10•3–1. Ideal LC resonance circuit as an electromagnetic harmonic oscillator.

of the two cases in the classical limit suggests an identical formal structure on the quantum level. We therefore *postulate* the following:

(a) The correct quantum-mechanical description of the *LC* oscillator is in terms of a state function Ψ, or, in Dirac notation, a state vector $|\Psi\rangle$. In a representation in which the charge q is the independent variable, $|\Psi(q,t)|^2\,dq$ is the probability that the oscillator has a charge between q and $q + dq$ on its upper plate at time t.

(b) The time evolution of the state function is governed by an equation of the form

$$i\hbar\frac{\partial}{\partial t}|\Psi\rangle = \hat{H}|\Psi\rangle, \qquad (10\text{•}3\text{–}9)$$

where $\hat{H}$ is the Hermitian operator

$$\hat{H} = \frac{1}{2M}\hat{p}^2 + \frac{1}{2}M\omega^2\hat{q}^2, \qquad (10\text{•}3\text{–}10)$$

obtained from (10•3–6) by replacing the variables p and q with the Hermitian operators $\hat{p}$ and $\hat{q}$.

(c) The operators $\hat{p}$ and $\hat{q}$ form a canonically conjugate pair satisfying the commutation relation

$$[\hat{q}, \hat{p}] = i\hbar. \qquad (10\text{•}3\text{–}11)$$

(d) We obtain the "charge representation" referred to under (a) if we implement (10•3–11) by setting

$$\hat{q} = q, \qquad \hat{p} = -i\hbar\frac{\partial}{\partial q}. \qquad (10\text{•}3\text{–}12\text{a,b})$$

Formally, all of this is completely straightforward. What is new is the application of this previously derived formalism to electromagnetic rather than mechanical oscillations. We see from the formalism that the energy quanta of the electromagnetic field—the photons—arise formally as the equal-energy steps between the energy eigenstates of a harmonic oscillator.

The quantization of the *LC* circuit is the most elementary example of the quantization of an electromagnetic system. It illustrates the central idea of field quantization with a minimum of formal encumbrances. This central idea and its mathematical formalism are readily generalized to arbitrary electromagnetic systems, including specifically electromagnetic waves in free space. Such an extension leads ultimately to a full-blown **quantum electrodynamics**, an advanced topic beyond the scope of the present text.[2]

---

[2] An excellent introduction to this topic, written for readers with backrounds in engineering and applied physics, is found in D. Marcuse, *Engineering Quantum Electrodynamics* (New York: Academic, 1980).

◆ PROBLEM TO SECTION 10.3

#10•3-1: Voltage and Current Uncertainties

Give expressions for the voltage and current uncertainties ΔV and ΔI of the ideal *LC circuit*, as well as for the uncertainty product $\Delta V \Delta I$. Express the results as functions of the frequency ω and of the ratio

$$Z = \sqrt{L/C}, \tag{10•3–13}$$

which has the dimension of an impedance. Give numerical values, assuming that $\omega/2\pi = 1$ GHz and

$$Z = \sqrt{\mu_0/\varepsilon_0} \approx 377\ \Omega, \tag{10•3–14}$$

the characteristic impedance of free space.

10.4 PROBABILITIES, PHOTONS, AND PHASE

10.4.1 Interpretation of the Expansion Coefficients

It is evident that the set $\{c_n\}$ of expansion coefficients in (10•2–1) fully specifies the state $|\Psi\rangle$. The inverse is also true: If we take, on both sides of (10•2–1), the inner product with one of the energy eigenstates $|n\rangle$, orthogonality causes all terms on the right-hand side except the corresponding $|n\rangle$-term to vanish, and we obtain

$$c_n = \langle n \mid \Psi\rangle \cdot \exp(i\omega_n t). \tag{10•4–1}$$

This specifies the expansion coefficients in terms of $|\Psi\rangle$. Hence, the set of expansion coefficients contains exactly the same information as the wave function Ψ; it may therefore serve as an alternative *representation* of the state $|\Psi\rangle$, in exactly the same way in which the position and momentum representations are alternative representations. Because the coefficients c_n refer to the energy eigenfunctions of the harmonic oscillator, we call the new representation the **energy representation** of the harmonic oscillator.

Because the harmonic oscillator energy eigenvalues form a discrete set, the representation based on them is a **discrete** representation, in contrast to the continuous position and momentum representations. There is nothing unusual about this; mathematically, it is similar to the way in which the discrete set of Fourier coefficients of a continuous periodic function forms a discrete representation of that function.

We recall that $\langle\Psi\mid\Psi\rangle$ is a sum over probabilities. We therefore interpret $\Sigma|c_n|^2$ also as a sum over probabilities, analogously to the way we interpreted $\langle\Psi\mid\Psi\rangle$ and $\langle\Phi\mid\Phi\rangle$ as alternative sums over probabilities. Specifically, we postulate that

$$P(n) = |c_n|^2 \tag{10•4–2}$$

is the probability that the harmonic oscillator is "found" in the state $|n\rangle$ with the energy $\mathcal{E}_n$ if an energy measurement is performed. For the specific case of the quasi-classical oscillating state, insertion of the expansion coefficients from (10•2–32) yields

$$P(n) = \frac{\langle n \rangle^n}{n!} \exp(-\langle n \rangle), \tag{10•4–3}$$

which is the Poisson distribution we encountered already in chapter 1 in the context of the photon statistics in a light beam. We will comment on this coincidence presently.

There is one important difference between the probabilities $|c_n|^2$ and the probabilities for position and momentum. Because the states $|n\rangle$ form a discrete rather than a continuous set, the $|c_n|^2$ represent probabilities for sharply defined discrete properties, rather than probability densities within a continuous probability distribution. Implied in our interpretation of the c_n is therefore the assertion that the probability of finding energy values *other* than one of the discrete $\mathcal{E}_n$ is zero.

To illustrate this important point, consider again an electromagnetic oscillator with frequency ω. We saw that such an oscillator is, quantum-mechanically, just another harmonic oscillator: The energy quanta $\hbar\omega$ of the electromagnetic field—the photons—are simply the energy increments between adjacent stationary states of a harmonic oscillator! Applied to an electromagnetic harmonic oscillator, the statement that an energy measurement can lead only to discrete energy values is simply a restatement of the fact that, if the photons contained in a given field are counted, their number is necessarily an integer—there are no fractional photons. It is this non-existence of fractional photons that causes photons to be indivisible objects, in the sense of our discussion in chapter 1.

The new alternative description of photons as the harmonic oscillator quanta of a field has proven immensely fruitful. Modern quantum field theory views *all* of the various indivisible objects of nature as quanta of an appropriate field, with different kinds of objects (e.g., photons, electrons, etc.) belonging to different kinds of fields. Conversely, associated with every field is a set of quanta that act as discrete, indivisible objects, which have, by virtue of being discrete and indivisible, particle-like as well as wave-like properties.

In the light of these comments, consider now the probability distribution (10•4–3). In chapter 1, we discussed the counting of photons contained in a classically specified amount of energy intercepted from a light beam. We concluded that the numbers actually found in different interception experiments vary statistically from experiment to experiment, and we *postulated* that the distribution is that for randomly distributed events—the Poisson distribution. Only the *average* number of photons that materialize is equal to the number corresponding to the classically calculated energy. Our result (10•4–3) now shows that this point of view of randomly distributed individual objects is completely consistent with the point of view that the intercepted light represents a harmonic oscillator in a quasi-classical oscillating state.

A final comment is due concerning the zero-point energy $\mathcal{E}_0 = \hbar\omega/2$ of the harmonic oscillator, which does not appear to show up in the photon description. This is simply a result of picking a different zero of the energy scale in the two descriptions: In the photon description, the actual ground state of the system, with zero photons, serves as the natural zero of the energy scale, whereas in the quasi-mechanical harmonic oscillator description, the zero of the potential energy is used as the zero of the energy scale.

10.4.2 Uncertainty Relation for Photon Number and Oscillation Phase

In chapter 9, we gave a rigorous interpretation of the energy-time uncertainty relation

$$\Delta\mathcal{E}\Delta t \geq \hbar/2 \tag{10•4–4}$$

in terms of the uncertainty of the energy of a system and the uncertainty in the time at which any time-dependent observable would pass through a specified value. We apply this interpretation here to the harmonic oscillator and express it as an uncertainty relation between the quantum photon number n and the phase α of the oscillation.

With $\mathcal{E} = (n + 1/2)\hbar\omega$, we evidently have

$$\Delta\mathcal{E} = \hbar\omega \cdot \Delta n. \tag{10•4–5}$$

If we express any of the oscillating quantities in the form $\cos(\omega t - \alpha)$, then the uncertainty Δt in the occurrence of a specified event may be expressed as a phase uncertainty via

$$\Delta t = \frac{1}{\omega}\Delta\alpha. \tag{10•4–6}$$

Inserting (10•4–6) and (10•4–5) into (10•4–4) yields

$$\boxed{\Delta n \Delta\alpha \geq \tfrac{1}{2}.} \tag{10•4–7}$$

If we apply this to a quasi-classical state, we may replace Δn in (10•4–7) by the value (1•4–22) appropriate for Poisson-distributed events:

$$\Delta n = \sqrt{\langle n \rangle}. \tag{10•4–8}$$

This leads to a phase uncertainty for quasi-classical states,

$$\Delta\alpha \geq \frac{1}{2\sqrt{\langle n \rangle}}. \tag{10•4–9}$$

Evidently, a large number of photons must be present before the phase uncertainty is reduced to a small value.

Chapter 11

COMPOSITE SYSTEMS

11.1 CONFIGURATION-SPACE FORMALISM

11.1.1 The Problem

Up to this point, we have always studied the motion of a single object—usually an electron—in an *externally* given potential. In nature, these external potentials are, in turn, generated by other objects. For example, in the hydrogen atom, the electron moves in the electric field of the proton. But the proton also moves in the electric field of the electron. *Both* motions must be treated quantum-mechanically. Conceptually, the problem is nearly symmetric; the only difference is the much higher mass of the proton.

Evidently, we need a generalization of the single-particle Schroedinger equation to systems of several objects. Which form must such a generalization take? Perhaps the most basic difference between classical and quantum mechanics is the occurrence of a wave function *and* its statistical interpretation. The latter states that the physical meaning of the wave function is related to a *probability density;* specifically, $|\psi|^2 d^3r$ is the probability of "finding" the single object in the small volume d^3r. This probabilistic interpretation must be expected to carry over to multi-particle systems.

We consider specifically the simplest case, that of two *distinguishable* particle-like objects, say, an electron and a proton. Let us refer to them simply as particles 1 and 2. If the particles interact, the probability of finding particle 1 in a volume element d^3r_1, near $\mathbf{r} = \mathbf{r}_1$, must depend on the position $\mathbf{r}_2$ of particle 2. But particle 2 has a probability distribution of its own. Accordingly, the only meaningful question is in regard to the **joint probability** of finding particle 1 in a d^3r_1-vicinity of $\mathbf{r}_1$ and, *simultaneously,* particle 2 in a d^3r_2-vicinity of $\mathbf{r}_2$. We denote this joint probability here by

$$P(\mathbf{r}_1, d^3r_1|\mathbf{r}_2, d^3r_2). \tag{11•1–1}$$

Presumably, two-particle quantum mechanics and, by extension, multi-particle quantum mechanics, must be interpretable in terms of such joint probabilities. This suggests a formalism in terms of a two-particle wave function $\psi(\mathbf{r}_1, \mathbf{r}_2)$ with the statistical interpretation

$$P(\mathbf{r}_1, d^3r_1|\mathbf{r}_2, d^3r_2) = |\psi(\mathbf{r}_1, \mathbf{r}_2)|^2 d^3r_1 d^3r_2. \tag{11•1–2}$$

In single-particle quantum mechanics, the wave function evolves with time through a time evolution equation of the general form

$$i\hbar\frac{\partial}{\partial t}\psi = \hat{H}\psi, \tag{11•1–3}$$

where the operator $\hat{H}$ is a Hermitian operator—the Hamiltonian—that plays a dual role as time evolution operator and as operator for the total energy. In fact, we saw in chapters 8 and 10 that all of classical Newtonian dynamics follows from this formulation, as well as the dynamics of purely electromagnetic oscillators.

We therefore postulate that this overall form persists in the quantum mechanics of coupled systems; in our present case of two particles, we have

$$i\hbar\frac{\partial}{\partial t}\psi(\mathbf{r}_1, \mathbf{r}_2) = \hat{H}\psi(\mathbf{r}_1, \mathbf{r}_2), \tag{11•1–4}$$

with a Hamiltonian of the form

$$\hat{H} = \hat{H}_1 + \hat{H}_2 + \hat{H}_{12}. \tag{11•1–5}$$

Here $\hat{H}_1$ is the single-particle Hamiltonian of particle 1 alone, $\hat{H}_2$ that of particle 2 alone, and $\hat{H}_{12}$ the interaction energy between the two particles, serving as an interaction Hamiltonian, which is responsible for the effects of the particle interaction on the time evolution of the overall state of the system.

The specific form of the interaction Hamiltonian $\hat{H}_{12}$ depends on the nature of the interaction. In the case of a pure electrostatic Coulomb interaction between two charged particles with charges Q_1 and Q_2, the interaction

Hamiltonian is simply the *classical* mutual potential energy of the particles:

$$\hat{H}_{12} = V(\mathbf{r}_1, \mathbf{r}_2) = \frac{Q_1 Q_2}{4\pi\epsilon_0} \cdot \frac{1}{|\mathbf{r}_1 - \mathbf{r}_2|}. \tag{11•1–6}$$

Note that this potential energy depends only on the *relative* coordinates of the two particles, not on their absolute positions in space, nor on their momenta. We will encounter other forms of interaction Hamiltonians later, including interaction Hamiltonians that depend on the momentum of the object.

The above formalism is called the **configuration space formalism** of multi-particle quantum mechanics: The two-particle wave functions, which depend on the space coordinates of *both* particles, are functions, not in ordinary three-dimensional space, but in a six-dimensional **configuration space**, in which each point refers to the real-space positions of two particles. The extension to an arbitrary number of particles is obvious: For N particles, the N-particle wave function, written as $\psi(\mathbf{r}_1, \ldots, \mathbf{r}_N)$, is a function in a $3N$-dimensional configuration space. In such a system, there are $\frac{1}{2}N(N-1)$ different interacting particle pairs, each with an interaction energy of its own. Hence, the overall Hamiltonian is of the form

$$\hat{H} = \sum_{\nu=1}^{N} \hat{H}_\nu + \sum_{\mu<\nu}^{N} \hat{H}_{\mu\nu}, \tag{11•1–7}$$

containing N single-particle terms and $\frac{1}{2}N(N-1)$ mutual interaction terms.

11.1.2 The Limit of Non-Interacting Particles

In the limit of non-interacting particles, the Schroedinger wave equation

$$i\hbar\frac{\partial}{\partial t}\psi(\mathbf{r}_1, \mathbf{r}_2) = (\hat{H}_1 + \hat{H}_2)\psi(\mathbf{r}_1, \mathbf{r}_2) \tag{11•1–8}$$

can always be separated into two single-particle problems by setting

$$\psi(\mathbf{r}_1, \mathbf{r}_2) = \psi_a(\mathbf{r}_1)\psi_b(\mathbf{r}_2), \tag{11•1–9}$$

where each of the two single-particle wave functions obeys a single-particle equation,

$$i\hbar\frac{\partial}{\partial t}\psi_a(\mathbf{r}_1) = \hat{H}_1\psi_a(\mathbf{r}_1) \quad \text{and} \quad i\hbar\frac{\partial}{\partial t}\psi_b(\mathbf{r}_2) = \hat{H}_2\psi_b(\mathbf{r}_2). \tag{11•1–10a,b}$$

We have added the subscripts a and b to specify the particular single-particle state of each particle.

This procedure evidently reduces the non-interacting-particle problem to a set of problems in "ordinary" single-particle quantum mechanics.

The product wave function (11•1–9) implies that the joint probability in (11•1–2) can also be written as a product,

$$P(\mathbf{r}_1, d^3r_1|\mathbf{r}_2, d^3r_2) = P_a(\mathbf{r}_1, d^3r_1) \cdot P_b(\mathbf{r}_2, d^3r_2), \tag{11•1–11}$$

where $P_a(\mathbf{r}_1, d^3r_1)$ and $P_b(\mathbf{r}_2, d^3r_2)$ are the two single-particle probabilities associated with the single-particle states a and b. The product form (11•1–11) means that the probability distributions of the two individual particles are statistically independent. In the absence of the interaction Hamiltonian, the motion of the two particles is uncorrelated, and the probability of finding particle 1 in a d^3r_1-vicinity of $\mathbf{r}_1$ is independent of where particle 2 is and vice versa.

11.2 CENTER-OF-MASS MOTION VS. INTERNAL DYNAMICS

11.2.1 Separation of Variables

The case of principal interest in the present chapter is a system of several particles in free space, in which all potentials are due to particle-particle interactions, with no *external* potential acting on the particles. In this case, the terms in the first sum in (11•1–7) are simply the various kinetic energy terms,

$$\hat{H}_\nu = \hat{T}_\nu = \frac{\hat{p}_\nu^2}{2M_\nu}, \tag{11•2–1}$$

while the interaction terms in the second sum can depend only on the coordinate *differences* of the two particles involved in each interaction term:

$$\hat{H}_{\mu\nu} = V_{\mu\nu}(\mathbf{r}_\mu - \mathbf{r}_\nu). \tag{11•2–2}$$

The center-of-mass motion of the system can then always be split off, and the problem can be reduced to one of the internal dynamics of a system of particles relative to one another. To see this, consider the operator for the total momentum, defined as the sum over all single-particle momenta:

$$\hat{\mathbf{P}} \equiv \sum_{\nu=1}^{N} \hat{\mathbf{p}}_\nu. \tag{11•2–3}$$

Every term in the sum in (11•2–3) commutes with every single-particle kinetic energy term of the form (11•2–1). Furthermore, the non-vanishing commutators of the form $[\hat{\mathbf{p}}, V]$ cancel in pairs:

$$[\hat{\mathbf{P}}_\nu, V_{\mu\nu}(\mathbf{r}_\nu - \mathbf{r}_\mu)] = -[\hat{\mathbf{p}}_\mu, V_{\mu\nu}(\mathbf{r}_\nu - \mathbf{r}_\mu)]. \tag{11•2–4}$$

Hence, the operator for the total momentum commutes with the overall Hamiltonian:

$$[\hat{\mathbf{P}}, \hat{H}] = 0. \tag{11•2–5}$$

But this means that the energy eigenfunctions can always be chosen to be eigenfunctions of the total momentum as well. The expectation value of total momentum is then time-independent, even for states that are not momentum eigenstates. We have thus recovered the classical law of conservation of momentum.

Furthermore, if the energy eigenfunctions are chosen as eigenfunctions of the total momentum, then they are also eigenfunctions of the translational kinetic energy operator associated with the center of-mass motion

$$\hat{T}_{\text{com}} \equiv \frac{\hat{\mathbf{P}}^2}{2M}, \tag{11•2–6}$$

where

$$M = \sum_{\nu=1}^{N} M_\nu \tag{11•2–7}$$

is the total mass of the system. But this means that the overall Hamiltonian can be split into two parts:

$$\hat{H} = \hat{T}_{\text{com}} + \hat{H}_{\text{int}}, \tag{11•2–8}$$

Here $\hat{T}_{\text{com}}$ describes the translational center-of-mass motion and $\hat{H}_{\text{int}}$ describes the internal dynamics. Any eigenstate of both the total Hamiltonian and the translational Hamiltonian must then also be an eigenstate of the internal Hamiltonian.

Mathematically, this is simply a separation-of-variables attempt. To accomplish it, we must first replace the N independent variables $\mathbf{r}_\nu$ by a combination of one variable that describes the center-of-mass motion, plus a new set of $N - 1$ variables that describe the *internal* motion of the system.

The natural *position* variable to describe the center-of-mass motion is of course the center-of-mass position itself,

$$\mathbf{R} \equiv \frac{1}{M} \sum_{\nu=1}^{N} M_\nu \mathbf{r}_\nu. \tag{11•2–9}$$

Exercise: Show that the Cartesian components $\hat{X}_u$ and $\hat{P}_v$ of the operators $\hat{\mathbf{R}}$ and $\hat{\mathbf{P}}$ form canonically conjugate pairs, in the sense that they satisfy commutation relations of the form

$$[\hat{X}_u, \hat{P}_v] = i\hbar\delta_{uv}, \quad (u, v = x, y, z), \tag{11•2–10}$$

analogous to the commutation relations (7•3–8) for the canonically conjugate pairs for the single-particle components of position and momentum,

$$[\hat{x}_u, \hat{p}_v] = i\hbar\delta_{uv}, \quad (u, v = x, y, z). \tag{11•2–11}$$

From (11•2–11), and from the fact that single-particle operators for different particles always commute, show that (11•2–10) is satisfied.

The choice of the proper independent position variables for the internal dynamics of the system is less obvious, except in the simplest case, a two-particle system. In that limit, there is only a single inter-particle potential present, which is of the form $V(\mathbf{r}_2 - \mathbf{r}_1)$, depending only on the inter-particle separation vector

$$\mathbf{r} = \mathbf{r}_2 - \mathbf{r}_1. \tag{11•2–12}$$

Evidently, $\mathbf{r}$ is the natural choice for the *internal* position variable.

Given the choice (11•2–12), the next problem is that of determining the internal *momentum* operator that is canonically conjugate to $\mathbf{r}$. We proceed by first subtracting the translational kinetic energy from the total kinetic energy, to obtain the *internal* kinetic energy:

$$\hat{T}_{\text{int}} = \left(\frac{\hat{\mathbf{p}}_1^2}{2M_1} + \frac{\hat{\mathbf{p}}_2^2}{2M_2}\right) - \frac{(\hat{\mathbf{p}}_1 + \hat{\mathbf{p}}_2)^2}{2M} = \frac{1}{2M} \cdot \frac{[M_1\hat{\mathbf{p}}_2 - M_2\hat{\mathbf{p}}_1]^2}{M_1M_2}. \tag{11•2–13}$$

The last form suggests introducing

$$\hat{\mathbf{p}} \equiv \frac{1}{M} \cdot [M_1\hat{\mathbf{p}}_2 - M_1\hat{\mathbf{p}}_2] \tag{11•2–14}$$

as the **internal momentum**, leading to

$$\hat{T}_{\text{int}} = \frac{\hat{\mathbf{p}}^2}{2M_r}, \tag{11•2–15}$$

where M_r is a **reduced mass**, defined via

$$\frac{1}{M_r} \equiv \frac{M}{M_1M_2} = \frac{1}{M_1} + \frac{1}{M_2}. \tag{11•2–16}$$

The form (11•2–15) is evidently of the desired form of a single-particle kinetic energy operator, albeit with a mass different from that of the two particles involved. It is left to the reader to show that the internal momentum, thus defined, is indeed canonically conjugate to the internal position.

By combining $\hat{T}_{\text{int}}$ and the interaction potential $V(\mathbf{r})$, we obtain the internal Hamiltonian

$$\hat{H}_{\text{int}} = \frac{\hat{\mathbf{p}}^2}{2M_r} + V(\mathbf{r}), \tag{11•2–17}$$

which is formally identical to the Hamiltonian of a single particle of mass M_r in the potential $V(\mathbf{r})$. Evidently, we have achieved our goal of separating the overall Hamiltonian into two parts, as in (11•2–8).

Example: Hydrogen Atom. Applied to the hydrogen atom, our treatment implies a reduced mass defined by

$$\frac{1}{m_r} \equiv \frac{1}{m_e} + \frac{1}{m_p} = \frac{m_H}{m_e m_p}, \tag{11•2–18}$$

where $m_H = m_e + m_P$ is the mass of the entire hydrogen atom. In chapter 3—see (3•2–33)—we claimed that there was a need for the introduction of such a reduced mass into the hydrogen problem, and we pointed out that the net effect on the energy levels would be to replace the Rydberg constant R_∞ with the hydrogen Rydberg constant

$$R_H = \frac{m_p}{m_H} R_\infty, \tag{11•2–19}$$

a small but important correction. The above treatment now gives a rigorous justification for these earlier claims.

The eigenfunctions of the overall Schroedinger equation may be written in the form

$$\psi(\mathbf{R}, \mathbf{r}) = \psi_{\text{com}}(\mathbf{R})\, \psi_{\text{int}}(\mathbf{r}), \tag{11•2–20}$$

where ψ_{com} and ψ_{int} obey the equations

$$\hat{H}_{\text{com}}\psi_{\text{com}} = \mathcal{E}_{\text{com}}\psi_{\text{com}} \quad \text{and} \quad H_{\text{int}}\psi_{\text{int}} = \mathcal{E}_{\text{int}}\psi_{\text{int}}. \tag{11•2–21a,b}$$

The overall energy is

$$\mathcal{E} = \mathcal{E}_{\text{com}} + \mathcal{E}_{\text{int}}, \tag{11•2–22}$$

the sum of the (unquantized) translational kinetic energy $\mathcal{E}_{\text{com}}$ of the entire system and the (quantized) internal energy $\mathcal{E}_{\text{int}}$.

11.2.2 Wave Properties of Composite Objects

Our treatment implies that the center-of-mass motion of composite objects exhibits wave properties just like that of elementary objects, a point already alluded to in Section 1.4.

Because the center-of-mass position vector $\mathbf{R}$ and the total momentum $\mathbf{P}$ are canonically conjugate variables, we may pick a representation in which $\mathbf{R}$ is one of the independent variables, in which case the total-momentum operator $\hat{\mathbf{P}}$ may be written in the form

$$\hat{\mathbf{P}} = -i\hbar\nabla_{\mathbf{R}}. \tag{11•2–23}$$

The Schroedinger equation (11•2–21a) for the center-of-mass motion then becomes the ordinary free-space wave equation for particles with mass M,

$$-\frac{\hbar^2}{2M}\nabla^2\psi_{\text{com}} = \mathcal{E}_{\text{com}}\psi_{\text{com}}, \tag{11•2–24}$$

with plane-wave-like eigenfunctions of the form

$$\psi_{\text{com}}(\mathbf{R}) \propto \exp(i\mathbf{K} \cdot \mathbf{R}). \tag{11•2–25}$$

The eigenvalues for the total momentum and the center-of-mass energy are

$$\mathbf{P} = \hbar\mathbf{K}, \tag{11•2–26}$$

and

$$\mathcal{E}_{\text{com}} = \frac{\hbar^2 K^2}{2M} = \frac{P^2}{2M}. \tag{11•2–27}$$

The wave functions (11•2–25) imply a wave-like propagation of the overall system, with wavelengths

$$\lambda_{\text{com}} = \frac{2\pi}{K} = \frac{2\pi\hbar}{P}, \tag{11•2–28}$$

just as for elementary objects.

Diffraction effects with composite objects are readily observed, most easily with thermal helium atomic beams. The translational kinetic energy of a thermal He atom is insufficient to raise the atom from its internal ground state to any of its internal higher energy eigenstates by being somehow converted into internal energy. As a result, the behavior of the atom during any interaction with a stationary object, such as a diffraction grating, becomes indistinguishable from that of an indivisible object. In fact, diffraction effects with beams of thermal helium atoms are utilized to supplement electron diffraction in the study of the surface structure of crystals.

◆ PROBLEM TO SECTION 11.2

#11•2-1: Vibrations and Rotations of a Diatomic Molecule

Consider two particles with equal mass M having an equilibrium distance a and interacting via a harmonic oscillator potential for deviations from that equilibrium distance, with the interaction Hamiltonian

$$\hat{H}_{12} = \tfrac{1}{2}K \cdot (|\mathbf{r}_2 - \mathbf{r}_1| - a)^2, \tag{11•2–29}$$

where K is a "spring constant," as in section 2.3.

In the absence of any rotation ($l = 0$), the molecule would be a simple harmonic oscillator, oscillating about the equilibrium inter-atomic distance a with a frequency

$$\omega_0 = \sqrt{K/M_r} = \sqrt{2K/M}. \tag{11•2–30}$$

The situation changes somewhat for states with finite angular momentum. The addition of the centrifugal potential alters the shape of the overall effective potential. To the first order, the overall effective potentials for different values of $l > 0$ may, in the vicinity of their minima, still be approximated as pure harmonic oscillator potentials, but with slightly increased equilibrium distances and—more importantly—slightly higher effective spring constants and oscillation frequencies. Using such a treatment, determine the resulting oscillation frequencies and their dependence on the angular momentum quantum number l.

The experimental study of the vibration-rotation spectra of molecules is a powerful tool for elucidating the internal structure of molecules.

11.3 BEYOND TWO PARTICLES: NORMAL MODES IN COUPLED HARMONIC OSCILLATOR SYSTEMS

11.3.1 The Problem

When the number of particles exceeds 2, the number of different particle-particle interaction terms in the Hamiltonian, $\frac{1}{2}N(N - 1)$, exceeds the number of independent position variables, $N - 1$, available to describe the interactions. As a rule, the problem can then no longer be reduced to a set of single-particle problems. For example, consider the helium atom, one of the simplest real three-particle problems, consisting of one nucleus and two electrons. With the He nucleus being much heavier than the electrons, the natural independent internal variables are the two electron position vectors relative to the nucleus,

$$\mathbf{r}_1 = \mathbf{r}_{e1} - \mathbf{r}_n \quad \text{and} \quad \mathbf{r}_2 = \mathbf{r}_{e2} - \mathbf{r}_n. \tag{11•3–1}$$

Such an assignment handles the electron-nucleus interaction well enough, but the electron-electron interaction continues to involve the difference between two independent variables, $\mathbf{r}_{e2} - \mathbf{r}_{e1} = \mathbf{r}_2 - \mathbf{r}_1$, making it impossible to treat the overall dynamics as the linear superposition of two single-particle problems. Such problems invariably cannot be solved exactly in closed form, but require approximation or numerical techniques. The specific case of the ground state of the He atom will in fact be our prime example for discussing the variational approximation method, developed in the next chapter.

11.3.2 Normal-Mode Oscillations in Coupled Harmonic Oscillator Systems

There is, however, one class of problems of practical importance for which separation of variables can always be achieved, namely, when all particle-particle interaction potentials are pure harmonic oscillator-like potentials. The internal vibrations of the different atoms in a molecule or in a crystal can, to the first order, be described by such a superposition-of-harmonic-oscillators model.

In classical dynamics, purely harmonic systems are capable of undergoing purely sinusoidal oscillations, with all particles oscillating in synchronism at the same frequency, one of a set of certain common characteristic frequencies. Ignoring three linearly independent pure translational motions that do not leave the center of mass of the system invariant and two linearly independent non-oscillatory pure rotations, there are $3N - 5$ distinguishable *modes* of oscillation, called the **normal modes** of the system, each with a characteristic oscillation pattern and a characteristic frequency. All overall oscillations of the system can be described as linear superpositions of those normal modes, with an appropriate amplitude and phase for each mode. The characteristic frequencies of different modes need not all be different, but different modes always have different oscillation patterns, even if they have the same characteristic frequency.

In quantum mechanics, each of the classical normal modes acts as a separate harmonic oscillator, and in the absence of any rotational kinetic energy, the internal Hamiltonian can always be separated into a linear superposition of $3N - 5$ one-dimensional harmonic oscillator Hamiltonians, each depending only on its own independent variable. When rotations are present, certain complications occur, as in Problem 11•2–1, but not of interest to us at this point.

We illustrate the procedure here, via an example that is sufficiently simple that the classical normal modes can be determined by inspection. We consider a system of three particles, of equal mass M, confined in their motion to the x-axis. We assume that the mutual potential energies between the particles are pure harmonic oscillator potentials, with potential minima such that the system has its potential energy minimum when the three particles are equidistantly spaced with spacing a (**Fig. 11•3–1**).

We assume that the center-of-mass motion has already been separated out and that the minimum-potential positions of the three particles are $x = -a$, 0, and $+a$. We describe the dynamics of the system in terms of the *displacements* x_1, x_2, and x_3 of the three particles from their equilibrium positions. Note that, for a fixed center of mass, the three positions are not independent of each other, but must satisfy

$$x_1 + x_2 + x_3 = 0. \tag{11•3–2}$$

We assume further that there are harmonic oscillator interactions only between particles 1 and 2 and between 2 and 3, both with the same spring constant. To keep the problem as simple as possible, we assume that there is no *direct* interaction between particles 1 and 3, leaving the inclusion of such an interaction as an exercise to the reader. We may then write the sum of the two interaction Hamiltonians as

$$\hat{H}_{12} + \hat{H}_{23} = \tfrac{1}{2}M\omega_0^2 \cdot [(x_1 - x_2)^2 + (x_3 - x_2)^2], \tag{11•3–3}$$

where ω_0 is the frequency with which particles 1 and 3 would oscillate *if* particle 2 were kept fixed in space.

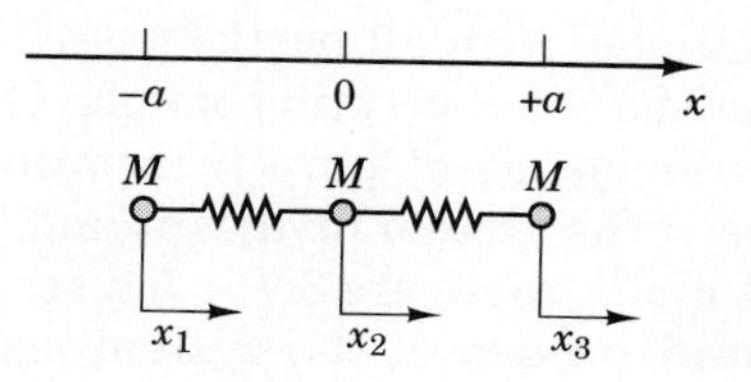

Figure 11•3–1. Tri-atomic linear molecule.

The kinetic energy part of the overall Hamiltonian is evidently

$$\hat{T} = \frac{1}{2M} \cdot (\hat{p}_1^2 + \hat{p}_2^2 + \hat{p}_3^2), \tag{11•3–4}$$

where the three $\hat{p}$'s are the momentum operators canonically conjugate to the three x's, obeying the canonical commutation relations

$$[\hat{x}_1, \hat{p}_1] = [\hat{x}_2, \hat{p}_2] = [\hat{x}_3, \hat{p}_3] = i\hbar. \tag{11•3–5}$$

All position and momentum operators belonging to *different* particles commute:

$$[\hat{x}_\nu, \hat{p}_\mu] = 0 \text{ for } \nu \neq \mu. \tag{11•3–6}$$

Separating out the center-of-mass motion implies that the three momentum operators are not independent of each other, but must satisfy

$$\hat{p}_1 + \hat{p}_2 + \hat{p}_3 = 0, \tag{11•3–7}$$

similar to (11•3–2).

The symmetric three-particle system has two normal modes. One of these is symmetric or even, about the plane $x = 0$, and the other is antisymmetric, or odd (**Fig. 11•3–2**).

For the even mode, we have

$$x_3 = -x_1, \qquad x_2 = 0, \tag{11•3–8a,b}$$

while for the odd mode,

$$x_3 = +x_1, \qquad x_2 = -(x_1 + x_3). \tag{11•3–9a,b}$$

We may describe the state of the system in terms of two mode amplitudes

$$x_+ \equiv \tfrac{1}{2}(x_1 - x_3) \quad \text{and} \quad x_- \equiv \tfrac{1}{3}(x_1 + x_3 - x_2), \tag{11•3–10a,b}$$

where x_+ measures the amplitude of the even-mode part of any oscillation, while x_- measures the amplitude of the odd-mode part. Note that for the even mode, we have $x_- = 0$, and for the odd mode, $x_+ = 0$.

We may similarly describe the momenta in terms of two new mode momenta

$$\hat{p}_+ \equiv \hat{p}_1 - \hat{p}_3 \quad \text{and} \quad \hat{p}_- \equiv \hat{p}_1 + \hat{p}_3 - \hat{p}_2. \tag{11•3–11a,b}$$

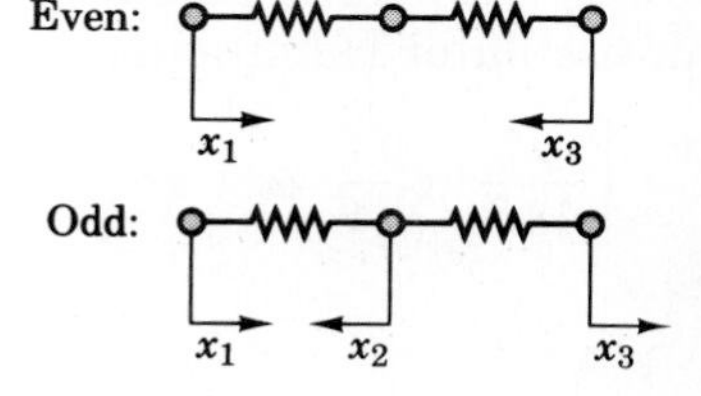

Figure 11•3–2. Even and odd normal modes of the tri-atomic molecule.

It is left to the reader to show that both of the new position-momentum pairs are again canonically conjugate pairs, satisfying

$$[\hat{x}_+, \hat{p}_+] = [\hat{x}_-, \hat{p}_-] = i\hbar; \qquad [\hat{x}_+, \hat{p}_-] = [\hat{x}_-, \hat{p}_+] = 0. \tag{11•3–12a,b}$$

Using (11•3–11a,b), the particle momenta $\hat{p}_1, \hat{p}_2$, and $\hat{p}_3$ may be expressed as functions of the mode momenta $\hat{p}_+$ and $\hat{p}_-$. When this is done, and the results are inserted into the expression (11•3–4) for the total kinetic energy, we obtain

$$\hat{T} = \frac{1}{2M} \cdot (\hat{p}_1^2 + \hat{p}_2^2 + \hat{p}_3^2) = \frac{1}{2M} \cdot \left[\frac{1}{2}\hat{p}_+^2 + \frac{3}{8}\hat{p}_-^2\right]. \tag{11•3–13}$$

This may be brought into the form of two single-particle kinetic energies, by defining two "effective" masses

$$M_+ \equiv 2M \quad \text{and} \quad M_- \equiv 8M/3, \tag{11•3–14}$$

which leads to the final form,

$$\hat{T} = \frac{\hat{p}_+^2}{2M_+} + \frac{\hat{p}_-^2}{2M_-}. \tag{11•3–15}$$

Similarly, by using (11•3–10a,b), we may express the particle positions $\hat{x}_1$, $\hat{x}_2$, and $\hat{x}_3$ as functions of the mode amplitudes $\hat{x}_+$ and x_-. When this is done, and the results are inserted into the expression (11•3–3) for the total potential energy, we obtain

$$\hat{H}_{\text{int}} = \frac{M}{2}\omega_0^2 \cdot \left[2x_+^2 + \frac{81}{8}x_-^2\right]. \tag{11•3–16}$$

This, too, may be brought into a more useful form, by replacing M with M_+ and M_- from (11•3–14), and by defining two mode frequencies via

$$\omega_+^2 = \omega_0^2 \quad \text{and} \quad \omega_-^2 = \frac{3^5}{2^6}\omega_0^2. \tag{11•3–17a,b}$$

This leads to

$$\hat{H}_{\text{int}} = \frac{M_+}{2}\omega_+^2 x_+^2 + \frac{M_-}{2}\omega_-^2 x_-^2. \tag{11•3–18}$$

By adding the kinetic energies from (11•3–13), we obtain the overall Hamiltonian for the internal dynamics of the three-particle system in the form of a sum of two separated harmonic oscillator Hamiltonians:

$$\hat{H} = \hat{H}_+ + \hat{H}_- = \left[\frac{\hat{p}_+^2}{2M_+} + \frac{M_+}{2}\omega_+^2 \cdot x_+^2\right] + \left[\frac{\hat{p}_-^2}{2M_-} + \frac{M_-}{2}\omega_-^2 \cdot x_-^2\right]. \tag{11•3–19}$$

Exercise: Inclusion of a Third Spring. Generalize the treatment of the tri-atomic linear molecule by adding an interaction between particles 1 and 3, of the form

$$\hat{H}_{13} = \tfrac{1}{2}M\omega_1^2 \cdot (x_3 - x_1)^2, \qquad (11\text{•}3\text{–}20)$$

where ω_1 is another characteristic frequency.

11.4 INDISTINGUISHABLE PARTICLES: EXCHANGE CORRELATIONS

Throughout the foregoing, we had interpreted the positions $\mathbf{r}_1$ and $\mathbf{r}_2$ as the positions of particles 1 and 2, respectively. This is a *physically* meaningful distinction only if the two particles are in fact distinguishable, like an electron and a proton, and at the beginning of section 1 we had therefore assumed explicitly that we are dealing with distinguishable particles.

Assume now that the two particles are indistinguishable, like two electrons. The joint probability $P(\mathbf{r}_1, d^3r_1|\mathbf{r}_2, d^3r_2)$ is then simply the probability that both d^3r_1 and d^3r_2 contain one particle each. The distinction between the two position variables becomes a pure bookkeeping device, to account for the fact that we are looking for one particle each at two different positions. But no physical meaning can be attached to any questions as to which particle is which: The particles do not have individual identities of their own, and an interchange of the two particles in the wave function cannot lead to a different physical state; it is simply a different mathematical description of the same state. All this means that, for indistinguishable particles, the joint probability in (11•1–1) for any physical state must be invariant under particle interchange:

$$P(\mathbf{r}_2, d^3r_2|\mathbf{r}_1, d^3r_1) = P(\mathbf{r}_1, d^3r_1|\mathbf{r}_2, d^3r_2). \qquad (11\text{•}4\text{–}1)$$

In terms of the configuration-space wave functions, this implies that only those wave functions are **physically admissible** that meet the condition

$$|\psi(\mathbf{r}_2, \mathbf{r}_1)|^2 = |\psi(\mathbf{r}_1, \mathbf{r}_2)|^2 \qquad (11\text{•}4\text{–}2)$$

Any wave functions not meeting this condition are not physically admissible in multi-particle quantum mechanics, just as non-normalizable wave functions are not physically admissible in single-particle quantum mechanics.

The condition (11•4–2) does *not* require the wave function itself to be what is called **symmetric** under particle exchange,

$$\psi(\mathbf{r}_2, \mathbf{r}_1) = +\psi(\mathbf{r}_1, \mathbf{r}_2). \qquad (11\text{•}4\text{–}3a)$$

A wave function that is antisymmetric under particle exchange,

$$\psi(\mathbf{r}_2, \mathbf{r}_1) = -\psi(\mathbf{r}_1, \mathbf{r}_2), \qquad (11\text{•}4\text{–}3b)$$

would satisfy (11•4–2) just as well.

Given a two-particle wave function $\psi(\mathbf{r}_1, \mathbf{r}_2)$ that does *not* satisfy either of these conditions, it is always possible to construct from it another function that does, by a linear superposition of one of the two following forms,

$$\psi_\pm(\mathbf{r}_1, \mathbf{r}_2) = C \cdot [\psi(\mathbf{r}_1, \mathbf{r}_2) \pm \psi(\mathbf{r}_2, \mathbf{r}_1)], \tag{11•4–4}$$

where C is an appropriate normalization factor.

A full discussion of this topic must wait until chapter 22. We quote here only the result for the simplest case, a system of two electrons. In this case, for parallel spins, only those wave functions turn out to be physically admissible that are antisymmetric under particle exchange, as in (11•4–3b), while for antiparallel spins, the admissible solutions must be symmetric, as in (11•4–3a).[1]

The conditions (11•4–3a,b) imply a built-in correlation between the motion of the particles, especially for the case of states that are antisymmetric under particle exchange: The antisymmetry condition (11•4–3b) requires that for $\mathbf{r}_2 = \mathbf{r}_1$, the wave function be equal to its own negative, which means that it must be zero:

$$\psi(\mathbf{r}_1, \mathbf{r}_1) = 0. \tag{11•4–5}$$

In words, the probability per unit volume of finding two electrons—or any other two identical fermions—within the same infinitesimally small volume is zero: Parallel-spin electrons "avoid each other." This is nothing other than our first encounter with the **Pauli exclusion principle**. We will say more about it in chapter 22. The correlations between antiparallel-spin electrons are of the opposite kind, as if the electrons attracted each other. We will consider them later, too.

Note that these **exchange correlations** are *not* caused by the Coulomb repulsion between electrons; they would exist even if the interaction Hamiltonian were zero. Any correlation due to the Coulomb repulsion between electrons is *in addition* to the exchange correlations.

◆ PROBLEM TO SECTION 11.4

#11•4-1: Correlated States of Two Particles in a Square Well

Suppose that a one-dimensional square well of length L with infinitely high walls contains two non-interacting indistinguishable particles, one in single-particle state m, the other in the different single-particle state n, using the terminology of section 2.2. The physically admissible energy eigenfunctions for this combination are of one of the two following forms

$$\psi_\pm(x_1, x_2) = [\psi_{m,n}(x_1, x_2) \pm \psi_{n,m}(x_1, x_2)]/\sqrt{2}, \tag{11•4–6}$$

---

[1] This statement is an over-simplification, correct only if the wave function is written as a scalar function of the space coordinates alone, not including the spin into the wave function itself. See chapters 21 and 22 for a complete treatment that is free of these restrictions.

where

$$\psi_{m,n}(x_1, x_2) = (2/L)\sin(m\pi x_1/L)\sin(n\pi x_2/L), \tag{11•4–7a}$$

$$\psi_{n,m}(x_1, x_2) = (2/L)\sin(n\pi x_1/L)\sin(m\pi x_2/L). \tag{11•4–7b}$$

(a) Show that the two superpositions in (11•4–6) correspond to states in which the motion of the two particles is correlated, meaning that the probability distribution of each of the particles is dependent upon the position of the other particle, rather than being the same for all positions of the other particle.

(b) Assume that the two occupied single-particle states are $m = 1$ and $n = 2$. Demonstrate the correlation by giving closed-form expressions for the probability density distribution $\rho_2(x_2)$ for particle 2 under two different conditions:

(1) Averaged over all possible locations of particle 1.

(2) For only those cases when particle 1 is in a narrow interval around $x_1 = L/3$.

(c) Make a graph of $\rho_2(x_2)$ for both cases, for both superpositions given above. Give a discussion expressing in words why the different distributions imply correlated motions, and describe qualitatively the nature of the correlations.

(d) If there existed a repulsive force between the two particles, such as the Coulomb repulsion, which of the two linear superpositions would have the lower expectation value of the energy? Why?

11.5 NORMALIZATION, EXPECTATION VALUES, AND SUCH

The normalization of multi-particle state functions is analogous to that of the single-particle case. If $|\Psi(\mathbf{r}_1, \mathbf{r}_2)|^2$ represents a joint probability density, then the two integrals

$$\rho(\mathbf{r}_1) = \int |\Psi(\mathbf{r}_1, \mathbf{r}_2)|^2 d^3r_2 \quad \text{and} \quad \rho(\mathbf{r}_2) = \int |\Psi(\mathbf{r}_1, \mathbf{r}_2)|^2 d^3r_1 \tag{11•5–1}$$

are the probability densities for particles 1 and 2, each averaged over all locations of the "other" particle. These single-particle probability densities must integrate to unity:

$$\iint |\Psi(\mathbf{r}_1, \mathbf{r}_2)|^2 d^3r_1 d^3r_2 = 1. \tag{11•5–2}$$

Generally, for N particles,

$$\int\cdots\int |\Psi(\mathbf{r}_1, \cdots \mathbf{r}_N)|^2 d^3r_1 \cdots d^3r_N = 1. \tag{11•5–3}$$

We will, as a rule, find it convenient not to write out the multiple integrals in (11•5–2) and (11•5–3), but to denote these normalization integrals again by the more compact Dirac notation

$$\langle\Psi|\Psi\rangle = 1, \tag{11•5–4}$$

regardless of the number of particles involved. The Dirac brackets stand for integration—or summation—over *all* variables, whatever and however many they are.

By considerations entirely analogous to those in chapter 6, it follows again that two multi-particle state functions Φ and Ψ corresponding to *different* eigenvalues of some multi-particle Hermitian operator are again orthogonal, in the sense of the multiple integral

$$\int \ldots \int \Phi^*(\mathbf{r}_1, \ldots, \mathbf{r}_N)\, \Psi\, (\mathbf{r}_1, \ldots, \mathbf{r}_N)\, d^3r_1 \ldots d^3r_N = 0. \tag{11•5–5}$$

For this type of expression, too, we retain the notation

$$\langle\Phi|\Psi\rangle = 0. \tag{11•5–6}$$

The extension to matrix elements of a multi-particle operator $\hat{A}$ should be obvious.

Chapter 12

VARIATIONAL PRINCIPLE

12.1 VARIATIONAL THEOREM

We now turn to the derivation of the variational theorem we asserted, without proof, in chapter 2. More specifically, we show the following: Of all physically admissible wave functions, the one with the lowest energy expectation value $\langle H \rangle$ is an eigenfunction of the Hamiltonian. The proof will subsequently serve both as the basis for an approximation method for solving the Schroedinger equation and as the foundation for a full development of a theory of the properties of eigenfunctions of Hermitian operators.

Assume that the—as yet unknown—normalized wave function with the lowest value of $\langle H \rangle$ is called ψ_0. We can then construct other wave functions with higher $\langle H \rangle$-values by the prescription

$$\psi = \psi_0 + \alpha\phi_0, \tag{12•1–1}$$

where α is a real scale factor and ϕ_0 is any arbitrary (normalized) function, subject only to the constraint that ψ shall remain in the hermiticity domain of $\hat{H}$. The expectation value of the energy can then be written

$$\langle H \rangle \equiv \frac{\langle \psi | \hat{H}\psi \rangle}{\langle \psi | \psi \rangle}$$

$$= \frac{\langle\psi_0|\hat{H}\psi_0\rangle + \alpha[\langle\psi_0|\hat{H}\phi_0\rangle + \langle\phi_0|\hat{H}\psi_0\rangle] + \alpha^2\langle\phi_0|\hat{H}\phi_0\rangle}{1 + \alpha[\langle\psi_0|\phi_0\rangle + \langle\phi_0|\psi_0\rangle] + \alpha^2}, \qquad (12{\bullet}1\text{–}2)$$

where the denominator $\langle\psi|\psi\rangle$ has been added to re-normalize ψ.

If $\langle H\rangle$ is to be a minimum for $\alpha \to 0$, then ψ_0 must be such that

$$\frac{d}{d\alpha}\langle H\rangle = 0 \quad \text{for} \quad \alpha = 0. \qquad (12{\bullet}1\text{–}3)$$

Applied to (12•1–2), this leads to

$$[\langle\psi_0|\hat{H}\phi_0\rangle + \langle\phi_0|\hat{H}\psi_0\rangle] - [\langle\psi_0|\phi_0\rangle + \langle\phi_0|\psi_0\rangle]\langle\psi_0|\hat{H}\psi_0\rangle = 0. \qquad (12{\bullet}1\text{–}4)$$

If we use the symbol $\mathcal{E}_0$ for the expectation value of the energy for ψ_0,

$$\mathcal{E}_0 = \langle\psi_0|\hat{H}\psi_0\rangle, \qquad (12{\bullet}1\text{–}5)$$

and if we utilize the hermiticity of $\hat{H}$ in the first term in (12•1–4), we can re-write (12•1–4) in the form

$$\langle(\hat{H}\psi_0 - \mathcal{E}_0\psi_0)|\,\phi_0\rangle + \langle\phi_0\,|(H\psi_0 - \mathcal{E}_0\psi_0)\rangle = 0. \qquad (12{\bullet}1\text{–}6)$$

But this can be true for an essentially arbitrary ϕ_0 only if

$$\hat{H}\psi_0 = \mathcal{E}_0\psi_0, \qquad (12{\bullet}1\text{–}7)$$

that is, if ψ_0 is an eigenfunction of $\hat{H}$, with $\mathcal{E}_0$ itself as the energy eigenvalue.

Equation (12•1–7) represents the **variational theorem** in its simplest form. It is readily extended in two directions, to higher energy eigenstates, and to Hermitian operators other than $\hat{H}$.

Higher Energy States

The energy eigenfunctions for higher energy states must be orthogonal to the ground state. Consider, therefore, the energy expectation value for that class of functions that are orthogonal to ψ_0:

$$\langle\psi|\psi_0\rangle = 0. \qquad (12{\bullet}1\text{–}8)$$

The entire energy minimization procedure can then be repeated almost verbatim: Amongst those functions ψ that satisfy (12•1–8), there will be one, called ψ_1, with the lowest energy expectation value $\langle H\rangle$. (If there is more than one such function, as for a degenerate energy level, let ψ_1 be any one of those.) We can then construct other wave functions with higher (or equal) $\langle H\rangle$-values by the prescription, similar to (12•1–1),

$$\psi = \psi_1 + \alpha\phi_1, \qquad (12{\bullet}1\text{–}9)$$

where ϕ_1 is subject to the same constraints as ϕ_0, plus the *additional* constraint imposed by (12•1–8),

$$\langle\phi_1|\psi_0\rangle = 0. \qquad (12{\bullet}1\text{–}10)$$

By repeating the same argument as for ψ_0, one finds that ψ_1 is also an eigenfunction of $\hat{H}$:

$$\hat{H}\psi_1 = \mathcal{E}_1\psi_1, \tag{12•1–11}$$

where $\mathcal{E}_1$ is the second-lowest energy eigenvalue.

The process can be repeated ad infinitum, at each step restricting the class of competing functions further to those that are orthogonal to all previous eigenfunctions. If one of the energy eigenvalues is N-fold degenerate, the process will automatically generate a set of N mutually orthogonal—and hence, linearly independent—energy eigenfunctions with this common eigenvalue, before proceeding to the next higher energy eigenvalue. In this way, all energy eigenfunctions can—at least in principle—be generated, in the order of increasing energy eigenvalues:

$$\mathcal{E}_0 \leq \mathcal{E}_1 \leq \mathcal{E}_2 \leq \ \ldots \ \leq \mathcal{E}_n \leq \ \ldots . \tag{12•1–12}$$

The equals signs in (12•1–12) stand for the case of degenerate energy levels.

Other Hermitian Operators

Although we assumed at the beginning that $\hat{H}$ is the Hamilton operator, we have nowhere utilized this assumption, except implicitly by assuming that the operator $\hat{H}$ does in fact have a *minimum* eigenvalue, rather than having eigenvalues ranging from $-\infty$ to $+\infty$. Our derivation therefore holds automatically for any Hermitian operator that has a bounded set of eigenvalues. (The eigenvalues may have a maximum rather than a minimum, or even both.)

The process can be readily extended to any Hermitian operator $\hat{A}$ whose eigenvalue spectrum ranges from $-\infty$ to $+\infty$, by considering the operator $\hat{A}^2$, which is itself a Hermitian operator whose eigenvalues are simply the squares A_n^2 of the eigenvalues A_n of $\hat{A}$ itself.[1] By minimizing the expectation value of $\hat{A}^2$, one automatically generates the set of eigenfunctions of $\hat{A}^2$ in the order of increasing absolute value of the eigenvalues of $\hat{A}$:

$$|A_0| \leq |A_1| \leq |A_2| \leq \ \ldots \ \leq |A_n| \leq \ \ldots . \tag{12•1–13}$$

PROBLEM TO SECTION 12.1

#12•1-1: Eigenvalues and Eigenfunctions of the Square of a Hermitian Operator

Suppose that $|\psi\rangle$ is an eigenstate of the operator $\hat{A}^2$, with the nonzero eigenvalue A^2, but that $|\psi\rangle$ is *not* an eigenstate of the operator $\hat{A}$.

[1] It can be shown that $\hat{A}^2$ has neither additional eigenvalues nor additional eigenfunctions. For details see problem #12•1-1 at the end of the section.

Show that in this case there must exist two non-vanishing states

$$|+A\rangle \equiv (\hat{A} - A)|\psi\rangle \quad \text{and} \quad |-A\rangle \equiv (\hat{A} + A)|\psi\rangle \tag{12•1–14a,b}$$

that *are* eigenstates of the operator $\hat{A}$, with the eigenvalues $\pm A$, that is,

$$\hat{A}|+A\rangle = +A|+A\rangle \quad \text{and} \quad \hat{A}|-A\rangle = -A|-A\rangle. \tag{12•1–15a,b}$$

Evidently, under the initial premise, both $+A$ and $-A$ must be eigenvalues of the operator $\hat{A}$.

Show further that the original state $|\psi\rangle$ is a linear superposition of the eigenstates $|+A\rangle$ and $|-A\rangle$, namely,

$$|\psi\rangle = \frac{1}{2A}(|-A\rangle - |+A\rangle). \tag{12•1–16}$$

Hence, there are no additional eigenvalues.

12.2 VARIATIONAL APPROXIMATION METHOD

12.2.1 The Idea

The variational theorem is often used to estimate the energy of the ground state of a quantum-mechanical system, particularly when the system is difficult to treat rigorously. To this end, one starts out with a "plausibly-looking" ad hoc wave function that contains one or more adjustable parameters. The energy expectation value thereby becomes a function of these parameters. The parameters are subsequently determined by differentiating $\langle H\rangle$ with respect to each of them and requesting that the derivatives vanish.

If the wave functions are determined accurately enough, the next higher energy level can also be obtained by restricting the competition to wave functions that are orthogonal to the ground state. If the problem contains symmetries, higher states can frequently be obtained even without an accurate knowledge of the ground state. For example, in one-dimensional problems with inversion symmetry, the wave functions must be either even or odd. By restricting the competition to either class of functions, both the lowest even-parity and the lowest odd-parity state can be determined in this fashion.

Because the actual energy of the ground state must always be lower than the energy estimated from the variational principle, the latter is often employed to determine whether the ground state of a quantum-mechanical system is a bound state. As an example, consider once again the unsymmetric one-dimensional potential well of Fig. 2•4–1, repeated here in **Fig. 12•2–1**.

As we saw in chapter 2, if such a well is shallow enough, it need not have a bound state. But if an approximate wave function can be found that leads to an energy expectation value in the range $\mathcal{E} \leq V_\infty$, then the existence of a bound state is assured, no matter how crude that approximate wave function might be. If no such function can be found, the question remains open, but often it is

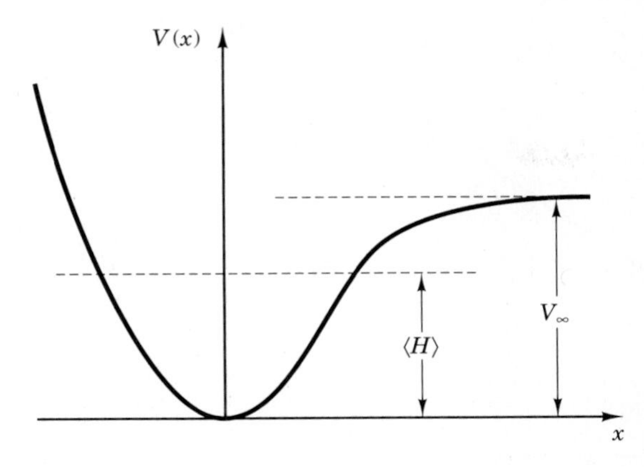

Figure 12•2–1. Simple potential well.

possible in such a way to prove the existence of a bound state. A famous example of this approach occurs in the BCS theory of superconductivity.

12.2.2 Example: An Electron in a Uniform Electric Field

We illustrate here the variational approximation principle by returning to the problem of an electron that is driven against an infinitely high abrupt potential wall by an electric field E (**Fig. 12•2–2**).

We treated this problem already in chapter 6, as an illustration of the WKB approximation for the overall potential energy

$$V(x) = \begin{Bmatrix} \infty \\ eEx \end{Bmatrix} \quad \text{for} \quad \begin{Bmatrix} x < 0 \\ x > 0 \end{Bmatrix}. \tag{12•2–1}$$

In the present chapter, we use the variational approximation principle to obtain an approximate value for the ground-state energy, to be compared with both the exact value and the WKB value.

The energy eigenfunction of the ground state must have the qualitative shape shown in Fig. 12•2–2, going to zero linearly for $x \to 0$, and in an exponential-like fashion for $x \to \infty$. Probably the simplest function exhibiting this qualitative behavior is a product of x and an exponential, which may be written as

$$\psi(x) = Cx \cdot \exp(-\tfrac{1}{2}\gamma x), \tag{12•2–2}$$

where γ is a suitable adjustable parameter and C is a normalization constant. We shall use (12•2–2) here as a trial wave function and shall adjust γ in such a way that the energy expectation value becomes a minimum.

We first normalize ψ:

$$\int_0^\infty \psi^2 \, dx = \frac{2C^2}{\gamma^3} = 1. \tag{12•2–3}$$

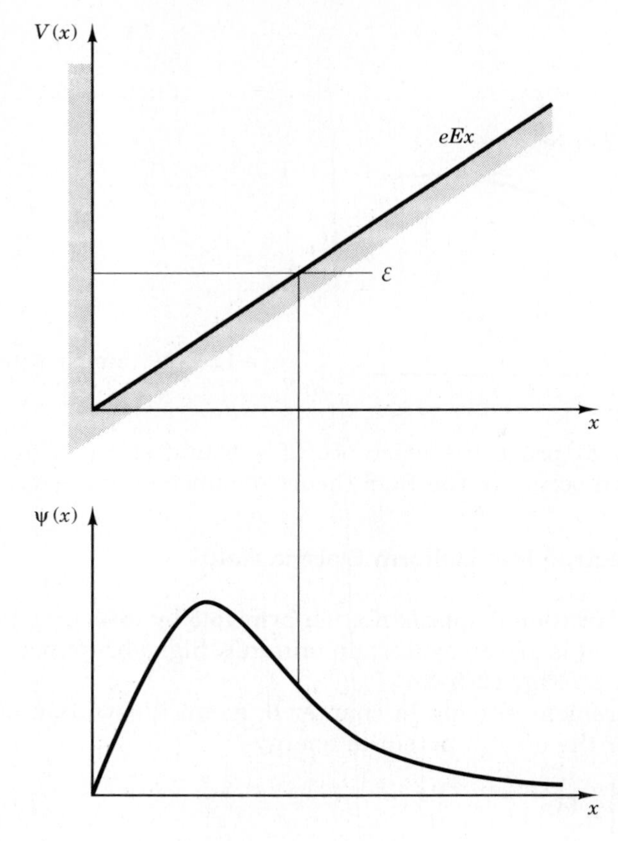

Figure 12•2–2. Potential energy for electron in a uniform electric field at a potential wall, and qualitative shape of the ground-state wave function.

This gives the normalization coefficient C as a function of the variational parameter γ.

It is left to the reader to show that the expectation values of the potential and kinetic energy are given by

$$\langle \mathcal{E}_{\text{pot}} \rangle = eE \int_0^\infty \psi^2 x \, dx = \frac{3eE}{\gamma}, \tag{12•2–4}$$

and

$$\langle \mathcal{E}_{\text{kin}} \rangle = -\frac{\hbar^2}{2m_e} \int_0^\infty \psi \frac{d^2\psi}{dx^2} \, dx = \frac{\hbar^2 \gamma^2}{8m_e}. \tag{12•2–5}$$

With (12•2–4) and (12•2–5), the expectation value of the total energy is

$$\langle \mathcal{E} \rangle = \frac{3eE}{\gamma} + \frac{\hbar^2 \gamma^2}{8m_e}. \tag{12•2–6}$$

We see that the potential energy decreases inversely with γ, while the kinetic energy increases quadratically with γ.

Exercise: Make a qualitative plot of the relation (12•2–6).

The total energy has a minimum for

$$\gamma = \left(6eE \cdot \frac{2m_e}{\hbar^2}\right)^{1/3} = \frac{\sqrt[3]{6}}{L}, \tag{12•2–7}$$

where

$$L = \left(\frac{\hbar^2}{2m_e \cdot eE}\right)^{1/3} \tag{12•2–8}$$

is the natural unit of length for the problem we introduced in chapter 6—see (6•3–8)—along with the natural unit of energy (6•3–6),

$$\mathcal{E}_0 = eE \cdot L = \left[\frac{\hbar^2(eE)^2}{2m_e}\right]^{1/3}. \tag{12•2–9}$$

The energy at that minimum is

$$\langle \mathcal{E} \rangle = \frac{9}{2\sqrt[3]{6}} \cdot \left[\frac{\hbar^2(eE)^2}{2m_e}\right]^{1/3} = \frac{9\mathcal{E}_0}{2\sqrt[3]{6}} = 2.48\mathcal{E}_0, \tag{12•2–10}$$

about 6% higher than the exact value, $2.338\mathcal{E}_0$, from (6•3–16a).

Considering the crudeness of the approximation, the agreement is surprisingly good, but still not as good as the WKB approximation, which yielded a value $2.321\mathcal{E}_0$—see (6•3–15a)—below the exact value, rather than above it. The accuracy can be readily improved by using a trial wave function that approximates more closely the asymptotic behavior of the true wave function for $x \to \infty$. The simple falloff like $\exp(-\frac{1}{2}\gamma x)$ in our trial wave function, with an exponent that is *linear* in x, would have been appropriate for penetration into a barrier of *constant* height, that is, for a potential energy that levels out at infinity, as in the hydrogen atom. For a monotonically increasing potential, the falloff must be faster, but clearly not as fast as for the harmonic oscillator, for which the potential increases quadratically and the wave function falls off with an exponent that is quadratic in x. A simple study of the asymptotic behavior of the wave function for large x, similar to our study for the hydrogen atom and the harmonic oscillator, shows that an exponent proportional to the $\frac{3}{2}$ power of x is appropriate, and this is in fact the form of the exponent in the WKB approximation. This suggests an improved trial wave function of the form

$$\psi(x) = Cx \cdot \exp[-\tfrac{1}{2}(\gamma x)^{3/2}]. \tag{12•2–11}$$

Exercise: Show that, if the wave function is written in the form

$$\psi(x) = C \cdot \exp[-f(x)], \tag{12•2–12}$$

the dominant term in the exponent for large values of x must be a $\frac{3}{2}$-power term.

The actual determination of the ground-state energy using this trial wave function (12•2–11) is left to the problems at the end of the section.

PROBLEMS TO SECTION 12.2

#12•2-1: Improved Ground State for the Triangular Well

Estimate the ground-state energy for an electron in an electric field, using the improved trial wave function (12•2–11). Express your final answer in the form $\langle \mathcal{E} \rangle_{\min} = \alpha \cdot \mathcal{E}_0$. Give a *closed-form algebraic* expression for the proportionality factor α first, without inserting decimal values for any fractions, roots, or gamma function terms. Give a decimal value for α at the end.[2]

#12•2-2: Screened Coulomb Potential

In many problems in science, including solid-state physics,[3] one encounters a potential that differs from the Coulomb potential between an electron and a proton by an attenuation factor of the form $\exp(-\kappa r)$, namely,

$$V(r) = -\frac{e^2}{4\pi\epsilon_0 r} \cdot \exp(-\kappa r), \tag{12•2–13}$$

called the **screened Coulomb potential**. The quantity κ is called the **screening parameter**. For sufficiently large values of κ, all bound states will have been squeezed out of the spherical potential well.

(a) Estimate the maximum value of κ for which the existence of at least one bound state is guaranteed. Express your answer in terms of a multiple of $1/a_0$, where a_0 is the Bohr radius.

(b) Estimate the binding energy, in terms of the Rydberg energy R_∞, for the case $\kappa a_0 = 1/2$.

#12•2-3: Bound State of Delta Function "Dipole"

We recall that a single **negative** delta function of strength $-S$ always has a single bound state. Suppose we add a **positive** delta function to the potential, a distance D away, with a strength $+S$ of the same absolute magnitude, but with opposite sign, as shown in **Fig. 12•2–3**.

This addition makes a positive contribution to the energy of the bound state, thus pushing that state up and weakening its binding energy. It is not obvious whether the bound state can get completely pushed out of the negative δ-well and, if so, under what conditions. Use the variational principle to show that a bound state *always* remains. Do this by calculating the expectation value of the energy for the extremely crude trial wave

---

[2] You will need $\Gamma(5/3) = 0.902745$, a value not contained in all tables.

[3] See, for example, Kittel, *Introduction to Solid State Physics,* 6th ed. (New York: Wiley, 1986), p. 266.

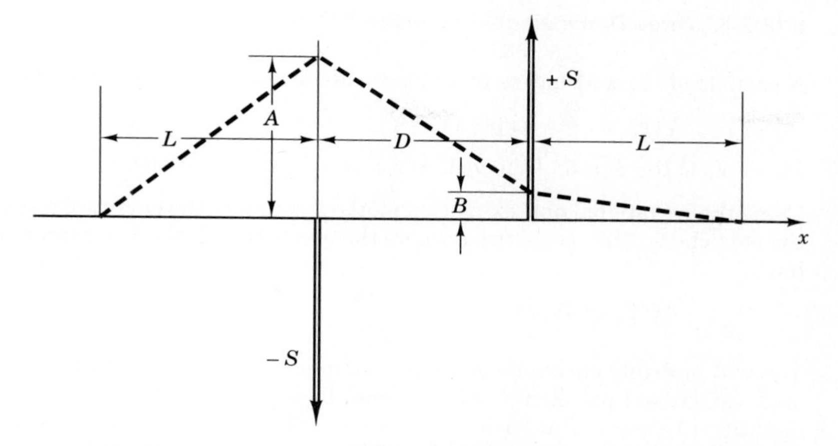

Figure 12•2–3. Potential energy diagram and trial wave function for a delta function "dipole" potential.

function consisting of three straight lines, superimposed as a dashed line on the potential energy diagram in Fig. 12•2–3. Do not overlook the *kinetic* energy associated with the two kinks in the calculation of the expectation value of the energy!

By adjusting the three quantities A, B, and L, show that it is always possible to obtain a negative expectation value of the *total* energy. Explain why this proves that a bound state must exist.

#12•2-4: Can the Last Bound State Be Pushed Out?

Figure 12•2–4 shows an elaboration on the simple potential of the preceding problem: a potential consisting of a square barrier adjacent to a square well, with the height and the width of the barrier being the same as the depth and the width of the well. This potential always has bound states if either V_0 or a is large enough, and there are no bound states left when the well itself vanishes. Show that there will always remain at least one bound state, as long as both V_0 and a remain *nonzero,* no matter how small.

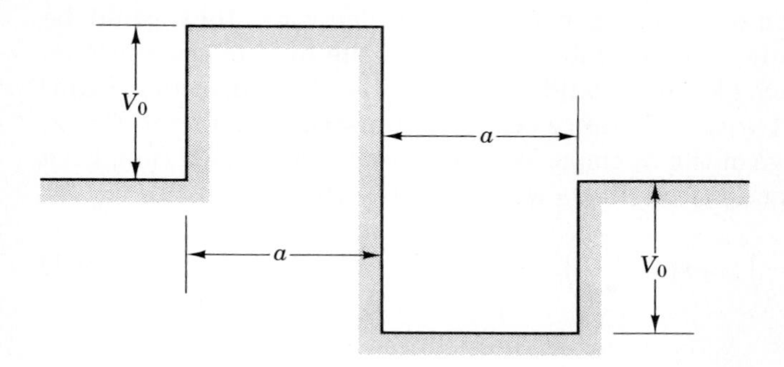

Figure 12•2–4. Well-and-barrier combination.

#12•2-5: Three-Dimensional Exponential Well

A particle of mass M moves in a spherically symmetric potential well of the form

$$V(\mathbf{r}) = -V_0 \cdot \exp(-\Gamma r), \tag{12•2–14}$$

where V_0 is the depth of the well and Γ is a decay parameter.

Using the variational approximation method, estimate the energy of the ground state of the particle, for the specific case where the quantities V_0 and Γ are inter-related according to

$$4\hbar^2\Gamma^2 = 3V_0 M. \tag{12•2–15}$$

You will probably encounter a certain complicated higher-order algebraic equation for your variational parameter, which would be hard to solve for arbitrary values of the quantities V_0 and Γ. But for the specific case where the two quantities are inter-related according to (12•2–15), you should be able to guess the variational parameter by inspection of the algebraic equation for it.

Express the final energy estimate in terms of the ratio $\langle \mathcal{E} \rangle / V_0$.

12.3 GROUND STATE OF THE HELIUM ATOM

With its two electrons, the neutral helium atom is the simplest multi-electron system. The theoretical understanding of the ground-state energy of the He atom has historically played an important role in the evolution of quantum mechanics. Initial attempts to calculate that energy had failed. When quantum mechanics based on the Schroedinger equation was able to calculate the ground-state energy with remarkable accuracy, this was a major triumph of the new mechanics.

The method employed was the variational approximation. We present the treatment here because it is an instructive example both of the handling of multi-particle wave functions and of the application of the variational method to a non-trivial problem.

In the ground state, the two electrons in the He atom have opposite spin. *If* there were no Coulomb repulsion between the electrons, they could be treated as non-interacting, moving independently in the field of the nucleus. The wave function for each electron would then simply be the hydrogen ground-state wave function, but with a distance scale compressed by a factor $Z = 2$, due to the doubled charge of the nucleus. With the proper normalization, both single-electron wave functions might be written in the common form

$$\psi(r) = \frac{1}{\sqrt{\pi}}\left(\frac{Z}{a_0}\right)^{3/2} \exp\left(-\frac{Zr}{a_0}\right), \tag{12•3–1}$$

with $Z = 2$.

The two-electron wave function would then be of the simple form (11•1–9) of a product of two single-electron wave functions, re-written here as

$$\psi(\mathbf{r}_1, \mathbf{r}_2) = \psi_\uparrow(\mathbf{r}_1)\psi_\downarrow(\mathbf{r}_2). \tag{12•3–2}$$

Note that, with the same single-particle wave function chosen for both electrons, the product wave function (12•3–2) satisfies the exchange symmetry condition (11•4–3a) for two electrons with antiparallel spins.

The central idea behind the variational approximation to the He atom problem is to retain the forms (12•3–1) and (12•3–2) of the wave function, but to treat Z as an adjustable parameter, to be chosen in such a way as to minimize the expectation value of the total energy. This is not a purely ad hoc choice; such a wave function may be justified as a good first-order approximation by the following physical argument.

At any distance r, part of the nuclear charge seen by each electron is always screened by those parts of the charge of the other electron that are farther inside than r. Effectively, each electron sees only a (distance-dependent) fraction of the total nuclear charge, approaching the full nuclear charge $2e$ close to the nucleus, but decaying to e at large distances. Our wave function ignores this distance dependence of the effective nuclear charge and treats the atom *as if* both electrons were still independent, but each seeing only a reduced nuclear charge $f \cdot e$.

The effect of the distance scale factor Z on the expectation value of the kinetic energy is the same as if we replaced the Bohr radius in (3•2–5) by a_0/Z. This yields the kinetic energy *per electron,*

$$\langle \mathcal{E}_{\mathrm{kin}} \rangle = Z^2 \frac{\hbar^2}{2m_e a_0^2} = Z^2 R_\infty, \tag{12•3–3}$$

where, in the second equality, we have drawn on the definition (3•2–9) of the Rydberg energy.

The potential energy in the Coulomb field of the nucleus requires more care. The compression of the distance scale implied by the substitution $a_0 \to a_0/Z$ brings each electron closer to the nucleus and would cause the potential energy per electron to increase by a factor Z, even without a change in the nuclear charge. The doubling of the nuclear charge contributes another factor of 2, leading to a potential energy per electron given by

$$\langle \mathcal{E}_{e\text{-}n} \rangle = -2Z \frac{e^2}{4\pi\epsilon_0 a_0} = -4ZR_\infty. \tag{12•3–4}$$

The task that is left is to calculate the expectation value for the Coulomb repulsion energy between the electrons, which is given by

$$\langle \mathcal{E}_{e\text{-}e} \rangle = \frac{e^2}{4\pi\epsilon_0} \cdot \iint \frac{|\psi(r_1, r_2)|^2}{|\mathbf{r}_2 - \mathbf{r}_1|} d^3r_1 d^3r_2. \tag{12•3–5}$$

The factor preceding the integral equals $2R_\infty a_0$. If we substitute this value, insert the wave function (12•3–1), and make the variable substitution

$$2Z\mathbf{r}/a_0 = \mathbf{s}, \tag{12•3–6}$$

we may re-arrange (12•3–5) to read as follows:

$$\langle \mathcal{E}_{e\text{-}e} \rangle = \frac{R_\infty Z}{16\pi^2} \int \exp(-s_1) \cdot \left[\int \frac{\exp(-s_2)}{|\mathbf{s}_2 - \mathbf{s}_1|} d^3 s_2 \right] d^3 s_1. \tag{12•3–7}$$

The complicated double volume integral can be evaluated in closed form (see below), yielding the value $20\pi^2$ and leading to

$$\langle \mathcal{E}_{e\text{-}e} \rangle = \frac{5}{4} Z R_\infty. \tag{12•3–8}$$

From (12•3–3), (12•3–4), and (12•3–8), we obtain the expectation value for the total energy:

$$\langle \mathcal{E}_{\text{tot}} \rangle = 2\langle \mathcal{E}_{\text{kin}} \rangle + 2\langle \mathcal{E}_{e\text{-}n} \rangle + \langle \mathcal{E}_{e\text{-}e} \rangle = \left[2Z^2 - \frac{27}{4} Z \right] R_\infty. \tag{12•3–9}$$

This has a minimum for

$$Z = Z_0 \equiv \frac{27}{16}, \tag{12•3–10}$$

corresponding to the variational estimate for the ground-state energy

$$\mathcal{E}_0 \approx \langle \mathcal{E}_{\text{tot}} \rangle_{\text{min}} = -2\left(\frac{27}{16}\right)^2 R_\infty = -\frac{729}{128} R_\infty. \tag{12•3–11}$$

Experimentally, the quantity that is known accurately is the **first ionization energy** I, defined as the energy required to detach the first electron. If one electron is removed, the remaining electron is a hydrogen-like electron in a Coulomb potential with $Z = 2$, with a ground-state energy of $-4R_\infty$. According to our theory, we should therefore have

$$I = -\mathcal{E}_0 - 4R_\infty \approx \frac{217}{128} R_\infty \approx 23.07 \text{ eV}. \tag{12•3–12}$$

The experimental value is about 24.5 eV. Considering the relative crudeness of our approximation scheme, the agreement is remarkably good. By using trial wave functions that reflect better the radial dependence of the mutual screening effects of the two electrons upon each other, an agreement with the experimental value well within the uncertainty of the latter is obtainable.

Mathematical Detail: The Integral in (12•3–7)

The volume integral over $\mathbf{s}_2$ inside the square brackets in (12•3–7) may be evaluated by introducing polar coordinates for $\mathbf{s}_2$, with the direction of $\mathbf{s}_1$ as the polar axis:

$$\left[\int \ldots d^3\mathbf{s}_2\right] = 2\pi \int_0^{\infty}\left[\left[\int_0^{\pi} \frac{\sin\theta\, d\theta}{\sqrt{s_1^2 + s_2^2 - 2s_1s_2\cos\theta}}\right]\right] \exp(-s_2)s_2^2 ds_2. \tag{12•3–13}$$

The integration over θ is readily executed:

$$\int_0^{\pi} \ldots d\theta = \frac{1}{s_1s_2}[|s_1 + s_2| - |s_1 - s_2|] = \begin{Bmatrix} 2/s_1 & \text{if} & s_2 < s_1 \\ 2/s_2 & \text{if} & s_2 > s_1 \end{Bmatrix}. \tag{12•3–14}$$

If this is inserted into (12•3–13), we obtain

$$\left[\int \ldots d^3s_2\right] = \frac{4\pi}{s_1}[-(s_1^2 + 2s_1)e^{-2s_1} + 2], \tag{12•3–15}$$

and if this in turn is inserted into the outer integral over $\mathbf{s}_1$ in (12•3–7), we obtain the result claimed earlier:

$$\int \ldots d^3s_1 = 20\pi^2. \tag{12•3–16}$$

◆ PROBLEM TO SECTION 12.3

#12•3-1: Improved He Ground State

The trial wave function (12•3–2), based on two single-particle wave functions of the form (12•3–1) with the same value of Z, assumes that the two electrons have the same radial density distribution. An improved ground state can be obtained by allowing for the possibility that the two electrons have different radial distributions. The electron closer to the nucleus would, on the average, see a larger fraction of the total nuclear charge, while the electron farther away would see a smaller fraction, due to the screening by the other electron. The desired refinement can be achieved by using a trial wave function of the form

$$\psi(\mathbf{r}_1, \mathbf{r}_2) = \frac{C}{2\pi} \cdot \left(\frac{Z_\alpha Z_\beta}{a_0^2}\right)^{3/2}\left[\exp\left(-\frac{Z_\alpha r_1 + Z_\beta r_2}{a_0}\right) + \exp\left(-\frac{Z_\beta r_1 + Z_\alpha r_2}{a_0}\right)\right], \tag{12•3–17}$$

with *two* variational parameters Z_α and Z_β.

Because the two electrons are indistinguishable, two terms are necessary in (12•3–17), to meet the exchange condition (11•4–3b) for antiparallel spins. Note that for $Z_\alpha = Z_\beta$, (12•3–17) reduces to the earlier form, with $C = 1$.

Carry out the energy minimization using the trial wave function (12•3–17). Warning: The calculations are fairly tedious.

Chapter 13

EXPANSION PRINCIPLE AND MATRIX FORMULATION

13.1 EXPANSION THEOREM: EIGENFUNCTIONS AS COMPLETE ORTHOGONAL SETS

13.1.1 Derivation of the Theorem

We come now to a central theorem, the **expansion theorem**, that forms the mathematical basis for much of what is to follow in the remaining chapters. Roughly speaking, it states that any state function that is likely to occur in any quantum-mechanical calculations can be expanded as a linear superposition of the eigenfunctions of *any* Hermitian operator, similar to the way in which it can be expanded as a linear superposition of trigonometric functions in Fourier analysis.

Our derivation draws on the variational theorem. We derive the expansion theorem here first for expansion by eigenfunctions of the Hamiltonian and generalize it later to other operators.

The discussion is best carried out in Dirac notation. We denote the eigenstates of the Hamiltonian by the ket vectors $|n\rangle$, such that

$$\hat{H}|n\rangle = \mathcal{E}_n|n\rangle, \; n = 1, 2, 3, \ldots . \qquad (13{\bullet}1\text{–}1)$$

Without loss in generality, we may assume that the zero of energy has been chosen such that all energy eigenvalues are positive and that they are ordered by increasing energy:

$$0 < \mathcal{E}_1 \leq \mathcal{E}_2 \leq \mathcal{E}_3 \leq \ldots . \qquad (13{\bullet}1\text{–}2)$$

We further assume that the eigenvalues are *discrete,* that is, that there is only a finite number of eigenstates in any finite energy interval. We will comment on the generalization to continuous eigenvalues in section 13.4.

Finally, we assume that all eigenstates of $\hat{H}$ are normalized and that any eigenstates belonging to the same degenerate eigenvalue are chosen to be orthogonal,

$$\langle n|m\rangle = \delta_{nm}. \qquad (13{\bullet}1\text{–}3)$$

Consider now a state $|\Psi\rangle$ of which we assume only that it belongs to the hermiticity domain of $\hat{H}$ and has a finite energy expectation value

$$\langle \mathcal{E}\rangle = \langle\Psi|\hat{H}|\Psi\rangle. \qquad (13{\bullet}1\text{–}4)$$

Suppose that $|\Psi\rangle$ is expanded into the *finite* series

$$|\Psi\rangle = \sum_{n=1}^{N} a_n|n\rangle + |R_N\rangle, \qquad (13{\bullet}1\text{–}5)$$

where the a_n are as yet unspecified expansion coefficients and $|R_N\rangle$ is a remainder. Because we have said nothing yet about $|R_N\rangle$, (13•1–5) can always be satisfied. What we prove here is the following: If the a_n are chosen as

$$a_n = \langle n|\Psi\rangle, \qquad (13{\bullet}1\text{–}6)$$

and if N is increased so that higher and higher energy eigenstates are included, then the norm of the remainder $|R_N\rangle$ goes to zero:

$$\text{As } \mathcal{E}_N \to \infty : \langle R_N|R_N\rangle \to 0. \qquad (13{\bullet}1\text{–}7)$$

It is in this sense that the $|n\rangle$ form a complete set.

If the function $\Psi(\mathbf{r})$ is a non-pathological function, free from unphysical discrete singularities, the remainder $|R_N\rangle$ will itself be a continuous function, not containing any discrete singularities, in which case a vanishing norm means that the remainder itself goes to zero for $N \to \infty$,

$$\lim_{N\to\infty} |R_N\rangle = 0. \qquad (13{\bullet}1\text{–}8)$$

We can then write (13•1–5) in the form

$$\boxed{|\Psi\rangle = \sum_{n=1}^{\infty} a_n|n\rangle,} \qquad (13{\bullet}1\text{–}9)$$

with the a_n given by (13•1–6).

To prove (13•1–7), we form the inner product of (13•1–5) with each of the $|n\rangle$ and draw on the orthonormality relation (13•1–3) to obtain

$$\langle n|\Psi\rangle = a_n + \langle n|R_N\rangle,\ n = 1, 2, \ldots, N. \tag{13•1–10}$$

If we now select a_n according to (13•1–6), it follows that $|R_N\rangle$ is orthogonal to all the $|n\rangle$:

$$\langle n|R_N\rangle = 0,\ n = 1,2, \ldots, N. \tag{13•1–11}$$

We next calculate the expectation value of the energy for $|\Psi\rangle$, again using (13•1–3):

$$\begin{aligned}\langle \mathcal{E}\rangle &= \langle\Psi|\hat{H}|\Psi\rangle \\ &= \sum_{n=1}^{N} a_n^* a_n \mathcal{E}_n + \sum_{n=1}^{N}[a_n^* \langle n|H|R_N\rangle + a_n\langle R_N|\hat{H}|n\rangle] + \langle R_N|H|R_N\rangle.\end{aligned} \tag{13•1–12}$$

Because of (13•1–1) and (13•1–11), the terms in the square bracket vanish, and the expression may be re-written as

$$\langle \mathcal{E}\rangle = \sum_{n=1}^{N}|a_n|^2\mathcal{E}_n + \langle R_N|R_N\rangle\langle\mathcal{E}_R\rangle, \tag{13•1–13}$$

where $\langle\mathcal{E}_R\rangle$ is the expectation value of the energy for the remainder $|R_N\rangle$ *after* $|R_N\rangle$ has been re-normalized.

From (13•1–11) and the variational theorem, it follows that

$$\langle\mathcal{E}_R\rangle \geq \mathcal{E}_N. \tag{13•1–14}$$

Thus,

$$0 \leq \langle R_N|R_N\rangle \leq \frac{1}{\mathcal{E}_N}\left(\langle\mathcal{E}\rangle - \sum_{n=1}^{N}|a_n|^2\,\mathcal{E}_n\right) \xrightarrow[\mathcal{E}_N\to\infty]{} 0. \tag{13•1–15}$$

Because of (13•1–2), the sum contains only positive terms; hence, the numerator in (13•1–15) can only *decrease* with increasing N, whereas the denominator increases without limit. This implies (13•1–7), the relation we wanted to prove.

An essential point in the proceding derivation was our assumption that the state $|\Psi\rangle$ has a finite energy expectation value. When this assumption is not fulfilled, the expansion will in general not be possible. For example, the eigenfunctions of a one-dimensional box with infinitely high walls vanish outside the box and hence cannot be used to expand any function $\Psi(x)$ that has finite values outside the box. And indeed, such a function would have an infinite expectation value for the potential energy, using the box Hamiltonian. On the other hand, it would be possible to expand the box eigenfunctions in terms of, say, the harmonic oscillator eigenfunctions.

The expansion theorem is readily generalized to other Hermitian operators. Although we had assumed that $\hat{H}$ is the Hamilton operator, only the hermiticity of $\hat{H}$ was actually used in our derivation—just as with the variational theorem—together with the fact that its eigenvalues have a *minimum* value. Obviously, our proof applies automatically to *any* Hermitian operator whose eigenvalues exhibit a minimum and, by simple sign reversal, to those whose eigenvalues exhibit a maximum rather than a minimum.

In order to extend the proof to Hermitian operators $\hat{A}$ whose eigenvalues extend from $-\infty$ to $+\infty$, it is merely necessary to expand by eigenstates of the operator $\hat{A}^2$. As we stated in section 12.1, the eigenstates of $\hat{A}^2$ are the same as those of $\hat{A}$ itself, and by ordering these eigenstates in the order of increasing squares of eigenvalues, A_n^2, our earlier proof can be applied. The only non-trivial change is in the condition for the expandability of the state $|\Psi\rangle$. Instead of requesting a finite expectation value $\langle A\rangle = \langle\Psi|\hat{A}|\Psi\rangle$, we must now request that

$$\langle A^2\rangle = \langle\Psi|\hat{A}^2|\Psi\rangle = \langle\hat{A}\Psi|\hat{A}\Psi\rangle \tag{13•1–16}$$

remain finite.

All the information contained in the original state $|\Psi\rangle$ is evidently contained in the set of expansion coefficients, which may therefore be used as an alternative representation of the state $|\Psi\rangle$, fully equivalent to, say, the position representation in terms of a state function $\Psi(\mathbf{r})$. We commented on this point previously in chapter 10, in the context of the interpretation of the expansion coefficients for the non-stationary states of a harmonic oscillator. It is common practice to name the representation by the name of the Hermitian operator whose eigenstates serve as the basis for the expansion. If, as assumed here, these eigenvalues form a discrete set, the representation is a discrete one, as in the earlier harmonic oscillator case.

13.1.2 Normalization and Inner Product

Given an expansion of the form (13•1–5) that satisfies the limit (13•1–7), we obtain a simple general relation between the norm of the state $|\Psi\rangle$ and its expansion coefficients:

$$\langle\Psi|\Psi\rangle = \sum_{n=1}^{N}|a_n|^2 + \langle R_N|R_N\rangle \underset{N\to\infty}{\longrightarrow} \sum_{n=1}^{\infty}|a_n|^2. \tag{13•1–17}$$

If $|\Psi\rangle$ is normalized,

$$\boxed{\langle\Psi|\Psi\rangle = \sum_{n=1}^{\infty}|a_n|^2 = 1.} \tag{13•1–18}$$

The relation (13•1–17) is readily generalized to the inner product of two state vectors. Let $|\Psi\rangle$ be given by the expansion (13•1–9) and $|\Phi\rangle$ by

$$|\Phi\rangle = \sum_{n=1}^{\infty} b_n |n\rangle. \tag{13•1–19}$$

We then obtain immediately

$$\langle\Psi|\Phi\rangle = \sum_{n=1}^{\infty} a_n^* b_n. \tag{13•1–20}$$

If $|\Phi\rangle$ and $|\Psi\rangle$ are orthogonal,

$$\langle\Psi|\Phi\rangle = \sum_{n=1}^{\infty} a_n^* b_n = 0. \tag{13•1–21}$$

13.1.3 Physical Interpretation of the Expansion Coefficients as Measurement Probability Amplitudes

We saw earlier that an expectation value $\langle A\rangle = \langle\Psi|\hat{A}|\Psi\rangle$ of the observable 'A' can correspond to a *sharp* measurement value only if $\langle A\rangle$ is an eigenvalue A_n of the operator $\hat{A}$ and if the state $|\Psi\rangle$ is an eigenstate $|n\rangle$ of $\hat{A}$ with that eigenvalue A_n. The expansion theorem enables us now to turn to the question regarding the distribution of the actual measurement values when the state $|\Psi\rangle$ is *not* an eigenstate of $\hat{A}$, and to relate the *probabilities* that certain measurement values of 'A' will occur to the expansion coefficients of the state $|\Psi\rangle$ in any expansion of $|\Psi\rangle$ by the eigenstates of the operator $\hat{A}$. The problem is a direct generalization of the interpretation of the expansion coefficients for a time-dependent superposition state of the harmonic oscillator, discussed in section 7.4.

With $\hat{A}$ being a Hermitian operator, we can expand the state $|\Psi\rangle$ as in (13•1–9), except that now the states $|n\rangle$ are eigenstates of $\hat{A}$ rather than $\hat{H}$:

$$\hat{A}|n\rangle = A_n|n\rangle. \tag{13•1–22}$$

We can then express the expectation value $\langle A\rangle$ in terms of the expansion coefficients a_n. From the form (13•1–6) of the expansion coefficients, we obtain[1]

$$\begin{aligned}\langle A\rangle \equiv \langle\Psi|\hat{A}|\Psi\rangle &= \sum_m\sum_n a_m^* a_n \langle m|\hat{A}|n\rangle = \sum_m\sum_n a_m^* a_n A_n \langle m|n\rangle \\ &= \sum_n |a_n|^2 A_n.\end{aligned} \tag{13•1–23}$$

On the other hand, the expectation value, being defined as a probabilistic average, is also related to the probabilities $P(A_m')$ for *measuring* the value A_m'

[1] For simplicity, we omit the summation limits from now on. Unless specifically stated otherwise, all sums go over all values of the summation index, usually 1 to ∞.

according to the fundamental relation

$$\langle A \rangle = \sum_n P(A'_n)A'_n, \tag{13•1–24}$$

where the sum goes over all possible *measurement* values. We have designated the *measurement* values by the primed symbol A' rather than the unprimed symbol A chosen for the *eigenvalues,* in order to keep the *conceptual* distinction between the two clear. However, the analogy between (13•1–23) and (13•1–24) is obvious, and we will in fact argue presently that the two sets are indeed the same.

The probabilities must of course add up to unity:

$$\sum_n P(A'_n) = 1. \tag{13•1–25}$$

This is evidently the analog of the normalization condition (13•1–18).

A comparison of (13•1–23) with (13•1–24) and of (13•1–18) with (13•1–25) leads to the interpretation of the quantities $|a_n|^2$ as the probabilities that, during a measurement of A, the system materializes in the state $|n\rangle$. If the state $|n\rangle$ has a non-degenerate eigenvalue of $\hat{A}$, this statement is equivalent to the statement that $|a_n|^2$ is the probability $P(A_n)$ that a measurement of A yields the eigenvalue A_n:

$$P(A') = |a_n|^2 \quad \text{if} \quad A' = A_n \quad \text{(non-degenerate)}. \tag{13•1–26a}$$

If the eigenvalue A_n is degenerate, we must sum the probabilities over all degenerate states to obtain the probability that a measurement of A yields the eigenvalue A_n:

$$P(A') = \sum_{A_n = A'} |a_n|^2 \quad \text{if} \quad A' = A_n \text{ (degenerate)}. \tag{13•1–26b}$$

Contained in this interpretation is the far-reaching claim that the probability for any other value is zero:

$$P(A) = 0 \quad \text{if} \quad A \neq A_n \quad \text{for all } n. \tag{13•1–27}$$

Or, put differently:

> In any measurement of the value of an observable 'A' for any system, the only values that can occur are the different eigenvalues A_n of the Hermitian operator $\hat{A}$ that corresponds to the observable 'A'.

The relations (13•1–26) and (13•1–27) are often called the **theorem of measurement probabilities**.

Our argument, based on a simple comparison of the expansion sums with the measurement sums, does not constitute a rigorous proof that our interpretation of the expansion coefficient is the *only* interpretation consistent with the identification of the expectation values calculated by the quantum-mechanical formalism with actual measurement averages. It is possible to give such a proof, but we shall not do so here.

The most interesting part about the theorem is the emphatic exclusion of non-eigenvalues from the set of possible measurement values. We commented on this point earlier, in our discussion of non-stationary states of the harmonic oscillator in section 10.4. The present discussion simply extends these considerations to *all* observables of *all* systems. If a deviation from this prediction were ever found, it would constitute a refutation of the probabilistic interpretation of quantum mechanics itself, at least in its present form.

◆ PROBLEM TO SECTION 13.1

#13•1-1: Displaced Harmonic Oscillator

A one-dimensional harmonic oscillator is displaced by x_0 along the x-axis, with the Hamiltonian

$$\hat{H} = \frac{\hat{p}^2}{2M} + \frac{M}{2}\omega^2 \cdot (x - x_0)^2. \tag{13•1–28}$$

It must be possible to write the energy eigenstates of the displaced oscillator as linear superpositions of those of the undisplaced oscillator. Determine the expansion coefficients for the *ground* state of the displaced oscillator.

The following procedure is suggested:

(a) Re-write the Hamiltonian (13•1–28) in terms of the raising and lowering operators $\hat{a}^+$ and $\hat{a}^-$ for the undisplaced oscillator, and re-arrange the result into an expression of the form

$$\hat{H} = \hbar\omega \cdot (\hat{b}^+\hat{b}^- + \tfrac{1}{2}), \tag{13•1–29}$$

where $\hat{b}^+$ and $\hat{b}^-$ are two new operators that are functions of the old operators $\hat{a}^+$ and $\hat{a}^-$ and that obey the same commutation relation.

(b) Drawing on the form (13•1–29), determine what mathematical condition the displaced ground state must satisfy. Re-express this condition in terms of a condition involving the original operators $\hat{a}^+$ and/or $\hat{a}^-$.

(c) Proceed to write the displaced ground-state wave function as an explicit superposition of the undisplaced energy eigenfunctions. The procedure bears some formal similarity to the problem of the quasi-classical oscillating state. Explain why such a similarity is to be expected.

13.2 STATE VECTORS AND OPERATOR MATRICES

13.2.1 States as Vectors in Hilbert Space

In our introduction of Dirac notation in section 7.2, we noted the formal analogy between the inner product of two states in quantum mechanics and the inner product of two complex vectors in a generalized multi-dimensional space. In fact, much of the terminology of Dirac notation was built on this analogy, and we introduced the term **state vector** (without, however, using it very much subsequently). The analogy becomes complete with the introduction of the expansion of states by the eigenstates of Hermitian operators.

Consider the inner product $\langle\Psi|\Phi\rangle$ of two states, now written in the form (13•1–20), as a relation between the expansion coefficients of the two states. The right-hand side of this relation has exactly the same form as the inner product of two vectors $\mathbf{a}^*$ and $\mathbf{b}$ whose Cartesian components are the a_n^* and b_n. The only difference is that the space in which these vectors "live" is not the ordinary three-dimensional "real" space, but an infinite-dimensional generalized mathematical space, called **Hilbert space**. The expansion of a state vector $|\Psi\rangle$ in terms of the eigenstates of the Hermitian operator $\hat{A}$ is mathematically equivalent to setting up a system of Cartesian coordinate axes in that space, such that the eigenstates of $\hat{A}$ act as the mutually orthogonal unit vectors that define the axes. The expansion coefficients a_n^* and b_n in (13•1–20) and elsewhere are the components of the respective state vectors relative to the axis basis vectors.

The normalization condition (13•1–18) is the equation of a generalized sphere in Hilbert space. In this space, all normalized state vectors are unit vectors, and different states are represented by vectors with different "directions."

13.2.2 Matrix Representations of Operators

The representation of the state $|\Psi\rangle$ as a vector in Hilbert space may be expressed mathematically by writing the Dirac ket vector $|\Psi\rangle$ as an (infinite) single-column matrix:

$$|\Psi\rangle = \begin{bmatrix} a_1 \\ a_2 \\ a_3 \\ \vdots \end{bmatrix} = \begin{bmatrix} \langle 1|\Psi\rangle \\ \langle 2|\Psi\rangle \\ \langle 3|\Psi\rangle \\ \vdots \end{bmatrix}. \tag{13•2–1a}$$

The bra $\langle\Psi|$ dual to $|\Psi\rangle$ is the Hermitian conjugate (or adjoint) single-row matrix

$$\langle\Psi| = (a_1^*, a_2^*, a_3^*, \ldots). \tag{13•2–1b}$$

This representation of the states leads to the representation of operators by matrices. Consider the operator equation

$$|\Phi\rangle = \hat{A}|\Psi\rangle. \tag{13•2–2}$$

Assume that $|\Phi\rangle$ is expanded by the basis states $|n\rangle$, as in (13•1–19), with expansion coefficients b_n. If we insert (13•1–19) into (13•2–2) and take, on both sides, the inner product with the bra vector $\langle m|$, we obtain

$$b_m = \sum_n A_{mn} a_n \quad (m = 1,2,3, \ldots), \tag{13•2–3}$$

where the quantities

$$A_{mn} = \langle m|\hat{A}|n\rangle \tag{13•2–4}$$

are the **matrix elements** of the operator $\hat{A}$. Equations (13•2–3) form a set of equations, one for each value of m. They can be lumped together in the single matrix equation

$$\begin{bmatrix} b_1 \\ b_2 \\ b_3 \\ \vdots \end{bmatrix} = \begin{bmatrix} A_{11} & A_{12} & A_{13} & \cdots \\ A_{21} & A_{22} & A_{23} & \cdots \\ A_{31} & A_{32} & A_{33} & \cdots \\ \vdots & \vdots & \vdots & \ddots \end{bmatrix} \cdot \begin{bmatrix} a_1 \\ a_2 \\ a_3 \\ \vdots \end{bmatrix}. \tag{13•2–5}$$

The two column vectors in (13•2–5) are the representations of the two state vectors $|\Phi\rangle$ and $|\Psi\rangle$, in the basis of the $|n\rangle$'s in Hilbert space. The matrix

$$\hat{A} = \begin{bmatrix} A_{11} & A_{12} & A_{13} & \cdots \\ A_{21} & A_{22} & A_{23} & \cdots \\ A_{31} & A_{32} & A_{33} & \cdots \\ \vdots & \vdots & \vdots & \ddots \end{bmatrix} \tag{13•2–6}$$

is the representation of the operator $\hat{A}$ in that basis. The action of an operator on a state vector is represented by ordinary matrix multiplication of a matrix with a column vector. This is, of course, the reason for the name **matrix elements** for the quantities $\langle m|\hat{A}|n\rangle$.

The representation of operators as matrices has proven to be a very fruitful concept. The reason for this is the following. Consider the product of two operators,

$$\hat{C} = \hat{A}\hat{B}, \tag{13•2–7}$$

namely, the operator that consists of first applying the operator $\hat{B}$ to a state vector and then applying the operator $\hat{A}$ to the result. The operator $\hat{C}$ has a matrix representation itself, and one finds easily that the matrix elements of $\hat{C}$ are given in terms of those of $\hat{A}$ and $\hat{B}$, via

$$C_{mn} = \sum_l A_{ml} B_{ln}. \tag{13•2–8}$$

The proof is left to the reader. The expression (13•2–8) is, of course, the conventional rule for matrix multiplication. Thus, we see that the product of two operators is represented in any matrix representation by the conventional matrix product of the corresponding matrices.

13.2.3 Schroedinger Equation in Matrix Form

One of the most important applications of the matrix formalism is to the Schroedinger equation. In a matrix representation, the latter assumes the form

$$\begin{bmatrix} H_{11} & H_{12} & H_{13} & \cdots \\ H_{21} & H_{22} & H_{23} & \cdots \\ H_{31} & H_{32} & H_{33} & \cdots \\ \vdots & \vdots & \vdots & \ddots \end{bmatrix} \begin{bmatrix} a_1 \\ a_2 \\ a_3 \\ \vdots \end{bmatrix} = \mathcal{E} \begin{bmatrix} a_1 \\ a_2 \\ a_3 \\ \vdots \end{bmatrix}. \tag{13•2–9}$$

The matrix on the left-hand side is the matrix representation of the hamiltonian $\hat{H}$ in an *arbitrary* system of basis states $|n\rangle$. The matrix elements of $\hat{H}$ are, of course,

$$H_{mn} = \langle m|H|n\rangle. \tag{13•2–10}$$

The formalism becomes simplest when the basis states are the eigenstates of the Hamiltonian itself, as in (13•1–1). In that case, all off-diagonal matrix elements of $\hat{H}$ vanish,

$$H_{mn} = \mathcal{E}_n \delta_{mn}, \tag{13•2–11}$$

and the matrix representation of $\hat{H}$ becomes a simple diagonal matrix. The Schroedinger equation (13•2–9) then simplifies to

$$\begin{bmatrix} \mathcal{E}_1 & 0 & 0 & \cdots \\ 0 & \mathcal{E}_2 & 0 & \cdots \\ 0 & 0 & \mathcal{E}_3 & \cdots \\ \vdots & \vdots & \vdots & \ddots \end{bmatrix} \begin{bmatrix} a_1 \\ a_2 \\ a_3 \\ \vdots \end{bmatrix} = \mathcal{E} \begin{bmatrix} a_1 \\ a_2 \\ a_3 \\ \vdots \end{bmatrix}. \tag{13•2–12}$$

But in this case, the Schroedinger equation is already solved: For each n, we have a solution of the form

$$a_m = \delta_{mn}, \quad \mathcal{E} = \mathcal{E}_n. \tag{13•2–13}$$

The task of solving the Schroedinger equation is therefore mathematically equivalent to finding a way to get from (13•2–9) to (13•2–12). Utilizing the matrix formalism itself to accomplish this goal forms the basis for one of the methods for actually solving the equation, known under the name **degenerate perturbation theory**. It will be the subject of chapter 14.

13.3 DIRAC NOTATION

13.3.1 Outer Product

The mathematical manipulation of operator matrices and matrix elements is greatly facilitated by an extension of the Dirac notation and by the introduction of a second kind of symbolic product between a bra vector and a ket vector, called the **outer product** or **operator product**.

Suppose we insert the expression (13•1–5) for the expansion coefficient into the expansion expression (13•1–9):

$$|\Psi\rangle = \sum_n |n\rangle\langle n|\Psi\rangle. \tag{13•3–1}$$

Consider now one of the individual terms in the sum on the right-hand side, written here as

$$|\Phi\rangle = |n\rangle\langle n|\Psi\rangle, \tag{13•3–2}$$

which is a special case of the more general expression

$$|\Phi\rangle = |m\rangle\langle n|\Psi\rangle. \tag{13•3–3}$$

We may view the last expression in two different ways:

(a) As the result of first forming the inner product of $\langle n|$ and $|\Psi\rangle$, which is a number, and then multiplying the state vector $|m\rangle$ by this number, by ordinary algebraic multiplication.

(b) As a result of operating with a new kind of linear operator $\hat{P}_{mn}$ on the state vector $|\Psi\rangle$:

$$|\Phi\rangle = \hat{P}_{mn}|\Psi\rangle. \tag{13•3–4}$$

This is a perfectly legitimate operator, whose action is fully defined for every admissible state vector $|\Psi\rangle$ for which the inner product $\langle n|\Psi\rangle$ exists, namely, by the requirement that the results of (13•3–3) and (13•3–4) be identical, regardless of $|\Psi\rangle$. We may express this *definition* by writing, symbolically,

$$\hat{P}_{mn}|\Psi\rangle = [|m\rangle\langle n|]|\Psi\rangle, \tag{13•3–5}$$

where, on the right-hand side, we have broken the inner product $\langle n|\Psi\rangle$ into its two factors $\langle n|$ and $|\Psi\rangle$ and have added grouping brackets around the ket-bra combination $|m\rangle\langle n|$.

Because (13•3–5) must hold for every admissible $|\Psi\rangle$, me may go one step further and omit $|\Psi\rangle$ from the expression altogether and write (13•3–5) as a relation between the operator $\hat{P}_{mn}$ itself and the content of the grouping brackets, i.e.,

$$\boxed{\hat{P}_{mn} = |m\rangle\langle n|.} \tag{13•3–6}$$

We view this combination of a ket vector $|m\rangle$ to the left with a bra vector $\langle n|$ to the right as a second kind of product of a bra and a ket vector, called the **outer product** (or operator product) of $|m\rangle$ and $\langle n|$. The outer product is evidently not a number, but a linear operator, whose action is *defined* by the requirement that

$$[|m\rangle\langle n|]|\Psi\rangle \equiv |m\rangle[\langle n|\Psi\rangle], \tag{13•3–7}$$

for every admissible $|\Psi\rangle$.

The identity of the two different sequences in (13•3–7) means that it does not matter whether the inner or the outer product operation is executed first; hence, we may continue to omit the square grouping brackets in (13•3–7) altogether. In terms of mathematical axiomatics, the two kinds of products satisfy the *distributive law* with respect to each other.

The idea of two different kinds of products between two state vectors is superficially similar to the idea of two different kinds of products between two vectors in ordinary vector algebra: their scalar product and their vector product. However, this analogy is of limited value: While the inner product of two state vectors is indeed analogous to the scalar product of two ordinary vectors, the outer product of two state vectors cannot be readily compared to the vector product of two ordinary vectors.

13.3.2 Projection Operators, the Unit Operator, and the Closure Relation

The simplest kind of outer product is that of a basis bra vector $|n\rangle$ with its dual ket $\langle n|$:

$$\hat{P}_{nn}|\Psi\rangle = |n\rangle\langle n|. \tag{13•3–8}$$

This operator evidently extracts from $|\Psi\rangle$ the contribution $a_n|n\rangle$ of the basis state vector $|n\rangle$ in the eigenvector expansion (13•1–9). In vector language, we may say that it *projects* the state vector $|\Psi\rangle$ onto the basis vector $|n\rangle$. It is therefore called a **projection operator**.

With the aid of the various projection operators, the expansion (13•3–1) of the state vector $|\Psi\rangle$ may be re-written as

$$|\Psi\rangle = \sum_n |n\rangle\langle n|\Psi\rangle = \left[\sum_n |n\rangle\langle n|\right]|\Psi\rangle, \tag{13•3–9}$$

where, on the far right, we have *first* grouped the projection operators and *then* operated with their sum on the state $|\Psi\rangle$. The result of the operation, on any state $|\Psi\rangle$, with the sum over all projection operators, is evidently the original state $|\Psi\rangle$ itself. The sum over all projection operators is therefore simply the **unit operator**:

$$\boxed{\sum_n |n\rangle\langle n| = \sum_n \hat{P}_{nn} = \hat{1}.} \tag{13•3–10}$$

The form (13•3–10) of the unit operator is often found useful in quantum-mechanical calculation in two different ways:

(a) If a mathematical expression can be re-arranged in such a way that the left-hand side of (13•3–10) occurs as a factor embedded between other factors, that factor can be omitted. A simple example would be the sequence

$$\sum_n \langle n|\Phi\rangle\langle\Psi|n\rangle = \sum_n \langle\Psi|n\rangle\langle n|\Phi\rangle = \langle\Psi|\left[\sum_n |n\rangle\langle n|\right]|\Phi\rangle = \langle\Psi|\Phi\rangle, \tag{13•3–11}$$

which is nothing other than the relation (13•1–20) between the expansion coefficients of $|\Phi\rangle$ and $|\Psi\rangle$.

(b) Conversely, it is sometimes useful to insert the dummy operator $\hat{1}$ as a factor and then employ (13•3–10) in reverse order. An elementary example would be a simple re-derivation of the matrix multiplication rule (13•2–7) for the matrix elements of product operators. If $\hat{C} = \hat{A}\hat{B}$, we may write

$$\begin{aligned} C_{mn} &\equiv \langle m|\hat{C}|n\rangle = \langle m|\hat{A}\hat{B}|n\rangle = \langle m|\hat{A}\,\hat{1}\hat{B}|n\rangle \\ &= \sum_l \langle m|\hat{A}|l\rangle\langle l|\hat{B}|n\rangle = \sum_l A_{ml}B_{lm}. \end{aligned} \tag{13•3–12}$$

But this is of course the same as (13•2–8).

We will frequently use these kinds of manipulations in the chapters that follow.

Closure Relation

It is useful to re-express the relation (13•3–9) involving the unit operator in the notation of conventional state functions in the position representation, by substituting

$$|\Psi\rangle \rightarrow \Psi(\mathbf{r}), \tag{13•3–13a}$$

$$|n\rangle \rightarrow \psi_n(\mathbf{r}), \tag{13•3–13b}$$

and by re-expressing the inner products as conventional integrals. This leads to

$$\begin{aligned} \Psi(\mathbf{r}) &= \sum_n \psi_n(\mathbf{r}) \int \psi_n^*(\mathbf{r}')\Psi(\mathbf{r}')\, d^3\mathbf{r}' \\ &= \int \left[\sum_n \psi_n^*(\mathbf{r}')\psi_n(\mathbf{r})\right]\Psi(\mathbf{r}')\, d^3\mathbf{r}', \end{aligned} \tag{13•3–14}$$

where, in the last expression, we have interchanged summation and integration.

Just like (13•3–9), the relation (13•3–14) expresses $\Psi(\mathbf{r})$ in terms of itself, inside and outside the integral, for any $\Psi(\mathbf{r})$ that can be expanded according to (13•1–9). But (13•3–14) can be true for an arbitrary $\Psi(\mathbf{r})$ only if the square bracket in it is a three-dimensional Dirac delta function,

$$\sum_{n=1}^{\infty} \psi_n^*(\mathbf{r}')\psi_n(\mathbf{r}) = \delta(\mathbf{r} - \mathbf{r}'). \tag{13•3–15}$$

Here $\delta(\mathbf{r} - \mathbf{r}')$ is defined by the two requirements (see appendix A) that

$$\int_{\Omega} \delta(\mathbf{r} - \mathbf{r}')\, d^3\mathbf{r} = 0 \quad \text{if } \Omega \text{ does not contain } \mathbf{r}' \tag{13•3–16a}$$

and

$$\int_{\Omega} \delta(\mathbf{r} - \mathbf{r}')\, d^3\mathbf{r} = 1 \quad \text{if } \Omega \text{ contains } \mathbf{r}'. \tag{13•3–16b}$$

The relation (13•3–15) is commonly called the **closure relation** for the eigenfunctions $\psi_n(\mathbf{r})$. The name indicates that the relation (13•3–14) is valid for arbitrary $\Psi(\mathbf{r})$ only if the set of ψ_n's is "closed," meaning complete.

In our derivation of the closure relation, we have considered only the simplest case, that of a set of eigenfunctions in the position representation for one particle, defined in ordinary three-dimensional space. It is readily generalized to other representations, such as the momentum representation, and to representations with more than three independent variables, such as occur in systems with more than one particle. We ignore these generalizations here.

13.3.3 Generalization to Arbitrary Operators

As a generalization of (13•3–6), we adopt the convention that any numerical multiplier of $\hat{P}_{mn}$ may be pulled between the state vectors $|m\rangle$ and $\langle n|$:

$$|m\rangle\, a \langle n| \equiv a\hat{P}_{mn}. \tag{13•3–17}$$

This generalization permits us to apply the outer product notion to arbitrary linear operators operating on the state vectors of quantum mechanics. The generalization is most easily obtained by employing the unit operator $\hat{1}$ defined in (13•3–10), and it is in fact a good demonstration of the utility of this operator. Let $\hat{A}$ be the operator we wish to express in outer product notation. By multiplying $\hat{A}$ on both sides with $\hat{1}$, we may write

$$\begin{aligned} \hat{A} = \hat{1}\hat{A}\hat{1} &= \left[\sum_m |m\rangle\langle m|\right]\hat{A}\left[\sum_n |n\rangle\langle n|\right] \\ &= \sum_m\sum_n |m\rangle\langle m|\hat{A}|n\rangle\langle n| = \sum_m\sum_n |m\rangle A_{mn}\langle n|. \end{aligned} \tag{13•3–18}$$

In going from the first line in (13•3–18) to the second line, we have used the distributive law (13•3–7) for inner and outer products and then dropped the square brackets. The second line in (13•3–18) may be viewed as three operators applied in succession, first $|n\rangle\langle n|$, then $\hat{A}$, and finally, $|m\rangle\langle m|$, followed by summations over m and n. But it can also be viewed as in the last equality, as a single outer product of the form (13•3–17), with $a = A_{mn}$, followed by the same double summation.

We see from (13•3–18) that *any* linear operator $\hat{A}$ can be written as the sum over all the outer products that can be formed between the basis vectors, each one multiplied by the appropriate matrix element.

13.3.4 Hermitian Operators and Hermitian Conjugate Operator Pairs

Let the operator $\hat{A}$ now be a Hermitian operator. From the hermiticity condition (7•4–3), one finds immediately that the matrix elements of $\hat{A}$ satisfy the condition

$$A_{nm}^* = (\langle n|\hat{A}|m\rangle)^* = \langle m|\hat{A}|n\rangle = A_{mn}. \qquad (13•3–19)$$

Matrices with this property are called **Hermitian matrices**. In fact, the concept of Hermitian *matrices* is historically much older than that of Hermitian *operators;* the latter were given their name precisely because their matrix representations are Hermitian matrices.

If two operators $\hat{a}$ and $\hat{a}^\dagger$ form a Hermitian conjugate pair, then instead of (13•3–19), we have

$$(a_{nm})^* = (\langle n|\hat{a}|m\rangle)^* = \langle m|\hat{a}^\dagger|n\rangle = a_{mn}^\dagger. \qquad (13•3–20)$$

Exercise: Show that the raising and lowering operators for the harmonic oscillator may be written

$$\hat{a}^+ = \sum_n |n+1\rangle\sqrt{n+1}\langle n| \qquad (13•3–21a)$$

and

$$\hat{a}^- = \sum_n |n\rangle\sqrt{n+1}\langle n+1|. \qquad (13•3–21b)$$

Show that these two forms satisfy (13•3–20). Derive the commutation relation (10•1–4) from (13•3–21a,b).

13.4 CONTINUOUS EIGENVALUES

At the beginning of this chapter, we made the simplifying assumption that the eigenstates that are used to expand some general state vector belong to a *discrete* set of eigenvalues of a Hermitian operator. This enabled us to write the eigenstate expansion as the simple sums (13•1–5) and (13•1–9), and it also simplified the theorem of measurement probabilities.

The simplifying assumption of discrete eigenvalues is an idealization. For example, while the eigenvalues of the Hamilton operator for the bound states of a system are indeed discrete, those for states above the binding energy are continuous. In those energy ranges, the sum in (13•1–5) and (13•1–9) must be replaced by an integral. The most general way to write (13•1–9) is then

$$|\Psi\rangle = \sum a_d(A)|A\rangle_d + \int a_c(A)|A\rangle_c\, dA, \qquad (13•4–1)$$

where the sum goes over all discrete eigenvalues of $\hat{A}$ and the integral over all continuous ones, with

$$\hat{A}|A\rangle = A|A\rangle \tag{13•4–2}$$

in both regimes.

In the orthogonality and closure relations (13•1–21) and (13•3–15), too, the sums must be replaced by the combination of a sum and an integral.

From a practical point of view, the separate formal treatment of the continuous portion of the eigenvalue spectrum is a nuisance. Fortunately, at least for the most important case, the expansion by *energy* eigenfunctions, the eigenvalue spectrum is easily converted into a dense but discrete spectrum by simply enclosing the system in a large but finite volume. Consider the energy eigenvalues of a particle in a potential well of finite depth. By changing the problem from a well in infinite space to a well inside a very large but finite box (**Fig. 13•4–1**), we convert the previously free continuum states to bound states inside the large outer box. The energy eigenvalue spectrum then becomes very dense, but discrete. By making the box sufficiently large, the properties of the individual state remain essentially unchanged. One might even argue that in many cases the enclosed problem is more realistic than the unbounded one. The procedure has the additional advantage that it permits one to count the number of quasi-continuum states per unit volume of the box, an important requirement in many problems.

The mathematical implementation of the boxing-in takes the form of suitable boundary conditions at the surface of the fictitious box. The simplest form of such boundary conditions would be to set $\psi = 0$ at the surface, corresponding to infinitely high potential walls. However, such hard-wall boundary conditions lead to eigenfunctions that do not carry any translational current, which is undesirable in the quasi-continuum range, because it means that linear superpositions of box eigenstates are needed to describe current-carrying states, which are what is usually of interest in that range. It is therefore often preferable to use **periodic boundary conditions**: One assumes a cube-shaped box with edges of length L parallel to the axes of a Cartesian coordinate system and wave function that is periodic in space, with periods L:

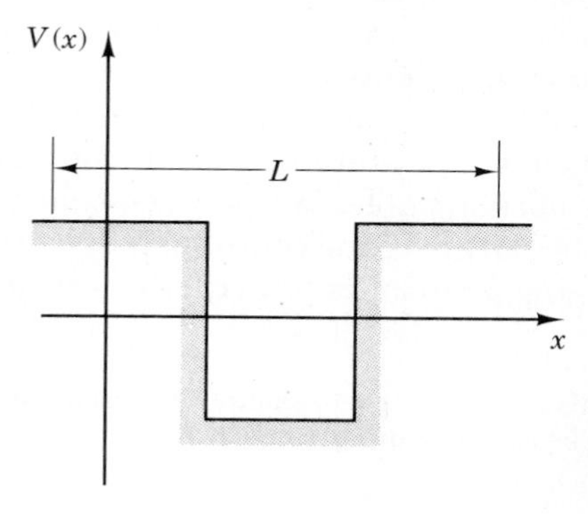

Figure 13•4–1. Discretization of states by enclosing the system in a large box.

$$\psi(x + L, y, z) = \psi(x, y + L, z) = \psi(x, y, z + L) = \psi(x, y, z). \tag{13•4–3}$$

Although it would be possible to choose a confinement volume with a more complicated shape, there is usually no incentive to do so.

In one dimension, periodic boundary conditions may be viewed as representing the motion of an object along a closed loop. In two- and three-dimensional problems, no such simple visualization is possible.

Unfortunately, this discretization of the eigenvalue spectrum cannot always be done. For some operators—notably, the position operator—the eigenvalues are inherently continuously distributed. We must therefore consider continuous eigenvalues, at least to a minimal extent.[2]

The expansion coefficients $a_d(A)$ in the discrete portion of the expansion (13•4–1) continue to be given by (13•1–6), which may also be written as

$$a_d(A) = \langle A|\Psi\rangle_d. \tag{13•4–4a}$$

To determine the expansion coefficients $a_c(A)$ in the continuous portion of the spectrum, we take the inner product of (13•4–1) with the bra $\langle A|$ taken from the continuous spectrum:

$$\langle A|\Psi\rangle_c = \int a_c(A')\,\langle A|A'\rangle_c\,dA'. \tag{13•4–5}$$

Because of orthogonality—-which is valid regardless of the nature of the eigenvalue spectrum—the integrand vanishes everywhere except for $A = A'$. In order that both the entire integral and the expansion coefficient be finite, the norm $\langle A|A'\rangle_c$ must be proportional to a Dirac δ-function. If we make the simplest possible choice,

$$\langle A|A'\rangle_c = \delta(A - A'), \tag{13•4–6}$$

the relation (13•4–5) reduces to (13•4–4a), except for the replacement of the subscripts d and c. We can then lump both relations together by writing

$$a_{c,d}(A) = \langle A|\Psi\rangle_{c,d}. \tag{13•4–4b}$$

The normalization condition (13•4–6) expresses the principal difference between the eigenstates of the discrete and of the continuous portions of the eigenvalue spectrum: To be usable as bases for an eigenstate expansion, the former *must* be normalizable, whereas the latter must *not*! We noticed earlier that in those cases in which we encountered continuous eigenvalue spectra, the eigenfunctions were not normalizable. We now see that this was not an accident, but is inherent in the nature of the continuous spectrum. We also see that this property, far from being a nuisance, is actually desirable. Finally, (13•4–6)

---

[2] The remaining material of this subsection is not essential for understanding anything that follows prior to chapter 22, and may be skipped during a first reading.

goes beyond the statement that the continuous spectrum eigenfunctions are not normalizable, by stating the quantitative measure of non-normalizability that is the exact equivalent of a unit norm for the discrete spectrum eigenfunctions.

The change in normalization implies a change in the interpretation of the continuous spectrum expansion coefficients in terms of measurement probabilities. The probability of measuring exactly any specific ("infinitely sharp") value in a continuum of possible values is, in general, zero. Therefore, $|a_c(A)|^2$ cannot be interpreted as the probability of the occurrence of the *exact* measurement value A. Instead, $|a_c(A)|^2\,dA$ is the probability of a measurement value occurring in the *interval* $(A, A + dA)$. That is, $|a_c(A)|^2$ is a **probability density**:

$$\rho(A) = |a_c(A)|^2. \tag{13•4–7}$$

13.5 EIGENVALUES AS A UNITARY TRANSFORMATION PROBLEM

13.5.1 Transformation to a Different Basis: Unitary Operators

As we stated in section 13.2, the task of solving the Schroedinger equation is equivalent to finding a way to get from the general non-diagonal form (13•2–9) to the diagonal-matrix form (13•2–12). From the point of view of linear algebra, the problem is equivalent to finding the matrix transformation that diagonalizes the Hamiltonian matrix. If we designate the non-diagonal form in (13•2–9) by the symbol $(\hat{H})$ and the diagonal form in (13•2–12) by $(\hat{H})'$, we may write the transformation as

$$(\hat{H})' = \hat{U}^{-1}(\hat{H})\hat{U}, \tag{13•5–1}$$

where $\hat{U}$ is a certain matrix and $\hat{U}^{-1}$ is its inverse. The problem is to find $\hat{U}$.

This is a special case of a task encountered in many quantum-mechanical problems: changing from one set of basis vectors to a different set. We denote the old set by unprimed bra and ket vectors, such as $|l\rangle$, $|m\rangle$, $\langle n|$, and the new one by primed ones, $|l'\rangle$, $|m'\rangle$, $\langle n'|$, etc. We assume that both sets are orthonormal.

Each of the new basis vectors may be expressed as a finite linear superposition of the old ones,

$$|n'\rangle = \sum_m |m\rangle\langle m|n'\rangle, \tag{13•5–2a}$$

and vice versa:

$$|n\rangle = \sum_{m'} |m'\rangle\langle m'|n\rangle. \tag{13•5–2b}$$

The expansion coefficients $\langle m|n'\rangle$ in (13•5–2a) form the elements of a matrix

$$\hat{U} = \begin{bmatrix} \langle 1|1'\rangle & \langle 1|2'\rangle & \langle 1|3'\rangle & \dots \\ \langle 2|1'\rangle & \langle 2|2'\rangle & \langle 2|3'\rangle & \dots \\ \langle 2|1'\rangle & \langle 2|2'\rangle & \langle 2|3'\rangle & \dots \\ \vdots & \vdots & \vdots & \ddots \end{bmatrix}. \tag{13•5–3a}$$

Similarly, the coefficients $\langle m'|n\rangle$ in (13•5–2b) also form a matrix, namely,

$$\hat{U}^{\dagger} = \begin{bmatrix} \langle 1'|1\rangle & \langle 1'|2\rangle & \langle 1'|3\rangle & \dots \\ \langle 2'|1\rangle & \langle 2'|2\rangle & \langle 2'|3\rangle & \dots \\ \langle 3'|1\rangle & \langle 3'|2\rangle & \langle 3'|3\rangle & \dots \\ \vdots & \vdots & \vdots & \ddots \end{bmatrix}. \tag{13•5–3b}$$

Because of the relation $(\langle m|n'\rangle)^* = \langle n'|m\rangle$, the matrix $\hat{U}^{\dagger}$ is the **Hermitian conjugate** of $\hat{U}$, *defined* as complex conjugate of the transpose of $\hat{U}$. Hence its designation as $\hat{U}^{\dagger}$, a notation we introduced earlier (chapter 7) for Hermitian conjugate pairs of operators. It actually originates from the linear algebra properties of the matrices representing the operators. In linear algebra, Hermitian conjugate pairs of matrices and operators are also often called **adjoint** pairs; we do not employ this terminology here.

The key point now is that $\hat{U}^{\dagger}$ is not only the Hermitian conjugate of $\hat{U}$, it is also the inverse of $\hat{U}$. To show that, we form the matrix elements of the matrix product $\hat{U}\hat{U}^{\dagger}$:

$$(\hat{U}\hat{U}^{\dagger})_{mn} = \sum_{l'} \langle m|l'\rangle\langle l'|n\rangle = \langle m|n\rangle = \delta_{mn}. \tag{13•5–4}$$

Here the first equality follows from the rules of matrix multiplication, the second from the unit operator property (13•3–6) of the sum over all projection operators. Although we had derived the latter property in section 3 with the sum going over the complete set of all state vectors, the terms outside a finite-dimensional sub-space may be dropped from the sum if the operator is never applied to states outside this sub-space. This constraint evidently applies in our case.

Eq. (13•5–4) shows that the product $\hat{U}\hat{U}^{\dagger}$ is indeed the unit matrix. Hence we have

$$\boxed{\hat{U}^{\dagger} = \hat{U}^{-1} \quad \text{or} \quad \hat{U}^{\dagger}\hat{U} = 1.} \tag{13•5–5}$$

Operators that satisfy this condition are called **unitary operators**, and the corresponding matrices are **unitary matrices**.

Consider now a state vector $|\Psi\rangle$, expanded in terms of both the old and the new set:

$$|\Psi\rangle = \hat{1}|\Psi\rangle = \sum_m |m\rangle\langle m|\Psi\rangle = \sum_{n'} |n'\rangle\langle n'|\Psi\rangle. \tag{13•5–6}$$

The old expansion coefficients $\langle n|\Psi\rangle$ are readily expressed in terms of the new ones:

$$\langle m|\Psi\rangle = \langle m|\hat{1}|\Psi\rangle = \sum_{n'} \langle m|n'\rangle\langle n'|\Psi\rangle. \tag{13•5–7}$$

This is a set of equations, one for each value of m. The equations may be lumped together into a single matrix equation,

$$\begin{bmatrix} \langle 1|\Psi\rangle \\ \langle 2|\Psi\rangle \\ \langle 3|\Psi\rangle \\ \vdots \end{bmatrix} = \hat{U} \begin{bmatrix} \langle 1'|\Psi\rangle \\ \langle 2'|\Psi\rangle \\ \langle 3'|\Psi\rangle \\ \vdots \end{bmatrix}, \tag{13•5–8}$$

where $\hat{U}$ is again the matrix defined in (13•5–3a). We express (13•5–8) more compactly by writing

$$|\Psi\rangle = \hat{U}|\Psi'\rangle, \tag{13•5–9}$$

where $|\Psi\rangle$ and $|\Psi'\rangle$ denote, not different states, but *the same state in different representations.*

Thus, we see that the same operator $\hat{U}$ is involved in the transformation both of the basis vectors and of the state vectors. Notice, however, that in (13•5–2) and in (13•5–9) the transformation goes in opposite directions. In (13•5–2), the new basis vectors are expressed as functions of the old basis vectors; symbolically,

$$|\text{new basis}\rangle = \hat{U}|\text{old basis}\rangle\ . \tag{13•5–10a}$$

But in (13•5–9), the old state vector components are expressed as functions of the new state vector components:

$$|\text{old states}\rangle = \hat{U}|\text{new states}\rangle\ . \tag{13•5–10b}$$

This subtle but important distinction is easily overlooked, creating mathematical difficulties.

The operators $\hat{U}$ and $\hat{U}^\dagger$ and their matrix elements differ from the matrix operators considered earlier in a conceptually important way: They are not operators that transform one state vector into a different state vector, both expressed relative to the same set of basis vectors; instead, they leave the state vectors unchanged, but transform the basis vectors and, with them, the expansion coefficients of the unchanged state vectors relative to the changed basis.

This distinction between the two kinds of operations is the Hilbert space analog of a distinction that occurs already in the problem of the relative rotation of a vector and a system of coordinate axes describing that vector in ordinary space. Such a rotation may be viewed either as a rotation of the vector relative to a fixed coordinate system, or as a rotation of the coordinate system relative to a fixed vector. The rotation axes and the magnitudes of the rotation angles are the same in both views, but the directions of rotation are opposite. The interchange of the operator direction in (13•5–10a,b) is entirely analogous.

13.5.2 Transformation of Operators

A change of the basis used to express the state vectors leads to a change in the matrix elements of any matrix operators that act upon the state vectors. Consider, for example, the operator equation

$$|\Phi\rangle = \hat{A}|\Psi\rangle. \tag{13•5–11}$$

Assume that both $|\Phi\rangle$ and $|\Psi\rangle$ are expressed in a new basis by means of the unitary transformation (13•5–9) and its equivalent for $|\Phi\rangle$:

$$\hat{U}|\Phi'\rangle = \hat{A}\hat{U}|\Psi'\rangle. \tag{13•5–12}$$

Operating on both sides with $\hat{U}^\dagger$ and making use of the unitarity relation $\hat{U}^\dagger\hat{U} = 1$ leads to

$$|\Phi'\rangle = \hat{U}^\dagger\hat{A}\hat{U}|\Psi'\rangle = \hat{A}'|\Psi'\rangle. \tag{13•5–13}$$

Here,

$$\boxed{(\hat{A})' = \hat{U}^\dagger(\hat{A})\hat{U}} \tag{13•5–14}$$

is the matrix representing the operator that transforms the state $|\Psi'\rangle$ into $|\Phi'\rangle$, with both state vectors expressed in the new basis. Because $|\Phi'\rangle$ and $|\Psi'\rangle$ are simply new representations of the states $|\Phi\rangle$ and $|\Psi\rangle$, rather than new states, $(\hat{A})'$ and $(\hat{A})$ are simply different representations of the same physical operator $\hat{A}$, rather than physically different operators. We have put the symbol $(\hat{A})$ in parentheses, to indicate that we are dealing with different matrix representations for the same physical operator, not with different operators.

The matrix elements $\langle m'|\hat{A}|n'\rangle$ of $\hat{A}$ in the new representation are readily expressed as functions of the old matrix elements:

$$\langle m'|\hat{A}|n'\rangle = \langle m'|\hat{1}\hat{A}\hat{1}|n'\rangle = \sum_m\sum_n \langle m'|m\rangle\langle m|\hat{A}|n\rangle\langle n|n'\rangle. \tag{13•5–15}$$

The transformation formalism applies specifically when $\hat{A}$ is the Hamiltonian $\hat{H}$. If we choose as the new set of basis states the states in which the Hamiltonian matrix is diagonal and as the old set the set of unperturbed basis states, then the operator $\hat{U}$ defined by (13•5–3a) using this assignment is evidently the transformation operator that diagonalizes the Hamiltonian.

Note that (13•5–3a) implies that the eigenvectors of the Hamiltonian are in fact known. Hence, the transformation formalism is of little use for actually finding the eigenvectors and eigenvalues, except in rare cases in which the transformation matrix can be guessed. Instead, the value of the formalism lies in the knowledge that there *is* such a matrix, because from this fact, a number of conclusions can be drawn that would otherwise not be apparent.

13.5.3 Transform Invariants

Several quantities remain invariant under a change of basis. The most important of these are the determinant of the matrix corresponding to any Hermitian operator and the trace of that matrix

$$\mathrm{Tr}(\hat{A}) \equiv \sum_n A_{nn}. \tag{13•5–16}$$

The invariance of the determinant follows because the determinant of any product matrix equals the product of the determinants of the factor matrices. From (13•5–14) and (13•5–5), it then follows that

$$\det(\hat{A}') = \det(\hat{U}^\dagger) \cdot \det(\hat{A}) \cdot \det(\hat{U}) = \det(\hat{U}^\dagger \hat{U}) \cdot \det(\hat{A}) = \det(\hat{A}). \tag{13•5–17}$$

To prove the invariance of the trace, we set $m' = n'$ in (13•5–15) and sum over n':

$$\begin{aligned} \mathrm{Tr}(\hat{A}') &= \sum_{n'} \langle n'|\hat{A}|n'\rangle = \sum_{n'}\sum_{m}\sum_{n} \langle n'|m\rangle \langle m|\hat{A}|n\rangle \langle n|n'\rangle \\ &= \sum_{m}\sum_{n}\sum_{n'} \langle n|n'\rangle \langle n'|m\rangle \langle m|\hat{A}|n\rangle \\ &= \sum_{m}\sum_{n} \langle n|m\rangle \langle m|\hat{A}|n\rangle = \sum_{n} \langle n|\hat{A}|n\rangle = \mathrm{Tr}(\hat{A}). \end{aligned} \tag{13•5–18}$$

If we choose as basis the eigenvectors of $\hat{A}$, i.e.,

$$\hat{A}|n\rangle = A_n|n\rangle, \tag{13•5–19}$$

we obtain

$$\mathrm{Tr}(\hat{A}) = \sum_n \langle n|\hat{A}|n\rangle = \sum_n A_n \langle n|n\rangle = \sum_n A_n. \tag{13•5–20}$$

In words, the trace of a matrix is the sum of the eigenvalues of the operator represented by that matrix. Because the trace is invariant under the transformation (13•5–14), the result (13•5–20) must also be true in any basis into which the eigenvectors of $\hat{A}$ can be transformed.

13.5.4 Orthogonality Relations for Unitary Operators

We return to the matrix representation of the unitary operator $\hat{U}$, as given in (13•5–3a). According to (13•2–1a), each of the columns of this matrix is simply a vector representation of one of the new basis vectors $|n'\rangle$ in the old basis $|n\rangle$. But the different new basis vectors form, of course, an orthonormal set. Hence, the different columns in (13•5–3a) must themselves be orthonormal vectors, obeying the **column orthonormality** relation

$$\sum_l U_{ml}^* U_{nl} = \delta_{mn}. \tag{13•5–21a}$$

Similarly, each of the rows in (13•5–3a) is simply a vector representation of one of the old basis vectors $|n\rangle$ in the new basis $|n'\rangle$. But because the different old basis vectors must also form an orthonormal set, the different rows in (13•5–3a) must also be orthonormal vectors, obeying the **row orthonormality** relation

$$\sum_l U_{lm}^* U_{ln} = \delta_{mn}. \tag{13•5–21b}$$

Although our argument was based here on the idea that the matrix operator $\hat{U}$ is a transformation matrix between two orthonormal sets of basis vectors, the two orthonormality relations hold for any unitary matrix.

Exercise: Show that the orthonormality relations (13•5–21) are obeyed by the matrix elements of *any* unitary matrix $\hat{U}$ and that any matrix $\hat{U}$ obeying these relations is a unitary matrix.

Chapter 14

PERTURBATION THEORY, I: "DEGENERATE" PERTURBATION THEORY

14.1 WHAT IS PERTURBATION THEORY?

Only a small fraction of quantum-mechanical problems can be solved rigorously in closed form. The particle-in-a-box problem, the harmonic oscillator, and the hydrogen atom are three of them, and there are a handful of others. Even these rigorously solvable cases are already idealizations: A box with infinitely hard walls does not occur in nature. In many "harmonic" oscillators, the actual restoring force is not exactly proportional to the displacement. And in the hydrogen atom, we had to neglect a number of small but real corrections to the simple non-relativistic Coulomb Hamiltonian.

It is only the idealization, achieved by neglecting the complications that are actually present, that makes these problems "rigorously" tractable. In many real cases, such nearly exact idealizations are not possible, and approximations are required to obtain any solutions at all. A good example is the motion of the electrons in the periodic potential inside a crystal, which will in fact serve as a standard example throughout this chapter and the following one. In other cases, the *deviations* from an exactly solvable idealized situation are the essence of the problem, like the splitting of the l-degeneracy in the hydrogen atom under the influence of various corrections to the Hamiltonian. In one form or another, similar problems arise in almost all actual quantum-mechanical calculations. This confronts us with the need to develop a method for the approximate solution of the Schroedinger equation in such cases.

We encountered earlier two techniques for approximate solutions: the *WKB approximation* and the *variational approximation.* But these are special techniques of restricted applicability, and a more general and more systematic approach is called for.

In many problems of practical importance, the actual Hamiltonian is in some sense "close" to an idealized problem that *is* rigorously tractable in closed form. In such problems, it is useful to split the actual Hamiltonian $\hat{H}$ of the problem into two terms,

$$\hat{H} = \hat{H}^{(0)} + \hat{W}, \tag{14•1–1}$$

where $\hat{H}^{(0)}$ is the **unperturbed** Hamiltonian, for which the Schroedinger equation *can* be solved, and $\hat{W}$ is a **perturbation** that modifies the solution. Because of the expansion theorem, it is possible to expand the eigenfunctions $|\psi\rangle$ of the *actual* Hamiltonian in terms of the eigenfunctions $|n\rangle$ of the *unperturbed* Hamiltonian. That is, we set

$$|\psi\rangle = \sum_n a_n |n\rangle, \tag{14•1–2}$$

where the $|n\rangle$ obey

$$\hat{H}^{(0)}|n\rangle = \mathcal{E}_n^{(0)}|n\rangle \tag{14•1–3}$$

and where the expansion coefficients may be expressed in terms of $|\psi\rangle$ by the relation (13•1–6),

$$a_n = \langle n|\Psi\rangle. \tag{14•1–4}$$

The $\mathcal{E}_n^{(0)}$ in (14•1–3) are the energy eigenvalues of the unperturbed Hamiltonian.

The problem of solving the Schroedinger equation then reduces to the determination of the values of the expansion coefficient a_n in (14•1–2), in such a way that $|\psi\rangle$ satisfies the Schroedinger equation

$$[\hat{H}^{(0)} + \hat{W}]|\psi\rangle = \mathcal{E}|\psi\rangle. \tag{14•1–5}$$

By making the expansion coefficients in (14•1–2) time-dependent, i.e., $a_n = a_n(t)$, this expansion technique can equally well be used to determine the solutions of the time-dependent Schroedinger wave equation

$$i\hbar\frac{\partial}{\partial t}|\Psi\rangle = [\hat{H}^{(0)} + \hat{W}]|\Psi\rangle. \tag{14•1–6}$$

So far, the above is an *exact* re-formulation of the problem, rather than an approximation. What makes it an approximation is that, in actual problem-solving practice, the expansion (14•1–2) can contain only a finite number N of terms:

$$|\psi\rangle = \sum_{N\ \text{terms}} a_n |n\rangle. \tag{14•1–7}$$

Often N will be a small number. For simplicity, we will, as a rule, number the N terms included in (14•1–7) from $n = 1$ through $n = N$.

The various approximation methods based on this central idea are lumped together under the name **perturbation theory**. Depending on the nature of the problem, perturbation theory may be **stationary** or **time-dependent**. Stationary perturbation theory concerns itself with the energy eigenvalues and eigenfunctions in the presence of a time-*independent* perturbation operator $\hat{W}$. Time-dependent perturbation theory concerns itself with the time evolution of a quantum-mechanical system under the influence of a perturbation. This chapter and the following one deal with stationary perturbation theory. However, the distinction is not a completely sharp one: It is possible to view stationary perturbation theory as a limiting case of the time-dependent theory, by treating the perturbation $\hat{W}$ as a time-dependent perturbation that was absent at $t = -\infty$ and that has been "turned on" infinitesimally slowly since then.

Within stationary perturbation theory, two quite different and complementary methods stand out. The first of these, treated in this chapter, is based on the matrix form (13•2–9) of the Schroedinger equation, but now written as a finite-matrix equation of the form

$$\begin{bmatrix} H_{11} & H_{12} & \cdots & H_{1N} \\ H_{21} & H_{22} & \cdots & N_{2N} \\ \vdots & \vdots & & \vdots \\ H_{N1} & H_{N2} & \cdots & H_{NN} \end{bmatrix} \begin{bmatrix} a_1 \\ a_2 \\ \vdots \\ a_N \end{bmatrix} = \mathcal{E} \begin{bmatrix} a_1 \\ a_2 \\ \vdots \\ a_N \end{bmatrix}, \tag{14•1–8}$$

with the matrix elements (13•2–10),

$$H_{mn} = \langle m | H | n \rangle. \tag{14•1–9}$$

In this method, all terms in the expansion (14•1–7) are treated on an equal footing. In effect, we study the interaction of a selected group of initially unperturbed states *with each other*, under the influence of the perturbation. This approach is particularly suited to the problem of how an energy degeneracy of several states may be split up by a perturbation; hence, it is often referred to as **degenerate perturbation theory**. The name is somewhat misleading, though: As we shall see, the method is just as useful in many non-degenerate problems.

In the second method, treated in the next chapter, one term in the expansion (14•1–7) is singled out and assumed dominant. In effect, we study the way the selected state is perturbed by its interaction with all other states when the perturbation is "turned on." This approach is particularly useful when the selected state is *not* degenerate with the perturbing states; hence, it is often referred to as **non-degenerate perturbation theory**. Again, the name does not reflect the full utility of the method.

Regardless of which of the two methods is used, one of the first decisions we must make is which states to include in the expansion (14•1–7) and which to omit. Usually, we are interested only in a limited subset of states, their eigenvalues, and—maybe—their eigenfunctions. The determination of the properties of this *set of interest* usually requires the inclusion of states that are not of interest by themselves, but which perturb the states of interest by a non-negligible amount. But there is nothing to be gained by including in the perturbation treatment more states than are necessary to achieve the desired accuracy of the set of interest, even if the computations required to do so would not strain the available computational resources. This evidently calls for a criterion for which states to include in the expansion and which to omit.

The strength of the mutual perturbation of two states $|m\rangle$ and $|n\rangle$ is governed by two quantities:

(a) The energetic separation $|\mathcal{E}_m - \mathcal{E}_n|$ between the two states. The larger this separation, the weaker will be the perturbation.

(b) The magnitude of the "mixed" matrix elements or "coupling" matrix elements of the perturbation,

$$W_{mn} = \langle m|\hat{W}|n\rangle, \tag{14•1–10}$$

taken between the two states. The smaller this matrix element, the weaker will be the effect of the perturbation.

Taken together, these two facts yield the first rule that any state degenerate with one of the states of interest and having a nonzero mixed matrix element with that state, *must* be included in the expansion.

For non-degenerate interacting states, we need an estimate of the error $\Delta\mathcal{E}_{ix}$ made in the energy of one the states of interest $|i\rangle$ by *excluding* the state $|x\rangle$ from the expansion. We shall see later, during the derivation of (15•2–2a), that, *to the first order,* this error is

$$\Delta\mathcal{E}_{ix} = \frac{|W_{ix}|^2}{\mathcal{E}_x - \mathcal{E}_i}. \tag{14•1–11}$$

The qualifier "to the first order" refers to the following: The energies and matrix elements in each term in (14•1–11) are the values taken from the two *unperturbed* states $|i\rangle$ and $|x\rangle$. As a result of the interaction of $|i\rangle$ and $|x\rangle$ with other states, these energies and the matrix elements change, and it is really these changed—but unknown—values that *should* be used in (14•1–11), rather than the known original values.

As a result, it is sometimes necessary to include states in the expansion that have a negligible *first-order* effect on the states of interest, including even states that have zero mixed matrix elements with the states of interest, but which nevertheless change the states of interest *indirectly*, through their effect on other states. We will encounter examples of such *indirect interactions* later.

14.2 "DEGENERATE" PERTURBATION THEORY: THE PRINCIPLE

14.2.1 The Schroedinger Equation in Finite Matrix Form

As stated earlier, degenerate perturbation theory is based on the finite matrix form (14•1–8) of the Schroedinger equation. Note that the contribution from the unperturbed Hamiltonian is by definition purely diagonal and that all off-diagonal elements arise from the perturbation. Often the perturbation has no diagonal elements.

The relation (14•1–8) represents a system of N homogeneous, linear, simultaneous equations for the N unknowns a_1 through a_N, which may be written as

$$\sum_{n=1}^{N}[H_{mn} - \mathcal{E}\delta_{mn}]a_n = 0 \quad \text{for} \quad m = 1, 2, \ldots, N. \tag{14•2–1}$$

Such a system has a non-trivial solution only if its characteristic determinant vanishes:

$$\begin{vmatrix} H_{11} - \mathcal{E} & H_{12} & \ldots & H_{1N} \\ H_{21} & H_{22} - \mathcal{E} & \ldots & H_{2N} \\ \vdots & \vdots & & \vdots \\ H_{N1} & H_{N2} & \ldots & H_{NN} - \mathcal{E} \end{vmatrix} = 0. \tag{14•2–2}$$

This is an algebraic equation for $\mathcal{E}$, of degree N, commonly called the **secular equation**.[1] It has N roots—all real, but not necessarily all different—which are the energy eigenvalues of the $N \times N$ Hermitian matrix in (14•2–1). These eigenvalues can, at least *in principle*, always be determined by numerical techniques. Once the eigenvalues have been found, the eigenvector corresponding to each eigenvalue could be determined from (14•2–1) by standard linear algebra techniques—if that is still of interest. Often the eigenvalues are all that is wanted, and the re-formulation (14•2–2) of the perturbation problem as a matrix eigenvalue problem offers the possibility of obtaining them without ever having to determine the eigenvectors themselves.

By including enough states in the expansion, the perturbed eigenvalues may be determined, at least in principle, to whatever accuracy is needed. Depending on the nature of the problem, this may or may not lead to large

[1] The name dates back to an equation in the perturbation theory of 18th century celestial mechanics, describing the way the mutual gravitational attraction of the planets perturbed their orbits, leading to long-term changes in the orbits. (*Saeculum* is Latin for *century*.)

matrices. However, as we shall see, determining the eigenvalues of even large Hermitian matrices by numerical methods is an entirely practical approach to quantum-mechanical problem solving, and it is in fact widely employed. We will discuss this point further in section 14.5, including appropriate computational strategies.

14.2.2 An Alternative Point of View: Degenerate Perturbation Theory as a Variational Problem

The secular equation (14•2–2) is also obtained if we view the finite expansion (14•1–7) as a variational trial function, with the expansion coefficients as variational parameters, to be varied until a minimum expectation value of the energy has been found.

The energy expectation value for the function $|\psi\rangle$ follows readily as

$$\langle \mathcal{E} \rangle = \sum_m^N \sum_n^N a_m^* a_n H_{mn} \Big/ \sum_n^N a_n^* a_n. \tag{14•2–3}$$

The denominator has been added in order not to need the usual normalization constraints on the expansion coefficients.

To find the stationary values of $\langle \mathcal{E} \rangle$, we differentiate $\langle \mathcal{E} \rangle$ with respect to each of the a_n and the a_n^*, treating a_n and a_n^* as independent variables for the purposes of the differentiation, similar to the way x and y are independent *spatial* variables.

Exercise: To show that this is legitimate, consider the functions

$$f(z, z^*) = F(x, y), \text{ where } z = x + iy \quad \text{and} \quad z^* = x - iy. \tag{14•2–4a,b}$$

Show that the conditions

$$\left(\frac{\partial f}{\partial z}\right)_{z^*} = \left(\frac{\partial f}{\partial z^*}\right)_z = 0 \quad \text{and} \quad \left(\frac{\partial F}{\partial x}\right)_y = \left(\frac{\partial F}{\partial y}\right)_x = 0 \tag{14•2–5a,b}$$

are equivalent to each other.

We choose to differentiate with respect to the various a^*'s:

$$\frac{\partial \langle \mathcal{E} \rangle}{\partial a_l^*} = 0 \quad \text{for} \quad l = 1, 2, \ldots, N. \tag{14•2–6}$$

We have given the differentiation index the new name l, to avoid confusion with the summing indices m and n in (14•2–3). From (14•2–3) and (14•2–6), we obtain the set of conditions

$$\left(\sum_n^N a_n H_{ln}\right)\left(\sum_n^N a_n^* a_n\right) - a_l \sum_m^N \sum_n^N a_m^* a_n H_{mn} = 0 \quad \text{for} \quad l = 1, 2, \ldots, N. \tag{14•2–7}$$

If we divide this by the non-vanishing sum $\sum a_n^* a_n$, then re-insert $\mathcal{E} = \langle \mathcal{E} \rangle$ from (14•2–3), and finally re-name the index l as m, we obtain exactly the relation (14•2–1). This proves the claimed equivalence of degenerate perturbation theory to a variational treatment.

The lowest root of the secular equation (14•2–2) is evidently the lowest energy expectation value that can be obtained for the state function expansion (14•1–7). According to the variational theorem, the *exact* energy eigenvalue of the state corresponding to this lowest root can only be even lower. Conversely, the highest root of (14•2–2) corresponds to a state whose *exact* energy can only be higher. Hence, (14•2–2) gives an inner limit to the range of exact eigenvalues for the states included in the expansion (14•1–7).

Furthermore, because the diagonal elements of the eigenvalue equation are energy expectation values for the set of basis functions employed in the expansion (14•1–7), the lowest root of (14•2–2) cannot be above the lowest diagonal element in (14•2–2), nor the highest root below the highest diagonal element:

$$\mathrm{Min}(\mathcal{E}) \leq \mathrm{Min}(H_{nn});\ n = 1, \ldots, N; \tag{14•2–8a}$$

$$\mathrm{Max}(\mathcal{E}) \geq \mathrm{Max}(H_{nn});\ n = 1, \ldots, N. \tag{14•2–8b}$$

A quantity that is often of considerable interest in perturbation problems is the *mean value* $\bar{\mathcal{E}}$ of the different energy eigenvalues, along with its changes under the influence of the perturbation. We denote this mean value with $\bar{\mathcal{E}}$ rather than $\langle \mathcal{E} \rangle$, to distinguish it from the energy expectation value for a wave function. We recall from chapter 13 that the sum of the eigenvalues of a matrix is equal to the sum of the diagonal elements of the matrix; hence, the mean eigenvalue is given by

$$\bar{\mathcal{E}} \equiv \frac{1}{N} \sum_{n=1}^{N} \mathcal{E}_n = \frac{1}{N} \mathrm{Tr}(\hat{H}), \tag{14•2–9}$$

where

$$\mathrm{Tr}(\hat{H}) \equiv \sum_{n=1}^{N} H_{nn} \tag{14•2–10}$$

is the **trace** of the matrix $\hat{H}$. Note that if the perturbation does not have any diagonal matrix elements, the mean value of the energy expectation values will remain unchanged by it.

Finally, we state—without proof—that the wave functions corresponding to *different* roots of the secular equation are automatically orthogonal to each other, and that the eigenfunctions belonging to any multiple roots may always be chosen to be a mutually orthogonal set.

14.2.3 On the Distribution of Energy Eigenvalues

While Eqs. (14•2–8) specify a minimum spread of the perturbed energy eigenvalues, they establish neither a maximum spread nor an average spread, both of which are sometimes of interest. The most convenient measure for the average spread is the *standard deviation* $\Delta\mathcal{E}$ of the roots of the secular equation about their mean value $\bar{\mathcal{E}}$:

$$\Delta\mathcal{E} \equiv \sqrt{\overline{(\mathcal{E}^2)} - (\bar{\mathcal{E}})^2}. \tag{14•2–11}$$

To evaluate (14•2–11), we need the mean square of the eigenvalues,

$$\overline{\mathcal{E}^2} \equiv \frac{1}{N}\sum_n \mathcal{E}_n^2 = \frac{1}{N}\,\mathrm{Tr}(\hat{H}^2), \tag{14•2–12}$$

where the last form holds because the $\mathcal{E}_n^2$ are simply the eigenvalues of the matrix $\hat{H}^2$. But this may be written

$$\begin{aligned}\mathrm{Tr}(\hat{H}^2) &= \sum_n \langle n|\hat{H}^2|n\rangle = \sum_n \langle n|\hat{H}\hat{1}\hat{H}|n\rangle \\ &= \sum_n\sum_m \langle n|\hat{H}|m\rangle\langle m|\hat{H}|n\rangle = \sum_m\sum_n |H_{mn}|^2.\end{aligned} \tag{14•2–13}$$

Hence,

$$\bar{\mathcal{E}}^2 = \frac{1}{N}\sum_m\sum_n |H_{mn}|^2. \tag{14•2–14}$$

A knowledge of $\bar{\mathcal{E}}$ and $\Delta\mathcal{E}$ automatically places an *outer* limit on the roots of the secular equations, as well as an additional inner limit. It is left to the reader to show that

$$\bar{\mathcal{E}} - \Delta\mathcal{E}\cdot\sqrt{N-1} \le \mathrm{Min}(\mathcal{E}_n) \le \bar{\mathcal{E}} - \Delta\mathcal{E}/\sqrt{N-1} \tag{14•2–15a}$$

and

$$\bar{\mathcal{E}} + \Delta\mathcal{E}/\sqrt{N-1} \le \mathrm{Max}(\mathcal{E}_n) \le \bar{\mathcal{E}} + \Delta\mathcal{E}\cdot\sqrt{N-1}. \tag{14•2–15b}$$

14.3 SIMPLE TWO-STATE DEGENERATE PERTURBATION THEORY

14.3.1 Introduction

The simplest degenerate perturbation problem is the splitting of a two-fold degenerate energy level by a perturbation. To the first order, the problem may be treated by including only the two degenerate states themselves in the expansion (14•1–7). This leads to the secular equation

$$\begin{vmatrix} H_{11} - \mathcal{E} & H_{12} \\ H_{21} & H_{22} - \mathcal{E} \end{vmatrix} = 0, \tag{14•3–1}$$

which is a quadratic equation in $\mathcal{E}$ with the two roots

$$\mathcal{E}_\pm = \tfrac{1}{2}(H_{11} + H_{22}) \pm \sqrt{\tfrac{1}{4}(H_{11} - H_{22})^2 + |H_{12}|^2}. \tag{14•3–2}$$

The unperturbed Hamiltonian is by definition diagonal; hence, the off-diagonal matrix elements H_{12} and H_{21} are due to the perturbation. The diagonal matrix elements include the contributions of both the unperturbed Hamiltonian and the perturbation.

Once the energy eigenvalues have been determined, the associated eigenfunctions may be found. To this end, recall that the secular equation (14•3–1) corresponds to two simultaneous equations for the expansion coefficients a_1 and a_2 in the expansion (14•1–7):

$$[H_{11} - \mathcal{E}]a_1 + H_{12}a_2 = 0, \qquad (14•3–3a)$$

$$H_{21}a_1 + [H_{22} - \mathcal{E}]a_2 = 0. \qquad (14•3–3b)$$

If $|\psi\rangle$ is to be normalized, we must further have

$$|a_1|^2 + |a_2|^2 = 1. \qquad (14•3–4)$$

From (14•3–3a) and (14•3–4), we obtain

$$a_1 = \frac{H_{12}}{D} \quad \text{and} \quad a_2 = \frac{E - H_{11}}{D}, \qquad (14•3–5a)$$

where

$$|D|^2 = |H_{12}|^2 + (\mathcal{E} - H_{11})^2. \qquad (14•3–5b)$$

This formulation is unsymmetric in the subscripts '1' and '2'. By starting with (14•3–3b) rather than (14•3–3a), we obtain a mathematically equivalent form in which the subscripts '1' and '2' have been interchanged everywhere in (14•3–5a,b). It is possible to symmetrize the expressions (14•3–5), but the result is unwieldy.

14.3.2 Example: Electron in a Simple Periodic Potential

The Problem

We illustrate two-state degenerate perturbation theory by applying it to a simple one-dimensional periodic potential (**Fig. 14•3–1**) of the form

$$V(x) = -2\,U_G \cos Gx, \quad \text{where } G = 2\pi/a. \qquad (14•3–6)$$

We treat this potential as a perturbation acting on free electrons with the unperturbed plane-wave eigenfunctions

$$\psi_k(x) = C\exp(ikx). \qquad (14•3–7)$$

This problem is one of the simplest possible models for the electron propagation through the periodic potential in a crystal, and we will use it extensively as a standard example throughout our treatment of perturbation theory.

Plane waves are not normalizable over infinite space. We therefore convert the problem into one over a finite space by imposing the periodic boundary

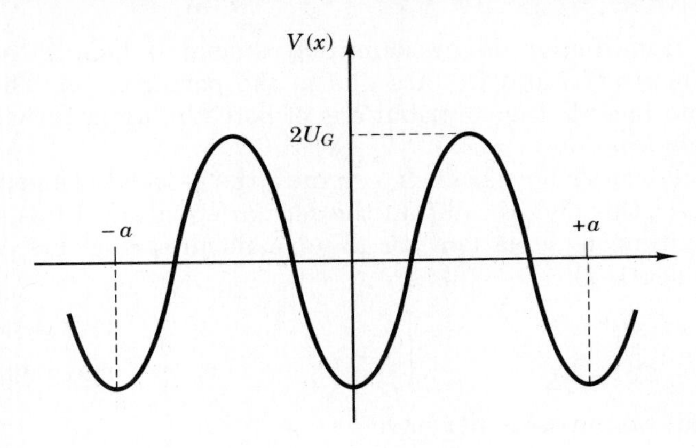

Figure 14•3–1. Simple cosine potential.

condition

$$\psi_k(x + L) = \psi_k(x), \qquad (14\bullet3\text{–}8a)$$

where

$$L = Na \qquad (14\bullet3\text{–}8b)$$

is the length of an array of N potential wells, the length of our "crystal." This leads to

$$|k\rangle \mathrel{\hat{=}} \psi_k(x) = \frac{1}{\sqrt{L}} e^{ikx}. \qquad (14\bullet3\text{–}9)$$

The boundary conditions (14•3–8) restrict the allowed wave numbers k to the discrete set

$$k = \frac{n}{N} G \qquad (n = 0, \pm1, \pm2, \ldots, \pm\infty). \qquad (14\bullet3\text{–}10)$$

The reader is asked to show that wave functions corresponding to different allowed values of k are orthogonal.

The unperturbed energy as a function of k is given by

$$\mathcal{E}_0(k) = \frac{\hbar^2 k^2}{2m_e}. \qquad (14\bullet3\text{–}11)$$

The mixed matrix elements $H_{12} = \langle k_1 | \hat{H} | k_2 \rangle = \langle k_1 | \hat{W} | k_2 \rangle$ between two arbitrary plane-wave states $|k_1\rangle$ and $|k_2\rangle$ are given by

$$H_{12} = -\frac{U_G}{L} \int_0^L \{\exp[i(k_2 - k_1 + G)x] + \exp[i(k_2 - k_1 - G)x]\}\, dx. \qquad (14\bullet3\text{–}12)$$

For plane waves that are allowed by the criterion (14•3–10), the matrix elements vanish, unless

$$k_2 - k_1 = \pm G, \qquad (14\bullet3\text{–}13)$$

in which case they are equal to $-U_G$. Hence,

$$\boxed{H_{12} = -U_G \cdot \delta(k_2 - k_1, \pm G).} \tag{14•3–14}$$

Thus, we see that the matrix elements connect every unperturbed state $|k\rangle$ with only two other unperturbed states.

If k has a value in the vicinity of $\pm G/2$, one of the latter two states will have an energy close to that of $|k\rangle$, and the interaction between those two nearly degenerate states then dominates the problem. This is the case we consider here, neglecting the effect of the higher energy states that are coupled to this pair. It then becomes convenient to re-write the two plane-wave numbers included in the perturbation treatment in the form

$$k_1 = k_- = -G/2 + k', \qquad k_2 = k_+ = +G/2 + k', \tag{14•3–15}$$

where $k' \ll G$. The overall perturbed wave function may then be written as

$$\psi(x) = \exp(ik'x) \cdot [a_+ \exp(+iGx/2) + a_- \exp(-iGx/2)]. \tag{14•3–16}$$

The Limit k = ±G/2

The problem becomes trivially simple when the two wave numbers are *exactly* equal to $\pm G/2$. In that case, the two unperturbed plane waves $|k_-\rangle$ and $|k_+\rangle$ are degenerate, with the unperturbed energy

$$F \equiv \mathcal{E}_0(\pm G/2) = \frac{\hbar^2 G^2}{8m_e}, \tag{14•3–17}$$

which we will use as the **natural unit of energy** for our problem.

According to (14•3–14), the perturbation (14•3–6) makes no contribution to the diagonal matrix elements in the secular equation, which are therefore given by

$$H_{11} = H_{22} = \mathcal{E}_0(G/2) = F. \tag{14•3–18}$$

The off-diagonal matrix elements coupling the two plane waves $|+G/2\rangle$ and $|-G/2\rangle$ are

$$H_{21} = H_{12} = \langle -G/2|\hat{W}|+G/2\rangle = -U_G. \tag{14•3–19}$$

Insertion of (14•3–18) and (14•3–19) into (14•3–2) yields the two simple roots

$$\mathcal{E} = F \pm U_G. \tag{14•3–20}$$

The Vicinity of k = ±G/2

We extend this simple result to states *in the vicinity* of $\pm G/2$, with a small nonzero value of k' (**Fig. 14•3–2**).

According to (14•3–14), the mixed matrix elements between the two states remain unchanged: $H_{21} = H_{12} = -U_G$. But now, the two diagonal matrix ele-

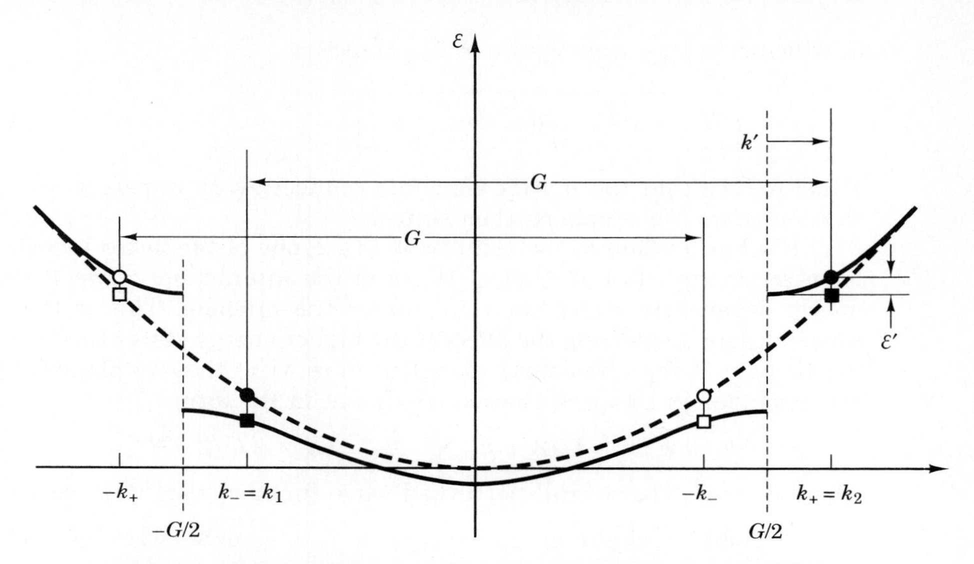

Figure 14•3–2. Plane waves involved in the simplest two-fold degenerate perturbation treatment of the states in the vicinity of $k = \pm G/2$.

ments are no longer equal to each other. If $k' > 0$, we may write

$$H_{11} = \frac{\hbar^2}{2m_e}(-\tfrac{1}{2}G + k')^2 = F - 2\sqrt{F\mathcal{E}'} + \mathcal{E}', \tag{14•3–21a}$$

$$H_{22} = \frac{\hbar^2}{2m_e}(+\tfrac{1}{2}G + k')^2 = F + 2\sqrt{F\mathcal{E}'} + \mathcal{E}'. \tag{14•3–21b}$$

Here F is again given by (14•3–18), and

$$\mathcal{E}' = \frac{\hbar^2(k')^2}{2m_e} \tag{14•3–22}$$

is the kinetic energy of a free electron with the wave number k'. If $k' < 0$, the right-hand sides in (14•3–21a,b) must be interchanged.

When these values are inserted into (14•3–2), we obtain the two energies

$$\mathcal{E}_{\pm} = F + \mathcal{E}' \pm \sqrt{4F\mathcal{E}' + U_G^2}. \tag{14•3–23}$$

The two roots correspond to two different linear superpositions of the two unperturbed plane-wave states $|k_1\rangle$ and $|k_2\rangle$. It is often convenient to view each of the two perturbed states *as if* it had arisen from a particular one of the two participating unperturbed plane waves, with the other plane wave acting as the perturbing wave. For $U_G \to 0$, the energies $\mathcal{E}_+$ and $\mathcal{E}_-$ in (14•3–23) go over into the different unperturbed energies for the participating plane waves, with

$\mathcal{E}_+$ going over into the higher of the two unperturbed energies and $\mathcal{E}_-$ into the lower one:

$$\mathcal{E}_\pm \to \mathcal{E}_0(k_\pm), \text{ where } k_\pm \equiv \pm G/2 + |k'|. \qquad (14\text{•}3\text{–}24)$$

We may therefore identify the perturbed state with the energy $\mathcal{E}_+$ as the perturbed state for $k = k_+$ and that with $\mathcal{E} = \mathcal{E}_-$ as the perturbed state for $k = k_-$, as indicated by the squares in Fig. 14•3–2. The perturbed energies may then be represented as a function $\mathcal{E}(k)$ of the wave number k, as shown in the figure. In solid-state theory, this description is called the **extended-zone** description. Note that $\mathcal{E}(k)$ depends only on the magnitude of k, not on its sign; hence, along with each state, its "mirror image in k-space" is automatically included. We therefore show two pairs of perturbed states; the pair indicated by filled circles corresponds to $k' > 0$, and the pair indicated by empty circles corresponds to $k' < 0$.

Effective Mass

For $k' = 0$, we have $\mathcal{E}' = 0$, and we obtain again $\mathcal{E} = F \pm U_G$. For sufficiently small finite values of k', $\mathcal{E}'$ will also be small, and the square root in (14•3–23) can be expanded:

$$\mathcal{E}_\pm \approx F + \mathcal{E}' \pm U_G\left(1 + \frac{2F\mathcal{E}'}{U_G^2}\right) = (F \pm U_G) + \frac{\hbar^2(k')^2}{2m^*}. \qquad (14\text{•}3\text{–}25)$$

Here we have defined an **effective mass** m^* via

$$\frac{m_e}{m^*} = \left(1 \pm \frac{2F}{U_G}\right). \qquad (14\text{•}3\text{–}26)$$

Thus, we see that in the vicinity of $k = \pm G/2$, $\mathcal{E}(k)$ varies parabolically with the distance k' from the singularity. This is the same as the behavior of a *free* electron with wave number k', but with an effective mass m^*, rather than the free-electron mass m_e. The effective mass in the upper of the two energy bands is always positive and smaller than the free-electron mass. The mass in the lower band is negative for $U_G < 2F$.

Mathematically, the negative effective mass simply expresses the fact that the energy in the lower band has a *maximum* at $\pm G/2$. However, the meaning of the negative effective mass goes far beyond this formal analogy: We will see in chapter 17 that an electron in this energy range is accelerated by an external force *as if* it had indeed a negative mass m^*. This negative-mass concept is central to the theory of electron transport in solids, and it is discussed extensively in texts on solid-state physics. To pursue it here would go beyond the scope of a text on quantum mechanics.

Note that both the energy gap and the effective mass scale with the potential amplitude U_G. Hence, a narrow energy gap implies a small effective mass. This result is another fundamental aspect of the effective mass theory of semiconductors.

For $U_G > 2F$, Eq. (14•3–26) predicts that the energy at $k = \pm G/2$ for the lower band turns into a relative minimum rather than a maximum. However, for such large values of U_G, two-state degenerate perturbation theory, employing only the two unperturbed state functions $|k_+\rangle$ and $|k_-\rangle$, is no longer a good approximation. In that case, it becomes necessary to include higher energy plane waves in the expansion, and when this is done, the lower band energy retains a relative maximum at $\pm G/2$. We shall not attempt such a much more complicated four-state degenerate calculation here.

Wave Functions and Charge Distribution

We conclude our treatment by calculating the perturbed wave functions. To this end, recall that the secular equation (14•3–1) corresponds to the two simultaneous equations for the expansion coefficients a_- and a_+ in (14•3–16). From (14•3–5a,b), (14•3–19), and (14•3–23), we find, after some elementary but tedious manipulations, that

$$a_\pm^2 = \frac{1}{2} \pm \frac{\sqrt{F\mathcal{E}'}}{\sqrt{4F\mathcal{E}' + U_G^2}}. \tag{14•3–27}$$

The relation (14•3–27) does not specify the signs of the coefficients; it is not difficult to show that the following sign rule holds:

$$a_- a_+ \gtrless 0 \quad \text{for} \quad \mathcal{E} \gtrless F. \tag{14•3–28}$$

Exercise: Derive (14•3–27) and (14•3–28).

The perturbed wave functions (14•3–16) no longer have a uniform probability density. From (14•3–15), (14•3–16), (14•3–27), and (14•3–28), we find readily that

$$\rho = |\psi|^2 = \frac{1}{L}(1 + A \cdot \cos Gx), \tag{14•3–29a}$$

where

$$A = 2a_- a_+ = \pm\frac{U_G}{\sqrt{4F\mathcal{E}' + U_G^2}} \quad \text{for} \quad \mathcal{E} \gtrless F. \tag{14•3–29b}$$

This means that the electron distribution in the states below the energy gap gets more concentrated in the potential troughs—as one might expect—while in the states above the gap, it gets more concentrated in the regions of high potential energy. In fact, the lowering and raising of the energies of the states below and above the gap are due precisely to this charge re-distribution.

Exercise: Show that in the limit $\mathcal{E}' \ll F$, the square-root term in (14•3–23) is equal to the change in the expectation value of the potential energy for the charge distribution specified by (14•3–29).

Note that for $k' = 0$, Eqs. (14•3–29) reduce to

$$\rho_c = \frac{1}{L}(1 + \cos Gx) \quad \text{and} \quad \rho_s = \frac{1}{L}(1 - \cos Gx), \tag{14•3–30a,b}$$

where the subscripts c and s stand for *cosine* and *sine*. Clearly, the (even) cosine function has its density maxima in the regions of the potential minima, with the density actually going to zero at the potential maxima, leading to a lowered energy. The (odd) sine function has the opposite density distribution, leading to a raised energy.

◆ PROBLEMS TO SECTION 14.3

#14•3-1: Potential Well with Deformed Bottom

A one-dimensional potential well with infinitely high walls has a sinusoidally deformed bottom with the potential

$$V(x) = \begin{Bmatrix} -A\sin(\pi\kappa/2a) & \text{for } |x| < a \\ \infty & \text{for } |x| > a \end{Bmatrix}, \tag{14•3–31}$$

shown in **Fig. 14•3–3.**

(a) Calculate the approximate energy eigenvalues for the two lowest energy eigenstates in this potential, using a linear superposition of the two lowest states of the *unperturbed* (flat-bottom) well as the trial wave function. Give closed-form analytic expressions for the two energies, as functions of A.

(b) Assume that $A = \mathcal{E}_1$, where $\mathcal{E}_1$ is the lowest energy eigenstate for the *unperturbed* (flat-bottom) well. For this specific case, indicate the energies of the two states

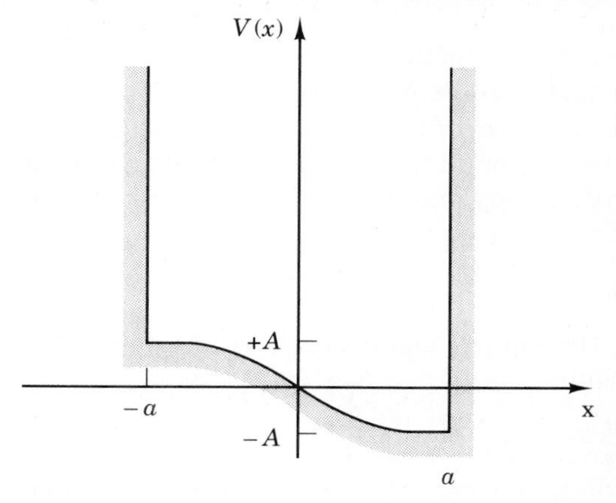

Figure 14•3–3. Potential well with sinusoidal bottom.

on the graph of the potential in Fig. 14•3–3. Give numerical values for the two energies, in eV, assuming that the object has the mass of an electron and that $a = 1$ nm. Make a quantitative plot of the wave functions for the two states.

#14•3-2: Anharmonic Oscillator

A one-dimensional harmonic oscillator is perturbed by a small perturbation that is proportional to x^3 which may always be written in the form

$$W(x) = \alpha \cdot \hbar\omega \cdot (x/L)^3, \tag{14•3–32}$$

where $\alpha(\ll 1)$ is a numerical parameter characterizing the strength of the perturbation, and ω and L are the natural frequency and the natural unit of length of the oscillator.

(a) With the help of a sketch of the perturbed potential, discuss why this is, strictly speaking, no longer a bound-state problem at all, but a problem of quasi-bound states, in the sense of section 2.6. Up to which quantum number $n = n_{\max}$ would you expect perturbation theory to give reasonable values for the energy of the quasi-bound states?

(b) Use two-state degenerate perturbation theory to estimate the effect of the perturbation on the energy of the two lowest states of the oscillator, as a function of α. Whatever the validity of the two-state procedure for the perturbation of the ground-state energy, why is this procedure a very poor one for the perturbation of the second state, even for small values of α?

(c) Now do a three-state perturbation treatment, assuming $\alpha = 1/10$, solving the cubic secular equation numerically. Compare the result with the analytical result obtained under (b), for the same value of α.

14.4 FACTORIZABLE HIGHER ORDER PROBLEMS

14.4.1 The Principle

The simplicity of a closed-form algebraic solution for two-state perturbation problems disappears for most problems requiring more than two states—unless the matrix elements happen to be simple enough that the equation can *somehow* be factored into two or more tractable equations of lower degree.

As an example, consider any one-dimensional potential exhibiting inversion symmetry:

$$V(-x) = V(x). \tag{14•4–1}$$

As we saw in chapter 2, the energy eigenfunctions are then either symmetric or anti-symmetric functions:

$$\psi(-x) = \pm\psi(-x). \tag{14•4–2}$$

Examination of the off-diagonal matrix elements for such a potential shows that *all* of these elements between states with opposite parity are zero. Hence, the two sets of states do not perturb each other. The secular equation can then be factored into a product of two separate equations, one for the even-parity states and one for the odd-parity states. To see this, we arrange the ordering of all the states in such a way that all the even-parity states are listed first, followed by all the odd-parity states. The secular determinant then assumes the form

$$D = \begin{vmatrix} D_+ & \vdots & 0 \\ \cdots & \cdot & \cdots \\ 0 & \vdots & D_- \end{vmatrix} = 0. \tag{14•4–3}$$

Here, D_+ and D_- are secular determinants for the even- and odd-parity states alone, and the 0's stand for null matrices. But (14•4–3) is of course equivalent to the two separate equations

$$D_+ = 0 \quad \text{and} \quad D_- = 0. \tag{14•4–4a,b}$$

Such factorizations or similar simplifications are possible surprisingly often, almost always as the consequence of some form of symmetry that is present in the Hamiltonian, as in our simple example. If the factored equations are quadratic, the problem then becomes solvable in closed form, but even a purely numerical solution is greatly simplified by breaking down a secular equation of a high degree into a more easily solved set of equations of lower degrees. Furthermore, this factorization is exact, not just an approximation. Evidently, the utilization of symmetries is a powerful tool, of such importance that we shall devote an entire chapter (16) to it later.

14.4.2 Example: Refinement of the Energy Gap of the Cosine Potential

Secular Equation

Suppose we wished to improve the accuracy of the calculation of the energy gap at $k = \pm G/2$ for large values of the potential amplitude U_G by including the two additional plane waves with $k = \pm 3G/2$ into the perturbation expansion, each of which couples to one of the plane-wave basis states $|\pm G/2\rangle$. The unperturbed energies of those additional plane waves are

$$\mathcal{E}_0(\pm 3G/2) = 9\hbar^2 G^2/8m_e = 9F, \tag{14•4–5}$$

which is above the two gap edges.

A naive use of four plane waves for the perturbation expansion is easily found to lead to the secular determinant

$$\begin{vmatrix} 9F - \mathcal{E} & -U_G & 0 & 0 \\ -U_G & F - \mathcal{E} & -U_G & 0 \\ 0 & -U_G & F - \mathcal{E} & -U_G \\ 0 & 0 & -U_G & 9F - \mathcal{E} \end{vmatrix} = 0, \tag{14•4–6}$$

an unfactored fourth-degree equation.

However, there is no need to use *propagating* plane waves as basis set for the eigenfunction expansion; suppose instead that we had used the four *standing* plane waves

$$|c,1\rangle = \frac{1}{\sqrt{2}}(|+G/2\rangle + |-G/2\rangle) \triangleq \sqrt{2/L} \cdot \cos(Gx/2), \tag{14•4–7a}$$

$$|c,3\rangle = \frac{1}{\sqrt{2}}(|+3G/2\rangle + |-3G/2\rangle)$$
$$\triangleq \sqrt{2/L} \cdot \cos(3Gx/2), \tag{14•4–7b}$$

$$|s,1\rangle = \frac{1}{\sqrt{2}i}(|+G/2\rangle - |-G/2\rangle) \triangleq \sqrt{2/L} \cdot \sin(Gx/2), \tag{14•4–7c}$$

$$|s,3\rangle = \frac{1}{\sqrt{2}i}(|+3G/2\rangle - |-3G/2\rangle)$$
$$\triangleq \sqrt{2/L} \cdot \sin(3Gx/2). \tag{14•4–7d}$$

These are simply linear superpositions of the original propagating plane waves, belonging to the same two energies. Because the sine and cosine functions have opposite parity, there will be no mixed matrix elements between them, and the 4×4 secular determinant factors into two 2×2 determinants, one for the even-parity states and one for the odd-parity ones. The remaining off-diagonal matrix elements remain unchanged; for example,

$$\begin{aligned}&\langle c,1|\hat{W}|c,3\rangle \\ &= \tfrac{1}{2}(\langle +G/2|\hat{W}|+3G/2\rangle + \langle -G/2|\hat{W}|+3G/2\rangle \\ &\quad + \langle +G/2|\hat{W}|-3G/2\rangle + \langle -G/2|\hat{W}|-3G/2\rangle) \\ &= \tfrac{1}{2}(-U_G + 0 + 0 - U_G) = -U_G,\end{aligned} \tag{14•4–8a}$$

where, in the last line, we have utilized the relation (14•3–14) for the plane-wave matrix elements. Similarly,

$$\langle s,1|\hat{W}|s,3\rangle = -U_G. \tag{14•4–8b}$$

With the new basis functions, the diagonal matrix elements of the perturbation $\hat{W}$ are no longer all zero; it is left to the reader to show that

$$\langle c,1|\hat{W}|c,1\rangle = -U_G, \langle s,1|\hat{W}|s,1\rangle = +U_G, \tag{14•4–9a,b}$$

$$\langle c,3|\hat{W}|c,3\rangle = \langle s,3|\hat{W}|s,3\rangle = 0. \tag{14•4–9c,d}$$

With these matrix elements, we obtain the secular equation for the even-parity (cosine) states,

$$\begin{vmatrix} F - U_G - \mathcal{E} & -U_G \\ -U_G & 9F - \mathcal{E} \end{vmatrix} = 0, \tag{14•4–10a}$$

which has the roots

$$\mathcal{E}_{c,\pm} = 5F - \tfrac{1}{2}U_G \pm \sqrt{(4F + \tfrac{1}{2}U_G)^2 + U_G^2}. \tag{14•4–11a}$$

Similarly, for the odd-parity (sine) states,

$$\begin{vmatrix} F + U_G - \mathcal{E} & -U_G \\ -U_G & 9F - \mathcal{E} \end{vmatrix} = 0, \tag{14•4–10b}$$

$$\mathcal{E}_{s,\pm} = 5F + \tfrac{1}{2}U_G \pm \sqrt{(4F - \tfrac{1}{2}U_G)^2 + U_G^2}. \tag{14•4–11b}$$

We see readily that in the limit of a vanishing perturbation, $U_G \to 0$, the *lower* two of the four roots, $\mathcal{E}_{c,-}$ and $\mathcal{E}_{s,-}$, go over into the unperturbed energy at $k = \pm G/2$, $\mathcal{E}^{(0)} = F$, while the higher roots, $\mathcal{E}_{c,+}$ and $\mathcal{E}_{s,+}$, go over into the unperturbed energy at $k = \pm 3G/2$, $\mathcal{E}^{(0)} = 9F$. Hence, the two lower roots represent the energies of the energy gap at $\pm G/2$, whereas the higher roots represent a new gap at $\pm 3G/2$, as shown in **Fig. 14•4–1**.

Energy Gap at $\pm G/2$

If we expand the square roots in (14•4–11a,b) by powers of U_G and drop all terms higher than linear, we recover the result (14•3–20) of the two-wave treatment of section 14.2. In that section, we treated the energy gap at $\pm G/2$ for $U_G = F/2$ by two-state perturbation theory. If we now insert $U_G = F/2$ into (14•4–11a,b), we obtain the result

$$\left.\begin{aligned} \mathcal{E}_{s,-} &= 1.4668F = F + 0.9336U_G, \\ \mathcal{E}_{c,-} &= 0.4707F = F - 1.0586U_G. \end{aligned}\right\} \tag{14•4–12}$$

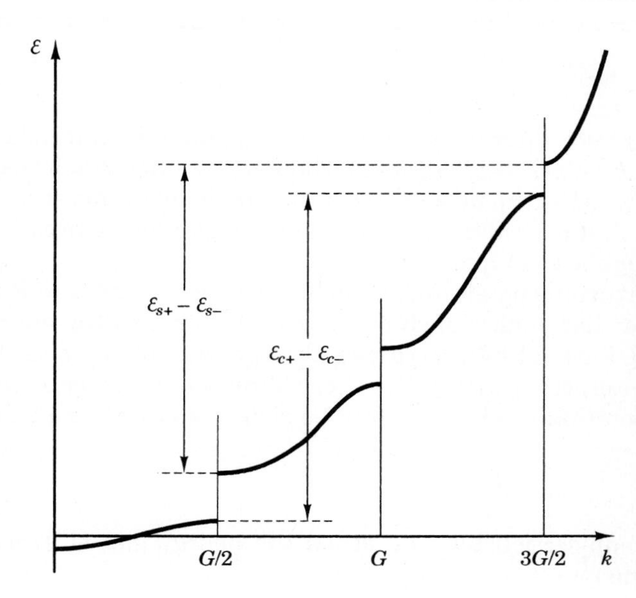

Figure 14•4–1. Additional energy gaps at $k = 3G/2$ and $k = G$. The new gap at $k = 3G/2$ arises as a result of the coupling of the plane waves with $k = G/2$ to the plane waves (14•4–7b) and (14•4–7d) with $k = 3G/2$. The other new gap, at $k = G$, arises as a result of the coupling of the plane waves $|\pm G\rangle$ with the state at $k = 0$; it is the object of one of the problems at the end of this section.

Comparing this to the earlier result (14•3–20), we see that both energies have been pushed down by about $0.06U_G$, with almost no change in the energy gap.

Energy Gap at ±3G/2

More interesting is the occurrence of a second energy gap at $k = \pm 3G/2$, shown in Fig. 14•4–1. Even though there is no direct matrix element between the unperturbed plane-wave states $|-3G/2\rangle$ and $|+3G/2\rangle$, the two states interact *indirectly*, via the path

$$|-3G/2\rangle \leftrightarrow |-G/2\rangle \leftrightarrow |+G/2\rangle \leftrightarrow |+3G/2\rangle, \tag{14•4–13}$$

involving two intermediate states. We call this a **second-order indirect interaction**.

If we again expand the square roots in (14•4–11a,b) by powers of U_G, but now for the two higher roots, $\mathcal{E}_{c,+}$ and $\mathcal{E}_{s,+}$, the linear terms cancel. We find a common upward shift of both roots, which is quadratic in U_G, and an energy gap proportional to U_G^3:

$$\mathcal{E}_{s,+} - \mathcal{E}_{c,+} \cong \frac{F}{32}\left(\frac{U_G}{F}\right)^3. \tag{14•4–14}$$

Exercise: Derive (14•4–14). If you obtain a different pre-factor, you probably overlooked a term in the expansion of the square roots. Also, evaluate the common quadratic upward shift of the two roots, and show that it is equal in magnitude to the quadratic terms in the downward shift of the two lower roots.

Other Gaps

Similar gaps due to indirect interactions occur for all integer multiples of $k = \pm G/2$, including at $k = \pm G$. (Fig. 14•4–1). The gap at $\pm mG/2$ is proportional to U_G^m. The case $m = 1$ is the task of one of the problems to this section. Unfortunately, for $m > 2$, the secular equation can no longer be factored into equations that are no higher than quadratic.

Note that, if the perturbing potential had not been a pure cosine potential, but had contained higher harmonics, such as cos $2Gx$ and/or cos $3Gx$, energy gaps at $2G/2$ and/or $3G/2$ would have occurred already without the need for higher-order indirect interactions. Usually, these direct-coupling gaps will dominate for realistic potentials, which tend to be rich in spatial harmonics.

Limitations

Our treatment in this sub-section has concerned the energy gap at *exactly* $\pm G/2$. Unfortunately, the four-wave secular equation away from $\pm G/2$ can no longer be factored into two quadratic equations. Although it would be possible to write the unperturbed states as standing waves of the form

$$\cos[(G/2 + k')x] \qquad \text{and} \qquad \sin[(G/2 + k')x], \tag{14•4–15a}$$

the two plane-wave components contained in these forms, $\exp[\pm i(G/2 + k')x]$, are no longer coupled by the perturbation to each other, but to the separate pair $\exp[\pm i(G/2 - k')x]$. This would make it necessary to include the additional standing waves

$$\cos[(G/2 - k')x] \qquad \text{and} \qquad \sin[(G/2 - k')x], \tag{14•4–15b}$$

along with two pairs in which $G/2$ has been replaced by $3G/2$. Once this is done, one obtains an 8×8 secular determinant that factors into two 4×4 rather than 2×2 determinants—which now require a numerical solution. In this case, it is simpler to continue to work with propagating plane waves.

14.4.3 Identical Perturbation Matrix Elements: a Class of Exactly Solvable Special Cases

The Principle

One sometimes encounters situations in which an N-fold degeneracy of arbitrary order is split by a perturbation, all matrix elements of which are the same,

$$W_{mn} = W. \tag{14•4–16}$$

This leads to a secular equation of the form

$$\begin{vmatrix} W - \Delta\mathcal{E} & W & W & \cdots & W \\ W & W - \Delta\mathcal{E} & W & \cdots & W \\ W & W & W - \Delta\mathcal{E} & \cdots & W \\ \vdots & \vdots & \vdots & \ddots & \vdots \\ W & W & W & \cdots & W - \Delta\mathcal{E} \end{vmatrix} = 0, \tag{14•4–17}$$

where $\Delta\mathcal{E}$ is the energy shift due to the perturbation. This problem can be factored exactly, regardless of order, by elementary row-and-column manipulation of the determinant, of the kind that leaves the value of the determinant unchanged.

As a first step, we subtract the first column from all other columns:

$$\begin{vmatrix} W - \Delta\mathcal{E} & +\Delta\mathcal{E} & +\Delta\mathcal{E} & \cdots & +\Delta\mathcal{E} \\ W & -\Delta\mathcal{E} & 0 & \cdots & 0 \\ W & 0 & -\Delta\mathcal{E} & \cdots & 0 \\ \vdots & \vdots & \vdots & \ddots & \vdots \\ W & 0 & 0 & \cdots & -\Delta\mathcal{E} \end{vmatrix} = 0. \tag{14•4–18a}$$

We next add rows 2 through N to the first row:

$$\begin{vmatrix} NW - \Delta\mathcal{E} & 0 & 0 & \cdots & 0 \\ W & -\Delta\mathcal{E} & 0 & \cdots & 0 \\ W & 0 & -\Delta\mathcal{E} & \cdots & 0 \\ \vdots & \vdots & \vdots & \ddots & \vdots \\ W & 0 & 0 & \cdots & -\Delta\mathcal{E} \end{vmatrix} = 0. \tag{14•4–18b}$$

All elements above the diagonal now vanish, and the value of the determinant is the product of its diagonal elements, leading to the energy eigenvalue shifts

$$\Delta\mathcal{E} = \begin{Bmatrix} NW & \text{single} \\ 0 & (N-1)-\text{fold} \end{Bmatrix}. \tag{14•4–19}$$

Evidently, one state is split off, by N-times the value of the perturbation matrix element; the remaining $N - 1$ states remain degenerate and unshifted.

A variation on this theme is the case that all off-diagonal matrix elements have one common value W, and all diagonal elements have a different common value $\Delta\mathcal{E}_0$, which is often zero. It is left to the reader to show that in this case we obtain

$$\Delta\mathcal{E} = \begin{Bmatrix} \Delta\mathcal{E}_0 + (N-1)W & \text{single} \\ \Delta\mathcal{E}_0 - W & (N-1)-\text{fold} \end{Bmatrix}, \tag{14•4–20}$$

instead of (14•4–19).

Example: Cube-Shaped Potential Well with a Central Perturbation

To illustrate the foregoing, we consider here one of the six-fold degenerate energy levels of a particle in a cube-shaped well of volume $\Omega = L^3$. From section 2.2, we recall that the normalized energy eigenfunctions of this system are of the form

$$|l, m, n\rangle \mathrel{\hat{=}} \psi_{lmn}(x, y, z) = \left(\frac{2}{L}\right)^{3/2} \sin\frac{l\pi x}{L} \cdot \sin\frac{m\pi x}{L} \cdot \sin\frac{n\pi x}{L}, \tag{14•4–21}$$

with the energy levels

$$\mathcal{E} = (l^2 + m^2 + n^2)\mathcal{E}_1. \tag{14•4–22}$$

Here

$$\mathcal{E}_1 = \frac{\hbar^2\pi^2}{2\,ML^2} \tag{14•4–23}$$

is the ground-state energy of a particle in a one-dimensional well of width L (see (2•2–9)), serving as the natural unit of energy of both the square well and the cubic well problems.

We consider specifically that subset of states for which all three quantum numbers l, m, and n are odd numbers, like 1, 3, 5. From (14•4–21), we see that the wave functions for those states do not vanish at the center of the box, but have the common value

$$\psi(\tfrac{1}{2}L, \tfrac{1}{2}L, \tfrac{1}{2}L) = \left(\frac{2}{L}\right)^{3/2}. \tag{14•4–24}$$

Hence, if we add a δ-function of strength S at the center of the well, all matrix elements of this perturbation have the same value,

$$W = S \cdot \left(\frac{2}{L}\right)^3 = \frac{8S}{\Omega}. \tag{14•4–25}$$

If the three quantum numbers l, m, and n are all different (and if there are no other states with the same energy), then the unperturbed energy level is six-fold degenerate, and the perturbation will split this degeneracy into a single state shifted by $6W$ and 5 states remaining degenerate at their unperturbed energy.

We will return to this example in chapter 16, in the context of the question to what extent degeneracies are a consequence of the symmetry of the potential and to what extent they are "accidental." We will also encounter perturbations of this type in chapter 17, in conjunction with a first-order perturbation model of the energy band structure of electrons in a three-dimensional periodic potential.

Exercise: Investigate the splitting of the energy level $83\mathcal{E}_1$ by a central δ-function.

◆ PROBLEMS TO SECTION 14.4

#14•4-1: Energy Gap at $k = \pm G$ for the Cosine Potential

Give a full and detailed treatment of three-wave degenerate perturbation theory involving the two degenerate plane-wave states $|\pm G\rangle$ and their interaction with the energetically separate state $|0\rangle$. Re-arrange the plane-wave states in such a way that the secular equation factors from the beginning.

Calculate the depression of the energy at $k = 0$ by the interaction. Show that in the limit of small values of the potential amplitude, the depression may be expanded into a Taylor series, the first term of which is quadratic in U_G. Also, calculate the energy gap at $k = \pm G$. Show that for small values of U_G, its width is quadratic in U_G.

#14•4-2: Stark Effect of Hydrogen

Calculate the splitting of the four-fold degenerate level of the hydrogen atom, with $n = 2$, under the influence of a uniform electric field $E = E_z$. Show that enough matrix elements vanish that the fourth-degree secular equation factors into one quadratic and two linear equations. Give numerical values, in Rydbergs, for $E_z = 10^4$ V/cm.

Note: Watch out with the integrations for the non-vanishing matrix elements. They are in three dimensions!

#14•4-3: Cube-Shaped Well with Center Perturbation: Perturbed Energy Eigenstates

Determine the perturbed energy eigenstates for the initially six-fold degenerate energy level $35\mathcal{E}_1$ of the perturbed cube-shaped well treated in the text. All these states must be linear superpositions of the form

$$|\Psi\rangle = a_{135}|1, 3, 5\rangle + a_{531}|5, 3, 1\rangle + \cdots, \qquad (14\text{•}4\text{–}26)$$

with six different sets of coefficients.

(a) Which conclusions about the sets of a's for the five unshifted states can you draw merely from the fact that their energies remain unchanged?

(b) You should be able to guess the set of a's for the single split-off state. Show that this guess is in fact a state with the correct energy. Show that the conclusions drawn under (a) imply orthogonality of all unshifted states to the split-off state.

(c) Using systematic trial and error, and drawing on the requirement that the unshifted states must not only be orthogonal to the split-off state, but also to each other, find sets of expansion coefficients for the five unshifted states. Note that these sets are not unique and that some of the expansion coefficients for any given set may be zero.

(d) Give an estimate of how strong the perturbing δ-function may be before overlap with other states not included in your perturbation treatment may introduce errors on the order of $\mathcal{E}_1$.

14.5 COMPUTATIONAL ISSUES

Once a degenerate perturbation problem can no longer be factored into quadratic equations, it becomes a numerical computer problem, regardless of whether the secular determinant is a 3×3 or a $1{,}000 \times 1{,}000$ determinant, with surprisingly little dependence of the computational strategies on size, once the latter exceeds 3×3 or, at most, 4×4.

Such problems should not be viewed as mathematically intractable, in either program complexity or required computer time. A treatment of suitable numerical methods falls outside the scope of a text on quantum mechanics; the purpose of the present section is to provide minimal guidance to those readers who are not familiar with such methods. Much of the presentation draws on the book by Press et al.,[2] hereafter referred to as PFTV, and that by Acton.[3]

---

[2] W. H. Press, B. P. Flannery, S. A. Teukolsky, and W. T. Vetterling, *Numerical Recipes: The Art of Scientific Computing* (Cambridge, England: Cambridge University Press, 1986). See especially chapter 11.

[3] F. S. Acton, *Numerical Methods that Work* (New York: Harper & Row, 1970).

The science—and art—of diagonalizing large matrices is very highly developed,[4,5] and readers will not find it necessary to write their own programs for this task, any more than they must write their own programs for extracting square roots. In fact, readers are explicitly advised *against* writing their own programs: The matrix eigenvalue problem is one where seemingly obvious pencil-and-paper approaches suffer badly from inefficiency and round-off errors if applied to large matrices, and the best approaches are non-obvious. Those wishing to ignore this advice may do so at their own peril, but such attempts are more likely to qualify as mathematical recreation than as quantum-mechanical problem solving. Although these warnings do not apply to small matrices, there is usually no incentive to treat small matrices differently.

Centralized computer installations almost always provide programs for matrix diagonalization. For stand-alone personal computers and workstations, many commercial numerical analysis packages are available that contain excellent matrix diagonalization capabilities. Readers should consult up-to-date sources for what is available; the field is changing too rapidly to give meaningful references in a textbook. Complete source code listings of several of the most widely used matrix diagonalization routines have been published for the most common computer languages.[6] They are even available at low cost in machine-readable form,[7,8] to eliminate the tedium of re-entering computer code by hand—possibly incorrectly.

All practically important methods for diagonalizing large Hermitian matrices with arbitrary matrix elements are based on matrix transformation *sequences* of the form

$$(\hat{H})' = \hat{T}^{-1}(\hat{H})\hat{T}, \tag{14•5–1}$$

with the goal that the transformed matrix $(\hat{H})'$ be closer to diagonal form than the old matrix $(\hat{H})$. Inasmuch as a transformation accomplishing exact diagonalization in a *single* step is not known beforehand, the strategy is to employ a *sequence* of cleverly chosen *simple* transformations that gradually reduce the off-diagonal matrix elements. Different techniques differ in their choice of transformation matrices—and in their computational efficiency and ease of use.

---

[4] J. H. Wilkinson, *The Algebraic Eigenvalue Problem* (Oxford: Clarendon Press, 1965).

[5] J. H. Wilkinson and C. Reinsch, *Linear Algebra Handbook for Automatic Computation*, Vol. II (Berlin: Springer Verlag, 1971).

[6] E. Anderson, Z. Bai, C. Bishof, J. Demmel, J. Dongarra, J. Du Croz, A. Greenbaum, S. Hamarling, A. McKenney, S. Oustouchov, and D. Sorenson, *LAPACK Users' Guide*, (Philadelphia: Society for Industrial and Applied Mathematics (SIAM), 1992).

[7] ACM Algorithms Distribution Service, c/o IMSL Inc., 2500 Park West Tower One, 2500 West Boulevard, Houston, TX.

[8] The algorithms in the book by Press et al. are obtainable on diskettes from the publisher.

For most problems likely to be encountered by most readers, computer time is not a dominant issue, and the ease of learning and using the method tends to be more important. Nevertheless, it is desirable to have understanding of at least the order of magnitude of the computer work loads involved, if only to be able to judge at what point computer time *might* become important. The efficiency of the algorithms is most easily compared by comparing the number M of floating-point multiplications required: In an *efficient* matrix diagonalization program, the total computer time spent tends to scale with the time spent doing such multiplications, usually within a factor of about 2, except for very small matrices ($N \lessdot 10$), for which the size-independent computational overhead tends to dominate. Transformations of the form (14•5–1) involve matrix multiplications. A simple product of two $N \times N$ matrices with *arbitrary* real coefficients requires N^3 multiplications, and all general-purpose matrix diagonalizations scale with N^3 for large N. In general, we may write

$$M = M_3 N^3 + 0(N^2), \tag{14•5–2}$$

where $0(N^2)$ designates unspecified terms of order N^2 or lower. For large values of N, the cubic term dominates. The values of M_3 vary over a very wide range, depending on the method and on the nature of the problem.

The simplest of those methods that are in widespread practical use is the **Jacobi transformation** method, discussed extensively in the relevant literature. It is relatively inefficient: PFTV quote a value of $M_3 \sim 12$ to 20, depending on the matrix and on the accuracy required. This estimate assumes that only the eigenvalues are required, but no eigenvectors, and that all matrix elements are real. For, say, a 100×100 matrix, PFTV's upper estimate amounts to about 2×10^7 multiplications. Assuming a 100% overhead for other (necessary) operations, and another 100% for inefficiencies in the computer code generated by the compiler, we estimate a total computer time of about 6×10^7 multiplication cycles. On a computer that can execute, say, 10^6 (double-precision) floating-point multiplications[9] per second, the computing time should be about 1 minute. If the eigenvectors are required along with the eigenvalues, the time required for Jacobi's method increases by about 50%.

Often, more efficient methods or faster computers will be available to the user for little or no additional effort. But if they are not readily available, there is little incentive to go to lengthy efforts to seek them out, unless much bigger matrices than 100×100 must be handled or a very large number of such computations must be performed.

The most efficient techniques first reduce the matrix to a symmetric **tri-diagonal** matrix[10] and then extract the eigenvalues (and possibly the

---

[9] For all but the smallest matrices, all variables should be at least double-precision variables with 64-bit word lengths, preferably 80-bit extended-precision variables. If the variables are complex, both the real and the imaginary parts should have the required precision.

[10] A tri-diagonal matrix is one in which nonzero elements occur only on the diagonal itself and on the two sub-diagonals adjacent to the main diagonal. An example of a particularly simple real, symmetric tri-diagonal form is the 4×4 secular determinant (14•4–6) we encountered earlier.

eigenvectors) from the latter. The reason for this apparent detour is that *rigorous* tri-diagonalization of an arbitrary Hermitian matrix can be achieved with a finite number ($N - 2$) of simple transformations, and the extraction of eigenvalues from a tri-diagonal matrix is itself a fairly simple and computationally efficient process. The method used for the tri-diagonalization is usually the so-called **Householder** method. If only the eigenvalues are needed, the coefficient M_3 has the remarkably low value $M_3 = 2/3$, a factor of 30 below the upper-limit estimate for the Jacobi transformation! If the eigenvectors are also needed, the additional computations double M_3. These estimates again assume real matrix elements.

Once a symmetric tri-diagonal matrix has been achieved, many methods are available to extract the eigenvalues from it. *All* of these are necessarily iterative, meaning that a mathematically exact diagonalization for arbitrary coefficients would require an infinite number of iteration cycles. In practice, the better methods converge rapidly to any reasonable finite accuracy that is within the capability of the hardware. PFTV recommend the so-called **QL algorithm** as the general-purpose method of default, to be used whenever there are no specific reasons to choose something else. If only the eigenvalues are needed, the QL work loads are linear in N and hence negligible compared to the Householder work load for large matrices, giving this overall approach a speed advantage up to 30:1 over the Jacobi method, at a relatively small cost in ease of use. Some of the commercial packages obtain additional improvements by coding the algorithm in more efficient machine language.

Extracting the eigenvectors along with the eigenvalues from a symmetric tri-diagonal matrix involves work loads that scale again with N^3, but with smaller coefficients than the Jacobi transformation. PFTV estimate $M_3 \sim 3$ for the QL algorithm, giving the combination Householder-plus-QL an overall coefficient of about 4.3, still 5 to 7 times faster than the Jacobi transformation route.

Throughout the above, we have assumed that the matrix elements are real. Each complex multiplication requires four real multiplications, increasing the computer time considerably, but somewhat less than a full factor of four. More important than the cost in efficiency is often the fact that diagonalization programs capable of handling complex Hermitian matrices tend to be much harder to find than similar programs for real symmetric matrices, except as raw code in FORTRAN, PASCAL, or C. Readers wishing to use algorithms for complex Hermitian matrices are urged to make sure beforehand that both the real and the imaginary parts of all complex variables are handled in double precision or extended precision. Such considerations evidently place a premium on the use of basis functions for which all matrix elements are real. This is usually possible, but not always. Alternative strategies are found in the numerical analysis literature—for example, in PFTV.

Chapter 15

PERTURBATION THEORY, II: "NONDEGENERATE" PERTURBATION THEORY

15.1 RE-FORMULATION OF THE SCHROEDINGER EQUATION

As we stated at the beginning of chapter 14, degenerate perturbation theory treats all states included in the expansions (14•1–2) or (14•1–7) on an equal footing. We now turn to the opposite approach, that of singling out one particular state as dominant and studying the way this state is perturbed by its interaction with all other states.

Our point of departure is again the Schroedinger equation in the form (14•1–5),

$$[\hat{H}^{(0)} + \hat{W}]|\psi\rangle = \mathcal{E}|\psi\rangle. \tag{15•1–1}$$

We designate the selected unperturbed basis state by $|u\rangle$ and its unperturbed energy by $\mathcal{E}_u$:

$$\hat{H}^{(0)}|u\rangle = \mathcal{E}_u|u\rangle, \tag{15•1–2}$$

and we assume that $|u\rangle$ is normalized:

$$\langle u|u\rangle = 1. \tag{15•1–3}$$

The perturbed state may always be written

$$|\psi\rangle = |u\rangle + |\phi\rangle, \tag{15•1–4}$$

where $|\phi\rangle$ is the perturbation. If we are willing to forgo normalization of the *perturbed* state function $|\psi\rangle$, the perturbation $|\phi\rangle$ may be chosen in such a way that it is orthogonal to the unperturbed state $|u\rangle$,

$$\langle u|\phi\rangle = 0. \tag{15•1–5}$$

In this case, $|\psi\rangle$ is no longer normalized to unity; instead,

$$\langle\psi|\psi\rangle = 1 + \langle\phi|\phi\rangle, \tag{15•1–6}$$

and

$$\langle u|\psi\rangle = 1. \tag{15•1–7}$$

These are the normalizations we use here.

If the perturbation $|\phi\rangle$ were known, the perturbed energy eigenvalue $\mathcal{E}$ in the Schroedinger equation (15•1–1) could be obtained by simply taking the inner product of (15•1–1) with $\langle u|$:

$$\mathcal{E} = \mathcal{E}\langle u|\psi\rangle = \langle u|\hat{H}^{(0)}|\psi\rangle + \langle u|\hat{W}|\psi\rangle. \tag{15•1–8}$$

Because of the hermiticity of $\hat{H}^{(0)}$, and with the help of (15•1–7), the first term on the far right is simply $\mathcal{E}_u$:

$$\langle u|\hat{H}^{(0)}|\psi\rangle = \mathcal{E}_u\langle u|\psi\rangle = \mathcal{E}_u. \tag{15•1–9}$$

With this, (15•1–8) becomes

$$\mathcal{E} = \mathcal{E}_u + \langle u|\hat{W}|\psi\rangle. \tag{15•1–10}$$

This result is exact.

If the perturbation is sufficiently weak, the contribution $|\phi\rangle$ to $|\psi\rangle$ will be small. If we neglect this contribution on the right-hand side of (15•1–10) altogether, and replace $|\psi\rangle$ by its unperturbed limit $|u\rangle$, the relation (15•1–10) reduces to the **first-order approximation** for the energy,

$$\mathcal{E} \approx \mathcal{E}_u + \Delta\mathcal{E}^{(1)}, \qquad \text{where} \qquad \Delta\mathcal{E}^{(1)} = \langle u|\hat{W}|u\rangle. \tag{15•1–11a,b}$$

The correction term $\Delta\mathcal{E}^{(1)}$ is called the **first-order energy correction**; it is simply the expectation value of the perturbation, taken for the unperturbed state.

Similar first-order energy corrections apply to all other eigenstates of the unperturbed Hamiltonian. If the perturbation had no off-diagonal matrix elements, the first-order energy corrections would account fully for the effect of the perturbation. In particular, there would be no effect on the state functions; the changes to the latter are entirely due to the off-diagonal matrix elements of the perturbation.

This observation suggests splitting the perturbation into a diagonal and an off-diagonal part,

$$\hat{W} = \hat{D} + \hat{W}', \tag{15•1–12}$$

and splitting the overall Hamiltonian $\hat{H}$, not as in (15•1–1), but into a diagonal term

$$\hat{H}^{(d)} = \hat{H}^{(0)} + \hat{D} \tag{15•1–13}$$

and the off-diagonal part $\hat{W}'$ of the perturbation. The new split is accomplished by writing all operators as sums over Dirac outer products, namely,

$$\hat{D} = \sum_n |n\rangle\langle n|\hat{W}|n\rangle\langle n| = \sum_n |n\rangle W_{nn}\langle n|, \tag{15•1–14}$$

$$\hat{W}' = \sum_m \sum_{n\neq m} |m\rangle\langle m|\hat{W}|n\rangle\langle n|, \tag{15•1–15}$$

and

$$\hat{H}^{(d)} = \sum_n |n\rangle(\mathcal{E}_n + \langle n|\hat{W}|n\rangle)\langle n| = \sum_n |n\rangle\mathcal{E}_n'\langle n|, \tag{15•1–16}$$

where we have defined the first-order-corrected energy eigenvalues

$$\mathcal{E}_n' = \mathcal{E}_n + \langle n|\hat{W}|n\rangle. \tag{15•1–17}$$

The eigenstates of $\hat{H}^{(d)}$ are the same as those of $\hat{H}^{(0)}$, obeying

$$\hat{H}^{(d)}|n\rangle = E_n'|n\rangle, \tag{15•1–18}$$

but with the new eigenvalues $\mathcal{E}_n'$.

The Schroedinger equation (15•1–1) and the energy relation (15•1–10) now become

$$[\hat{H}^{(d)} + \hat{W}']|\psi\rangle = \mathcal{E}|\psi\rangle, \tag{15•1–19}$$

and

$$\boxed{\mathcal{E} = \mathcal{E}_u' + \langle u|\hat{W}|\phi\rangle.} \tag{15•1–20}$$

Note that the latter is still exact.

To obtain the perturbation corrections to the state function itself, we first re-arrange (15•1–19) as follows:

$$[\mathcal{E} - \hat{H}^{(d)}]|\psi\rangle = (\mathcal{E} - \mathcal{E}_u')|u\rangle + [\mathcal{E} - \hat{H}^{(d)}]|\phi\rangle = \hat{W}'|\psi\rangle. \tag{15•1–21}$$

So far, no eigenfunction expansion has been used. We next re-write the operators occurring in (15•1–21) as sums over Dirac outer products, using (15•1–15) and (15•1–16):

$$\sum_n |n\rangle(\mathcal{E} - \mathcal{E}'_n)\langle n|\psi\rangle = \sum_m \sum_{n\neq m} |m\rangle\langle m|\hat{W}|n\rangle\langle n|\psi\rangle. \tag{15•1–22}$$

Note that all sums so far include the unperturbed state $|u\rangle$ itself.

We now make the central assumption that gives the technique the name **non-degenerate perturbation theory**, namely, that the actual (perturbed) energy eigenvalue $\mathcal{E}$ does *not* coincide with any of the first-order-corrected *unperturbed* energy eigenvalues, except possibly with $\mathcal{E}'_u$:

$$\mathcal{E} \neq \mathcal{E}'_n \quad \text{for all} \quad n \neq u. \tag{15•1–23}$$

We can then define the operator

$$\hat{G} = \sum_{m\neq u} |m\rangle \frac{1}{\mathcal{E} - \mathcal{E}'_m} \langle m|, \tag{15•1–24}$$

where the term with $m = u$ is omitted from the sum. We apply this operator to both sides of (15•1–22). On the left-hand side, we obtain

$$\sum_{m\neq u} \sum_n |m\rangle \frac{1}{\mathcal{E} - \mathcal{E}'_m} \langle m|n\rangle(\mathcal{E} - \mathcal{E}'_n)\langle n|\psi\rangle$$
$$= \sum_{m\neq u} |m\rangle\langle m|\psi\rangle = |\phi\rangle. \tag{15•1–25}$$

Here the first equality utilizes the ortho-normality of the different basis states, and the second equality utilizes the condition (15•1–5) for the orthogonality of the perturbation $|\phi\rangle$ to the unperturbed state $|u\rangle$. Evidently, (15•1–25) isolates the perturbation to the wave function—which is the whole purpose of the otherwise non-obvious introduction of the operator $\hat{G}$.

The right-hand side of (15•1–22) becomes

$$\sum_{l\neq u} \sum_m \sum_{n\neq m} |l\rangle \frac{1}{\mathcal{E} - \mathcal{E}'_l} \langle l|m\rangle\langle m|\hat{W}|n\rangle\langle n|\psi\rangle$$
$$= \sum_{m\neq u} \sum_{n\neq m} |m\rangle \frac{\langle m|\hat{W}|n\rangle}{\mathcal{E} - \mathcal{E}'_m} \langle n|\psi\rangle \,. \tag{15•1–26}$$

By equating (15•1–25) to (15•1–26) and replacing $|\phi\rangle$ with $|\psi\rangle - |u\rangle$, we obtain finally

$$\boxed{|\psi\rangle = |u\rangle + \sum_{m\neq u} \sum_{n\neq m} |m\rangle \frac{\langle m|\hat{W}|n\rangle}{\mathcal{E} - \mathcal{E}'_m} \langle n|\psi\rangle.} \tag{15•1–27}$$

This is a recursion relation that expresses $|\psi\rangle$ in terms of itself, via the matrix elements of the perturbation, and the various energies. Like (15•1–10) and (15•1–20), this relation is also exact *if* both sums go over the complete set of states, except those explicitly excluded.

Taken together, (15•1–20) and (15•1–27) form a set of coupled equations that permit—at least in principle—an iterative numerical determination of both the energy eigenvalues and the associated eigenfunctions, to whatever accuracy is desired.

15.2 SECOND-ORDER PERTURBATION THEORY

15.2.1 The Formalism

In many cases of practical interest, it is not necessary to carry out the full-blown iteration of both energy eigenvalues and eigenfunctions. The double sum on the right-hand side of (15•1–27) is the correction $|\phi\rangle$ to the wave function. If the perturbation is sufficiently weak, this correction will be small. In that case, the correction may itself be approximated by replacing $|\psi\rangle$ inside the correction term with the unperturbed state function $|u\rangle$:

$$|\phi\rangle \approx |\phi^{(1)}\rangle = \sum_{m\neq u} |m\rangle \frac{\langle m|\hat{W}|u\rangle}{\mathcal{E} - \mathcal{E}_m'}. \tag{15•2–1}$$

This is called the **first-order correction to the state function**. If we insert it into (15•1–4) and the result into (15•1–20), we obtain the second-order approximation for the energy,

$$\boxed{\mathcal{E} \approx \mathcal{E}_u' + \sum_{m\neq u} \frac{|\langle m|\hat{W}|u\rangle|^2}{\mathcal{E} - \mathcal{E}_m'} = \mathcal{E}_u' + \Delta\mathcal{E}^{(2)},} \tag{15•2–2a}$$

where

$$\Delta\mathcal{E}^{(2)} = \sum_{m\neq u} \frac{|\langle u|\hat{W}|m\rangle|^2}{\mathcal{E} - \mathcal{E}_m'} \tag{15•2–2b}$$

is the **second-order energy correction**.

The relations (15•2–2a,b) form the backbone of what is called **second-order perturbation theory**, the most common approximation in perturbation theory. Note that the state function no longer shows up explicitly in this formulation, similar to the way it no longer shows up in the secular equation of degenerate perturbation theory.

Each term in $\Delta\mathcal{E}^{(2)}$ has a sign equal to the sign of $\mathcal{E} - \mathcal{E}_m'$. This means that states with energies below the energy of the perturbed state tend to raise the energy of the perturbed state, while states with energies above the perturbed state tend to depress the energy of that state: The interaction between different states is **repulsive**. This is one of the central general properties of second-order perturbation theory.

In the form (15•2–2), the formal expression for $\Delta\mathcal{E}^{(2)}$ still depends explicitly on the perturbed energy $\mathcal{E}$ itself. Hence, (15•2–2a) is really not an explicit expression for the perturbed energy, but rather a recursive iterative algorithm for the energy, as will be explained below. However, the energy corrections $\Delta\mathcal{E}^{(1)}$ and $\Delta\mathcal{E}^{(2)}$ are often small compared to the separations of the unperturbed energy $\mathcal{E}_u$ from all other unperturbed energy levels that interact with $|u\rangle$,

$$|\Delta\mathcal{E}^{(1)} + \Delta\mathcal{E}^{(2)}| \approx |\mathcal{E} - \mathcal{E}'_u| \ll |\mathcal{E}_m - \mathcal{E}_u|, \tag{15•2–3}$$

for all $m \neq u$ for which $\langle u|\hat{W}|m\rangle \neq 0$. In this case—and *only* in this case—we may make a *second approximation* beyond replacing $|\psi\rangle$ by $|u\rangle$ in (15•1–27): We may replace the *true* energy $\mathcal{E}$ in the denominators in (15•2–1) and (15•2–2) by the *unperturbed* energy $\mathcal{E}_u$. This yields the simplified first-order perturbation to the state function

$$|\phi^{(1)}\rangle = \sum_{m\neq u} |m\rangle \frac{\langle m|\hat{W}|u\rangle}{\mathcal{E}_u - \mathcal{E}_m}, \tag{15•2–4}$$

and the simplified second-order energy correction

$$\boxed{\Delta\mathcal{E}^{(2)} = \sum_{m\neq u} \frac{|\langle m|\hat{W}|u\rangle|^2}{\mathcal{E}_u - \mathcal{E}_m},} \tag{15•2–5}$$

instead of (15•2–1) and (15•2–2b).

This is the simplest form of second-order non-degenerate perturbation theory, and it is the most widely quoted form, sometimes referred to as (second-order) **Rayleigh-Schroedinger** (RS) **perturbation theory**, to distinguish it from the recursive iteration approach implied by (15•2–2), which is referred to as (second-order) **Brillouin-Wigner** (BW) **perturbation theory**. In fact, whenever the term "perturbation theory" is used in the literature, without a further qualifier, it usually refers to the RS approximation, carried either to the first or to the second order.

The reason for this apparent preference for the RS formalism, especially in textbooks, is precisely the existence of the simple closed-form expression (15•2–5), whereas BW perturbation theory is a numerical algorithm that does not lead to closed-form analytical expressions for either the perturbed energy or the perturbed state function. The BW formalism is therefore rarely quoted in purely formal deliberations. But it is well suited—and indeed preferable—for computer implementation in numerical calculations, where its accuracy is invariably higher than that of RS perturbation, with sometimes startlingly superior convergence properties near degeneracies.

It is evident from (15•2–4) and (15•2–5) that the RS approximation diverges if the unperturbed state $|u\rangle$ is degenerate with one of the perturbing states $|m\rangle$, unless the matrix element $\langle u|\hat{W}|m\rangle$ vanishes. This result is clearly incorrect; it arises from replacing, in the denominators in (15•2–1) and (15•2–2), the true energy $\mathcal{E}$ by the unperturbed energy $\mathcal{E}_u$. It is this breakdown of the approximations (15•2–4) and (15•2–5) that has given the name *non-degenerate* perturbation theory to the entire method.

One alternative in such situations is the use of degenerate perturbation theory, discussed in chapter 14. But this is by no means the only possibility. The divergence of non-degenerate perturbation theory is easily avoided by *not* replacing $\mathcal{E}$ *with* $\mathcal{E}_u$, instead treating (15•2–2) as the basis of an iterative recursion scheme. We will discuss such iteration schemes in the next section.

Mathematically, the second-order Rayleigh-Schroedinger approximation (15•2–5) may be viewed as the third term of a Taylor series expansion of the energy in terms of powers of the perturbation $\hat{W}$. The RS series expansion can be carried to higher order, but the higher-order terms are rarely used. They are unwieldy, and the convergence properties of the series are usually poor: Power series expansions are a mathematical straightjacket ill suited to many quantum-mechanical perturbation problems. In cases in which the second-order Rayleigh-Schroedinger result gives insufficient accuracy, and even more when it diverges, it is almost always preferable to change the method, rather than carrying the series to a higher order.

15.2.2 Example: The Cosine Potential Re-Visited

We illustrate second-order perturbation theory by applying it to the simple one-dimensional cosine potential discussed extensively in section 14.3 (see (14•3–6) and Fig. 14•3–1):

$$W(x) = V(x) = -2\,U_G \cos Gx, \qquad G = 2\pi/a. \tag{15•2–6}$$

As in section 14.3, we treat this potential as acting as a perturbation on free electrons, with the unperturbed plane-wave eigenfunctions of the form (14•3–9).

We shall designate the specific basis state whose perturbation we study as $|K\rangle$, with a capital K, retaining the designation $|k\rangle$ with a lower-case k for unperturbed plane waves. The matrix elements of the perturbation were calculated in (14•3–14).

$$\boxed{\langle K|\hat{W}|k\rangle = -U_G\delta(k, K \pm G).} \tag{15•2–7}$$

From (15•2–7), it follows that the expectation value of the perturbation vanishes. Hence, there is no first-order energy correction:

$$\Delta\mathcal{E}^{(1)} = \langle K|\hat{W}|K\rangle = 0. \tag{15•2–8}$$

Because of this, we need not distinguish between the "primed" and "unprimed" energies introduced in section 15.1, and we will omit the prime everywhere.

Because every plane-wave state $|k\rangle$ interacts only with the two other states $|k \pm G\rangle$, the sum in the expression (15•2–2a) for the second-order energy correction contains only two terms. If we write $|K\rangle$ for $|u\rangle$ and insert into (15•2–2) the expressions (15•2–7) and (15•2–8) for the matrix elements, we obtain, after minor manipulations,

$$\mathcal{E} = \mathcal{E}_0(K) + \frac{U_G^2}{\mathcal{E} - \mathcal{E}_0(K - G)} + \frac{U_G^2}{\mathcal{E} - \mathcal{E}_0(K + G)}, \tag{15•2–9a}$$

which may also be written

$$\mathcal{E} = \mathcal{E}_0(K) + 2U_G^2 \cdot \frac{\mathcal{E} - \mathcal{E}_0(K) - 4F}{[\mathcal{E} - \mathcal{E}_0(K) - 4F]^2 - 16\mathcal{E}_0(K) \cdot F}. \tag{15•2–9b}$$

where we have again inserted the natural unit of energy $F = \mathcal{E}_0(G/2)$ from (14.3–17).

The result (15•2–9b) still contains the energy $\mathcal{E}$ on the right-hand side, as is appropriate for second-order Brillouin-Wigner perturbation theory. If we make the transition to Rayleigh-Schroedinger perturbation theory by setting

$$\mathcal{E} = \mathcal{E}_0(K), \tag{15•2–10}$$

on the right-hand side of (15•2–9b), we obtain

$$\Delta\mathcal{E}^{(2)}(K) = -\frac{U_G^2}{2[F - \mathcal{E}_0(K)]}. \tag{15•2–11}$$

We discuss here specifically the case $K = 0$, deferring the discussion of the more general case to later. At $K = 0$, we have $\mathcal{E}_0(K) = 0$; hence,

$$\Delta\mathcal{E}^{(2)}(0) = -U_G^2/2F. \tag{15•2–12}$$

Evidently, the ground state is depressed by the repulsive interaction with the two states $|\pm G\rangle$.

We would expect (15•2–12) to be a good approximation for a weak perturbation, but it clearly cannot be valid for a strong perturbation: The energy of the ground state cannot be lower than the minimum of the potential energy, which implies that

$$\mathcal{E} > -2U_G, \tag{15•2–13}$$

and (15•2–12) violates this condition unless $|U_G| < 4F$. As $|U_G|$ approaches this limit, RS perturbation theory must become invalid, leading to the validity condition

$$|U_G| \ll 4F. \tag{15•2–14}$$

Evidently, Rayleigh-Schroedinger perturbation theory behaves quite differently than degenerate perturbation theory, which is equivalent to a variational approximation. In the latter, the approximate ground state can never drop below the true ground state. But in RS perturbation theory, the energy is calculated, not as an expectation value, but from the relation (15•1–20), which contains the perturbation to the wave function in a different way. If the first-order correction to the wave function overshoots, as the result of an incorrect energy denominator in (15•2–4), the RS energy correction will similarly overshoot. Such an overshoot occurs whenever the magnitude of the true energy denominator $\mathcal{E} - \mathcal{E}_m$ in (15•2–1) is significantly larger than that of the unperturbed RS energy denominator $\mathcal{E}_u - \mathcal{E}_m$ in (15•2–4). This is precisely what happens in our case as the potential amplitude increases.

We obtain a much better approximation by retaining the full Brillouin-Wigner form (15•2–9a) and setting $K = 0$. Normally, BW perturbation theory does not lead to closed-form expressions for the energy, but our example is one of the few exceptions, leading to the simple quadratic equation

$$\mathcal{E}^2 - 4\mathcal{E}F, = 2U_G^2, \qquad (15•2–15)$$

with the solutions

$$\mathcal{E} = 2F \pm \sqrt{4F^2 + 2U_G^2}. \qquad (15•2–16)$$

A little reflection shows that it is the lower of the two roots that corresponds to the perturbed state at $K = 0$, and that this root clearly satisfies the condition $\mathcal{E} > -2U_G$. In fact, the result is identical to what we would have obtained *if* we had treated the perturbation of this state by three-wave degenerate perturbation theory. More accurate values could be obtained by going beyond second-order BW perturbation theory, using the full BW iteration discussed in section 15.3.

15.2.3 Divergences near Degeneracies

The over-correction in the first-order perturbed wave function in RS perturbation theory leads to a catastrophic divergence near degeneracies, as shown in **Fig. 15•2–1** for the specific case of our cosine potential, when $K \to \pm G/2$.

In this limit the unperturbed state $|K\rangle$ becomes degenerate with the unperturbed state $|-K\rangle$, and the repulsive interaction between those two states dominates the perturbation of the state $|K\rangle$. We assume here that K is positive and consider the limit $K \to +G/2$. If K approaches $G/2$ from below, the repelling state $|K - G\rangle$ has an energy above that of $|K\rangle$, leading to a lowering of the energy of $|K\rangle$. If K approaches $G/2$ from above, the repelling state $|K - G\rangle$ has an energy below that of $|K\rangle$, leading to a raising of the energy of $|K\rangle$. In the RS approximation, both the lowering and the raising of the energy are calculated from the *unperturbed* energy differences between the two states,

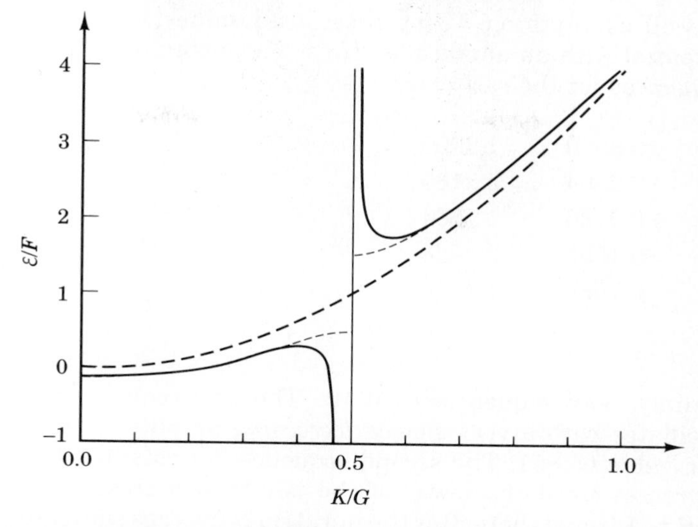

Figure 15•2–1. Energy of an electron in a periodic cosine potential with an amplitude $2U_G = F$. The solid line represents the perturbed energy, calculated from the Rayleigh-Schroedinger approximation, while the dashed line is the result of a Brillouin-Wigner iteration.

rather than from the final perturbed energy differences. Because the unperturbed energy differences vanish as $K \to \pm G/2$, the calculated energy corrections diverge in this limit, as shown by the solid curve in Fig. 15•2–1.

The divergence may be eliminated by going back to the BW formulation (15•2–9b) of the problem and solving that equation numerically. There are several methods available to do this. The simplest, requiring the least amount of programming, is to view (15•2–9b) as a recursive re-insertion algorithm of the form

$$\mathcal{E} \leftarrow f(\mathcal{E}), \tag{15•2–17}$$

where $f(\mathcal{E})$ is the right-hand side of (15•2–9b). This method usually converges rapidly for roots far away from any degeneracies, but it tends to be unstable near degeneracies.

We demonstrate this instability here for our example of the energy gap at $K = G/2$. The unperturbed energy eigenvalue for $K = G/2$ is $\mathcal{E} = F$. If the iteration of (15•2–9b) were started with this initial value of $\mathcal{E}$, the denominator on the right-hand side would vanish and the iteration would blow up immediately. But there is no need to start the iteration with the unperturbed energy values. We know that the two perturbed values of the energy straddle the unperturbed value. This calls for two starting values of $\mathcal{E}$ on opposite sides of $\mathcal{E} = F$, but the exact numbers do not matter; $\mathcal{E} = 0.8F$ and $\mathcal{E} = 1.2F$ should do as well as anything. Using those seed values, and assuming again a cosine potential with an amplitude $2U_G = F$, we obtain the following two numerical sequences for the ratio $\mathcal{E}/F$:

0.8000	1.2000
−0.2804	2.2180
+0.7780	1.1684
−0.1556	2.4528
0.7564	1.1340
⋮	⋮

Plainly, both sequences oscillate. The first sequence is just barely stable, it oscillates with a very slowly *decreasing* amplitude, eventually converging to the value 0.4851. The second sequence is unstable: Its oscillation amplitude increases until the lower of the two values crosses the unperturbed value $\mathcal{E}/F = 1.0$, and thereafter the iteration converges slowly toward the same value 0.4851 as the first sequence, rather than toward the desired value *above* 1.0.

Such oscillations and instabilities are the bane of many iteration techniques. If only a small number of computations is required, the slowness of a particular convergence may be acceptable. But instabilities are not. Fortunately, it is always possible in such cases to *force* convergence by inserting, on the left-hand side of the basic BW energy relation (15•2–2a), not the most recently calculated value $\mathcal{E}_{\text{new}}$ of the energy, but a weighted average of this value and the preceding value $\mathcal{E}_{\text{old}}$,

$$\mathcal{E} = w\mathcal{E}_{\text{old}} + (1 - w)\mathcal{E}_{\text{new}}, \tag{15•2–18}$$

with a suitable value for the weighting factor w. In our example, setting $w = 1/2$ will lead to a rapid convergence of both iteration sequences, but it will often be preferable to switch to an *inherently* more stable method of solving the BW energy equation (15•2–2a).

15.2.4 A Stable Alternative Approach: Newton's Method

Newton's method for finding roots of equations is well suited to the specific structure of the BW perturbation problem. We treat (15•2–2a) as an equation of the form

$$f(\mathcal{E}) = 0, \tag{15•2–19}$$

with

$$f(\mathcal{E}) = \mathcal{E} - \mathcal{E}_u - \sum_{m \neq u} \frac{|\langle u|\hat{W}|m\rangle|^2}{\mathcal{E} - \mathcal{E}_m}. \tag{15•2–20}$$

The desired root of (15•2–19) is found by picking a suitable *seed* value for $\mathcal{E}$—often $\mathcal{E}_u$ itself—and homing in on the correct energy via the iteration sequence

$$\mathcal{E} \leftarrow \mathcal{E} - \frac{f(\mathcal{E})}{f'(\mathcal{E})}, \tag{15•2–21}$$

where

$$f'(\mathcal{E}) = 1 + \sum_{m \neq u} \frac{|\langle u|\hat{W}|m\rangle|^2}{[\mathcal{E} - \mathcal{E}_m]^2}. \tag{15•2–22}$$

If $\mathcal{E}_u$ is degenerate with another state, it cannot be used as a seed value, because the sum in (15•2–20) will then still contain another term with a pole at $\mathcal{E} = \mathcal{E}_u$, even though the state $|u\rangle$ has been omitted from the sum. By using as a seed value an energy just below or just above $\mathcal{E}_u$, far away from whatever additional poles might be present, the sequence (15•2–22) usually converges rapidly toward the nearest root in that direction, without any numerical instability problems.

In our example, from (15•2–9a), with $\mathcal{E}_u = \mathcal{E}_0(K)$,

$$f(\mathcal{E}) = \mathcal{E} - \mathcal{E}_0\,(K) - \frac{U_G^2}{\mathcal{E} - \mathcal{E}_0\,(K - G)} - \frac{U_G^2}{\mathcal{E} - \mathcal{E}_0\,(K + G)}, \tag{15•2–23}$$

and, from this,

$$f'(\mathcal{E}) = 1 + \frac{U_G^2}{[\mathcal{E} - \mathcal{E}_0\,(K - G)]^2} + \frac{U_G^2}{[\mathcal{E} - \mathcal{E}_0\,(K + G)]^2}. \tag{15•2–24}$$

Using these expressions, and the same seed values as earlier, both $\mathcal{E}/F$ sequences converge very rapidly:

0.8000	1.2000
0.6510	1.3403
0.5210	1.4567
0.4851	1.4836
0.4851	1.4836
⋮	⋮

If we re-express the two final values as energies, we obtain, with $2U_G = F$,

$$\mathcal{E} = \begin{Bmatrix} 1.4836F = F + 0.9673U_G \\ 0.4815F = F - 1.0298U_G \end{Bmatrix}, \tag{15•2–25}$$

which is roughly halfway between the earlier result (14•3–20) obtained by two-state degenerate perturbation theory and the more accurate subsequent result (14•4–12) obtained by four-state degenerate perturbation theory. The reason for the difference is the following. In the second-order BW treatment, we have not included the plane wave $|-3G/2\rangle$ amongst the waves perturbing the state $|+G/2\rangle$, because there was no matrix element coupling the two states *directly*. This was a consequence of not treating the two plane-wave states $|+G/2\rangle$ and $|-G/2\rangle$ on an equal footing, but treating $|+G/2\rangle$ as the unperturbed state and $|-G/2\rangle$ as part of the perturbation. With no direct coupling between $|+G/2\rangle$ and $|-3G/2\rangle$, the coupling proceeded only indirectly, via the plane-wave state $|-G/2\rangle$, which was itself treated as part of the perturbation. But in second-order BW perturbation theory, interactions amongst the perturbing states are neglected. If we had performed a higher-order BW iteration including $|-3G/2\rangle$, we would have obtained a result similar to that of the degenerate four-wave treatment.

By using appropriate values of $\mathcal{E}_0(K \pm G)$ in (15•2–9b), together with suitable seed values, the BW iteration is readily applied to values of K away from $G/2$. The broken line in Fig. 15•2–1 was obtained by such a calculation, using as seed value for every K the converged energy from the preceding iteration closer to $G/2$. Outside the range $0.4G < K < 0.6G$, the BW result merges smoothly into the RS result, but the unphysical divergence of RS perturbation theory is removed.

Our example shows the power of Brillouin-Wigner perturbation theory in a degenerate case in which Rayleigh-Schroedinger perturbation theory fails dismally. But it also shows that convergence, or convergence toward the energy eigenvalue of interest, is not automatic, but requires attention to the stability of the iteration algorithm. In the absence of any degeneracies or near-degeneracies, simple re-insertion iteration has the advantage of requiring a minimum of programming. If degeneracies are present, a more stable algorithm is needed, and Newton's method *usually* provides this stability, along with faster convergence. Its main drawback is that it requires more programming, but once that programming has been done, the method is advantageously used for non-degenerate energy ranges as well.

Remaining Convergence Problems

Although *usually* faster and more stable than re-insertion iteration, Newton's method is not free of convergence problems of its own. In most perturbation problems, the unperturbed state interacts with other states of higher energy. As a result of the repulsive nature of the second-order energy correction, in *second-order* BW perturbation theory the correct perturbed state can never get pushed beyond the next higher unperturbed state. Yet the iteration step (15•2–21) can cause the iteration to jump beyond the latter state and then converge to an incorrect energy. This can also happen if there is an interacting state below the unperturbed state of interest, and if the coupling matrix element to the lower energy state is much larger than that to the higher-energy state. In such cases the Newton algorithm must be started with a seed value just below the upper state, where the repulsion by the upper state is stronger. Problem 3 at the end of this section gives an example.

◆ PROBLEMS TO SECTION 15.2

#15•2-1: Energy Gap of a Square-Wave Periodic Potential

Using second-order Brillouin-Wigner perturbation theory, calculate the energy gap of a square-wave potential with the amplitude $\Delta V = F$, where F is again given by (14•3–17).

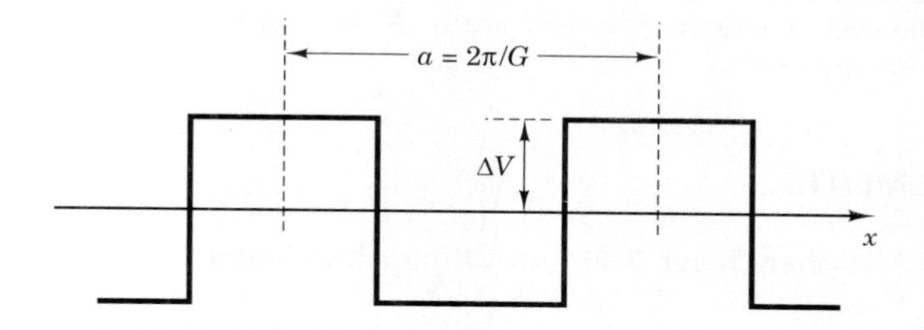

#15•2-2: BW Iteration Divergence as a Feedback Instability

The iterative solution of a relation such as the basic BW relation (15•2–2a) by re-insertion is basically a mathematical feedback loop. Let $\delta\mathcal{E}_i$ be the error in the energy after the ith iteration. We can then define a "loop gain" Γ via

$$\delta\mathcal{E}_{i+1} = \Gamma \cdot \delta\mathcal{E}_i. \tag{15•2–26}$$

Like other feedback systems, this iteration will be stable only if the absolute magnitude of Γ is less than unity;

$$|\Gamma| < 1 \quad \text{for stability.} \tag{15•2–27}$$

(a) Show that the loop gain for the re-insertion algorithm is

$$\Gamma = \sum_{m \neq 0} \frac{|\langle m|\hat{W}|u\rangle|^2}{[\mathcal{E} - \mathcal{E}_m]^2}, \tag{15•2–28}$$

where $\mathcal{E}$ is the actual energy of the state. Apply this expression to the energy gap example treated in the text, and show that the upper of the two roots is indeed just barely unstable, while the lower root is just barely stable.

(b) Show that the loop gain for Newton's method is

$$\Gamma = \frac{f(\mathcal{E}) \cdot f''(\mathcal{E})}{[f'(\mathcal{E})]^2}, \tag{15•2–29}$$

where $f(\mathcal{E})$ is the function (15•2–20). Note that this gain is strongly energy-dependent, going to zero as $f(\mathcal{E}) \to 0$, implying increasingly rapid conversion as the energy approaches the true root. Note also that the convergence is instantaneous for a vanishing second derivative of $f(\mathcal{E})$.

#15•2-3: Delta Dipole Perturbation of Particle in a Square Well

The second energy eigenstate of a particle in a square well of width L is perturbed by a pair of Dirac δ-functions of strength $+S$ and $-S$, placed at $x = L/2 \pm d$. Assume that $S = \mathcal{E}_1 L$. Use second-order BW perturbation theory with Newton's-method iteration to determine the perturbation of the energy of the state $|2\rangle$, as a function of the displacement d of the δ-functions from the center of the well. Show that, for d-values in the vicinity of $L/6$, Newton's method does not converge to an energy below $\mathcal{E}_3$ if the unperturbed energy $\mathcal{E}_2$ is used as a seed value. Show that convergence is restored by starting sufficiently close to $\mathcal{E}_3$. Add a suitable test-and-reset algorithm to Newton's method to handle such divergences automatically. Make a plot of the function $f(\mathcal{E})$ over the relevant energy range.

15.3 REFINEMENTS

15.3.1 Higher-Order Brillouin-Wigner Iterations

Second-order BW iteration is still not exact: Even though the energy is iterated, the iteration continues to operate with the approximate perturbed wave function (15•2–1), rather than with the exact form (15•1–27).

When the perturbation is sufficiently strong, the accuracy of second-order BW iteration may still be insufficient, and even some *qualitative* features of the true solution may be missing. For example, we saw in chapter 14 that the simple cosine potential exhibits energy gaps not only at $K = \pm G/2$, but also at multiples of $G/2$. These additional gaps for a pure cosine potential are inherently impossible to obtain in second-order perturbation theory, which uses only a first-order correction, (15•2–1) or (15•2–4), to the state function.

We therefore return now to the solution of the original set (15•1–10) and (15•1–27) of coupled equations, and convert this set into a set of equations for the expansion coefficients a_n in a *finite* expansion of the energy eigenfunction $|\psi\rangle$, of the form (14•1–7),

$$|\psi\rangle = \sum_{n=1}^{N} a_n\,|n\rangle. \tag{15•3–1}$$

The expansion coefficients are related to $|\psi\rangle$ according to (14•1–4):

$$a_n = \langle m\,|\,\psi\rangle. \tag{15•3–2}$$

If we insert (15•3–1) and (15•3–2) into (15•1–10) and (15•1–27), and also make the terminology more compact by writing

$$W_{mn} \equiv \langle m\,|\,\hat{W}\,|\,n\rangle, \tag{15•3–3}$$

we obtain a set of recursion relations for the expansion coefficients that may be written in the form of the following three sets of equations:

$$\mathcal{E} = \mathcal{E}_u + \sum_{n}^{N} W_{un} a_n; \tag{15•3–4}$$

$$b_m = \sum_{n}^{N} W_{mn} a_n, \quad \text{for all values of } m \neq u; \tag{15•3–5}$$

$$a_m = \frac{b_m}{\mathcal{E} - \mathcal{E}_m}, \quad \text{for all values of } m \neq u. \tag{15•3–6}$$

This is a very simple set of equations for both the energy and the wave function. One of the easiest ways to solve these equations is by re-insertion iteration. One starts out with a plausible ad hoc set of initial expansion coefficients, which may be simply

$$a_n = \delta_{nu}, \tag{15•3–7}$$

although it will often be advantageous to include contributions from states with $n \neq u$ from the beginning. Whatever the initial set, one next calculates an initial energy estimate from (15•3–4) and a set of intermediate coefficients b_m from (15•3–5). A revised set of expansion coefficients is then obtained from (15•3–6), which are again inserted into (15•3–4), to calculate a new and presumably improved value for the energy $\mathcal{E}$.

If desired, the loop (15•3–5) → (15•3–6) → (15•3–4) may be executed again for further improvement, and if the process converges, the cycle may be repeated until the desired accuracy has been achieved. Given such convergence, this BW iteration algorithm is much simpler than the diagonalization algorithms of degenerate perturbation theory. It is also computationally more efficient if only a small number of states are of interest.

Note that, in accordance with (15•1–4), the amplitude a_u of the original unperturbed state never gets updated, thus resulting in a non-normalized perturbed state, as given in (15•1–6). If desired, normalization may of course be restored after completion of the iteration.

The computation time of the higher-order BW algorithm is dominated by the set of N equations represented by (15•3–5). Executing this set requires N^2 multiplications per iteration loop; the other two equations have linear work loads. If there are no degeneracies or near-degeneracies present, the iteration tends to converge rapidly, leading to total work loads per state on the order of $n_i N^2$, where n_i is the number of iterations needed for convergence. If only one state or a small number of states are of interest, this kind of iteration is far more economical than extracting those states by degenerate perturbation theory—not to mention the much greater simplicity of the algorithm.

The principal difficulty are remaining convergence problems caused by small energy denominators in (15•3–6), similar to the convergence problems in second-order Brillouin-Wigner iteration. The cure is basically the same: Conversion of the algorithm into one that works, not by re-insertion iteration, but by Newton's method. The reader confronted with such problems should have little difficulties working out the details, by extension of the simpler scenario of the use of Newton's method in second-order BW perturbation theory, discussed in section 15.2.

PROBLEMS TO SECTION 15.3

#15•3-1: Energy Gap at $K = \pm G$ for the Cosine Potential

(a) Show that, for the cosine potential (15•2–6), the BW recursion relations (15•3–4) and (15•3–5) reduce to

$$\mathcal{E} = \mathcal{E}_0(K) - U_G \cdot (a_{-1} + a_{+1}), \tag{15•3–8}$$

$$b_m = -U_G \cdot (a_{m-1} + a_{m+1}), \quad \text{for all } m \neq 0, \tag{15•3–9}$$

with (15•3–6) remaining unchanged. Here

$$a_m = \langle K + mG \,|\, \psi \rangle. \tag{15•3–10}$$

(b) Use these relations to compute the energy gap at $K = G$ for a potential with the amplitude $2U_G = F$. As basis set, use the 11 plane waves from $|-5G\rangle$ to $|+5G\rangle$. But rather than assuming that all amplitudes other than that of the unperturbed state $|+G\rangle$ are initially zero, include also the amplitude a_{-2} of the state $|-G\rangle$. Run the iteration with the two sets of starting values $a_{-2} = \pm a_0$, with all other coefficients initially zero. Explain why this is a good choice.

Compare the resulting energy gap with the gap at $K = G/2$.

#15•3-2: The Anharmonic Oscillator by Higher-Order BW Iteration

Consider once more the anharmonic oscillator of problem 14.3–2, containing a small perturbation of the form

$$W(x) = \alpha \cdot \hbar\omega \cdot (x/L)^3, \tag{15•3–11}$$

where α ($\ll 1$) is a numerical parameter characterizing the strength of the perturbation and ω and L have their previous meanings.

In contrast to the treatment in chapter 14, write the wave function as a five-term expansion

$$|\psi\rangle = \sum_{n=0}^{4} a_n |n\rangle. \tag{15•3–12}$$

Set up a computer program for a BW recursion algorithm. Run that algorithm for two different set of initial conditions:

(1) Perturbation of the ground state:

$$a_n = \delta_{n0}, \qquad \mathcal{E} = \tfrac{1}{2}\hbar\omega. \tag{15•3–13a}$$

(2) Perturbation of the first excited state:

$$a_n = \delta_{n1}, \qquad \mathcal{E} = \tfrac{3}{2}\hbar\omega. \tag{15•3–13b}$$

Assuming that $\alpha = 0.1$, run the iteration until the energies have converged to $\pm 0.001\hbar\omega$. Try iteration by simple re-substitution first, switching to more complicated algorithms only if re-substitution should not converge. List the final values of the energy and the final sets of expansion coefficients.

15.4 ADIABATIC PERTURBATIONS

At the beginning of chapter 14, we commented that stationary perturbation theory may be viewed as a limiting case of the time-dependent theory, by treating the perturbation $\hat{W}$ as a time-dependent perturbation that was absent at $t = -\infty$ and that has been "turned on" infinitesimally slowly. We now consider this case on the border to time-dependent perturbation theory in more detail, except that we assume that the "turning on" of the perturbation takes place over a finite time, from $t = 0$ to $t = T$. We show the following: If the system was initially in one of the sharp eigenstates of its Hamiltonian, it will, under certain circumstances, remain in a sharp eigenstate of the *changed* Hamiltonian, provided that the change in the Hamiltonian was performed slowly enough. This is the basic content of the **adiabatic theorem**.

The problem is best treated by considering the Hamiltonian as not depending on the time *explicitly*, but only *implicitly*, through another parameter λ that in turn depends on time. That is, we set

$$\hat{H} = \hat{H}\,[\lambda(t)]. \tag{15•4–1}$$

Our problem is then to solve the time-dependent Schroedinger wave equation,

$$i\hbar\frac{\partial}{\partial t}|\Psi(t)\rangle = \hat{H}[\lambda(t)]\,|\Psi(t)\rangle. \tag{15•4–2}$$

We assume that the change in the Hamiltonian is monotonic. Without loss of generality, we may then always define the parameter λ in such a way that it, too, changes monotonically from $\lambda = 0$ to $\lambda = 1$, and that this change is linear in t, over the time T:

$$\lambda(t) = t/T. \tag{15•4–3}$$

In this case, we may convert the Schroedinger wave equation (15•4–2) into an equation in λ,

$$\frac{i\hbar}{T}\frac{\partial}{\partial\lambda}|\Psi(\lambda)\rangle = \hat{H}(\lambda)\,|\Psi(\lambda)\rangle, \tag{15•4–4}$$

with the initial condition

$$|\Psi(0)\rangle = |n,0\rangle. \tag{15•4–5}$$

Here $|n, 0\rangle$ stands for the initial eigenfunctions of $\hat{H}[0]$, for $\lambda = 0$:

$$\hat{H}[0]\,|n,0\rangle = \mathcal{E}_n[0]\,|n,0\rangle. \tag{15•4–6}$$

For every value of λ, there exists a complete orthonormal set of eigenfunctions of $\hat{H}[\lambda]$, which depend on λ, but not *explicitly* on the time. If we denote the elements of this set by $|m, \lambda\rangle$, the entire set satisfies

$$\hat{H}[\lambda]\,|m, \lambda\rangle = \mathcal{E}_m[\lambda]\,|m, \lambda\rangle, \tag{15•4–7}$$

where the $\mathcal{E}_m[\lambda]$ are the corresponding energy eigenvalues. It is then possible to expand the complete time-dependent wave function $|\Psi(\lambda)\rangle$ in terms of the $|m, \lambda\rangle$, with λ-dependent coefficients,

$$|\Psi(\lambda)\rangle = \sum_m a_m(\lambda)|m, \lambda\rangle, \tag{15•4–8}$$

and with the initial condition

$$a_m(0) = \delta_{mn}. \tag{15•4–9}$$

We will find it useful to split off from $a_m(\lambda)$ a time-dependent phase factor by writing

$$\begin{aligned} a_m(\lambda) &= c_m(\lambda) \cdot \exp\left\{-\frac{iT}{\hbar}\int_0^\lambda \mathcal{E}_m(\lambda')d\lambda'\right\} \\ &= c_m(\lambda) \cdot \exp\left\{-iT\int_0^\lambda \omega_m(\lambda')d\lambda'\right\}. \end{aligned} \tag{15•4–10}$$

In the second line, we have introduced the frequencies ω_m, defined via

$$\hbar\omega_m(\lambda) \equiv \mathcal{E}_m(\lambda). \tag{15•4–11}$$

Instead of (15•4–9), we now have

$$c_m(0) = \delta_{mn}. \tag{15•4–12}$$

We insert (15•4–10) into the expansion (15•4–8) and the result into the Schroedinger wave equation (15•4–2). Because of the eigenvalue relations (15•4–6), many terms cancel out, and we obtain, after some manipulation,

$$\sum_m \left\{\left[\frac{dc_m}{d\lambda}\cdot|m, \lambda\rangle + c_m \cdot \frac{d}{d\lambda}|m, \lambda\rangle\right]\cdot \exp\left[-iT\int_0^\lambda \omega_m d\lambda'\right]\right\} = 0. \tag{15•4–13}$$

The proof is left to the reader.

We next take the inner product of (15•4–13) with the bra

$$\exp\left[+iT\int_0^\lambda \omega_l d\lambda'\right]\cdot\langle l, \lambda|. \tag{15•4–14}$$

Because of the orthonormality of the $|m, \lambda\rangle$ belonging to the same λ, this leads to

$$\begin{aligned} \frac{dc_l}{d\lambda} &= -c_l\langle l, \lambda|\frac{d}{d\lambda}|l, \lambda\rangle \\ &\quad - \sum_{m\neq l} c_m \cdot \langle l, \lambda|\frac{d}{d\lambda}|m, \lambda\rangle \cdot \exp\left[+iT\int_0^\lambda \omega_{lm} d\lambda'\right], \end{aligned} \tag{15•4–15}$$

where we have taken the term with $m = l$ out of the sum and have introduced the energy difference frequencies

$$\omega_{lm} = \omega_l - \omega_m. \tag{15•4–16}$$

The set (15•4–15) of differential equations describes the evolution of the various eigenstate amplitudes as the parameter λ increases. Consider now specifically a state $|l, \lambda\rangle$ that is *not* degenerate with *any* other state $|m, \lambda\rangle$, for *any* value of λ. In this case, because of (15•4–11) and (15•4–16), it follows that

$$\omega_{lm} \neq 0 \quad \text{for} \quad m \neq l. \tag{15•4–17}$$

All terms in the split-off sum in (15•4–15) are then oscillating functions of λ, whose rate of oscillations in λ increases with increasing T. Assume specifically that

$$\omega_{lm}(\lambda) \gg \frac{1}{T}, \tag{15•4–18}$$

for all values λ and of $m \neq n$. Upon integration of (15•4–15), the contributions from different phases of the oscillation then tend to cancel; in the limit $T \to \infty$, the cancellation will be complete. Therefore, if the total change of λ from 0 to 1 is spread out over a sufficiently long time T, we may neglect the sum terms in (15•4–15). Multiplying the remainder by c_n^* and adding the result to its own complex conjugate leads to

$$\begin{aligned} \frac{d}{d\lambda}|c_l|^2 &= -|c_l|^2\left[\langle l, \lambda|\frac{d}{d\lambda}|l, \lambda\rangle + \text{c.c.}\right] \\ &= -|c_l|^2 \frac{d}{d\lambda}\langle l, \lambda|l, \lambda\rangle = 0, \end{aligned} \tag{15•4–19}$$

where the final zero results from the λ-independence of the normalization of the eigenstates $|l, \lambda\rangle$. But (15•4–19) means that the occupation probability for any state that is not degenerate with any other state will remain constant in the limit of a sufficiently slowly varying Hamiltonian, even if the accumulated total variation is large.

If all the states remain non-degenerate for all values of λ, the occupation probabilities of *all* states remain the same. If some of the states are degenerate with one another for some or all values of λ, then transitions amongst the degenerate states can take place, but not to or from other states. Applied specifically to the case where the system is initially in the non-degenerate sharp state $|n, 0\rangle$, and assuming that the state $|n, \lambda\rangle$ remains non-degenerate for all values of λ, we obtain the result that the system will remain in the sharp state and will follow the changes of that state itself with λ. This is the **adiabatic theorem**.

The cancellation condition (15•4–18) states that the Hamiltonian must change sufficiently slowly that it does not contain any frequency components ω large enough that the associated quantum energy $\hbar\omega$ could cause transitions between the states. This is not a severe requirement. For energy differences corresponding to optical transition frequencies, and for "man-made" changes in the Hamiltonian—such as a change in a macroscopic external field—this condition is usually well-satisfied.

On the other hand, if the energy level $\mathcal{E}_n[\lambda]$ crosses that for $\mathcal{E}_m[\lambda]$ for some value of λ (**Fig. 15•4–1**), the system may at that point make transitions to the state $|m, \lambda\rangle$. The case then becomes one of full-fledged time-dependent perturbation theory.

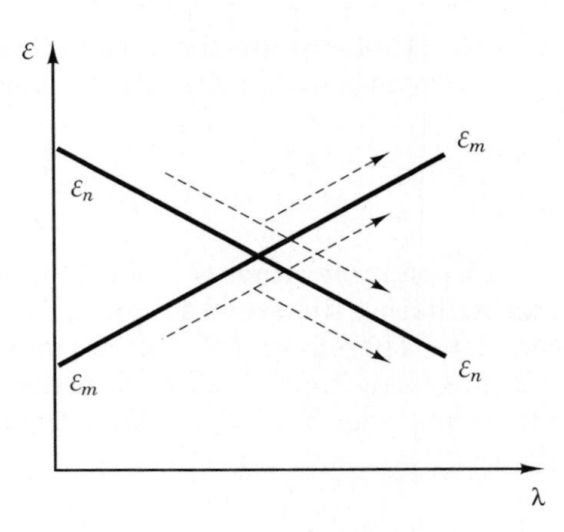

Figure 15•4–1. When the energy levels cross, the system may make transitions between the two crossing levels, even for an adiabatic perturbation.

Chapter 16

SYMMETRY

16.1 SYMMETRY AND SYMMETRY OPERATORS

16.1.1 Introduction: Symmetry Degeneracies

In our treatment of degenerate perturbation theory in chapter 14, we saw that the presence of a simple inversion symmetry in the potential caused selected matrix elements to vanish, thereby making some perturbation problems separable. In the present chapter, we follow up on the importance of symmetry in quantum mechanics, on degeneracies caused by symmetries, and on the utilization of symmetry arguments in simplifying quantum-mechanical problems.

Many important quantum-mechanical problems have Hamiltonians that are invariant under *multiple* symmetry operations. For example, the Hamiltonian of a particle in a cube-shaped well, discussed in section. 2.2, is invariant under all operations that turn a cube into itself. Similarly, the Hamiltonian of the hydrogen atom is invariant under all rotations, by arbitrary angles, about any axis that goes through the center of the atom, and it is also invariant under inversion. And the periodic potential inside crystals (away from the surface) is invariant under all translations that transform the atomic arrangement inside the crystals into itself, and usually, the potential will also be invariant under certain rotations and reflections. The list could easily be extended.

Hamiltonians that are invariant under *multiple* symmetries often show degeneracies of at least some of their energy levels, meaning that there are two or more linearly independent eigenstates that belong to the same energy eigenvalue.[1] As a simple example, which will serve us as a standard example to illustrate many of the points of interest, consider the states $|n_x, n_y\rangle$ of the two-dimensional cylindrical harmonic oscillator (CHO), introduced in section 2.3. The Hamiltonian of the cylindrical harmonic oscillator is invariant under rotations by arbitrary angles about the z-axis and also under reflection at any plane that contains that axis. If we ignore the motion along the z-direction, the energy eigenvalues are, from (2•3–48a),

$$\mathcal{E}_n = \hbar\omega \cdot (n + 1), \quad \text{with } n \equiv n_x + n_y, \tag{16•1–1}$$

where n_x and n_y are non-negative integers. For a given value $n > 0$, there are $n + 1$ different pairs n_x, n_y belonging to the same energy.

Some of these degeneracies are **symmetry degeneracies**, caused by the symmetry invariance properties of the potential alone, *independently of the specific shape of that potential*, as long as the potential retains its symmetry invariances. Others are so-called **accidental degeneracies**, caused by the specific shape of the potential and capable of being split by a suitable perturbation, even if the latter retains the symmetry invariances of the unperturbed potential. Degeneracies higher than two-fold often involve both accidental and symmetry degeneracies side by side. One of our tasks will be to understand the distinction between the two kinds.

Example: Degeneracies in the Hydrogen Atom Probably the best example of an accidental degeneracy is the l-degeneracy of the states $|n, l\rangle$ of the hydrogen atom with a common value of n but different values of l. It is an accidental degeneracy, caused by the simplicity of the Coulomb potential. It will be split by any deviation of the potential $V(r)$ from a simple $1/r$-dependence, as well as by relativistic effects in electron dynamics. Because such splittings can be measured to a much higher accuracy than the energy levels themselves, a study of the splittings is one of the most sensitive tests not only of any deviations from a simple non-relativistic Coulomb potential, but of quantum mechanics itself.

On the other hand, the $(2l + 1)$-fold m-degeneracy of the set of states with a common value of l and n is a symmetry degeneracy: We saw in chapter 2 that this degeneracy occurs regardless of the radial dependence of $V(r)$.

The l-degeneracy manifests itself already, in a somewhat hidden form, in non-relativistic classical dynamics, where any spherically symmetric potential that varies with distance purely as $1/r$ leads to closed elliptical orbits whose orientation in space is time-independent. This behavior suggests that there exists a quantum-mechanical operator that somehow "measures" the orientation of the elliptical classical orbit in space

---

[1] The terms *degenerate* and *degeneracy* must always be understood in the context of the eigenvalues of an operator; they do not refer to the functions per se. If used without reference to a specific operator, they usually refer to a degeneracy of the energy eigenvalue, the case of greatest interest.

and that commutes with the Hamiltonian. Such an operator can indeed be constructed,[2] and the commuting of the Hamiltonian with it is related to the l-degeneracy. Accidental degeneracies that are due to such special *dynamical* properties of the Hamiltonian are sometimes referred to as **dynamical degeneracies**, and the underlying invariances as **dynamical invariances**, as opposed to purely geometrical invariances.

One of the simplest examples of a true symmetry degeneracy is the two-fold degeneracy of the second-lowest energy level of the cylindrical harmonic oscillator. The two linearly independent eigenfunctions for the energy level with $n = n_x + n_y = 1$, obtained by separation of variables in Cartesian coordinates, are of the form

$$|1, 0\rangle \mathrel{\hat{=}} \psi_{10}(x, y) = \psi_1(x) \cdot \psi_0(y) = \sqrt{2}\, A_0^2 \frac{x}{L} \exp\left(-\frac{x^2 + y^2}{2L^2}\right) \tag{16•1–2a}$$

$$|0, 1\rangle \mathrel{\hat{=}} \psi_{01}(x, y) = \psi_0(x) \cdot \psi_1(y) = \sqrt{2}\, A_0^2 \frac{y}{L} \exp\left(-\frac{x^2 + y^2}{2L^2}\right), \tag{16•1–2b}$$

where we have drawn upon the appropriate one-dimensional eigenfunctions from section 2.3. The two functions ψ_{01} and ψ_{10} are shown in perspective surface views in **Fig. 16•1–1**. Neither of the functions exhibits the $\pm 90°$ rotational invariance of the cylindrical harmonic oscillator potential itself. Instead, a

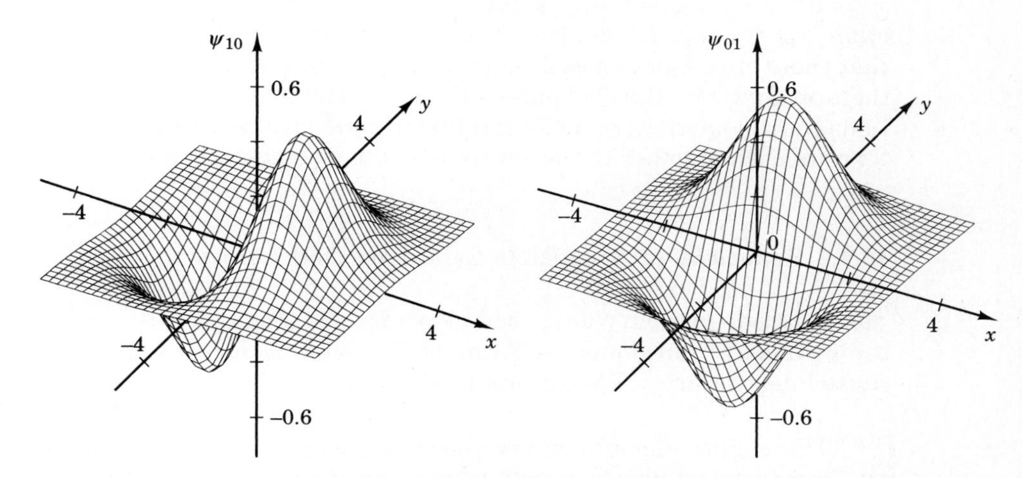

Figure 16•1–1. Perspective surface views of the two wave functions $\psi_{10}(x, y)$ (left) and $\psi_{01}(x, y)$ (right) of the cylindrical harmonic oscillator.

---

[2] See, for example, section 30 in the text by Schiff, listed in the Appendix G.

$\pm 90°$ rotation about the z-axis turns the functions into each other, except for irrelevant sign changes.

The harmonic oscillator potential itself is invariant under such a rotation. Hence, if one function is an energy eigenfunction, the other one must be one, too. This degeneracy is therefore a consequence of the $\pm 90°$ rotational invariance of the cylindrical harmonic oscillator, independently of the special form of the potential otherwise.

Exercise: Suppose we perturb the cylindrical harmonic oscillator by adding a perturbation of the form

$$W(x, y) = w(x) + w(y), \tag{16•1–3}$$

where $w(x)$ and $w(y)$ are identical unspecified *even* functions of their arguments. A simple example would be

$$W(x, y) = c \cdot (x^4 + y^4). \tag{16•1–4}$$

Although such perturbations destroy the rotational invariance of the cylindrical harmonic oscillator Hamiltonian by *arbitrary* angles, they retain the invariance under all operations under which a square would be transformed into itself, namely, all rotations and reflections that interchange the Cartesian axes or reverse the direction of one or both of the axes. By separation of variables in Cartesian coordinates—which remains possible for the form (16•1–3)—show that, even though all one-dimensional energy eigenvalues may change, the degeneracy of the two states $|1, 0\rangle$ and $|0, 1\rangle$ remains.

If we were given one of the two degenerate eigenfunctions without the other, we could conclude, from the reduced symmetry of that function alone, that there *must* exist an additional function, obtained by a $\pm 90°$ rotation from the first function, that belongs to the same energy eigenvalue: The two functions $\psi_{10}(x, y)$ and $\psi_{01}(x, y)$ form a linearly independent set, wherein the existence of each member of the set implies the existence of the other.

16.1.2 Reducible vs. Irreducible Symmetry Degeneracies

The reduced symmetry described above for the second-lowest energy level of the cylindrical harmonic oscillator is *the* defining feature of *all* symmetry-caused degeneracies. We express it—for now—as follows:

> The energy eigenfunctions associated with a symmetry-degenerate energy eigenvalue *necessarily* have a symmetry lower than the full symmetry invariance of the Hamiltonian, and the various eigenfunctions in the degenerate set can be transformed into each other (or into linear combinations of each other) by symmetry operations under which the Hamiltonian is invariant.

However, a warning is in order immediately: The inverse of this statement is not necessarily true! Given a set of linearly independent degenerate energy eigenfunctions that *can* be transformed into each other by symmetry operations under which the Hamiltonian is invariant, it does *not* follow that the *entire* degeneracy is due to symmetry; it *may* be *partially* accidental. We call such degeneracies **reducible symmetry degeneracies**, to distinguish them from true symmetry degeneracies, which we call **irreducible symmetry degeneracies**.

The Six-Fold Degeneracies of a Particle in a Cube

A prime example of a set of reducible symmetry degeneracies is the set of degeneracies of the energy levels of a particle in a cube-shaped square well with infinitely high walls, given in (2•2–15),

$$\mathcal{E}(n_x, n_y, n_z) = \frac{\hbar^2 \pi^2}{2ML^2} \cdot \left(n_x^2 + n_y^2 + n_z^2\right) = \mathcal{E}_1 n^2, \tag{16•1–5}$$

where $\mathcal{E}_1$ is the natural unit of energy of the problem, specified in (2•2–9).

If the three quantum numbers are different, there will be $3! = 6$ different permutations of those numbers and, hence, six linearly independent states, all with the same energy eigenvalue, leading to (at least) a six-fold degeneracy of those eigenvalues. This degeneracy *appears* to be due to the invariance of the Hamiltonian under permutation of the axes. The six corresponding energy eigenfunctions obtained by separation of variables in Cartesian coordinates all have a symmetry lower than that of the Hamiltonian, but are converted into each other by rotations or other symmetry operations that interchange the axes. Furthermore, it is easily shown that the six-fold degeneracies are not split by perturbations that can be written in the form

$$W(x, y, z) = w(x) + w(y) + w(z), \tag{16•1–6}$$

under which the Hamiltonian remains separable in Cartesian coordinates. The proof is left to the reader.

Nevertheless, these degeneracies are partially accidental, and they can be split by suitable perturbations that *cannot* be written in the separable form (16•1–6). To show that this is so, consider the situation when the three quantum numbers in (16•1–5) are three different odd numbers, like 1, 3, and 5. We saw in section 14.4 that in this case a perturbation by a three-dimensional δ-function at the center of the well—which cannot be written in the separable form (16•1–6)—will split the six-fold degeneracy into a single level and a remaining five-fold degenerate level.

What happens here is that the energy eigenfunctions obtained by separation of variables in Cartesian coordinates are ill adapted to the description of a case in which this separability has been destroyed by the perturbation and the accidental degeneracies are obscured by also-present symmetry degeneracies of lower order. However, it is always possible to re-arrange the set of six basis functions into two or more symmetry-decoupled sets. Consider, for example, the equal-weight linear superposition of the six Cartesian basis states

$$|\Psi\rangle = \frac{1}{\sqrt{6}}[|3, 1, 5\rangle + |1, 5, 3\rangle + |5, 3, 1\rangle + |3, 1, 5\rangle + |1, 5, 3\rangle + |5, 3, 1\rangle]. \tag{16•1–7}$$

Inspection shows that this state is invariant under all operations that convert a cube into itself; hence, it violates the necessary condition stated above for being part of a symmetry degeneracy.

If we split off the state $|\Psi\rangle$ from the original six-fold degenerate set, the remaining five states will of course also change, in such a way that they are orthogonal to $|\Psi\rangle$ and to each other. We shall return to this task later, after having acquired the necessary mathematical tools from formal group theory. We shall find that the symmetry of a cube (in the absence of spin) has no irreducible degeneracies higher than three-fold,[3] and lower degeneracies are common. Hence, the five-fold degeneracy remaining in our cube problem after splitting off the state $|\Psi\rangle$ must itself be reducible, and we shall find that it can in turn be reduced to a pair of irreducible two-fold symmetry degeneracies and another single state.

The CHO States with n = 2

We return once more to the cylindrical harmonic oscillator (CHO) and consider the three-fold degenerate next-higher level, $n = 2$. The three energy eigenfunctions obtained by separation of variables on Cartesian coordinates are

$$|1, 1\rangle = 2A_0^2 \frac{xy}{L^2} \exp\left(-\frac{x^2 + y^2}{2L^2}\right), \tag{16•1–8a}$$

$$|2, 0\rangle = \frac{A_0^2}{\sqrt{2}}\left(2\frac{x^2}{L^2} - 1\right) \exp\left(-\frac{x^2 + y^2}{2L^2}\right), \tag{16•1–8b}$$

$$|0, 2\rangle = \frac{A_0^2}{\sqrt{2}}\left(2\frac{y^2}{L^2} - 1\right) \exp\left(-\frac{x^2 + y^2}{2L^2}\right), \tag{16•1–8c}$$

where we have again drawn upon the appropriate one-dimensional eigenfunctions from section 2.3.

Evidently, all three states have a lower rotational symmetry than the potential itself, thus meeting the necessary conditions stated above for belonging to a symmetry degeneracy. The two states $|2, 0\rangle$ and $|0, 2\rangle$ are transformed

[3] This is true only if spin is ignored; for spin-1/2 objects, inclusion of the spin raises the maximal irreducible degeneracy to four-fold.

into each other by $\pm 90°$ rotations, but the state $|1, 1\rangle$ superficially does not appear to be related by symmetry to the other two. On closer inspection, however, the situation is more complex.

Consider the two alternative states

$$|2, \pm\rangle \equiv \frac{1}{\sqrt{2}} \cdot (|2,0\rangle \pm |0,2\rangle), \tag{16•1–9}$$

obtained from the Cartesian states $|2,0\rangle$ and $|0,2\rangle$ by linear superposition and providing an equally legitimate representation of that degenerate pair of states:

$$|2, +\rangle = A_0^2\left(\frac{x^2 + y^2}{L^2} - 1\right) \cdot \exp\left(-\frac{x^2 + y^2}{2L^2}\right), \tag{16•1–10a}$$

$$|2, -\rangle = A_0^2\left(\frac{x^2 - y^2}{L^2}\right) \cdot \exp\left(-\frac{x^2 + y^2}{2L^2}\right). \tag{16•1–10b}$$

The two new states have very different properties from each other, in both their radial and their angular dependence. The state $|2,+\rangle$ has a nonzero value at the coordinate origin and does not depend on the direction at all. Evidently, this state does not satisfy the necessary condition stated earlier for being part of a true symmetry degeneracy; hence, our degeneracy is a reducible degeneracy, and the state $|2,+\rangle$ will be split off by a suitable symmetry-preserving perturbation. As in the case of the cube, a δ-function placed at the coordinate origin will do the job.

The other new state, $|2,-\rangle$, has a null at the origin and changes sign under rotation by $\pm 90°$, but this an irrelevant change, because it does not create a new *physical* state: The rotated function continues to represent the same state. However, on closer inspection, the state $|2,-\rangle$ may be transformed into the state $|1,1\rangle$ by a $\pm 45°$ rotation, one of the symmetry operations under which the CHO Hamiltonian is invariant.

Exercise: Demonstrate the transformation properties of the states $|1,1\rangle$ and $|2,-\rangle$ by considering the contours $f(x, y) = constant$ of the polynomial pre-factors in (16•1–8a) and (16•1–10b). Show that in both cases the contours are simple hyperbolas that are rotated relative to each other by 45°.

The two states $|1,1\rangle$ and $|2,-\rangle$ thus satisfy the necessary condition for a symmetry-degenerate pair, and the remaining two-fold degeneracy is indeed irreducible under any perturbation that is invariant under rotations by $\pm 45°$—which implies that it is also invariant under rotations by integer multiples of these angles. Accordingly, this two-fold symmetry degeneracy will not be split. However, a perturbation that is invariant only under rotations by $\pm 90°$, but not by $\pm 45°$, will, in general split the degeneracy.

Exercise: Splitting of Three-Fold Degeneracy. Sets of strategically placed δ-functions tend to be excellent tools to "probe" degenerate states for their reducible nature. Determine the splitting of the CHO energy level with $n = 2$ under the influence of four two-dimensional δ-functions of equal strength S placed along the two Cartesian axes, equidistant from the origin. For which placements will an accidental degeneracy of one of the states (16•1–10) with the state $|1,1\rangle$ remain?

16.1.3 Elementary Symmetry Transformations

To incorporate the intuitive concept of symmetry invariance into the mathematical formalism of quantum mechanics, we need the concept of **symmetry operators**. We discuss here symmetry operators for four different kinds of elementary symmetry operation: rotations, translations, reflections, and three-dimensional inversions. More complicated symmetry operations can be built up from these four elementary operations. For example, a rotation may be followed (or preceded) by a translation, etc.

Rotations

For simplicity, we pick a cylindrical polar coordinate system (ρ, ϕ, z), with the z-axis chosen as the rotation axis. For much of what follows, it is irrelevant whether or not the z-dimension is present; we may therefore omit the z-coordinate, as if the problem were a two-dimensional one, of rotations within a plane.

Let $f(\rho, \phi)$ be an arbitrary function defined in the plane of rotation. We can then *define* a **rotation operator** $\hat{C}(\alpha)$ by requesting that, for *any* $f(\rho, \phi)$,

$$\hat{C}(\alpha)f(\rho, \phi) \equiv f(\rho, \phi - \alpha), \qquad (16•1–11)$$

where α is the rotation angle. The result of the operation (16•1–11) is a new "rotated function"

$$f_{\text{new}}(\rho, \phi) \equiv \hat{C}(\alpha)f(\rho, \phi) \equiv f(\rho, \phi - \alpha). \qquad (16•1–12)$$

This function has the same value, at every point (ρ, ϕ) in the plane of rotation, that the "old" function $f(\rho, \phi)$ had at the "old" point $(\rho, \phi - \alpha)$ "where (ρ, ϕ) came from," rotated *backward* from (ρ, ϕ) by the angle α (**Fig. 16•1–2**). But this means that the contours of $f_{\text{new}}(\rho, \phi)$ are rotated *forward* by the angle α relative to the contours of the original function $f(\rho, \phi)$.

The relation (16•1–12) is a special case of a more general principle: With every rigid coordinate transformation $\mathbf{r}_{\text{old}} \to \mathbf{r}_{\text{new}}$, we associate a **function transformation operator** $\hat{U}$ that transforms every "old" function $f_{\text{old}}(\mathbf{r})$ into a "new" function $f_{\text{new}}(\mathbf{r})$,

$$f_{\text{new}}(\mathbf{r}) = \hat{U}f_{\text{old}}(\mathbf{r}), \qquad (16•1–13)$$

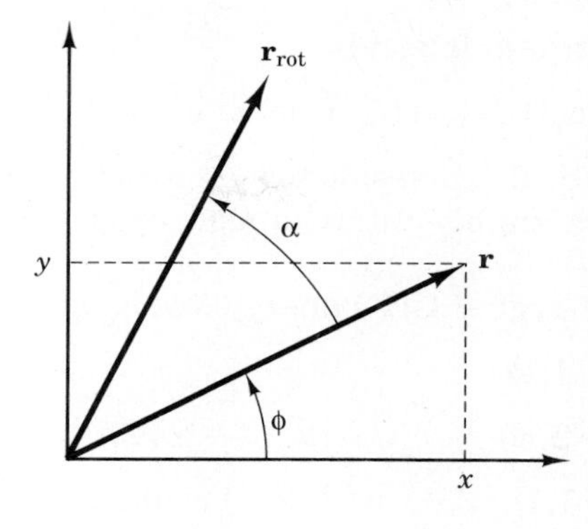

Figure 16•1–2. The rotation operator is defined in such a way that the new rotated function $f_{\text{rot}}(\mathbf{r})$ has the same value at $\mathbf{r} = \mathbf{r}_{\text{rot}}$ that the original function $f(\mathbf{r})$ has at the original $\mathbf{r}$, $f_{\text{rot}}(\mathbf{r}_{\text{rot}}) = f(\mathbf{r})$.

such that

$$\boxed{f_{\text{new}}(\mathbf{r}_{\text{new}}) = f_{\text{old}}(\mathbf{r}_{\text{old}}).} \tag{16•1–14}$$

This is the general idea behind all symmetry operators we will define.

The definition (16•1–11) of the rotation operator may be applied to functions $f(x, y)$ that are expressed in Cartesian coordinates by setting

$$x = \rho \cdot \cos \phi \qquad \text{and} \qquad y = \rho \cdot \sin \phi, \tag{16•1–15a,b}$$

and then replacing ϕ with $\phi - \alpha$. This leads to the substitutions

$$\begin{aligned} x \to \rho \cdot \cos(\phi - \alpha) &= \rho \cdot \cos \phi \cdot \cos \alpha + \rho \cdot \sin \phi \cdot \sin \alpha \\ &= x \cdot \cos \alpha + y \cdot \sin \alpha, \end{aligned} \tag{16•1–16a}$$

and

$$\begin{aligned} y \to \rho \cdot \sin(\phi - \alpha) &= \rho \cdot \sin \phi \cdot \cos \alpha + \rho \cdot \cos \phi \cdot \sin \alpha \\ &= y \cdot \cos \alpha - x \cdot \sin \alpha. \end{aligned} \tag{16•1–16b}$$

Insertion of these into (16•1–11) yields

$$\begin{aligned} &\hat{C}(\alpha) f[x, y] \\ &\equiv f[(x \cdot \cos \alpha + y \cdot \sin \alpha), (y \cdot \cos \alpha - x \cdot \sin \alpha)]. \end{aligned} \tag{16•1–17}$$

Often, the rotation angle will be an integer fraction of 2π, i.e., $\alpha = 2\pi/n$, in which case it is common to use the simpler notation $\hat{C}_n$ for the rotation operator:

$$\hat{C}_n \equiv \hat{C}(2\pi/n). \tag{16•1–18}$$

We are particularly interested in the $\pm 90°$ rotations, i.e., $n = \pm 4$, for which (16•1–17) reduces to

$$\hat{C}_{\pm 4} f(x, y) \equiv f(\pm y, \mp x). \tag{16•1–19}$$

Application to our CHO example yields

$$\hat{C}_{\pm 4}\,|\,1{,}0\rangle = \pm|0{,}1\rangle, \qquad \hat{C}_{\pm 4}\,|\,0{,}1\rangle = \mp|\,1{,}0\rangle, \qquad (16\text{•}1\text{–}20\text{a,b})$$

giving a formal mathematical expression to our earlier observation that the two states $|1{,}0\rangle$ and $|0{,}1\rangle$ are simply interchanged by a $\pm 90°$ rotation, with or without a sign change.

Turning to the next-higher CHO energy level, we find that

$$\hat{C}_{\pm 4}|\,1{,}1\rangle = -|\,1{,}1\rangle, \qquad (16\text{•}1\text{–}21)$$

$$\hat{C}_{\pm 4}|\,2,+\rangle = +|\,2,+\rangle, \qquad \hat{C}_{\pm 4}|\,2,-\rangle = -|\,2,-\rangle. \qquad (16\text{•}1\text{–}22\text{a,b})$$

Each of the three states $|\,1{,}1\rangle$, $|\,2,+\rangle$, and $|\,2,-\rangle$ evidently turns into itself under $\pm 90°$ rotation, again with or without an irrelevant sign change. We say that these three states are **symmetry eigenstates** of the symmetry operator pair $\hat{C}_{\pm 4}$, with the **symmetry eigenvalues** ± 1. This concept of a symmetry eigenstate will turn out to be an important concept in the formal discussion of symmetry degeneracies.

Exercise: Arbitrary Rotation Angles. Show that, for an *arbitrary* rotation angle α, the harmonic oscillator eigenstates $|\,1{,}0\rangle$ and $|\,0{,}1\rangle$ transform according to

$$\hat{C}(\alpha)|\,1{,}0\rangle = +\cos\alpha\cdot|\,1{,}0\rangle + \sin\alpha\cdot|\,0{,}1\rangle, \qquad (16\text{•}1\text{–}23\text{a})$$

$$\hat{C}(\alpha)|\,0{,}1\rangle = -\sin\alpha\cdot|\,1{,}0\rangle + \cos\alpha\cdot|\,0{,}1\rangle. \qquad (16\text{•}1\text{–}23\text{b})$$

That is, each transformed function is a linear superposition of the original functions. Show that the two transformed functions remain normalized and orthogonal to each other.

Derive the set of transform equations analogous to (16•1–23) for a 45° rotation ($\alpha = \pi/4$; $\hat{C}(\alpha) = \hat{C}_8$), applied to the three Cartesian CHO states $|\,1{,}1\rangle$, $|\,2{,}0\rangle$, and $|\,0{,}2\rangle$ defined in (16•1–8a–c). Show that the result of the transformation of $|\,1{,}1\rangle$ may be re-expressed as

$$\hat{C}_8|\,1{,}1\rangle = -|\,2,-\rangle. \qquad (16\text{•}1\text{–}24)$$

Our discussion concerning rotations in a plane is easily extended to rotations in three-dimensional space, about axes with arbitrary orientations. For example, in the discussion of problems with full cubic symmetry, there are operators for 23 different rotations, distributed over four different classes: (1) six $\pm 90°$ rotations about the three cubic axes; (2) three 180° rotations about the three cubic axes; (3) eight $\pm 120°$ rotations about the four body diagonals of the cube; and (4) six 180° rotations about the six axes that connect the midpoints of opposite edges of the cube. The corresponding operators are readily constructed following the general principle expressed formally in (16•1–13) and (16•1–14), and the interested reader should have no difficulties doing so. We shall discuss some of these operators later.

Reflections at a Plane

Let $f(x, y)$ be a function of the Cartesian coordinates x and y. (We again ignore the z-coordinate.) We can then *define* an operator $\hat{\sigma}_x$ by requesting that, for every function $f(x, y)$,

$$\hat{\sigma}_x f(x, y) = f(-x, y). \tag{16•1–25}$$

This operator evidently interchanges the function values on opposite sides of the **mirror plane** $x = 0$; that is, it performs a **reflection** of the function contours at that plane. We can similarly define a reflection operator $\hat{\sigma}_y$ performing a reflection at the plane $y = 0$ via the analogous definition

$$\hat{\sigma}_y f(x, y) \equiv f(x, -y). \tag{16•1–26}$$

Reflections at the planes $x = 0$ and $y = 0$ are two special cases out of an infinite set of possible reflection planes, just as $\pm 90°$ rotations are only special cases of rotations by arbitrary angles. The simplest alternative examples are the two reflection operators defined via

$$\hat{\sigma}_{d+} f(x, y) \equiv f(y, x), \qquad \hat{\sigma}_{d-} f(x, y) \equiv f(-y, -x). \tag{16•1–27a,b}$$

These evidently perform reflections at the two diagonals $y = \pm x$.

Exercise: Study the effects of the four reflection operators defined above on the various CHO energy eigenstates we discussed earlier. Determine which states are also symmetry eigenstates of one or more of the operators. For later use—see (16•2–6) and (16•2–7)—determine also the transformation rules for the states $|1,0\rangle$ and $|0,1\rangle$.

The generalization of our treatment to mirror planes with an arbitrary inclination relative to the x- and y-axes is the subject of one of the problems at the end of the section.

As was the case for rotations, our discussion for reflections within the xy-plane is easily extended to reflections in three-dimensional space. In the case of cubic symmetry, there are nine different reflections, distributed over two different classes: (a) three reflections at the planes perpendicular to the cubic axes and (b) six reflections at the diagonal planes.

Three-Dimensional Inversion

Another important symmetry operator is the three-dimensional inversion operator $\hat{I}$, defined by the requirement that

$$\hat{I} f(\mathbf{r}) = f(-\mathbf{r}), \tag{16•1–28}$$

for any function $f(\mathbf{r})$. In Cartesian coordinates,

$$\hat{I} f(x, y, z) = f(-x, -y, -z). \tag{16•1–29}$$

Note that this is not the same as the inversion in *two* dimensions, which is simply a 180° rotation:

$$\hat{C}_2 f(x, y) = f(-x, -y). \qquad (16{\bullet}1\text{–}30)$$

Translations

The periodic potentials inside a crystal are translationally invariant under displacement by entire sets of vectors. Let $\mathbf{a}$ be such a vector and $f(\mathbf{r})$ an arbitrary function. Following the general principle contained in (16•1–13) and (16•1–14) for the definition of symmetry operators, we can then *define* a **translation operator** $\hat{T}(\mathbf{a})$ by stipulating that, for any $f(\mathbf{r})$,

$$\hat{T}(\mathbf{a}) f(\mathbf{r}) \equiv f(\mathbf{r} - \mathbf{a}). \qquad (16{\bullet}1\text{–}31)$$

Note that, with the minus sign on the right-hand side, this definition corresponds to a *forward* displacement of the contours of $f(\mathbf{r})$ by the vector $+\mathbf{a}$.

16.1.4 Commutation Properties of Symmetry Operators

As with other sets of operators, among the most important properties of the symmetry operators are their commutation properties, both with the Hamiltonian and with each other.

Commutation with the Hamiltonian

Let $\hat{U}$ be any of the symmetry operators under which the potential is invariant. A central property of all such symmetry operators is that they commute with the Hamiltonian whose symmetries they represent;

$$\boxed{\hat{U}\hat{H} = \hat{H}\hat{U}, \quad \text{or} \quad [\hat{U}, \hat{H}] = 0.} \qquad (16{\bullet}1\text{–}32)$$

We view this commutation relation as the basic formal expression for the invariance of $\hat{H}$ under the symmetry operation $\hat{U}$.

To derive (16•1–32), we note first that invariance of the potential under a symmetry operation may be written as

$$\hat{U} V(\mathbf{r}) = V(\mathbf{r}). \qquad (16{\bullet}1\text{–}33)$$

From the definitions of the symmetry operators in terms of coordinate transformations, it follows next that the result of applying the operator $\hat{U}$ to the product of two functions is obtained by operating with $\hat{U}$ on each function separately:

$$\hat{U}[f(\mathbf{r})\, g(\mathbf{r})] = [\hat{U} f(\mathbf{r})][\hat{U} g(\mathbf{r})]. \qquad (16{\bullet}1\text{–}34)$$

If we apply this to the function pair $V(\mathbf{r})\Psi(\mathbf{r})$ and draw on (16•1–33), we obtain

$$\begin{aligned} \hat{U} V(\mathbf{r})\Psi(\mathbf{r}) &\equiv \hat{U}[V(\mathbf{r})\Psi(\mathbf{r})] \\ &= [\hat{U} V(\mathbf{r})][\hat{U}\Psi(\mathbf{r})] = V(\mathbf{r})\hat{U}\Psi(\mathbf{r}). \end{aligned} \qquad (16{\bullet}1\text{–}35)$$

This must be true for any $\Psi(\mathbf{r})$; hence, the operator $\hat{U}$ commutes with $V(\mathbf{r})$:

$$\hat{U}V = V\hat{U}, \quad \text{or} \quad [\hat{U}, V] = 0. \tag{16•1–36}$$

The same is true for the Laplace operator:

$$\hat{U}\nabla^2 = \nabla^2\hat{U}, \quad \text{or} \quad [\hat{U}, \nabla^2] = 0. \tag{16•1–37}$$

To see this, recall that all the symmetry operations we are considering are *rigid* transformations without any change in shape of the transformed function.[4] But if $f(\mathbf{r})$ undergoes a rigid motion expressed by $\hat{U}f(\mathbf{r})$, then the function $g(\mathbf{r}) \equiv \nabla^2 f(\mathbf{r})$ simply moves along, and the new value of $g(\mathbf{r})$ after the transformation can be expressed both as $\hat{U}g(\mathbf{r}) = \hat{U}[\nabla^2 f(\mathbf{r})]$ and as $\nabla^2[\hat{U}f(\mathbf{r})]$. But this invariance implies (16•1–37). Combining (16•1–36) and (16•1–37) then yields the claimed commutation relation (16•1–32).

Non-Commuting Symmetries

Although the symmetry operators all commute with the Hamiltonian, they do not necessarily commute with each other. The situation is similar to what we encountered in chapter 7 in our discussion of the angular momentum operators, where we saw that the angular momentum component operators all commuted with $\hat{L}^2$ and, hence, with any spherically symmetric Hamiltonian, but not with each other.

As an example, consider the successive action of the operators $\hat{\sigma}_x$ and $\hat{C}_4$, introduced earlier to describe one of the rotations and one of the reflections in the xy-plane. It is instructive to visualize their action graphically, in terms of the transformation of the small unsymmetric triangle labeled '1', placed onto the square in **Fig. 16•1–3**. We may view this triangle as one of the contours of

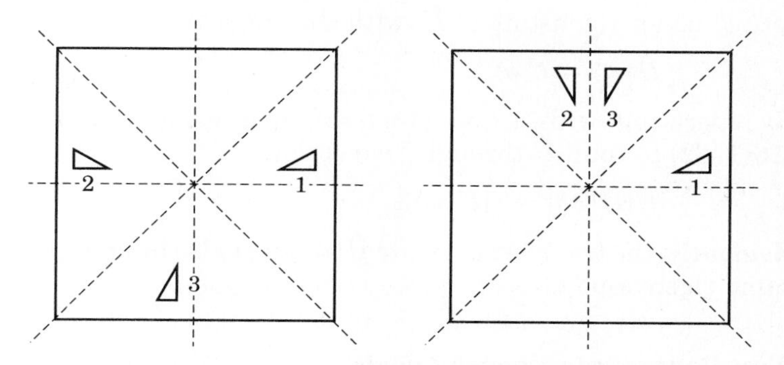

Figure 16•1–3. Sequence-dependence of symmetry operations converting a square into itself.

[4] Ignoring, of course, the trivial changes associated with inversions and reflections, which are irrelevant to the present argument.

a function $f(x, y)$ that we wish to transform. In the left-hand panel, we show what happens if we first execute the reflection $\hat{\sigma}_x$ at the plane $x = 0$ and then the 90° counterclockwise rotation $\hat{C}_4$; the triangle moves first to '2' and then to '3.' The overall result is the same as if we had performed a reflection $\hat{\sigma}_{d-}$ at the diagonal $y = -x$. This equivalence implies the operator product relation

$$\hat{C}_4\hat{\sigma}_x = \hat{\sigma}_{d-}. \tag{16•1–38a}$$

If we invert the order of the two operations, the different transformation in the right-hand panel results, which is the same as if we had performed a reflection $\hat{\sigma}_{d+}$ at the diagonal $y = +x$:

$$\hat{\sigma}_x\hat{C}_4 = \hat{\sigma}_{d+}. \tag{16•1–38b}$$

These results may also be obtained purely formally from the definitions (16•1–19) and (16•1–27a,b) of the various operators involved:

$$\hat{C}_4\hat{\sigma}_x f(x, y) = \hat{C}_4 f(-x, +y) = f(-y, -x) = \hat{\sigma}_{d-}f(x, y), \tag{16•1–39a}$$

$$\hat{\sigma}_x\hat{C}_4 f(x, y) = \hat{\sigma}_x f(+y, -x) = f(+y, +x) = \hat{\sigma}_{d+}f(x, y). \tag{16•1–39b}$$

Evidently, the rotation by $\pi/2$ and the reflection at the plane $x = 0$ do not commute.

The existence of such non-commuting operator pairs is common. As we shall see presently, it is a *necessary* condition for the existence of irreducible symmetry degeneracies.

16.1.5 Symmetry Degeneracies and Non-Commuting Symmetry Operators

Given that $\hat{U}$ is a symmetry operator that commutes with the Hamiltonian $\hat{H}$, let $|\psi\rangle$ be an eigenstate of $\hat{H}$, with the eigenvalue $\mathcal{E}$:

$$\hat{H}|\psi\rangle = \mathcal{E}|\psi\rangle. \tag{16•1–40}$$

By operating with $\hat{U}$ on (16•1–40) and using the commutation relation (16•1–32) to "pull $\hat{U}$ through," we obtain

$$\hat{H}[\hat{U}|\psi\rangle] = \mathcal{E}[\hat{U}|\psi\rangle]. \tag{16•1–41}$$

Evidently, the transformed state $\hat{U}|\psi\rangle$ must also be an eigenstate of $\hat{H}$, with the same eigenvalue $\mathcal{E}$.

Non-Degenerate Energy Levels

If the energy eigenvalue $\mathcal{E}$ is non-degenerate, then $\hat{U}|\psi\rangle$ can differ from the original state $|\psi\rangle$ only by a proportionality factor U:

$$\hat{U}|\psi\rangle = U|\psi\rangle \tag{16•1–42}$$

Put differently, in this case $|\psi\rangle$ must be an eigenstate not only of the Hamiltonian $\hat{H}$, but of $\hat{U}$ as well. States that are eigenstates of a symmetry operator, as in (16•1–42), are called **symmetry eigenstates**.

Degenerate Energy Levels

Suppose now that the energy eigenvalue $\mathcal{E}$ is degenerate. It is then still true that the transformed state $\hat{U}|\psi\rangle$ must also be an eigenstate of $\hat{H}$, with the same eigenvalue $\mathcal{E}$ as $|\psi\rangle$, but we can no longer be sure that $\hat{U}|\psi\rangle$ is simply proportional to the original state $|\psi\rangle$. All we can say is that in this case $\hat{U}|\psi\rangle$ must be a linear superposition of the states belonging to the degenerate set with energy $\mathcal{E}$. The key point now is that the set of degenerate energy eigenstates can always be re-arranged in such a way that all states in the set are eigenstates not only of $\hat{H}$, but of $\hat{U}$ as well. We saw in our discussion of the expansion theorem in chapter 13 that this is true if the two operators involved are both Hermitian, but the result also holds for the symmetry operators.[5]

Therefore, if *all* symmetry operators commute both with the Hamiltonian *and with each other*, then the set of degenerate energy eigenstates can always be arranged in such a way that all states in the set are simultaneously eigenstates of all symmetry operators. But in that case, there are no longer any symmetry operators left that transform the different states into each other. Hence, by definition, the degeneracy cannot be a symmetry degeneracy, but must be accidental. If the Hamiltonian is sufficiently complex that all accidental degeneracies have been split, there will be no degeneracies left at all.

This conclusion cannot be drawn when the symmetry operators present do not all commute with each other. In that case, it will in general no longer be possible to select the degenerate energy eigenstates as simultaneous eigenstates of all symmetry operators. Instead, there may then be some symmetry operators that transform the states amongst each other, the hallmark of a symmetry degeneracy. Thus, we conclude that

> For symmetry degeneracies to occur, it is *necessary* that the set of symmetry operators of the system contain operators that do not commute with each other.

We noted earlier that, in our CHO example, the two energy eigenstates $|1,0\rangle$ and $|0,1\rangle$ are symmetry eigenstates of *some* of the operators we discussed, but not of *all* of them. This is a direct consequence of the fact that the operators that transform a square into itself do not all commute with each other—as we saw in Fig. 16•1–3.

[5] The proof of this assertion will emerge later in the form of a procedure by which the symmetry eigenstates may actually be constructed from a given initial set.

◆ PROBLEMS TO SECTION 16.1

#16•1-1: Reflection at an Arbitrarily Inclined Plane

Generalize the definition of reflection operators to reflection at *any* plane perpendicular to the xy-plane that contains within itself the z-axis, but with an arbitrary inclination angle β between the reflection plane and the x-axis. (**Fig. 16•1–4**). Write the definition in polar coordinates. Show that the corresponding **reflection operator** $\hat{\sigma}(\beta)$ is defined by stipulating that, for any $f(\rho, \phi)$,

$$\hat{\sigma}(\beta)f(\rho, \phi) \equiv f(\rho, 2\beta - \phi). \tag{16•1–43}$$

Transform (16•1–43) to Cartesian coordinates for arbitrary angles β. Show that, for the specific angles $\beta = 0$, $\pi/2$, and $\pm\pi/4$, the general form reduces to the previous expressions for $\hat{\sigma}_y$, $\hat{\sigma}_x$, and $\hat{\sigma}_{d\pm}$.

Also, consider the case $\beta = \pi/8$. Apply the resulting operator to our harmonic oscillator example states, and compare the results with those obtained by applying the 45° rotation operator to these states.

#16•1-2: The *l*-Degeneracy of the Hydrogen Atom as an Accidental Degeneracy

Investigate the way the three hydrogen atom states $|n, l, m\rangle = |2, 1, m\rangle$ are transformed amongst each other by reflection at the plane $y = z$ and by a clockwise 90° rotation about the y-axis (i.e., clockwise looking *along* the $+y$-axis). In all cases, write the result of the transformation as a linear superposition of the three original states.

16.2 GROUP THEORY FOR PEDESTRIANS

16.2.1 Symmetry Groups

When a Hamiltonian is invariant under multiple symmetries, the set of all such symmetries forms a **symmetry group**, in the precisely defined sense in which the word *group* is used in mathematical group theory.

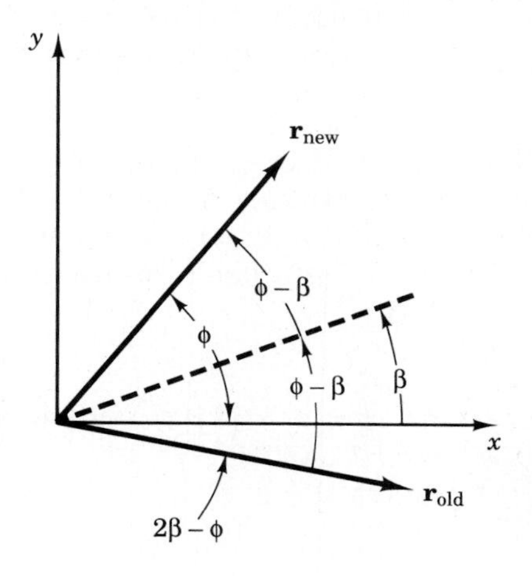

Figure 16•1–4. On the definition of the oblique reflection operator $\hat{\sigma}(\beta)$.

What makes this set of operations a group in the sense of mathematical group theory are the following properties:[6]

(a) If any two operations in a group are executed in succession, the result of the combination is itself an element of the group; that is, there exists an operation within the group that leads to the same result in a single step.
(b) There exists a "do-nothing" *identity operation* $\hat{E}$ that is explicitly considered an element of the group.[7]
(c) For every element $\hat{U}$ of the group, the group also contains its inverse $\hat{U}^{-1}$, not necessarily different from $\hat{U}$, defined by the requirement

$$\hat{U}\hat{U}^{-1} = \hat{U}^{-1}\hat{U} = \hat{E}(= 1). \tag{16•2–1}$$

The central condition is (a).

To illustrate these properties, we consider here the **group of a square**, that is, the group of all those symmetry operations under which anything with the rotation and reflection symmetry of a simple square is left invariant. This is the symmetry group of our CHO example under a perturbation of the form (16•1–3).

Given an initial orientation of a square, and *including* the do-nothing identity operation, there must be eight symmetry operations that transform that square into itself. To see this, suppose we number the four edges of the square from 1 to 4, proceeding either clockwise or counterclockwise around the square. The edge labeled 1 may then be in any one of four different equivalent positions, distinguishable only by the ad hoc labeling. Wherever edge 1 is located, we may arrange the remaining edges proceeding either clockwise or counterclockwise, leading to a doubling of the total number of possibilities, to eight. The number of operations contained in the group, including the identity operation, is called the **order of the group**.

The eight symmetry operations transforming a square into itself fall into five distinct intuitive **classes**:

1. The identity operation $\hat{E}$, corresponding to no rotation or reflection at all.
2. One rotation, $\hat{C}_2$, by $\pm\pi$. Note that a rotation by $-\pi$ leads to the same result; hence, the rotation by $\pm\pi$ must be counted as a single operation.
3. Two rotations, $\hat{C}_4$ and $\hat{C}_{-4}$, by $+\pi/2$ and $-\pi/2$.
4. Two reflections, $\hat{\sigma}_x$ and $\hat{\sigma}_y$, at the lines (or planes, in a three-dimensional view) perpendicular to the x- and y-axes.

[6] For a general background in group theory, see, for example, V. Heine, *Group Theory in Quantum Mechanics* (New York: Pergamon Press, 1960), or M. Tinkham, *Group Theory and Quantum Mechanics* (New York: McGraw Hill, 1964).

[7] The notation $\hat{E}$ for the unit or identity operator, commonly used in group theory, originates from the German 'Eins' (= 1).

5. Two additional reflections, $\hat{\sigma}_{d+}$ and $\hat{\sigma}_{d-}$, at the diagonals $y = +x$ and $y = -x$.

We saw earlier—see (16•1–38a,b)—that executing $\hat{C}_4$ and $\hat{\sigma}_x$ in succession leads to either $\hat{\sigma}_{d+}$ or $\hat{\sigma}_{d-}$, depending on the order of execution. By an analogous procedure, we can determine the results of successive operation for all possible operator pairs. This leads to Table 16•2–1, called the **group multiplication table** for the group of a square. Each cell contains the operator product $\hat{T}\hat{L}$—in that order—where $\hat{T}$ is the operator at the *top* of each column and $\hat{L}$ the operator at the *left* end of each row. Note that this implies applying first the operator $\hat{L}$ in the leftmost column and then the operator $\hat{T}$ in the topmost row.

Inspection of the table shows that it contains many non-commuting pairs. The only rotation that commutes with any of the reflections is the 180° rotation; no other rotations commute with any of the reflections. Among the reflections themselves, only those reflections commute with one another whose planes are perpendicular to each other; the two diagonal reflections do not commute with the Cartesian reflections. Recall that the existence of such non-commuting pairs of operators is a *necessary* condition for the occurrence of symmetry degeneracies.

Inverting vs. Non-Inverting Operations

Symmetry operations may be divided into inverting and non-inverting operations. Suppose we label the corners of the small unsymmetric triangle in Fig. 16•1–3 in some order, say, by increasing angle. Inverting operations invert a clockwise order into a counterclockwise one and vice versa, while non-inverting operations retain the original order. All reflections are inverting

TABLE 16•2–1. Group multiplication table for the group of a square.

$\hat{E}$	$\hat{C}_2$	$\hat{C}_4$	$\hat{C}_{-4}$	$\hat{\sigma}_x$	$\hat{\sigma}_y$	$\hat{\sigma}_{d+}$	$\hat{\sigma}_{d-}$
$\hat{C}_2$	$\hat{E}$	$\hat{C}_{-4}$	$\hat{C}_4$	$\hat{\sigma}_y$	$\hat{\sigma}_x$	$\hat{\sigma}_{d-}$	$\hat{\sigma}_{d+}$
$\hat{C}_4$	$\hat{C}_{-4}$	$\hat{C}_2$	$\hat{E}$	$\hat{\sigma}_{d+}$	$\hat{\sigma}_{d-}$	$\hat{\sigma}_y$	$\hat{\sigma}_x$
$\hat{C}_{-4}$	$\hat{C}_4$	$\hat{E}$	$\hat{C}_2$	$\hat{\sigma}_{d-}$	$\hat{\sigma}_{d+}$	$\hat{\sigma}_x$	$\hat{\sigma}_y$
$\hat{\sigma}_x$	$\hat{\sigma}_y$	$\hat{\sigma}_{d-}$	$\hat{\sigma}_{d+}$	$\hat{E}$	$\hat{C}_2$	$\hat{C}_{-4}$	$\hat{C}_4$
$\hat{\sigma}_y$	$\hat{\sigma}_x$	$\hat{\sigma}_{d+}$	$\hat{\sigma}_{d-}$	$\hat{C}_2$	$\hat{E}$	$\hat{C}_4$	$\hat{C}_{-4}$
$\hat{\sigma}_{d+}$	$\hat{\sigma}_{d-}$	$\hat{\sigma}_x$	$\hat{\sigma}_y$	$\hat{C}_4$	$\hat{C}_{-4}$	$\hat{E}$	$\hat{C}_2$
$\hat{\sigma}_{d-}$	$\hat{\sigma}_{d+}$	$\hat{\sigma}_y$	$\hat{\sigma}_x$	$\hat{C}_{-4}$	$\hat{C}_4$	$\hat{C}_2$	$\hat{E}$

operations, while all rotations and the identity operation are non-inverting. In three dimensions, inverting operations interchange left-handed and right-handed screws, while non-inverting operations leave what is called the **helicity** of the screw unchanged.

Exercise: Derive Table 16•2–1.

Subgroups

Very often, a subset of the elements of a group forms a smaller group in its own right. For example, the set of all non-inverting operations in a symmetry group forms a subgroup of the original group. It can be shown that the order of any subgroup must be an integer divisor of the original group.

Exercise: The group of a square has six subgroups of order 2 and three subgroups of order 4. Identify them.

Cyclic Operators

For any finite group, repeated application of the same operator must ultimately lead to the identity operator $\hat{E}$; that is,

$$\hat{U}^M = \hat{E}, \tag{16•2–2}$$

where M is an integer. The operator is said to be **cyclic** of order M. All reflections and all rotations by $\pm 180°$ are cyclic of order 2, as is three-dimensional inversion. In the case of the group of a square, all elements except $\hat{C}_4$ and $\hat{C}_{-4}$ are cyclic of order 2; $\hat{C}_4$ and $\hat{C}_{-4}$, representing rotations by $\pm 360°/4$, are cyclic of order 4. Rotations by $\pm 120°$ are cyclic of order 3. All symmetry operations that transform a cube into itself are cyclic of order 2, 3, or 4. Translational operators can be made into cyclic operators by imposing periodic boundary conditions.

Given a group element $\hat{U}$, the different powers of $\hat{U}$, from $\hat{U}$ to $\hat{U}^M = \hat{E}$, form a subgroup of order M, implying that M must be an integer fraction of the order of the group.

Classes

In our enumeration of the symmetry operations contained in the group of a square, we introduced the concept of *classes* of symmetry operators purely intuitively. In formal group theory, this intuitive concept is put on a rigorous basis: Two operations $\hat{A}$ and $\hat{B}$ in a group are said to belong to the same class if they can be transformed into each other by a relation of the form

$$\hat{C}\hat{B} = \hat{A}\hat{C}, \tag{16•2–3}$$

where $\hat{C}$ is a third operation also contained in the group.

In terms of the group multiplication table, this condition compares the row $\hat{C}$ with the column $\hat{C}$. If the element at the nth position in the row differs from the element at the nth position in the column, the two elements belong to the same class. The number of classes contained in a group plays an important role in formal group theory in determining what kind of symmetry degeneracies might occur.

16.2.2 Matrix Representations of Symmetry Operators and Groups

One of the central and most powerful concepts of formal group theory—especially as applied to quantum mechanics—is that of the **irreducible representations** of a symmetry group. It flows directly from the concept of an irreducible symmetry degeneracy. The central idea is simple: Like other operators in quantum mechanics, the symmetry operators and their algebraic properties can be represented by matrices. These matrices obey the same multiplication table as the group; hence, the set of matrices associated with the elements of a group may be used to represent the group itself.

To demonstrate the concept, we return again to our CHO example. Consider the relations (16•1–20a,b), which show the way the two 90° rotation operators $\hat{C}_4$ and $\hat{C}_{-4}$, transform the two degenerate states $|1,0\rangle$ and $|0,1\rangle$ into each other. These two pairs of relations may be combined into a pair of 2 × 2 matrix equations:

$$\hat{C}_{\pm 4}\begin{pmatrix}|1,0\rangle\\|0,1\rangle\end{pmatrix}=\begin{pmatrix}\pm|0,1\rangle\\\mp|1,0\rangle\end{pmatrix}=\begin{pmatrix}0 & \pm 1\\\mp 1 & 0\end{pmatrix}\begin{pmatrix}|1,0\rangle\\|0,1\rangle\end{pmatrix}. \qquad (16•2–4)$$

The 2 × 2 matrix on the far right-hand side is a representation of the operator pair $\hat{C}_4$ and $\hat{C}_{-4}$:

$$\hat{C}_{+4}=\begin{pmatrix}0 & +1\\-1 & 0\end{pmatrix},\qquad \hat{C}_{-4}=\begin{pmatrix}0 & -1\\+1 & 0\end{pmatrix}. \qquad (16•2–5a,b)$$

It is left to the reader to show—see the exercise following (16•1–27)—that the four reflection operators obey

$$\hat{\sigma}_x=\begin{pmatrix}-1 & 0\\0 & +1\end{pmatrix},\qquad \hat{\sigma}_y=\begin{pmatrix}+1 & 0\\0 & -1\end{pmatrix}, \qquad (16•2–6a,b)$$

$$\hat{\sigma}_{d+}=\begin{pmatrix}0 & +1\\+1 & 0\end{pmatrix},\qquad \hat{\sigma}_{d-}=\begin{pmatrix}0 & -1\\-1 & 0\end{pmatrix}. \qquad (16•2–7a,b)$$

The two remaining operators of the symmetry group of the square have the matrix representations

$$\hat{E}=\begin{pmatrix}+1 & 0\\0 & +1\end{pmatrix},\qquad \hat{C}_2=\begin{pmatrix}-1 & 0\\0 & -1\end{pmatrix}. \qquad (16•2–8a,b)$$

Note that all transformation matrices are unitary matrices, as must be the case for transformation matrices converting one orthonormal set of states into another.

If two operations are executed in succession, the combination is represented by the product of the two relevant matrices by ordinary matrix multiplication. Inasmuch as these matrices are simply another description of the symmetry operations, their ordinary matrix multiplication must obey the same multiplication table as the operators they represent. The reader is asked to confirm that this is indeed the case.

If the energy level had been three-fold degenerate, like the CHO energy level with $n = 2$, we would have obtained a set of 3×3 matrices, etc.

If the degeneracy is an irreducible symmetry degeneracy, the representation is called an **irreducible representation**; this is the case of interest in group theory. The matrices associated with a given irreducible degeneracy are not unique, but depend on the choice of the basis set of energy eigenfunctions. For example, rather than choosing the two states $|1{,}0\rangle$ and $|0{,}1\rangle$ to discuss the degeneracy of the second-lowest energy level of the cylindrical harmonic oscillator, we could have chosen the set

$$|\pm i\rangle \equiv \frac{1}{\sqrt{2}}[|1{,}0\rangle \pm i|0{,}1\rangle], \tag{16•2–9}$$

which is equivalent to $|1{,}0\rangle$ and $|0{,}1\rangle$. In fact, it is the set we would have obtained if we had solved the CHO eigenstate problem in cylindrical polar coordinates. Now, if we investigate the transformation properties of that set, we obtain a different set of 2×2 matrices representing the group. The two sets, along with all other sets similarly obtained, are considered **equivalent** representations, differing from each other only by a simple unitary transformation.

Exercise: Determine the transformation matrices associated with the basis function pair (16•2–9). Show that they have the same set of traces as the matrices in (16•2–5) through (16•2–8), as must be the case for matrices that differ from each other only by a unitary transformation.

If a state is non-degenerate, the 1×1 transformation "matrices" reduce to a set of numbers, namely, the symmetry eigenvalues of the state for the various operators. As an example, consider the CHO state $|1{,}1\rangle$. It is a symmetry eigenstate of all operators in the group of the square, with symmetry eigenvalues that are either $+1$ or -1, distributed as follows:

$$+1 : \hat{E},\ \hat{C}_2,\ \hat{\sigma}_{\pm d}; \qquad -1 : \hat{C}_{\pm 4},\ \hat{\sigma}_x,\ \hat{\sigma}_y. \tag{16•2–10}$$

If we replace each of the operators in the multiplication table with its eigenvalue for the state $|1{,}1\rangle$, the multiplication table is again satisfied, albeit in a highly redundant way. Hence, this set of eigenvalues is a valid one-dimensional representation of the group, even though it contains only two distinct elements. In fact, the simplest representation of every group, the so-called **identity representation**, represents every group element by the number +1: Regardless of the multiplication table of a group, this table remains trivially satisfied if *all* symmetry operators are replaced by the "matrix" +1.

16.2.3 Two Theorems

Formal group theory contains a rich set of powerful theorems about the properties of irreducible representations of a group, all but indispensable in the analysis of irreducible symmetry degeneracies in quantum mechanics. We simply state here, without proof,[8] the two most central of these theorems:

1. The theorem on the number of irreducible representations states that

> The number of distinct non-equivalent irreducible representations of a group is equal to the number N_C of classes contained in that group.

2. Let l_i be the dimensionality of the ith irreducible representation of a group, and let h be the order of the group. Then the **dimensionality theorem** states that

$$\boxed{\sum_{i=1}^{N_C} l_i^2 = h.} \qquad (16\text{•}2\text{–}11)$$

There probably does not exist a more powerful set of tools for predetermining the levels of possible (irreducible) symmetry degeneracy than these two theorems. More often than not, the two equations yield a unique set of l_i-values, without the need for invoking additional considerations.

As a first example, consider again the group of the square. As we saw earlier, this group has eight elements and five classes. The reader can confirm easily that the only solution of (16•2–11) for $N_C = 5$ and $h = 8$ is

$$1 \cdot 2^2 + 4 \cdot 1^2 = 8. \qquad (16\text{•}2\text{–}12)$$

This means that there can be no irreducible degeneracies higher than two-fold: All higher degeneracies must be either reducible or accidental. The group has one two-dimensional representation and four non-equivalent one-dimensional representations. The two-dimensional representation is the one given by the matrices derived above; we will say more about some of the one-dimensional representations later.

---

[8] A proof of these theorems lies outside of the scope of a text on quantum mechanics, and the interested reader must be referred to texts on group theory per se, such as the texts by Heine and by Tinkham cited earlier.

As a second example, consider the group of all rotations that convert a cube into itself, without inverting operations. There are $4 \times 6 = 24$ ways to orient a cube; hence, the cube rotation group must have 24 elements, distributed over five classes, namely, the identity class, plus the four classes of rotations enumerated earlier, following Eq. (16•1–24). The dimensionality theorem again yields a unique solution:

$$2 \cdot 3^2 + 1 \cdot 2^2 + 2 \cdot 1^2 = 24. \qquad (16\text{•}2\text{–}13)$$

There are two non-equivalent three-dimensional representations, one two-dimensional representation, and two one-dimensional representations. Hence, there cannot be any irreducible degeneracies higher than three-fold. We shall see later that this remains true even if the inverting operations are added. Evidently, the (at least) six-fold degeneracies of the energy levels of a particle inside a cubic well with infinitely high walls are all reducible.

16.2.4 Character Tables

We return to our earlier observation that for a given irreducible symmetry degeneracy, we may construct several equivalent irreducible matrix representations, differing from each other only by a simple unitary transformation. Under such a transformation, the trace of each matrix remains unchanged. The set of traces associated with a given representation is called the set of **characters** of the representation; it is *the* characteristic feature of that representation, independent of the set of basis functions used. It plays a central role in group theory; the full set of matrices is rarely needed.

For the one-dimensional representations, for which the basis states are symmetry eigenstates of all symmetry operators, the characters are of course the symmetry eigenvalues themselves.

From the unitarity of the matrices and the definition (16•2–3) of the classes of operators, it follows that symmetry operators belonging to the same class have the same character; hence, there are only as many independent characters associated with an irreducible representation as there are classes in the group. In our CHO example for $n = 2$, the characters for most classes are zero, the exceptions being the two single-operator classes involving $\hat{E}$ and $\hat{C}_2$, for which $\chi(\hat{E}) = +2$ and $\chi(\hat{C}_2) = -2$.

The set of all characters of a group may be written as a square **character table** with N_C rows and columns. For the group of a square, the character table, derived in Appendix F, is a 5×5 table, shown as the central portion of Table 16•2–2.

The top row lists the five different classes, in the form of a number indicating the number of elements in the class and a representative element identifying the class. Thus, the notation $2\hat{C}_4$ denotes the two-element class containing $\hat{C}_4$ and $\hat{C}_{-4}$. The absence of a numerical factor indicates that a class contains only one element, such as $\hat{E}$ and $\hat{C}_2$. The subsequent rows list the characters for the five irreducible representations.

TABLE 16•2–2. Character table of the group of a square.

	$\hat{E}$	$\hat{C}_2$	$2\hat{C}_4$	$2\hat{\sigma}_x$	$2\hat{\sigma}_d$	examples
A_1	+1	+1	+1	+1	+1	$c, x^2 + y^2$
A_2	+1	+1	+1	−1	−1	$xy(x^2 - y^2)$
B_1	+1	+1	−1	+1	−1	$x^2 - y^2$
B_2	+1	+1	−1	−1	+1	xy
E	+2	−2	0	0	0	x, y

The leftmost column gives the name commonly used for the representation in formal group theory; the rightmost column shows examples of simple polynomial functions that transform according to the representation. The notation E for the 2×2 representation must not be confused with the notation $\hat{E}$ for the identity operator. We would have preferred a less confusing name for the representation E, but the letter E is deeply entrenched as the name for two-dimensional representations.

To see the power and utility of such a character table, consider the three degenerate states $|1,1\rangle$, $|2,0\rangle$, and $|0,2\rangle$ of the cylindrical harmonic oscillator. We can immediately draw a number of conclusions from the character table alone, without ever considering the three eigenfunctions in detail. The character table of the square lists no three-dimensional representation; hence, the three-fold degeneracy must be reducible. Inasmuch as $|2,0\rangle$ and $|0,2\rangle$ are evidently transformed into each other by several of the operators in the symmetry group of a square, we might expect that they are symmetry degenerate with each other. However, a look at the character table reveals the reducible nature of this remaining degeneracy without going through the considerations in section 16.1 that led to the alternative functions in (16•1–10a,b): The only irreducible 2×2 representation of the group of a square is the representation E in Table 16•2–2, which has the character 0 under all reflections. But the two states $|2,0\rangle$ and $|0,2\rangle$ are even functions in both x and y and, hence, are symmetry eigenstates of $\hat{\sigma}_x$ with the eigenvalue $+1$, a value incompatible with the value $\chi(\hat{\sigma}_x) = 0$ in the representation E. This means that there *must* be a way to re-arrange the two states into a set of two new states, each of which transforms according to one of the four one-dimensional representations. Investigating the behavior of the states $|2,0\rangle$ and $|0,2\rangle$ for small values of x and y, and comparing that behavior with the polynomials in the rightmost column of Table 16•2–2, readily identifies one linear superposition that transforms according to A_1, namely $|2,+\rangle$ from (16•1–10a), and one that transforms according to B_1, namely $|2,-\rangle$ from (16•1–10b).

Character Orthogonality Relations

In more complicated situations, elementary considerations such as the preceding might not suffice to break up a reducible degeneracy into its irreducible constituents, and a more systematic approach might be needed, calling on an important set of additional properties of the character table: the two **character orthogonality relations**.

Let the notation $\chi^{(i)}(C_k)$ designate the character associated with the kth class in the ith irreducible representation. The **column orthogonality relation** then states that the table *columns* behave like orthogonal vectors in the N_C-dimensional space of representations; that is,

$$N_k \sum_{i=1}^{N_C} [\chi^{(i)}(C_k)]^* \, \chi^{(i)}(C_l) = h\delta_{kl}, \tag{16•2–14}$$

where h is the order of the group and N_k the number of elements in the kth class.

Similarly, the **row orthogonality relation** states that the table *rows* associated with the irreducible representations behave like orthogonal vectors in the h-dimensional space of all group elements. Because the characters within a given class are the same, this orthogonality is usually expressed as a sum over all classes, with a weighting factor equal to the number N_k of elements in each class:

$$\sum_{k=1}^{N_C} N_k \cdot [\chi^{(i)}(C_k)]^* \chi^{(j)}(C_k) = h\delta_{ij}. \tag{16•2–15}$$

Inspection of Table 16•2–2 shows that both the column and the row orthogonality relations are indeed satisfied. For a proof of these relations, we again refer the reader to texts on formal group theory.

The two orthogonality relations are very powerful tools both in the construction of character tables (see appendix F) and in the use of those tables in determining the nature of symmetry degeneracies.

16.2.5 Decomposing Reducible Degeneracies

Suppose we encountered an M-fold degeneracy and we wished to determine whether or not it was an irreducible degeneracy and, if not, which irreducible degeneracies might be contained within it. By inspecting the transformation properties of the M orthonormal energy eigenstates under the influence of the various symmetry classes of the group, we can determine the character vector $\chi(C_k)$ associated with the full set of degenerate states. If this character vector coincides with that of one of the irreducible representations, the degeneracy is an irreducible one. If there is no such coincidence, we have a reducible degeneracy, for which the characters must be the sum over the characters of the participating irreducible representations:

$$\chi(C_k) = \sum_{i=1}^{N_C} n_i \cdot \chi^{(i)}(C_k). \tag{16•2–16}$$

Here n_i is the number of times the ith irreducible representation occurs. If we multiply by $N_k \cdot [\chi^{(i)}(C_k)]^*$, sum over k, and draw on the row orthogonality relation (16•2–15), we obtain

$$n_j = \frac{1}{h} \sum_{k=1}^{N_C} N_k \cdot [\chi^{(j)}(C_k)]^* \chi(C_k). \tag{16•2–17}$$

The procedure is best illustrated by a simple example that is not quite as trivial as the earlier CHO case with $n = 2$. We move up two steps in the CHO energy eigenvalue ladder and consider the five-fold CHO energy level with $n = 4$. We then construct the two-part Table 16•2–3.

Consider first the left-hand part of the table, designed to extract the character vector. The top row of this part is the same as the top row of the character table; it contains one selected symmetry operator per column from each of the five classes of the group of the square. (It does not matter which operator is selected.) The leftmost column contains the different basis states associated with the energy level. Each of the inner fields of the table is associated with one selected basis state $|\psi\rangle$ and one selected symmetry operator $\hat{U}$. There are now three possibilities: (*i*) If the basis state is an eigenstate of the symmetry operator, we enter the symmetry eigenvalue into the field. (*ii*) If the result of operating with $\hat{U}$ on $|\psi\rangle$ is orthogonal to $|\psi\rangle$ itself, we enter zero. (*iii*) In all other cases (which are rare), we enter the expansion coefficient with which $|\psi\rangle$ itself appears in the expansion of $\hat{U}|\psi\rangle$ in terms of the states in the

TABLE 16•2–3. Character extraction table and representation count table for the five-fold degenerate CHO energy level with $n = 4$.

	$\hat{E}$	$\hat{C}_2$	$2\hat{C}_4$	$2\hat{\sigma}_x$	$2\hat{\sigma}_{d+}$
$\|4, 0\rangle$	+1	+1	0	+1	0
$\|3, 1\rangle$	+1	+1	0	−1	0
$\|2, 2\rangle$	+1	+1	+1	+1	+1
$\|1, 3\rangle$	+1	+1	0	−1	0
$\|0, 4\rangle$	+1	+1	0	+1	0
	+5	+5	−1	+1	+1

A_1	A_2	B_1	B_2
1/2	0	1/2	0
0	1/2	0	1/2
1	0	0	0
0	1/2	0	1/2
1/2	0	1/2	0
2	1	1	1

leftmost column, namely,

$$\langle\psi|\hat{U}|\psi\rangle. \tag{16•2–18}$$

Each column in the inner table is then simply the set of diagonal matrix elements of the 5×5 matrix representing the operator at the top. Summing each column gives the corresponding character listed in the bottom row.

Exercise: Prove this assertion.

Inserting the character vector from the last row into (16•2–17), with the characters of the irreducible representations taken from the character table, yields immediately

$$n(\mathrm{A}_1) = 2, \qquad n(\mathrm{A}_2) = n(\mathrm{B}_1) = n(\mathrm{B}_2) = 1, \qquad n(\mathrm{E}) = 0. \tag{16•2–19}$$

This result—especially the absence of any two-dimensional irreducible representation—could hardly have been guessed, and the systematic procedure used is more efficient than trial and error.

The right-hand part of the table has as its columns the various irreducible representations actually present. The bottom row contains the representation count from (16•2–19). The rows in between show the contributions to that count by the various basis states, obtained by inserting each row in the left-hand part of the table separately into (16•2–17).

Evidently, the state $|2{,}2\rangle$ is one of the two final states transforming according to the irreducible representation A_1; we express this by writing

$$|\mathrm{A}_1{,}1\rangle = |2{,}2\rangle. \tag{16•2–20a}$$

The second A_1-type state, $|\mathrm{A}_1{,}1\rangle$, is plainly a superposition of $|4{,}0\rangle$ and $|0{,}4\rangle$. By drawing on the rightmost column in the character table itself, or by working out the character vector, we find easily that

$$|\mathrm{A}_1{,}2\rangle = \frac{1}{\sqrt{2}}(|4{,}0\rangle + |0{,}4\rangle). \tag{16•2–20b}$$

Similarly, we find the remaining three irreducible-representation states:

$$|\mathrm{A}_2\rangle = \frac{1}{\sqrt{2}}(|3{,}1\rangle + |1{,}3\rangle), \tag{16•2–21}$$

$$|\mathrm{B}_1\rangle = \frac{1}{\sqrt{2}}(|4{,}0\rangle - |0{,}4\rangle), \qquad |\mathrm{B}_2\rangle = \frac{1}{\sqrt{2}}(|3{,}1\rangle - |1{,}3\rangle). \tag{16•2–22a,b}$$

Exercise: Show that the four-fold degenerate energy level with $n = 3$ can be reduced to two irreducible two-fold degeneracies.

Character Table of a Cube Rotation Group

Table 16•2–4, gives the character table for the cube rotation group, that is the group of rotations transforming a cube into itself, without inverting operations.[9] Here, the class $8\hat{C}_3$ designates the $\pm 120°$ rotations about the four body diagonals, $3\hat{C}_{2z}$ designates the $180°$ rotations about the Cartesian axes (which are the squares of the $\pm 90°$ rotations $6\hat{C}_{4z}$ about those axes), and $6\hat{C}_{2xy}$ represents the $180°$ rotations about the axes going through the midpoints of opposing edges of the cube. We will draw on this table later.

16.2.5 Inversion, Reflections, and Improper Rotations

If the inversion is added to a three-dimensional symmetry group that contains only rotations, it also adds to the group the product of the inversion with every rotation in the group, doubling the number of operations. The newly added elements, including the inversion itself, are called **improper rotations**.

For each class of the old group, this will add a new class, containing the same number of elements, but with each proper rotation replaced by its corresponding improper rotation. Evidently, both the number of classes and the number of irreducible representations double.

It is shown in formal group theory that in such cases the correct set of representations is obtained by simply doubling the number of irreducible representations for every dimensionality, and that the character table is obtained from the original character table by simply replacing every character in the

TABLE 16•2–4. Character table for the cube rotation group.

	$\hat{E}$	$8\hat{C}_3$	$3\hat{C}_{2z}$	$6\hat{C}_{2xy}$	$6\hat{C}_{4z}$
A_1	+1	+1	+1	+1	+1
A_2	+1	+1	+1	−1	−1
E	+2	−1	+2	0	0
T_1	+3	0	−1	−1	+1
T_2	+3	0	−1	+1	−1

[9] Strictly speaking, this is a subgroup of the full group of a cube.

original table with a 2×2 matrix, via the substitution

$$\chi \rightarrow \begin{pmatrix} +\chi & +\chi \\ +\chi & -\chi \end{pmatrix}. \tag{16•2–23}$$

Just as each newly added class contains the same number of elements as the corresponding old class, each newly added irreducible representation, represented by the second row in (16•2–23), has the same dimensionality as the corresponding old irreducible representation, without introducing any additional dimensionalities or levels of degeneracy.

The dimensionality theorem remains satisfied by doubling the number of times each dimensionality occurs. For the case of a cube, it now reads

$$4 \cdot 3^2 + 2 \cdot 2^2 + 4 \cdot 1^2 = 48. \tag{16•2–24}$$

Note that two alternative options for a group consisting of 48 elements distributed over 10 classes, namely,

$$\begin{aligned} &1 \cdot 6^2 + 1 \cdot 2^2 + 8 \cdot 1^2 = 48 \quad \text{and} \\ &2 \cdot 4^2 + 1 \cdot 3^2 + 7 \cdot 1^2 = 48, \end{aligned} \tag{16•2–25a,b}$$

must be ruled out because they do not exhibit the one-to-one correspondence between the dimensionalities of the old and the new irreducible representations.

Reflections as Improper Rotations

If the rotation part of an improper rotation is a two-fold rotation, the resulting improper rotation has the simple geometric meaning of a reflection at a plane perpendicular to the rotation axis and containing the center of inversion. In the case of cubic symmetry, there are nine such reflections: three at the three planes $x, y, z = 0$, given by

$$\hat{\sigma}_x = \hat{C}_{2x}\hat{I}, \text{ etc.}, \tag{16•2–26}$$

and six at the six diagonal mirror planes, namely,

$$\hat{\sigma}_{xy} \equiv \hat{C}_{2xy}\hat{I}, \text{ etc.} \tag{16•2–27}$$

Those operations in which the inversion is combined with rotations higher than two-fold have no simplified geometric interpretation in terms of ordinary reflections; they are symmetry operations of a more general kind, usually designated by the symbol $\hat{S}$. In the group of a cube, there are 14 such operations, 6 deriving from the $\pm 90°$ rotations about the Cartesian axes, such as

$$\hat{S}_{\pm x} \equiv \hat{C}_{\pm 4x}\hat{I}, \tag{16•2–28}$$

and 8 deriving from the $\pm 120°$ rotations about the body diagonals of the cube, like

$$\hat{S}_3 \equiv \hat{C}_3\hat{I}. \tag{16•2–29}$$

◆ PROBLEM TO SECTION 16.2

#16•2-1: Reduction of Assorted Degeneracies in a Cube

Determine the irreducible representations that are present in the following energy levels of a particle in a cube-shaped well with infinitely high walls:

$$\mathcal{E} = 4\mathcal{E}_1,\ 9\mathcal{E}_1,\ 14\mathcal{E}_1,\ 29\mathcal{E}_1,\ \text{and } 56\mathcal{E}_1, \qquad (16{\bullet}2\text{–}30)$$

where $\mathcal{E}_1$ has the same meaning as in (16•1–5).

16.3 MORE ON SYMMETRY OPERATORS AND THEIR EIGENSTATES

16.3.1 Symmetry Operators as Unitary Operators

All the symmetry operations we have discussed are rigid coordinate transformations, without any change in shape of the transformed function. This implies that normalization and inner product integrals are invariant under these transformations,

$$\int (\hat{U}\Phi)^*(\hat{U}\Psi)\, d^3r = \int \Phi^*\Psi\, d^3r, \qquad (16{\bullet}3\text{–}1a)$$

or, in Dirac notation,

$$\boxed{\langle \hat{U}\Phi \,|\, \hat{U}\Psi \rangle = \langle \Phi \,|\, \Psi \rangle.} \qquad (16{\bullet}3\text{–}1b)$$

This relation can be converted into an equivalent alternative form involving Hermitian conjugates. From the definition of Hermitian conjugate operator pairs in chapter 9, we obtain

$$\langle \hat{U}\Phi \,|\, \hat{U}\Psi \rangle = \langle \Phi \,|\, \hat{U}^\dagger\hat{U}\Psi \rangle = \langle \Phi \,|\, \Psi \rangle. \qquad (16{\bullet}3\text{–}2)$$

But this can be true for essentially arbitrary Ψ and Φ only if

$$\boxed{\hat{U}^\dagger\hat{U} = 1 \quad \text{or} \quad \hat{U}^\dagger = \hat{U}^{-1}.} \qquad (16{\bullet}3\text{–}3a,b)$$

This is nothing other than the definition (13•5–6) of **unitary operators**. Hence, the symmetry operators are unitary operators.

A unitary operator will also be Hermitian if it is equal to its own inverse. The symmetry operators for the 180° rotation, for the inversion, and for the reflection at a plane are three examples.

16.3.2 Eigenvalues and Orthogonality of Symmetry Eigenstates

If we designate the various eigenstates of a symmetry operator $\hat{U}$ as $|m\rangle$, $|n\rangle$, etc., we may write (16•1–42) in the form

$$\hat{U}|n\rangle = U_n|n\rangle. \tag{16•3–4}$$

The Hermitian conjugate of this is

$$\langle n|\hat{U}^\dagger = \langle n|U_n^*. \tag{16•3–5}$$

If we replace n by m in (16•3–5) and take the inner product with (16•3–4), we obtain

$$\langle m|\hat{U}^\dagger\hat{U}|n\rangle = U_m^* U_n\langle m|n\rangle. \tag{16•3–6}$$

But because of the unitarity condition (16•3–3), we have, on the left-hand side,

$$\langle m|\hat{U}^\dagger\hat{U}|n\rangle = \langle m|n\rangle. \tag{16•3–7}$$

The expressions (16•3–6) and (16•3–7) are compatible only if either $U_m^* U_n = 1$ or $\langle m|n\rangle = 0$.

If $m = n$, we must have $\langle n|n\rangle \neq 0$; hence,

$$U_n^* U_n = 1. \tag{16•3–8}$$

This property reflects again the fact that the symmetry operations are rigid transformations that do not change the normalization of the state.

Assume now that $U_m \neq U_n$. Then, because of (16•3–8),

$$U_m^* U_n = U_n/U_m \neq 1, \tag{16•3–9}$$

and we conclude that

$$\boxed{\langle m|n\rangle = 0 \quad \text{if} \quad U_m \neq U_n.} \tag{16•3–10}$$

In words, symmetry eigenstates with *different* eigenvalues are orthogonal, just as are eigenstates of Hermitian operators.

For the remainder of this chapter, we restrict ourselves to finite symmetry groups, in which case all symmetry operators are cyclic of some *finite* order M that is an integer fraction of the order of the symmetry group.

If $\hat{U}$ is cyclic of order M, it follows from (16•2–2) and (16•1–42) that

$$\hat{U}^M|\psi\rangle = U^M|\psi\rangle = |\psi\rangle; \tag{16•3–11}$$

hence,

$$U^M = 1. \tag{16•3–12}$$

This means that U must be of the form

$$\boxed{U_m = \exp(2\pi i m/M),} \tag{16•3–13}$$

where the m's are integers, positive, negative, or zero. Because there are only M independent values for U, the labels m may be chosen to be the integers from 0 to $M - 1$:

$$m = 0, 1, \ldots, M - 1. \tag{16•3–14}$$

If plotted in the complex U-plane, the different roots occur equidistantly distributed along the unit circle, as illustrated in **Fig. 16•3–1** for $M = 3$. One of the roots will always be $U = 1$. If M is an even number, -1 will be another root. For $M > 2$, all but at most two of the roots will be complex, and those roots always occur in complex conjugate pairs: If U_m is one root, then U_m^* is another.

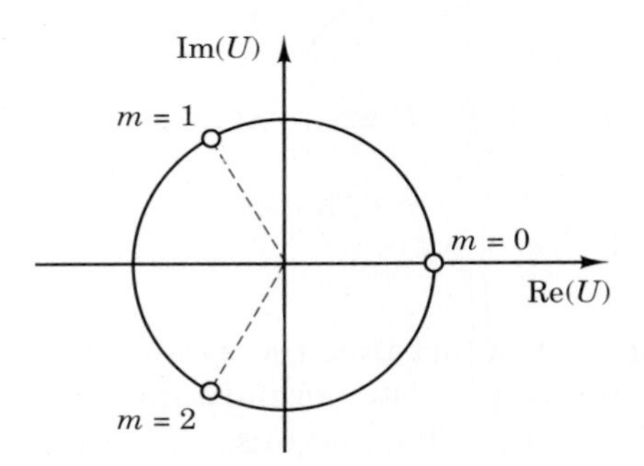

Figure 16•3–1. Location of the symmetry eigenvalues U on the unit circle in the complex plane; shown for a cyclic unitary operator of order 3.

Exercise: Show that, if $0 < n < M$, then

$$\sum_{m=0}^{M-1} (U_m)^n = 0, \tag{16•3–15a}$$

and if $0 < m < M$, then

$$\sum_{n=0}^{M-1} (U_m)^n = 0. \tag{16•3–15b}$$

16.3.3 Projection onto Symmetry Eigenstates

We now return to our earlier claim that the eigenstates of one symmetry operator can always be chosen to be eigenstates of all symmetry operators that commute with that operator, just as for commuting *Hermitian* operators. We show this by actually giving a procedure that leads to the desired arrangement.

The Principle: Cyclic Operators of Order 2

Consider first a symmetry operator $\hat{U}$ that is cyclic of order 2, like any of the reflection operators, any 180° rotation, or the three-dimensional inversion. Assume that $|\psi\rangle$ is one of the energy eigenstates belonging to an N-fold symmetry-degenerate energy eigenvalue, but is *not* a symmetry eigenstate of $\hat{U}$:

$$\hat{U}|\psi\rangle \neq \pm|\psi\rangle. \tag{16•3–16}$$

We can then construct two new states from $|\psi\rangle$ by the prescription

$$|\pm\rangle \equiv \tfrac{1}{2}(1 \pm \hat{U})|\psi\rangle. \tag{16•3–17}$$

Note that the states $|+\rangle$ and $|-\rangle$ are not necessarily normalized. In fact, if $|\psi\rangle$ had already been a symmetry eigenstate of $\hat{U}$, then one of the two new states would vanish. Inspection shows that $|+\rangle$ and $|-\rangle$, if nonzero, are symmetry eigenstates of $\hat{U}$:

$$\hat{U}|+\rangle = +|+\rangle, \qquad \hat{U}|-\rangle = -|-\rangle. \tag{16•3–18}$$

The operators on the right-hand side of (16•3–17),

$$\hat{P}(\pm 1) \equiv \tfrac{1}{2}(1 \pm \hat{U}), \tag{16•3–19}$$

evidently act as **projection operators**, in the sense of chapter 13: They "project" from $|\psi\rangle$ the two symmetry eigenstate contributions $|+\rangle$ and $|-\rangle$.

Because $\hat{U}$ commutes with the Hamiltonian, the projection operators $P_\pm$ do the same, and if $|\psi\rangle$ is an energy eigenstate of the Hamiltonian, then $|+\rangle$ and $|-\rangle$ are also energy eigenstates with the same energy eigenvalue as $|\psi\rangle$.

Finally, the original state $|\psi\rangle$ is simply the sum of the two projected states,

$$|\psi\rangle = |+\rangle + |-\rangle, \tag{16•3–20}$$

without any remainder. Hence, the projection procedure constitutes an expansion of the original state by symmetry eigenstates.

If the original degeneracy was no higher than two-fold, and both $|+\rangle$ and $|-\rangle$ are nonzero, the sorting into symmetry eigenstates of the operator $\hat{U}$ is complete, except possibly for the need to re-normalize. Otherwise, the procedure may be applied to additional states of the original degenerate set, until N linearly independent states have been generated. Because the total number of linearly independent states generated in this way cannot be larger (or smaller) than the original number, the result of one-half of all possible projection operations must be redundant, in that they either are zero or are linearly dependent on states already obtained earlier during preceding projection operations. Such redundant states must be discarded.

For degeneracies higher than two-fold, two or more states must have the same symmetry eigenvalue; but the states will then differ in other ways that make them linearly independent of each other. In particular, if the original set of states was a fully orthogonal set, the set of all projected states will also be fully orthogonal, but usually not normalized.

If there is another symmetry operator $\hat{U}'$ present that commutes with the original operator $\hat{U}$, then the new operator defines a pair of projection operators in its own right, which commutes with $\hat{U}$ and its projection operators. This new operator pair may then be applied to the N states resulting from the application of the preceding pair, leading to N new states that are now symmetry eigenstates of *both* $\hat{U}$ and $\hat{U}'$.

The procedure may evidently be repeated as long as there are additional symmetry operators left that commute not only with the Hamiltonian, but also with all previously utilized symmetry operators. The procedure ends when there are no more symmetry operators that commute with the set previously applied, at which point the final set of N states consists of states that are simultaneous symmetry eigenstates of all mutually commuting symmetry operators present.

Cyclic Operators of Order 3

The projection procedure is readily extended to cyclic operators of higher order by defining appropriate projection operators. The underlying principle emerges by considering cyclic operations of order 3, like the operator $\hat{C}_3$ for the 120° rotation about the body diagonals of a cube. The three possible symmetry eigenvalues are

$$1,\ \gamma = \exp(+2\pi i/3), \quad \text{and} \quad \gamma^2 = \gamma^* = \exp(-2\pi i/3). \tag{16•3–21}$$

Note that

$$1 + \gamma + \gamma^* = 0. \tag{16•3–22}$$

We claim that the projection operators for the three eigenvalues are

$$\hat{P}(1) = \frac{1}{3}[1 + \hat{C}_3 + \hat{C}_3^2], \tag{16•3–23a}$$

$$\hat{P}(\gamma) = \frac{1}{3}[1 + \gamma^*\hat{C}_3 + \gamma\hat{C}_3^2], \tag{16•3–23b}$$

$$\hat{P}(\gamma^*) = \frac{1}{3}[1 + \gamma\hat{C}_3 + \gamma^*\hat{C}_3^2]. \tag{16•3–23c}$$

Exercise: Show that, for an essentially arbitrary state $|\psi\rangle$,

$$\hat{C}_3[\hat{P}(1)|\psi\rangle] = \hat{P}(1)|\psi\rangle, \tag{16•3–24a}$$

$$\hat{C}_3[\hat{P}(\gamma)|\psi\rangle] = \gamma\hat{P}(\gamma)|\psi\rangle, \tag{16•3–24b}$$

$$\hat{C}_3[\hat{P}(\gamma^*)|\psi\rangle] = \gamma^*\hat{P}(\gamma^*)|\psi\rangle. \tag{16•3–24c}$$

Also, show that

$$|\psi\rangle = \hat{P}(1)|\psi\rangle + \hat{P}(\gamma)|\psi\rangle + \hat{P}(\gamma^*)|\psi\rangle, \tag{16•3–25}$$

which constitutes the desired expansion of the state $|\psi\rangle$ in terms of the symmetry eigenstates of $\hat{C}_3$.

Exercise: Generalization to Arbitrary Cyclic Order. Let $|\psi\rangle$ be an arbitrary state, let $\hat{U}$ be a cyclic unitary symmetry operator of finite order M, and let U_m be one of the M different eigenvalues of $\hat{U}$. Consider then the operator

$$\hat{P}_m \equiv \frac{1}{M}[1 + U_m^*\hat{U} + (U_m^*\hat{U})^2 + \ldots +(U_m^*\hat{U})^{M-1}]. \tag{16•3–26}$$

Show the following:

(a) The state $|\psi_m\rangle$ obtained by operating with $\hat{P}_m$ on $|\psi\rangle$,

$$|\psi_m\rangle = \hat{P}_m|\psi\rangle, \tag{16•3–27}$$

either is zero or is itself an eigenstate of the symmetry operator $\hat{U}$ with the eigenvalue U_m:

$$\hat{U}|\psi_m\rangle = U_m|\psi_m\rangle. \tag{16•3–28}$$

(b) The original state $|\psi\rangle$ is simply the sum over all the $|\psi_m\rangle$:

$$|\psi\rangle = \sum_{m=0}^{M-1} |\psi_m\rangle. \tag{16•3–29}$$

This provides the desired expansion of the original state in terms of eigenstates of $\hat{U}$.

16.3.4 Example: The Six-Fold Degeneracy of the Cube States

To demonstrate the power of the projection operator formalism, we re-visit the problem of the (at least) six-fold degeneracy of the unperturbed energy levels of a particle in a cubic well. We pointed out in section 16.1 that, in the case of three different odd quantum numbers, one state could readily split off from what had looked like a symmetry degeneracy, leaving a five-fold degeneracy of uncertain pedigree behind. From the dimensionality theorem, we know that a five-fold degeneracy must be reducible, but the exact splitting is not clear from inspection of the original Cartesian basis set.

In such cases it is often useful to switch to a different set of basis functions, chosen to be symmetry eigenfunctions of a different set of symmetry operators. Now, one of the most important symmetry invariances of a cube is the invariance under a 120° rotation $\hat{C}_3$ about one of the body diagonals of the cube. Applied to the original Cartesian states, the operator $\hat{C}_3$ has the properties

$$\hat{C}_3|k, l, m\rangle = |l, m, k\rangle, \qquad \hat{C}_3^2|k, l, m\rangle = |m, k, l\rangle. \tag{16•3–30a,b}$$

Note that, according to the general definition (16•1–13) of the action of symmetry operators, $\hat{C}_3$ represents a rotation about that cube diagonal for which $x = y = z$, with a counterclockwise sense of rotation if looking outwards from the origin into the direction of positive xyz-values: The rotated state $|l, m, k\rangle$ has the same behavior along the $+x$-axis that the "old" state $|k, l, m\rangle$ had along the $+y$-axis, etc.

We use the relations (16•3–30) here to project the symmetry eigenstates of $\hat{C}_3$ from the six Cartesian basis states associated with the low energy level for which the three Cartesian quantum numbers k, l, and m are three different odd numbers. The basis states are then even functions of all three coordinates, and we saw in section 14.4 that in this case a δ-function perturbation applied at the center of the well will split the six-fold degeneracy into a single level and a remaining five-fold degenerate level. Our objective is to reduce this remaining five-fold degeneracy to its irreducible constituents. To be specific, we assume that the triplet of quantum numbers is the triplet 1, 3, and 5.

If we insert (16•3–30a,b) into the projection operator $\hat{P}(1)$ from (16•3–23a), apply the resulting form to the state $|1,3,5\rangle$, and normalize the result, we obtain

$$
\begin{aligned}
|1,\Uparrow\rangle &\equiv \sqrt{3}\cdot\hat{P}(1)|1,3,5\rangle \\
&= \frac{1}{\sqrt{3}}[|1,3,5\rangle + |3,5,1\rangle + |5,1,3\rangle],
\end{aligned}
\tag{16•3–31a}
$$

where the factor $\sqrt{3}$ restores normalization. The notation $|1,\Uparrow\rangle$ indicates the eigenvalue and the fact that the quantum numbers 1, 3, and 5 are listed in "upward" cyclic order $1 \Rightarrow 3 \Rightarrow 5 \Rightarrow 1$. If we apply $\hat{P}(1)$ to either $|3,5,1\rangle$ or $|5,1,3\rangle$, we obtain the same result, demonstrating the redundancy of the projection process. On the other hand, starting with one of the remaining three basis states, such as $|3,1,5\rangle$, yields

$$
\begin{aligned}
|1,\Downarrow\rangle &\equiv \sqrt{3}\cdot\hat{P}(1)|3,1,5\rangle \\
&= \frac{1}{\sqrt{3}}[|3,1,5\rangle + |1,5,3\rangle + |5,3,1\rangle].
\end{aligned}
\tag{16•3–31b}
$$

If we turn next to the symmetry eigenstates with the symmetry eigenvalue γ, we obtain

$$
\begin{aligned}
|\gamma,\Uparrow\rangle &\equiv \sqrt{3}\cdot\hat{P}(\gamma)|1,3,5\rangle \\
&= \frac{1}{\sqrt{3}}[|1,3,5\rangle + \gamma^*|3,5,1\rangle + \gamma|5,1,3\rangle],
\end{aligned}
\tag{16•3–32a}
$$

$$
\begin{aligned}
|\gamma,\Downarrow\rangle &\equiv \sqrt{3}\cdot\hat{P}(\gamma)|3,1,5\rangle \\
&= \frac{1}{\sqrt{3}}[|3,1,5\rangle + \gamma^*|1,5,3\rangle + \gamma|5,3,1\rangle].
\end{aligned}
\tag{16•3–32b}
$$

The symmetry eigenstates $|\gamma^*, \Uparrow\rangle$ and $|\gamma^*, \Downarrow\rangle$ belonging to γ^* are simply the complex conjugates of those for γ:

$$|\gamma^*,\Uparrow\rangle = \frac{1}{\sqrt{3}}[|1,3,5\rangle + \gamma|3,5,1\rangle + \gamma^*|5,1,3\rangle] = (|\gamma,\Uparrow\rangle)^*, \tag{16•3–33a}$$

$$|\gamma^*,\Downarrow\rangle = \frac{1}{\sqrt{3}}[|3,1,5\rangle + \gamma|1,5,3\rangle + \gamma^*|5,3,1\rangle] = (|\gamma,\Downarrow\rangle)^*. \tag{16•3–33b}$$

The only symmetry operator that commutes with $\hat{C}_3$, other than $\hat{C}_{-3}$, is the inversion. However, inasmuch as all basis orbitals were already inversion eigenstates with the common eigenvalue $+1$, all projected states, too, are even under inversion. Hence, the question of inversion symmetry never comes up, and the irreducible representations contained in our energy level are those of the cube rotation group, without inversion operations.

For the same reasons, the six projected states are also symmetry eigenstates of the three 180° rotations about the Cartesian axes and of the three reflections at the Cartesian planes. This is true even though those operators do not commute with $\hat{C}_3$. Finally, all six projection states are also eigenstates of the remaining seven $\pm 120°$ rotation operators about the various body diagonals of the cube.

All remaining operators interchange two of the three axes, with or without a reversal of one or more axis directions. Because all states are even functions of all Cartesian coordinates, any interchange of axis directions is irrelevant, and the effect of all these operators is the same as that of the reflection operator $\hat{\sigma}_{xy}$ that interchanges the x- and y-axes. We find easily that

$$\hat{\sigma}_{xy}|1,\Uparrow\rangle = |1,\Downarrow\rangle, \qquad \hat{\sigma}_{xy}|1,\Downarrow\rangle = |1,\Uparrow\rangle, \tag{16•3–34}$$

$$\hat{\sigma}_{xy}|\gamma,\Uparrow\rangle = |\gamma^*,\Downarrow\rangle, \qquad \hat{\sigma}_{xy}|\gamma^*,\Downarrow\rangle = |\gamma,\Uparrow\rangle, \tag{16•3–35a}$$

$$\hat{\sigma}_{xy}|\gamma,\Downarrow\rangle = |\gamma^*,\Uparrow\rangle, \qquad \hat{\sigma}_{xy}|\gamma^*,\Uparrow\rangle = |\gamma,\Downarrow\rangle. \tag{16•3–35b}$$

From the first of these relations, we see immediately that the implied degeneracy is reducible, by constructing two new linear superpositions of the form

$$|1,\pm\rangle \equiv \frac{1}{\sqrt{2}}[|1,\Uparrow\rangle \pm |1,\Downarrow\rangle] = \frac{1}{\sqrt{6}}[(|3,1,5\rangle + |1,5,3\rangle + |5,3,1\rangle) \pm (|3,1,5\rangle + |1,5,3\rangle + |5,3,1\rangle)]. \tag{16•3–36}$$

This procedure does not work with the remaining two interchange pairs, because here the interchanges are between states belonging to different symmetry eigenvalues of $\hat{C}_3$. Instead, both pairs, $(|\gamma, \Uparrow\rangle, |\gamma^*, \Downarrow\rangle)$ and $(|\gamma, \Downarrow\rangle, |\gamma^*, \Uparrow\rangle)$, represent two-fold *irreducible* degeneracies.

Exercise: Confirm the preceding conclusions by constructing the character vectors for the three pairs of states, using the procedure given in section 16.2. Show that the state $|1,+\rangle$ belongs to the irreducible representation A_1 of the group O, while $|1,-\rangle$ belongs to A_2. Show that the character vector of both remaining pairs is that of the irreducible representation E.

Thus, we see, at long last, that the six-fold apparent symmetry degeneracy of the energy level $\mathcal{E} = 35\mathcal{E}_1$ can be split, by a suitable symmetry-preserving perturbation, into to two non-degenerate levels and two doubly degenerate levels. The reader should have no difficulty constructing perturbations that split both single levels from each other and from the two two-fold degenerate levels. Splitting the latter two requires extra thought: Because the two pairs of states are simply complex conjugates of each other, they cannot be split by a real scalar perturbing potential. Being complex, the states involved are current-carrying states, and the two complex conjugate sets differ from each other only by a reversal in current *direction*. The degeneracy between such states can be split by a suitable magnetic field perturbation, represented by a suitable vector potential **A**, such that the energy is shifted in opposite directions for opposite directions of current flow.

◆ PROBLEM TO SECTION 16.3

#16•3-1: Irreducible Representations for the Cubic-Well Level $\mathcal{E} = \mathcal{E}_1 \cdot (2^2 + 4^2 + 6^2)$ and Associated Eigenstates

Resolve the reducible degeneracy for the cubic-well level $\mathcal{E} = \mathcal{E}_1 \cdot (2^2 + 4^2 + 6^2)$, and determine the sets of eigenstates associated with the various irreducible representations that are present.

16.4 SYMMETRY IN PERTURBATION THEORY

16.4.1 Symmetry Factorization

There is no more powerful tool for the simplification of perturbation problems than the utilization of symmetry invariances in the Hamiltonian. Such invariances are not always present, and even when they are, they do not always suffice to make the problem tractable. But they provide rigorous rather than approximate simplifications, and if they can be applied at all, they should be applied *before* any approximation methods are employed. Usually, they will simplify the problem greatly.

Symmetry arguments are applicable whenever *both* the unperturbed Hamiltonian $\hat{H}^{(0)}$ and the perturbation $\hat{W}$ are invariant under a symmetry operator $\hat{U}$, that is, when both commute with $\hat{U}$. Such symmetry invariances are very common in quantum mechanics. The central idea behind utilizing these symmetries is a generalization of the idea presented already in section 11.4, where we factored a secular equation into two smaller equations by utilizing the inversion symmetry of the simple cosine potential.

"Simple" Symmetries

Suppose we are dealing with a perturbation theory problem in which the unperturbed Hamiltonian $\hat{H}^{(0)}$ commutes with the symmetry operator $\hat{U}$. It is then possible to choose the unperturbed eigenstates of $\hat{H}^{(0)}$ in such a way that they are also eigenstates of $\hat{U}$. Let $|m\rangle$ and $|n\rangle$ be two such eigenstates of both $\hat{H}^{(0)}$ and $\hat{U}$, but belonging to *different* eigenvalues U_m and U_n of $\hat{U}$:

$$\hat{U}|m\rangle = U_m|m\rangle, \qquad \hat{U}|n\rangle = U_n|n\rangle. \tag{16•4–1a,b}$$

Assume now that a perturbation $\hat{W}$ is present that is also invariant under the symmetry operation $\hat{U}$,

$$\hat{U}\hat{W} = \hat{W}\hat{U}. \tag{16•4–2a,b}$$

Consider, then, the matrix element

$$W_{mn} = \langle m|\hat{W}|n\rangle. \tag{16•4–3}$$

Because $\hat{U}$ is unitary, $\hat{U}\hat{U}^\dagger = 1$, and we may write

$$\langle m|\hat{W}|n\rangle = \langle m|\hat{U}^\dagger\hat{U}\hat{W}|n\rangle = \langle m|\hat{U}^\dagger\hat{W}\hat{U}|n\rangle = U_m^* U_n\langle m|\hat{W}|n\rangle. \tag{16•4–4}$$

Now, if $U_m \neq U_n$, then $U_m^* U_n \neq 1$, from (16•3–9). But in this case, the first and the last expression can be equal to each other only if *both* are zero,

$$\langle m|\hat{H}|n\rangle = \langle m|\hat{W}|n\rangle = 0. \tag{16•4–5}$$

This relation forms the basis for the application of symmetry arguments to perturbation theory.

Suppose we arrange the different eigenvalues of the symmetry operator $\hat{U}$ in a convenient order, U_1, U_2, U_3, . . . , etc., and then order all wave functions similarly, beginning with all wave functions corresponding to U_1, followed by all wave functions corresponding to U_2, etc. (The ordering *within* each group need not concern us.) If this ordering is done, the $N \times N$ secular equation (14•2–2) assumes the block form

$$\begin{vmatrix} D_1 & & & \\ & D_2 & & \\ & & D_3 & \\ & & & \text{etc.} \end{vmatrix} = 0 \tag{16•4–6}$$

where the different D_m stand for smaller secular determinants and the empty off-diagonal blocks stand for blocks of zeros. Each of the D_m involves only eigenstates of both $\hat{H}^{(0)}$ *and* $\hat{U}$ that belong to the same eigenvalue U_m of $\hat{U}$. The relation (16•4–6) represents, of course, a factoring of the Nth-degree secular equation into simpler equations of lower degree,

$$D_1 = 0, \qquad D_2 = 0, \qquad D_3 = 0, \text{ etc.} \tag{16•4–7}$$

Our argument is independent of how many states have been included in the secular equation; it is exact. In particular, it also applies to non-degenerate perturbation theory.

Exercise: Translational Invariance. Any one-dimensional periodic potential with the period a—such as our simple cosine potential of the two preceding chapters—is invariant under the translational operators $\hat{T}$ defined by

$$\hat{T}f(x) \equiv f(x + a), \tag{16•4–8}$$

corresponding to a displacement by $-a$, that is, a displacement to the left.

All unperturbed plane waves $|k\rangle$ are eigenstates of this operator, with the symmetry eigenvalues

$$T = e^{+ika}. \tag{16•4–9}$$

Because the periodic potential commutes with $\hat{T}$, only those unperturbed eigenstates can have a finite matrix element between each other that belong to the same symmetry eigenvalue T. Show that this implies that their wave numbers must differ by an integer multiple of $2\pi/a = G$:

$$\langle k_1|\hat{W}|k_2\rangle = 0, \quad \text{unless} \quad k_2 - k_1 = mG, \qquad (m = 0, \pm 1, \pm 2, \ldots, \text{etc.}). \tag{16•4–10}$$

The selection rule (14•3–14) found earlier for the simple cosine potential is a special case of (16•4–10), showing that for a potential whose Fourier expansion contains only a *single* spatial frequency, without higher harmonics, the set (16•4–10) must be further restricted to $m = 1$. The general form (16•4–10) holds for any arbitrary one-dimensional periodic potential with the period $a = 2\pi/G$.

Multiple Symmetries

If there is an additional symmetry operator $\hat{U}'$ present that commutes with both $\hat{H}^{(0)}$ and $\hat{U}$, then it is often possible to factor the individual secular equations in (16•4–6) and (16•4–7) further, by arranging the energy eigenstates within each secular equation by the different eigenvalues of $\hat{U}'$. The procedure can evidently be repeated as long as additional symmetry operators can be found that commute both with $\hat{H}^{(0)}$ and with all previously used symmetry operators. In fact, it is sometimes possible to obtain further factorization with the help of symmetry operators that do *not* commute with all previously used symmetry operators.

16.4.2 Example: Splitting of the Nine-Fold Degenerate Hydrogen Atom Level with $n = 3$ by a Quadrupole Perturbation

Symmetry Invariances

To demonstrate the application of symmetry arguments—particularly the utilization of multiple simultaneous symmetries—and to show the full repertoire of possibilities, we discuss here a fairly complicated problem, the splitting of the nine-fold degenerate hydrogen atom energy level with $n = 3$, under the influence of a so-called *quadrupole perturbation*, that is, a perturbation caused by four charges of equal magnitude but alternating sign, placed symmetrically in the same plane as the H-atom, all at the same distance from the atom, as in **Fig. 16•4–1**. Such a perturbation is invariant under the following symmetry operations:

$\hat{\sigma}_x, \hat{\sigma}_y, \hat{\sigma}_z$: Reflections at the planes $x = 0, y = 0, z = 0$;

$\hat{C}_{2x}, \hat{C}_{2y}, \hat{C}_{2z}$: Rotations by π about the x-, y-, and z-axes;

$\hat{I}$: Inversions about the coordinate origin.

All these are operations under which the unperturbed Hamiltonian is also invariant.

All seven symmetry operators listed above commute with one another. Hence, we expect that all true symmetry degeneracies get split by the perturbation, and only accidental degeneracies can remain.

The different symmetry operations are not independent of each other. We find readily that

$$\hat{I} = \hat{\sigma}_x \hat{C}_{2x} = \hat{\sigma}_y \hat{C}_{2y} = \hat{\sigma}_z \hat{C}_{2z} \tag{16•4–11}$$

and

$$\hat{\sigma}_x \hat{\sigma}_y = \hat{C}_{2z}, \qquad \hat{\sigma}_y \hat{\sigma}_z = \hat{C}_{2x}, \qquad \hat{\sigma}_z \hat{\sigma}_x = \hat{C}_{2y}. \tag{16•4–12}$$

This means that a consideration of, say, $\hat{C}_{2z}$, $\hat{\sigma}_z$, and $\hat{\sigma}_y$ is sufficient to obtain the symmetry properties of the unperturbed eigenstates under all symmetry operations.

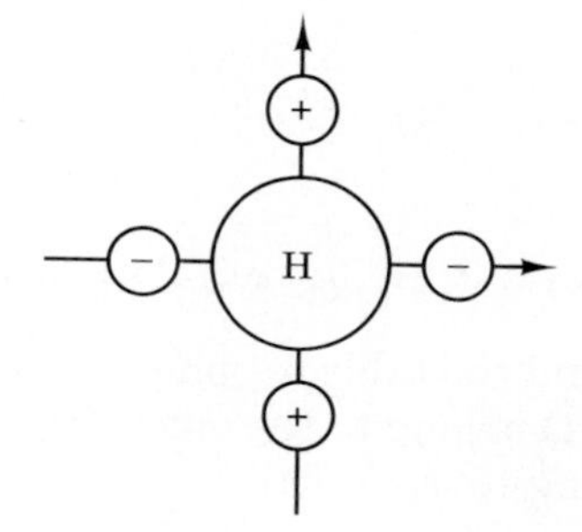

Figure 16•4–1. Hydrogen atom under the influence of a quadrupole perturbation.

Exercise: Confirm (16•4–11) and (16•4–12).

Note that the perturbation W by the four charges is *not* invariant under 90° rotation. Instead, we have

$$\hat{C}_{4z}W = -W, \tag{16•4–13}$$

that is, W is an eigenstate of $\hat{C}_{4z}$, with the eigenvalue -1. This property is not a geometrical symmetry *invariance* in the conventional sense of a symmetry operator *commuting* with $\hat{W}$. Rather, we view it as a **generalized invariance**, under a slightly more general class of transformations, namely, a geometrical invariance combined with a sign reversal. We shall see that such generalized invariances can cause additional simplifications and are therefore important in their own right.

Symmetry Properties of Basis States

The unperturbed eigenstates are of the form

$$|l, m\rangle = Y_l^m(\theta, \phi) \cdot \chi_{3,l}(r), \tag{16•4–14}$$

where Y_l^m are the spherical harmonics and the $\chi_{3,l}(r)$ are the radial functions. Because the symmetry properties depend only on the angular parts of $|l, m\rangle$, we ignore the radial functions $\chi_{3,l}(r)$ here.

We start our considerations with the rotation $\hat{C}_{2z}$. We have

$$\hat{C}_{2z} Y_l^m(\theta, \phi) = Y_l^m(\theta, \phi - \pi). \tag{16•4–15}$$

The ϕ-dependence of the spherical harmonics—see (3•1–21)—is simply

$$Y_l^m(\theta, \phi) \propto e^{im\phi}. \tag{16•4–16}$$

From this and (16•4–15), it follows that

$$\hat{C}_{2z} Y_l^m = (-1)^m Y_l^m, \tag{16•4–17}$$

which implies that the perturbation will not couple states with even values of m (including $m = 0$) to those with odd m.

Consider next the reflection $\hat{\sigma}_z$. Expressed in polar coordinates, it has the property

$$\hat{\sigma}_z f(\theta, \phi) = f(\pi - \theta, \phi). \tag{16•4–18}$$

Specifically,

$$\hat{\sigma}_z \sin\theta = \sin\theta, \qquad \hat{\sigma}_z \cos\theta = -\cos\theta. \tag{16•4–19a,b}$$

From these properties, and the table of spherical harmonics in appendix C, we find that $|1, 0\rangle$ and $|2, \pm 1\rangle$ belong to the eigenvalue $\sigma_z = -1$, with the remainder of the states belonging to $\sigma_z = +1$.

By drawing on the preceding two symmetry properties, we can sort the nine unperturbed states into four non-interacting subsets, as shown in the three leftmost columns of **Table 16•4–1**. Because of (16•4–11), the inversion does not split the four subsets further.

Consider next the reflection $\hat{\sigma}_y$:

$$\hat{\sigma}_y f(\theta, \phi) = f(\theta, -\phi), \tag{16•4–20a}$$

$$\hat{\sigma}_y e^{\pm im\phi} = e^{\mp im\phi}. \tag{16•4–20b}$$

The relation (16•4–20b) implies that the spherical harmonics with $m \neq 0$ are not eigenstates of $\hat{\sigma}_y$, but are interchanged by it. But in that case, we can construct a new set of cosine- and sine-like functions by linear superposition of the original functions:

$$|l, cm\rangle = +\frac{1}{\sqrt{2}}[|l, m\rangle + |l, -m\rangle] \propto \cos m\phi, \tag{16•4–21a}$$

$$|l, sm\rangle = -\frac{1}{\sqrt{2}}[|l, m\rangle - |l, -m\rangle] \propto \sin m\phi. \tag{16•4–21b}$$

TABLE 16•4–1. Transformation properties of the various unperturbed energy eigenstates of the hydrogen atom energy level with $n = 3$ under the influence of various symmetry operators. Horizontal lines separate different sets of symmetry eigenvalues. The last column shows the transformation by the 90° rotation operator, under which the perturbed Hamiltonian is *not* invariant.

$\|l, m\rangle$	C_{2z}	σ_z	$\|l, \ldots\rangle$	σ_y	$\hat{C}_{4z}\|l, \ldots\rangle$
$\|0, 0\rangle$	+1	+1	$\|0, 0\rangle$	+1	$+\|0, 0\rangle$
$\|2, 0\rangle$	+1	+1	$\|2, 0\rangle$	+1	$+\|2, 0\rangle$
$\|2, -2\rangle$	+1	+1	$\|2, c2\rangle$	+1	$-\|2, c2\rangle$
$\|2, +2\rangle$	+1	+1	$\|2, s2\rangle$	−1	$-\|2, s2\rangle$
$\|1, 0\rangle$	+1	−1	$\|1, 0\rangle$	+1	$+\|1, 0\rangle$
$\|1, -1\rangle$	−1	+1	$\|1, c1\rangle$	+1	$+\|1, s1\rangle$
$\|1, +1\rangle$	−1	+1	$\|1, s1\rangle$	−1	$-\|1, c1\rangle$
$\|2, -1\rangle$	−1	−1	$\|2, c1\rangle$	+1	$+\|2, s1\rangle$
$\|2, +1\rangle$	−1	−1	$\|2, s1\rangle$	−1	$-\|2, c1\rangle$

Because of (16•4–20a) and (16•4–21), the new functions are eigenstates of $\hat{\sigma}_y$,

$$\hat{\sigma}_y|l,cm\rangle = +|l,cm\rangle, \qquad \hat{\sigma}_y|l,sm\rangle = -|l,sm\rangle, \tag{16•4–22}$$

thus leading to the fourth and fifth columns in Table 16•4–1. It will therefore be useful to use these functions instead of the original ones.

Because of (16•4–12), $\hat{\sigma}_x$ is not independent of $\hat{\sigma}_y$ and $\hat{C}_{2z}$, and its application does not split the one remaining subset of three states further. Neither does the application of $\hat{C}_{2x}$ or $\hat{C}_{2y}$.

Finally, the last column in Table 16•4–1 shows the result of applying the 90° rotation operator $\hat{C}_{4z}$—which is *not* a symmetry operator of the potential—to the functions in the fourth column. The first five functions are eigenstates of that operator, with the eigenvalues ± 1. In the last four rows, the operation leads to a pairwise interchange of states, as indicated. We will draw on these properties shortly.

From the table, it is clear that the 9×9 secular equation can be factored immediately into six linear equations and one cubic equation.

The roots of the six linear equations are equal to the following six diagonal matrix elements, shown here in three groups:

$$\langle 2,s\,2|\hat{W}|2,s\,2\rangle, \qquad \langle 1,0|\hat{W}|1,0\rangle; \tag{16•4–23a,b}$$

$$\langle 1,c\,1|\hat{W}|1,c\,1\rangle, \qquad \langle 1,s\,1|\hat{W}|1,s\,1\rangle; \tag{16•4–23c,d}$$

$$\langle 2,c\,1|\hat{W}|2,c\,1\rangle, \qquad \langle 2,s\,1|\hat{W}|2,s\,1\rangle. \tag{16•4–23e,f}$$

The remaining cubic equation is of the form

$$\begin{vmatrix} \langle 0,0|\hat{W}|0,0\rangle - \Delta\mathcal{E} & \langle 0,0|\hat{W}|2,0\rangle & \langle 0,0|\hat{W}|2,c\,2\rangle \\ \langle 2,0|\hat{W}|0,0\rangle & \langle 2,0|\hat{W}|2,0\rangle - \Delta\mathcal{E} & \langle 2,0|\hat{W}|2,c\,2\rangle \\ \langle 2,c\,2|\hat{W}|0,0\rangle & \langle 2,c\,2|\hat{W}|2,0\rangle & \langle 2,c\,2|\hat{W}|2,c\,2\rangle - \Delta\mathcal{E} \end{vmatrix} = 0, \tag{16•4–24}$$

where $\Delta\mathcal{E} = \mathcal{E} - \mathcal{E}^{(0)}$ is the energy shift from the unperturbed value.

16.4.3 Effect of the Generalized 90° Rotation Invariance

We now turn to the utilization of the generalized invariance property (16•4–13). Let $|\Psi\rangle$ be an arbitrary admissible state vector, and let us apply the operator $\hat{C}_{4z}$ to $\hat{W}|\Psi\rangle$. From the property (16•1–34) of symmetry operators and the property (16•4–13) of $\hat{W}$, we obtain

$$\hat{C}_{4z}\hat{W}|\Psi\rangle = (\hat{C}_{4z}\hat{W}) \cdot \hat{C}_{4z}|\Psi\rangle = -\hat{W} \cdot \hat{C}_{4z}|\Psi\rangle. \tag{16•4–25}$$

Because this must be true for any admissible $|\Psi\rangle$, we may drop $|\Psi\rangle$ from the equation and write (16•4–25) in the form

$$\hat{C}_{4z}\hat{W} = -\hat{W}\hat{C}_{4z}. \tag{16•4–26}$$

The operators $\hat{C}_{4z}$ and $\hat{W}$ anti-commute, in contrast to the commutation relation (16•4–2b) for the other operators.

In applying this property to the various matrix elements in our problem, we must distinguish between two cases, depending upon whether the basis states $|m\rangle$ and $|n\rangle$ are eigenstates of $\hat{C}_{4z}$ or whether they are interchanged by $\hat{C}_{4z}$.

Symmetry Eigenstates

Let $|m\rangle$ and $|n\rangle$ be eigenstates of $\hat{C}_{4z}$, with the eigenvalues $C_{4z}(m)$ and $C_{4z}(n)$. By utilizing the unitarity of $\hat{C}_{4z}$, along with the anti-commutation relation (16•4–26), the matrix element $\langle m|\hat{W}|n\rangle$ may be transformed as follows:

$$\begin{aligned}\langle m|\hat{W}|n\rangle &= \langle m|\hat{C}_{4z}^{\dagger}\hat{C}_{4z}\hat{W}|n\rangle = -\langle m|\hat{C}_{4z}^{\dagger}\hat{W}\hat{C}_{4z}|n\rangle \\ &= C_{4z}^{*}(m)\cdot C_{4z}(n)\cdot\langle m|\hat{W}|n\rangle. \end{aligned} \tag{16•4–27}$$

But this implies

$$\boxed{\langle m|\hat{W}|n\rangle = 0, \quad \text{unless} \quad C_{4z}^{*}(m)\cdot C_{4z}(n) = -1.} \tag{16•4–28}$$

If we apply (16•4–28) to our quadrupole perturbation problem, we find that all diagonal matrix elements for the first five functions in Table 16•4–1 vanish. This includes the two matrix elements in (16•4–23a,b) and all diagonal matrix elements in the 3×3 secular equation (16•4–24).

In addition, two pairs of off-diagonal matrix elements also vanish:

$$\langle 0,0|\hat{W}|2,c2\rangle = \langle 2,c2|\hat{W}|0,0\rangle = 0, \tag{16•4–29a}$$

$$\langle 2,0|\hat{W}|2,c2\rangle = \langle 2,c2|\hat{W}|2,0\rangle = 0. \tag{16•4–29b}$$

With these simplifications, the secular equation (16•4–24) factors further into

$$\begin{vmatrix} -\Delta\mathcal{E} & 0 & \langle 0,0|\hat{W}|2,c2\rangle \\ 0 & -\Delta\mathcal{E} & \langle 2,0|\hat{W}|2,c2\rangle \\ \langle 2,c2|\hat{W}|0,0\rangle & \langle 2,c2|\hat{W}|2,0\rangle & -\Delta\mathcal{E} \end{vmatrix}$$

$$= \Delta\mathcal{E}\cdot[|\langle 0,0|\hat{W}|2,c2\rangle|^2 + |\langle 2,0|\hat{W}|2,c2\rangle|^2 - (\Delta\mathcal{E})^2] = 0. \tag{16•4–30}$$

This has the three roots

$$\Delta\mathcal{E} = \begin{cases} 0 \\ \pm\sqrt{|\langle 0,0|\hat{W}|2,c2\rangle|^2 + |\langle 2,0|\hat{W}|2,c2\rangle|^2}. \end{cases} \tag{16•4–31a,b}$$

Symmetry-Interchanged States

We turn next to the other case, where the basis states $|m\rangle$ and $|n\rangle$ are interchanged by $\hat{C}_{4z}$,

$$\hat{C}_{4z}|m\rangle = \pm|n\rangle, \tag{16•4–32}$$

and to the effect of this interchange property on the *diagonal* matrix elements of $|m\rangle$ and $|n\rangle$. In this case, we obtain, by a manipulation similar to that in (16•4–27),

$$\begin{aligned}\langle m|\hat{W}|m\rangle &= \langle m|\hat{C}_{4z}^{\dagger}\hat{C}_{4z}\hat{W}|m\rangle \\ &= -\langle m|\hat{C}_{4z}^{\dagger}\hat{W}\hat{C}_{4z}|m\rangle = -\langle n|\hat{W}|n\rangle. \end{aligned} \tag{16•4–33}$$

Applied to the four states in the lower right corner of Table 16•4–1, this means that the energy shifts of these states come in pairs that differ only by a sign, namely,

$$\langle 1,c\,1|\hat{W}|1,c\,1\rangle = -\langle 1,s\,1|\hat{W}|1,s\,1\rangle \tag{16•4–34a}$$

and

$$\langle 2,c\,1|\hat{W}|2,c\,1\rangle = -\langle 2,s\,1|\hat{W}|2,s\,1\rangle \tag{16•4–34b}$$

Final Level Splittings

The overall result of our manipulations is shown in **Fig. 16•4–2**: The originally nine-fold degenerate energy level splits up into a three-fold degenerate unshifted energy level and three symmetrically split pairs. Two of the unshifted states arise from the vanishing of the diagonal matrix elements in (16•4–23a,b), and the third from the vanishing root (16•4–31a) of the 3×3 secular equation. Of the three symmetrically split pairs, two arise from the sign reversals in (16•4–34a,b), and the third from the symmetric pair of roots (16•4–31b).

The actual calculation of the splittings requires the knowledge of the specific analytical form of $\hat{W}$. Such calculations tend to be tedious, and we forgo them here; our purpose was to illustrate the power of symmetry arguments in *avoiding* having to calculate matrix elements, not to calculate those that remain.

The remaining three-fold degeneracy is a consequence of the quadrupole symmetry of the perturbation. Yet it is not a simple symmetry degeneracy in

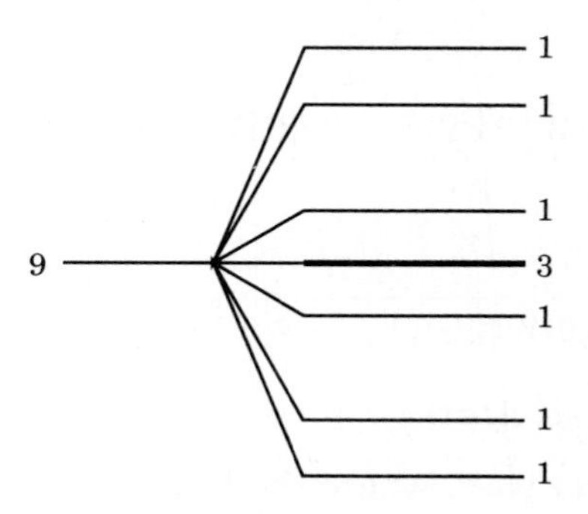

Figure 16•4–2. Splitting of $n = 3$ energy level of the hydrogen atom under the influence of the quadrupole perturbation shown in Fig. 16•4–1. The originally nine-fold degenerate level splits into six non-degenerate levels and one three-fold degenerate level. The magnitudes of the splittings shown are qualitative.

the narrow sense defined earlier, where it was possible to obtain the different degenerate wave functions from each other by applying one or more symmetry operators to one of the functions. Nor is it a purely accidental degeneracy, unrelated to any symmetry properties. We call it a **generalized symmetry degeneracy**.

Exercise: The simple cosine potential reverses its sign under translation by one-half the lattice period, $a/2$:

$$\hat{U}V(x) = V(x + a/2) = -V(x). \tag{16•4–35}$$

Show that this property alone implies that the difference between the interacting wave numbers k_1 and k_2 must be an *odd* multiple of G.

16.4.4 Non-Commuting Symmetries

Our quadrupole perturbation example did not involve any non-commuting operators amongst the set of operators under which both the Hamiltonian and the perturbation were invariant.[10] When non-commuting symmetries are present, it is no longer possible to select the unperturbed basis functions to be symmetry eigenstates of *all* the symmetry operators, and the question arises how to choose these basis functions. To obtain the maximum factorization, the basis states should be selected so as to transform according to the irreducible representation of the symmetry group. For details, we must again refer the reader to the specialized literature on group theory.

In practice, it is often possible to factor the problem by elementary methods, without invoking formal group theory. As a rule, some, but not all, of the symmetry operators in a symmetry group commute with one another. A good point of departure is then to select a subset of mutually commuting symmetry operators from the full group and to pick the basis states to be symmetry eigenstates of that subset. If the subset was well chosen—which may require some trial and error—a large degree of factorization is likely to result. Further simplifications can then often be obtained by going over the factor determinants, one at a time, and investigating the way the functions left within each of those determinants transform under one of more of the symmetry operations that were left outside the initial subset. If this transformation is strictly among the functions within the same factor determinant, further factorization is often possible.

---

[10] Note, however, that the 90° rotation operator invoked at the end does not commute with all the other operators.

◆ PROBLEM TO SECTION 16.4

#16•4-1: A Ten-Fold Degeneracy Split by a Cubic-Symmetry Potential

Consider the ten-fold degenerate energy level of the *spherical* harmonic oscillator, with the quantum number

$$n_r = n_x + n_y + n_z = 3. \tag{16•4–36}$$

Calculate quantitatively the splitting of this energy level under the influence of the perturbation

$$\hat{W} = \alpha \cdot \frac{4\hbar\omega}{L^4}(x^2y^2 + y^2z^2 + z^2x^2), \tag{16•4–37}$$

where α is a dimensionless strength parameter, assumed to be $\ll 1$, and L is the natural unit of length for the harmonic oscillator.

Chapter 17

ELECTRONS IN PERIODIC CRYSTAL POTENTIALS[1]

17.1 k-SPACE

17.1.1 Bloch's Theorem

The dominant feature of the physics of electrons in crystals is the translational periodicity of their Hamiltonian. The periodic arrangement of the atoms in a perfect crystal means that it is possible to define three vectors $\mathbf{a}_1$, $\mathbf{a}_2$, and $\mathbf{a}_3$, called **primitive translation vectors**, such that the atomic arrangement inside the crystal and, hence, the Hamiltonian of the electrons is invariant under translations by each of these vectors,

$$\hat{H}(\mathbf{r} + \mathbf{a}_1) = \hat{H}(\mathbf{r} + \mathbf{a}_2) = \hat{H}(\mathbf{r} + \mathbf{a}_3) = \hat{H}(\mathbf{r}). \qquad (17{\bullet}1\text{–}1)$$

[1] The purpose of this chapter is to provide the reader with some of the quantum-mechanical background drawn upon in texts and/or courses on solid-state physics, especially semiconductor physics. It is to *supplement* such texts and courses, not to *substitute* for the treatment given in them.

More generally,

$$\hat{H}(\mathbf{r} + \mathbf{T}) = \hat{H}(\mathbf{r}), \tag{17•1–2a}$$

where $\mathbf{T}$ is *any* vector of the form

$$\mathbf{T} = u_1\mathbf{a}_1 + u_2\mathbf{a}_2 + u_3\mathbf{a}_3, \tag{17•1–2b}$$

in which the three u's are independent arbitrary integers, positive, negative, or zero. Any displacement vector of this form is called a **lattice translation**.

The set of endpoints of all vectors of the form (17•1–2b) with integer u's is called the **point lattice** of the crystal structure, usually referred to simply as the **lattice**. It is a periodic array of mathematical points that serves as the mathematical frame of reference for the atoms themselves. For example, the crystal structure of most of the important semiconductors, such as silicon (Si) and gallium arsenide (GaAs), have as their underlying point lattice the so-called **face-centered cubic** lattice, shown in **Fig. 17•1–1**.

Note that the three orthogonal edges of the cube shown in the figure are not *primitive* translation vectors: There is no way we can get to the lattice points at the centers of the cube faces by stringing such vectors together. The true primitive translation vectors for the face-centered cubic lattice are the vectors pointing from one of the corners of the cube to the lattice points at the *centers* of each of the three adjacent faces. By stringing the latter vectors together, every lattice point can be reached, never leading to a point that is not a lattice point.

Associated with each lattice point is a set of atoms, usually more than one, called the **basis** of the crystal structure. For example, in GaAs, there is a basis of one Ga and one As atom per lattice point. The Ga atoms by themselves form a face-centered cubic (fcc) lattice, as do the As atoms,[2] but these two fcc **sublattices** are displaced relative to each other by the vector

$$\mathbf{d} = \tfrac{1}{4}(\mathbf{a}_1 + \mathbf{a}_2 + \mathbf{a}_3). \tag{17•1–3}$$

In Si, the same two sublattices occur, but each is occupied by the same atomic species. The point lattice of the structure may be chosen to coincide with one of the atomic sublattices, but it does not have to be.

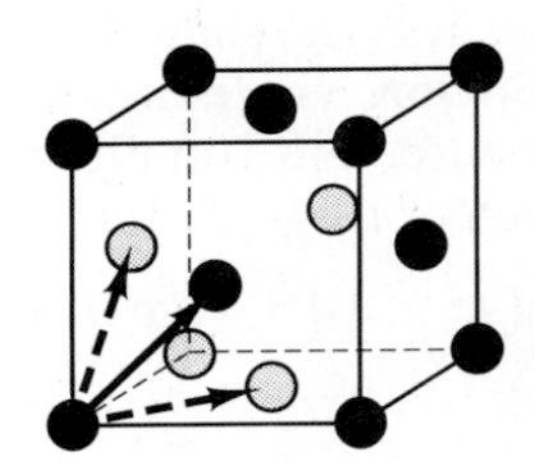

Figure 17•1–1. Face-centered cubic lattice and its primitive translation vectors.

[2] See section 17.4 for details.

The three primitive translation vectors may be viewed as the edges of a parallelepiped with the volume

$$\Omega_0 = \mathbf{a}_1 \cdot \mathbf{a}_2 \times \mathbf{a}_3. \tag{17•1–4}$$

The entire crystal may be viewed as being made up of a periodic repetition of such **primitive cells**, with neither gaps nor overlaps at the cell boundaries. In order for a set of translation vectors to be considered primitive, it is essential that the volume Ω_0 defined by them be the smallest possible volume from which the crystal can be constructed by periodic repetition of identical primitive cells without gaps or overlaps. It is an exercise in elementary geometry to show that the primitive cell volume of the fcc lattice has a volume one-quarter of the volume spanned by the cube edges in Fig. 17•1–1.

Strictly speaking, only infinitely large and perfect crystals would be truly invariant under translation by the lattice translation vectors (17•1–2a,b). Real crystals are finite and have surfaces and defects, so that (17•1–1) and (17•1–2) are not valid near the surfaces. However, the *bulk* properties of electrons deep inside a macroscopic crystal are governed by the *local* periodicity of the Hamiltonian, and for the study of these bulk properties, translational invariance is essential.

In our discussion of a one-dimensional periodic potential in chapters 14 and 15, we eliminated the irrelevant surface altogether, by pretending that the finite crystal formed an endless ring of circumference Ma, where M was the number of cells. The wave function in such an endless ring satisfies the **periodic boundary condition**

$$\psi(x + Ma) = \psi(x). \tag{17•1–5}$$

By imposing this boundary condition on the wave functions, we can account for the finite size of the crystal while retaining the translational periodicity of the Hamiltonian. The idea to treat the position coordinate as a cyclical coordinate may be extended to a three-dimensional periodic Hamiltonian as well, even though it can then no longer be visualized geometrically as a crystal bent to terminate upon itself. Hence, we generalize (17•1–5) to

$$\psi(\mathbf{r} + M\mathbf{a}_1) = \psi(\mathbf{r} + M\mathbf{a}_2) = \psi(\mathbf{r} + M\mathbf{a}_3) = \psi(\mathbf{r}), \tag{17•1–6}$$

where, for simplicity, we have assumed that the period in each of the three primitive translation directions contains the same number M of cells, corresponding to a crystal volume of M^3 cells.

We next associate with the three primitive translations three function transformation operators defined by[3]

$$\hat{T}_1\psi(\mathbf{r}) \equiv \psi(\mathbf{r} + \mathbf{a}_1), \tag{17•1–7a}$$

---

[3] Note that according to the general definition (16•1–12) of the function transformation operators, the three operators $\hat{T}$ correspond to translations by $-\mathbf{a}_1$, $-\mathbf{a}_2$, and $-\mathbf{a}_3$.

$$\hat{T}_2\psi(\mathbf{r}) \equiv \psi(\mathbf{r} + \mathbf{a}_2), \tag{17•1–7b}$$

$$\hat{T}_3\psi(\mathbf{r}) \equiv \psi(\mathbf{r} + \mathbf{a}_3). \tag{17•1–7c}$$

Because of the periodic boundary conditions (17•1–6), these operators act as cyclic operators of order M, and all our considerations in chapter 16 about cyclical operators apply.

The net result of successive translations does not depend on their order; therefore, the three operators commute. But this means that it must be possible to choose *all* energy eigenfunctions in such a way that they are also eigenfunctions of *all three* translation operators:

$$\hat{T}_1\psi(\mathbf{r}) = T_1\psi(\mathbf{r}), \text{ etc.} \tag{17•1–8}$$

We apply this result to a general lattice translation $\mathbf{T}$ of the form (17•1–2b). Because the wave function must be an eigenfunction of all three translation operators, we must have

$$\psi(\mathbf{r} + \mathbf{T}) = (T_1)^{u_1}(T_2)^{u_2}(T_3)^{u_3}\psi(\mathbf{r}). \tag{17•1–9}$$

All three symmetry eigenvalues must be of the form $\exp(i\alpha)$—see (16•3–13). Given three eigenvalues of this form, it is always possible to find a vector $\mathbf{k}$, called the **Bloch wave vector,** such that

$$\exp(i\mathbf{k} \cdot \mathbf{a}_1) = T_1, \tag{17•1–10a}$$

$$\exp(i\mathbf{k} \cdot \mathbf{a}_2) = T_2, \tag{17•1–10b}$$

$$\exp(i\mathbf{k} \cdot \mathbf{a}_3) = T_3. \tag{17•1–10c}$$

If these are inserted into (17•1–9), we obtain **Bloch's theorem,**

$$\boxed{\psi(\mathbf{k}, \mathbf{r} + \mathbf{T}) = \exp(i\mathbf{k} \cdot \mathbf{T})\psi(\mathbf{k}, \mathbf{r}),} \tag{17•1–11}$$

where we have added the designator $\mathbf{k}$ to the argument list of the wave function.

In words, under any displacement $\mathbf{T}$ that leaves the Hamiltonian invariant, the energy eigenfunctions are simply multiplied by phase factors of the form $\exp(i\mathbf{k} \cdot \mathbf{T})$, where $\mathbf{k}$ is a wave vector that is a characteristic property of that particular wave function.

The Bloch theorem can be brought into an alternative form by defining a **Bloch lattice function** $u(\mathbf{k}, \mathbf{r})$ via

$$\boxed{\psi(\mathbf{k}, \mathbf{r}) = \exp(i\mathbf{k} \cdot \mathbf{r})u(\mathbf{k}, \mathbf{r}),} \tag{17•1–12a}$$

where $u(\mathbf{k}, \mathbf{r})$ is periodic with the lattice periodicity,

$$u(\mathbf{k}, \mathbf{r} + \mathbf{T}) = u(\mathbf{k}, \mathbf{r}). \tag{17•1–12b}$$

Conversely, (17•1–11) follows from (17•1–12a) and (17•1–12b). Hence, (17•1–11) and (17•1–12) are equivalent formulations of Bloch's theorem.

17.1.2 k-Space

The mathematical language appropriate to the description of periodic functions is the language of Fourier analysis. The generalization of Fourier analysis to three-dimensional periodicity with oblique translation axes leads to the concept of a **reciprocal lattice** associated with each "direct" lattice.

As a preliminary, consider a simple plane wave of the form

$$f(\mathbf{r}) = \exp(i\mathbf{G} \cdot \mathbf{r}). \tag{17•1–13}$$

Such a plane wave will have lattice periodicity if the product $\mathbf{G} \cdot \mathbf{T}$ is an integer multiple of 2π for *all* lattice translations $\mathbf{T}$. If we write $\mathbf{T}$ in the form (17•1–2b), this condition becomes

$$\mathbf{G} \cdot \mathbf{T} = u_1\mathbf{G} \cdot \mathbf{a}_1 + u_2\mathbf{G} \cdot \mathbf{a}_2 + u_3\mathbf{G} \cdot \mathbf{a}_3 = 2\pi n, \tag{17•1–14}$$

where n is *any* integer. Inasmuch as the three integers u_1, u_2, and u_3 are independent of each other, each of the three products $\mathbf{G} \cdot \mathbf{a}_i$ must itself be an integer multiple of 2π,

$$\mathbf{G} \cdot \mathbf{a}_i = 2\pi v_i (i = 1, 2, 3), \tag{17•1–15}$$

where the v_i's are another set of independent integers.

We bring (17•1–15) into a different form by defining three vectors $\mathbf{b}_1$, $\mathbf{b}_2$, and $\mathbf{b}_3$, called the **reciprocal lattice vectors**, such that[4]

$$\mathbf{a}_i \cdot \mathbf{b}_j = 2\pi\delta_{ij}. \tag{17•1–16}$$

In words, each of the three **b**-vectors is perpendicular to the plane formed by *two* of the three **a**-vectors, and its length is such that the scalar product with the *third* **a**-vector is simply 2π. If $\mathbf{a}_1$, $\mathbf{a}_2$, and $\mathbf{a}_3$ are perpendicular to each other, the vectors $\mathbf{b}_1$, $\mathbf{b}_2$, and $\mathbf{b}_3$ will also be mutually perpendicular, and each will be parallel to one of the ordinary lattice vectors. But usually that will not be the case.

With the help of the reciprocal lattice vectors, we may express the condition (17•1–15) as the requirement that the vector $\mathbf{G}$ be of the form of a linear superposition of the three **b**-vectors, with integer coefficients,

$$\mathbf{G} = v_1\mathbf{b}_1 + v_2\mathbf{b}_2 + v_3\mathbf{b}_3. \tag{17•1–17}$$

This condition evidently defines a periodic lattice of points in a space defined by the three vectors $\mathbf{b}_1$, $\mathbf{b}_2$, and $\mathbf{b}_3$, similar to the way the three prim-

---

[4] Sometimes—especially in X-ray crystallography—the reciprocal lattice vectors are defined without the factor 2π.

itive translation vectors $\mathbf{a}_1$, $\mathbf{a}_2$, and $\mathbf{a}_3$ define a lattice in real space. Because the dimensions of all vectors in this new space are length^{-1}, this space is called **reciprocal space**, and the lattice of the **G**'s is called the **reciprocal lattice** associated with the real-space lattice. The reciprocal lattice vectors $\{\mathbf{b}_i\}$ are the primitive translation vectors of the reciprocal lattice.

In terms of the reciprocal lattice vectors, the three-dimensional generalization of Fourier analysis is of the form

$$f(\mathbf{r}) = \sum_{\mathbf{G}} A_{\mathbf{G}} \exp(i\mathbf{G} \cdot \mathbf{r}), \tag{17•1–18}$$

where the sum goes over all reciprocal lattice vectors and the $A_{\mathbf{G}}$ are suitable (complex) amplitudes.

The Bloch wave vector **k** is a evidently a vector with the dimension length^{-1}, and may therefore be viewed as a vector in reciprocal space. Hence, the latter is also often referred to as **k-space**, and that is the terminology we will use here.

According to Bloch's theorem, each stationary state of the electrons is represented by a specific point in **k**-space. The various physical properties of the electrons are therefore functions of the location of the state in **k**-space. Of particular importance are the eigenvalue of the energy and the expectation value of the velocity:

$$\mathcal{E} = \mathcal{E}(\mathbf{k}) \quad \text{and} \quad \langle \mathbf{v} \rangle = \mathbf{v}(\mathbf{k}). \tag{17•1–19a,b}$$

As we shall see, an understanding of the **k**-space behavior of electrons is the key to understanding the electron dynamics in real space.

17.1.3 Allowed k-Vectors

Bloch's theorem makes no statement about which **k**-vectors actually occur. In a finite crystal, the possible **k**-vectors are restricted by the requirement that the wave function satisfy the periodic boundary conditions (17•1–6). If we insert the Bloch form (17•1–12) of the wave function into these conditions, we obtain

$$\exp(iM\mathbf{k} \cdot \mathbf{a}_1) = \exp(iM\mathbf{k} \cdot \mathbf{a}_2) = \exp(iM\mathbf{k} \cdot \mathbf{a}_3) = 1. \tag{17•1–20}$$

This means that the three exponents in (17•1–20) must be integer multiples of $2\pi i$:

$$\mathbf{k} \cdot \mathbf{a}_i = \frac{2\pi}{M} n_i \qquad (i = 1, 2, 3), \tag{17•1–21}$$

where the three n's are independent arbitrary integers, positive, negative or zero.

With the aid of the reciprocal lattice vectors $\mathbf{b}_1$, $\mathbf{b}_2$, and $\mathbf{b}_3$, the condition (17•1–21) means that the Bloch wave vector **k** must be of the form

$$\mathbf{k} = (n_1\mathbf{b}_1 + n_2\mathbf{b}_2 + n_3\mathbf{b}_3)/M. \tag{17•1–22}$$

In words, the allowed states are uniformly distributed in **k**-space, with as many states per primitive cell of the *reciprocal* lattice as there are *real-space* primitive cells in the entire crystal. This is the first of several properties that make the **k**-space concept a simplifying central formal concept in solid-state theory.

17.1.4 The Reduced Zone

The three eigenvalues T_1, T_2, and T_3 in (17•1–10) and the condition (17•1–20) do not fully specify the vector **k**: It is always possible to add to **k** any of the reciprocal lattice vectors **G**: From (17•1–15), it follows that, if **k** satisfies (17•1–10) and (17•1–20), then **k** + **G** also does.

To eliminate this redundancy, we adopt the convention that the vector **k** be chosen as the *shortest* vector that satisfies Eqs. (17•1–10). Such a vector is called a **reduced** Bloch wave vector. The portion of **k**-space occupied by the reduced Bloch wave vectors is called the **reduced zone** of **k**-space, or the **first Brillouin zone**. If the three primitive translation vectors are perpendicular to each other, the reduced zone is simply the rectangular parallelepiped defined by

$$-\pi/a_x < k_x \le \pi/a_x, \tag{17•1–23a}$$

$$-\pi/a_y < k_y \le \pi/a_y, \tag{17•1–23b}$$

$$-\pi/a_z < k_z \le \pi/a_z, \tag{17•1–23c}$$

where we have chosen a Cartesian coordinate system in **k**-space.

For non-perpendicular primitive translation vectors, the reduced zone is a more complicated polyhedron in **k**-space, as is illustrated in **Fig. 17•1–2** for a simple two-dimensional example involving oblique axes. For the reduced zones of three-dimensional lattices, the reader is referred to texts on solid-state physics.

The requirement that the Bloch wave vector be the shortest vector possible that satisfies (17•1–10) removes the redundancy, unless **k** lies exactly on the surface of the reduced zone. In this case, there exists an alternative vector

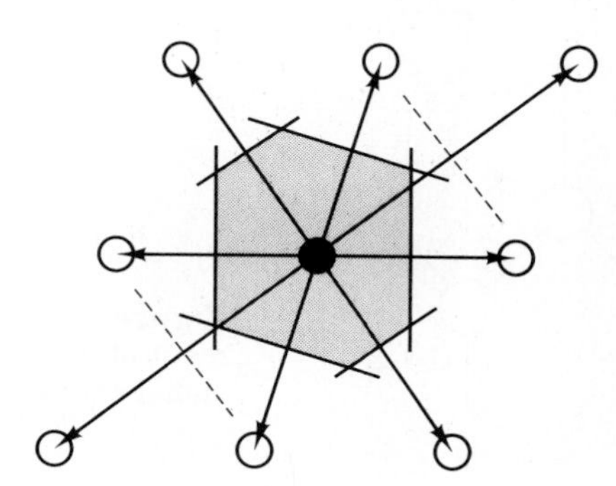

Figure 17•1–2. Reduced zone for a two-dimensional reciprocal lattice with oblique axes. The boundaries are obtained by drawing perpendicular bisectors to all nearby reciprocal lattice vectors and excluding all states outside the innermost set of bisectors.

k′ on the opposite surface, corresponding to the same state and differing from **k** by a reciprocal lattice vector. In (17•1–23), we have eliminated this remaining redundancy by including any allowed **k**-vectors on one of the two opposite surfaces, but not on both.

◆ PROBLEM TO SECTION 17.1

#17•1-1: Direct and Reciprocal Lattice and Reduced Zone, for a Simple Two-Dimensional Atomic Arrangement

Figure 17•1–3 shows the atomic arrangement of a fictitious two-dimensional "crystal."

(a) Using the "atom" shown in black as the origin of the lattice, indicate the primitive translation vectors of this lattice. (There are several possible choices; use the simplest one, and explain.) How many atoms are there per lattice point?

(b) Determine the reciprocal lattice vectors corresponding to this lattice, and make a quantitative graph of the reciprocal lattice. Assume that the grid shown in the figure is a 1Å grid. For the reciprocal lattice graph, use a scale such that the distance $2\pi/1$Å in reciprocal space shows as 3 inches on your drawing.

(c) Construct the reduced zone, and indicate it on your reciprocal-lattice plot.

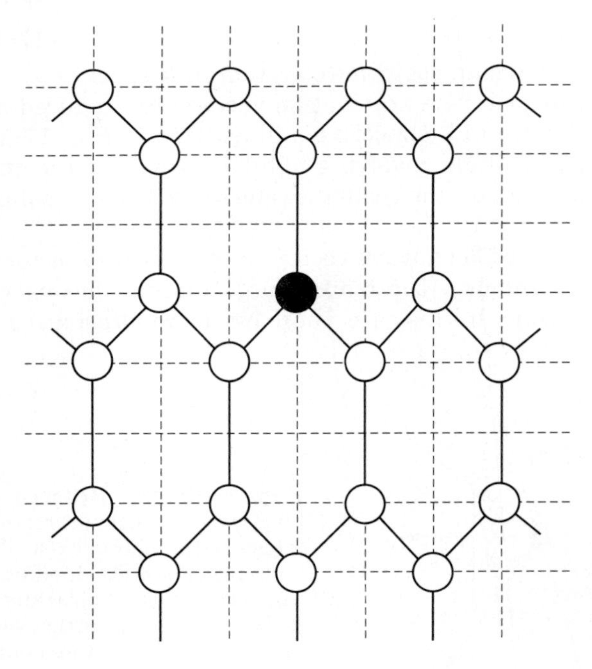

Figure 17•1–3. Atomic arrangement of a fictitious two-dimensional crystal.

17.2 ENERGY BANDS

17.2.1 An Example: The Cosine Potential Re-Visited

The most important consequence of the periodic potential is that the allowed energy eigenvalues of such a system occur in bands, with the possibility of forbidden gaps between the allowed bands. We observed this consequence already for the Kronig-Penney potential in chapter 6 and for the cosine potential in chapters 14 and 15. The overall result of the periodic potential in the cosine case was shown in Fig. 14•4–1, where we represented the allowed energies as a function of the full (unreduced) plane wave number k. In **Fig. 17•2–1**, we re-plot those earlier results in the more fundamental reduced-zone representation.[5]

In the second and third band, the heavy-line sections have been moved over by $-G$ from the extended-zone range $G/2 < k \leq 3G/2$, and the bottom of the fourth band has been moved over by $-2G$. Each of these sections has a mirror image, with the same energy but reversed k-value, shown as broken lines in the figure. These mirror images originate from the negative-k mirror images of the original extended-zone bands, omitted from Fig. 14•4–1.

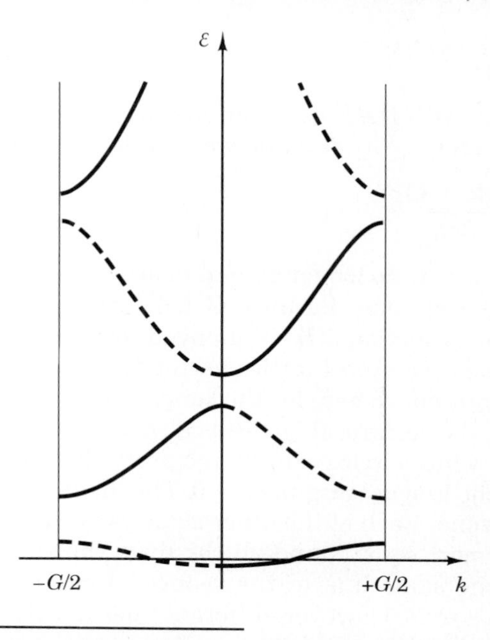

Figure 17•2–1. Energy bands for the cosine potential from Fig. 14•3–1, re-plotted in the reduced-zone representation.

[5] The earlier representations in terms of an unreduced free-electron plane wave vector are commonly called **extended-zone** representations. From our translational-invariance-based point of view, they are less fundamental than the reduced-zone representation; they are useful only in cases where the Bloch waves are treated as perturbed plane waves.

17.2.2 The Empty-Lattice Approximation

When dealing with two- or three-dimensional periodic potentials, it is useful to carry out the transformation to the reduced-zone representation *before* embarking on any perturbation calculations, as if the periodic potential had been turned off. This is called the **empty-lattice** approximation.

The electron wave functions are then simply plane waves of the form

$$\psi(\mathbf{r}) = \frac{1}{\sqrt{\Omega}} \exp(i\mathbf{K} \cdot \mathbf{r}), \tag{17•2–1}$$

where Ω is the volume of the crystal and where we have designated the wave vector of the free electron as $\mathbf{K}$, to distinguish it from the reduced Bloch wave vector $\mathbf{k}$. The plane waves are converted to Bloch wave form by splitting the free-electron wave vector into a reciprocal lattice vector $\mathbf{G}$ and a reduced Bloch wave vector "remainder,"

$$\mathbf{K} = \mathbf{G} + \mathbf{k}, \tag{17•2–2}$$

where $\mathbf{G}$ is chosen in such a way that $\mathbf{k}$ falls inside the reduced zone. If this is substituted into (17•2–1), we obtain the Bloch wave form (17•1–12), with

$$u(\mathbf{k}, \mathbf{r}) = \frac{1}{\sqrt{\Omega}} \exp(i\mathbf{G} \cdot \mathbf{r}), \tag{17•2–3}$$

which is clearly periodic with the lattice periodicity.

The energy of each empty-lattice plane-wave state may then be written as

$$\mathcal{E}(\mathbf{k}, \mathbf{G}) = \frac{\hbar^2(\mathbf{k} + \mathbf{G})^2}{2m_e}. \tag{17•2–4}$$

The Bloch wave vectors $\mathbf{k}$ now no longer extend to arbitrarily large values, but are confined to the reduced zone. Because $\mathbf{G}$ belongs to a discrete set, the original *single* continuous function $\mathcal{E}(\mathbf{K})$ is changed into a discrete *set* of functions $\mathcal{E}(\mathbf{k}, \mathbf{G})$, one for each reciprocal lattice vector $\mathbf{G}$.

We illustrate this procedure here for the simple case of a two-dimensional square lattice for which the reciprocal lattice is also a square lattice, shown in **Fig. 17•2–2**. The small white circles are the reciprocal lattice points, and the black circle represents the $\mathbf{k}$-space origin, $\mathbf{G} = 0$. The innermost square, shown in gray, is the reduced zone. Each of the other squares is shifted as a whole to coincide with the innermost square, and all the lines shown as double-ended arrows correspond to the same line in the reduced zone, namely, the cross section taken along the k_x-axis. However, different lines in extended space may correspond to different energies.

If we plot the energy along the double-arrow cross section as a function of reduced $\mathbf{k}$ for each of the nine squares shown, we obtain the plot of **Fig. 17•2–3**, showing nine parabolic sections. The sections shown as double lines are two-fold degenerate, representing pairwise equal energies for some of the squares in Fig. 17•2–2, at least along the cross section shown. It is left as an exercise

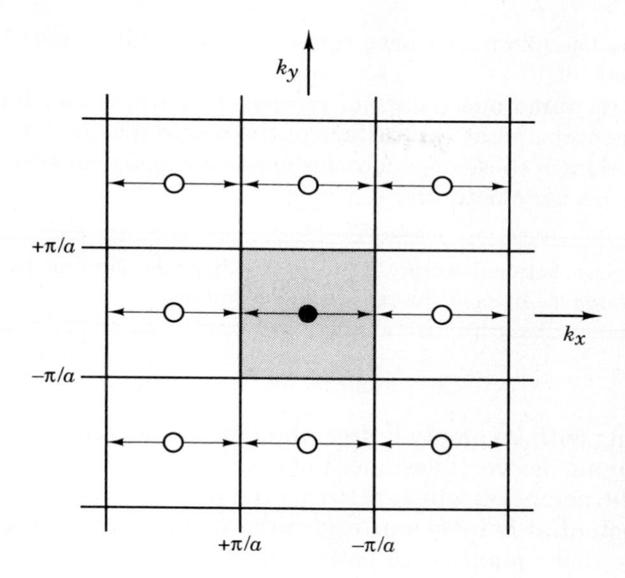

Figure 17•2–2. Two-dimensional extended **k**-space for a simple square lattice. The reduced zone is shown shaded. All extended **k**-space cross sections shown as double-ended arrows are projected onto the same cross section of the reduced zone.

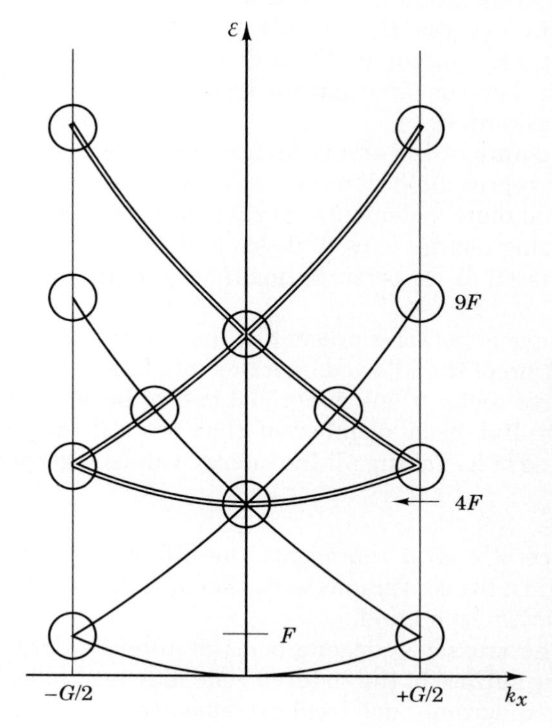

Figure 17•2–3. Cross-section along the k_x axis ($k_y = 0$) through the empty-lattice band structure for a simple square lattice. Only the lowest-energy branches are shown. The energy F is the quantity defined in (14•3–17). The degeneracies at the points circled are accidental degeneracies that will be split by a suitable periodic potential. The lines shown as heavier lines represent two states that are accidentally degenerate along the entire k_x axis.

for the reader to identify the extended-space square(s) from which each of the parabolic sections arises.

Note that there are numerous points of crossover of the empty-lattice bands, not only at the center and at the surface of the reduced zone, but also at intermediate points. Most of these degeneracies are accidental degeneracies, which will be split by a periodic potential.

Exercise: Construct a cross-sectional empty-lattice $\mathcal{E}(k)$ diagram similar to Fig. 17•2–3, taken along the *diagonal* lines of the reduced-zone square.

17.2.3 Energy Gaps

As long as we are dealing with an empty lattice, the reduced-zone representation is simply a bookkeeping device. It assumes physical significance as soon as we "turn on" the periodic potential, which perturbs the free-electron energies. Because the periodic potential is invariant under the translation operations introduced earlier, only those plane-wave pairs can perturb each other that belong to the same set of three eigenvalues of the three lattice translation operators; plane waves differing in one or more eigenvalues do not have mixed matrix elements and hence are not coupled by the potential. But all states having the same set of translation eigenvalues belong to the same reduced Bloch wave vector $\mathbf{k}$ and vice versa. Hence, only planes waves having the same reduced Bloch wave vector $\mathbf{k}$ can couple. Whether or not they will in fact couple depends on whether or not the appropriate spatial Fourier component is present in the periodic potential.

In particular, the points in the empty-lattice $\mathcal{E}(\mathbf{k})$ diagram where two or more bands cross over represent degeneracies between states that may be coupled by nonzero mixed matrix elements. In that case, the degeneracies will in general be split, causing energy gaps to develop at the points of crossover, similar to the gaps discussed in our perturbation theory treatments of chapters 14 and 15.

The splitting of the degeneracies present in the empty-lattice limit leads naturally to the separation of the allowed energies into bands. For each value of the reduced Bloch wave vector, there is not just one value of the energy, but an infinite discrete set. But each member of that set still depends quasi-continuously on the value of $\mathbf{k}$. Taking all the energy values together, we have

$$\mathcal{E} = \mathcal{E}_n(\mathbf{k}). \qquad (17\text{•}2\text{–}5)$$

The discrete dependence of $\mathcal{E}$ on n represents the different allowed **energy bands** of crystalline solids; the continuous variation of $\mathcal{E}_n(\mathbf{k})$ with $\mathbf{k}$ represents the energy variation *within* each band.

The splitting of the crossover degeneracies at intermediate points between the center and the surface of the reduced zone may lead to local minima and maxima in the $\mathcal{E}(\mathbf{k})$ diagram. Such local extrema are a common feature in the energy band structure of semiconductors, with important practical consequences. For example, silicon does not have a single conduction band minimum at $\mathbf{k} = 0$; rather, it has a set of six conduction band minima at a set of intermediate points along the cubic axes of **k**-space. Note that such crossover points cannot occur for one-dimensional periodic potentials, which *must* have their band extrema either at the center of the reduced zone or at its edge.

The quantitative determination of the energy band structure is an exercise in perturbation theory. In the immediate vicinity of the crossovers, the interaction between the crossing bands usually dominates the entire interaction, and the effect of other bands at the same **k**, but energetically farther away, tends to be much smaller and may be neglected in first-order calculations. If there are only two bands involved in the crossover, the perturbation treatment of the band splitting then reduces to the simple two-fold degenerate perturbation treatment discussed extensively in chapter 14. For higher order degeneracies, such as the one shown in Fig. 17•2–3 at $\mathcal{E} = 4F$, higher order degenerate perturbation theory may or may not be required, depending upon the nature of the degeneracy. Often the higher order secular determinant can be factored, as a result of rotation or reflection invariances that are present in addition to the translational invariances. Some of the degeneracies may be symmetry degeneracies, which remain unsplit.

Exercise: The "Bed-of-Nails" Potential. The simplest potential with which to illustrate the at least partial splitting of all degeneracies in the empty-lattice approximation is a weak (inverted) "bed-of-nails" potential in which a delta-function perturbation with strength $-S$ is placed at every lattice point, as a simple model of an atomic core potential. Treat the degeneracies by degenerate perturbation theory, including only those states that are actually degenerate with each other. Show that, right at each degeneracy, this treatment results in a formalism as in section 14.4.3, in which all matrix elements of the perturbation are the same, leading to the splitting-off of one state from the degeneracy. Make a qualitative sketch of how you would expect these splittings to change the cross-sectional $\mathcal{E}(k_x)$ diagram of Fig. 17•2–3 for the square lattice. Note that some of the lines in this figure represent degenerate pairs of states, which will themselves be split by the perturbation; these splittings must be included. Explain why the procedure breaks down away from most of the crossover points in the figure; make a best guess of which lines remain connected to each other at those points.

◆ PROBLEMS TO SECTION 17.2

#17•2-1: Splitting of Four-Fold Degeneracy of the Empty Square Lattice at $\mathcal{E} = 4F$

Drawing on the symmetry properties of the empty square lattice, re-arrange the four empty-lattice energy eigenfunctions for $\mathcal{E} = 4F$ into symmetry eigenfunctions, such that all four functions have different sets of symmetry eigenvalues. Which of the four states is split off immediately by the "bed-of-nails" perturbation, without having to invoke interactions with energetically lower or higher states? Identify the irreducible representations of the group of the square to which the three remaining states belong, and determine whether or not part of the remaining three-fold degeneracy at $\mathcal{E} = 4F$ is a symmetry degeneracy. Investigate the splitting of that remaining three-fold degeneracy by interaction with energetically lower or higher states, assuming an inverted "bed-of-nails" potential.

#17•2-2: Splitting of Crossover Degeneracy of the Empty Square Lattice at an Intermediate Point in k-Space

Investigate the splitting, by the inverted "bed-of-nails" potential, of the three-fold degeneracy at the energy at the crossover point about halfway between the zone center and the zone boundary shown in Fig. 17•2–3, including the vicinity of the point of exact degeneracy.

#17•2-3: Wannier Functions

In the theory of electrons in perturbed periodic potentials, it is often useful to expand the perturbed electron wave function as a linear superposition, not of the Bloch waves, but of the **Wannier functions**, which are defined as

$$w_n(\mathbf{T}, \mathbf{r}) = \frac{1}{\sqrt{N}} \sum_{\mathbf{k}} e^{-i\mathbf{k} \cdot \mathbf{T}} \psi_n(\mathbf{k}, \mathbf{r}). \tag{17•2–6}$$

Here, $\mathbf{T}$ is one of the lattice points of the crystal, N is the number of distinct lattice points (that is, the number of primitive cells in the crystal), n is the band index, and the sum goes over all allowed $\mathbf{k}$-values in the *reduced* zone. There is a different set of Wannier functions associated with each band, with one Wannier function per lattice point in each set. For every value of $\mathbf{k}$, the bands are counted strictly in order of increasing energy. Hence, when two free-electron branches cross each other in the empty-lattice approximation, the sections of a given free-electron branch on opposite sides of the crossing point get allocated to different bands.

(a) Construct and plot the Wannier functions for an arbitrary lattice point T for the lowest two bands of a one-dimensional crystal in the empty-lattice approximation, and show that those functions are localized about the point $x = T$.

(b) Show that the Wannier functions have the following properties (not just for an empty-lattice crystal, but generally):

(1) The Wannier functions associated with the lattice point $\mathbf{T}$ differ from those for $\mathbf{T} = 0$ only by a displacement by the vector $\mathbf{T}$.
(2) Wannier functions belonging to different lattice points or to different bands are orthogonal.
(3) The set of all Wannier functions forms a complete set, just like the set of Bloch waves.

17.3 ELECTRON DYNAMICS

17.3.1 Expectation Value of the Velocity

The function $\mathcal{E}(\mathbf{k})$ is the dominant quantity of the theory of electrons in crystals. Many of the properties of the electrons can be calculated from it, without further reference to the energy eigenfunctions.

Consider the expectation value of the velocity for Bloch waves. We recall that for linear superpositions of *ordinary* plane waves, the group velocity of the waves may be written in the form (1•3–1),

$$\mathbf{v} = \nabla_{\mathbf{k}}\omega \equiv \left(\frac{\partial\omega}{\partial k_x}, \frac{\partial\omega}{\partial k_y}, \frac{\partial\omega}{\partial k_z}\right), \tag{17•3–1}$$

where $\nabla_{\mathbf{k}}$ is the gradient operator in **k**-space. With the help of the PEdB relation $\mathcal{E} = \hbar\omega$, (17•3–1) may also be written as

$$\boxed{\mathbf{v} = \frac{1}{\hbar}\nabla_{\mathbf{k}}\mathcal{E}(\mathbf{k}).} \tag{17•3–2}$$

The latter form remains valid for the expectation value of the velocity for Bloch waves, provided that **k** is understood to be the Bloch wave vector, rather than an ordinary plane-wave vector. The simplicity of the form (17•3–2) for the group velocity of the energy eigenstates is a second reason for the importance of **k**-space in the physics of electrons in crystals.

To derive (17•3–2), we note first that the expectation value of the velocity for a normalized state with the wave function ψ is simply the integral of the probability current density, taken over all space:

$$\mathbf{v} \equiv \langle\mathbf{v}\rangle = \int \mathbf{j}\, d^3r = -\frac{i\hbar}{2m_e}\int(\psi^*\nabla\psi - \psi\nabla\psi^*)\, d^3r. \tag{17•3–3}$$

This expression may be transformed into the form (17•3–2) by the following trick. We write the Schroedinger equation for Bloch waves in the form

$$-\frac{\hbar^2}{2m_e}\nabla^2\psi = [\mathcal{E}(\mathbf{k}) - V]\psi \tag{17•3–4}$$

and take the gradient in **k**-space. The result is of the form

$$-\frac{\hbar^2}{2m_e}\nabla^2\boldsymbol{\phi} = [\mathcal{E}(\mathbf{k}) - V]\boldsymbol{\phi} + [\nabla_{\mathbf{k}}\mathcal{E}(\mathbf{k})]\psi, \tag{17•3–5}$$

where $\boldsymbol{\phi}$ is defined as

$$\boldsymbol{\phi} = \nabla_{\mathbf{k}}[\exp(i\mathbf{k}\cdot\mathbf{r})u(\mathbf{k},\mathbf{r})] = i\mathbf{r}\psi + \boldsymbol{\chi}, \tag{17•3–6a}$$

with

$$\chi = \exp(i\mathbf{k}\cdot\mathbf{r})\nabla_{\mathbf{k}}u(\mathbf{k},\mathbf{r}). \tag{17•3–6b}$$

For compactness, we have omitted the dependence on $\mathbf{k}$ and $\mathbf{r}$ in the notation for ψ, $\boldsymbol{\phi}$, and $\boldsymbol{\chi}$. Note that χ is itself of the form of a Bloch wave.

If we insert (17•3–6a) on both sides of (17•3–5), we obtain

$$-\frac{\hbar^2}{2m_e}\cdot[2i\nabla\psi + i\mathbf{r}\nabla^2\psi + \nabla^2\boldsymbol{\chi}] = [\mathcal{E}(\mathbf{k}) - V][i\mathbf{r}\psi + \boldsymbol{\chi}] + [\nabla_{\mathbf{k}}\mathcal{E}(\mathbf{k})]\psi. \tag{17•3–7}$$

Because ψ is a solution of (17•3–4) with the eigenvalue $\mathcal{E}(\mathbf{k})$, the troublesome $i\mathbf{r}\psi$-terms[6] cancel out, and the remaining terms may be re-arranged to read

$$-\frac{i\hbar^2}{m_e}\cdot\nabla\psi = [\mathcal{E}(\mathbf{k}) - \hat{H}]\boldsymbol{\chi} + [\nabla_{\mathbf{k}}\mathcal{E}(\mathbf{k})]\psi. \tag{17•3–8}$$

We next multiply with ψ^* and integrate:

$$-\frac{i\hbar^2}{m_e}\cdot\int\psi^*\nabla\psi\,d^3r = \int\psi^*[\mathcal{E}(\mathbf{k}) - \hat{H}]\boldsymbol{\chi}\,d^3r + [\nabla_{\mathbf{k}}\mathcal{E}(\mathbf{k})]\int\psi^*\psi\,d^3r. \tag{17•3–9}$$

By letting the Hermitian operator $\hat{H}$ operate to the left, and drawing once more on the eigenfunction properties of ψ, the first integral on the right-hand side is seen to vanish. The second integral is simply the normalization integral. We thus obtain

$$-\frac{i\hbar^2}{m_e}\cdot\int\psi^*\nabla\psi\,d^3r = \nabla_{\mathbf{k}}\mathcal{E}(\mathbf{k}). \tag{17•3–10}$$

Insertion of this result and its complex conjugate into (17•3–3) shows that (17•3–2) indeed remains valid.

The relation (17•3–10) may be re-written in terms of the Bloch lattice functions $u(\mathbf{k},\mathbf{r})$, rather than the full Bloch waves $\psi(\mathbf{k},\mathbf{r})$:

$$\frac{m_e}{\hbar}\mathbf{v} = \frac{m_e}{\hbar^2}\nabla_{\mathbf{k}}\mathcal{E}(\mathbf{k}) = \mathbf{k} - i\int u^*\nabla u\,d^3r. \tag{17•3–11}$$

The integral represents the deviation from the behavior of an ordinary plane wave with wave vector $\mathbf{k}$.

17.3.2 Newton's Law in k-Space

With the help of the PEdB relation $\mathbf{p} = \hbar\mathbf{k}$, Newton's second law for classical particles in free space,

---

[6] The product $\mathbf{r}\psi_{\mathbf{k}}$ does not satisfy periodic boundary conditions and hence does not belong to the hermiticity domain of the crystal Hamiltonian.

$$\frac{d\mathbf{p}}{dt} = \mathbf{F}, \tag{17•3–12}$$

may be also written in the form

$$\boxed{\hbar \frac{d\mathbf{k}}{dt} = \mathbf{F}.} \tag{17•3–13}$$

Again, this second form remains valid—in a certain sense—for the time evolution of Bloch waves under the influence of a uniform force, provided that **k** is understood to be the reduced Bloch wave vector. More specifically, we show here the following: If $\Psi(\mathbf{r}, t)$ starts out as a function belonging to a sharp value of the reduced Bloch wave vector **k**, then it will remain a function belonging to a sharp value of the reduced Bloch wave vector **k**, but this **k**-vector itself evolves with time according to (17•3–13), until it reaches the edge of the reduced zone.

We consider a periodic potential with Hamiltonian $\hat{H}_0$ to which a uniform external force **F** has been added. The total Hamiltonian is then

$$\hat{H} = \hat{H}_0 - \mathbf{F} \cdot \mathbf{r}, \tag{17•3–14}$$

and the problem of determining the electron dynamics is that of solving the Schroedinger wave equation

$$i\hbar \frac{\partial \Psi}{\partial t} = [\hat{H}_0 - \mathbf{F} \cdot \mathbf{r}]\Psi. \tag{17•3–15}$$

We proceed by considering the time derivatives of the expectation values of the three lattice translation operators $\hat{T}_1$, $\hat{T}_2$, and $\hat{T}_3$, defined in Eqs.(17•1–7). From the law for the time derivatives of expectation values, which remains valid for non-Hermitian operators such as the three $\hat{T}$'s, we find, for $\langle T_1 \rangle$,

$$\frac{d}{dt}\langle T_1 \rangle = \frac{i}{\hbar}\langle [\hat{H}, \hat{T}_1] \rangle. \tag{17•3–16}$$

On the right-hand side, the periodic part $\hat{H}_0$ of the Hamiltonian commutes with the translation operator. Without loss of generality, we may choose the direction of the primitive displacement vector $\mathbf{a}_1$ as the x-direction, in which case the commutator reduces to

$$[\hat{H}, \hat{T}_1] = -F_x[x, \hat{T}_1] = F_x a_1 \hat{T}_1. \tag{17•3–17}$$

Hence, (17•3–16) becomes simply

$$\frac{d}{dt}\langle T(t) \rangle = \frac{iF_x a_1}{\hbar}\langle T(t) \rangle. \tag{17•3–18}$$

This is a differential equation for $T(t)$, the most general solution of which may be written in the form

$$\langle T(t) \rangle = T_0 \exp[ik_x(t)a_1], \tag{17•3–19}$$

where $k_x(t)$ satisfies (17•3–13).

Equation (17•3–19) describes a motion of the expectation value of $\hat{T}_1$ along a circle in the complex plane, with the time-independent radius $|T_0|$. Suppose now that at $t = 0$ the state of the electron was a pure Bloch wave from a single band, with the wave vector component $\mathbf{k}_x = \mathbf{k}_{x0}$. Such a Bloch wave is an eigenstate of $\hat{T}_1$, with the eigenvalue $\exp(ik_{x0}a)$, located on the unit circle in the complex T-plane, corresponding to $|T_0| = 1$. But in that case, (17•3–19) states that the expectation value of $\hat{T}_1$ will remain on the unit circle. Analogous arguments apply to the other components of $\mathbf{k}$.

The reader can easily confirm that a linear superposition of Bloch waves with *different* reduced Bloch wave vectors has an expectation value of $\hat{T}_1$ located *inside* the unit circle in the complex T-plane. All states with T-values *on* the unit circle are either pure Bloch waves from a single band or linear superpositions of Bloch waves from different bands, but all with the same reduced Bloch wave vector. Hence, (17•3–19) means that (17•3–13) is an *exact* law for the dynamics of the state of the electron *within* each of the participating bands. The only complication that can arise consists of transitions into other bands. Even when such transitions take place, however, they are always into Bloch states with the same (time-dependent) reduced Bloch wave vector as in the original band.

A study of these field-induced interband transitions, which lies outside the scope of this text, shows that the transition rate remains negligibly small, except for very large fields—typically on the order of 10^6 V/cm. To the extent that such transitions into other bands can be ignored, a very simple picture emerges from our calculations: An electron initially in a state with the sharp Bloch wave vector $\mathbf{k}(0)$ retains a sharp wave vector, but this wave vector is itself time-dependent, moving with the constant speed $\mathbf{F}/\hbar$ through $\mathbf{k}$-space. We have here a third example of the important role that $\mathbf{k}$-space arguments play in the theory of crystalline solids: the $\mathbf{k}$-space motion is much simpler than the motion in real space.

Because of (17•3–13), the quantity

$$\mathbf{P} \equiv \hbar\mathbf{k} \tag{17•3–20}$$

is often called the **crystal momentum**.

What happens when the electron reaches the edge of the reduced zone? Recall that points located perpendicularly opposite to each other on the edges of the reduced zone and differing from each other by a reciprocal lattice vector $\mathbf{G}$ really represent merely two different descriptions of the same state with the same wave function. But this means that an electron leaving the reduced zone at one point simply re-enters it at the point perpendicularly opposite. Such processes are called **umklapp processes**;[7] they play an important role in the description of electron dynamics in solids.

[7] From the German verb *umklappen,* to flip over.

The uniform speed of motion through **k**-space, together with the uniform distribution of states in **k**-space, implies that the entire electron distribution moves rigidly. Questions of more than one electron being squeezed into one state and thereby violating the Pauli exclusion principle never arise.

17.3.3 Effective Mass

By combining (17•3–2) with (17•3–13), we obtain a relation for the acceleration in real space. For the x-component of the acceleration,

$$\begin{aligned}\frac{d}{dt}v_x &= \frac{1}{\hbar}\frac{d}{dt}\left(\frac{\partial \mathcal{E}}{\partial k_x}\right)\\ &= \frac{1}{\hbar}\left[\frac{\partial^2 \mathcal{E}}{\partial k_x^2}\frac{dk_x}{dt} + \frac{\partial^2 \mathcal{E}}{\partial k_x \partial k_y}\frac{dk_y}{dt} + \frac{\partial^2 \mathcal{E}}{\partial k_x \partial k_z}\frac{dk_z}{dt}\right]\\ &= \frac{1}{\hbar^2}\left[\frac{\partial^2 \mathcal{E}}{\partial k_x^2}F_x + \frac{\partial^2 \mathcal{E}}{\partial k_x \partial k_y}F_y + \frac{\partial^2 \mathcal{E}}{\partial k_x \partial k_z}F_z\right].\end{aligned} \tag{17•3–21}$$

The last line may be written

$$\frac{d}{dt}v_x = \frac{F_x}{m_{xx}} + \frac{F_y}{m_{xy}} + \frac{F_z}{m_{xz}}, \tag{17•3–22}$$

where the quantities $1/m_{vw}$, defined by

$$\frac{1}{m_{vw}} = \frac{1}{\hbar^2}\frac{\partial^2 \mathcal{E}}{\partial k_v \partial k_w} \qquad (v,\, w = x,\, y,\, z), \tag{17•3–23}$$

are three components of a tensor, called the **effective mass tensor**.

Expressions cyclically analogous to (17•3–22) are obtained for the y- and z-components of the acceleration. The three components can be lumped together in the form

$$\frac{d\mathbf{v}}{dt} = \left\|\frac{\mathbf{1}}{\mathbf{m}}\right\|\mathbf{F}, \tag{17•3–24}$$

where

$$\left\|\frac{\mathbf{1}}{\mathbf{m}}\right\| \equiv \left\|\begin{matrix}\frac{1}{m_{xx}} & \frac{1}{m_{xy}} & \frac{1}{m_{xz}}\\ \frac{1}{m_{yx}} & \frac{1}{m_{yy}} & \frac{1}{m_{yz}}\\ \frac{1}{m_{zx}} & \frac{1}{m_{zy}} & \frac{1}{m_{zz}}\end{matrix}\right\| \tag{17·3–25}$$

is the effective mass tensor.

Like the group velocity of the Bloch electrons, the elements (17•3–23) of the effective mass tensor may be expressed in terms of integrals involving the

Bloch lattice functions $u(\mathbf{k},\mathbf{r})$. By differentiating (17•3–11), we find, for m_{xx},

$$\frac{m_e}{m_{xx}} = \frac{m_e}{\hbar^2}\frac{\partial^2 \mathcal{E}}{\partial k_x^2} = i\int\left(\frac{\partial u^*}{\partial x}\frac{\partial u}{\partial k_x} - \frac{\partial u}{\partial x}\frac{\partial u^*}{\partial k_x}\right) d^3r, \tag{17•3–26}$$

with analogous expressions for the other diagonal elements. The off-diagonal elements, such as m_{xy}, may be expressed in two different ways,

$$\frac{m_e}{m_{xy}} = \frac{m_e}{\hbar^2}\frac{\partial^2 \mathcal{E}}{\partial k_x \partial k_y} = i\int\left(\frac{\partial u^*}{\partial x}\frac{\partial u}{\partial k_y} - \frac{\partial u}{\partial x}\frac{\partial u^*}{\partial k_y}\right) d^3r$$

$$= i\int\left(\frac{\partial u^*}{\partial y}\frac{\partial u}{\partial k_x} - \frac{\partial u}{\partial y}\frac{\partial u^*}{\partial k_x}\right) d^3r, \tag{17•3–27}$$

again with cyclic equivalents. The proof is left to the reader.

17.3.4 Effective Mass and Ehrenfest's Theorem

Our above treatment of the effective-mass concept was purely formal, giving little insight into the underlying physics. A better insight is obtained by viewing the acceleration of a crystal electron from the point of view of Ehrenfest's theorem (7•6–18). There are two forces present: the crystal force exerted by the periodic crystal potential $V(\mathbf{r})$,

$$\mathbf{F} = -\nabla V, \tag{17•3–28}$$

and the force $-e\mathbf{E}$ due to the externally applied electric field, assumed here to be spatially uniform. This leads to Ehrenfest's theorem for the x-component of the acceleration,

$$m_e \frac{d}{dt}\langle v_x\rangle = \langle\psi|F_x - eE_x|\psi\rangle = \langle\psi|F_x|\psi\rangle - eE_x, \tag{17•3–29}$$

where $|\psi\rangle$ is the state of the electron *in the presence of the external field.* Expressions analogous to (17•3–29) hold for the other acceleration components.

The key point is now that the state $|\psi\rangle$ is not the unperturbed state $|\psi_0\rangle$ that would be a solution of the Schroedinger wave equation in the absence of the external field, but is a state that has been "polarized" by the field, as evidenced by the non-unity dielectric constant of the crystal.

In the absence of any external field, and hence, in the absence of any polarization, the acceleration is of course zero:

$$\langle\psi_0|\mathbf{F}|\psi_0\rangle = 0. \tag{17•3–30}$$

This means that, if polarization were absent in the presence of an external field, (17•3–29) would reduce to Newton's law for a free electron,

$$m_e \frac{d}{dt} \langle \mathbf{v} \rangle = -e\mathbf{E}. \tag{17•3–31}$$

Any deviations from this behavior are caused by the polarization of the wave function and the interaction of the polarization with the periodic crystal potential.

In the spirit of non-degenerate perturbation theory, we expand the state $|\psi\rangle$ into a Taylor series of the form,

$$|\psi\rangle = |\psi_0\rangle - [|\chi_x\rangle eE_x + |\chi_y\rangle eE_y + |\chi_z\rangle eE_z] + O(E^2), \tag{17•3–32}$$

where the three $|\chi\rangle$'s are measures of the polarizability of the electron wave function by the three Cartesian field components. By writing (17•3–32) in terms of three different $|\chi\rangle$'s we make allowance for anisotropy of the polarization. An expansion of the form (17•3–32) should certainly be valid in the limit of small fields.

If we insert (17•3–32) into (17•3–29) and drop all terms higher than linear in the components of $\mathbf{E}$, we obtain

$$\begin{aligned} m_e \frac{d}{dt} \langle v_x \rangle = & -[1 + \langle \psi_0 | F_x | \chi_x \rangle + \langle \chi_x | F_x | \psi_0 \rangle] eE_x \\ & -[\langle \psi_0 | F_x | \chi_y \rangle + \langle \chi_y | F_x | \psi_0 \rangle] eE_y \\ & -[\langle \psi_0 | F_x | \chi_z \rangle + \langle \chi_z | F_x | \psi_0 \rangle] eE_z. \end{aligned} \tag{17•3–33}$$

This relation is of the form of Newton's law with an effective mass tensor $\|1/m^*\|$ if we set

$$\frac{m_e}{m_{xx}} = 1 + \langle \psi_0 | F_x | \chi_x \rangle + \langle \chi_x | F_x | \psi_0 \rangle, \tag{17•3–34a}$$

$$\frac{m_e}{m_{xy}} = \langle \psi_0 | F_x | \chi_y \rangle + \langle \chi_y | F_x | \psi_0 \rangle, \tag{17•3–34b}$$

$$\frac{m_e}{m_{xz}} = \langle \psi_0 | F_x | \chi_z \rangle + \langle \chi_z | F_x | \psi_0 \rangle. \tag{17•3–34c}$$

Analogous relations hold for the other acceleration directions.

By executing the perturbation treatment implied in (17•3–32), it is possible to show that the different components of the inverse-mass tensor are indeed given by the relations (17•3–26) and (17•3–27). The derivation draws on procedures developed later, in section 17.5; see also the problems to that section.

Our presentation shows that the physics behind the effective mass is really that of an *effective force* that is *proportional* to the externally applied force, but not *equal* to it, nor necessarily in the same direction.

◆ PROBLEM TO SECTION 17.3

#17•3-1: Collision-Less Electron Dynamics in k-Space and Real Space

The function $\mathcal{E}(\mathbf{k})$ for the electrons in the lowest band of a fictitious two-dimensional crystal with a square lattice (with lattice constant a) has the simple form

$$\mathcal{E}(k_x, k_y) = -\mathcal{E}_0 \cdot [\cos k_x a + \cos k_y a]. \tag{17•3–35}$$

(a) Make a semi-quantitative graph of the contours of constant energy in the reduced zone, for at least the three energies $\mathcal{E} = 0, \pm\mathcal{E}_0$.

(b) An electron is initially located at $\mathbf{k} = 0$ in $\mathbf{k}$-space and at $\mathbf{r} = 0$ in real space. At $t = 0$, a force $\mathbf{F}$ is turned on which points in an oblique direction, such that $F_x = 2F_y$. Show the trajectory of the electron through the reduced zone, including umklapp processes, for the time interval $0 \le t \le 4T$, where

$$T = \frac{\pi\hbar}{aF_x}. \tag{17•3–36}$$

(c) Calculate and plot the x- and y-components of the velocity and of the position of the electron, all as functions of time, also for $0 \le t \le 4T$.

(d) Make a graph of the trajectory of the electron in the x-y plane of real space.

17.4 THE SYMMETRY OF k-SPACE

17.4.1 Inversion Symmetry

In the absence of vector potentials, the time-independent Schroedinger equation is a real equation. If $\psi(\mathbf{r})$ is a solution, then $\psi^*(\mathbf{r})$ must also be a solution, with the same energy eigenvalue $\mathcal{E}$. But according to Bloch's theorem, if $\psi(\mathbf{r})$ belongs to the Bloch wave vector $\mathbf{k}$, then $\psi^*(\mathbf{r})$ belongs to $-\mathbf{k}$. Thus, we see that, if $\mathcal{E} = \mathcal{E}(\mathbf{k})$ is the energy of a state with the Bloch wave vector $\mathbf{k}$, there must exist another state with the opposite Bloch wave vector but the same energy,

$$\boxed{\mathcal{E}(-\mathbf{k}) = \mathcal{E}(\mathbf{k}),} \tag{17•4–1}$$

and with the corresponding Bloch lattice functions,

$$u(-\mathbf{k}, \mathbf{r}) = u^*(\mathbf{k}, \mathbf{r}). \tag{17•4–2}$$

The properties (17•4–1) and (17•4–2) are sometimes referred to as **inversion symmetry** of $\mathbf{k}$-space. Note that this symmetry is present even if the crystal potential itself does not have a center of inversion.

17.4.2 Rotational Symmetry

Almost all crystal structures contain rotational invariances in addition to their translational periodicities. As an example, consider the atomic arrangement in a GaAs crystal. The basic building block is a tetrahedron, formed by, say, a Ga atom at the center and its four nearest As neighbors at the corners (or vice versa), as shown in **Fig. 17•4–1**. All tetrahedra have the same orientation in space. In silicon, the two kinds of atoms are the same. The four outer atoms may be viewed as occupying four of the eight corners of the cube shown. The cubes themselves are arranged into a half-filled face-centered cubic lattice (**Fig. 17•4–2**), in such a way that the corner atoms in Fig. 17•4–1 are shared by the four cubes that come together at that corner.

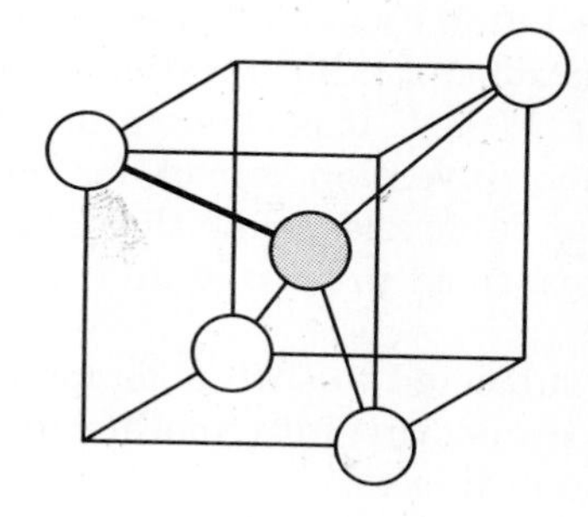

Figure 17•4–1. Structural building block of the crystal structure of semiconductors such as Si and GaAs. Each atom forms four bonds with its nearest neighbors. In III-V compounds such as GaAs, alternate atoms, shown as white and gray, belong to the third and fifth columns of the periodic table; in Si, the atoms are of the same kind.

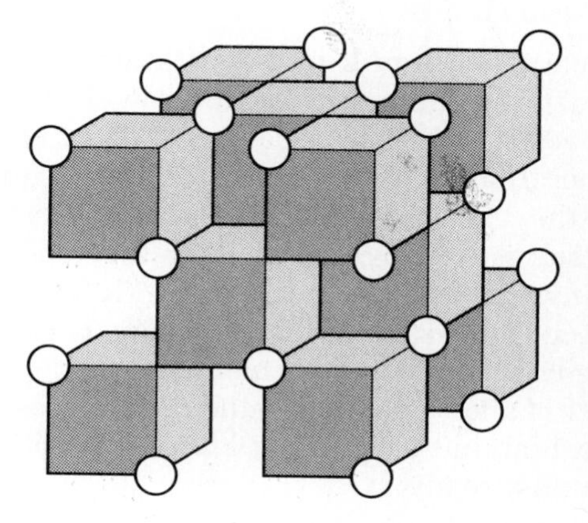

Figure 17•4–2. Overall crystal structure, made up from the building blocks shown in Fig. 17•4–1. The tetrahedra themselves are arranged in a face-centered cubic lattice. Only the atoms at each cube corner are shown, each shared by the four cubes coming together at that corner.

Let us single out one of these tetrahedra. Any rotation of this particular tetrahedron into itself also turns the atomic configuration of the entire crystal into itself. Hence, it is a symmetry operation under which the Hamiltonian is invariant. There are two classes of such rotations:

(a) Rotations by $\pm 120^\circ$ about any of the four body diagonals of the cube in Fig. 17•4–1a.

(b) Rotations by 180° (but not by 90°!) about any of the three axes that go through the center of the cube and that are perpendicular to the cube faces.

Any energy eigenfunction rotated by one of these invariance operations must again be an energy eigenfunction with the same energy eigenvalue. Applied to the Bloch waves, this means the following: If the Bloch wave vector $\mathbf{k}$ is not parallel to the axis of rotation, rotating the wave function will lead to another Bloch wave with a different wave vector $\mathbf{k}'$, obtained from the original wave vector by this rotation. But this means that the energy function $\mathcal{E}_n(\mathbf{k})$ must be invariant under all rotations under which the crystal is invariant.

But this is not all: In addition to simple rotations, many crystal structures contain **compound symmetry operations** that combine a rotation with either an inversion or a (non-primitive) translation, in such a way that neither the rotation nor the translation by itself is a symmetry operation. We consider here specifically combinations of a 90° rotation with the inversion about the center of the cube in Fig. 17•4–1. It is clear from inspection that the tetrahedron is *not* invariant under inversion, nor under rotations by $\pm 90°$ about any of the three perpendicular cubic axes going through the cube center. However, if *both* operations are performed in succession (in either order), the structure turns into itself.

If a Bloch wave is subjected to such a compound operation, the periodic part $u(\mathbf{k},\mathbf{r})$ of the wave transforms into another periodic function. The factor $\exp(i\mathbf{k} \cdot \mathbf{r})$ transforms according to

$$\exp(i\mathbf{k} \cdot \mathbf{r}) \rightarrow \exp(i\mathbf{k}' \cdot \mathbf{r}), \tag{17•4–3}$$

where $\mathbf{k}'$ is the vector obtained from $\mathbf{k}$ by the rotation alone. Evidently, the entire Bloch wave is transformed into another function of Bloch wave form, but with the rotated wave vector $\mathbf{k}'$. Because the energy must be invariant under the overall symmetry operation, this means that the energy function $\mathcal{E}(\mathbf{k})$ must also be invariant under the $\pm 90°$ rotation that leads from $\mathbf{k}$ to $\mathbf{k}'$, even though the crystal structure itself is not invariant under that operation alone. Together with the earlier invariances, this means that $\mathcal{E}(\mathbf{k})$ is invariant under *all* rotations that convert a cube into itself, which is a higher rotational symmetry than that of the tetrahedral atomic arrangement itself.

There are a number of additional compound symmetries under which the overall structure is invariant, but they do not lead any additional invariances of $\mathcal{E}(\mathbf{k})$ beyond those found already.

17.5 k · p THEORY

17.5.1 Schroedinger-Bloch Equation

If we insert the energy eigenfunctions in Bloch form (17•1–12a) into the Schroedinger equation and multiply the result by $\exp(-i\mathbf{k} \cdot \mathbf{r})$, we obtain an equation for the Bloch lattice functions,

$$\hat{B}(\mathbf{k})u(\mathbf{k}, \mathbf{r}) = \mathcal{E}(\mathbf{k})u(\mathbf{k}, \mathbf{r}), \tag{17•5–1a}$$

where

$$\hat{B}(\mathbf{k}) \equiv \exp(-i\mathbf{k}\cdot\mathbf{r})\hat{H}\exp(+i\mathbf{k}\cdot\mathbf{r})$$

$$= \hat{H} + \frac{\hbar}{m_e}\mathbf{k}\cdot\mathbf{p} + \frac{\hbar^2k^2}{2m_e} = \hat{H} - \frac{i\hbar^2}{m_e}\mathbf{k}\cdot\nabla + \frac{\hbar^2k^2}{2m_e}. \qquad (17\text{•}5\text{–}1\text{b})$$

We call (17•5–1) the **Schroedinger-Bloch equation**. The reader is asked to supply the missing steps in (17•5–1b) and to show that the operator $\hat{B}(\mathbf{k})$ is Hermitian.

The requirement that the Bloch lattice functions $u(\mathbf{k}, \mathbf{r})$ be periodic with the lattice periodicity makes the eigenvalue problem contained in (17•5–1) a boundary value problem, just as if we had requested that the wave function vanish at the surface of each cell (but of course, with different eigenfunctions and eigenvalues). For a given value of $\mathbf{k}$, (17•5–1) has a periodic solution only for selected values $\mathcal{E}_n$ of the energy $\mathcal{E}$. As we saw already from the empty-lattice limit, for each value of $\mathbf{k}$ there is an infinite number of such energy eigenvalues.

Let us single out a vector $\mathbf{k}$ and keep it fixed. We consider then the infinite set of functions $u_n(\mathbf{k}, \mathbf{r})$, viewed as functions of $\mathbf{r}$ alone. As eigenfunctions of a Hermitian operator, these functions form an orthogonal set, and they can be normalized. Furthermore, they form a *complete* set, in the restricted sense that any function with lattice periodicity can be expanded in terms of the Bloch lattice functions for any selected $\mathbf{k}$. If we designate the set of functions $u_n(\mathbf{k}, \mathbf{r})$ for a common value of $\mathbf{k}$ by the simple ket symbol $|n\rangle$, this completeness means that, for operations within any set of lattice-periodic functions, the projection operators $|n\rangle\langle n|$ obey the closure relation familiar from section 13.3:

$$\sum_n |n\rangle\langle n| = \hat{1}. \qquad (17\text{•}5\text{–}2)$$

In particular, the Bloch lattice functions for one value of $\mathbf{k}$ may be expanded in terms of those for another value of $\mathbf{k}$. Expansions in terms of the set for $\mathbf{k} = 0$ are of particular interest, because for $\mathbf{k} = 0$ the Bloch waves and the Bloch lattice functions are the same. Such expansions are widely used in the theory of solids.

17.5.2 k · p Perturbation Theory

The Bloch operator has the form of an unperturbed Hamiltonian $\hat{H}$ plus a perturbation

$$\hat{W} = \frac{\hbar}{m_e}\mathbf{k}\cdot\mathbf{p} + \frac{\hbar^2k^2}{2m_e} = -\frac{i\hbar^2}{m_e}\mathbf{k}\cdot\nabla + \frac{\hbar^2k^2}{2m_e}. \qquad (17\text{•}5\text{–}3)$$

This implies that the Bloch waves *in the vicinity* of the point $\mathbf{k} = 0$ may be obtained by perturbation theory from those *right at* $\mathbf{k} = 0$. The k^2-term in (17•5–3) causes a trivial energy shift; the non-trivial term is the $\mathbf{k}\cdot\mathbf{p}$ term. Because of its form, this kind of perturbation treatment is commonly called

k · p theory. Inasmuch as the point $\mathbf{k} = 0$ is a point where the energy bands tend to have maxima or minima, $\mathbf{k} \cdot \mathbf{p}$ theory gives important information about the vicinity of the band edges, including information specifically about the effective masses at and near the edges.

We restrict ourselves here to the simplest case, in which the periodic potential has removed all degeneracies at $k = 0$. In that case, we may treat the two perturbation terms in (17•5–3) by second-order non-degenerate Brillouin-Wigner perturbation theory. The first-order correction to the energy of band n is evidently

$$\Delta\mathcal{E}^{(1)} = W_{nn} = \frac{\hbar}{m_e}\mathbf{k} \cdot \mathbf{p}_{nn} + \frac{\hbar^2 k^2}{2m_e}. \tag{17•5–4}$$

If the degeneracy at $\mathbf{k} = 0$ has been removed, as we assumed, the inversion symmetry of $\mathbf{k}$-space requires that $\mathbf{k} = 0$ corresponds to an energy minimum or maximum and that the term linear in $\mathbf{k}$ vanishes. In that case, only the k^2-term in (17•5–3) makes a contribution, which is simply equal to the kinetic energy of a free electron with the wave number k.

Exercise: Show that in the absence of a degeneracy at $\mathbf{k} = 0$, the diagonal matrix element of the momentum operator does indeed vanish:

$$\mathbf{p}_{nn} = 0. \tag{17•5–5}$$

On the other hand, the k^2-term in (17•5–3) has no mixed matrix elements and, hence, does not contribute to the second-order correction to the energy, which arises purely from the first term in (17•5–3):

$$\Delta\mathcal{E}^{(2)} = \sum_{m\neq n} \frac{|W_{mn}|^2}{\mathcal{E}_n(\mathbf{k}) - \mathcal{E}_m(0)} = \frac{\hbar^2}{m_e^2} \sum_{m\neq n} \frac{|\mathbf{k} \cdot \mathbf{p}_{mn}|^2}{\mathcal{E}_n(\mathbf{k}) - \mathcal{E}_m(0)}. \tag{17•5–6}$$

We next restrict ourselves further to cubic crystals, in which case the $\mathcal{E}_n(\mathbf{k})$ relation must have cubic symmetry in $\mathbf{k}$-space. This means that the numerators in the sum in (17•5–4) must themselves have cubic symmetry. But for a quadratic form in the components of $\mathbf{k}$, this means that the numerators must be simply proportional to k^2,

$$|\mathbf{k} \cdot \mathbf{p}_{mn}|^2 = k^2 |p_{mn}|^2, \tag{17•5–7}$$

with an isotropic off–diagonal momentum matrix element as the proportionality factor. If (17•5–7) is inserted into (17•5–4), we obtain

$$\mathcal{E}_n(\mathbf{k}) - \mathcal{E}_n(0) = \frac{\hbar^2 k^2}{2m_e}\left[1 + \frac{2}{m_e} \sum_{m\neq n} \frac{|p_{mn}|^2}{\mathcal{E}_n(\mathbf{k}) - \mathcal{E}_m(0)}\right]. \tag{17•5–8}$$

17.5.3 Effective Masses, Oscillator Strengths, and Their Sum Rules

Consider the limit of k-values that are sufficiently small that the energy denominators in the sum may be replaced by their unperturbed limits,

$$\mathcal{E}_n(\mathbf{k}) - \mathcal{E}_m(0) \approx \mathcal{E}_n(0) - \mathcal{E}_m(0), \tag{17•5–9}$$

thus making the transition to Rayleigh-Schroedinger perturbation theory. In this limit, the right-hand side of (17•5–8) becomes simply proportional to k^2, and may be viewed as the k^2-term of a Taylor expansion of the dispersion relation $\mathcal{E}(\mathbf{k})$ about $\mathbf{k} = 0$,

$$\mathcal{E}(\mathbf{k}) = \mathcal{E}(0) + \left(\frac{\partial \mathcal{E}}{\partial k}\right)_0 k + \frac{1}{2} \cdot \left(\frac{\partial^2 \mathcal{E}}{\partial k^2}\right)_0 k^2 + \cdots, \tag{17•5–10}$$

where the notation $(\ldots)_0$ indicates that the derivatives are to be taken at $\mathbf{k} = 0$. Under our assumptions, the term linear in k vanishes, and we may re-write (17•5–8) as

$$\boxed{\frac{m_e}{\hbar^2}\left(\frac{\partial^2 \mathcal{E}_n}{\partial k^2}\right)_0 = 1 + \frac{2}{m_e}\sum_{m \neq n} \frac{|p_{mn}|^2}{\mathcal{E}_n(0) - \mathcal{E}_m(0)}.} \tag{17•5–11}$$

But because of (17•3–23), the left-hand side is simply the ratio m_e/m_n^* of the free-electron mass to the effective mass of band n at $k = 0$. Hence, we may re-write (17•5–11) further as

$$\frac{m_e}{m_n^*} = 1 + \frac{2}{m_e}\sum_{m \neq n} \frac{|p_{mn}|^2}{\mathcal{E}_n(0) - \mathcal{E}_m(0)}. \tag{17•5–12}$$

If the point $k = 0$ is an energy minimum, the effective mass there is positive; if it is a maximum, it is negative. Because of the repulsive nature of second-order perturbation theory, any bands below the band of interest will push the energy up, whereas bands above will push it down. The strongest effect will usually be exerted by the bands nearest to the band of interest. This explains why effective masses within a given semiconductor family correlate with the energy gaps, in such a way that the narrower-gap semiconductors tend to have lower effective masses than the wider gap materials.

Effective-Mass Sum Rule

Each term in the sum in (17•5–12) shows up in the effective-mass relation for both bands coupled by that term, but with opposite signs. If we average over the inverse masses for all bands, the terms in the sum cancel out, and we obtain the **effective mass sum rule**

$$\overline{\left[\frac{m_e}{m_n^*}\right]} = 1, \tag{17•5–13}$$

where the overbar implies averaging over all bands. Although we derived this relation here under restrictive assumptions, it holds generally.

The quantities inside the sum in (17•5–11) play a central role also in the different context of optical transitions between different bands: As we shall see in chapter 19, the strength of optical transitions between two isotropic $\mathbf{k} = 0$ states is directly proportional to these quantities. If we include the factor $2/m_e$ preceding the sum, the individual terms are what is commonly called the **oscillator strengths** involving the two states. Dropping the restriction to $\mathbf{k} = 0$ and to isotropic matrix elements, the general definition of the oscillator strength for x-polarization is

$$F_x^{m,n} \equiv \frac{2}{m_e} \frac{|\langle m, \mathbf{k}|\hat{p}_x|n, \mathbf{k}\rangle|^2}{\mathcal{E}_m(\mathbf{k}) - \mathcal{E}_n(\mathbf{k})}, \tag{17•5–14a}$$

with cyclic equivalents for other polarizations. Note that the states $|m, \mathbf{k}\rangle$ and $|n, \mathbf{k}\rangle$ in the momentum matrix element are full Bloch waves containing the plane-wave factor $\exp(i\mathbf{k} \cdot \mathbf{r})$, not just the Bloch lattice functions.

It can be shown that the oscillator strengths for transitions from a state in band n to states in other bands, but with the same $\mathbf{k}$, obey the ***F*-sum rule**

$$\sum_{m \neq n} F_x^{m,n} = 1 - \frac{m_e}{\hbar^2}\left(\frac{\partial^2 \mathcal{E}}{\partial k_x^2}\right), \tag{17•5–14b}$$

again with obvious cyclic equivalents. This is an exact relation, not an approximation.

Our earlier result (17•5–12) is simply a special case of this more general result, and we may in fact view our $\mathbf{k} \cdot \mathbf{p}$ perturbation treatment as a derivation of the sum rule for the specific case of $\mathbf{k} = 0$ and isotropic matrix elements. It is possible to extend the $\mathbf{k} \cdot \mathbf{p}$ perturbation derivation of the F-sum rule to the more general case (see the problems).

17.5.4 Non-Parabolicity

In many semiconductors, the energy gap is sufficiently narrow that the Rayleigh-Schroedinger approximation (17•5–9) quickly becomes a poor approximation for energies away from the band edge. The $\mathcal{E}_n(\mathbf{k})$ relation then ceases to be parabolic.

We consider here only the simplest case, that of a single conduction band perturbed by a single valence band. We select the bottom of the conduction band as the zero of the energy, $\mathcal{E}_C(0) = 0$, in which case the top of the valence band is at the energy $\mathcal{E}_v = -\mathcal{E}_G$. With this notation, (17•5–8) may be written as

$$\mathcal{E}_C(k) = \frac{\hbar^2 k^2}{2m_e}\left[1 + \frac{2}{m_e} \cdot \frac{|p_{cv}|^2}{\mathcal{E}_C(k) + \mathcal{E}_G}\right]. \tag{17•5–15}$$

The effective mass at the bottom of the conduction band is evidently given by

$$\frac{m_e}{m_C^*} = \left[1 + \frac{2}{m_e} \cdot \frac{|p_{cv}|^2}{\mathcal{E}_G}\right]. \tag{17•5–16}$$

Often, the effective mass is known experimentally. It is then useful to express (17•5–15) in terms of the effective mass rather than the momentum matrix element:

$$\mathcal{E}_C(k) = \frac{\hbar^2 k^2}{2m_e}\left[1 + \frac{m_e - m_C^*}{m_C^*} \cdot \frac{\mathcal{E}_G}{\mathcal{E}_C(k) + \mathcal{E}_G}\right]. \tag{17•5–17}$$

This is a quadratic equation in $\mathcal{E}_C(k)$, which may be written as

$$\left[\mathcal{E}_C(k) - \frac{\hbar^2 k^2}{2m_e}\right] \cdot [\mathcal{E}_C(k) + \mathcal{E}_G] = \frac{\hbar^2 k^2}{2m_C^*} \cdot \frac{m_e - m_C^*}{m_e} \cdot \mathcal{E}_G. \tag{17•5–18}$$

In narrow-gap semiconductors, in which non-parabolicity is important, the effective mass is usually small compared to the free-electron mass,

$$m_C^* \ll m_e. \tag{17•5–19}$$

In this case, (17•5–18) may be further simplified to

$$\boxed{\mathcal{E}_C(k) \cdot [\mathcal{E}_C(k) + \mathcal{E}_G] = \frac{\hbar^2 k^2}{2m_C^*} \cdot \mathcal{E}_G.} \tag{17•5–20}$$

It is in this form that **k** · **p** theory is most often used in semiconductor theory.

Note that for sufficiently large values of k, the $\mathcal{E}(k)$ relation becomes linear,

$$\mathcal{E}_C(k) = \hbar k \cdot \sqrt{\mathcal{E}_G/2m_C^*}, \tag{17•5–21}$$

with a limiting velocity

$$v_{\text{lim}} = \sqrt{\mathcal{E}_G/2m_C^*}. \tag{17•5–22}$$

Exercise: Show that this behavior is *formally* identical to that of a relativistic particle, with m_C^* playing the role of the rest mass M and $\mathcal{E}_G/2$ that of the rest mass energy Mc^2.

Our treatment has neglected a number of complications. First of all, in many cases of interest, not all degeneracies are lifted by the periodic potential. For example, in Si, the III-V compounds, and other semiconductors of similar crystal structure, the valence bands remain partially degenerate at $k = 0$. It then becomes necessary to include more than two bands in the perturbation treatment. If the effect of the perturbation on the degenerate valence band itself is to be studied, it becomes preferable to treat the problem by degenerate perturbation theory, although the Brillouin-Wigner formalism will work. Finally, if degeneracies are present *and* the crystal does not have a center of inversion, then the diagonal momentum matrix element may become nonzero, leading to (small) linear terms in the $\mathcal{E}(\mathbf{k})$ relation of the degenerate bands.

Exercise: Show that (17•5–6) remains valid if the crystal has a center of inversion, even if band n is degenerate at $k = 0$.

PROBLEMS TO SECTION 17.5

#17•5-1: "Off-Center" k · p Perturbation Theory

Replace the **k**-terms in (17•5–1b) via

$$\mathbf{k} = \mathbf{k}_0 + \mathbf{k}', \tag{17•5–23}$$

and redo the $\mathbf{k} \cdot \mathbf{p}$ perturbation treatment, but treating $\mathbf{k}_0$ as part of the unperturbed problem and $\mathbf{k}'$ as the perturbation. Use this treatment to generalize the sum rule (17•5–11) to arbitrary values of **k**.

#17•5-2: Ehrenfest's Theorem Completed

Using perturbation theory, determine the perturbation terms of the electron wave function in (17•3–32), and transform the result into such a form that the sum rule (17•5–11) can be used to obtain the basic result (17•3–23) for the effective mass. For simplicity, restrict yourself to a one-dimensional treatment.

Chapter 18

ROTATIONAL INVARIANCE AND ANGULAR MOMENTUM

18.1 OPERATOR ALGEBRA AND EIGENVALUES

18.1.1 Spherical Symmetry and Angular Momentum: A Review

One of the most common and most important symmetry invariances occurring in nature is **full spherical symmetry**, that is, invariance under all rotations and reflections that convert a sphere into itself. This includes, in particular, all rotations by infinitesimally small angles about *any* axis that goes through the center of the sphere.

Consider specifically rotations by a small angle $\delta\alpha$ about the polar z-axis in a system of spherical polar coordinates. Let $f(r, \theta, \phi)$ be an essentially arbitrary function expressed in that system. The rotation operator $\hat{C}_z(\delta\alpha)$ for such a rotation is then defined, from (16•1–11), by the requirement that

$$\hat{C}_z(\delta\alpha)f(r, \theta, \phi) = f(r, \theta, \phi - \delta\alpha). \tag{18•1–1}$$

For an infinitesimally small $\delta\alpha$, we may expand the right-hand side into a truncated Taylor series,

$$f(r, \theta, \phi - \delta\alpha) = f - \frac{\partial f}{\partial \alpha} \cdot \delta\alpha = \left[1 + \frac{i\delta\alpha}{\hbar} \cdot \hat{L}_z\right]f, \tag{18•1–2}$$

where, in the last equality, we have expressed the angular derivative in terms of the operator

$$\hat{L}_z = -i\hbar\frac{\partial}{\partial\phi}. \tag{18•1–3}$$

Any Hamiltonian with full spherical symmetry must evidently commute with this operator. But the operator is nothing other than the polar-coordinate representation (7•5–4c) of the angular momentum operator $\hat{L}_z$. Thus, we see that the angular momentum operator $\hat{L}_z$ is basically a differential symmetry operator for infinitesimal rotations. Inasmuch as it is arbitrary which direction is chosen as the z-direction, the argument holds for the other components of $\hat{\mathbf{L}}$ as well.

Exercise: Work out the details for the other Cartesian components $\hat{L}_x, \hat{L}_y$ of the angular momentum operator $\hat{\mathbf{L}}$, by drawing on the Cartesian form (16•1–17) of the rotation operator and its cyclic equivalents.

Hence, we are led again to the result already obtained in chapter 7 by a different line of reasoning—that in the case of a spherically symmetric potential $V(r)$, the angular momentum operator $\hat{\mathbf{L}}$ and its Cartesian component operators $\hat{L}_x$, $\hat{L}_y$, and $\hat{L}_z$ all commute with the Hamiltonian:

$$[\hat{L}_x, \hat{H}] = [\hat{L}_y, \hat{H}] = [\hat{L}_z, \hat{H}] = 0. \tag{18•1–4}$$

Comment: Symmetry and the Classical Law of Conservation of Angular Momentum We pointed out in section 8.1 that the commutation relations (18•1–4) imply that the expectation value $\langle\mathbf{L}\rangle$ and its Cartesian components are time-invariant, and that this is the quantum-mechanical origin of the law of conservation of angular momentum in an isotropic force field. Inasmuch as the properties of these operators follow from the rotational invariance of the Hamiltonian, we see that the dynamical law of conservation of angular momentum is a consequence of rotational invariance.

The purpose of the present chapter is to investigate the properties of the angular momentum operators in more detail, drawing on the full operator formalism that we have developed since chapter 7, in roughly the same spirit in which we re-examined the harmonic oscillator in chapter 10. As in that earlier case, we shall find new and important insights. In particular, we shall find that the properties of the angular momentum operators alone, in the absence of additional constraints, permit not only integer values for the quantum numbers l and m, but half-integer values as well, thereby setting the stage for the later incorporation of the electron spin into the angular momentum formalism.

In our earlier treatment, the claim that l and m must be integers had followed, *not* from the properties of the angular momentum operators themselves, but from the *additional* constraint that each eigenfunction be a single-valued scalar analytic function that turns into itself under azimuthal rotation by 2π. But we shall see later (chapter 21) that the spin properties of the electron call for a representation in terms of *two* coupled wave functions, which are *interchanged* under rotation by 2π, rather than being invariant. The conclusion that l and m must be integers can then no longer be drawn. One of the principal tasks of the present chapter is to determine the constraints imposed on the angular momentum eigenvalues by the operator algebra alone.

The central property of this operator algebra is that components of $\hat{\mathbf{L}}$ do not commute with each other, but obey the commutation relations (7•5–13),

$$\boxed{[\hat{L}_x, \hat{L}_y] = i\hbar\hat{L}_z; \quad [\hat{L}_y, \hat{L}_z] = i\hbar\hat{L}_x; \quad [\hat{L}_z, \hat{L}_x] = i\hbar\hat{L}_y.} \tag{18•1–5a,b,c}$$

Recall from chapter 16 that the existence of non-commuting symmetry operators is a necessary condition for the occurrence of symmetry degeneracies; the above commutation relations are what permits the occurrence of the m-degeneracy in spherically symmetric potentials.

We also recall that, even though the components of $\hat{\mathbf{L}}$ do not commute with each other, they all commute with the square of $\hat{\mathbf{L}}$—see (7•5–14):

$$[\hat{L}^2, \hat{L}_x] = [\hat{L}^2, \hat{L}_y] = [\hat{L}^2, \hat{L}_z] = 0, \tag{18•1–6}$$

$$\hat{L}^2 = \hat{L}_x^2 + \hat{L}_y^2 + \hat{L}_z^2. \tag{18•1–7}$$

In fact, we saw in chapter 7 that the relations (18•1–6) are necessary for (18•1–4) to be valid.

From the set of commutation relations, we concluded in chapter 7 that the energy eigenstates of a spherically symmetric Hamiltonian may always be chosen to be eigenstates of $\hat{L}^2$ and of *one* of the angular momentum component operators, invariably selected to be $\hat{L}_z$. We expressed this joint eigenvalue problem in the form (7•5–15),

$$\boxed{\hat{L}^2|l, m\rangle = \hbar^2 l(l + 1)|l, m\rangle,} \tag{18•1–8}$$

$$\boxed{\hat{L}_z|l, m\rangle = \hbar m|l, m\rangle,} \tag{18•1–9}$$

where we have written the eigenvalues of $\hat{L}^2$ and of $\hat{L}_z$ in the forms $\hbar^2 l(l + 1)$ and $\hbar m$, and where we have designated the joint eigenstates as $|l, m\rangle$. Without loss of generality, we may continue to use this notation here, even if we drop the restriction that l and m be integers: The eigenvalues of $\hat{L}_z$ can always be split into a product $\hbar m$ with a real value of m. Also, with $\hat{L}^2$ being the sum of three squares of Hermitian operators, it cannot have negative eigenvalues, and any non-negative eigenvalue can always be written as $\hbar^2 l(l+1)$, with a suitable non-negative value of l.

18.1.2 Stepping Operators

Because the $|l, m\rangle$ are eigenstates not only of $\hat{L}_z$, but also of $\hat{L}_z^2$, with the eigenvalues $\hbar^2 m^2$, the eigenvalue problem (18•1–8) is equivalent to the slightly simpler problem

$$(\hat{L}_x^2 + \hat{L}_y^2)|l, m\rangle = \hbar^2[l(l+1) - m^2]|l, m\rangle. \tag{18•1–10}$$

The occurrence of the sum of the squares of two non-commuting Hermitian operators on the left-hand side of (18•1–10) is reminiscent of the Hamiltonian of the harmonic oscillator, and it suggests the introduction of the Hermitian conjugate pair of operators

$$\hat{L}_+ = \hat{L}_x + i\hat{L}_y, \qquad \hat{L}_- = \hat{L}_x - i\hat{L}_y, \tag{18•1–11a,b}$$

similar to the introduction of the raising and lowering operators $\hat{a}^+$ and $\hat{a}^-$ in the harmonic oscillator case. By expressing the operator $\hat{L}^2$ in terms of $\hat{L}_\pm$ and $\hat{L}_z$, rather than in terms of $\hat{L}_x$, $\hat{L}_y$, and $\hat{L}_z$, the solution of the joint eigenvalue problem (18•1–8) and (18•1–9) is readily possible. We proceed in several steps.

(a) From the commutation relations (18•1–5) for the components of $\hat{\mathbf{L}}$, we find

$$\hat{L}_+\hat{L}_- = \hat{L}_x^2 + \hat{L}_y^2 - i[\hat{L}_x, \hat{L}_y] = \hat{L}_x^2 + \hat{L}_y^2 + \hbar\hat{L}_z = \hat{L}^2 - \hat{L}_z(\hat{L}_z - \hbar), \tag{18•1–12a}$$

where, in the last equality, we have drawn on (18•1–7). Similarly, for the inverse order of factors on the left-hand side of (18•1–12a),

$$\hat{L}_-\hat{L}_+ = \hat{L}_x^2 + \hat{L}_y^2 - \hbar\hat{L}_z = \hat{L}^2 - \hat{L}_z(\hat{L}_z + \hbar) \tag{18•1–12b}$$

If we apply the two product operators $\hat{L}_+\hat{L}_-$ and $\hat{L}_-\hat{L}_+$ to $|l, m\rangle$, and use (18•1–8) and (18•1–10), we obtain

$$\hat{L}_+\hat{L}_-|l, m\rangle = \hbar^2[l(l+1) - m(m-1)]|l, m\rangle. \tag{18•1–13a}$$

$$\hat{L}_-\hat{L}_+|l, m\rangle = \hbar^2[l(l+1) - m(m+1)]|l, m\rangle, \tag{18•1–13b}$$

We see that the states $|l, m\rangle$ are eigenstates also of the operators $\hat{L}_+\hat{L}_-$ and $\hat{L}_-\hat{L}_+$, with the eigenvalues shown on the right-hand sides of (18•1–13a,b).

Because the operators $\hat{L}_+$ and $\hat{L}_-$ form a Hermitian conjugate pair, the eigenvalues of $\hat{L}_+\hat{L}_-$ and $\hat{L}_-\hat{L}_+$ cannot be negative, and we obtain from (18•1–13) the important dual restriction

$$l(l+1) \geq m(m \pm 1), \tag{18•1–14}$$

which implies

$$-l \leq m \leq +l, \tag{18•1–15}$$

a result we already obtained in (7•5–16) by a slightly different line of reasoning.

(b) We next consider the commutation relations for $\hat{L}_{\pm}$. Becaue $\hat{L}_x$, $\hat{L}_y$, and $\hat{L}_z$ commute with $\hat{L}^2$, the same is true for $\hat{L}_{\pm}$,

$$[\hat{L}^2, \hat{L}_{\pm}] = 0, \tag{18•1–16}$$

but otherwise $\hat{L}_+$ and $\hat{L}_-$ commute neither with each other nor with the components of $\hat{L}$. In particular,

$$[\hat{L}_z, \hat{L}_{\pm}] = [\hat{L}_z, \hat{L}_x] \pm i[\hat{L}_z, \hat{L}_y] = i\hbar\hat{L}_y \pm \hbar\hat{L}_x = \pm\hat{L}_{\pm}. \tag{18•1–17}$$

These relations are of the general form of the commutation relations (9•1–21) for stepping operators,

$$[\hat{A}, \hat{S}^{\pm}] = \pm(\Delta A)\hat{S}^{\pm}. \tag{18•1–18}$$

Applied to our present case, the relations (18•1–17) imply that the operators $\hat{L}_+$ and $\hat{L}_-$ act as stepping operators for quanta of the z-component of the angular momentum, similar to the way the operators $\hat{a}^+$ and $\hat{a}^-$ were stepping operators for the quanta of energy of the harmonic oscillator. More specifically, the result of the operation $\hat{L}_{\pm}|l, m\rangle$ must either be zero or represent another eigenstate of $\hat{L}_z$, with the eigenvalue $\hbar(m + 1)$ for $\hat{L}_+|l, m\rangle$ and $\hbar(m - 1)$ for $\hat{L}_-|l, m\rangle$.

At the same time, because $\hat{L}_{\pm}$ commute with $\hat{L}^2$, the states $\hat{L}_{\pm}|l, m\rangle$—if they are nonzero—remain eigenstates of $\hat{L}^2$ with the same eigenvalues as before:

$$\hat{L}^2(\hat{L}_{\pm}|l, m\rangle) = \hbar^2 l(l + 1)(\hat{L}_{\pm}|l, m\rangle). \tag{18•1–19}$$

Taken together, these two facts mean that $\hat{L}_+|l, m\rangle$ and $\hat{L}_-|l, m\rangle$ are proportional to the states $|l, m \pm 1\rangle$:

$$\hat{L}_+|l, m\rangle = c_m^+|l, m + 1\rangle, \tag{18•1–20a}$$

$$\hat{L}_-|l, m\rangle = c_m^-|l, m - 1\rangle, \tag{18•1–20b}$$

where c_m^+ and c_m^- are proportionality factors, to be determined next.

(c) We restrict ourselves to the (+)-case, leaving the (−)-case to the reader. Recall that any operator operating on a ket vector to its right is equivalent to the Hermitian conjugate operator acting on the dual bra vector to its left, with the Hermitian conjugate result. Because $\hat{L}_-$ is the Hermitian conjugate of $\hat{L}_+$, we can therefore write

$$\langle l, m|\hat{L}_- = (c_m^{\pm})^*\langle l, m + 1|. \tag{18•1–20c}$$

We take the inner product of (18•1–20b) and (18•1–20c):

$$\langle l, m|\hat{L}_-\hat{L}_+|l, m\rangle = |c_m^+|^2\langle l, m + 1|l, m + 1\rangle. \tag{18•1–21}$$

But as we saw in (18•1–13a), the state $|l, m\rangle$ is an eigenstate of $\hat{L}_-\hat{L}_+$, with the eigenvalue given in (18•1–13a). Thus, the left-hand side of (18•1–21) can be written

$$\hbar^2[l(l+1) - m(m+1)]\langle l, m | l, m\rangle. \tag{18•1–22}$$

If both $|l, m\rangle$ and $|l, m+1\rangle$ represent normalized states, the Dirac brackets in (18•1–22) and on the right-hand side of (18•1–21) have the value 1, and we obtain

$$|c_m^+|^2 = \hbar^2 \cdot [l(l+1) - m(m+1)]. \tag{18•1–23a}$$

Similarly, one finds that for the (−)-case, (18•1–20b),

$$|c_m^-|^2 = \hbar^2 \cdot [l(l+1) - m(m-1)]. \tag{18•1–23b}$$

Insertion into (18•1–20) yields

$$\hat{L}_+ |l, m\rangle = \hbar \cdot \sqrt{l(l+1) - m \cdot (m+1)}\,|l, m+1\rangle, \tag{18•1–24a}$$

$$\hat{L}_- |l, m\rangle = \hbar \cdot \sqrt{l(l+1) - m \cdot (m-1)}\,|l, m-1\rangle. \tag{18•1–24b}$$

(d) We saw in (18•1–15) that there cannot exist any eigenstates $|l, m\rangle$ for which m falls outside the range $\pm l$. From (18•1–24), we now see that the repeated application of the raising operator $\hat{L}_+$ or of the lowering operator $\hat{L}_-$ does indeed ultimately lead to a vanishing result: Upward truncation takes place for $m = +l$, and downward truncation for $m = -l$,

$$\hat{L}_+ |l, l\rangle = \hat{L}_- |l, -l\rangle = 0. \tag{18•1–25}$$

Our treatment still leaves open which values of m and l can actually occur. Because the individual possible m-values between $m = -l$ and $m = +l$ are separated in unit steps, the difference $2l$ must be an integer. This implies that l must be either an integer itself or a half-integer. In the first case, the possible values of m are also integers, including zero:

$$l = \text{integer:} \qquad m = -l, -l+1, \ldots, -1, 0, +1, \ldots, l-1, l. \tag{18•1–26}$$

In the second case, the possible values of m are also half-integers:

$$l = \text{half-integer:} \qquad m = -l, -l+1, \ldots, -\tfrac{1}{2}, +\tfrac{1}{2}, \ldots, l-1, l. \tag{18•1–27}$$

18.1.3 Restriction to Integer Quantum Numbers for the *Orbital* Angular Momentum

It is only at this point that we must draw on additional information to select between the preceding two possibilities. The restriction to integer values of m, and hence of l, is a consequence of the specific form (18•1–3) of the operator $\hat{L}_z$, which has eigenfunctions of the form (7•5–11),

$$\Psi(r, \phi, \theta) = f(r, \theta)\exp(im\phi). \tag{18•1–28}$$

But Ψ must turn into itself under rotation about the polar axis by 2π, and it is *this* requirement that calls for integer values of m, and hence of l.

The states belonging to different values of l are usually not degenerate. As we commented in chapter 16, the Coulomb potential with its simple form $V(r) \propto 1/r$ is one of the exceptions to this rule.

Regardless of $V(r)$, the stepping operator formalism shows that, for a given value of l, *all* values of m between $-l$ and $+l$ must occur,

$$m = 0, \pm 1, \pm 2, \ldots, \pm l, \tag{18•1–29}$$

leading to a $(2l + 1)$-fold degenerate set of eigenfunctions of the Hamiltonian and to the geometric representation of the angular momentum eigenstates given earlier, in Fig.7•5–2.

The angular momentum eigenvalue m also occurs in problems of invariance under continuous rotations about a *single* axis. However, in this case, there does not exist a suitable stepping operator formalism, and we can no longer draw the conclusion that all m-values in a certain range must occur. For example, it can be shown that in the case of the cylindrical harmonic oscillator, only alternate values of m occur amongst the states belonging to the same energy eigenvalue.

18.1.4 Looking Ahead: Electron Spin as a Case of Half-Integer Angular Momentum Eigenvalues

As claimed earlier, the angular momentum operators $\hat{\mathbf{L}}$ and $\hat{L}^2$ introduced in chapter 7 are not the only kinds of operators satisfying the set of commutation relations (18•1–5) and (18•1–6) that have formed the mathematical basis for our discussion in the present section. In our discussion of the electron spin in chapter 21, we will be led to define a set of three 2×2 matrix operators of the form

$$\hat{S}_x = \frac{\hbar}{2}\begin{pmatrix} 0 & 1 \\ 1 & 0 \end{pmatrix}, \quad \hat{S}_y = \frac{\hbar}{2}\begin{pmatrix} 0 & -i \\ i & 0 \end{pmatrix}, \quad \hat{S}_z = \frac{\hbar}{2}\begin{pmatrix} 1 & 0 \\ 0 & -1 \end{pmatrix}, \tag{18•1–30}$$

along with the operator

$$\hat{S}^2 = \hat{S}_x^2 + \hat{S}_y^2 + \hat{S}_z^2 = \frac{3\hbar^2}{4}\begin{pmatrix} 1 & 0 \\ 0 & 1 \end{pmatrix}. \tag{18•1–31}$$

It is an elementary exercise in matrix algebra to show that this set of operators has the same commutation relations as the set of orbital angular momentum operators and that the operator eigenvalues are $\pm\hbar/2$ for $\hat{S}_z$ and $3\hbar^2/4$ for $\hat{S}^2$. It is also evident that such operators are meaningful only if they operate, not on scalar state functions, but on 2×1 column matrices of the form

$$\overline{\Psi} = \begin{pmatrix} \psi_1 \\ \psi_2 \end{pmatrix}. \tag{18•1–32}$$

As we shall see in chapter 21, the properties of the electron spin will lead us to the introduction of precisely this kind of state functions, appropriately called **spinors**.

◆ PROBLEMS TO SECTION 18.1

#18•1-1: Total Angular Momentum in Composite Systems

Consider a system of N interacting particles in free space, similar to the systems that were the topic of section 11.2. Show that the angular momentum operators of the individual particles do not commute with the Hamiltonian, and hence, the individual angular momenta are not conserved. Assume that all potentials are due to *isotropic* particle-particle interactions depending only on the coordinate *differences* of the two particles involved in each interaction term, with no *external* potential acting on the particles. Show that in this case the operator for the total angular momentum,

$$\hat{\mathbf{L}} \equiv \sum_{\nu=1}^{N} \hat{\mathbf{L}}_\nu = \sum_{\nu=1}^{N} (\hat{\mathbf{r}} \times \hat{\mathbf{p}})_\nu, \qquad (18\bullet1\text{–}33)$$

commutes with the Hamiltonian and that the total angular momentum *is* conserved. Show further that the operator algebra for the various total angular momentum operators is the same as for the single-particle case in a spherically rotationally invariant potential.

#18•1-2: Angular Momentum in the Cylindrical Harmonic Oscillator

The potential energy for a cylindrical harmonic oscillator may be written in Cartesian coordinates as

$$V(x, y) = \tfrac{1}{2} M\omega^2 (x^2 + y^2), \qquad (18\bullet1\text{–}34)$$

and the energy eigenvalues may be obtained by separation of variables in these coordinates, leading to a set of energy eigenstates $|n_x, n_y\rangle$ that may be represented as a set of points in an (n_x, n_y) plane.

(a) Draw such a representation, indicating which states belong to the same energy and the degree of degeneracy of each energy level.

(b) Define *two* sets of raising and lowering operators, $\hat{a}_x^{\pm}$ and $\hat{a}_y^{\pm}$, for the two orthogonal oscillation components. Write both the Hamiltonian and the operator $\hat{L}_z$ in terms of these four operators.

(c) Use the new form of $\hat{L}_z$ and the (n_x, n_y) plane diagram to show that none of the $|n_x, n_y\rangle$ are eigenstates of $\hat{L}_z$ except $|0, 0\rangle$.

(d) Construct the Hermitian conjugate pair of operators

$$\hat{S}^{\pm} = \hat{a}_x^{+} \pm i\hat{a}_y^{+}. \qquad (18\bullet1\text{–}35)$$

What is their action on the state $|n_x, n_y\rangle$, with regard to both the energy and the angular momentum of the resulting state?

(e) Show that the operators $\hat{S}^{\pm}$ may be used to generate the full set of all eigenstates in such a way that every state generated is an eigenstate of both the Hamiltonian and $\hat{L}_z$. Give a rule that states which m-values occur for a given energy. Actually construct the states for the second- and third-lowest energy eigenvalue.

#18•1-3: Spherical Harmonic Oscillator

The potential energy of a spherical harmonic oscillator may be written in polar coordinates as

$$V(r) = \tfrac{1}{2}M\omega^2 r^2. \tag{18•1–36}$$

We know from the solution of the problem by separation of variables in Cartesian coordinates that the energy eigenvalues of this system are of the form

$$\mathcal{E} = \hbar\omega \cdot (n + \tfrac{3}{2}) \tag{18•1–37}$$

where

$$n = n_x + n_y + n_z \geq 0 \tag{18•1–38}$$

is a non-negative integer. The three n's describe each state in terms of its Cartesian product wave functions, but in polar coordinates the energies must be expressed in terms of the quantum numbers l, m, and a radial quantum number n_r that is related to the number of nulls in the radial part of the wave function.

(a) Out of the total set of eigenvalues, which ones occur amongst the states that have $l = 0$?

(b) For nonzero values of l, estimate the *lowest* energy eigenvalue as a function of l. *Note:* The correct procedure requires including the quantum-mechanical zero-point energy of the one-dimensional equivalent effective radial potential. This energy can be estimated by approximating the vicinity of the effective potential minimum as a harmonic oscillator in its own right.

18.2 ANGULAR MOMENTUM EIGENFUNCTIONS

18.2.1 Spherical Harmonics Re-Visited

Having put the eigenvalues of the operators $\hat{L}^2$ and $\hat{L}_z$ on a firm mathematical foundation, we return briefly to the eigenfunctions and place their determination into the framework of the angular momentum operators themselves. As we saw in section 3.1, the energy eigenfunctions for a spherically symmetric potential may always be obtained by separation of variables, in the form (3•1–21),

$$\psi(r, \theta, \phi) = R(r) \cdot Y_l^m(\theta, \phi), \tag{18•2–1}$$

where the functions

$$Y_l^m(\theta, \phi) = \frac{1}{\sqrt{2\pi}} f(\theta) e^{im\phi} \tag{18•2–2}$$

are the **spherical harmonics**. Because the angular momentum operators involve only the angular variables, the radial term $R(r)$ in (18•2–1) is irrelevant for our present purposes. We therefore omit here the radial factor and identify the ket symbol $|l, m\rangle$ with the spherical harmonics:

$$|l, m\rangle \triangleq Y_l^m(\theta, \phi). \tag{18•2–3}$$

Similar to the case of the harmonic oscillator, the angular momentum eigenfunctions are most conveniently generated recursively with the help of the stepping operators $\hat{L}_\pm$. If we insert the polar coordinate forms (7•5–4a,b) of the operators $\hat{L}_x$ and $\hat{L}_y$ into the definition (18•1–11) of the stepping operators, we obtain

$$\hat{L}_\pm = \hat{L}_x \pm i\hat{L}_y = \hbar e^{\pm i\phi} \cdot \left[\pm\frac{\partial}{\partial\theta} + i \cot\theta\frac{\partial}{\partial\phi}\right]. \tag{18•2–7}$$

Applied to the eigenstates of $\hat{L}_z$, the ϕ-derivative may be replaced by im, leading to

$$\hat{L}_\pm = \hbar e^{\pm i\phi} \cdot \left[\pm\frac{\partial}{\partial\theta} - m \cdot \cot\theta\right]. \tag{18•2–8}$$

The cases $m = \pm l$ are of particular interest, because the application of $\hat{L}_+$ to $|l, l\rangle$ and of $\hat{L}_-$ to $|l, -l\rangle$ must yield zero. Inserting the forms (18•2–8) into (18•1–25) leads to the differential equation for the θ-dependence of both $Y_l^l(\theta, \phi)$ and $Y_l^{-l}(\theta, \phi)$:

$$\frac{\partial}{\partial\theta}Y_l^{\pm l}(\theta, \phi) = l \cdot \cot\theta \cdot Y_l^{\pm l}(\theta, \phi). \tag{18•2–9}$$

It has the solutions

$$Y_l^{\pm l}(\theta, \phi) = C_l e^{\pm il\phi} \cdot \sin^l\theta, \tag{18•2–10a}$$

where C_l is a normalization factor, and where we have inserted the known ϕ-dependence from (18•2–2). In Cartesian coordinates,

$$|l, \pm l\rangle = C_l\frac{(x \pm iy)^l}{r^l}. \tag{18•2–10b}$$

***

Exercise: Show that the normalization integral for a given value of l can be reduced to that for $l - 1$ via an integration by parts, and that the values for the normalization factors may be obtained through the recursion relation

$$C_l^2 = C_{l-1}^2 \cdot \frac{2l + 1}{2l}, \quad \text{with} \quad C_0^2 = \frac{1}{4\pi}. \tag{18•2–11}$$

***

Given the set of the $|l, l\rangle$ or of the $|l, -l\rangle$, the remaining spherical harmonics are readily determined by repeated application of the upward or downward stepping relations (18•1–20a,b), with the normalization coefficients given in (18•1–23b). A table of the lowest-order functions, mentioned already in chapter 3, is given in appendix C.

A detailed study of the spherical harmonics is an exercise in mathematics rather than physics. This includes particularly the relation of the term $f(\theta)$ in (18•2–2) to the **associated Legendre functions**, usually denoted as $P_l^m(\cos\theta)$. Because these and other properties are discussed in every conceivable detail in numerous texts on mathematical physics, we refer the reader to that literature.[1]

18.2.2 Angular Momentum and Parity

Full spherical symmetry implies invariance not only under arbitrary *proper* rotations, but under inversion as well. The inversion commutes with all rotations about any axis that goes through the center of inversion. As a result, the angular momentum eigenfunctions are also eigenfunctions of the inversion operator.

Specifically, we claim that

$$\hat{I}\,|l, m\rangle = (-1)^l \cdot |l, m\rangle. \qquad (18•2–12)$$

In words, the parity depends only on the quantum number l, independently of the quantum number m, in such a way that states with even values of l have even parity and those with odd l odd parity. This result is easily obtained by first considering the special case of $|l, l\rangle$, in the Cartesian form (18•2–10b): Under inversion, x and y go over into $-x$ and $-y$, and $|l, l\rangle$ evidently satisfies (18•2–12). Hence, we see that the parity of the eigenfunctions with $m = l$ is even for even l and odd for odd l. The eigenfunctions for $m \neq l$ can now be obtained from those for $m = l$ by successive applications of the lowering operator $\hat{L}_-$. Because these operators represent pure rotations, which commute with $\hat{I}$, the parity of a wave function does not change in the process, and all wave functions with the same l have the same parity.

18.2.3 More Commutators

In many problems, we need the commutators not only between the angular momentum operators, but also between the latter and some of the position coordinates. In the following table, we give the values of selected commutators

[1] See, for example, the book by Morse and Feshbach, cited earlier, as well as many quantum mechanics texts.

of the form $[\hat{L}, \hat{r}]$, where $\hat{L}$ is one of the angular momentum operators in the leftmost column of the table and $\hat{r}$ is one of the position operators in the top row.

	$\hat{z}$	$\hat{r}_+ = \hat{x} + i\hat{y}$	$\hat{r}_- = \hat{x} - i\hat{y}$	
$\hat{L}_z$	0	$\hbar(\hat{x} + i\hat{y})$	$-\hbar(\hat{x} - i\hat{y})$	(18•2–13a,b,c)
$\hat{L}_+ = \hat{L}_x + i\hat{L}_y$	$-\hbar(\hat{x} + i\hat{y})$	0	$+2\hbar\hat{z}$	(18•2–14a,b,c)
$\hat{L}_- = \hat{L}_x - i\hat{L}_y$	$+\hbar(\hat{x} - i\hat{y})$	$-2\hbar\hat{z}$	0	(18•2–15a,b,c)

We give the x- and y-positions not by themselves, but in the form of the Hermitian conjugate pair $\hat{x} + i\hat{y}$ and $\hat{x} - i\hat{y}$, because this is the form in which they are often needed. The derivation of the results in the table is left to the reader.

Note that the two commutators $[\hat{L}_z, \hat{r}_+]$ and $[\hat{L}_z, \hat{r}_-]$ are again of the stepping operator form (18•1–18); that is, the operators $\hat{r}_+$ and $\hat{r}_-$ raise or lower the azimuthal angular momentum eigenvalues by $\pm\hbar$. This is of course simply the consequence of the presence of the factor $e^{\pm i\phi}$ in the polar coordinate form of these operators. However, in contrast to the earlier stepping operators $\hat{L}_+$ and $\hat{L}_-$, the operators $\hat{r}_+$ and $\hat{r}_-$ do not commute with $\hat{L}^2$ and hence do not leave the orbital quantum number l invariant.

18.2.4 Simple Matrix Elements

Just as in the case of the harmonic oscillator, matrix elements of the form

$$\langle l', m' | \hat{W} | l, m \rangle \tag{18•2–16}$$

may often be evaluated by drawing on the stepping operators and on the commutation relations for the operators, without needing to know the actual eigenfunctions. The central idea is to express the eigenfunctions $|l, m\rangle$ as results of repeated application of the raising operator $\hat{L}_+$ to $|l, -l\rangle$ or of the lowering operator $\hat{L}_-$ to $|l, +l\rangle$:

$$|l, m\rangle = C_m^+ (\hat{L}_+)^{l+m} |l, -l\rangle \quad \text{or} \quad |l, m\rangle = C_m^- (\hat{L}_-)^{l-m} |l, +l\rangle. \tag{18•2–17a,b}$$

Here C_m^+ and C_m^- are two nonzero real factors obtainable from the factors in (18•1–20); their exact values do not concern us at this point (see the problems). The Hermitian conjugate expressions are

$$\langle l, m| = C_m^+ \langle l, -l| (\hat{L}_-)^{l+m} \quad \text{and} \quad \langle l, m| = C_m^- \langle l, +l| (\hat{L}_+)^{l-m}. \tag{18•2–18a,b}$$

As a simple example, we consider the matrix elements for

$$\hat{W} = x \pm iy = r \cdot e^{\pm i\phi} \sin\theta \tag{18•2–19}$$

and

$$\hat{W} = z = r \cdot \cos\theta, \tag{18•2–20}$$

which we will need in the next chapter, where they play an important role in the theory of optical transition probabilities. We claim that

$$\langle l', m' | x + iy | l, m \rangle = 0 \quad \text{unless} \quad m' = m + 1 \text{ and } l' = l \pm 1, \tag{18•2–21a}$$

$$\langle l', m' | x - iy | l, m \rangle = 0 \quad \text{unless} \quad m' = m - 1 \text{ and } l' = l \pm 1. \tag{18•2–21b}$$

and

$$\langle l', m' | z | l, m \rangle = 0 \quad \text{unless} \quad m' = m \quad \text{and} \quad l' = l \pm 1. \tag{18•2–22}$$

The conditions for m follow directly from the polar coordinate form of $\hat{W}$ in (18•2–19) and (18•2–20); otherwise the implied integrations over ϕ will lead to zero. Also, because both $x \pm iy$ and z have odd parity, the states $|l', m'\rangle$ and $|l, m\rangle$ must have opposite parity, which implies that l' and l must differ by an odd number. What remains to be shown is that in all cases we must have specifically

$$l' = l \pm 1. \tag{18•2–23}$$

We consider the matrix elements of $x \pm iy$ first. With the help of (18•2–18b) and (18•2–16), the matrix element in (18•2–21a) may be written as

$$(C_{m'}^{-})^* C_m^{+} \cdot \langle l', +l' | (\hat{L}_+)^{l'-m'} (x + iy) (\hat{L}_+)^{l+m} | l, -l \rangle. \tag{18•2–24}$$

Because $\hat{L}_+$ and $x + iy$ commute, and because $m' - m = 1$, the Dirac bracket may be re-arranged in two different ways, either as

$$\langle l', +l' | (x + iy)(\hat{L}_+)^{l+l'-1} | l, -l \rangle \tag{18•2–25a}$$

or as

$$\langle l', +l' | (\hat{L}_+)^{l+l'-1} (x + iy) | l, -l \rangle. \tag{18•2–25b}$$

But because of the termination of the upward stepping sequence according to (18•1–25), the form (18•2–25a) can be nonzero only if

$$l' - 1 \le l \le l' + 1. \tag{18•2–26a}$$

Similarly, because of the termination of the downward stepping sequence, (18•2–25b) can be nonzero only if

$$l - 1 \le l' \le l + 1. \tag{18•2–26b}$$

But for $l' \neq l$, these two relations are compatible only if (18•2–23) is satisfied. By analogous reasoning, drawing on the operators $\hat{L}_-$ rather than $\hat{L}_+$, the reader may easily confirm that (18•2–23) also must be satisfied for the (−)-case.

Given the matrix elements for $x + iy$ and $x - iy$, those for x and y follow by addition and subtraction. We evidently have

$$\langle l', m' | x | l, m \rangle = \langle l', m' | y | l, m \rangle = 0,$$
$$\text{unless} \quad m' = m \pm 1 \quad \text{and} \quad l' = l \pm 1. \tag{18•2–27}$$

The matrix elements for z may be reduced to those for $x + iy$ and $x - iy$ by re-expressing z in terms $x + iy$ and $x - iy$ via the commutation relations (18•2–14c) or (18•2–15b):

$$z = +\frac{1}{2\hbar}[\hat{L}_+, (x - iy)] = -\frac{1}{2\hbar}[\hat{L}_-, (x + iy)]. \tag{18•2–28}$$

It is left to the reader to show that this leads to (18•2–22).

◆ PROBLEMS TO SECTION 18.2

#18•2-1: Values of Non-Vanishing Matrix Elements for $x \pm iy$ and z

(a) Show that the coefficients C_m^+ and C_m^- in (18•2–17a,b) have the values

$$C_m^\pm = \sqrt{\frac{(l \pm m)!}{(2l)!(l \pm m)!}}. \tag{18•2–29}$$

(b) Drawing on this result, determine the actual value of the non-vanishing matrix elements in (18•2–21) and (18•2–22).

#18•2-2: Commutator for Angular Momentum Operators

Evaluate the commutator

$$[(\hat{L}_x + i\hat{L}_y), F(\mathbf{r})], \tag{18•2–30}$$

where $F(\mathbf{r})$ is given in spherical polar coordinates:

$$F(\mathbf{r}) = \tfrac{1}{2}r^2 \sin^2 \theta \cdot \sin 2\phi. \tag{18•2–31}$$

Express the final answer both in Cartesian coordinates and in spherical polar coordinates.

18.3 SPLITTING OF THE *m*-DEGENERACY IN A MAGNETIC FIELD

As an application of the angular momentum operator formalism, we determine here the splitting of the m-degeneracy of the energy levels of an electron in a spherically symmetric potential $V(r)$ by a uniform magnetic field $\mathbf{B} = \mathbf{B}_z$. The magnetic field destroys the full spherical symmetry, but it retains single-axis rotational invariance about the direction of the field—provided we choose the gauge of the vector potential $\mathbf{A}$ in such a way that it retains the latter invariance, namely, as the circular gauge shown in Fig. 8•2–1, given by (8•2–2):

$$\mathbf{A} = \frac{1}{2}\begin{Bmatrix} -By \\ +Bx \\ 0 \end{Bmatrix} = \frac{1}{2}\mathbf{B} \times \mathbf{r}. \tag{18•3–1}$$

With this choice, the Hamiltonian assumes the form

$$\hat{H} = -\frac{\hbar^2}{2m_e}\nabla^2 - \frac{i}{2}\hbar\omega_L \cdot \left(x\frac{\partial}{\partial y} - y\frac{\partial}{\partial x}\right) + \frac{1}{8}m_e\omega_L^2 \cdot (x^2 + y^2) + V(r). \tag{18•3–2}$$

The second term on the right-hand side of (18•3–2) may be written

$$-\frac{i}{2}\hbar\omega_L \cdot \left(x\frac{\partial}{\partial y} - y\frac{\partial}{\partial x}\right) = \frac{1}{2}\omega_L \hat{L}_z, \tag{18•3–3}$$

where $\hat{L}_z$ is the operator for the z-component of the angular momentum. Because the entire Hamiltonian (18•3–2) is rotationally invariant about the z-axis, its eigenstates may be chosen also to be eigenstates of $\hat{L}_z$, that is, $|\psi\rangle = |m\rangle$, with

$$\hat{L}_z|m\rangle = m\hbar|m\rangle, \tag{18•3–4}$$

where the magnetic quantum number m must be an integer. In fact, it is this occurrence of the quantum number m in magnetic problems that has given the name **quantum number** to m. By drawing on (18•3–4), the Schroedinger equation for the eigenstates $|m\rangle$ may be simplified by replacing the operator $\hat{L}_z$ inside (18•3–2) by its eigenvalue, leading to a form that may be written as

$$[\hat{H}_0 + \tfrac{1}{8}m_e\omega_L^2 \cdot (x^2 + y^2)]|m\rangle = \mathcal{E}'_m|m\rangle, \tag{18•3–5}$$

where $\hat{H}_0$ is the Hamiltonian in the absence of any magnetic field, and

$$\mathcal{E}'_m = \mathcal{E}_m - \tfrac{1}{2}m\hbar\omega_L. \tag{18•3–6}$$

The overall Hamiltonian on the left-hand side of (18•3–5) combines the potential $V(r)$ with a *cylindrical* harmonic oscillator potential with an oscillation frequency $\omega = \omega_L/2$, rather than ω_L.

Most potentials $V(r)$ of actual interest confine the electron very much more strongly than the magnetic harmonic oscillator potential. This is readily seen by considering the natural unit of length for this oscillator, the quantity

$$L = \sqrt{\frac{\hbar}{m_e\omega}} = \sqrt{\frac{2\hbar}{m_e\omega_L}} = \sqrt{\frac{2\hbar}{eB}}. \tag{18•3–7}$$

Numerically, for a field of 1 Tesla, $L \approx 36$ nm, decreasing inversely with the square root of the field. By comparison, most potentials $V(r)$ of actual interest confine the electron to a fraction of a nanometer. This means that the magnetic harmonic oscillator term is at most a very weak perturbation, shifting the energy levels upward by an amount that is small compared to the energy $\hbar\omega_L$. To the first order, this term may therefore be neglected, to be treated by perturbation theory, if necessary (see the problems). For now, we simply neglect the harmonic oscillator term in (18•3–5); that is, we write

$$\hat{H}_0|m\rangle = \mathcal{E}'_m|m\rangle, \tag{18•3–8}$$

which is *mathematically* identical to the Schroedinger equation in the absence of a magnetic field and which, therefore, has the same eigenvalues. What has changed is the meaning of these eigenvalues: They no longer represent the *actual* energy eigenvalues, but are related to the latter via (18•3–6). This means that the actual energy eigenvalues—that is, those of the original Schroedinger equation (18•3–4)—have changed from their zero-field values $\mathcal{E}(0)$ to

$$\boxed{\mathcal{E}_m(B) = \mathcal{E}_m(B) + \tfrac{1}{2}m\hbar\omega_L.} \tag{18•3–9a}$$

In the absence of a magnetic field, those states that differ only in their magnetic quantum number are degenerate in energy. Equation (18•3–9a) shows that this degeneracy is removed by the magnetic field. Because we have $2l + 1$ different values of m, from $m = -l$ to $m = +l$, the $(2l + 1)$-fold degenerate level splits up into $2l + 1$ equidistant magnetic sub-levels (**Fig. 18•3–1**), with a level separation

$$\Delta\mathcal{E} = \tfrac{1}{2}\hbar\omega_L = \frac{e\hbar}{2m_e}B. \tag{18•3–10a}$$

The energy-level splittings given by (18•3–9a) and (18•3–10a) have a simple physical interpretation: They are those associated with the potential energy $-\mu_L \cdot \mathbf{B}$ of a magnetic moment

$$\boldsymbol{\mu}_L = -\frac{e\hbar}{2m_e}\mathbf{L}, \tag{18•3–11}$$

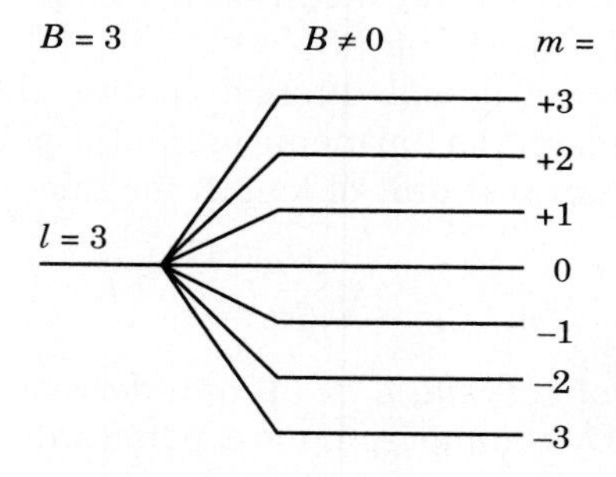

Figure 18•3–1. Magnetic level splitting in a spherically invariant potential, for $l = 3$.

proportional in magnitude to the orbital angular momentum **L**, but opposite in direction. But the projection of the angular momentum on the symmetry axis—that is, on the **B**-field—is quantized, with the values $m\hbar$. This leads to the quantization of the z-component of the magnetic moment and hence of the potential energy, (18•3–9).

The quantity

$$\mu_B = \frac{e\hbar}{2m_e} = 9.27408 \times 10^{-24}\ \mathrm{J/T}, \tag{18•3–12}$$

occurring in (18•3–11) is called the **Bohr magneton**; it is the natural unit of the magnetic moment in the quantum mechanics of electrons. The relations (18•3–9a) and (18•3–10a) are often written in the form

$$\mathcal{E}(B) = \mathcal{E}(0) + m\mu_B B \tag{18•3–9b}$$

and

$$\Delta\mathcal{E} = \mu_B B. \tag{18•3–10b}$$

It should be obvious that there must be a magnetic moment associated with the angular momentum: An angular momentum corresponds to a ring current. Furthermore, the proportionality constant in (18•3–11),

$$\gamma_L = \mu_L/L = e/2m_e, \tag{18•3–13}$$

is what we would expect from simple classical considerations: Consider a charge e of mass m_e moving with velocity v on a circular orbit of radius r. Such a motion corresponds to a circular current $I = ev/2\pi r$ and hence to a magnetic moment

$$\mu_L = I \cdot \pi r^2 = \tfrac{1}{2}evr. \tag{18•3–14a}$$

The corresponding angular momentum is

$$L = m_e v r, \tag{18•3–14b}$$

leading to the ratio (18•3–13).

We return to the energy-level splittings (18•3–10) themselves. Note first, by comparing (18•3–10a) with (8•2–12) and (18•3–4), that the splittings in a rotationally invariant confining potential are only one-half as large as those in free space. It is useful to develop some feeling for the numerical magnitudes of the splittings. One finds, from (18•3–10b) and (18•3–12),

$$\Delta\mathcal{E}/B = 5.79 \times 10^{-5}\ \mathrm{eV/T}. \tag{18•3–15a}$$

For readily realizable laboratory fields ($B \sim 10$ Tesla), this is a very small energy compared to, say, the near-1 eV energies of transitions in the optical range (or even compared to room-temperature $k_B T$, about 0.0258 eV). A second useful measure of the splitting is the equivalent frequency,

$$\nu/B = \Delta\mathcal{E}/2\pi\hbar B = 14.0\ \mathrm{GHz/T}, \tag{18•3–15b}$$

indicating the microwave-type frequencies involved in transitions between the magnetic sub-levels belonging to the same originally degenerate level.

These magnetic energy splittings are the principal mechanism of the **Zeeman effect**, that is, of the splitting of spectral lines in a magnetic field. On closer inspection, it turns out that our treatment is incomplete, and that the real Zeeman effect is more complicated because of the existence of an *intrinsic* magnetic moment of the electron due to the electron spin, discussed later. But the qualitative picture, and the numerical orders of magnitude involved, remain close to our simple description.

We close our discussion by noting that the essential identity of the effective Schroedinger equation (18•3–8) with the zero-field case means that, to the order considered here, the energy eigenfunctions do not change at all.

◆ PROBLEMS TO SECTION 18.3

#18•3-1: Atomic Diamagnetism

Investigate the effects of the $(x^2 + y^2)$ term in (18•3–5), subsequently omitted in (18•3–8) as being a very weak peturbation. Show that this term raises all energy levels by an amount proportional to B^2. Explain why such a term represents a diamagnetic contribution to the magnetic susceptibility of the atoms.

#18•3-2: Precession of Angular Momentum

An electron moving in a spherically symmetric potential perturbed by a magnetic field is in a state $|\Psi\rangle$ that is not an angular momentum eigenstate; hence, the expectation value $\langle \mathbf{L} \rangle$ of the angular momentum vector may be time-dependent. Show that $\langle \mathbf{L} \rangle$ precesses about the magnetic field, as illustrated in **Fig. 18•3–2**. Compare this behavior with the precession of a classical magnetic dipole with angular momentum $\mathbf{L}$ and magnetic moment μ, under the influence of the torque exerted by the magnetic field on the magnetic dipole.

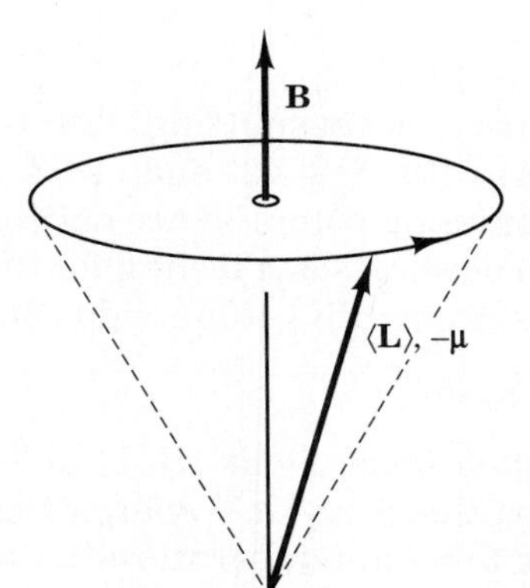

Figure 18•3–2. Precession of an angular momentum about a magnetic field.

Chapter 19

TIME-DEPENDENT PERTURBATION THEORY

19.1 INTRODUCTION

In this chapter, we turn to the time evolution of a quantum system under the influence of an external perturbation. We assume that the system was at time $t = 0$ in the unperturbed state $|1\rangle$ of the unperturbed Hamiltonian $\hat{H}^{(0)}$, with the energy $\mathcal{E}_1$:

$$\hat{H}^{(0)}|1\rangle = \mathcal{E}_1|1\rangle. \tag{19•1–1}$$

At $t = 0$, an external perturbation $\hat{W}$ is turned on, which may itself be time-dependent. The total Hamiltonian for $t > 0$ is then

$$\hat{H} = \hat{H}^{(0)} + \hat{W}(t). \tag{19•1–2}$$

We are interested in the time evolution of the state $|\Psi(t)\rangle$ of the system under the influence of the perturbation, via the Schroedinger wave equation

$$i\hbar\frac{\partial}{\partial t}|\Psi\rangle = [\hat{H}^{(0)} + \hat{W}]|\Psi\rangle. \tag{19•1–3}$$

The prototype of such a process would be the transitions between the stationary states of a quantum system—say, an atom[1]—under the influence of an electromagnetic wave. This will in fact be the case of greatest interest to us.

As in stationary perturbation theory, we express the state function $|\Psi(t)\rangle$ as an expansion in terms of the eigenfunctions $|n\rangle$ of the unperturbed Hamiltonian $\hat{H}^{(0)}$:

$$|\Psi(t)\rangle = \sum_n a_n(t)|n\rangle, \tag{19•1–4}$$

except that now the expansion coefficients a_m are time-dependent.

The $|n\rangle$ in (19•1–4) satisfy

$$\hat{H}^{(0)}|n\rangle = \mathcal{E}_n|n\rangle, \tag{19•1–5}$$

and we assume, as always, that the unperturbed eigenfunctions have been chosen as an orthonormal set:

$$\langle m|n\rangle = \delta_{mn}. \tag{19•1–6}$$

If we insert the expansion (19•1–4) into the Schroedinger wave equation (19•1–3) and form the inner product with $\langle m|$, we obtain

$$i\hbar\frac{d}{dt}a_m(t) = \sum_n a_n(t)H_{mn} = a_m(\mathcal{E}_m + W_{mm}) + \sum_{n\neq m} a_n(t)W_{mn}. \tag{19•1–7}$$

Here H_{mn} and W_{mn} are the matrix elements of the total Hamiltonian and of the perturbation, as in chapters 13 and 14.

Although simple, this formulation has the drawback that part of the time dependence of each expansion coefficient is due to the time evolution of each energy eigenstate itself, rather than to transitions between states. It is often convenient to split off the intrinsic time evolution of each state. If we assume that the diagonal matrix elements are time-independent, this is accomplished by the transformation

$$a_n(t) = c_n(t)\cdot\exp\left(-\frac{i}{\hbar}H_{nn}t\right). \tag{19•1–8}$$

If this is inserted into (19•1–7), the result can be re-arranged in the form

$$i\hbar\frac{d}{dt}c_m(t) = \sum_{n\neq m} c_n(t)\exp(i\omega_{mn}t)W_{mn}, \tag{19•1–9}$$

[1] We use the word "atom" here as a generic term, including molecules, electrons in crystals, etc.

where we have introduced the **coupling frequencies** ω_{mn} defined by

$$\hbar\omega_{mn} \equiv H_{mm} - H_{nn} = (\mathcal{E}_m - \mathcal{E}_n) + (W_{mm} - W_{nn}). \tag{19•1–10}$$

Note that the c-coefficients would be time-independent if there were no off-diagonal matrix elements, that is, in the absence of any coupling between the states. Such a formalism is said to be in the **interaction picture** of the perturbation, as opposed to the a-coefficient formalism, which is said to be in the **Schroedinger picture**. We use here the interaction picture formalism.

The expansion coefficients have a simple operational interpretation as probability amplitudes: If the perturbation is turned off after a time t, the quantity

$$P_n(t) = |c_n(t)|^2 = |a_n(t)|^2 \tag{19•1–11}$$

is the probability that the system can subsequently be "found" in the stationary state $|n\rangle$ of the unperturbed Hamiltonian $\hat{H}^{(0)}$. As initial condition, we assume that the system is in the state $|1\rangle$ at $t = 0$,

$$c_n(0) = a_m(0) = \delta_{n1}. \tag{19•1–12}$$

The probabilistic interpretation of the expansion coefficients contained in (19•1–11) and (19•1–12) hinges on a subtle but important assumption we have made: that the perturbation is not turned on until after the initial condition (19•1–12) is established, and that it is turned off before a state-determining measurement is performed. According to the theorem of measurement probabilities of chapter 12, a measurement of the energy of the system at any time must yield an eigenvalue of whatever Hamiltonian is actually present at the time of the measurement. If we wish to place the system into the stationary state $|1\rangle$ of the unperturbed Hamiltonian, or if we wish to measure in which state $|n\rangle$ of this Hamiltonian the system is to be found, we must make sure that the Hamiltonian at the time of the state preparation and at the time of the measurement is indeed the unperturbed Hamiltonian. Otherwise, we prepare or measure a state of the perturbed Hamiltonian.

These are somewhat restrictive conditions, particularly if we wish to consider the effect of perturbations that can inherently not be turned off, as would be the case with the interaction of an atom with the zero-point vibrations of the electromagnetic field surrounding it. In such cases, the initial condition can itself be given only as a probability distribution. It is possible to extend time-dependent perturbation theory to such cases, by the introduction of so-called **density matrices**. However, such a treatment would go beyond the scope of this text, and the interested reader is referred to more advanced or more specialized texts.[2]

The two sets (19•1–7) and (19•1–9) of equations are infinite sets of equations for an infinite number of variables, just as in stationary perturbation

---

[2] See, for example, Dietrich Marcuse, *Engineering Quantum Electrodynamics* (New York: Academic, 1980).

theory. There is, however, a major difference: (19•1–7) and (19•1–9) are first-order linear differential equations in time t, rather than algebraic equations. This is, of course, what we would expect for a time-evolution rather than stationary problem.

As in a stationary perturbation theory, the infinite number of equations in (19•1–7) and (19•1–9) will force us again into approximation techniques that assume that $\hat{W}$ is only a weak perturbation, relative to the dominant unperturbed Hamiltonian $\hat{H}^{(0)}$. Again, two distinctly different approximations stand out:

(a) In the first approximation, reminiscent of stationary degenerate perturbation theory, only a few selected states are included in the expansions (19•1–4) or (19•1–9). All transitions amongst those states are considered, including repeated back-and-forth transitions, but all transitions to or from other states are ignored. The first approximation is considered in sections 19.2 and 19.3, for the specific case of a simple two-state system.

(b) In the second approximation, we assume that the transition takes place into a *continuum* of an infinite number of states, but only the initial transitions are considered; all subsequent transitions within the continuum and especially back to the initial state are ignored. The second approximation is considered in section 19.4.

19.2 A STEP PERTURBATION ACTING ON A TWO-LEVEL SYSTEM

19.2.1 The Formalism

The simplest and most fundamental of all perturbations is one that, once it has been turned on, remains constant with time until it is turned off:

$$\frac{\partial \hat{W}}{\partial t} = 0 \quad \text{for} \quad t > 0. \tag{19•2–1}$$

We call such a perturbation a **step perturbation**. It is a highly idealized case, but its mathematical treatment forms the basis for the treatment of more complicated perturbations, including specifically perturbations by an electromagnetic wave.

In the present section, we simplify the problem further by considering only transitions between just two interacting states, ignoring all other states, and leaving it open for now under which conditions and to what extent such a neglect is in fact justified. If we designate the two states as $|1\rangle$ and $|2\rangle$, the time evolution equations (19•1–9) for the two coefficients c_1 and c_2 may be written

$$i\hbar \frac{d}{dt} c_1(t) = c_2(t) W_{12} \exp(-i\delta\omega t), \tag{19•2–2a}$$

$$i\hbar \frac{d}{dt} c_2(t) = c_1(t) W_{21} \exp(+i\delta\omega t), \tag{19•2–2b}$$

where we have introduced an **energy mismatch frequency** $\delta\omega$ defined by

$$\delta\omega \equiv \omega_{21} = \frac{1}{\hbar}(H_{22} - H_{11}) = \frac{1}{\hbar}[(\mathcal{E}_2 - \mathcal{E}_1) + (W_{22} - W_{11})]. \tag{19•2–3}$$

The system (19•2–2) is readily solved by setting

$$c_1(t) = b_1 \exp[i(\omega - \tfrac{1}{2}\delta\omega)t], \tag{19•2–4a}$$

$$c_2(t) = b_2 \exp[i(\omega + \tfrac{1}{2}\delta\omega)t]. \tag{19•2–4b}$$

where ω is an as-yet unknown frequency. Insertion into (19•2–2) leads to a pair of algebraic equations, which may be written

$$b_1\hbar(\omega - \tfrac{1}{2}\delta\omega) + b_2 W_{21} = 0, \tag{19•2–5a}$$

$$b_1 W_{21} + b_0\hbar(\omega + \tfrac{1}{2}\delta\omega) = 0. \tag{19•2–5b}$$

They have solutions only if their determinant vanishes. This leads to the condition on the unknown frequency ω in (19•2–4):

$$\boxed{\omega = \pm\Omega, \quad \text{where} \quad \Omega^2 = \frac{1}{\hbar^2}|W_{12}|^2 + (\tfrac{1}{2}\delta\omega)^2.} \tag{19•2–6}$$

The most general solution of (19•2–4) consists of linear superpositions of the special solutions for the two allowed frequencies $\omega = \pm\Omega$:

$$c_1(t) = [b_1^+ e^{+i\Omega t} + b_1^- e^{-i\Omega t}]\exp(-\tfrac{i}{2}\delta\omega t), \tag{19•2–7a}$$

$$c_2(t) = [b_2^+ e^{+i\Omega t} + b_2^- e^{-i\Omega t}]\exp(+\tfrac{i}{2}\delta\omega t). \tag{19•2–7b}$$

We are interested in the solution that satisfies the initial condition (19•1–12), which requires

$$c_1(0) = b_1^+ + b_1^- = 1, \tag{19•2–8a}$$

$$c_2(0) = b_2^+ + b_2^- = 0. \tag{19•2–8b}$$

From (19•2–5b), (19•2–6), and (19•2–8), we obtain

$$b_1^\pm = \frac{1}{2} \pm \frac{\delta\omega}{4\Omega}, \tag{19•2–9}$$

and with these values, (19•2–7a) and (19•2–7b) assume the final form

$$c_1(t) = \left[\cos\Omega t + \frac{i\delta\omega}{2\Omega}\sin\Omega t\right] \cdot \exp(-\tfrac{i}{2}\delta\omega t), \tag{19•2–10a}$$

$$c_2(t) = -\frac{iW_{21}}{2\Omega} \sin \Omega t \cdot \exp(+\tfrac{i}{2}\,\delta\omega t). \tag{19•2–10b}$$

The associated probabilities $|c_1|^2$ and $|c_2|^2$ are

$$|c_1(t)|^2 = \cos^2 \Omega t + \left(\frac{\delta\omega}{2\Omega}\right)^2 \sin^2 \Omega t = 1 - C^2 \sin^2 \Omega t, \tag{19•2–11a}$$

and

$$|c_2(t)|^2 = C^2 \sin^2 \Omega t, \tag{19•2–11b}$$

where

$$C^2 = 1 - \left(\frac{\delta\omega}{2\Omega}\right)^2 = \frac{|W_{21}|^2}{\hbar^2\Omega^2} = \frac{|W_{21}|^2}{(\frac{1}{2}\hbar\,\delta\omega)^2 + |W_{21}|^2}. \tag{19•2–12}$$

19.2.2 Interpretation

Final-State Probability Oscillations

According to our discussion in section 19.1, the quantity $|c_2(t)|^2$ is the *transition probability* from state $|1\rangle$ to state $|2\rangle$, defined as the probability that the system materializes in state $|2\rangle$ if the perturbation is turned off after having acted for a time t. Equation (19•2–11b) states that, for a two-level system, the transition probability oscillates with time between zero and a peak value C^2, as shown in **Fig. 19•2–1**. The oscillation period is $T = 2\pi/\Omega$. If the perturbation acts for an exact integer multiple of this period, the system is guaranteed to be "found" in its original state $|1\rangle$, even though a shorter action would have led to a nonzero transition probability.

The peak transition probability C^2 occurs at all half-integer multiples of T. Its magnitude depends on the energy mismatch frequency $\delta\omega$ defined in (19•2–3). According to (19•2–12), this peak transition probability reaches unity for $\delta\omega = 0$, implying that in this case—and for t being a half-integer multiple of T—a transition from state $|1\rangle$ to state $|2\rangle$ is guaranteed. With increasing

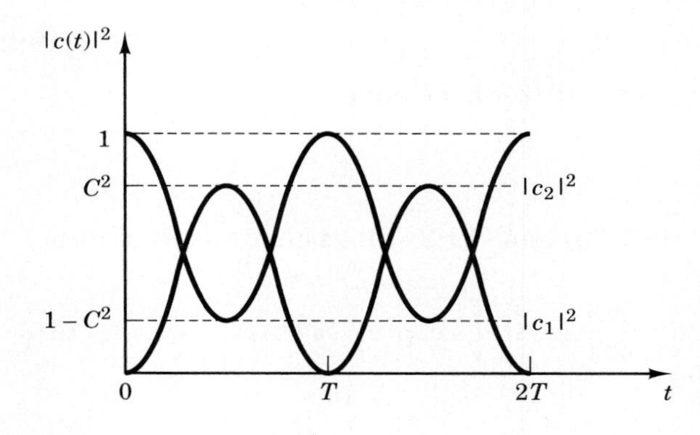

Figure 19•2–1. Oscillations of the transition probability $|c_2(t)|^2$ from state $|1\rangle$ to state $|2\rangle$ and of the probability $|c_1(t)|^2$ that the system remains in state $|1\rangle$.

energy mismatch $\hbar\delta\omega$ between the two states, the peak transition probability C^2 decreases, as shown in **Fig. 19•2–2**. We evidently have a resonance-like behavior for the transition probability oscillations, with the maximum occurring for $\hbar\delta\omega = 0$.

Resonances of the form (19•2–12) are common; they are called **Lorentzian** resonances, to distinguish them from resonances with other shapes, such as a Gaussian. For a fixed strength of the perturbation, the maximum transition probability has dropped to 1/2 for

$$\hbar\delta\omega = H_{22} - H_{11} = \pm 2|W_{21}|. \tag{19•2–13}$$

The interval between the two points $\pm\hbar\delta\omega$,

$$\boxed{\Delta\mathcal{E} = \hbar\Delta\omega = 4|W_{21}|,} \tag{19•2–14}$$

is commonly used as a measure for the energetic width of the resonance. It is called the **f**ull **w**idth at **h**alf **m**agnitude, usually abbreviated FWHM.

Transition Resonance and Energy Conservation

To understand this resonant behavior better, we first note that, whenever an external perturbation causes a system to make a transition from one state to another with different energy, the energy difference is due to an energy exchange between the perturbed system of interest and a perturbing external system. If the two energies $\mathcal{E}_1$ *and* $\mathcal{E}_2$ are different, net work must be done by the outside perturbation to make up that difference. A little reflection shows that the process of turning the perturbation on or off involves work itself.

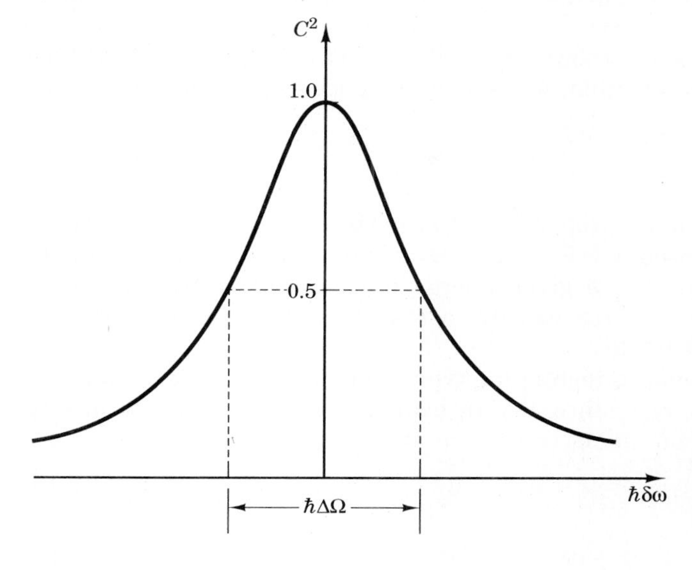

Figure 19•2–2. Lorentzian resonant nature of the maximum transition probability C^2 as a function of the energy mismatch $\hbar\delta\omega$.

Suppose the system is initially in the sharp state $|1\rangle$, and the perturbation is turned on abruptly. At first, the wave function will still be that of the original state $|1\rangle$, and hence the expectation value of the *kinetic* energy will also remain initially unchanged. But if the process of turning on the perturbation places the state into a different potential, the potential energy may change; the term W_{11} contained in (19•2–3) is simply the amount of work that must be done by the perturbation to accomplish this change.

When the perturbation is subsequently turned off—an essential assumption in our interpretation given earlier—work is again done, which may or may not recover the work done during turn-on. The term W_{22} is the amount of work that *would* have to be done by the perturbation to turn *on* the perturbation *if* the system were initially in state $|2\rangle$. This means that W_{22} would be the amount of work *recovered* when the perturbation is turned off, *if* the state function of the system at the instant of turning-off were in fact that of state $|2\rangle$. Usually, that last condition will not be satisfied, and the amount of work recovered will depend on the state into which the system is forced by the perturbation.

If the system of interest ends up in a mixed state with an unsharp energy, this simply means that we cannot predict in which one of two or more possible energy eigenstates the system may be found, and hence the work done during turnoff remains unknown, until a state-determining measurement is done, which determines not only the final state, but the work done during turning off the perturbation.

Given this interpretation, the resonance relation (19•2–12) simply states that a high transition probability will occur only if the energy difference $\mathcal{E}_2 - \mathcal{E}_1$ between the two *unperturbed* states matches the net work $W_{11} - W_{22}$ that must be done to turn the step perturbation on and off, so that the balance of work to be done by the perturbation *during* its action stays within a resonance range on the order of $\pm 2|W_{12}|$.

Under no circumstances should the finite resonance width be interpreted as a lack of energy conservation, or as energy conservation merely "on the average."

Resonance Width

The resonance width is itself proportional to the strength of the perturbation. Thus, we have a sharp resonance for weak perturbations and a broad resonance for strong perturbations. For a given energy mismatch $\hbar\delta\omega$, the maximum transition probability $|C|^2$ increases with increasing strength of the perturbation, as shown in **Fig. 19•2–3**.

It is useful to develop a feeling for typical magnitudes of the energetic width of the transition resonance. As an example, consider the action of a uniform electric field E on an electron in an atom. We then have

$$\hat{W} = eEx; \qquad (19•2–15)$$

hence,

$$W_{12} = eE \cdot \langle 1|x|2\rangle = eEx_{12}. \qquad (19•2–16)$$

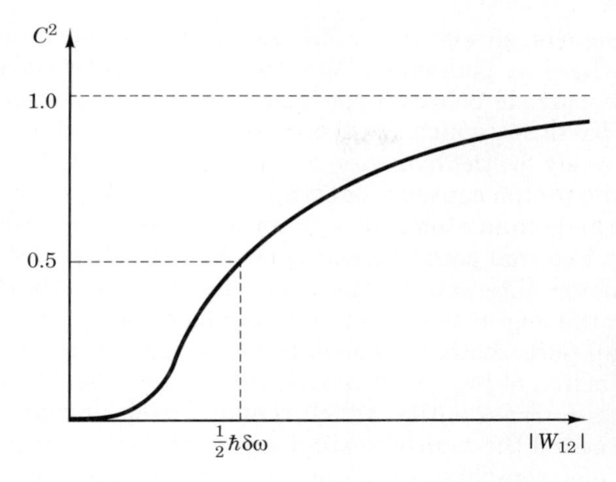

Figure 19•2–3. Dependence of the maximum transition probability on the strength of the perturbation, in the presence of a fixed energy mismatch $\hbar\delta\omega$.

The matrix element x_{12} is at most on the order of the physical size of the atom; often it will be much smaller. Thus, we may say that in an atom perturbed by a uniform field, transitions can take place only between states whose energies differ at most by the potential energy difference in the field, taken over a distance on the order of the size of the atom. For man-made macroscopic fields ($E \leq 10^6$ V/cm), this is invariably a small energy ($\leq 10^{-2}$ eV). Applied to the hydrogen atom, our result means that such fields will not cause transitions between states with different principal quantum numbers n, although they may cause transitions between the degenerate states with different values of l belonging to the same n.

19.3 PERTURBATION BY AN ELECTROMAGNETIC WAVE

19.3.1 The Problem

Probably the most important problem of time-dependent perturbation theory in low-energy quantum mechanics is the absorption and emission of photons during the interaction of an electron quantum system with an electromagnetic wave. For simplicity, we assume that only a single wave frequency ω_0 is present, as might be the situation inside a high-Q electromagnetic resonator. We refer to such a single-frequency field as a **mode** of the resonator.

Because of the quantization of electromagnetic energy, the electromagnetic field mode is itself a quantum system, which is in turn perturbed by the electron system. Quantum mechanically, each field mode is a harmonic oscillator, similar to the way we treated the LC circuit as a quantum-mechanical harmonic oscillator in chapter 10. Strictly speaking, we are therefore dealing, not with the perturbation of an electron quantum system by an *external* classical field, but with the *internal* interactions within a larger quantum system that consists of two interacting sub-systems perturbing each other.

The situation is somewhat similar to our discussion of composite quantum systems in chapter 11, where we pointed out that the external potentials $V(\mathbf{r})$ showing up in the single-particle Schroedinger equation are really caused by interactions with other particles, which are themselves subject to the laws of quantum mechanics. Already for the hydrogen atom, we found that the action of the electron back on the proton caused a downward shift of all energy levels by a common factor; in the helium atom, the system could not be described at all in terms of a common external potential acting on the individual electrons. The case of electron-photon interaction falls somewhat between those two extremes, depending on the degree of excitation of the field oscillator.

We treat the overall perturbation problem in two stages. In the present chapter, we neglect the action of the electron back onto the field and treat the field as a purely classical *external* entity, which remains itself unchanged by the interaction. This is called the **semiclassical approximation**. It gives a valid description of the electron-photon interaction in the limit of waves that are sufficiently strong that the number of photons contained in each mode of interest is very large. In that limit, the change of field energy by one photon has a very small effect. In fact, all waves that can be described classically by an amplitude *and a phase* automatically meet this criterion: As we saw in chapter 10, it is possible to assign a well-defined classical phase to a wave only if the number n of quanta contained in the wave has a large uncertainty $\Delta n \gg 1$, which requires that the expectation value $\langle n \rangle = (\Delta n)^2$ of the number of quanta contained in the wave is even larger.

The semi-classical approximation does not give a satisfactory *quantitative* account of the most common and important transition process of all, the *spontaneous* emission of a light quantum into a field mode that is initially empty. In that limit, the action of the electron on the field dominates the problem, and a fully quantized treatment of the field mode and the electron-field interaction becomes essential to obtain quantitatively correct results. We will give such a treatment in the next chapter.

19.3.2 The Semiclassical Interaction Hamiltonian

The electric field in an electromagnetic wave is not curl-free and, hence, cannot be expressed as the gradient of an electrostatic potential. Instead, the field enters the Hamiltonian via its vector potential. We showed in chapter 8 that in the presence of a vector potential $\mathbf{A}$, the Hamiltonian for an object of mass m_e and charge $-e$ is of the form (8•1–16):

$$\hat{H} = \frac{1}{2m_e}[\hat{\mathbf{p}} + e\mathbf{A}]^2 + V(\mathbf{r}). \qquad (19\bullet3\text{–}1)$$

Although our arguments in chapter 8 leading to this form assumed static fields, we postulate that (19•3–1) remains valid for rapidly time-varying vector potentials.

The electric field is easily obtained from the vector potential via the fundamental relation

$$\mathbf{E}(\mathbf{r}, t) = -\frac{\partial}{\partial t}\mathbf{A}(\mathbf{r}, t). \tag{19•3–2}$$

We also recall from electromagnetic theory that the vector potential may always be chosen divergence-free:

$$\text{div}\,\mathbf{A} = 0. \tag{19•3–3}$$

It is left to the reader to show that this implies a vector potential that commutes with the momentum operator:

$$\hat{\mathbf{p}} \cdot \mathbf{A} = \mathbf{A} \cdot \hat{\mathbf{p}}. \tag{19•3–4}$$

With the help of (19•3–4), we may re-write (19•3–1) as

$$\hat{H} = \frac{\hat{p}^2}{2m_e} + \frac{e}{m_e}\mathbf{A} \cdot \hat{\mathbf{p}} + \frac{e^2 A^2}{2m_e} + V(\mathbf{r}). \tag{19•3–5}$$

We view the first and last terms in (19•3–5) as the unperturbed Hamiltonian of our problem,

$$\hat{H}^{(0)} = \frac{\hat{p}^2}{2m_e} + V(\mathbf{r}), \tag{19•3–6}$$

and the remaining terms as a perturbation by the electromagnetic field.

Most electromagnetic wave fields of practical interest are sufficiently weak that the $\mathbf{A}^2$-term in (19•3–5) will be small compared to the linear term. In fact, the $\mathbf{A}^2$-term may be viewed as part of the reaction of the electron back onto the field and taken out of the electron Hamiltonian altogether, to be lumped with a field Hamiltonian. Consistent with our semiclassical point of view of ignoring such reaction effects, we simply neglect the quadratic term here and write, for the perturbation,

$$\boxed{\hat{W} = \frac{e}{m_e}\mathbf{A} \cdot \hat{\mathbf{p}}.} \tag{19•3–7}$$

The occurrence of the momentum operator reflects the fact that the energy exchange between charges and fields depends on the ability of the field to induce a *current,* not merely on the presence of static charges: Without a current, no work is done.

To simplify the discussion further, we restrict ourselves (temporarily) to transitions between bound states of a system that is very small compared to the wavelength of the wave. This is called the **dipole approximation**, because it will turn out that the interaction of the field with the electron system then assumes the form of an interaction with a simple electron dipole. It is an excellent assumption for atomic systems perturbed by visible or infrared light, and even more so for longer wavelengths. We will discuss the opposite limit of electrons in a crystal later.

The assumption that the system is small compared to the wavelength means that we may neglect the position dependence of the both the electric field and the vector potential and treat both as time-dependent only. More specifically, we assume a linearly polarized and perfectly sinusoidal electric field of the form

$$\mathbf{E}(t) = \mathbf{E}_0 \sin(\omega_0 t). \tag{19•3–8}$$

From the relation (19•3–2) between electric field and vector potential, we have

$$\mathbf{A}(t) = \frac{\mathbf{E}_0}{\omega_0} \cdot \cos \omega_0 t. \tag{19•3–9}$$

If this is inserted into (19•3–7), we obtain

$$\begin{aligned} \hat{W} &= \frac{eE_0}{m_e \omega_0} \cdot \hat{p}_{\mathbf{E}} \cdot \cos \omega_0 t \\ &= \hat{W}' \cdot [\exp(+i\omega_0 t) + \exp(-i\omega_0 t)], \end{aligned} \tag{19•3–10}$$

where $\hat{p}_{\mathbf{E}}$ is the momentum component in the direction of $\mathbf{E}_0$ and

$$\hat{W}' = \frac{eE_0}{2m_e \omega_0} \cdot \hat{p}_{\mathbf{E}}. \tag{19•3–11}$$

We assume again, as in section 19.2, that the electron system may be treated as a pure two-level system. If we insert (19•3–10) into the basic time evolution equations (19•1–9), the result may be written in the form

$$i\hbar \frac{d}{dt} c_1(t) = c_2(t) W'_{12}[\exp(-i\delta\omega_+ t) + \exp(-i\delta\omega_- t)], \tag{19•3–12a}$$

$$i\hbar \frac{d}{dt} c_2(t) = c_1(t) W'_{21}[\exp(+i\delta\omega_+ t) + \exp(+i\delta\omega_- t)], \tag{19•3–12b}$$

where we have defined two new energy mismatch frequencies $\delta\omega_+$ and $\delta\omega_-$ via

$$\hbar\delta\omega_+ \equiv \hbar(\omega_{21} - \omega_0) = \mathcal{E}_2 - (\mathcal{E}_1 + \hbar\omega_0), \tag{19•3–13a}$$

$$\hbar\delta\omega_- \equiv \hbar(\omega_{21} + \omega_0) = (\mathcal{E}_2 + \hbar\omega_0) - \mathcal{E}_1. \tag{19•3–13b}$$

In writing (19•3–13a,b), we have also used the fact that for bound states the expectation value of the momentum must be zero, which implies a zero diagonal matrix element of the perturbation $\hat{W}$:

$$W_{11} = W_{22} = 0. \tag{19•3–14}$$

We note that (19•3–12a,b) are of a form similar to the earlier evolution equations (19•2–2a,b) for a step perturbation, except for the occurrence of *two* oscillating terms on the right-hand sides, with two different energy mismatch frequencies. To understand the roles of the two different-frequency terms in (19•3–10), it is instructive to treat them *as if* they represented two separate and independent perturbations.

Suppose, therefore, that we ignore the $\delta\omega_-$-terms in (19•3–12a,b) and consider the effect of the $\delta\omega_+$-terms alone. The relations (19•3–12a,b) then reduce exactly to the form (19•2–2a,b) found earlier for the step perturbation, with $\delta\omega_+$ taking on the role of the energy mismatch frequency $\delta\omega$ and W'_{12} taking on the role of W_{12}.

Evidently, the entire step perturbation formalism of section 19.2 may then be taken over verbatim, including the conclusion that strong transitions between the two states take place only if the energy mismatch frequency is sufficiently small to fall within a narrow resonance bandwidth on the order of $\pm 2|W'_{21}|/\hbar$, from (19•2–13). We shall see shortly that in almost all cases of practical interest, the energetic width $\Delta\mathcal{E} = \hbar\Delta\omega$ of the transition will indeed be very small compared to the photon energy $\hbar\omega_0$. With our re-definition (19•3–13a) of the energy mismatch frequency $\delta\omega_+$, this means that non-negligible transitions take place only if the energetic ordering of the states $|1\rangle$ and $|2\rangle$ is such that $\mathcal{E}_1 < \mathcal{E}_2$ *and* the photon energy of the electromagnetic wave matches the energy difference between the two electron states.

By a completely analogous argument, the $\delta\omega_-$-terms, taken alone, would also lead to a formalism of the form (19•2–2a,b), but with the different energy mismatch frequency $\delta\omega_-$. A little reflection shows that those terms cause non-negligible transitions only if the energetic ordering of the states $|1\rangle$ and $|2\rangle$ is reversed, such that $\mathcal{E}_1 > \mathcal{E}_2$ *and*, the photon energy of the electromagnetic wave again matches the energy difference between the two electron states.

The presence of two terms ensures that the transitions can take place regardless of which of the two electron states has the higher energy, but given a particular choice, only one of the two oscillating terms plays a role; the more rapidly oscillating term may be dropped. **Fig. 19•3–1** shows the retained term for the case of photon absorption; the reader is requested to draw the appropriate diagram for photon emission.

In either case, we conclude that a system initially in state $|1\rangle$ will exhibit an oscillating transition probability into state $|2\rangle$ if and only if the energy difference between the states is within the resonance bandwidth of the quantum energy of the electromagnetic wave. A maximum transition probability $C^2 = 1$ will occur only if the two energies are *exactly* equal. The energetic width (FWHM) of the resonance is again given by (19•2–14),

$$\Delta\mathcal{E} = \hbar\Delta\omega = 4|W'_{12}|. \qquad (19\bullet3\text{–}15)$$

Our earlier comments about energy conservation also apply again.

From our semiclassical point of view, dropping one of the two terms in (19•3–10) appears to be an approximation. But as we shall see in the next chapter, in a rigorous quantized-field formalism, the dropped term never shows

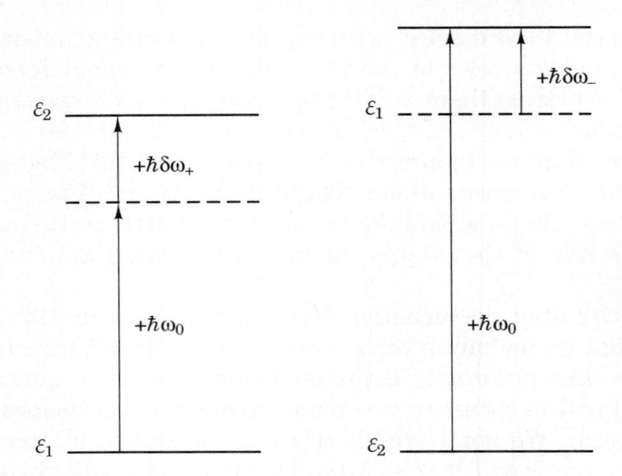

Figure 19•3–1. Definition of photon energy $\hbar\omega_0$ and energy mismatch $\hbar\delta\omega$ for absorptive transitions between two energy levels $\mathcal{E}_1$ and $\mathcal{E}_2$, for both kinds of energetic ordering of the two levels. The arrows indicate the directions in which the indicated quantities are counted positively.

up in the first place, and omitting it in the semiclassical treatment not only simplifies the latter, but brings it closer to an exact formalism. What this all means is that it we may view the interaction not as an *external oscillating* perturbation acting on the electron system alone, but as an *internal step* perturbation of the form (19•3–11), acting on the combined electron-plus-photon system, *with the unperturbed energies of this combined system containing the photon energies along with the electron energies.* The energy increase in the electron portion of the overall system during the absorption of a photon is accompanied by an energy decrease in the field portion in the overall system (**Fig. 19•3–2**).

Similarly, the energy decrease in an electron system during the emission of a photon is accompanied by an energy increase in the field system. In either case, only those transitions can take place for which the sum of the two energies

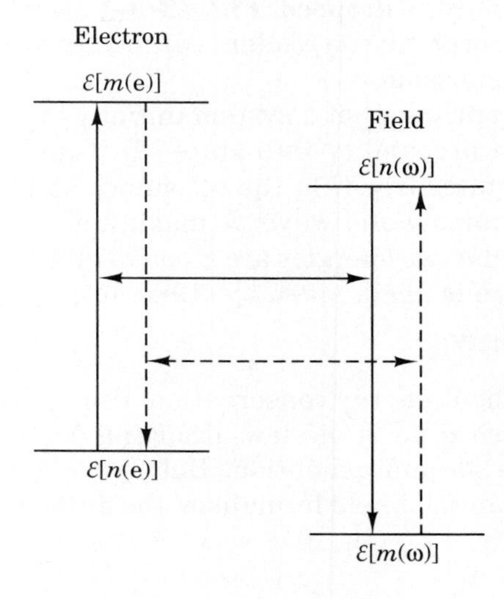

Figure 19•3–2. Radiative transitions. An electron system and a field oscillator make transitions simultaneously, in opposite energetic directions, in such a way that the sum of the two energies stays within the narrow resonance bandwidth about exact energy conservation.

stays within the narrow resonance bandwidth about exact energy conservation, just as for the step perturbation case discussed in section 19.2.

Only such transitions will take place for which the lower energy *electron* state—whichever it may be—is associated with a photon state containing one additional photon. By adding the energy of one photon to the lower energy electron state, the energy mismatch frequencies defined in (19•3–13a,b) become precisely the frequencies associated with the energy mismatch of the *combined* system before and after the transition.

This re-interpretation of the mathematical formalism of the semiclassical approximation leads us naturally to the coupled quantum system point of view espoused at the beginning of this section. All that is left is to determine the exact form of a suitable Hamiltonian for the photon portion of the combined quantum system and its relation to the classical vector potential. This will be the task of the next chapter. For the remainder of the present chapter, we will retain the semiclassical formalism, in the sense that we will continue to treat the vector potential as an external classical perturbation, *but with the augmentation that we will treat it as a non-oscillating step perturbation* given by (19•3–11), rather than as the original fully classical oscillating perturbation (19•3–10).

19.3.3 Resonance Width

Evidently, the entire transition behavior between two states depends on the matrix element

$$W'_{12} = \frac{|E_0|}{2m_e\omega_0}|p_{12}|, \tag{19•3–16}$$

where we have drawn on (19•3–11) and where $p_{12} = \langle 1|\hat{p}_{\mathbf{E}}|2\rangle$ is the momentum matrix element between the two states, with the momentum taken in the direction of $\mathbf{E}_0$. It is useful to replace the momentum matrix element by the position matrix element $x_{12} = \langle 1|\hat{x}|2\rangle$, by drawing on the general relation

$$\hat{p}_x = \frac{im_e}{\hbar}\cdot[\hat{H},\hat{x}], \tag{19•3–17}$$

derived in chapter 7—see (7•6–12). This relation is true for any electron Hamiltonian that does not contain a vector potential, including the unperturbed Hamiltonian $\hat{H}^{(0)}$ of our problem. With (19•3–17),

$$\langle 1|\hat{p}_x|2\rangle = \frac{im_e}{\hbar}\langle 1|\hat{H}\hat{x} - \hat{x}\hat{H}|2\rangle. \tag{19•3–18}$$

Because $\hat{H}$ is Hermitian, and $|1\rangle$ and $|2\rangle$ are eigenstates of $\hat{H} = \hat{H}^{(0)}$ with the eigenvalues $\mathcal{E}_1$ and $\mathcal{E}_2$, (19•3–18) is simply

$$\langle 1|\hat{p}_x|2\rangle = \frac{im_e}{\hbar}\cdot(\mathcal{E}_1 - \mathcal{E}_2)\cdot\langle 1|\hat{x}|2\rangle = im_e\omega_{12}x_{12}. \tag{19•3–19}$$

Inserting this into (19•3–16) yields

$$\boxed{W'_{12} = \frac{e|E_0|}{2} \cdot \frac{|\omega_{12}|}{\omega_0} \cdot |x_{12}| \approx \frac{e|E_0|}{2} \cdot |x_{12}|,} \tag{19•3–20}$$

where, in the last equality, we have replaced $|\omega_{12}|$ by ω_0 because only those transitions have a non-negligible probability for which $|\omega_{12}| \approx \omega_0$.

The product $e|x_{12}|$ has the form of an electric dipole moment. It is this circumstance that gives rise to the name **dipole approximation** to the approximation (19•3–9), in which any position dependence of the vector potential has been neglected.

It is useful to develop a feeling for typical magnitudes of the energetic width of the transition resonance. The matrix element x_{12} is at most on the order of the physical size of the system inside which the electron resides, such as an atom; often it will be much smaller. Hence, we may say that the energetic width of the transition range can be at most on the order of the potential energy difference in the field E_0, taken over a distance on the order of the size of the atom.

From electromagnetic theory, we know that the time-averaged energy flux density $-\mathbf{S}$ of a linearly polarized electromagnetic wave (the average magnitude of its Poynting vector) is given by

$$\overline{S} = \frac{E_0^2}{2\eta}, \tag{19•3–21}$$

where $\eta = (\mu_0/\epsilon_0)^{1/2} = 377\ \Omega$ is the wave impedance of free space. Most monochromatic light sources other than lasers are much weaker than sunlight. For sunlight, $\overline{S} \approx 10^3\ \mathrm{W/m^2}$; hence, $E_0 \approx 868\ \mathrm{V/m}$. Assuming $x_{12} \approx 10^{-10}$ m, we obtain $|W'_{12}| \approx 4.3 \cdot 10^{-8}$ eV. Evidently, the resonance width $4|W'_{12}|$ is indeed *very* small compared to typical photon energies of, say, 2 eV.

From a quantitative point of view, our estimate is only as valid as the assumptions that went into it, namely, that transitions caused by an incident wave are the only transitions present. This is an oversimplification. In most physical systems there are additional transitions present, besides the incident light itself, which can contribute greatly to the resonance broadening.

19.3.4 Critique of the Semiclassical Approximation

If one sets the electric field amplitude E_0 in our semiclassical treatment to zero, there is no perturbation left, and the treatment predicts that the system will remain indefinitely in its state, even if the state is the upper state. But it is, of course, an elementary fact of nature that almost any quantum system will decay to its ground state if it is permitted to radiate off its internal energy without receiving any energy inputs from the outside. In fact, most light in nature other than laser light is generated by this spontaneous emission process. Our treatment does not account for this elementary fact.

The resolution of the problem lies again in treating the transition in terms of two interacting quantum systems. The field system would then presumably be some sort of harmonic oscillator, and the matrix elements for the transition would be expressed, not in terms of electric field strengths, but in terms of harmonic oscillator quantum numbers. These may be expressed in terms of classical electric field amplitudes by equating the quantum energy $(n + 1/2)\hbar\omega_0$ to the total classical field energy contained in the particular field mode, which is proportional to E_0^2. We recall that the harmonic oscillator matrix elements all depended on the *higher* of the two quantum numbers involved in the transition, which is never zero. But that means that the electric field amplitude used to express the matrix elements must also be chosen to be the amplitude in the higher of the two field states. In a photon *absorption* event this is the field *before* the transition, and in a photon *emission* event it is the field *afterwards*. In either case, this upper-state field is never zero, and hence there is always a finite probability of photon emission. We shall elaborate on this point in the next chapter.

The spontaneous emission is simply a form of interaction between the atomic system and the field system, where the field system happens to be initially in its ground state. But this is not at all the same as neglecting the presence of the field system altogether, which is in effect what we do if we set $E_0 = 0$.

19.3.5 Removal of the Dipole Approximation: Interaction between Bloch Waves

We return once again to the original form (19•3–7) of the perturbation, but rather than neglecting the position-dependence of the vector potential, we now assume that the latter has the form of a traveling wave with the wave vector $\mathbf{k}$:

$$\mathbf{A}(\mathbf{r}, t) = \frac{\mathbf{E}_0}{\omega_0} \cdot \cos(\mathbf{k} \cdot \mathbf{r} - \omega_0 t). \tag{19•3–22}$$

The $\mathbf{A} \cdot \mathbf{p}$-term in the perturbation may then be written in the form

$$\frac{e}{m_e}\mathbf{A} \cdot \hat{\mathbf{p}} = \hat{W} \cdot \exp(+i\omega_0 t) + \hat{W}^\dagger \cdot \exp(-i\omega_0 t), \tag{19•3–23}$$

which differs from (19•3–10) in that $\hat{W}'$ has been replaced by the Hermitian conjugate pair

$$\hat{W} = \frac{eE_0}{m_e\omega_0} \cdot e^{-i\mathbf{k} \cdot \mathbf{r}} \cdot \hat{p}_\mathbf{E} \quad \text{and} \quad \hat{W}^\dagger = \frac{eE_0}{m_e\omega_0} \cdot e^{+i\mathbf{k} \cdot \mathbf{r}} \cdot \hat{p}_\mathbf{E} \cdot \tag{19•3–24a,b}$$

As before, one of the two terms in (19•3–23) should be omitted, and if the remainder of the calculation in sub-section 19.3.2 is repeated, one finds that the only difference is the occurrence of a factor $\exp(+i\mathbf{k}\cdot\mathbf{r})$ or $\exp(-i\mathbf{k}\cdot\mathbf{r})$ in the matrix element W_{12}. These factors account for the fact that the absorbed or emitted photon carries the momentum $\hbar\mathbf{k}$, in addition to the energy $\hbar\omega$, and they ensure that the overall transition process conserves momentum.

An important case of this kind is the absorption or emission of light in a crystalline solid. Here the energy eigenfunctions are Bloch waves. We find easily that transitions between Bloch waves can have a nonzero matrix element only if the electron Bloch wave vectors $\mathbf{k}_e$ and $\mathbf{k}_e'$ before and after the transition obey

$$\mathbf{k}_e' - \mathbf{k}_e = \pm\mathbf{k}_p, \tag{19•3–25}$$

where $\mathbf{k}_p$ is the wave vector of the photon. The upper sign holds for the absorption of a photon, the lower for its emission. The wavelengths of light of practical interest are much larger than the lattice parameters of crystals. As a result, the photon wave vectors of interest are much smaller than the reciprocal lattice vectors: Optical transitions between Bloch waves in crystals are very nearly vertical in an $\mathcal{E}(\mathbf{k})$ band diagram, and for many purposes, they may be treated as if they were exactly vertical. This rule of $\mathbf{k}$-conservation may be broken, however, if photons are not solely responsible for the transition. We will return to this point in section 19.6.

19.4 TRANSITIONS INTO A CONTINUUM OF STATES: FERMI'S GOLDEN RULE

19.4.1 The Problem

In the preceding sections of the chapter, we considered in detail the transitions between two *discrete* states of a quantum system. All other states of the system were assumed to be energetically so far removed that their participation could be neglected. We now turn to the opposite extreme, of transitions from an initial state $|1\rangle$ into a continuum of states,[3] as shown in **Fig. 19•4–1**. Such situations are very common.

We mention briefly three examples:

(a) The scattering of a free particle with an initial wave vector $\mathbf{k}$ by a potential $W(\mathbf{r})$. The potential acts as a time-independent perturbation causing transitions to other plane-wave states.

(b) The ionization of an atom by light (the photo effect). The light wave acts as a perturbation causing transitions of an electron from an initial discrete bound state into the free-electron continuum.

[3] The initial state might be a member of the continuum itself.

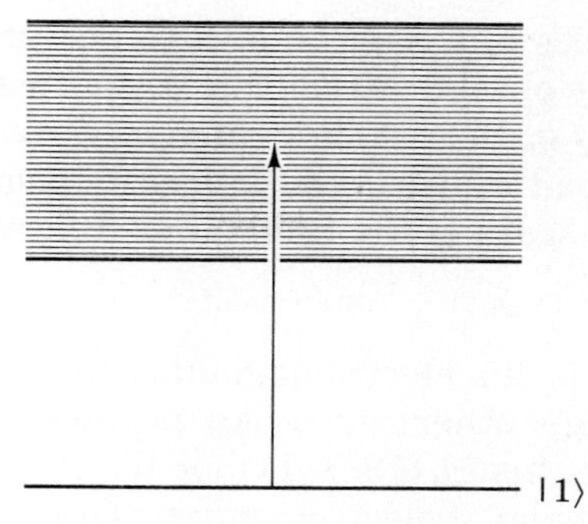

Figure 19•4–1. Transitions from an initial state $|1\rangle$ into a continuum.

(c) The transition between two discrete states of an electron in an atom under the influence of broadband rather than monochromatic electromagnetic radiation. Although the final state of the *electron* is not a member of a continuum, the final state of the electromagnetic field is, and as a result, the formalism of example (b) is applicable to this third case as well.

To treat these cases, we go back to the basic time evolution equation (19•1–9) for the expansion coefficients:

$$i\hbar \frac{d}{dt} c_m(t) = \sum_{n \neq m} c_n(t) \exp(i\omega_{mn}t) W_{mn}. \tag{19•4–1}$$

Our task is to integrate this set with the initial conditions

$$c_n(0) = \delta_{n1}. \tag{19•4–2}$$

The problem becomes readily tractable if the conditions are such that only the *primary* transitions from the initial state into one of the states of the continuum matter, and that all subsequent *secondary* transitions *within* the continuum may be neglected. This will certainly be the case for perturbations by an electromagnetic wave, if the continuum is narrow compared to the energy separation of the continuum from the initial state. It may or may not be true in other scenarios, calling for a case-by-case examination. But given this restriction, we can drastically simplify the set (19•4–1).

In the equation for $m = 1$, we must retain the sum over all states of the continuum:

$$i\hbar \frac{d}{dt} c_1(t) = \sum_{n \neq 1} c_n(t) \exp(i\omega_{1n}t) W_{1n}. \tag{19•4–3}$$

But in each of the equations with $m \neq 1$, all terms describing transitions *within* the continuum $(m, n, \neq 1)$ are neglected, leaving only the terms with $n = 1$, describing transitions from the initial state into the continuum:

$$i\hbar \frac{d}{dt} c_m(t) = c_1(t) \exp(i\omega_{m1}t) W_{m1}. \tag{19•4–4}$$

We treat here the case of a harmonic wave perturbation. (The case of a step perturbation is easily obtained from that for a harmonic perturbation by letting the wave frequency go to zero.) For simplicity, we again make the dipole approximation and assume that $W_{nn} = 0$; both restrictions are easily removed. The perturbation is then again of the form (19•3–10), which we write here as

$$\Delta\hat{H} = \hat{W} \cdot [\exp(+i\omega_0 t) + \exp(-i\omega_0 t)]. \tag{19•4–5}$$

If the perturbing wave is an electromagnetic wave, $\hat{W}$ is again given by (19•3–11). However, many other wave-like perturbations are of the form (19•4–5), and we shall not insert (19•3–11) for $\hat{W}$.

Just as in our discussion of electromagnetic transitions between two *discrete* electron states, we argue again that the two-term oscillating perturbation may be treated as a single-term step perturbation, provided we include the quantum energy $\hbar\omega_0$ of the oscillation in the definition of the energy mismatch frequencies. If the continuum is *above* the initial state, we have, in generalization of (19•3–13a),

$$\hbar\delta\omega_n \equiv \hbar(\omega_{n1} - \omega_0) = \mathcal{E}_n - (\mathcal{E}_1 + \hbar\omega_0); \tag{19•4–6a}$$

otherwise we obtain, in generalization of (19•3–13b),

$$\hbar\delta\omega_n \equiv \hbar(\omega_{n1} + \omega_0) = (\mathcal{E}_n + \hbar\omega_0) - \mathcal{E}_1. \tag{19•4–6b}$$

In either case, (19•4–3) and (19•4–4) become

$$i\hbar\frac{d}{dt}c_1(t) = \sum_{n\neq 1} c_n(t)\exp(-i\delta\omega_n t)W_{1n} \tag{19•4–7}$$

and

$$i\hbar\frac{d}{dt}c_m(t) = c_1(t)\exp(+i\delta\omega_m t)W_{m1}; \qquad (m \neq 1). \tag{19•4–8}$$

19.4.2 The "Golden Rule"

The set of Eqs. (19•4–7) and (19•4–8), while still forming a set of a large number of coupled differential equations, is much simpler than the original set (19•4–1). In particular, if somehow $c_1(t)$ were already known, all of the other c's could be obtained by integration of (19•4–8).

Now it is quite clear what the overall shape of $c_1(t)$ must be (**Fig. 19•4–2**). Starting with its initial value 1, $c_1(t)$ must gradually fall off, due to transitions of the system into other states, and if there is a very large number of such accessible states in the energetic vicinity of the initial state, this falloff should eventually lead to a very low final value of $c_1(t)$. In short, we would expect an exponential-like behavior of $c_1(t)$, of the form

$$c_1(t) = e^{-\gamma t}, \tag{19•4–9}$$

with a suitable coefficient γ that remains to be determined.

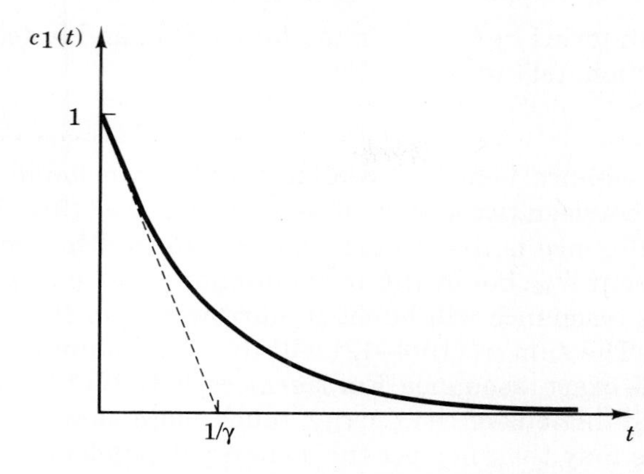

Figure 19•4–2. Exponential-like falloff of the probability amplitude $c_1(t)$ of the initial state $|1\rangle$.

If (19•4–9) is inserted into (19•4–8) and the equation is integrated, we obtain

$$c_m(t) = \frac{W_{m0}}{i\hbar} \cdot \frac{\exp[(-\gamma + i\delta\omega_m)t] - 1}{-\gamma + i\delta\omega_m}. \tag{19•4–10}$$

To determine whether (19•4–9) is indeed a good trial function—and if so, the value of γ—we insert both (19•4–9) and (19•4–10) into (19•4–7). On the left-hand side, this yields

$$-i\hbar\gamma e^{-\gamma t}. \tag{19•4–11}$$

The right-hand side is more complicated:

$$\sum_{n\neq 1} \frac{|W_{n1}|^2}{-i\hbar} \cdot \frac{\exp(-\gamma t) - \exp(-i\delta\omega_n t)}{\gamma - i\delta\omega_n}. \tag{19•4–12}$$

At first glance, the expressions (19•4–11) and (19•4–12) do not look at all similar. However, the difference is deceptive, and under certain conditions it is indeed possible to transform (19•4–12) into the form (19•4–11).

Inasmuch as the states of interest were assumed to form a quasi-continuum, we may replace the sum (19•4–12) by an integral, according to the standard prescription

$$\sum \ldots \Rightarrow \int \ldots D(\delta\mathcal{E})\, d(\delta\mathcal{E}), \tag{19•4–13}$$

where $D(\delta\mathcal{E})$ is the **density of states**, viewed here as a function of the energy mismatch $\delta\mathcal{E} = \hbar\delta\omega$ for the particular transition. Similarly, we may also view the matrix element W_{n0} as a function of the energy mismatch by writing $W(\delta\mathcal{E})$ instead of W_{n0}.

Consider now the $\delta\omega$-dependence of the two terms contained in (19•4–12) and, hence, in (19•4–13),

$$\frac{\exp(-\gamma t)}{\gamma - i\delta\omega} \quad \text{and} \quad \frac{\exp(-i\delta\omega t)}{\gamma - i\delta\omega}. \tag{19•4–14a,b}$$

The absolute value of both terms has a maximum for $\delta\omega = 0$, and it decreases with increasing ω proportionately to

$$(\gamma^2 + \delta\omega^2)^{-1/2}. \tag{19•4–15}$$

This is a Lorentzian resonance behavior similar to what we found in our discussion of transitions between two states, in section 19.2; see (19•2–12) and Fig. 19.2–2. The only difference is that the resonance bandwidth is given no longer by the matrix element W_{12}, but by the as-yet unknown rate parameter γ.

In many cases, this resonance will be much narrower than the width of the continuum of states. The sum in (19•4–12) will then be dominated by the terms in the vicinity of the exact resonance. Furthermore, both $W(\delta\mathcal{E})$ and $D(\delta\mathcal{E})$ are often slowly varying functions of the energy, much more slowly than the expression (19•4–15). We may then neglect their energy dependence over the narrow resonance and evaluate the integral by simply replacing both $W(\delta\mathcal{E})$ and $D(\delta\mathcal{E})$ with their values at the exact resonance, $\delta\mathcal{E} = 0$, where the remainder of the integrand has its maximum:

$$W(\delta\mathcal{E}) \to W(0); \qquad D(\delta\mathcal{E}) \to D(0) \tag{19•4–16a,b}$$

Furthermore, we will make only a negligible error if we extend the integration limits symmetrically to $\pm\infty$. With these substitutions, the sum (19•4–12) simplifies to

$$\sum \ldots \to i|W(0)|^2 D(0) \cdot e^{-\gamma t} \cdot \int_{-\infty}^{+\infty} \frac{1 - \exp[(\gamma - i\delta\omega)t]}{\gamma - i\delta\omega} d(\delta\omega). \tag{19•4–17}$$

The resulting integral can be evaluated in closed form, yielding the value $-\pi$, provided that $t > 0$. The (somewhat tedious) proof is left to the reader. If that value is inserted in (19•4–17), the resulting expression is simply

$$-i\pi|W(0)|^2 D(0)e^{-\gamma t}, \tag{19•4–18}$$

which is identical to (19•4–11) if we set

$$\gamma = \frac{\pi}{\hbar}|W(0)|^2 D(0). \tag{19•4–19}$$

Clearly, our trial function $c_1(t)$ of (19•4–9) will be a good approximation whenever the sum (19•4–12) can be approximated well by the integral (19•4–17).

According to (19•4–9), the probability that the overall system is still in its initial state is

$$|c_1(t)|^2 = e^{-2\gamma t}. \tag{19•4–20}$$

Evidently, 2γ is the (initial) transition probability rate. We therefore re-write (19•4–19) as

$$\boxed{2\gamma = \frac{2\pi}{\hbar}|W(0)|^2 D(0).} \tag{19•4–21}$$

This is Fermi's **Golden Rule**.

The probability that the system has made a transition to another (unspecified) state is

$$P(n \neq 0) = 1 - |c_1(t)|^2 = 1 - e^{-2\gamma t}. \tag{19•4–22}$$

For values $t \ll 1/2\gamma$, the transition probability $P(n \neq 0)$ increases linearly with time:

$$\boxed{P(n \neq 1) \cong 2\gamma t = \frac{2\pi}{\hbar}|W(0)|^2 D(0)t; \qquad (t \ll 1/2\gamma).} \tag{19•4–23}$$

It is in this "crippled" linearized form that the Golden Rule is often given.

Because of (19•4–20), the quantity

$$\boxed{\tau = 1/2\gamma} \tag{19•4–24}$$

has the meaning of the **lifetime** of the initial state in the presence of the perturbation.

In using the Golden Rule, one should always keep in mind that it is an approximation. This warning is necessary because, in the literature on various transition processes, one often finds Fermi's Golden Rule stated as an axiomatic point of departure, as though it were a rigorous law, without any discussion as to whether it is in fact applicable. Sometimes it is actually only a poor approximation. "Practicing the Golden Rule without a license" is a common offense, although it tends to pale compared to the misuse of the uncertainty relations.

19.4.3 The Energy Range of the Transition

We turn now to the question of the probability distribution of the system over the various final states. Once the transition probability rate constant γ has been determined, the actual probability distribution follows directly from (19•4–10), with $W_{m0} \to W(0)$:

$$|c_m(t)|^2 = \frac{|W(0)|^2}{\hbar^2}\frac{e^{-2\gamma t} - 2e^{-\gamma t}\cos\delta\omega_m t + 1}{\gamma^2 + \delta\omega_m^2}. \tag{19•4–25}$$

In the asymptotic limit $t \gg \tau$,

$$|c_m(t)|^2 \to \frac{|W(0)|^2}{\hbar^2} \frac{1}{\gamma^2 + \delta\omega_m^2} = \frac{1}{\pi\hbar D(0)} \frac{1}{\gamma^2 + \delta\omega_m^2}. \tag{19•4–26}$$

Viewed as a function of the energy difference $\hbar\delta\omega_m$ between the initial state and the final state, the final-state probabilities are again Lorentzian-distributed. The maximum transition probability occurs for those states that are energetically closest to exact resonance, $\delta\omega_m = 0$. With increasing energy separation, the transition probability falls off, reaching one-half its maximum value when $|\delta\omega_m| = \gamma$. As in sections 19.2 and 19.3, we can again define an FWHM transition width. Here, however, its value depends in a somewhat different way on the interaction matrix element than it did in the earlier sections, namely,

$$\hbar\Delta\omega = 2\hbar\gamma = 2\pi|W(0)|^2 D(0). \tag{19•4–27}$$

Note that the angular frequency $\Delta\omega$ associated with this width is equal to the inverse of the lifetime τ, of the initial state:

$$\Delta\omega = 1/\tau. \tag{19•4–28}$$

19.4.4 Step Perturbation

Although we have derived Fermi's Golden Rule for a harmonic wave perturbation that satisfies (19•4–3), our results remain valid for a step perturbation by simply setting $\omega_0 = 0$, provided that we restrict ourselves to sufficiently short elapsed times such that

$$t \ll \tau = 1/2\gamma. \tag{19•4–29}$$

This restriction arises as follows. For a step perturbation, we no longer can automatically neglect transitions within the continuum. However, as long as (19•4–29) is satisfied, the probability amplitudes c_n with $n \neq 0$ will still be sufficiently small that in each of the equations (19•4–1), for $m \neq 0$, the right-hand side will still be dominated by the contributions from the c_1-term. Hence, the initial transition probability continues to be given by (19•4–23). The extent to which the validity of our treatment extends to times $t \gtrsim \tau$ can be decided only on a case-by-case basis, by actually computing the intra-continuum matrix elements. Often these matrix elements vanish, in which case our wave perturbation treatment can be taken over in its entirety by simply setting $\omega_0 = 0$.

19.4.5 Transitions within a Two-Level System Induced by Broadband Electromagnetic Radiation

The time evolution of a two-level system under the influence of a single monochromatic wave is a highly idealized case, realizable in practice only under special circumstances. A much more common situation is that the electromagnetic radiation is broadband, containing many modes with many different frequencies in the vicinity of the exact transition frequency.

The most elementary case is that of a simple two-level atomic system interacting with broadband electromagnetic radiation. To be specific, we assume that the electromagnetic field is the field inside a resonant cavity, sufficiently large that there are many cavity modes within the resonance bandwidth of the atomic system. The field is then a linear superposition of many different cavity modes, each of which represents a separate harmonic oscillator. Hence, we have basically the problem of energy exchange between two systems, one a two-level system and the other a system and with a quasi-continuum of many levels. One of the two systems is an electronic one, the other a photonic one, but which is which is irrelevant to the basic formalism. Hence, we may take over the Golden Rule formalism without change, the only difference being that now $D(0)$ is not the density of *electronic* states at the resonance energy, but the density of *photonic* states, that is, the density of cavity modes. The extension of this line of reasoning to a broadband perturbation acting on a continuum of atomic levels is left to the reader.

19.5 OSCILLATOR STRENGTHS, SELECTION RULES, AND ANGULAR MOMENTUM OF PHOTONS

19.5.1 Oscillator Strengths

We saw in sections 19.3 and 19.4 that the central quantity that determines the strength of the radiative transitions between the different states of an atomic system is the dipole matrix element between the coupled states. For an electric field in the x-direction, this matrix element is proportional to

$$x_{mn} = \langle m | x | n \rangle. \tag{19•5–1}$$

For an arbitrary direction of the electric field $\mathbf{E}$, the quantity that matters is

$$\mathbf{E} \cdot \mathbf{r}_{mn}, \tag{19•5–2}$$

where

$$\mathbf{r}_{mn} = \langle m | \mathbf{r} | n \rangle. \tag{19•5–3}$$

The numerical values of these matrix elements may vary over a very wide range. It is useful to introduce a standard reference transition and to express other matrix elements relative to this reference. The simplest and most convenient reference is the transition between the ground state ($m = 0$) and the first excited state ($n = 1$) of the one-dimensional harmonic electron oscillator. From (10•1–7), we obtain, for this transition,

$$(x_{01}^2)_{\mathrm{HO}} = L^2 |\langle 0 | Q | 1 \rangle|^2 = \frac{1}{2} L^2 = \frac{\hbar}{2m_e \omega}, \tag{19•5–4}$$

where we have drawn on the relation $x = QL$ between the true position coordinate x, the dimensionless position coordinate Q, and on the natural unit of length $L = \sqrt{\hbar/m_e\omega}$ for the harmonic oscillator, from (2•3–7).

Using this matrix element as a reference standard, we can express the strength of a transition of interest relative to that of a harmonic oscillator with the same frequency, $\omega = \omega_{mn}$, by expressing the quantity $|x_{mn}|^2$ characterizing that strength as

$$|x_{mn}|^2 = F^x_{mn} \cdot (x^2_{01})_{\text{HO}} = F^x_{mn} \cdot \frac{\hbar}{2m_e\omega_{mn}}, \tag{19•5–5}$$

where the proportionality factor F^x_{mn} is called the **oscillator strength** of the transition for an electric field in the x-direction. Analogous definitions hold for F^y_{mn} and F^z_{mn}. We further introduce a **mean oscillator strength**

$$F_{mn} = \frac{1}{3}(F^x_{mn} + F^y_{mn} + F^z_{mn}), \tag{19•5–6}$$

which averages the oscillator strength over the possible polarization directions of the electric field.

It is useful to replace the energy $\hbar\omega_{mn}$ with $\mathcal{E}_m - \mathcal{E}_n$ and to re-write the definition (19•5–5) of the oscillator strength in the form

$$F^x_{mn} = |\langle m|x|n\rangle|^2 \cdot \frac{2m_e}{\hbar^2} \cdot (\mathcal{E}_m - \mathcal{E}_n). \tag{19•5–7}$$

Note that F^x_{mn} is not necessarily positive: We have $F_{mn} < 0$ if $\mathcal{E}_m < \mathcal{E}_n$. Defined in this way, the oscillator strengths obey the important **sum rule**

$$\sum_m F^x_{mn} = 1. \tag{19•5–8}$$

To derive this rule, consider the double commutator

$$[[x, \hat{H}_0], x] = 2x\hat{H}_0x - \hat{H}_0x^2 - x^2\hat{H}_0. \tag{19•5–9}$$

We take the expectation value of this operator for the state $|n\rangle$,

$$\langle n|[[x, \hat{H}_0], x]|n\rangle = 2\langle n|x\hat{H}_0x|n\rangle - 2\mathcal{E}_n\langle n|x^2|n\rangle, \tag{19•5–10}$$

then replace $\hat{H}_0$ on the right-hand side by

$$\hat{H}_0 = \sum_m |m\rangle\mathcal{E}_m\langle m|, \tag{19•5–11}$$

and finally re-write the term $\langle n|x^2|n\rangle$ by inserting the unit operator between the two x factors:

$$\langle n|x^2|n\rangle = \langle n|x\hat{1}x|n\rangle = \sum_m \langle n|x|m\rangle\langle m|x|n\rangle = \sum_m |\langle m|x|n\rangle|^2. \tag{19•5–12}$$

This leads to the result

$$\langle n\,|[[x, \hat{H}_0], x]|\,n\rangle = 2\sum_m |\langle m\,|\,x\,|\,n\rangle|^2\,(\mathcal{E}_m - \mathcal{E}_n). \tag{19•5–13}$$

On the other hand, if the unperturbed Hamiltonian is of the conventional form, the double commutator is easily evaluated directly, via

$$[x, \hat{H}_0] = \frac{i\hbar}{m_e}\hat{p}_x, \qquad [[x, \hat{H}_0], x] = \frac{\hbar^2}{m_e}, \tag{19•5–14a,b}$$

Hence, the left-hand side of (19•5–13) is simply $\hbar^2/m_e$, and if this is inserted, we may write

$$\frac{2m_e}{\hbar^2}\sum_m |\langle m\,|\,x\,|\,n\rangle|^2\,(\mathcal{E}_m - \mathcal{E}_n) = 1, \tag{19•5–15}$$

which is equivalent to (19•5–8).

19.5.2 Selection Rules: Introduction, and the Parity Selection Rule

There occur numerous cases in which certain of the oscillator strengths vanish exactly, due to symmetries in the Hamiltonian that lead to the vanishing of the dipole matrix elements between some of the system states. The transitions between the associated states are said to be **forbidden**. Whether a transition is allowed or forbidden may depend not only on the two states themselves; it may also depend on the polarization of the wave. The rules that specify which transitions are allowed—and for which polarizations—are called **selection rules**. We discuss in this section the selection rules for the two most important types of symmetry invariances: inversion symmetry and rotational symmetry. The selection rules for translational invariance were already covered in section 19.3; see (19•3–25).

The simplest and most pervasive selection rule is the **parity selection** rule. It applies to any system whose unperturbed Hamiltonian is invariant under the inversion operation

$$\hat{H}^{(0)}(-\mathbf{r}) = \hat{H}^{(0)}(+\mathbf{r}). \tag{19•5–16}$$

Recall that in such cases it is always possible to select the eigenfunctions of $\hat{H}^{(0)}$ such that they are also eigenfunctions of the inversion operator, with the eigenvalues ± 1:

$$\hat{I}\psi(\mathbf{r}) = \psi(-\mathbf{r}) = \pm\psi(\mathbf{r}). \tag{19•5–17}$$

Eigenfunctions that are even in $\mathbf{r}$ are said to have **even parity**, those that are odd, **odd parity**. The factor $\mathbf{r}$ in (19•5–3) is itself odd under inversion. Therefore, the product $\psi_m^*\mathbf{r}\psi_n$ will be an odd function if the two wave functions have the same parity and will be an even function if they have opposite parity. But only in the latter case can the integral over this product be non-vanishing. In this way, we arrive at the

> **Parity Selection Rule**: In systems with inversion symmetry, only those dipole transitions are allowed for which the parity changes during the transition.

The parity selection rule assumes a particularly simple form if the Hamiltonian has full spherical symmetry. We saw in (18•2–12) that in this case the parity of all states is equal to the parity of the orbital quantum number l, being even for even l, and odd for odd l. Hence, the orbital quantum number must change by an odd number during a dipole transition. More specifically, the Cartesian components of the vector matrix elements of the form $\langle l', m' | \mathbf{r} | l, m \rangle$ all vanish unless $\Delta l = \pm 1$. Accordingly, we obtain the following selection rule:

> In systems with full spherical symmetry, only those dipole transitions are allowed for which the orbital quantum number l changes by ± 1 during the transition:
>
> $$\boxed{\Delta l = \pm 1.} \tag{19•5–18}$$

19.5.3 Selection Rules for the Azimuthal Quantum Number

In the case of the *azimuthal* quantum number m for rotationally invariant systems, we must distinguish between polarization of the electric field parallel to the symmetry axis and polarization perpendicular to it. This is true even for full spherical symmetry, because of the preferential formal treatment of the z-axis relative to the other Cartesian axes; in effect, the z-axis is treated as if it were the only symmetry axis. We treat here specifically the case of full spherical symmetry, but the results are formulated in such a way that they also apply to rotational invariance about a single axis. The generalization of our derivations to the true single-axis case is left to the reader.

Polarization Parallel to the Rotation Axis

This case is trivial. The operators $\hat{L}_z$ and $\hat{z}$ commute, which means that both $|m\rangle$ and $\hat{z}|m\rangle$ are eigenfunctions of $\hat{L}_z$, with the same quantum number m. Clearly, the matrix element for z between two states can be finite only if both states have the same azimuthal quantum number,

$$\boxed{\Delta m = 0,} \tag{19•5–19}$$

a result already contained in (18•2–22).

Polarization Perpendicular to the Rotation Axis

Polarization within the xy-plane involves the evaluation of matrix elements of the form $\langle l', m'|x|l, m\rangle$ and $\langle l', m'|y|l, m\rangle$, where l and l' are subject to (19•5–18). From (18•2–27), we obtain the selection rule for the azimuthal quantum number m,

$$\boxed{\Delta m = \pm 1.} \tag{19•5–20}$$

Circularly Polarized Waves

Consider a wave propagating in the $+z$-direction, having two equally strong Cartesian electric field components that are given by

$$E_x(z, t) = +E_0 \cos(kz - \omega_0 t), \tag{19•5–21a}$$

and

$$E_y(z, t) = \pm E_0 \sin(kz - \omega_0 t), \tag{19•5–21b}$$

with a $\pm 90°$ phase shift relative to each other. Such waves are called **circularly polarized**. Now consider the wave with the (+)-sign in (19•5–21b). If the electric field is viewed as a function of t, in the fixed plane $z = 0$, the field vector rotates clockwise. Viewed as a function of z along the direction of propagation, at the fixed instant of time $t = 0$, the field vector describes a right-handed screw through space, as shown in **Fig. 19•5–1**. Such a wave is referred to either as being **right-handed circularly polarized** or as having **negative helic-**

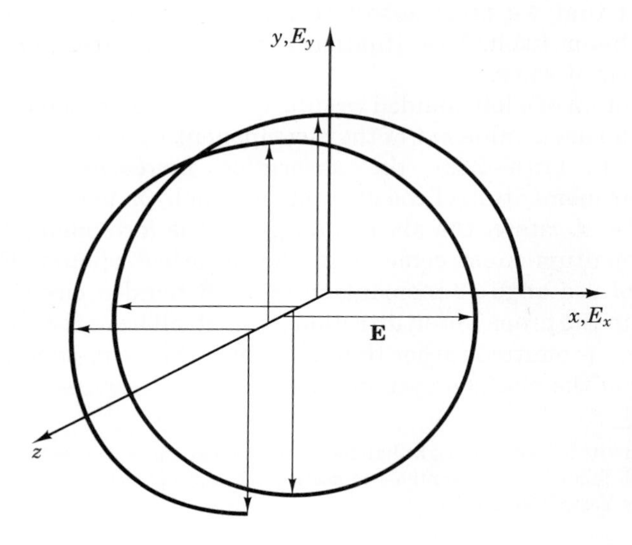

Figure 19•5–1. Right-handed screw-like pattern of the electric field vector for a right-hand circularly polarized wave (negative helicity).

ity.[4] Similarly, the wave with the $(-)$-sign rotates counterclockwise in time and corresponds to a left-handed helix in space; it is said to be **left-handed circularly polarized** or to have **positive helicity**.

Being linear superpositions of two transversely polarized waves, such waves obey the selection rule (18•2–20). But which of the two cases in (18•2–20) applies is now correlated with whether the transition involves the emission or absorption of a photon and with the sense of circular polarization. As an example, consider the absorption of a photon from a left-handed wave. It is left to the reader to show that the relevant matrix elements of x and y occur then in the combination $\langle l', m'|x + iy|l, m\rangle$. According to (18•2–21a), the latter matrix element vanishes unless $\Delta m = m' - m = +1$. Considering the remaining cases yields the following table of selection rules, which shows the value of $\Delta m = m' - m$, the change in angular momentum of the atomic system during an interaction event with a circularly polarized wave:

	Left-Handed Wave (Positive Helicity)	Right-Handed Wave (Negative Helicity)	
Emission	$\Delta m = -1$	$\Delta m = +1$	(19•5–22a,b)
Absorption	$\Delta m = +1$	$\Delta m = -1$	(19•5–22c,d)

19.5.4 The Angular Momentum of Photons

The selection rules show that the angular momentum of an electron system always changes by $\pm\hbar$ when a photon is either emitted or absorbed by that system. But this means that we must associate an angular momentum of magnitude $\hbar$ with the photon itself. The situation is most transparent in the case of a *circularly* polarized wave.

Consider the absorption of a left-handed circularly polarized photon by an electron in a state with a sharp value $m\hbar$ of the z-component L_z of its angular momentum. According to (19•5–22c), the absorption increases this z-component of the angular momentum of the electron system by $\hbar$, to $(m + 1)\hbar$. (If there is no such higher L_z state, the absorption process is forbidden.) The additional angular momentum must come from the absorbed photon. Evidently, the component of the angular momentum of a left-handed circularly polarized photon, taken in the propagation direction, must itself have the sharp value $+\hbar$. If such a photon is emitted rather than absorbed, the z-component of the angular momentum of the electron system must of course *decrease* by the

[4] The terminology on what constitutes left- and right-handed circular polarization is a rich source of confusion. Our presentation follows the conventions in section 7.2 of J. D. Jackson, *Classical Electrodynamics*, 2d Ed. (New York: Wiley, 1975).

sharp amount $\hbar$, to $(m - 1)\hbar$, in agreement with (19•5–22a). Again, if there is no such lower L_z state, the emission process is forbidden.

Similarly, one concludes that a right-handed circularly polarized photon has an angular momentum with a sharp z-component $-\hbar$, *opposite* to the propagation direction.

What about a linearly polarized wave? Such a wave must be viewed as a linear superposition of two equally strong circularly polarized waves; that is, it represents a mixed state with respect to the photon angular momentum, corresponding to photons whose angular momentum component in the propagation direction alternates randomly between $+\hbar$ and $-\hbar$.

The assignment of an integer multiple of $\hbar$ to the angular momentum of a photon, rather than a half-integer multiple, means that photons are bosons. That is, of course, nothing new: It follows already from the fact that the energy in any given field mode, while quantized, may contain an arbitrary number of quanta. Our present argument is more specific though, in giving the actual value of the angular momentum.

19.6 INDIRECT TRANSITIONS

19.6.1 Introduction

We have so far explicitly neglected the complications that occur when a system, after making a first transition from an initial state $|1\rangle$ to a second state $|2\rangle$, makes a further transition to a final third state $|3\rangle$. Such two-stage transitions become essential to the overall transition process when a *direct* transition between an initial state $|1\rangle$ and a final state $|3\rangle$ is *forbidden,* in the sense that the perturbation has a zero matrix element $\langle 1|\hat{W}|3\rangle$ between the two states, even though the transition would be *energetically* possible. In such cases, there may be *indirect transition* paths between $|1\rangle$ and $|3\rangle$, via one or more intermediate states $|2\rangle$ that *do* have nonzero matrix elements $\langle 1|\hat{W}|2\rangle$ and $\langle 2|\hat{W}|3\rangle$ with *both* the initial and the final state. Sometimes a single intermediate state is not sufficient, and two or more states in succession are needed.

An important and illustrative example of an indirect transition is the absorption of light in an indirect-gap semiconductor, in which the conduction band minimum occurs at a Bloch wave vector **k** different from the valence band maximum (**Fig. 19•6–1**). As we saw earlier, optical absorption processes in solids have a nonzero matrix element only between states with very nearly the same reduced Bloch wave vector **k**. Put differently, direct transition processes must be very nearly vertical in an $\mathcal{E}(\mathbf{k})$ diagram, as in Fig. 19•6–1. In an indirect-gap semiconductor, such a direct transition requires a photon energy $\hbar\omega = \mathcal{E}_d$ appreciably larger than the net energy gap $\mathcal{E}_g$. In reality, however, there is always an appreciable amount of absorption in the photon energy range $\mathcal{E}_g \leq \hbar\omega < \mathcal{E}_d$, due to **indirect transitions**, such as the one illustrated

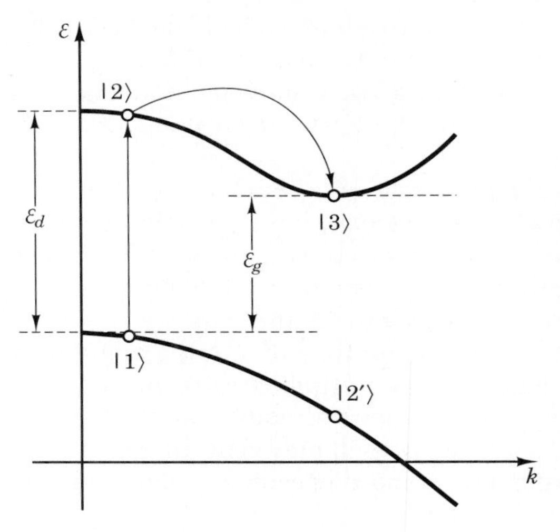

Figure 19•6–1. Absorption in an indirect-gap semiconductor. A photon raises an electron from the initial state $|1\rangle$ to the intermediate virtual state $|2\rangle$, from where it is scattered by a phonon, or quantum of vibrational energy, to the final state $|3\rangle$. Only the energy for the overall process is conserved; the partial transitions to and from the intermediate virtual state may violate energy conservation. The state labeled $|2'\rangle$ provides an alternative route.

in Fig. 19•6–1. What is shown there is a first (vertical) transition from an initial state $|1\rangle$ to an intermediate state $|2\rangle$, followed immediately by a second transition from $|2\rangle$ to a final state $|3\rangle$. During that second transition, a quantum $\hbar\omega_L$ of one of the lattice vibrations is either emitted or absorbed. Such quanta are called **phonons**.

The photon energy required for such an indirect transition is *not* the energy $\mathcal{E}_d$ for the direct transition alone, but is the energy required to conserve energy for the *overall* process, namely $\mathcal{E}_g \pm \hbar\omega_L$. Those processes that involve the *emission* of a phonon require a photon energy $\mathcal{E}_g + \hbar\omega$. For those in which an existing phonon is *absorbed*, the required photon energy is $\mathcal{E}_g - \hbar\omega$. In either case, because the phonon energy—typically around 0.03 eV—is usually much smaller than the difference $\mathcal{E}_d - \mathcal{E}_g$, the onset of phonon-assisted absorption is usually close to the *net* energy gap $\mathcal{E}_g$.

Inasmuch as the transitions to and from the intermediate state $|2\rangle$ do not conserve energy by themselves, the probability that the system will ever materialize in the intermediate state $|2\rangle$ is extremely small, as discussed earlier for the simple two-level system. Hence, the intermediate state is commonly called a **virtual state.** It is strictly a transient state, acting as a "catalyst" for the overall transition, but with an extremely low probability for itself. Exactly how low that probability is depends on the energy mismatch: The larger the mismatch, the lower the probability. The analysis to be presented will show this in more detail.

The process shown in Fig. 19•6–1 is not the only indirect route possible. An alternative process would be to lift an electron from the occupied state $|2'\rangle$ in the lower band, vertically below $|3\rangle$, and to re-occupy this state immediately from $|1\rangle$. The overall effect is one of an indirect transition from $|1\rangle$ to $|3\rangle$ via $|2'\rangle$.

19.6.2 The Formalism

We consider here only the simplest possible case of an indirect transition. We assume that there are only three participating states: $|1\rangle$, $|2\rangle$, and $|3\rangle$. To be specific, we assume that the energy ordering of the states is as shown in **Fig. 19.6–2**, with

$$\mathcal{E}_1 < \mathcal{E}_3 < \mathcal{E}_2, \tag{19•6–1}$$

the same as in the case of phonon-assisted absorption.

We treat the problem by the semiclassical approximation and assume that the perturbation Hamiltonian contains two oscillating terms of the form

$$\hat{A}[\exp(i\omega_A t) + \exp(-i\omega_A t)], \tag{19•6–2a}$$

$$\hat{B}[\exp(i\omega_B t) + \exp(-i\omega_B t)], \tag{19•6–2b}$$

which are jointly responsible for the transition from $|1\rangle$ to $|3\rangle$. We leave it open whether or not $\hat{A}$ and $\hat{B}$ correspond to different kinds of physical processes, as in phonon-assisted absorption processes, or whether they simply represent different frequencies of one common process. Our formalism will cover both cases.

We assume that the operator $\hat{A}$ has a matrix element $A_{12} = \langle 1|\hat{A}|2\rangle$ between the initial and the virtual state and that $\hat{B}$ has a matrix element $B_{23} = \langle 2|\hat{B}|3\rangle$ between the virtual state and the final state. For simplicity, all other matrix elements are assumed to be zero:

$$\langle 1|\hat{A}|1\rangle = \langle 2|\hat{A}|3\rangle = \langle 2|\hat{A}|2\rangle = 0, \tag{19•6–3a}$$

$$\langle 1|\hat{B}|1\rangle = \langle 1|\hat{B}|2\rangle = \langle 2|\hat{B}|2\rangle = 0. \tag{19•6–3b}$$

The wave function $|\Psi(t)\rangle$ may be written in the interaction picture as

$$|\Psi\rangle = c_1 \exp(-i\omega_1 t)\cdot|1\rangle + c_2 \exp(-i\omega_2 t)\cdot|2\rangle + c_3 \exp(-i\omega_3 t)\cdot|3\rangle, \tag{19•6–4}$$

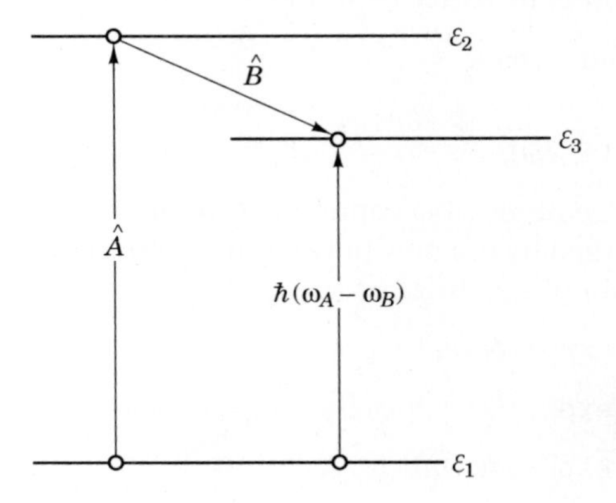

Figure 19•6–2. Energetic ordering assumed for the indirect transition discussed in the text.

where the three ω's are the frequency equivalents of the appropriate energies. Inserting (19•6–4) into the Schroedinger wave equations, and taking inner products with $\langle 1|$, $\langle 2|$, and $\langle 3|$, leads to the three equations

$$i\hbar\dot{c}_1 = c_2 H_{12} \exp(i\omega_{12}t), \tag{19•6–5a}$$

$$i\hbar\dot{c}_2 = c_1 H_{21} \exp(i\omega_{21}t) + c_3 H_{23} \exp(i\omega_{23}t), \tag{19•6–5b}$$

$$i\hbar\dot{c}_3 = c_2 H_{32} \exp(i\omega_{32}t). \tag{19•6–5c}$$

If the perturbation terms (19•6–2) are inserted into (19•6–5), each H-term in (19•6–5) leads to a pair of terms. As before, we retain only the less rapidly oscillating term of each pair and transform this oscillating-perturbation formalism into a step perturbation formalism with suitable energy mismatch frequencies, defined via

$$\hbar\delta\omega_A \equiv \hbar(\omega_{21} - \omega_A) = \mathcal{E}_2 - (\mathcal{E}_1 + \hbar\omega_A), \tag{19•6–6a}$$

$$\hbar\delta\omega_B \equiv \hbar(\omega_{32} + \omega_B) = (\mathcal{E}_3 + \hbar\omega_B) - \mathcal{E}_2. \tag{19•6–6b}$$

As in section 19.3, these definitions represent the inclusion of the quantum energies of the oscillating perturbation in the unperturbed energies of the overall Hamiltonian of the perturbation problem. Note that the signs are chosen as in Fig. 19•3–1, if we view the virtual state both as the final state of the first transition and as the initial state of the second.

Substituting (19•6–6) into (19•6–5) and dropping all more rapidly oscillating terms leads to

$$i\hbar\dot{c}_1 = c_2 A_{12} \exp(-i\delta\omega_A t), \tag{19•6–7a}$$

$$i\hbar\dot{c}_2 = c_1 A_{21} \exp(+i\delta\omega_A t) + c_3 B_{23} \exp(-i\delta\omega_B t), \tag{19•6–7b}$$

$$i\hbar\dot{c}_3 = c_2 B_{32} \exp(+i\delta\omega_B t). \tag{19•6–7c}$$

To strip the problem to its essence, we restrict ourselves to the most interesting case, where the various energies add up in such a way that energy for the *overall* transition is at exact resonance:

$$\mathcal{E}_3 - \mathcal{E}_1 = \hbar(\omega_A - \omega_B), \tag{19•6–8}$$

or

$$\delta\omega_A = -\delta\omega_B \; (= \delta\omega). \tag{19•6–9}$$

The energy mismatch frequency $\delta\omega$ common to both transitions (> 0 for the energetic ordering assumed by us) now plays a role similar to $\delta\omega$ in section 19.3. In this case, (19•6–7) simplifies further to

$$i\hbar\dot{c}_1 = c_2 A_{12} \exp(-i\delta\omega t), \tag{19•6–10a}$$

$$i\hbar\dot{c}_2 = c_1 A_{21} \exp(+i\delta\omega t) + c_3 B_{23} \exp(+i\delta\omega t), \tag{19•6–10b}$$

$$i\hbar\dot{c}_3 = c_2 B_{32} \exp(-i\delta\omega t). \tag{19•6–10c}$$

We need the solution that satisfies the initial conditions

$$c_1(0) = 1, \qquad c_2(0) = 0, \qquad c_3(0) = 0. \tag{19•6–11a–c}$$

In particular, we are interested in $c_3(t)$. The solution of this purely mathematical problem is tedious. We simply state the answer here, leaving its derivation to the problems at the end of the section.

It is useful first to introduce several characteristic frequencies, via the following definitions:

$$\hbar^2\Omega_0^2 = |A_{21}|^2 + |B_{32}|^2, \tag{19•6–12}$$

$$\Omega^2 = (\tfrac{1}{2}\,\delta\omega)^2 + \Omega_0^2, \tag{19•6–13}$$

and

$$\omega_\pm = \Omega \pm \tfrac{1}{2}\,\delta\omega = \sqrt{(\tfrac{1}{2}\,\delta\omega)^2 + \Omega_0^2} \pm \tfrac{1}{2}\,\delta\omega. \tag{19•6–14}$$

In terms of those frequencies, we find that the final-state probability amplitude is of the form

$$\boxed{c_3(t) = \frac{D}{2}\cdot[f_+\,\exp(+i\omega_-t) + f_-\,\exp(-i\omega_+t) - 1],} \tag{19•6–15}$$

where we have introduced the additional new quantities

$$D = \frac{2A_{21}B_{32}}{|A_{21}|^2 + |B_{32}|^2} \tag{19•6–16}$$

and

$$f_\pm = \frac{\omega_\pm}{2\Omega} = \frac{1}{2} \pm \frac{\delta\omega}{4\Omega}. \tag{19•6–17}$$

Note that

$$f_+ + f_- = 1. \tag{19•6–18}$$

To interpret the result (19•6–15), consider first the case $\delta\omega = 0$, corresponding to perfect energy conservation for *both* partial transitions, called a **resonant indirect transition**. From (19•6–15), with the help of (19•6–14) and (19•6–17), we obtain

$$c_3(t) = \frac{D}{2}(\cos\,\Omega_0 t - 1). \tag{19•6–19}$$

We evidently have a sinusoidal probability oscillation very similar to that discussed in section 19.2 for a simple step perturbation, except that the peak transition probability $|D|^2$ is no longer 100%, even at exact resonance, unless $|D| = 1$, which happens if and only if the strengths of the two perturbations are *matched* such that

$$|A_{21}| = |B_{32}|. \tag{19•6–20}$$

In general, we will have $|D|^2 < 1$: The absolute magnitude of D, which determines the maximum transition probability, is simply the ratio of the geometric average of $|A_{12}|^2$ and $|B_{23}|^2$ to their arithmetic average. This ratio cannot exceed unity, and it reaches unity only when the two quantities are equal:

$$|D|^2 \begin{cases} =1 & \text{if } |A_{21}| = |B_{32}|, \\ <0 & \text{otherwise.} \end{cases} \tag{19•6–21}$$

The off-resonance case $\delta\omega \neq 0$ is more complicated. The solution (19•6–15) evidently contains a superposition of two circular motions in the complex plane, offset from the origin of the plane. The radii and the frequencies of the two motions are inversely related. The motion along the larger circle (radius $f_+ \cdot |D|/2$) is counterclockwise for $\delta\omega > 0$ and has the lower frequency ($\omega = \omega_-$). The motion along the smaller circle (radius $f_- \cdot |D|/2$) is clockwise for $\delta\omega > 0$ and has the higher frequency ($\omega = \omega_+$). The circles are offset in such a way that $c_3 = 0$ for $t = 0$. The resulting trajectories in the complex plane are shown in **Fig. 19•6–3** for two values of the mismatch frequency, $\delta\omega = 0.5\Omega_0$ and $\delta\omega = 4\Omega_0$.

The motion is exactly the same as if we had initially marked the point $c_3 = 0$ on the smaller of the two circles contained in (19•6–15) and then rolled that circle along the *inside* of an *envelope circle* of diameter $|D|$, the center of which is at $-D/2$. Such rolling trajectories, known as *cycloids,* occur in many problems in physics. Note the periodic sharp cusps wherever the rolling circle touches the envelope circle.

If D is complex, the entire pattern is rotated by a fixed angle about the origin of the complex c-plane. In any event, $|c_3(t)|$ cannot exceed $|D|$, the

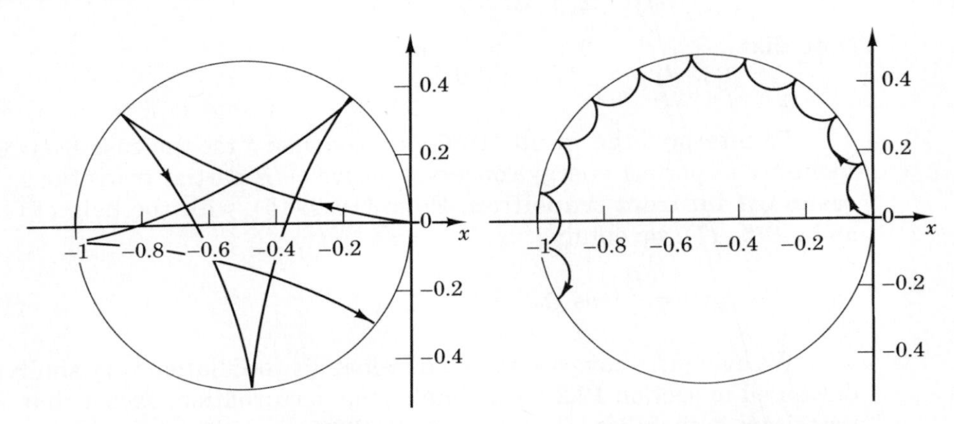

Figure 19•6–3. Cycloidal motion of the probability amplitude in the complex plane, for two values of the mismatch frequency, $\delta\omega = 0.5\Omega_0$ and $\delta\omega = 4\Omega_0$. Both curves assume that $D = 1$; both go from $t = 0$ to $t = 5\pi/\Omega_0$.

diameter of the envelope circle, which in turn cannot exceed unity, from (19•6–21).

Note, however, that the diameter of the envelope circle, and hence the peak transition probability, do not get changed by moving off resonance, as long as the overall transition conserves energy. In particular, if the strength of the two intermediate transitions is *matched,* in the sense of Eq. (19•6–21), the peak transition probability continues to be 100%, even if the two intermediate transitions do not conserve energy.

What *does* happen with increasing energy mismatch of the virtual state is that the *rate* decreases with which the transition probability reaches the peak probability. With increasing mismatch, the radius of the smaller circle decreases, and the overall transition behavior becomes dominated by the larger circle alone, with its frequency ω_-. In the limit of a large mismatch, $|\delta\omega| \gg \Omega_0$, the square root in (19•6–14) may be expanded, leading to

$$\omega_- = (\tfrac{1}{2}\,\delta\omega)\left[\sqrt{1 + \left(\frac{2\Omega_0}{\delta\omega}\right)^2} - 1\right] \approx (\tfrac{1}{2}\,\delta\omega)\cdot\tfrac{1}{2}\cdot\left(\frac{2\Omega_0}{\delta\omega}\right)^2$$

$$= \frac{\Omega_0^2}{\delta\omega} \qquad (19•6–22)$$

For a sufficiently large mismatch, it would simply take forever for a high final-state probability to build up—*if* there were no other perturbations acting on the system. In reality, other perturbations are inevitably present, and realistic probabilities for the final state build up only for a small mismatch.

◆ PROBLEM TO SECTION 19.6

#19•6-1: Derivation of the Final-State Probability Amplitude Expression (19•6–15)

The following procedure is suggested for deriving (19•6–15). By inserting (19•6–10a,c) into (19•6–10b), derive a second-order differential equation for $c_2(t)$ alone. Show that this equation has oscillatory solutions, with the frequencies $+\omega_+$ and $-\omega_-$. Use the initial condition (19•6–11b) to determine $c_2(t)$, except for an as-yet unknown factor. Insert the result into (19•6–10a) and (19•6–10c), and integrate with the initial conditions (19•6–11a,c). Finally, determine the unknown factor in $c_2(t)$, which is also contained in the expression for $c_1(t)$ and $c_3(t)$, by stipulating that (19•6–10b) be satisfied.

Chapter 20

ELEMENTS OF FIELD QUANTIZATION

20.1 THE FIELD HAMILTONIAN

20.1.1 The Classical Field Hamiltonian

In this chapter, we complete our discussion of radiative transitions by treating the electromagnetic field as a full-blown quantum system in its own right that happens to interact with the electron system. The central idea is simple: The electromagnetic field is viewed—still classically—as a linear superposition of single-frequency modes. Each of these modes represents a separate quantum harmonic oscillator, and the electron-field interaction is an interaction of the electron with this set of harmonic oscillators.

The problem is in many ways an elaboration of our earlier treatment of the LC-circuit as a quantum system in section 10.3. There we found that the circuit equations for the LC-circuit could be written in harmonic oscillator form by using the charge Q on the capacitor plates as the *position-like* dynamical variable that describes the momentary classical state of the oscillator. The product IL of the current and the inductance then served as the *momentum-like* dynamical variable, characterizing the classical state of motion of the oscillator.

We visualize the electromagnetic fields interacting with the electron as confined inside a resonant cavity of finite volume Ω, with suitable boundary conditions at the walls of the cavity. This represents no restriction: The free-space limit may always be achieved by letting all relevant cavity dimensions go to infinity. The purpose of the cavity is to make the modes discrete and to make it possible to count them. The assumption of a resonant cavity with boundary condition does not even restrict us to standing waves: By assuming that the cavity is a simple parallelepiped with periodic boundary conditions at the surfaces, the various field quantities may all be chosen as traveling rather than standing waves. This is what we will do here.

For any single-frequency mode, each of the various field quantities satisfies the same simple differential equation

$$\frac{\partial^2 F}{\partial t^2} = -\omega^2 F, \tag{20•1–1}$$

where the symbol F may stand for *any* of the components of any of the fields, such as the electric field $\mathbf{E}(\mathbf{r}, t)$, the magnetic field $\mathbf{B}(\mathbf{r}, t)$, the magnetic vector potential $\mathbf{A}(\mathbf{r}, t)$, etc., as well as the scalar electrostatic potential $\Phi(\mathbf{r}, t)$.

Each of the classical field quantities of every mode is still a function of both time and position. Different modes have different spatial field distributions. We will be interested in the interaction of an electron with a selected single mode, the spatial field distribution of which we may assume to be known. To specify the classical state of any specified single mode, it is then sufficient to know the (time-dependent) value of any single one of the various field quantities, taken at a suitable *fixed* point of reference. Given that local reference value, the position dependence of all fields is automatically known from the known spatial field distribution of that mode.

Without loss of generality, we may place the coordinate origin at the point of reference and introduce a set of position-independent *reference* field amplitudes, such as

$$\begin{gathered} \mathbf{E}_0(t) \equiv \mathbf{E}(\mathbf{r} = 0, t), \qquad \mathbf{B}_0(t) \equiv \mathbf{B}(\mathbf{r} = 0, t), \\ \mathbf{A}_0(t) \equiv \mathbf{A}(\mathbf{r} = 0, t), \quad \text{etc.} \end{gathered} \tag{20•1–2}$$

There is a different set of such reference field amplitudes for each mode. Any one of these different field amplitudes from the same set may be used as the independent position-like variable to describe the dynamics of the harmonic field oscillator for the particular mode represented by that set. Along with it, a second conjugate quantity is needed, to serve as the canonically conjugate instantaneous momentum-like variable. All this is similar to the way we used the charge Q on the capacitor plates as the position-like variable and the quantity IL as the momentum-like variable to describe the dynamics of the LC-oscillator.

It will be most convenient to work with the electric field $\mathbf{E}_0(t)$ and the vector potential $\mathbf{A}_0(t)$. Because of the fundamental relation between electric field and vector potential,

$$\frac{\partial \mathbf{A}(\mathbf{r}, t)}{\partial t} = -\mathbf{E}(\mathbf{r}, t), \tag{20•1–3}$$

we have

$$\frac{d\mathbf{A}_0}{dt} = -\mathbf{E}_0. \tag{20•1–4}$$

By differentiating this relation once more and drawing on (20•1–1), we also obtain

$$\frac{d\mathbf{E}_0}{dt} = \omega^2 \mathbf{A}_0. \tag{20•1–5}$$

The pair (20•1–4) and (20•1–5) looks suspiciously similar to a pair of classical Hamilton-Jacobi equations for a harmonic oscillator. The classical Hamiltonian for such an oscillator is of the form

$$H = \frac{1}{2M}p^2 + \frac{M}{2}\omega^2 q^2, \tag{20•1–6}$$

where M is the mass of the oscillating object, q its position coordinate, and p its momentum. The two canonically conjugate variables p and q satisfy the classical Hamilton-Jacobi equations (10•3–7) and (10•3–8):

$$\frac{dq}{dt} = +\left(\frac{\partial H}{\partial P}\right)_q = \frac{p}{M}, \qquad \frac{dp}{dt} = -\left(\frac{\partial H}{\partial q}\right)_p = -M\omega^2 q. \tag{20•1–7a,b}$$

The similarity of (20•1–4) and (20•1–5) to (20•1–7a,b) is evident. The occurrence of ω^2 in both (20•1–5) and (20•1–7b) suggests the use of the vector potential amplitude $\mathbf{A}_0(t)$ as the position-like dynamical variable, and a suitable quantity proportional to the electric field $\mathbf{E}_0(t)$ as the momentum-like variable.

To pursue this idea further, consider the quantity

$$U = \gamma \cdot (E_0^2 + \omega^2 A_0^2), \tag{20•1–8}$$

where γ is an as-yet unspecified multiplier. From (20•1–4) and (20•1–5), we see that U is time-independent:

$$\frac{dU}{dt} = 0. \tag{20•1–9}$$

Being quadratic in the fields, U is clearly related to the energy contained in the mode, and by a suitable choice of γ, it can be made equal to that energy.

Which value to choose for γ depends on the field configuration of the mode.

We consider here the simplest possible case, a linearly polarized propagating plane wave,

$$\mathbf{E}(\mathbf{r}, t) = \mathbf{E}_{00} \cos(\mathbf{k} \cdot \mathbf{r} - \omega t). \tag{20•1–10}$$

Such a wave has a uniform energy density

$$u = \tfrac{1}{2}\epsilon_0 \mathbf{E}_{00}^2. \tag{20•1–11a}$$

If the mode-defining resonator has the volume Ω, we obtain the total energy contained in the mode by simply multiplying with Ω, which leads to

$$U = \tfrac{1}{2}\epsilon_0 \mathbf{E}_{00}^2 \Omega. \tag{20•1–11b}$$

The quantity U in (20•1–8) is equal to this if we set

$$\gamma = \tfrac{1}{2}\epsilon_0 \Omega, \tag{20•1–12}$$

which leads to

$$U = \tfrac{1}{2}\epsilon_0 \Omega \cdot (\mathbf{E}_0^2 + \omega^2 \mathbf{A}_0^2). \tag{20•1–13}$$

This result is brought into the canonical form (20•1–6) by selecting

$$\mathbf{q} = \mathbf{A}_0(t), \tag{20•1–14}$$

$$\mathbf{p} = -\epsilon_0 \Omega \cdot \mathbf{E}_0(t), \tag{20•1–15}$$

and

$$M = \epsilon_0 \Omega. \tag{20•1–16}$$

Here p and q are canonically conjugate *momentum-like* and *position-like* variables characterizing the mode, and M is a *mass-like* variable, all three characterizing the electromagnetic mode.

20.1.2 Transition to Quantum Mechanics

So far, the above is a purely classical transformation of the properties of a plane wave to a harmonic oscillator formalism. However, just as in our quantization of an LC circuit in section 10.3, the identical formal structure of the electromagnetic wave to a harmonic oscillator in the classical limit must be the consequence of an identical formal structure on the quantum level. We therefore postulate:

> The correct quantum-mechanical description of each electromagnetic wave oscillator is in terms of a state $|\Psi\rangle$, the time evolution of which is governed by a Schroedinger wave equation-like equation of the form
>
> $$i\hbar \frac{\partial}{\partial t}|\Psi\rangle = \hat{H}|\Psi\rangle, \tag{20•1–17}$$
>
> where $\hat{H}$ is the Hermitian *operator*

$$\hat{H} = \frac{1}{2M}\hat{p}^2 + \frac{M}{2}\omega^2\hat{q}^2, \tag{20•1–18}$$

obtained from the Hamilton *function* (20•1–6) by replacing p and q with a pair of canonically conjugate Hermitian operators $\hat{p}$ and $\hat{q}$ that satisfy the commutation relation

$$[\hat{q}, \hat{p}] = i\hbar. \tag{20•1–19}$$

The energy eigenvalue spectrum of this field oscillator is of course of the familiar form

$$\mathcal{E}_n = \hbar\omega \cdot (n + \tfrac{1}{2}), \qquad n = 0, 1, 2, \ldots. \tag{20•1–20}$$

We could obtain the vector potential representation if we implemented (20•1–19) by setting

$$\hat{q} = q, \qquad \hat{p} = -i\hbar\frac{\partial}{\partial q}, \tag{20•1–21a,b}$$

which would lead to a standard harmonic oscillator Schroedinger equation. However, we will never actually need this equation. As in chapter 10, almost all harmonic oscillator properties of interest can be expressed in terms of a few matrix elements. All we need to take over the entire earlier formalism is to express the natural unit of "length" L of the field oscillator in terms of our variables M and ω:

$$L^2 = \frac{\hbar}{M\omega} = \frac{\hbar}{\epsilon_0\Omega\omega}. \tag{20•1–22}$$

We will need principally the matrix elements of $\mathbf{q} = \mathbf{A}_0$ itself:

$$\mathbf{A}_{mn} = \langle m(\omega)|\mathbf{A}_0|n(\omega)\rangle. \tag{20•1–23}$$

Here $|m(\omega)\rangle$ and $|n(\omega)\rangle$ refer to the eigenstates of a field mode with the frequency ω. These matrix elements are formally equivalent to the position matrix element $\langle m|\hat{x}|n\rangle$ of a mechanical harmonic oscillator. From (10•1–9),[1] for an oscillator with a natural unit of length L,

$$x_{mn} = \langle m|x|n\rangle = \begin{cases} L\sqrt{n_+/2} & \text{if } m = n \pm 1, \\ 0 & \text{otherwise,} \end{cases} \tag{20•1–24a}$$

where n_+ is the larger of the two quantum numbers m and n:

$$n_+ = \begin{cases} m & \text{if} \quad m = n + 1, \\ n & \text{if} \quad n = m + 1. \end{cases} \tag{20•1–24b}$$

---

[1] Note that the variable q used here is the formal equivalent of the variable x in chapter 10, not of the dimensionless variable Q used there, which is related to x via $x = QL$.

The field matrix elements are obtained from the matrix elements of a mechanical oscillator by simply replacing the quantity L with the value from (20•1–22):

$$A_{mn} = \begin{cases} \hbar\sqrt{\dfrac{n_+}{2\epsilon_0\,\Omega\hbar\omega}} & \text{if } m(\omega) = n(\omega) \pm 1, \\ 0 & \text{otherwise.} \end{cases} \tag{20•1–25}$$

Here, n_+ is defined as in (20•1–24b), with m and n replaced by $m(\omega)$ and $n(\omega)$.

20.2 RADIATIVE TRANSITIONS AS INTERACTIONS BETWEEN COUPLED QUANTUM SYSTEMS

We now apply the formalism developed in section 20.1 to the interaction between an electron and the electromagnetic field. For simplicity, we consider in the present section only the interaction with a single field mode; we will generalize to an arbitrary number of modes in section 20.3.

The Hamiltonian for the overall system can then be written

$$\hat{H} = \hat{H}(e) + \hat{H}(\omega) + \hat{W}, \tag{20•2–1}$$

where $\hat{H}(e)$ is the Hamiltonian of the electron system by itself, $\hat{H}(\omega)$ the Hamiltonian of the field by itself, and $\hat{W}$ the interaction Hamiltonian.

For simplicity, we assume that the electron system by itself is that of a single electron in an external potential $V(\mathbf{r})$, so that

$$\hat{H} = \frac{1}{2M}\hat{p}^2 + V(\mathbf{r}). \tag{20•2–2}$$

We also make again the dipole approximation, that is, we assume that the electron system is sufficiently small compared to the electromagnetic wavelength that we may neglect the variations of the electromagnetic fields over the electron system. Both assumptions could be easily removed if desired, but they simplify the mathematics greatly.

We showed in chapter 19 that the interaction between an electron and an electromagnetic wave is described by the operator

$$\hat{W} = \frac{e}{m_e}\hat{\mathbf{A}}\cdot\hat{\mathbf{p}}, \tag{20•2–3}$$

involving the vector potential of the wave.[2] In chapter 19, this operator occurred as an *external* perturbation acting on the electron system, and $\hat{\mathbf{A}}$ was an explicitly time-dependent *external* parameter, rather than a true operator. In

---

[2] As in chapter 19, we are ignoring again the term quadratic in the vector potential, which showed up in (19•3–5). See problem 20•2–1 for the effect of this term.

the present formalism, $\hat{W}$ represents an *internal* interaction between two parts of an overall quantum system, and $\hat{\mathbf{A}}$ is now an operator operating on the field portion of the state function, just as $\hat{\mathbf{p}}$ operates on the electron portion. We performed such a transition earlier as an ad hoc modification to the semiclassical treatment. The present treatment finally gives a rigorous justification for that procedure.

20.2.1 The Matrix Elements of the Interaction Hamiltonian

We treat the effect of the interaction Hamiltonian $\hat{W}$ by the methods of time-dependent perturbation theory. That is, we write the overall Hamiltonian (20•2–1) as the sum of an unperturbed Hamiltonian and a perturbation,

$$\hat{H} = \hat{H}^{(0)} + \hat{W}, \tag{20•2–4a}$$

where the unperturbed Hamiltonian,

$$\hat{H}^{(0)} = \hat{H}(e) + \hat{H}(\omega), \tag{20•2–4b}$$

is the sum of the unperturbed electron Hamiltonian and the unperturbed field Hamiltonian.

Because the unperturbed Hamiltonian consists of two terms with separated independent variables, its eigenfunctions are simply products of electron eigenfunctions and field eigenfunctions. We express this by writing *symbolically*

$$|n\rangle = |n(e), n(\omega)\rangle = |n(e)\rangle\,|n(\omega)\rangle, \tag{20•2–5}$$

where $|n(e)\rangle$ is the electron eigenstate and $|n(\omega)\rangle$ the field eigenstate that jointly form the unperturbed *overall* system eigenstate $|n\rangle$.

We need the matrix elements

$$W_{mn} = \langle m|\hat{W}|n\rangle = \frac{e}{m_e}\langle m(e), m(\omega)|\hat{\mathbf{A}}\cdot\hat{\mathbf{p}}|n(e), n(\omega)\rangle. \tag{20•2–6}$$

The operator $\hat{\mathbf{A}}\cdot\hat{\mathbf{p}}$ is the product of an operator $\hat{\mathbf{A}}$ that operates only on the field and an operator $\hat{\mathbf{p}}$ that operates only on the electron. Because the unperturbed state functions are themselves such products, the overall matrix elements can be factored into products of a field matrix element and an electron matrix element:

$$W_{mn} = \frac{e}{m_e}\mathbf{A}_{mn}\cdot\mathbf{p}_{mn}. \tag{20•2–7}$$

The electron matrix element is again simply the momentum matrix element

$$\mathbf{p}_{mn} = \langle m(e)|\hat{\mathbf{p}}|n(e)\rangle; \tag{20•2–8}$$

the field matrix element was given in (20•1–23) and (20•1–25).

20.2.2 Consequences

The properties (20•2–6) through (20•2–8) of the interaction matrix elements lead immediately to a number of important conclusions about the interaction of an electron system with a quantized electromagnetic wave:

(a) The matrix elements of the interaction are time-independent. This means that the interaction may be treated as a step perturbation acting on the *overall* quantum system, rather than as a wave perturbation acting on the electron subsystem alone. This is the final justification of our ad hoc step perturbation treatment in chapter 19.

Recall from chapter 19 that a step perturbation causes transitions only between states that fall within a certain resonance bandwidth of each other. Applied to our problem, this means that any change in the energy of the electron subsystem must be very nearly equal and opposite to any change in the energy of the field subsystem, as was illustrated already in Fig. 19•3–2.

(b) Only such transitions can take place during which the energy of the field changes by exactly one quantum $\hbar\omega$:

$$m(\omega) = n(\omega) \pm 1. \tag{20•2–9}$$

Together with the condition that the initial and the final state of the overall system fall within a resonance bandwidth of each other, this implies the famous Bohr relation between the radiation frequency and the energy difference separating the electron energy levels involved in the transition:

$$\hbar\omega \approx |\mathcal{E}_m - \mathcal{E}_n|. \tag{20•2–10}$$

The reasons why (20•2–10) need not be satisfied *exactly* are the same as those discussed already in chapter 19 in the same context. The comments made there about energy conservation apply again: Any energy difference represents an imbalance in external work required in placing the electron system into the wave field and removing it from the field.

(c) If the resonance condition (20•2–10) is satisfied, the electron subsystem can make a transition from an energetically higher state to an energetically lower state by emission of a quantum $\hbar\omega$, even if the field was initially in its ground state. In this case, simply $n_+(\omega) = 1$. This is the theoretical basis for the spontaneous emission of radiation of any excited electron system into "field-free" space. We put "field-free" in quotation marks because—as we have pointed out repeatedly—what is considered field-free in classical physics is quantum-mechanically simply a field system in its ground state.

It is instructive to re-express the field matrix element in a form that permits a quantitative comparison with the semiclassical matrix element (19•3–16). We first re-write (20•1–25) as

$$|\langle n(\omega) \pm 1|A(\omega)|n(\omega)\rangle|^2 = \frac{1}{2\omega^2} \cdot \frac{n_+\hbar\omega}{\epsilon_0\Omega} = \frac{1}{2\omega^2} \cdot \frac{u_+}{\epsilon_0} = \frac{E_+^2}{4\omega^2}, \tag{20•2–11}$$

where

$$u_+ = n_+ \cdot \hbar\omega/\Omega = \tfrac{1}{2}\epsilon_0 E_+^2 \tag{20•2–12}$$

is the classical electromagnetic energy density and E_+ the classical electric field amplitude associated with the presence of n_+ quanta of energy $\hbar\omega$ in the volume Ω. Note that u_+ and E_+ are the values associated with the number of quanta present either before or after the transition, *whichever is larger*. These values are therefore never zero. If (20•2–12) is inserted into (20•2–7), we may write

$$|W_{mn}|^2 = \frac{e^2E_+^2}{4m_e^2\omega^2} \cdot |p_{mn}|^2, \tag{20•2–13}$$

where the momentum is taken in the direction of **A**. The form (20•2–13) is exactly equal to the values we *would* have obtained from (19•3–16) *if* we had replaced the classical field E_0 there by the field E_+. For a large number of quanta the difference is negligible, but for a very small number it is decisive.

As we pointed out in section 19.3, it is often desirable to replace the momentum matrix element p_{mn} by the position or dipole matrix element r_{mn}, via the relation (19•3–19), which we write here in the form

$$|p_{mn}|^2 = m_e^2\omega_{mn}^2|r_{mn}|^2. \tag{20•2–14}$$

The matrix element r_{mn}, with or without the factor e, is often called the **dipole matrix element**, because it is proportional to the amplitude of an electric dipole constructed by linear superposition of the two states $|m(e)\rangle$ and $|n(e)\rangle$. When this substitution is made, (20•2–13) becomes

$$|W_{mn}|^2 = \frac{e^2n_+|\omega_{mn}|}{2\epsilon_0\Omega} \cdot |r_{mn}|^2 = \frac{1}{4}e^2E_+^2|r_{mn}|^2. \tag{20•2–15}$$

This form is usually simpler than the original form using the momentum matrix element.

Having established the quantitative connection between the fully quantized treatment of a wave perturbation and the semiclassical treatment of chapter 19, we may take over all the results of the former treatment by simply replacing $|W_{12}|^2$ in that treatment with $|W_{mn}|^2$ from (20•2–13) or (20•2–15).

No transitions will, of course, occur if the initial state of the electron system is the lower of the two states, and the field is initially also in its ground state, because the transition is then energetically impossible. But both the transition probability and the resonance width are nonzero for downward electron transitions, even for a field initially in its ground state.

◆ PROBLEM TO SECTION 20.2

#20•2-1: Effect of the A^2 Term in the Hamiltonian

Both in chapter 19 and in our present treatment, we neglected the term

$$\frac{e^2A^2}{2m_e}, \tag{20•2–16}$$

which showed up first in (19•3–5) in the electron Hamiltonian. Note that this term contains neither the position nor the momentum of the electron. Hence, if it is included in the treatment at all, it should be lumped into the wave Hamiltonian, where it changes the dispersion relation for the unperturbed waves in a way that depends on the number of electrons present. Analyze this effect in detail.

20.3 BROADBAND INTERACTIONS

20.3.1 The Hamiltonian and its Matrix Elements

For most radiative transition processes occurring in nature, the single-mode treatment given in section 20.2 is an oversimplification. Actual electromagnetic waves always have a finite bandwidth. In particular, any quantitative treatment of the spontaneous emission of radiation requires a consideration of *all* modes into which the emission can take place: As we saw, there is a finite emission probability into any initially empty field mode that falls within the resonance bandwidth of the transition. For an electron system in free space, there is always an infinite number of modes accessible within any finite resonance bandwidth. Although each mode makes only an infinitesimally small contribution to the overall interaction, the infinite number of contributions causes a finite overall interaction strength. The resonance bandwidth that governs this strength depends in turn on that strength. To determine both the overall strength of the spontaneous emission and its bandwidth, we need a theory that accounts quantitatively for the continuous nature of the electromagnetic spectrum.

Rather than working with a true continuum of interaction frequencies, we work with the quasi-continuum of the discrete resonance frequencies inside a very large cavity. If the cavity is sufficiently large, its size should not make any difference, and we shall indeed find that the dimensions of the cavity will drop out of the calculations in the end. We assume a cube-shaped cavity of volume $\Omega = L^3$ and non-reflecting periodic boundary conditions,

$$\begin{aligned} A(x, y, z, t) &= A(x + L, y, z, t) = A(x, y + L, z, t) \\ &= A(x, y, z + L, t). \end{aligned} \tag{20•3–1}$$

These conditions restrict the components of the allowed wave vectors **k** to values of the form

$$k_x = 2\pi n_x/L, \qquad k_y = 2\pi n_y/L, \qquad k_z = 2\pi n_z/L, \tag{20•3–2}$$

where n_x, n_y, and n_z are three arbitrary independent integers, including zero or negative values. The corresponding resonance frequencies are

$$\omega = \frac{2\pi c}{L}\sqrt{n_x^2 + n_y^2 + n_z^2}. \tag{20•3–3}$$

Associated with each triplet (n_x, n_y, n_z) are two transversely polarized modes, with polarization vectors that are perpendicular to each other and to the common wave vector **k**. The total number of modes in a narrow frequency interval $d\omega$ can be expressed as $D_\omega(\omega)\, d\omega$, where $D_\omega(\omega)$ is called the **mode density**, which can be shown to be given by

$$D_\omega(\omega) = \frac{\omega^2 \Omega}{\pi^2 c^3}. \tag{20•3–4}$$

Exercise: Derive (20•3–4).

Each of these modes represents a quantized harmonic oscillator with a Hamiltonian of the form (20•1–18). The total field Hamiltonian $\hat{H}(f)$ in (20•2–1) is the sum of the individual mode Hamiltonians:

$$\hat{H}(f) = \sum_\omega \hat{H}(\omega). \tag{20•3–5}$$

Similarly, the total interaction Hamiltonian is a sum over terms of the form (20•2–3), which we may write

$$\hat{W} = \frac{e}{m_e}\left[\sum_\omega \mathbf{A}(\omega)\right] \cdot \hat{\mathbf{p}}. \tag{20•3–6}$$

Finally, the field part $|n(e)\rangle$ of the unperturbed state function in (20•2–5) now becomes an infinite product of single-mode field eigenfunctions, one eigenfunction for each mode:

$$|n(f)\rangle = \prod_\omega |n(\omega)\rangle. \tag{20•3–7}$$

We need again the matrix elements of the perturbation $\hat{W}$ between the unperturbed overall system states. The electron matrix elements are not affected by going to an unlimited number of modes, except for the need to keep track of the direction in which the momentum is taken, relative to the different polarization directions of the various modes. We will say more about this aspect later.

The matrix elements $\mathbf{A}_{mn}$ of the vector potential in (20•2–7) now become sums of the form

$$\mathbf{A}_{mn}^{(\text{all})} = \sum_{\omega} \langle m(f) | \hat{\mathbf{A}}(\omega) | n(f) \rangle = \sum_{\omega} \mathbf{A}_{mn}(\omega). \tag{20•3–8}$$

Note that each of the vector potential operators acts on only one factor in the product (20•3–7). We may therefore write

$$\mathbf{A}_{mn}(\omega) = \langle m(\omega) | \hat{\mathbf{A}}(\omega) | n(\omega) \rangle \cdot \prod_{\omega' \neq \omega} \langle m(\omega') | n(\omega) \rangle, \tag{20•3–9}$$

where the product contains all modes ω' except the mode ω. Because of orthogonality, this product vanishes unless all modes contained in it retain their quantum state, in which case the product becomes unity. This means that only such interactions can take place during which *exactly* one mode changes its quantum state. The matrix elements $\mathbf{A}_{mn}(\omega)$ for the allowed transitions then reduce to the leading factor in (20•3–9), which is the same as in (20•2–7). Furthermore, as we saw in section 20.1, this matrix element will be nonzero only if the mode in question changes its quantum number $n(\omega)$ by exactly ± 1. The magnitude of the matrix element is then again given by (20•2–11).

20.3.2 Application of the Golden Rule

Consider now all those transitions in which the overall system was initially in a state $|n\rangle = |n(e), n(\omega)\rangle$ and during which the electron subsystem makes a transition $|n(e)\rangle \to |m(e)\rangle$. Associated with this electron transition are a very large number of possible field transitions, depending upon which of the field modes makes a transition along with the electron subsystem. Only those transitions will have a non-negligible probability for which the change in field energy very nearly or totally cancels the change in electron energy.

Energetically, these transitions form a continuum of final states, in exactly the same sense as in our Golden Rule treatment in chapter 19, where we discussed the transitions into a continuum of final states under the influence of a step perturbation. To apply the Golden Rule formalism to our problem, we need the energetic density of final states near exact resonance and the average absolute square of the perturbation matrix element, also near exact resonance.

Each field mode is associated with exactly one transition for which the matrix element is nonzero and for which the direction of the field energy change is opposite to the direction of the electron energy change. Therefore, the *energetic* density of final states near resonance is equal to the energetic density of modes near the resonance frequency $\omega_0 = |\omega_{mn}|$, or $1/\hbar$ times the *frequency* density of modes,

$$D_{\mathcal{E}}(0)\, d\mathcal{E} = D_{\mathcal{E}}(0)\, d(\hbar\omega) = D_{\omega}(\omega_0)\, d\omega. \tag{20•3–10}$$

From this and (20•3–4),

$$D_{\mathcal{E}}(0) = \frac{1}{\hbar} D_{\omega}(\omega_0) = \frac{\omega_0^2 \Omega}{\pi^2 \hbar c^3}. \tag{20•3–11}$$

In calculating the mean absolute square matrix element of the interaction matrix element, we must pay attention to the fact that this involves the scalar product $\mathbf{A} \cdot \mathbf{p}$ of two vectors. In section 20.1, we assumed specifically that the momentum $\mathbf{p}$ is taken in the direction of the vector potential $\mathbf{A}$. Because we are now dealing with vector potentials in different directions, we must write

$$|W_{mn}|^2 = \frac{e^2}{m_e^2} \cdot |\mathbf{A}_{mn} \cdot \mathbf{p}_{mn}|^2 = \frac{e^2}{m_e^2} \cdot |\mathbf{A}_{mn}|^2 |\mathbf{p}_{mn}|^2 \cdot \cos^2\theta, \tag{20•3–12}$$

where both $|\mathbf{A}_{mn}|$ and $|\mathbf{p}_{mn}|$ are understood to be the *magnitudes* of the vectors $\mathbf{A}_{mn}$ and $\mathbf{p}_{mn}$ and θ is the angle between these two vectors. With this change in the meaning of $\mathbf{p}_{mn}$, and with the addition of the cosine factor in (20•3–12), we can simply take over the value of $|W_{mn}|^2$ calculated in section 20.1. If we work with the form (20•2–15), we obtain

$$\overline{|W_{mn}|^2} = \frac{e^2 \hbar \omega_0 \overline{|r_{mn}|^2}}{2\epsilon_0 \Omega} \overline{[n_+ \cos^2\theta]}. \tag{20•3–13}$$

If this is inserted into the Golden Rule (19•4–21), together with (20•3–11), we obtain

$$\frac{1}{\tau} = 2\gamma = \frac{e^2 \omega_0^3 \overline{|r_{mn}|^2}}{\pi \hbar \epsilon_0 c^3} \overline{[n_+ \cos^2\theta]}. \tag{20•3–14}$$

Note that the volume of the cavity has canceled out.

20.3.3 The Spontaneous Emission of Radiation

We are now ready to treat the radiative decay of an electron system from a higher state $|n(e)\rangle$ to a lower state $|m(e)\rangle$ by the **spontaneous emission** of radiation, that is, in the absence of any already present external radiation. This corresponds to the case where all the modes falling within or near the resonance width of the transition are in their ground state, $n(\omega) = 0$. Because of (20•1–24b), this implies that $n_+ = 1$. Furthermore, because there are an equal number of equivalent modes polarized in each direction, the angular average is simply

$$\overline{\cos^2\theta} = \frac{1}{3}. \tag{20•3–15}$$

With these two results, (20•3–14) yields the final result,

$$\boxed{\frac{1}{\tau_0} = 2\gamma_0 = \frac{e^2\omega_0^3|r_{mn}|^2}{3\pi\hbar\epsilon_0 c^3}.} \tag{20•3–16}$$

The time τ_0 is the average time for which the electron system will reside in its upper state before having decayed. It is called the **spontaneous emission lifetime**.

The magnitude of this lifetime depends on the magnitude of the matrix element r_{mn}, which can vary over a very wide range. As a point of reference, we select the transition between the ground state and the first excited state of a one-dimensional harmonic oscillator of frequency ω_0 and mass m_e. As we saw in (19•5–4), in this case,

$$(x_{01}^2)_{\text{HO}} = \frac{\hbar}{2m_e\omega}. \tag{20•3–17}$$

Inserted into (20•3–16), this leads to

$$\boxed{\tau = \tau_{\text{HO}} = \frac{6\pi\epsilon_0 m_e c^3}{e^2\omega_0^2} = \frac{3\epsilon_0 m_e c\lambda_0^2}{2\pi e^2},} \tag{20•3–18}$$

where, in the last form, we have replaced the transition frequency by the corresponding electromagnetic wavelength, usually a more convenient quantity for optical transitions. For a wavelength in the near infrared, say, $\lambda_0 = 1\ \mu\text{m}$, corresponding to a frequency $\omega_0/2\pi$ of about 3×10^{14} Hz, (20•3–18) leads to $\tau_{\text{HO}} \approx 4.50 \times 10^{-8}$ s. More generally,

$$\tau_{\text{HO}} = 4.50 \times 10^{-8}\ \text{s} \cdot \left[\frac{\lambda_0}{1\ \mu\text{m}}\right]^2. \tag{20•3–19}$$

We see that, at least in this example, the spontaneous emission lifetime is very long compared to the oscillation period of the electromagnetic field involved in the transitions. For $\lambda = 1\ \mu\text{m}$, the lifetime exceeds 10^7 periods, and although this ratio decreases with decreasing wavelength, it remains large down to X-ray wavelengths.

The ratio of the spontaneous emission time to the oscillation period is directly related to the sharpness of the spontaneous emission frequency. As we saw in chapter 19, the FWHM energy range of the final states of the transitions into a continuum is given by $\hbar\Delta\omega = 2\hbar\gamma$, or

$$\frac{1}{\hbar}\Delta\mathcal{E} = \Delta\omega = 2\gamma = \frac{1}{\tau}. \tag{20•3–20}$$

This is the energy range of the emitted photons.

It is frequently useful, particularly in quantum electronics, to express the sharpness of the transition in terms of the Q-factor of the resonance,

$$Q = \frac{\mathcal{E}}{\Delta\mathcal{E}} = \frac{\omega_0}{\Delta\omega} = \omega_0\tau. \tag{20•3–21}$$

For our case of a harmonic oscillator, from (20•3–18),

$$Q = \frac{6\pi\epsilon_0 m_e c^3}{e^2\omega_0} = \frac{3\epsilon_0 m_e c\lambda_0}{e^2} = 8.47 \times 10^7 \cdot \left[\frac{\lambda_0}{1\ \mu\text{m}}\right]^2. \tag{20•3–22}$$

It is clear from this number that the spectral width of the emitted radiation is very narrow compared to the center frequency.

Comment: Relationship to the Radiation of a Classical Electrical Dipole Our harmonic oscillator result (20•3–18) no longer contains Planck's constant. Now, it is a good rule of thumb that the absence of Planck's constant from the result of a quantum calculation means that this result might also be obtainable in classical physics.[3] This is indeed the case here: The spontaneous emission lifetime τ_{HO} of a harmonic oscillator is nothing other than the damping time of a classical electrical dipole[4] consisting of a particle of charge e and mass m_e, bound elastically to a point in such a way that it can undergo harmonic oscillations with the frequency ω_0.

20.3.4 Stimulated Emission

If there is already a radiation field present within the bandwidth of the transition, we have $n_+ > 1$, leading to an enhanced emission probability rate

$$2\gamma = \frac{\overline{n_+}}{\tau_0}\overline{3\cos^2\theta}, \tag{20•3–23}$$

where τ_0 is the spontaneous emission lifetime. Because the additional transition rate is due to the radiation field already present, it is referred to as **stimulated emission**. It should not be viewed as a physical process separate from spontaneous emission: Both are parts of the same process; spontaneous emission is simply what is left over in the limit of vanishing driving field.

Nevertheless, from a *practical* point of view, spontaneous and stimulated emissions play quite different roles. As we have seen, spontaneous emission

---

[3] The rule has exceptions. The best known occurs in the theory of diamagnetic susceptibility, where Planck's constant cancels out in the final result, yet the result is fundamentally unobtainable from classical physics, if the laws of classical physics are applied consistently, without making hidden assumptions that inherently violate them.

[4] See, for example, Eqs. (17.3) and (17.57) in Jackson's text, cited in section 19.5. Jackson's terminology differs from ours: His Γ is the same as our $1/\tau$; his τ is $2e^2/3mc^3$.

takes place into *all* modes within the resonance width of the transition. The stimulated part of the emission, however, goes exactly into those modes which already contain one or more quanta, and the rate of stimulated emission into each mode is directly proportional to the number of quanta already present in that mode. Put differently, stimulated emission *amplifies* the radiation already present. This property forms the backbone of quantum electronics, and particularly of the most important quantum-electronic device, the laser, an acronym standing for *l*ight *a*mplifier by *s*timulated *e*mission of *r*adiation. In lasers, one or at most a few modes are made to build up in intensity to a very large number of photons, to the point that the stimulated emission into these few modes is much stronger than the combined spontaneous emission into all other modes. Under these circumstances, the averaging procedure over many modes, employed in the derivation of the Golden Rule, is no longer applicable, and the exact theory assumes a form intermediate between the purely oscillatory transition probabilities of a two-level system and the purely exponential transition probabilities of broadband transitions. The development of such a theory is a topic in quantum electronics.

We restrict ourselves here to stimulated emission driven by broadband radiation. In this case, it is possible to express the stimulated portion of the total transition rate in terms of the classical spectral energy density $u(\omega)$ of the radiation field that causes the stimulated emission. Assume that $n(\omega)$ is the average number of quanta *originally* contained in the modes with the frequency ω, that is, *before* the stimulated emission occurs. The average classical field energy contained in each mode is then $n(\omega) \cdot \hbar\omega$, and

$$u(\omega) = n(\omega) \cdot \hbar\omega D_\omega(\omega)/\Omega \tag{20•3–24}$$

is the classical energy density of the radiation field, per unit of (angular) frequency.

In order to apply this to (20•3–23), we recall that in the case of the emission of a photon by the electron system, the quantum number $n_+(\omega)$ is the number of quanta in the mode ω *after* the emitted quantum has been added to that mode. Hence,

$$n_+(\omega) = n(\omega) + 1. \tag{20•3–25}$$

From (20•3–23) through (20•3–25), and with the aid of (20•3–4) and (20•3–16), we obtain

$$2\gamma = \frac{1}{\tau_0} + \frac{\overline{n(\omega)}}{\tau_0} = \frac{1}{\tau_0} + B \cdot u(\omega_0), \tag{20•3–26}$$

where

$$B = \frac{\Omega}{\hbar\omega_0 D_\omega(\omega_0)} \cdot \frac{1}{\tau_0} = \frac{\pi^2}{\hbar}\left(\frac{c}{\omega_0}\right)^3 \frac{1}{\tau_0} = \frac{\lambda^3}{8\pi\hbar} \cdot \frac{1}{\tau_0} \tag{20•3–27}$$

is the **Einstein *B*-coefficient**. It was first introduced, on a purely ad hoc basis, in a famous 1917 paper in which Einstein postulated the concept of stimulated emission, to re-derive Planck's 1899 black-body radiation law from a photon and energy level point of view.[5] Equation (20•3–26) states that the stimulated portion of the total emission is proportional to the spectral energy density of the radiation field at the exact transition frequency.

The line width of the total emission—spontaneous plus stimulated—continues to be given by (20•3–20), that is,

$$\Delta\omega = \frac{1}{\tau} = \frac{1}{\tau_0} + B \cdot u(\omega_0). \tag{20•3–28}$$

20.3.5 Absorption

The process inverse to emission is the absorption of radiation by an electron system making an upward transition. The principal difference between the two processes is that now n_+ is the average quantum number in the interacting radiation field modes *before* the transition, not after it. That is, instead of (20•3–25), we now have

$$n_+(\omega) = n(\omega), \tag{20•3–29}$$

and instead of (20•3–26), we obtain the transition probability rate

$$2\gamma = Bu(\omega_0), \tag{20•3–30}$$

which is directly proportional to the spectral energy density at exact resonance and exactly the same as for the stimulated emission portion of the total emission probability rate. The absorption line width is not equal to the value 2γ given in (20•3–30), but is equal to the line width for *total* emission, (20•3–28). To obtain this result, it is necessary to modify the Golden Rule treatment by including secondary transitions within the continuum of final *field* states, during which the electron system first returns to its initial state by emitting a photon into a mode different from the one from which it absorbed a photon initially, followed by a new absorption of a third photon. Such a treatment goes beyond the scope of this text.

20.3.6 Planck's Black-Body Radiation Law

The expressions (20•3–26) and (20•3–30) for the transition probability rates for emission and absorption of a quantum and the relation (20•3–27) between the Einstein B-coefficient and the spontaneous emission lifetime permit us to derive an expression for the spectral energy density inside a cavity in thermal

---

[5] Reprinted, in translation, in ter Haar's book, quoted in the "General References." The coefficient actually used by Einstein was smaller by a factor 2π. This is because Einstein used the spectral energy density in terms of ν rather than $\omega = 2\pi\nu$. The two densities are related by $u_\omega\, d\omega = 2\pi u_\omega\, d\nu = u_\nu\, d\nu$.

equilibrium. To this end, consider a number of atoms inside a cavity whose walls are at a temperature T.

Suppose that there are N_1 atoms in the state $|1\rangle$ with energy $\mathcal{E}_1$ and N_2 in state $|2\rangle$ with energy $\mathcal{E}_2 > \mathcal{E}_1$. In thermal equilibrium, we must have

$$\frac{N^2}{N_1} = \exp\left(-\frac{\mathcal{E}_2 - \mathcal{E}_1}{k_B T}\right) = \exp\left(-\frac{\hbar\omega_0}{k_B T}\right). \qquad (20\text{•}3\text{–}31)$$

Furthermore, in equilibrium, the number of photon emission and absorption events must be equal:

$$N_1 B \cdot u(\omega_0) = N_2\left[\frac{1}{\tau_0} + B \cdot u(\omega_0)\right]. \qquad (20\text{•}3\text{–}32)$$

From (20•3–31), (20•3–32), and (20•3–27), we obtain

$$\boxed{u(\omega) = \frac{1}{B\tau_0}\frac{1}{N_1/N_2 - 1} = \frac{1}{\pi^2\hbar^2 c^3}\frac{(\hbar\omega)^3}{\exp(\hbar\omega_0/k_B T) - 1}.} \qquad (20\text{•}3\text{–}33)$$

This is Einstein's derivation of Planck's black-body radiation law, the law that gave birth to quantum mechanics.

20.4 CORRELATED PHOTON PAIRS

20.4.1 The Idea: Emission of Two Photons with Opposite Angular Momentum

In the physics of light emission from atoms, one often runs into situations in which selection rules prevent an atom in an excited state from making a direct radiative transition to the ground state, because both states have the same angular momentum, usually $l = 0$, in which case the transition is forbidden. In many such cases there exists a state with $l = 1$ at an intermediate energy, and the decay can take then place via this intermediate state through the emission of two photons, usually in very quick succession (**Fig. 20•4–1**).

We saw in chapter 19 that photons carry an angular momentum, related to their polarization. Inasmuch as the angular momentum of the atom reverts to its original value, the two photons must carry opposite angular momentum;

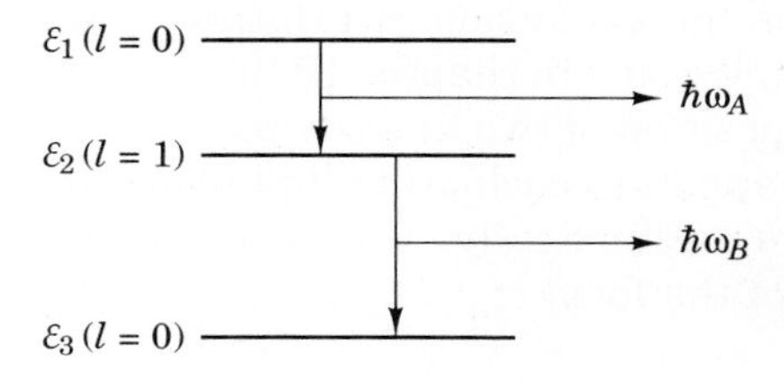

Figure 20•4–1. Two-photon cascade.

hence, there will be a strong correlation between the polarizations of the two photons.

For simplicity, assume that the two photons fly apart in exactly opposite directions and are circularly polarized. From chapter 19, a photon with positive helicity (= left-handed circular polarization) carries an angular momentum with a component $+\hbar$ pointing in the direction of photon propagation, while a photon with negative helicity (= right-handed circular polarization) carries the same angular momentum pointing "backwards." If the two photons fly apart in opposite directions, they must have the same helicity in order for the two angular momentum vectors to cancel. Whatever the helicity of the first photon, that of the second photon must be the same.

Although non-interacting, such correlated photons cannot be viewed as two independent objects, but must be treated as two constituents of a composite object, with a two-particle state function representing the two photons collectively. The formal description of a correlated photon pair with canceling angular momenta is in terms of the two-photon superposition state

$$|\Psi\rangle = \frac{1}{\sqrt{2}}[|+-\rangle + |-+\rangle]. \tag{20•4–1}$$

Here $|+-\rangle$ designates a two-photon state in which photon 1 has a z-component of the angular momentum of $+\hbar$, and photon 2 has a z-component of the angular momentum of $-\hbar$. In the other term, $|-+\rangle$, the two photons are interchanged. We explicitly ignore the specifications of all other properties of the two photons, such as the directions in which they fly off and, especially, their energies. None of these will play a role in the arguments that follow.

The superposition evidently gives equal weight to the two states, and the $(+)$-sign ensures that the superposition state itself is invariant under exchange of the two photons. The latter is not a matter of choice: We shall see in chapter 22 that invariance under exchange of any two photons is a necessary condition for multi-photon states.

Because photons do not interact, we may write each of the two-photon terms in (20•5–1) as the product of two one-photon states:

$$|\Psi\rangle = \frac{1}{\sqrt{2}}[|+\rangle_1|-\rangle_2 + |-\rangle_1|+\rangle_2]. \tag{20•4–2}$$

Although we have written the two-photon superposition state here in terms of one-photon states representing *circular* polarization, the latter simply serve as basis states for the expansion of arbitrary photon states, including *linearly* polarized states. We saw in chapter 19 that a circularly polarized wave is simply a linear superposition of two linearly polarized waves of equal amplitude, polarized at right angles to each other and phase shifted relative to each other by 90°. Quantum-mechanically, the corresponding superposition for *single*-photon states is of the form

$$|+\rangle = \frac{1}{\sqrt{2}}[|x\rangle + i|y\rangle] \quad \text{and} \quad |-\rangle = \frac{1}{\sqrt{2}}[|x\rangle - i|y\rangle], \qquad (20{\bullet}4\text{–}3\text{a,b})$$

where $|x\rangle$ and $|y\rangle$ are appropriate linearly polarized states, with mutually orthogonal polarizations and equal phases. If we insert (20•4–3a,b) into (20•4–2), the "mixed terms" cancel, and we obtain

$$|\Psi\rangle = \frac{1}{\sqrt{2}}[|x\rangle_1|x\rangle_2 + |y\rangle_1|y\rangle_2], \qquad (20{\bullet}4\text{–}4)$$

which implies that both photons must have the same *linear* polarization, whatever it may be.

20.4.2 A Simple Correlation Experiment

The above claim is an experimentally testable prediction, and it is useful to discuss here how such a test can be performed.

Consider the hypothetical experimental setup shown in **Fig. 20•4–2**. It consists of a light source emitting the photon pairs and two classical polarization-dependent beam splitters, labeled 'East' and 'West,' and placed along the $\pm\mathbf{z}$-directions, opposite from each other relative to the light source. Each beam splitter has two output channels, labeled here as 'X' and 'Y', and an incoming linearly polarized photon will be directed into one channel or the other, depending on the polarization state of the photon. Every output channel is followed by a photon counter.

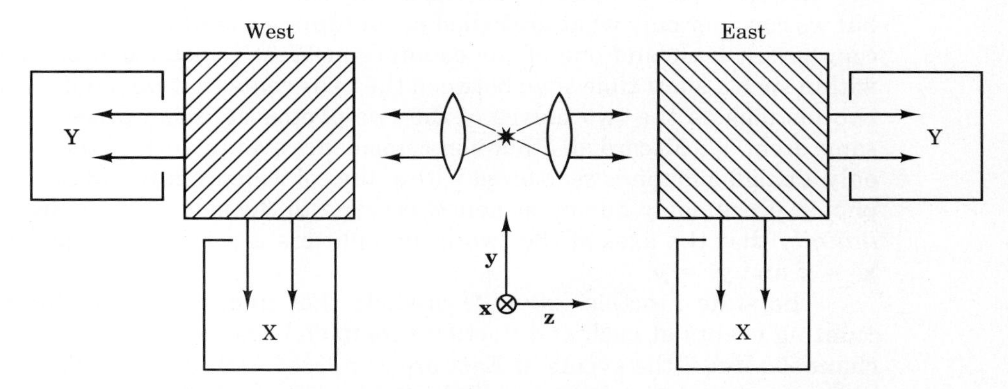

Figure 20•4–2. Photon correlation test setup. The two cross-hatched boxes are a schematic representation of polarization-sensitive beam splitters. Readers familiar with the polarization properties of so-called Brewster plates in classical optics may view the boxes as stacks of large numbers of ideal Brewster plates, in which case **y**-polarized photons pass through the stack, while **x**-polarized photons are deflected from the beam, as shown.

Polarization-dependent beam splitters are devices that are well known from classical wave optics. We need not be concerned here with how they are constructed; we simply quote their basic properties:

(a) Each beam splitter always has a pair of polarization axes, called here $\mathbf{x}$ and $\mathbf{y}$, perpendicular to each other and to the propagation direction $\pm\mathbf{z}$ of the incident beam.

(b) If the incoming beam is linearly polarized, with an electric vector $\mathbf{E}$ parallel to the $\pm\mathbf{x}$-axis, all photons will emerge in the same output channel of the beam splitter; we label this channel 'X.' Similarly, if the incident electric vector is parallel to the $\pm\mathbf{y}$-axis, all photons will emerge in the other output channel, labeled 'Y.' In those two limiting cases, the path of every photon is perfectly deterministic.

(c) If the electric vector $\mathbf{E}$ is at an angle θ relative to the $\mathbf{x}$-axis, the photons will be probabilistically distributed over the two output channels, with the probabilities

$$p_X = \cos^2\theta, \qquad p_Y = \sin^2\theta. \tag{20•4–5}$$

Returning to our experimental setup, we assume that the light source is weak and that the two photons belonging to the same pair are emitted in sufficiently quick succession—within a time interval Δt much shorter than the time between successive pairs—so that there is never any doubt as to which two photons registered by the counters at East and West belong to the same pair.

We record all photons registered by any one of the four quantum counters, but we consider only what are called **coincidence events**, in which one of the counters at East and one of the counters at West register one photon each, within such a short time span between the two events that we must ascribe the two photons as the two halves of the two-photon emission process from the same atom. We discard all other non-coincidence events, such as those in which only a single photon is registered within the coincidence interval Δt, the other photon presumably having somehow escaped to the side. Finally, we assume *initially* that the axes of the two beam splitters are aligned in parallel, i.e., $\mathbf{x}' = \mathbf{x}$ and $\mathbf{y}' = \mathbf{y}$.

The state function (20•4–4) predicts that under these conditions, the counting events at each end fluctuate completely randomly between the two channels. But if the events at East are compared with those at West, the two photons are registered either both by the two X counters or both by the two Y counters. Events in which one X counter and one Y counter respond do not occur. This is indeed what is found.

20.4.3 Conceptual Consequences

It is instructive to view the foregoing result from the point of view of the action of the photon counters on the state function.

Suppose that one of the photons materializes in one of the two X channels; it does not matter which one. We might say—loosely—that the direction of polarization of the photon has been "measured" to be the $\pm\mathbf{x}$-direction, but as we have seen repeatedly earlier, starting with chapter 1 in the context of a position "measurement," quantum mechanical measurements must be viewed, not in the classical sense of simply determining a pre-existing quantity, but as the *forcing* of the object into a state in which the variable has the sharp "measured" value: The measurement itself actively *creates* the measured value, rather than just passively observing it.

In our experiment, this means that out of the two terms in (20•4–4), the first term has materialized, and the second term must at this point be removed from the description of the state. The remaining photon must then also belong to the first term, that is, it must have the same polarization as the first intercepted photon. If, as we assumed, both beam splitters have the same axis orientation, the second photon is in what we call a **polarization eigenstate** of both beam splitters; it must then also materialize in one of the X channels, yielding the perfect correlation claimed earlier.

We are evidently dealing here with a collapse of the state function very similar to the collapse discussed in chapter 1, the only difference being that now *two* photons are involved, rather than just one, and the collapse is not one from a one-object state to a zero-object state, but from two objects to one: Once one of the counters has captured one photon, the state function collapses as in the case of a single photon capture, with the difference that it collapses, not to nothing, but to a single-photon state function with a polarization equal to that of the capture channel of the first photon. It is this "leftover" single-photon state that determines the probability distribution for the second photon capture event.

Suppose next that the two beam splitters are rotated relative to each other by an angle θ. In that case, the photon remaining after interception of the first photon by one of the beam splitters will not be in a polarization eigenstate of the other beam splitter. As a result, the response of the latter will no longer be fully deterministic, but will be probabilistically distributed over its two output channels, with the probabilities given in (20•4–5). The overall result is that, whatever the response of one of the two beam splitters, there is a probability

$$f_{\text{same}} = \cos^2 \theta, \qquad (20{\bullet}4\text{–}6)$$

that both counter pairs will record the two photons in the same channel.

The above argument nowhere assumes a spatial proximity of the two beam splitters. We commented already in chapter 1 that the collapse of the state function following a measurement event must be viewed as a *nonlocal* process taking place instantaneously throughout the full spatial extent of the state function, rather than propagating through space with a finite speed. It is what we might call *instantaneous action at a distance.* We also pointed out that this collapse could *not* be used to signal from one detector to another faster than with the speed of light and that, hence, there is no violation of the principles of the theory of relativity.

Our argument implies that the correlation should persist even if the two beam splitters are separated from each other by a large distance, and the polarization axes of the two beam splitters are rapidly and randomly switched between two or more different directions, in such a way that the final setting of each beam splitter is not decided until shortly before the photon arrives, long after both photons have been emitted by the atom. Under those circumstances, there is no way the setting of one of the beam splitters can be communicated to the other by any signal traveling with the speed of light, in time to influence the results there. This is a testable proposition, and it has been confirmed experimentally in what must be considered one of the great tests of quantum mechanics.[6]

20.4.4 The Demise of Hidden Variables: Bell Inequality

An essential ingredient in our line of reasoning above was the postulate that quantum mechanical measurements must be viewed as *forcing* the object into a state in which the variable has the sharp "measured" value, and that the measurement itself *creates* the measured value, rather than just passively observing a pre-existing value. This naturally brings us back to the question first raised in chapter 1, namely, whether the statistical nature of quantum mechanics might not be due to the presence of hidden variables. Such a claim was made in a famous 1935 paper by none other than Einstein, who found himself unable to accept not only the inherent indeterminacy claimed by orthodox quantum mechanics, but even more, what he called the "spooky action-at-a-distance" implied in the statistical interpretation, as applied to a correlated system.[7]

For much of the next 30 years, this remained a purely philosophical point, believed by many to be undecidable experimentally. It was not until 1964 that another famous paper, by John S. Bell, showed that *any* hidden-variable theory is subject to certain constraints as to the degree of correlations that may occur in systems of two correlated objects.[8] These constraints may be expressed in terms of certain inequalities, called **Bell inequalities**. Under suitable conditions, the predictions of quantum mechanics violate these constraints, thereby

---

[6] A. Aspect, J. Dalibard, and G. Roger, "Experimental Test of Bell's Inequalities Using Time-Varying Analyzers", *Phys. Rev. Lett.,* Vol. 25, Dec. 1982, pp. 1804–1807. The authors used a photon cascade in the calcium atom, involving photons with the wavelengths $\lambda_A = 511.3$ nm and $\lambda_B = 421.1$ nm.

[7] A. Einstein, B. Podolsky, and N. Rosen, "Can Quantum-Mechanical Description of Physical Reality be Considered Complete?" *Phys. Rev.* 47, May 1935, pp. 777–780. The paper is commonly referred to as the EPR paper.

[8] J. S. Bell, "On the Einstein Podolsky Rosen Paradox", *Physica,* Vol. 1, Nov./Dec. 1964, pp. 195–200.

opening up the possibility of deciding experimentally between the two interpretations.

A complete and rigorous discussion of Bell's theorem goes beyond the scope of this text;[9] we restrict ourselves here to the demonstration of an incompatibility of the quantum-mechanical predictions with the Bell inequality constraint in a simple example.[10]

We consider a hypothetical experimental scenario that happens to be particularly easy to analyze, namely, a setup as in Fig. 20•4–2, in which the beam splitters at each end are switched rapidly and randomly between just three axis orientations each—vertically and $\pm 120°$ from vertical, (**Fig. 20•4–3**)—an arrangement with a three-fold rotational invariance about the direction of the incident light beams.

We evidently have $3 \times 3 = 9$ different possible beam splitter combinations, occurring at random, but with equal frequency. We determine the probability that both photons materialize in equivalent counter channels, that is, either both in X channels or both in Y channels.

We first determine this probability under the rules of quantum mechanics. Consider the subset of events during which the counters at East register a photon in the X channel while the beam splitter is in position 1. If the beam splitter at West was also in position 1, "its" photon must then also emerge in the X channel. But if the West beam splitter was in one of the two $\pm 120°$ positions, the probability that the second photon will also materialize in the X channel will be only

$$p = \cos^2 120° = 1/4, \qquad (20•4–7)$$

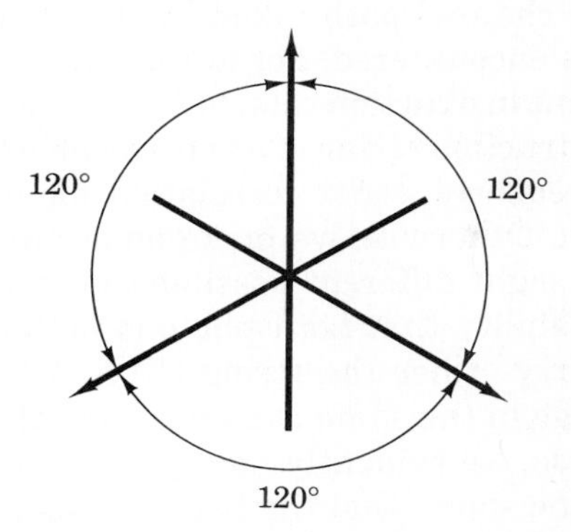

Figure 20•4–3. Three beam splitter orientations rotated from each other by $\pm 120°$, for a hypothetical photon correlation experiment.

[9] There exist a number of readable reviews of the topic in the "lighter" scientific literature, written specifically for the non-specialist. See, for example, B. d'Espagnat, "The Quantum Theory and Reality," *Scientific American,* Vol. 241, Nov. 1979, pp. 150–181; and especially D. Mermin, "Is the moon there when nobody looks? Reality and the quantum theory," *Physics Today,* Vol. 38, No. 4, April 1985, pp. 38–47. Both papers—especially Mermin's—contain extensive references to relevant original work.

[10] Our discussion is based on an adaptation of a simple argument given by Mermin, *op. cit.*

from (20•4–6). Averaging over the three equally probable positions of the East beam splitter, we obtain a probability

$$p = \frac{1}{3} \cdot \left(1 + 2 \cdot \frac{1}{4}\right) = \frac{1}{2} \tag{20•4–8}$$

that the East photon will also materialize in the X channel.

Our argument is symmetric in X and Y; by interchanging X and Y, we see readily that (20•4–8) also holds for the Y channel. Furthermore, because of the invariance of the entire setup under $\pm 120°$ rotations about the z-axis, the same probability results if the East beam splitter is in one of the $\pm 120°$ positions. Hence, quantum mechanics predicts that the overall probability that both photons materialize in the equivalent counter channels is exactly 1/2.

Consider now the alternative hypothesis, that the path of the photons through the beam splitters is not governed by quantum-mechanical probabilities, but that each photon carries with itself a set of hidden instructions that determine the path of the photon for each possible beam splitter setting, similar to the way each biological cell carries, within its DNA, its genetic code that controls the cell's evolution. With three different beam splitter positions to cover, we need a code of at least three bits to control the path. There are then eight different possible instructions, which we may write as

[X, X, X], [X, X, Y], [X, Y, X], [X, Y, Y], [Y, X, X], [Y, X, Y],

[Y, Y, X], [Y, Y, Y].

Here, the three positions correspond to the three beam splitter settings, and the two letters correspond to the channel path taken by the photon when the particular beam splitter setting is encountered. For uncorrelated photon pairs, the two photons may carry different instruction sets, but in our case of polarization correlation, whatever the instruction string of *one* of the photons, the other must carry the same string, or else the strict correlation for *parallel* beam splitter settings would be broken. Otherwise we make no assumptions about the frequency distribution of the eight different possible code strings.

Again, we ask for the probability that both photons materialize in the same channel. If the photons carry either the string [X, X, X] or the string [Y, Y, Y], both photons will emerge in the same channel, regardless of the two beam splitter settings; in that case, we evidently have $p = 1(>1/2)$. Consider next the string [X, X, Y]. Inspection shows that the two photons will emerge in the same channel for five of the nine possible beam splitter settings, namely, (1, 1), (2, 2), (3, 3), (1, 2), and (2, 1), while the two photons will emerge in different channels for the remaining four settings, (1, 3), (3, 1), (2, 3), and (3, 2). Evidently, for this code, $p = 5/9(>1/2)$. Inspection shows that the same channel distribution is obtained for the remaining five codes.

We may summarize these results by saying that the hidden-variable hypothesis implies a polarization correlation that obeys the inequality

$$p \geq \frac{5}{9}, \tag{20•4–9}$$

regardless of the frequency distribution of the hidden codes. If all codes occur with equal frequency, we obtain $p = (2 + 7/2)/9 = 11/18 > 1/2$. The inequality (20•4–9) is the Bell inequality for our case. The earlier quantum-mechanical prediction $p = 1/2$ evidently violates the Bell inequality, hence making the hypothesis of hidden variables experimentally testable.

In experimental practice, the physical rotation of beam splitters between three positions separated by $\pm 120°$ would be a slow "mechanical" process, making it impossible to obtain the fast switching rates shorter than the time of flight of the photons, which are required for the test of the predictions of quantum mechanics. What was done instead in the correlation experiments cited earlier was that the light beams were switched by opto-electronic means between just two fixed-orientation beam splitters at each end, rotated relative to one another by a suitable angle θ, chosen to yield a large discrepancy between the predictions of quantum mechanics and the corresponding Bell inequality. The analysis of the actual experiments is more complicated than that of our simple scenario discussed above, but the results unambiguously confirm the predictions of quantum mechanics and, hence, refute the hypothesis of hidden variables. In fact, to decide between the two contending possibilities was the whole purpose of those experiments.

Chapter 21

ELECTRON SPIN

21.1 SPIN AS AN INTERNAL DEGREE OF FREEDOM

21.1.1 Empirical Basis

Electrons, and most other elementary objects, have an **intrinsic angular momentum**, called their **spin**, that is not associated with any motion of the object through space, but is inseparable from the very existence of the object, just like its charge and mass. By *intrinsic*, we mean that this angular momentum must be included in the law of conservation of angular momentum, or else that law would be violated in certain interaction processes, similar to the way angular momentum conservation would be violated in electron-photon interactions if we neglected the angular momentum of photons, discussed in chapter 19. Even a plane wave would have such an intrinsic angular momentum.

Associated with the spin is an **intrinsic magnetic moment**, and many of the more obvious manifestations of spin are based on this magnetic moment more than on the (more fundamental) angular momentum itself. The conceptually simplest and most direct such manifestation is probably the Stern-Gerlach experiment.[1]

Consider a hydrogen atom in its ground state. This is an s-state with $l = 0$ and, hence, a state without *orbital* magnetic moment. If the electron has an *intrinsic* magnetic moment, this state will split in a magnetic field: If a beam of hydrogen atoms were directed through an *inhomogeneous* magnetic field perpendicular to the beam, as in **Fig. 21•1–1**, any atoms whose magnetic moment had a positive component in the direction of the magnetic field would be deflected toward stronger magnetic fields, while atoms with a negative component would be deflected toward weaker magnetic fields. Because the hydrogen atom is electrically neutral overall, no deflection due to the Lorentz force would obscure the effect. Experimentation with hydrogen *atomic* beams

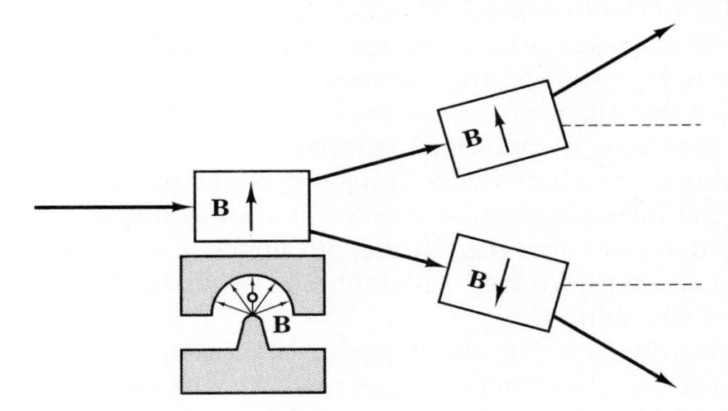

Figure 21•1–1. Idealized Stern-Gerlach experiment. A thin atomic beam is sent through a **spin polarizer**, consisting of an *inhomogeneous* magnetic field perpendicular to the beam direction. If both the field direction and the field gradient, taken on the beam axis, are in the plane of the drawing, the polarizer splits the beam into two beams within that plane. If the secondary beams are sent through a second set of polarizers, oriented the same way relative to their respective input beams, each of the secondary beams is deflected further in the same direction, but the secondary beams are not split again. If the secondary polarizers are rotated by 180° around the beam direction, the beams are deflected back to the original direction, as indicated by the broken lines, but again are not split further. If, however, the secondary polarizers are rotated out of the plane of the first polarizer, the secondary beams will be split again, into two tertiary beams, both in the plane of the secondary polarizers, out of the plane of the drawing. Stern and Gerlach did not yet have the secondary polarizers, which were added to the experiment later.

[1] For an excellent account of the details of the experiment and its historical background, see the text by French and Taylor referred to in the Appendix G.

is difficult—hydrogen comes in molecules—but atomic beams of alkali metal atoms will do equally well, as will atomic beams of a few other elements with a lone outer electron; Stern and Gerlach used silver. In all those atoms, the angular momenta and magnetic moments of the inner electrons cancel, leaving only the effect of the lone outer electron.

When this experiment is done, one finds that the beam is split into exactly two components, displaced symmetrically from the original beam direction, as shown in Fig. 21•1–1. *There is no undeflected component.* If one of the two components is sent through a second inhomogeneous field, also in the deflection plane, the beam is deflected again, but no additional splitting takes place. However, if the second field is at a right angle to the first, the beam is split again into two components of equal intensity, with a plane of splitting perpendicular to the first plane.

If we assume that the intrinsic magnetic momentum vector is proportional to the intrinsic angular momentum vector, then the Stern-Gerlach effect implies that the *projection* of the intrinsic angular momentum on the magnetic field direction is quantized, similar to the way the projection of an *orbital* angular momentum vector $\mathbf{L}$ upon a preferred axis may be quantized, *except* that the intrinsic angular momentum projection is capable of assuming only *two* values, of equal magnitude, but opposite sign. The case of zero projection does not occur! We view this inherent two-valuedness as the most basic property of the spin.

Purely formally, the behavior is the same *as if* the angular momentum quantum number l had the value $l = 1/2$, with the two m-states $m = \pm 1/2$. We saw in chapter 18 that the algebra of the angular momentum operator actually permits half-integer l-values; but they were ruled out for the *orbital* angular momentum by the additional requirement that the wave function be single-valued in the azimuthal angle ϕ. The spin angular momentum represents exactly that missing alternative of a half-integer angular momentum.[2]

If we take this view, the observations of the Stern-Gerlach experiment can be generalized into the following statement:

> The spinless quantum-mechanical description of the electron is incomplete. A complete description requires an additional *internal* degree of freedom that is not associated with the motion of the electron and that is capable of assuming exactly two independent sharp values. Each state of the spinless description represents a pair of states of the complete descrip-

---

[2] Historically, the experiment was performed several years before the existence of the electron spin had even been postulated by Uhlenbeck and Goudsmith, and Stern and Gerlach thought that they had confirmed an (incorrect!) prediction of the pre-quantum-mechanical Bohr atomic model according to which the ground state of hydrogen carried an orbital angular momentum and magnetic moment.

tion. In the absence of any magnetic fields or magnetic interactions, the two states of each pair have the same energy, and their dynamical behavior is identical and is correctly given by the spin-less quantum-mechanical formalism. Differences manifest themselves only in the presence of magnetic fields or interactions.

The above statement contains no reference to any specific magnitude of the angular momentum and of the associated magnetic moment. In fact, the statement about the two-valuedness of the additional degree of freedom does not even refer specifically to an angular momentum or a magnetic moment. It refers only to the far less specific fact that energetic and dynamic differences between the two spin states emerge only in the presence of magnetic fields. We shall see shortly how all the remaining properties of the spin and of its associated magnetic moment follow naturally from these very general basic postulates.

21.1.2 Spinor Wave Functions

To develop the mathematical formalism for the spin, we assume for the moment only that the electron *has* an angular-momentum-like directional property that we may call spin and that there are no more than two sharp values for the projection of this spin upon any arbitrarily selected reference direction, of equal magnitude but opposite sign. We leave open all quantitative details, such as the magnitude of either the angular momentum or the magnetic moment.

If the projection of the spin upon the reference direction has a sharp positive value, we call the corresponding state of the electron a **spin-up state**. If the spin projection has a sharp negative value, the corresponding state is called a **spin-down state**. Every possible state of the electron must be a linear superposition of these two states. We might write the state function for this linear superposition in a form such as

$$\Psi(\mathbf{r}) = \Psi\!\uparrow\!(\mathbf{r}) + \Psi\!\downarrow\!(\mathbf{r}). \tag{21•1–1}$$

But a problem arises immediately: In the absence of a magnetic field, the two states in (21•1–1) not only might be degenerate, but their *spatial* distributions, including their local momentum everywhere in space, might be identical. That is, they might have both the same magnitude and the same phase everywhere. Accordingly, we need to distinguish somehow between the state functions of the two physically distinct states, even if they have identical magnitudes and phases. This calls for an alternative formalism that automatically keeps track of the spin-up and spin-down portions of the superposition.

The simplest way to combine two independent scalar complex state functions into a single mathematical object is to write the state functions as components of a 2×1 column matrix with complex matrix coefficients,

$$\boxed{\overline{\Psi}(\mathbf{r}) = \begin{pmatrix}\psi_1\\0\end{pmatrix} + \begin{pmatrix}0\\\psi_2\end{pmatrix} = \begin{pmatrix}\psi_1\\\psi_2\end{pmatrix},} \tag{21•1–2}$$

a formalism first adopted by Wolfgang Pauli. Here the upper term represents the spin-up part of the superposition state, the lower term the spin-down part. Inasmuch as angular momentum and magnetic moment are vector properties, the reader might wonder why we do not replace the scalar wave function of spin-less quantum mechanics by a vector wave function, which we may write as a 3×1 column matrix:

$$\Psi = \begin{pmatrix}\psi_x\\\psi_y\\\psi_z\end{pmatrix}. \tag{21•1–3}$$

In fact, our electric field quantization formalism in chapter 20 was basically of this kind. However, such a formalism appears to be fundamentally incapable of giving a satisfactory account of the inherent two-valuedness of the spin; there is simply one component too many. This raises, of course, the question as to how we might associate a vector property with a mathematical description of a state in terms of two complex components. However, we shall see in the next section that this question has a simple answer.

Two-component column matrices of the form (21•1–2), with complex coefficients, are called **spinors**. To distinguish them from "ordinary" scalar wave functions, we shall often find it practical to place an overbar ($^-$) over them, as was done in (21•1–2) for $\overline{\Psi}$.

Associated with each spinor is an **adjoint spinor**, or **Hermitian conjugate spinor**, defined as the complex conjugate of the transposed spinor, a 2×1 row matrix of the form

$$\overline{\Psi}^\dagger(\mathbf{r}, t) = (\psi_1^*, \psi_2^*). \tag{21•1–4}$$

The product of each spinor with its Hermitian conjugate, taken in the order $\overline{\Psi}^\dagger\overline{\Psi}$, is a scalar:

$$\overline{\Psi}^\dagger\overline{\Psi} = \psi_1^*\psi_1 + \psi_2^*\psi_2. \tag{21•1–5}$$

In obvious generalization of the formalism for scalar Schroedinger wave functions, we interpret the first term on the right-hand side as the probability density for "finding" the electron with spin up, the second term as the probability density for "finding" it with spin down, and the sum as the overall probability density.

It is frequently useful to write spinors in terms of products of position-dependent scalar wave functions and suitable fixed spinors, as in

$$\overline{\Psi}(\mathbf{r}) = \begin{pmatrix}\psi_1(\mathbf{r})\\\psi_2(\mathbf{r})\end{pmatrix} = \psi_1(\mathbf{r}) \cdot \begin{pmatrix}1\\0\end{pmatrix} + \psi_2(\mathbf{r}) \cdot \begin{pmatrix}0\\1\end{pmatrix}. \tag{21•1–6}$$

The two simple spinors on the right-hand side have the property

$$(1, 0)\begin{pmatrix}1\\0\end{pmatrix} = (0, 1)\begin{pmatrix}0\\1\end{pmatrix} = 1; \tag{21•1–7}$$

hence, they are called **unit spinors**. We will later encounter other forms of unit spinors.

The notation may be made more compact by introducing the symbols $\uparrow$ and $\downarrow$ for the two unit spinors in (21•1–6),

$$\uparrow \equiv \begin{pmatrix}1\\0\end{pmatrix}, \qquad \downarrow \equiv \begin{pmatrix}0\\1\end{pmatrix}, \tag{21•1–8a,b}$$

leading to the quasi-scalar notation

$$\overline{\Psi}(\mathbf{r}) = \psi_1(\mathbf{r}) \cdot \uparrow + \psi_2(\mathbf{r}) \cdot \downarrow. \tag{21•1–9}$$

The Hermitian conjugate unit spinors are then written as

$$\uparrow^\dagger \equiv (1, 0), \qquad \downarrow^\dagger \equiv (0, 1). \tag{21•1–10a,b}$$

In Dirac notation, we might write

$$|\Psi\rangle = |\psi_1\rangle|\uparrow\rangle + |\psi_2\rangle|\downarrow\rangle. \tag{21•1–11}$$

Similar quasi-scalar notations, functionally equivalent to the above, but using symbols other than our $\uparrow$ and $\downarrow$ for the two unit spinors, are in widespread use.

21.1.3 The Pauli Spin Hamiltonian

If electrons must be represented by spinor wave functions, then all quantum-mechanical operators operating on full spinors must be of 2×2 matrix form, operating on the spinor components according to the rules of matrix multiplication:

$$\begin{pmatrix}\hat{A}_{11} & \hat{A}_{12}\\ \hat{A}_{21} & \hat{A}_{22}\end{pmatrix}\begin{pmatrix}\psi_1\\ \psi_2\end{pmatrix} = \begin{pmatrix}\hat{A}_{11}\psi_1 + \hat{A}_{12}\psi_2\\ \hat{A}_{21}\psi_1 + \hat{A}_{22}\psi_2\end{pmatrix}. \tag{21•1–12}$$

The matrix coefficients may themselves be operators, such as differential operators and even differential vector operators, such as $\hat{\mathbf{p}}$.

In particular, the Hamilton operator must be of the matrix form (21•1–12). This is equivalent to saying that instead of the single scalar Schroedinger wave equation, we must have one linear wave equation for each of the spinor components. The Stern-Gerlach experiment requires that the two equations be coupled, which means that the spinor Hamiltonian must contain off-diagonal terms, at least in the presence of magnetic fields. However, all experimental evidence indicates that in the *absence* of magnetic fields the spin-up state and the spin-down state behave identically. In that limit, we must therefore simply have two identical decoupled Schroedinger wave equations with scalar Hamiltonians of the "standard" scalar form

$$\hat{H}_{sc} = \frac{1}{2m_e}\hat{\mathbf{p}}^2 + V(\mathbf{r}), \tag{21•1–13}$$

one for the each of the two spinor components. In spin or operator notation, this implies a diagonal Hamiltonian of the form

$$\hat{H}_0 = \frac{1}{2m_e}\begin{pmatrix} \hat{p}^2 & 0 \\ 0 & \hat{p}^2 \end{pmatrix} + \begin{pmatrix} V(\mathbf{r}) & 0 \\ 0 & V(\mathbf{r}) \end{pmatrix} = \hat{I}\hat{H}_{sc}, \tag{21•1–14}$$

where

$$\hat{I} = \begin{pmatrix} 1 & 0 \\ 0 & 1 \end{pmatrix} \tag{21•1–15}$$

is the 2 × 2 unit matrix.

We go one step further: In spin-less quantum mechanics, the kinetic energy part of the Hamiltonian in the absence of a magnetic field is of the form of square of a simpler operator:

$$\hat{T} = \frac{1}{2m_e}\hat{p}^2. \tag{21•1–16}$$

It turns out that the kinetic energy part of the 2 × 2 matrix Hamiltonian (21•1–14) can be similarly written as the square of a simpler 2 × 2 matrix operator, namely,

$$\begin{pmatrix} \hat{p}_x^2 + \hat{p}_y^2 + \hat{p}_z^2 & 0 \\ 0 & \hat{p}_x^2 + \hat{p}_y^2 + \hat{p}_z^2 \end{pmatrix} = \begin{pmatrix} \hat{p}_z & \hat{p}_x - i\hat{p}_y \\ \hat{p}_x + i\hat{p}_y & \hat{p}_z \end{pmatrix}^2, \tag{21•1–17}$$

as is easily seen by executing the matrix multiplication. Note that the coefficients of the factor matrix are linear in the components of $\hat{\mathbf{p}}$ and that the matrix itself is Hermitian, two essential conditions if the splitting-up is to be physically meaningful.

The factor matrix on the right-hand side of (21•1–17) is not unique: The three coordinates x, y, and z may evidently be permuted, leading to the same squared matrix, and we may also interchange $\hat{p}_x + i\hat{p}_y$ with $\hat{p}_x - i\hat{p}_y$. All these forms simply represent different choices of the coordinate system; they are physically equivalent. We follow here the universal practice of working with the form given above.

Recall next, from chapter 8, that in the scalar Schroedinger formalism of spinless quantum mechanics the inclusion of magnetic fields was accomplished by replacing the momentum $\hat{\mathbf{p}}$ with the *kinetic* momentum $\hat{\boldsymbol{\pi}}$, according to

$$\boxed{\hat{\mathbf{p}} \rightarrow \hat{\boldsymbol{\pi}} \equiv \hat{\mathbf{p}} + e\mathbf{A},} \tag{21•1–18}$$

leading to the Hamiltonian

$$\hat{H}_{sc} = \frac{1}{2m_e}(\hat{\mathbf{p}} + e\mathbf{A})^2 + V(\mathbf{r}). \tag{21•1–19}$$

In the absence of any reason to the contrary, we would expect the same substitution also to account for the effects of magnetic fields in the presence of spin. However, at this point, an ambiguity arises: Recall from section 8.1 that the components of $\hat{\boldsymbol{\pi}}$ no longer commute, in contrast to those of $\hat{\mathbf{p}}$. Instead, by re-writing the commutation relations (8•1–11) in terms of the components of the kinetic momentum $\hat{\boldsymbol{\pi}}$, we obtain

$$[\hat{\pi}_x, \hat{\pi}_y] = -ie\hbar B_z, \ [\hat{\pi}_y, \hat{\pi}_z] = -ie\hbar B_x, \ [\hat{\pi}_z, \hat{\pi}_x] = -ie\hbar B_y. \tag{21•1–20a,b,c}$$

As a result, if we make the substitution (21•1–18) inside the factor matrix on the right-hand side of (21•1–17), and *then* square the matrix, we obtain

$$\begin{pmatrix} \hat{\pi}_z & \hat{\pi}_x - i\hat{\pi}_y \\ \hat{\pi}_x + i\hat{\pi}_y & \hat{\pi}_z \end{pmatrix}^2 = \begin{pmatrix} \hat{\pi}^2 & 0 \\ 0 & \hat{\pi}^2 \end{pmatrix} + e\hbar \begin{pmatrix} +B_z & B_x - iB_y \\ B_x + iB_y & -B_z \end{pmatrix}. \tag{21•1–21}$$

If we had made the substitution (21•1–18) *after* squaring the factor matrix, directly in the product matrix on the left-hand side of (21•1–17), we would have obtained only the diagonal first term in (21•1–21). Because the Stern-Gerlach experiment calls for a spinor Hamiltonian containing off-diagonal terms, we explicitly postulate that the substitution (21•1–18) must be made inside the factor matrix on the right-hand side of (21•1–17) *before* squaring that matrix.

For later use, we re-write (21•1–21) as

$$\begin{pmatrix} \hat{\pi}_z & \hat{\pi}_x - i\hat{\pi}_y \\ \hat{\pi}_x + i\hat{\pi}_y & \hat{\pi}_z \end{pmatrix}^2 = \hat{I} \cdot (\hat{\mathbf{p}} + e\mathbf{A})^2 + e\hbar(\hat{\sigma}_x B_x + \hat{\sigma}_y B_y + \hat{\sigma}_z B_z), \tag{21•1–22}$$

where

$$\boxed{\hat{\sigma}_x = \begin{pmatrix} 0 & 1 \\ 1 & 0 \end{pmatrix}, \quad \hat{\sigma}_y = \begin{pmatrix} 0 & -i \\ i & 0 \end{pmatrix}, \quad \hat{\sigma}_z = \begin{pmatrix} 0 & 0 \\ 1 & -1 \end{pmatrix}} \tag{21•1–23a,b,c}$$

are called the **Pauli spin matrices**.

If we substitute the expression (21•1–22) for the term $\hat{I} \cdot \hat{\mathbf{p}}^2$ in (21•1–14), we obtain a final spinor Hamiltonian consisting of two parts,

$$\hat{H} = \hat{H}_o + \hat{H}_s, \tag{21•1–24}$$

where

$$\hat{H}_o = \frac{1}{2m_e}\begin{pmatrix} (\hat{\mathbf{p}} + e\mathbf{A})^2 & 0 \\ 0 & (\hat{\mathbf{p}} + e\mathbf{A})^2 \end{pmatrix} + \begin{pmatrix} V(\mathbf{r}) & 0 \\ 0 & V(\mathbf{r}) \end{pmatrix} = \hat{I} \cdot \hat{H}_{sc} \tag{21•1–25}$$

is the quasi-scalar **orbital Hamiltonian** and

$$\hat{H}_s = \mu_B \cdot (\hat{\sigma}_x B_x + \hat{\sigma}_y B_y + \hat{\sigma}_z B_z) \quad (21\text{•}1\text{–}26)$$

is the **spin Hamiltonian** proper. In the last expression,

$$\boxed{\mu_B = \frac{e\hbar}{2m_e}} \quad (21\text{•}1\text{–}27)$$

is again the Bohr magneton already introduced in chapter 8.

In honor of Wolfgang Pauli, who first obtained the term $\hat{H}_s$—by different considerations—this term is often called the **Pauli spin Hamiltonian**. The quasi-scalar portion $\hat{H}_o$ of the overall Hamiltonian $\hat{H}$ is called the **orbital Hamiltonian** because it describes the energy of the motion of the electron in an external potential, a motion that, in the Bohr atomic model, was associated with classical particle orbits.

Exercise: Spin Matrix Algebra The Pauli spin matrices form the backbone of mathematical spin theory. As always, we are interested in the basic algebra of the operators. Derive the following properties:

(a) The square of each matrix is the unit matrix:

$$\hat{\sigma}_x^2 = \hat{\sigma}_y^2 = \hat{\sigma}_z^2 = \begin{pmatrix} 1 & 0 \\ 0 & 1 \end{pmatrix} = \hat{I}. \quad (21\text{•}1\text{–}28)$$

(b) The matrices anticommute, and the product of any two of the matrices is $\pm i$ times the third:

$$\hat{\sigma}_x \hat{\sigma}_y = -\hat{\sigma}_y \hat{\sigma}_x = i\hat{\sigma}_z, \quad (21\text{•}1\text{–}29a)$$

$$\hat{\sigma}_y \hat{\sigma}_z = -\hat{\sigma}_z \hat{\sigma}_y = i\hat{\sigma}_x, \quad (21\text{•}1\text{–}29b)$$

$$\hat{\sigma}_z \hat{\sigma}_x = -\hat{\sigma}_x \hat{\sigma}_z = i\hat{\sigma}_y. \quad (21\text{•}1\text{–}29c)$$

This immediately yields the all-important commutators

$$\boxed{[\hat{\sigma}_x, \hat{\sigma}_y] = 2i\hat{\sigma}_z, \quad [\hat{\sigma}_y, \hat{\sigma}_z] = 2i\hat{\sigma}_x, \quad [\hat{\sigma}_z, \hat{\sigma}_x] = 2i\hat{\sigma}_y.} \quad (21\text{•}1\text{–}30)$$

We will draw extensively on these properties throughout this chapter.

21.2 INTRINSIC MAGNETIC MOMENT

21.2.1 The Electron Energy in a Uniform Magnetic Field

The spin term $\hat{H}_s$ in the overall Hamiltonian represents the potential energy of a magnetic moment in the external magnetic field. Consider a spatially uniform magnetic field **B**. We can always choose the coordinate system in such a way that the z-axis points in the direction of **B**. In this case, (21•1–26) simplifies to

$$\hat{H}_s = \mu_B B \cdot \begin{pmatrix} 1 & 0 \\ 0 & -1 \end{pmatrix} = \mu_B B \hat{\sigma}_z. \tag{21•2–1}$$

Because the matrix $\hat{\sigma}_z$ commutes with the 2×2 identity matrix and the magnetic field is constant, the spin term commutes with the rest of the Hamiltonian, and all energy eigenstates can be chosen to be eigenstates of the spin term as well, and therefore, of $\hat{\sigma}_z$. Because $\hat{\sigma}_z^2 = \hat{I}$, the eigenvalues of $\hat{\sigma}_z$ are simply ± 1; hence, the effect of the spin term in the Hamiltonian is simply a splitting of every original energy eigenvalue $\mathcal{E}_0$ into two, according to

$$\boxed{\mathcal{E}_\pm = \mathcal{E}_0 \pm \mu_B B.} \tag{21•2–2}$$

The associated eigenstates, called **eigenspinors**, are easily seen to be the spin-up and spin-down states from (21•1–2),

$$\overline{\psi}_+ = \begin{pmatrix} \psi \\ 0 \end{pmatrix} = \psi \cdot \begin{pmatrix} 1 \\ 0 \end{pmatrix} \quad \text{and} \quad \overline{\psi}_- = \begin{pmatrix} 0 \\ \psi \end{pmatrix} = \psi \cdot \begin{pmatrix} 0 \\ 1 \end{pmatrix}, \tag{21•2–3a,b}$$

where ψ is a scalar energy eigenstate of the spinless Hamiltonian (21•1–19), obeying

$$\hat{H}_{\text{sc}} \psi = \mathcal{E}_0 \psi. \tag{21•2–4}$$

Note that these pure spin-up and spin-down states are indeed eigenspinors of the operator $\hat{\sigma}_z$, with the eigenvalues ± 1:

$$\hat{\sigma}_z \begin{pmatrix} 1 \\ 0 \end{pmatrix} = \begin{pmatrix} +1 & 0 \\ 0 & -1 \end{pmatrix} \begin{pmatrix} 1 \\ 0 \end{pmatrix} = +\begin{pmatrix} 1 \\ 0 \end{pmatrix}, \tag{21•2–5a}$$

$$\hat{\sigma}_z \begin{pmatrix} 0 \\ 1 \end{pmatrix} = \begin{pmatrix} +1 & 0 \\ 0 & -1 \end{pmatrix} \begin{pmatrix} 0 \\ 1 \end{pmatrix} = -\begin{pmatrix} 0 \\ 1 \end{pmatrix}. \tag{21•2–5b}$$

The simplest eigenspinors are the two unit spinors

$$\overline{\psi}_{+z} = \begin{pmatrix} 1 \\ 0 \end{pmatrix}, \qquad \overline{\psi}_{-z} = \begin{pmatrix} 0 \\ 1 \end{pmatrix}. \tag{21•2–6a,b}$$

More complicated eigenspinors may be written as a product of a scalar wave function and one of the unit eigenspinors $\overline{\psi}_{+z}$ or $\overline{\psi}_{-z}$.

The extra energy term $+\mu_B B$ has a very simple physical interpretation: It is the potential energy of a magnetic moment μ_B in a magnetic field B, with the moment directed opposite to the magnetic field.[3] More exactly, it is the energy of a magnetic moment whose *projection* upon the field direction has the magni-

---

[3] Our choice of signs anticipates that, for objects with negative charge, the magnetic moment should be expected to point opposite to the angular momentum.

tude $-\mu_B$. Similarly, $-\mu_B B$ is the potential energy of a magnetic moment whose projection upon the field direction has the magnitude $+\mu_B$. The occurrence of this extra energy term in (21•2–2) is therefore equivalent to the prediction that the electron possesses an intrinsic magnetic moment. Nor is this all: The fact that the extra term occurs only in two forms $\pm\mu_B B$ implies that this magnetic moment of an electron is quantized similar to the way angular momentum is quantized. This similarity to angular momentum is no accident: We will see shortly that the intrinsic magnetic moment of the electron is indeed associated with an intrinsic angular momentum that exists in addition to any orbital angular momentum. The only—but essential—difference between the quantization of the intrinsic angular momentum and that of the orbital angular momentum is that, for the intrinsic angular momentum, exactly two values of the projected momentum are allowed, while for the orbital angular momentum, the number of allowed projected momentum values is not restricted in such a way.

Thus, we see that our treatment of the overall Hamiltonian as a 2×2 matrix automatically leads to the conclusion, not only that there must be an intrinsic magnetic moment associated with the electron; but even that the magnitude μ_e of this intrinsic magnetic moment must be equal to the natural unit of *orbital* magnetic moment, μ_B. The latter prediction is readily testable experimentally; in fact, the ratio μ_e/μ_B can be measured with an accuracy that makes it one of one of the most accurately known constants in all of physics.[4] The best current value is[5]

$$\mu_e/\mu_B = 1.001\ 159\ 652\ 193 \pm 10 \times 10^{-12}. \tag{21•2–7}$$

The agreement with the prediction is good only to about 1 part in 1,000. However, this difference can be *quantitatively* accounted for by relativistic quantum electrodynamics, just like some of the discrepancies in the hydrogen spectrum. According to that theory, the value of μ_e/μ_B can be calculated by perturbation theory, leading to a power series in terms of powers of the so-called **fine-structure constant**

$$\alpha = \frac{e^2}{4\pi\epsilon_0 c\hbar} = 7.297\ 353\ 08 \times 10^{-3} = 1/137.035\ 996, \tag{21•2–8}$$

about which we will say more presently.

The first two terms in the perturbation expansion for μ_e/μ_B are

$$\mu_e/\mu_B = 1 + \alpha/2\pi + \dots, \tag{21•2–9}$$

leading to $\mu_e/\mu_B \sim 1.001\ 161$, within about 2 parts per million of the experimental value. By carrying the series expansion (21•2–9) to higher powers of α,

[4] For a very readable description of the actual experiment, see H. R. Crane, "The *g* Factor of the Electron," *Scientific American,* Vol. 218, No. 1, Jan. 1968, pp. 72–85.

[5] See E. R. Cohen and B. N. Taylor, "The 1986 adjustment of the fundamental physical constants," *Reviews of Modern Physics,* 59 (1987), pp. 1121–1148.

the agreement can be made so good that it is limited only by the accuracy with which the fundamental constants entering α are themselves known. In fact, the theory is used in reverse, to help in determining an accurate value for the fine-structure constant and, thereby, for the values of the fundamental constants.

Comment: The Fine-Structure Constant. The fine-structure constant first occurred in Sommerfeld's 1915 theory of the relativistic fine structure of the hydrogen atom, but its significance goes far beyond that. It is one of the truly fundamental constants of nature. Being dimensionless, it is independent of the system of units used. It usually shows up as the ratio of different natural units of energy or of length arising in different kinds of problems. For example, the natural unit of energy in relativistic electron dynamics is the energy $m_e c^2$, while the natural unit of energy in atomic physics is the Rydberg, R_∞, defined in (2•7–9). According to (2•7–9) and (21•2–8), the two energies are related via

$$R_\infty = \frac{\hbar^2}{2m_e a_0^2} = \frac{1}{2}\frac{e^2}{4\pi\epsilon_0 a_0} = \frac{m_e}{2}\left(\frac{e^2}{4\pi\epsilon_0\hbar}\right)^2 = \frac{\alpha^2}{2}\cdot m_e c^2. \qquad (21•2–10)$$

In effect, the fine-structure constant specifies the relative magnitude of the natural units of energy in quantum mechanics and in the theory of relativity. In modern quantum field theory, the speed of light and Planck's constant are considered more fundamental than the electron charge, which is viewed simply as a parameter specifying the coupling strength between two classes of objects, electrons and electromagnetic fields (= photons). From this point of view, the fine-structure constant simply expresses this coupling strength in terms of the more fundamental parameters c and $\hbar$. A theoretical prediction of the value of the fine-structure constant from more general principles is one of the as-yet unsolved challenges of theoretical physics.

Returning to (21•2–10), we recall that the expectation value of the kinetic energy in the ground state of the hydrogen atom is R_∞. If we write this as $m_e v^2/2$, where v is the RMS velocity of the electron, we see from (21•2–10) that $v = \alpha \cdot c$.

21.2.2 Spin as a Vector-like Property

In the argument that led us to postulate an additional term of the form (21•1–26) in the Hamiltonian, we assumed that the magnetic field was in the z-direction. Similar arguments can be made for magnetic fields in the x- and y-directions, except that the spin matrix $\hat{\sigma}_z$ must of course be replaced by $\hat{\sigma}_x$ or $\hat{\sigma}_y$.

Consider now a magnetic field $\mathbf{B}$ in an arbitrary direction, with the Cartesian components B_x, B_y, and B_z. We note that the combination of B's and $\hat{\sigma}$'s in (21•1–26) is formally the same as in the dot product of two vectors. This permits us to write (21•1–26) in the form

$$\hat{H}_s = \mu_B \mathbf{B} \cdot \hat{\boldsymbol{\sigma}}, \qquad (21•2–11)$$

where

$$\hat{\sigma} = \begin{pmatrix} \hat{\sigma}_x \\ \hat{\sigma}_y \\ \hat{\sigma}_z \end{pmatrix} \tag{21•2–12}$$

is a mathematical object that acts as a vector in real space, but the components of which are operators in spinor space. The idea of vectors whose components are operators is of course nothing new to us; what *is* new is that these operators operate in spinor space. What is meant by a vector whose Cartesian components are spin matrices is that this vector operator will extract, from any superposition of eigenspinors, the components of the spin vector associated with each term in the superposition, just as the momentum vector operator extracts, from any superposition of plane waves, the wave vector of each plane wave.

This line of reasoning now offers an answer to our earlier question, asked following Eq. (21•1–3), how we might associate a vector property with a mathematical description of a state in terms of two complex components. To answer this question, we consider the eigenspinors of the operator $\mathbf{B} \cdot \hat{\boldsymbol{\sigma}}$.

We characterize the direction of the magnetic field by the angles θ and ϕ in a system of spherical polar coordinates. In terms of such a system, the Cartesian field components are

$$B_x = B \sin\theta \cos\phi, \tag{21•2–13a}$$

$$B_y = B \sin\theta \sin\phi, \tag{21•2–13b}$$

$$B_z = B \cos\theta. \tag{21•2–13c}$$

With these, the product $\mathbf{B} \cdot \hat{\boldsymbol{\sigma}}$ in (21•1–26) and (21•2–11) may be written

$$\mathbf{B} \cdot \hat{\boldsymbol{\sigma}} = B\hat{\sigma}, \quad \text{where} \quad \hat{\sigma} = \begin{pmatrix} \cos\theta & e^{-i\phi}\sin\theta \\ e^{+i\phi}\sin\theta & -\cos\theta \end{pmatrix}. \tag{21•2–14a,b}$$

The requirement that a spinor of the form (21•1–2) be an eigenspinor of $\mathbf{B} \cdot \hat{\boldsymbol{\sigma}}$ may then be expressed as

$$\hat{\sigma}\bar{\psi} = \begin{pmatrix} \psi_1 \cos\theta + \psi_2 e^{-i\phi}\sin\theta \\ \psi_1 e^{+i\phi}\sin\theta - \psi_2\cos\theta \end{pmatrix} = \sigma\begin{pmatrix} \psi_1 \\ \psi_2 \end{pmatrix}, \tag{21•2–15}$$

where σ is the eigenvalue.

It is an elementary exercise in linear algebra to show that this pair of linear homogeneous equations for the two unknowns ψ_1 and ψ_2 has solutions only for the eigenvalues

$$\sigma = \pm 1, \tag{21•2–16}$$

the same values as found earlier. We also find easily that the components of the eigenspinors associated with these eigenvalues must obey the relations

$$\frac{\psi_2}{\psi_1} = +e^{-i\phi/2} \cdot \tan\theta/2 \quad \text{for } \sigma = +1 \tag{21•2–17a}$$

and

$$\frac{\psi_2}{\psi_1} = -e^{+i\phi/2} \cdot \cot\theta/2 \quad \text{for } \sigma = -1. \tag{21•2–17b}$$

The simplest spinors having this ratio are the two **symmetrized unit spinors**

$$\overline{\psi}_+ = \begin{pmatrix} +e^{-i\phi/2} \cdot \cos\theta/2 \\ +e^{+i\phi/2} \cdot \sin\theta/2 \end{pmatrix} \text{for } \sigma = +1, \tag{21•2–18a}$$

and

$$\overline{\psi}_- = \begin{pmatrix} +e^{-i\phi/2} \cdot \sin\theta/2 \\ -e^{+i\phi/2} \cdot \cos\theta/2 \end{pmatrix} \text{for } \sigma = -1. \tag{21•2–18b}$$

The full eigenfunctions of the Hamiltonian are then of the form

$$\overline{\Psi}(\mathbf{r}) = \psi(\mathbf{r})\overline{\psi}_\pm, \tag{21•2–19}$$

where $\psi(\mathbf{r})$ is a scalar function (or constant) and $\overline{\psi}_\pm$ is one of the unit spinors.

Evidently, for any direction characterized by the set of angles θ, ϕ, there is a pair of spinors of the form (21•2–17). Conversely, for any spinor with the components ψ_1 and ψ_2, the ratio of these components defines a set of angles θ, ϕ, via the relation (21•2–18a) and, hence, a direction. The alternative direction defined by (21•2–18b) is simply the opposite of that defined by (21•2–18a). Evidently, the absolute magnitude of the ratio ψ_2/ψ_1 defines the polar angle θ, while the phase defines the azimuthal angle ϕ.

The direction defined by those two angles is often loosely referred to as the *spin direction.* However, this is not strictly correct. We saw earlier that the spin matrices do not commute; hence, the different components of the vector of the intrinsic magnetic moment (and of the intrinsic angular momentum behind it) cannot all have sharp values simultaneously, and the direction defined by θ, ϕ is merely that direction upon which the *projection* of the total spin has a sharp value, not the—unsharp—direction of the spin vector itself. The situation is exactly the same as that encountered earlier for the orbital angular momentum. When necessary to avoid ambiguity, we shall therefore refer to the two kinds of directions as the **spin eigendirection** and the **spin vector direction**, respectively, rather than simply as the spin direction.

Exercise:

(a) Show that the normalized eigenspinors of $\hat{\sigma}_x$ and $\hat{\sigma}_y$ are of the forms

$$\overline{\psi}_{\pm x} = \frac{1}{\sqrt{2}}\begin{pmatrix} +1 \\ \pm 1 \end{pmatrix} \tag{21•2–20a}$$

and

$$\overline{\psi}_{\pm y} = \frac{1}{\sqrt{2}}\begin{pmatrix} +1 \\ \pm i \end{pmatrix}, \tag{21•2–20b}$$

except possibly for an irrelevant phase factor of unit magnitude.

(b) Reduce the spinor

$$\overline{\psi} = \begin{pmatrix} 2 + \sqrt{2} \\ 1 + i \end{pmatrix} \qquad (21{\bullet}2\text{–}21)$$

to the form (21•2–17), by splitting off an appropriate factor. What is the spin eigendirection, in terms of the angles θ, ϕ and of the Cartesian components of the eigendirection unit vector?

21.3 INTRINSIC ANGULAR MOMENTUM

21.3.1 Operator Algebra

The commutation relations (21•1–30) for the spin matrices are remarkably similar to those of the components of the (orbital) angular momentum operator $\hat{\mathbf{L}}$, derived in chapter 7:

$$[\hat{L}_x, \hat{L}_y] = i\hbar\hat{L}_z, \qquad [\hat{L}_y, \hat{L}_z] = i\hbar\hat{L}_x, \qquad [\hat{L}_z, \hat{L}_x] = i\hbar\hat{L}_y. \quad (21{\bullet}3\text{–}1)$$

We can make the similarity an identity by defining the **spin angular momentum operator**

$$\boxed{\hat{\mathbf{S}} = \tfrac{1}{2}\hbar\hat{\boldsymbol{\sigma}},} \qquad (21{\bullet}3\text{–}2)$$

with the commutation relations

$$\boxed{[\hat{S}_x, \hat{S}_y] = i\hbar\hat{S}_z, \quad [\hat{S}_y, \hat{S}_z] = i\hbar\hat{S}_x, \quad [\hat{S}_z, \hat{S}_x] = i\hbar\hat{S}_y,} \qquad (21{\bullet}3\text{–}3)$$

identical to (21•3–1), except for the names of the operators involved.

Note the factor 1/2 in the definition (21•3–2). Inasmuch as we know already that the Pauli spin matrices have the eigenvalues ± 1, it follows that the components of $\hat{\mathbf{S}}$ all have the eigenvalues $\pm\hbar/2$. Hence, we see that the spinor formalism naturally leads to a formalism in which an angular-momentum-like property shows up with half-integer eigenvalues for the angular momentum components, taken in units of $\hbar$. This is, of course, the intrinsic angular momentum anticipated earlier.

Furthermore, in analogy to the operator $\hat{L}^2$ in the theory of the orbital angular momentum, we may define an operator

$$\hat{S}^2 = \hat{S}_x^2 + \hat{S}_y^2 + \hat{S}_z^2, \qquad (21{\bullet}3\text{–}4)$$

whose general algebraic properties are essentially the same as those of $\hat{L}^2$. Recall that all properties of $\hat{L}^2$, except the restriction to integer values of l, arose purely from the commutation relations for the components of $\hat{\mathbf{L}}$. Inasmuch as the components of $\hat{\mathbf{S}}$ have the same commutation relations as those of $\hat{\mathbf{L}}$, it follows that the properties of $\hat{S}^2$ are the same as those of $\hat{L}^2$, except for the actual eigenvalues. Just as the eigenvalues $\hat{L}^2$ had to be of the form $\hbar^2 \cdot l(l+1)$, those of $\hat{S}^2$ must be of the form $\hbar^2 \cdot s(s+1)$, except that now $s = 1/2$, rather than an integer. Hence, the absolute magnitude of the intrinsic angular momentum is $\hbar \cdot \sqrt{3/4}$. The common reference to the electron as a spin-1/2 particle refers to the eigenvalues of the *projection* of the spin angular momentum onto a specified direction. Note also that the spin vector does not have a sharp direction; the reference to spin-up and spin-down states simply means that the up or down components have sharp values. The direction of the spin vector remains undefined; it may fall anywhere on a cone whose opening half-angle θ satisfies $\cos\theta = 1/\sqrt{3}$, about 54.7°.

What remains to be shown is that the spin angular momentum must be included in the law of conservation of angular momentum. We will show that later, in section 21.4.

21.3.2 Spin Precession of a Free Electron

The combination of an intrinsic angular momentum with an intrinsic magnetic moment implies a number of important and readily observable *spin precession* effects. Assume that a magnetic field $\mathbf{B}$ is present, in a direction different from that of the magnetic moment $\boldsymbol{\mu}$. Classically, this implies a torque $\boldsymbol{\mu} \times \mathbf{B}$ on the particle and, hence, a change in the angular momentum $\mathbf{S}$ of the particle, according to the classical equation of motion for angular moments,

$$\frac{d}{dt}\mathbf{S} = \boldsymbol{\mu} \times \mathbf{B}. \qquad (21\text{•}3\text{–}5)$$

If $\boldsymbol{\mu}$ and $\mathbf{S}$ are proportional to each other,

$$\boldsymbol{\mu} = -\gamma\mathbf{S}, \qquad (21\text{•}3\text{–}6)$$

then (21•3–5) becomes

$$\frac{d}{dt}\mathbf{S} = \gamma \cdot \mathbf{B} \times \mathbf{S}. \qquad (21\text{•}3\text{–}7)$$

The quantity γ is called the **gyromagnetic ratio** of the particle.

The torque in (21•3–7) is perpendicular to the angular momentum. It is an elementary exercise in classical mechanics to show that under those conditions, the axis of rotation of the particle, that is, the direction of the angular momentum, prescribes a precessional motion about the magnetic field, as shown in **Fig. 21•3–1**, with the (angular) precession frequency

$$\omega = \gamma B. \qquad (21\text{•}3\text{–}8)$$

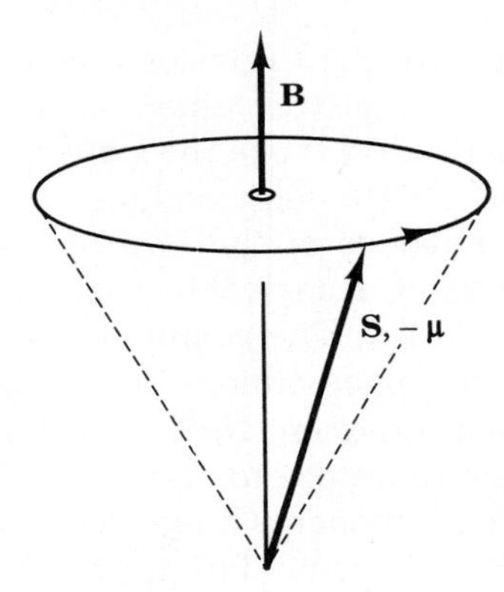

Figure 21•3–1. Precession of a classical magnetized gyro about a magnetic field.

Exercise: Show that the solution of (21•3–7) is indeed given by Fig. 21•3–1, with the frequency (21•3–8).

We show here, by actually solving the time-dependent Schroedinger wave equation, that the electron does indeed behave in this classical fashion. We consider a free electron moving in a uniform z-directed magnetic field $\mathbf{B} = \mathbf{B}_z$. The corresponding Schroedinger wave equation is

$$i\hbar\frac{\partial}{\partial t}\Psi = [\hat{H}_o + \hat{H}_s]\Psi, \tag{21•3–9}$$

where $\hat{H}_o$ is the quasi-scalar free-electron *orbital* Hamiltonian and $\hat{H}_s$ is the spin Hamiltonian (21•2–1). In our case,

$$\hat{H}_s = \frac{e\hbar}{2m_e}B\hat{\sigma}_z = \tfrac{1}{2}\hbar\omega_L\begin{pmatrix}+1 & 0\\ 0 & -1\end{pmatrix}, \tag{21•3–10}$$

where ω_L is again the Larmor frequency defined in chapter 8.

Because $\hat{H}_o$ does not depend on the spin and $\hat{H}_s$ does not depend on the position, we can separate variables by setting

$$\Psi(\mathbf{r}, t) = \Psi_o(\mathbf{r}, t)\overline{\Psi}(t) = \Psi_o(\mathbf{r}, t)\begin{pmatrix}\psi_1(t)\\ \psi_2(t)\end{pmatrix}, \tag{21•3–11}$$

where $\overline{\Psi}(t)$ is a time-dependent *unit* spinor. Insertion of (21•3–10) and (21•3–11) into (21•3–9) leads to the equation for $\overline{\Psi}(t)$:

$$i\hbar\frac{\partial}{\partial t}\overline{\Psi}(t) = \tfrac{1}{2}\omega_L\hat{\sigma}_z\overline{\Psi}(t). \tag{21•3–12}$$

Written in components, this becomes

$$i\frac{\partial}{\partial t}\psi_1 = +\tfrac{1}{2}\omega_L\psi_1, \qquad i\frac{\partial}{\partial t}\psi_2 = -\tfrac{1}{2}\omega_L\psi_2, \tag{21•3–13a,b}$$

with the solutions

$$\psi_1(t) = \psi_1(0) \cdot \exp\left(-\tfrac{i}{2}\,\omega_L t\right), \tag{21•3–14a}$$

$$\psi_2(t) = \psi_2(0) \cdot \exp\left(+\tfrac{i}{2}\,\omega_L t\right). \tag{21•3–14b}$$

If we assume that at $t = 0$ the unit spinor $\overline{\Psi}(t)$ corresponds to a spin eigendirection with the polar angles $\phi(0) = \phi_0$ and $\theta(0) = \theta_0$, then the solution of (21•3–13), subject to these initial conditions, becomes

$$\overline{\Psi}(t) = \begin{pmatrix} \exp[-\frac{i}{2}(\phi_0 + \omega_L t)]\cos\frac{1}{2}\theta_0 \\ \exp[+\frac{i}{2}(\phi_0 + \omega_L t)]\sin\frac{1}{2}\theta_0 \end{pmatrix}. \tag{21•3–15}$$

But according to our earlier discussion in section 21.2, this is a spinor whose polar angle ϕ advances linearly in time, corresponding precisely to the expected precessional motion, with an angular frequency

$$\omega = \omega_L. \tag{21•3–16}$$

Furthermore, this precession frequency is what one would expect from the classical equations (21•3–6) and (21•3–8): Setting $\omega = \omega_L$ in (21•3–8) implies an **intrinsic gyromagnetic ratio** for the electron of

$$\gamma \equiv \frac{\omega_L}{B} = \frac{e}{m_e} = \frac{\mu_B}{\hbar/2}. \tag{21•3–17}$$

But as we saw earlier, μ_B and $\hbar/2$ are the projections of the intrinsic magnetic moment and the intrinsic angular momentum upon B_z, for any eigenspinor of $\hat{\sigma}_z$.

Note that the ratio γ in (21•3–17) is twice the similarly defined ratio γ_L in (18•3–13) for the ratio of *orbital* magnetic moment to *orbital* angular momentum. Such gyromagnetic ratios γ that are different from the orbital gyromagnetic ratio γ_L occur quite frequently in atomic physics, and it is common to express them relatively to γ_L by defining a **g-factor**

$$\boxed{g \equiv \gamma/\gamma_L.} \tag{21•3–18a}$$

In this terminology, the electron has an **intrinsic *g*-factor** of 2:

$$g_e = \gamma_e/\gamma_L = 2. \tag{21•3–18b}$$

As we pointed out earlier, the intrinsic magnetic moment of the electron is about 1 part in 1000 larger than the Bohr magneton. Hence, the actual g-factor of the electron is slightly higher than 2:

$$g_e = 2\mu_e/\mu_B \sim 2.00232. \tag{21•3–19}$$

This is often referred to as the **anomalous (intrinsic) *g*-factor** of the electron, a terminology that dates back historically to the times before the deviation was understood quantitatively. As we saw, however, there is really nothing anomalous about it. We will neglect these relativistic corrections here.

Exercise: Treat the spin precession in a somewhat different way by considering the time derivative of the expectation value of the intrinsic angular momentum operator $\hat{\mathbf{S}}$. Draw on the general formalism of chapter 7 for the time derivative of any expectation value. Show that this leads to

$$\frac{d}{dt}\langle \mathbf{S} \rangle = -\gamma \langle \mathbf{S} \times \mathbf{B} \rangle, \tag{21•3–20a}$$

and, if the magnetic field is uniform,

$$\frac{d}{dt}\langle \mathbf{S} \rangle = -\gamma \langle \mathbf{S} \rangle \times \mathbf{B}, \tag{21•3–20b}$$

which is the same as (21•3–7) if we identify the classical vector $\mathbf{S}$ with $\langle \mathbf{S} \rangle$.

The spin precession frequency ω is related to the energy level difference between the spin-up and the spin-down state by the standard energy-frequency relationship

$$\Delta\mathcal{E} = 2\mu_B B = \frac{e\hbar}{m_e} = \hbar\omega. \tag{21•3–21}$$

This is, of course, no accident, and our derivation makes it clear why it must be that way: In the presence of a magnetic field, the solutions of the time-dependent Schroedinger wave equation can be expressed as a linear superposition of spin-up and spin-down eigenfunctions with time-dependent expansion coefficients. In the presence of that field, the two sets differ in energy by $\Delta\mathcal{E} = 2\mu_B B$, and this means that the time dependence of their expansion coefficients differs by a factor

$$\exp(-i\,\Delta\mathcal{E} \cdot t/\hbar) = \exp(-i\omega t/\hbar). \tag{21•3–22}$$

It is this difference in time dependence that leads to the spin precession about the magnetic field. The situation is quite similar to one we discussed in chapter 2 in conjunction with the harmonic oscillator, where we showed that non-stationary states of the harmonic oscillator correspond to a particle oscillating in space with the classical oscillation frequency associated with that oscillator.

◆ PROBLEM TO SECTION 21.3

#21•3-1: Spin Dynamics

An electron is initially in its spin-up state (relative to the z-direction) in zero magnetic field. At $t = 0$, a magnetic field is applied such that $B_x = B_y = B_z = B_0$. Describe the time evolution of the spin states of the electron. What is the maximum probability of ever finding the electron in its spin-down state (again, relative to the z-direction)?

21.4 SPIN-ORBIT INTERACTION AND TOTAL ANGULAR MOMENTUM

21.4.1 The Spin-Orbit Hamiltonian

The magnetic field seen by the intrinsic magnetic moment of an electron includes not only externally applied fields, but also any fields internal to the electron system. The most important source of internal magnetic fields are the internal currents in the system. Because these internal currents are due to what is called the **orbital** motion of the charged particles, the interaction between the resulting magnetic fields and the intrinsic magnetic moments is called the **spin-orbit interaction**.

The reference electron does not, of course, see its own magnetic field any more than it sees its own electric field. But if the electron moves, any external charge distribution moves relative to the electron, leading to a magnetic field in the frame of reference of the moving electron itself. This magnetic field is seen by the intrinsic magnetic moment of the electron, even in systems as simple as the hydrogen atom: If the electron moves relative to the proton in the hydrogen atom, the proton moves relative to the electron, causing a magnetic field to be seen by the electron.

Such a motion-induced magnetic field is related to the electric field $\mathbf{E}$ in the laboratory frame via a Lorentz transformation. Recall from the theory of relativity that the electric and magnetic fields seen by an observer depend upon the motion of the frame of reference of the observer. In particular, if in the laboratory frame of reference there exists a purely electric field $\mathbf{E}$, this electric field will cause a magnetic field to appear in any frame of reference that moves relative to the laboratory frame, unless the velocity $\mathbf{v}$ of relative motion is parallel to the electric field. Quantitatively, the Lorentz-transformed magnetic field is[6]

$$\mathbf{B}' = -\frac{(\mathbf{v} \times \mathbf{E})/c^2}{\sqrt{1 - (v/c)^2}} \xrightarrow[v \ll c]{} -\frac{(\mathbf{v} \times \mathbf{E})}{c^2}, \tag{21•4–1}$$

where c is the speed of light; the limit $v \ll c$ is the case of interest to us.

Despite its relativistic origin, this magnetic field is not negligible and can easily exceed typical laboratory fields, even when $v \ll c$, because the electric fields inside an atom can be very large. For example, inside a hydrogen atom, at a distance from the proton equal to the Bohr radius $r = a_0$, the electric field

---

[6] See, for example, J. D. Jackson, *Classical Electrodynamics* (New York: Wiley, 1962), particularly section 11.10; or A. Sommerfeld, *Electrodynamics,* Vol. 3 of *Lectures on Theoretical Physics* (New York: Academic Press, 1964), particularly section 28B.

is about 5.1×10^{11} V/m! The velocity of an electron at that distance from the proton, calculated classically, can be shown to be

$$v \approx \alpha c \approx 2.2 \times 10^6 \text{ m/s}, \tag{21•4–2}$$

where α is the fine-structure constant. Insertion of these values for the electric field and the electron velocity into (21•4–1) leads to a Lorentz-transformed magnetic field of about 12 Tesla, on the order of all but the strongest laboratory fields.

This argument suggests that in the presence of electric fields, the magnetic field $\mathbf{B}$ in the spin Hamiltonian (21•2–11) should be replaced by $\mathbf{B} - (\hat{\mathbf{v}} \times \mathbf{E})/c^2$, where $\hat{\mathbf{v}}$ is the velocity operator. However, the argument overlooks a second relativistic effect that is less widely known, but is of the same order of magnitude: In the presence of an electric field perpendicular to the electron velocity, a classical electron moves no longer along a straight line, but along a curved trajectory. In effect, the electron's own frame of reference is rotating. It can be shown[7] that this changes the interaction of the moving electron spin with the electric field and that the net result is that the spin-orbit interaction is cut in half. A rigorous derivation of this result requires a knowledge of some aspects of relativistic kinematics that, although not difficult, are probably unfamiliar to most readers. We therefore proceed here via a less rigorous but much simpler argument, by considering a slightly different special case in which the electron is *forced* to move along a straight line. This is easily accomplished by adding a magnetic field in the laboratory frame, at right angles to both the electric field and the electron velocity, with such a strength that the Lorentz force from the magnetic field exactly balances the electric force on the electron. In that case, there will be no effects due to a rotating frame of reference. If the electron velocity is in the x-direction, and the transverse part of the electric field in the y-direction, this cancellation will take place if the magnetic field is in the z-direction and has the strength

$$B_z = E_y/v_x. \tag{21•4–3}$$

If this *combination* of the two fields is Lorentz-transformed into the *uniformly* moving frame of the electron, the magnetic balancing field undergoes a Lorentz transformation, too, and the new magnetic field B_z' in the electron's own frame is[8]

$$B_z' = \frac{B_z - v_x E_y/c^2}{\sqrt{1 - (v/c)^2}}. \tag{21•4–4}$$

---

[7] See, for example, J. D. Jackson, *op. cit.*, section. 11.5.

[8] See again Jackson or Sommerfeld, *op. cit.*

If we expand this up to terms of the order v_x^2 and reexpress the term $v_x^2 B_z$ as $v_x E_y$, we may rewrite (21•4–4) as

$$B_z' = B_z \cdot \left(1 + \tfrac{1}{2}\frac{v_x^2}{c^2}\right) - \frac{v_x E_y}{c^2} = B_z - \tfrac{1}{2}\frac{v_x E_y}{c^2}. \tag{21•4–5}$$

This is the magnetic field that is seen by the intrinsic magnetic moment of the electron and that determines the potential energy of the moment in the presence of both electric and magnetic fields. The B_z-term is already included in the Pauli Hamiltonian; only the remainder enters a correction due to the spin-orbit interaction. Note that the $\mathbf{v} \times \mathbf{E}$-term differs from the naive value in (21•4–1) by the factor 1/2, commonly called the **Thomas precession factor**.

Our argument suggests that the magnetic field $\mathbf{B}$ in the spin Hamiltonian (21•2–11) should be replaced with an operator that is equivalent to

$$\mathbf{B}' = \mathbf{B} - \tfrac{1}{2}\frac{\mathbf{v} \times \mathbf{E}}{c^2} = \mathbf{B} + \tfrac{1}{2}\frac{\mathbf{E} \times \mathbf{v}}{c^2}. \tag{21•4–6}$$

The appropriate velocity operator is, of course, the operator

$$\hat{\mathbf{v}} = \frac{1}{m_e}(\hat{\mathbf{p}} + e\mathbf{A}) = \frac{1}{m_e}(-i\hbar\nabla + e\mathbf{A}). \tag{21•4–7}$$

Because the velocity is now an operator, and the electric field may be position dependent, we must investigate whether we are allowed to write simply $\hat{\mathbf{v}} \times \mathbf{E}$ or $-\mathbf{E} \times \hat{\mathbf{v}}$, or whether a more complicated symmetrized operator is required, as in the treatment of the Lorentz force in chapter 8. By inserting (21•4–7) into $\hat{\mathbf{v}} \times \mathbf{E}$ and applying the result to a wave function ψ, we obtain

$$\hat{\mathbf{v}} \times (\mathbf{E}\psi) = -\mathbf{E} \times (\hat{\mathbf{v}}\psi) - \frac{i\hbar}{m_e}(\nabla \times \mathbf{E})\psi, \tag{21•4–8}$$

where the curl operator on the far right operates only on the electric field, not on ψ. But for a *static* electric field—the only case of interest on the level of relativistic corrections—the curl vanishes,

$$\text{curl}\,\mathbf{E} = \nabla \times \mathbf{E} = 0, \tag{21•4–9}$$

and we find that we may use either of the two simple forms:

$$\hat{\mathbf{v}} \times \mathbf{E} = -\mathbf{E} \times \hat{\mathbf{v}}. \tag{21•4–10}$$

The $\mathbf{E} \times \hat{\mathbf{v}}$ form is the more convenient one. By inserting it into (21•4–6), and the result into (21•2–11), we finally obtain the overall spin Hamiltonian

$$\hat{H}_s = \frac{e\hbar}{2m_e}\hat{\boldsymbol{\sigma}} \cdot \left(\mathbf{B} + \tfrac{1}{2}\frac{\mathbf{E} \times \hat{\mathbf{v}}}{c^2}\right) = \hat{H}_{SB} + \hat{H}_{SO}, \tag{21•4–11}$$

where $\hat{H}_{SB}$ is the original spin Hamiltonian (21•2–11) and $\hat{H}_{SO}$ the **spin-orbit Hamiltonian**:

$$\hat{H}_{SB} = \frac{e\hbar}{2m_e}\hat{\boldsymbol{\sigma}} \cdot \mathbf{B}, \tag{21•4–12}$$

$$\hat{H}_{SO} = \frac{e\hbar}{4m_e c^2}\hat{\boldsymbol{\sigma}} \cdot (\mathbf{E} \times \hat{\mathbf{v}}) = \frac{\hbar}{4m_e c^2}\hat{\boldsymbol{\sigma}} \cdot [(\nabla V) \times \hat{\mathbf{v}}]. \tag{21•4–13}$$

The operator $\hat{\mathbf{v}}$ is again given by (21•4–7).

21.4.2 Spherical Symmetry and Total Angular Momentum

Of particular interest is the spin-orbit interaction in a potential of spherical symmetry, $V(\mathbf{r}) = V(r)$, in the absence of an *external* magnetic field. The product $\mathbf{E} \times \hat{\mathbf{v}}$ in the spin-orbit Hamiltonian (21•4–13) then assumes the form

$$\mathbf{E} \times \hat{\mathbf{v}} = \frac{E_r}{m_e r}(\hat{\mathbf{r}} \times \hat{\mathbf{p}}) = \frac{E_r}{m_e r}\hat{\mathbf{L}}, \tag{21•4–14}$$

where $E_r = E_r(r)$ is the radial component of the electric field and $\hat{\mathbf{L}}$ is the orbital angular momentum operator. The entire spin-orbit Hamiltonian then becomes

$$\hat{H}_{SO} = \frac{e\hbar}{4m_e^2 c^2}\frac{E_r}{r}(\hat{\boldsymbol{\sigma}} \cdot \hat{\mathbf{L}}) = \frac{e}{2m_e^2 c^2}\frac{E_r}{r}(\hat{\mathbf{S}} \cdot \hat{\mathbf{L}}), \tag{21•4–15}$$

where, in the second form, we have replaced $\hat{\boldsymbol{\sigma}}$ by $\hat{\mathbf{S}}$, via (21•3–2). Note that $\hat{H}_{SO}$ is still invariant under rotation, *if* the coordinate systems for $\mathbf{r}$ and $\mathbf{S}$ are rotated jointly. However, the product $\hat{\mathbf{S}} \cdot \hat{\mathbf{L}}$ and, hence, $\hat{H}_{SO}$ no longer commute with either $\hat{\mathbf{L}}$ or $\hat{\mathbf{S}}$. For example, for $\hat{L}_z$,

$$\begin{aligned}[\hat{L}_z, (\hat{\mathbf{S}} \cdot \hat{\mathbf{L}})] &= [\hat{L}_z, (\hat{S}_x\hat{L}_x + \hat{S}_y\hat{L}_y + \hat{S}_z\hat{L}_z)] \\ &= \hat{S}_x[\hat{L}_z, \hat{L}_x] + \hat{S}_y[\hat{L}_z, \hat{L}_y] \\ &= i\hbar(\hat{S}_x\hat{L}_y - \hat{S}_y\hat{L}_x) = +i\hbar(\hat{\mathbf{S}} \times \hat{\mathbf{L}})_z,\end{aligned} \tag{21•4–16}$$

where we have used the commutation relations (21•3–1) for the angular momentum operator components. Combined with its cyclically equivalent relations, we obtain

$$[\hat{\mathbf{L}}, (\hat{\mathbf{S}} \cdot \hat{\mathbf{L}})] = +i\hbar(\hat{\mathbf{S}} \times \hat{\mathbf{L}}). \tag{21•4–17}$$

By a similar procedure, we find that, for the commutator involving $\hat{\mathbf{S}}$,

$$[\hat{\mathbf{S}}, (\hat{\mathbf{S}} \cdot \hat{\mathbf{L}})] = -i\hbar(\hat{\mathbf{S}} \times \hat{\mathbf{L}}), \tag{21•4–18}$$

which differs from (21•4–17) by a change of sign on the right-hand side. This simple sign reversal suggests that we define an operator for the **total angular momentum**,

$$\boxed{\hat{\mathbf{J}} = \hat{\mathbf{L}} + \hat{\mathbf{S}},} \tag{21•4–19}$$

which evidently *does* commute with $\hat{\mathbf{S}} \cdot \hat{\mathbf{L}}$,

$$[\hat{\mathbf{J}}, (\hat{\mathbf{S}} \cdot \hat{\mathbf{L}})] = 0. \tag{21•4–20}$$

Because all constituents of $\hat{\mathbf{J}}$ commute with the rest of the Hamiltonian, $\hat{\mathbf{J}}$ itself commutes with the entire Hamiltonian.

The behavior expressed by the relations (21•4–17) through (21•4–20) has far-reaching consequences. Recall from chapter 7 that the classical law of conservation of angular momentum depends on the vanishing of the commutator of the angular momentum operator with the Hamiltonian. The relations (21•4–17) and (21•4–20) then imply that for objects with spin, the operator $\hat{\mathbf{L}}$ is no longer the correct operator for what is actually conserved. We must add the term $\hat{\mathbf{S}}$ to the orbital angular momentum to retain the law of conservation of angular momentum. This is what compels us, more than the mathematical properties of $\hat{\mathbf{S}}$, to view spin as a true angular momentum and $\hat{\mathbf{J}}$ as the operator for the *total* angular momentum.

The relations (21•4–17) through (21•4–20) imply that the eigenfunctions of the complete Hamilton operator can, in general, no longer be chosen such that they are also eigenfunctions of one of the components of $\hat{\mathbf{L}}$ and/or $\hat{\mathbf{S}}$; instead, they can now be chosen to be eigenstates of the total angular momentum operator $\hat{\mathbf{J}}$. However, just as with $\hat{\mathbf{L}}$ and $\hat{\mathbf{S}}$, they again cannot be simultaneous eigenstates of all $\hat{\mathbf{J}}$-components, because the latter once again do not commute: With $\hat{\mathbf{L}}$ operating only on the position variables and $\hat{\mathbf{S}}$ only on the spin, the commutators for the components of $\hat{\mathbf{J}}$ are identical to those for $\hat{\mathbf{L}}$ and $\hat{\mathbf{S}}$:

$$[\hat{J}_x, \hat{J}_y] = i\hbar\hat{J}_z, \qquad [\hat{J}_y, \hat{J}_z] = i\hbar\hat{J}_x, \qquad [\hat{J}_z, \hat{J}_x] = i\hbar\hat{J}_x. \tag{21•4–21}$$

Given these identical commutation relations, the entire mathematical formalism of the orbital angular momentum operators can be taken over for the operators of the total angular momentum, except for those changes that follow from the change from integer quantum numbers to half-integer ones.

In particular, we introduce the operator analogous to $\hat{L}^2$,

$$\hat{J}^2 = \hat{J}_x^2 + \hat{J}_y^2 + \hat{J}_z^2, \tag{21•4–22}$$

and write its eigenvalues analogously as $\hbar^2 \cdot j(j + 1)$, where j is now a half-integer, rather than an integer. Again, $\hat{J}^2$ commutes both with the Hamiltonian and with each of the components of $\hat{\mathbf{J}}$. Hence, we may choose the eigenstates of the Hamiltonian to be simultaneous eigenstates of both $\hat{J}^2$ and one of the components of $\hat{\mathbf{J}}$, invariably chosen to be $\hat{J}_z$. If we designate those eigenstates as $|j, m_j\rangle$, we may write

$$\hat{J}^2|j, m_j\rangle = \hbar^2 j(j+1)|j, m_j\rangle, \tag{21•4–23}$$

$$\hat{J}_z|j, m_j\rangle = \hbar m_j|j, m_j\rangle, \tag{21•4–24}$$

where the total azimuthal angular momentum quantum number m_j is now a half-integer rather than an integer, ranging from $-j$ to $+j$.

21.4.3 Example: Fine Structure of the Hydrogen Atom

As an example illustrating both the effects of spin-orbit interaction and the manipulation of spinor wave functions, we discuss here the splitting of the $n = 2$ energy level of the hydrogen atom under the influence of the spin-orbit interaction, also known as the **fine structure** of this level.

As a preliminary, we write out the product $\hat{\boldsymbol{\sigma}} \cdot \hat{\mathbf{L}}$ as a 2×2 matrix:

$$\begin{aligned}\hat{\boldsymbol{\sigma}} \cdot \hat{\mathbf{L}} &= \hat{\sigma}_x \hat{L}_x + \hat{\sigma}_y \hat{L}_y + \hat{\sigma}_z \hat{L}_z \\ &= \begin{pmatrix} 0 & +1 \\ +1 & 0 \end{pmatrix} \hat{L}_x + \begin{pmatrix} 0 & -i \\ +i & 0 \end{pmatrix} \hat{L}_y + \begin{pmatrix} +1 & 0 \\ 0 & -1 \end{pmatrix} \hat{L}_z \\ &= \begin{pmatrix} 0 & \hat{L}_- \\ \hat{L}_+ & 0 \end{pmatrix} + \begin{pmatrix} +\hat{L}_z & 0 \\ 0 & -\hat{L}_z \end{pmatrix}. \end{aligned} \tag{21•4–25}$$

In the last line, we have introduced the angular momentum stepping operators $\hat{L}_+$ and $\hat{L}_-$ defined in chapter 18. The interesting term is the off-diagonal matrix operator involving the stepping operators. We consider the effect of this operator on a pure spin-up state and a pure spin-down state, written symbolically as

$$|l, m, \uparrow\rangle = \begin{pmatrix} 1 \\ 0 \end{pmatrix} \cdot |l, m\rangle \quad \text{and} \quad |l, m, \downarrow\rangle = \begin{pmatrix} 0 \\ 1 \end{pmatrix} \cdot |l, m\rangle. \tag{21•4–26a,b}$$

For the spin-up state, we obtain

$$\begin{pmatrix} 0 & \hat{L}_- \\ \hat{L}_+ & 0 \end{pmatrix} \begin{pmatrix} 1 \\ 0 \end{pmatrix} \cdot |l, m\rangle = \begin{pmatrix} 0 \\ 1 \end{pmatrix} \hat{L}_+ |l, m\rangle \tag{21•4–27a}$$

$$= \begin{pmatrix} 0 \\ 1 \end{pmatrix} \hbar \sqrt{l(l+1) - m \cdot (m+1)}\, |l, m+1\rangle,$$

where we have drawn on the relation (18•1–25a) for the normalization coefficient for the upward stepping operator. The resulting state is evidently a spin-down state, but with its *orbital* angular momentum increased by one unit, thus keeping the *total* angular momentum constant. Similarly, for the action on the spin-down state,

$$\begin{pmatrix} 0 & \hat{L}_- \\ \hat{L}_+ & 0 \end{pmatrix} \begin{pmatrix} 0 \\ 1 \end{pmatrix} \cdot |l, m\rangle = \begin{pmatrix} 1 \\ 0 \end{pmatrix} \hbar \sqrt{l(l+1) - m(m-1)} \cdot |l, m-1\rangle. \tag{21•4–27b}$$

Turning now to the spin-orbit interaction in the hydrogen atom, we insert the Coulomb field into the spin-orbit Hamiltonian (21•4–15). The result may be written as

$$\hat{H}_{SO} = \frac{e\hbar}{4m_e^2c^2}\frac{e}{4\pi\epsilon_0 r^3}(\hat{\boldsymbol{\sigma}}\cdot\hat{\mathbf{L}}) = \tfrac{1}{2}\alpha^2 R_\infty\left(\frac{a_0}{r}\right)^3\frac{(\hat{\boldsymbol{\sigma}}\cdot\hat{\mathbf{L}})}{\hbar}, \tag{21•4–28}$$

where α is again the fine-structure constant and R_∞ the Rydberg energy. It is clear that the states with $l = 0$ remain unaffected by this perturbation, nor do they couple to the states with $l = 1$. We therefore only consider the splitting of the m-degeneracy of the latter by the perturbation.

In the absence of spin, we would have three basis states with $l = 1$; spin doubles that number. The states are listed in Table 21•4–1, in the order of decreasing value of $m_j = m + m_s$. Note that there are two basis states each associated with $m_j = \pm 1/2$.

Because $\hat{J}_z$ commutes with the perturbation, only terms belonging to the same value of m_j will couple. Hence, the 6×6 secular equation factors into two 1×1 and two 2×2 equations. The matrix elements for $m_j < 0$ are equal to those for $m_j > 0$, leading to a total of three diagonal and one off-diagonal matrix elements that must be considered. Each matrix element is a product of two factors, one due to the term $\hat{\boldsymbol{\sigma}}\cdot\hat{\mathbf{L}}$ and one due to the radial part of the perturbation.

We consider the radial part first. All states involved belong to the same quantum numbers n and l. Hence, they all have the same one-dimensional equivalent radial wave function,

$$\chi_{21}(s) = \frac{1}{2\sqrt{6a_0}}\cdot s^2\cdot e^{-s/2}, \tag{21•4–29}$$

taken from appendix D. We evidently, need the radial matrix element

$$\mathcal{E}_0 \equiv \tfrac{1}{2}\alpha^2 R_\infty \cdot \int_0^\infty \chi_{21}^2(r/a_0)\cdot(a_0/r)^3\,dr. \tag{21•4–30}$$

TABLE 21•4–1. Azimuthal quantum numbers for orbital angular momentum, spin angular momentum, and total angular momentum for the six basis states with $l = 1$.

	$\lvert +1, \uparrow\rangle$	$\lvert +1, \downarrow\rangle$	$\lvert 0, \uparrow\rangle$	$\lvert 0, \downarrow\rangle$	$\lvert -1, \uparrow\rangle$	$\lvert -1, \downarrow\rangle$
$m =$	$+1$	$+1$	0	0	-1	-1
$m_s =$	$+1/2$	$-1/2$	$+1/2$	$-1/2$	$+1/2$	$-1/2$
$m_j =$	$+3/2$	$+1/2$	$+1/2$	$-1/2$	$-1/2$	$-3/2$

The integral has the value 1/24, leading to

$$\mathcal{E}_0 = \frac{\alpha^2}{48} \cdot R_\infty. \tag{21•4–31}$$

We next turn to the $\hat{\boldsymbol{\sigma}} \cdot \hat{\mathbf{L}}$ contributions to the matrix elements. From our discussion about the action of this operator product, we see immediately that all diagonal contributions of $\hat{\boldsymbol{\sigma}} \cdot \hat{\mathbf{L}}$ are due to the $\pm\hat{L}_z$ matrix in (21•4–25) and are equal to $+m\hbar$ for the spin-up states and $-m\hbar$ for the spin-down states, whatever the value of m. The only nonzero off-diagonal contributions are the ones shown in Table 21•4–2; they all arise from the $\hat{L}_\pm$ matrix in (21•4–25), and they all have the value $\sqrt{2}\hbar$.

Table 21•4–2 is equivalent to two pairs of secular equations; the actual matrix elements differ from the entries in the table only by the factor $\mathcal{E}_0$. The two states with $m_j = \pm 3/2$ remain pure states, but are evidently shifted up by

$$\Delta\mathcal{E} = \mathcal{E}_0. \tag{21•4–32}$$

The remaining states become intermixed and split, with the energy shifts

$$\Delta\mathcal{E} = (-\tfrac{1}{2} \pm \tfrac{3}{2})\mathcal{E}_0 = +\mathcal{E}_0 \quad \text{and} \quad -2\mathcal{E}_0, \tag{21•4–33a,b}$$

each value representing two states.

Note that the two raised states remain degenerate with the two $m_j = \pm 3/2$ states, leading to the overall splitting pattern shown in **Fig. 21•4–1**.

What remains to be determined is the value of the total angular momentum quantum numbers j and m_j to which the various final states belong. By definition,

$$\hat{\mathbf{J}}^2 = (\hat{\mathbf{L}} + \hat{\mathbf{S}})^2 = \hat{\mathbf{L}}^2 + \hat{\mathbf{S}}^2 + 2\hat{\mathbf{L}} \cdot \hat{\mathbf{S}}. \tag{21•4–34}$$

TABLE 21•4–2. Matrix elements of the operator $\hat{\boldsymbol{\sigma}} \cdot \hat{\mathbf{L}}/\hbar$ for the six $l = 1$ states. Note that the lower right quarter of the table is simply the mirror image of the upper left quarter.

	$\lvert +1, \uparrow\rangle$	$\lvert +1, \downarrow\rangle$	$\lvert 0, \uparrow\rangle$	$\lvert 0, \downarrow\rangle$	$\lvert -1, \uparrow\rangle$	$\lvert -1, \downarrow\rangle$
$\lvert +1, \uparrow\rangle$	+1					
$\lvert +1, \downarrow\rangle$		−1	$\sqrt{2}$			
$\lvert 0, \uparrow\rangle$		$\sqrt{2}$	0			
$\lvert 0, \downarrow\rangle$				0	$\sqrt{2}$	
$\lvert -1, \uparrow\rangle$				$\sqrt{2}$	−1	
$\lvert -1, \downarrow\rangle$						+1

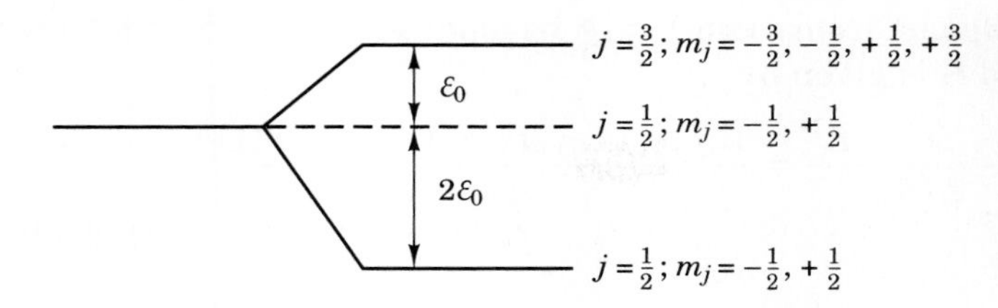

Figure 21•4–1. Fine-structure splitting of the $n = 2$ energy level of the hydrogen atom. The $l = 1$ states are split as shown by the solid lines; the $l = 0$ states remain unchanged (broken line), leading to an overall threefold split.

Evidently, those states whose energy was raised have a positive value of $\mathbf{L} \cdot \mathbf{S}$; hence, for them, we must have $J^2 > L^2$ and, therefore, $j > l$. Because j must be a half-integer, those states must belong to $j = 3/2$. Similarly, the states whose energy was lowered must belong to $j = 1/2$. The four states with $j = 3/2$ represent the states with different values of m_j, from $m_j = -3/2$ to $m_j = +3/2$, all four of which must be present.

21.4.4 Spin Precession Re-Visited: The Zeeman Effect

Semiclassical Vector Model

Even though the operators $\hat{\mathbf{L}}$ and $\hat{\mathbf{S}}$ no longer commute with the Hamiltonian in the presence of spin-orbit interaction, inspection shows that their squares $\hat{L}^2$ and $\hat{S}^2$ still commute. This means that it is possible to select the energy eigenstates in such a way that the *magnitudes* of both the orbital and the spin angular momentum still have sharp values, namely, $\sqrt{l(l+1)}$ and $\sqrt{s(s+1)}$. This leads to the semiclassical **vector model** for the total angular momentum shown in **Fig. 21•4–2**, in which the three relevant angular momentum vectors form a triangle in space. The lengths of the three vectors are such that the

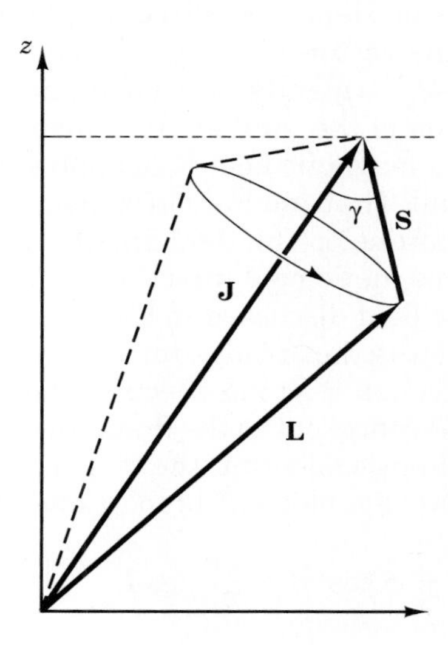

Figure 21•4–2. Semiclassical vector model of the total angular momentum. The drawing assumes that $m_j = j = 5/2$, $l = 2$, and $s = 1/2$. If viewed as made up from three classical vectors, the triangle made up by the three vectors rotates in time about the direction of **J**.

vectors cannot be collinear; from elementary trigonometry, we find that the angle γ between $\mathbf{J}$ and $\mathbf{S}$ is given by

$$\cos\gamma = \frac{J^2 + S^2 - L^2}{2JS} = \frac{j(j+1) + s(s+1) - l(l+1)}{2\sqrt{j(j+1)}\cdot\sqrt{s(s+1)}}. \tag{21•4–35}$$

In writing (21•4–35), we have deliberately *not* inserted $s = 1/2$, because the more general form (21•4–35) remains useful in the discussion of multi-electron systems, where s will be replaced by an analogous quantum number representing the *net* spin of *all* electrons, which will usually have a value $s \neq 1/2$.

It is left to the reader to show that, for given values of l and s, the quantum number j must obey

$$l - s \leq j \leq l + s. \tag{21•4–36}$$

For a single electron, $j = l \pm 1/2$.

Of the three classical vectors in Fig. 21•4–2, only $\mathbf{J}$ will have a time-independent direction. The other two will vary with time, in such a way that the triangle made up by the vectors rotates in space about the direction of $\mathbf{J}$, with a frequency equivalent to the contribution of the spin-orbit Hamiltonian to the total energy of the state.

Precession of the Total Magnetic Moment: The Zeeman Effect

An important consequence of the noncollinearity of $\mathbf{S}$ and $\mathbf{J}$ is that the total magnetic moment vector also is not collinear with $\mathbf{J}$, nor with $\mathbf{L}$ or $\mathbf{S}$: Recall that the ratio of magnetic moment to angular momentum for the spin is twice as large as the orbital ratio. Hence, the direction of the total magnetic moment is the same as that of the vector obtained by doubling the spin vector in Fig. 21•4–2—see **Fig. 21•4–3**. Evidently, the magnetic moment vector also precesses about the direction of the total angular momentum.

Suppose now that a magnetic field B_z is applied to an electron in a spherically symmetric potential. The combination of magnetic moment and magnetic field will again exert a torque on the total angular momentum, and the latter will precess about the magnetic field direction, just like the precession of the spin about the magnetic field discussed in section 21.1. What matters for the *net* torque and precession is the steady-state component $\boldsymbol{\mu}_J$ of the total magnetic moment in the direction of $\mathbf{J}$; the effects of the component perpendicular to $\mathbf{J}$ and rotating about $\mathbf{J}$ cancel out to the first order. From Fig. 21•4–2 and the relation (21•4–35), one finds easily that the magnetic moment component parallel to $\mathbf{J}$ may be written as a moment proportional to $\mathbf{J}$ itself,

$$\boldsymbol{\mu}_J = \frac{e}{2m_e}(\mathbf{J} + \mathbf{S}\cos\gamma) = \frac{e}{2m_e}g_J\mathbf{J}, \tag{21•4–37}$$

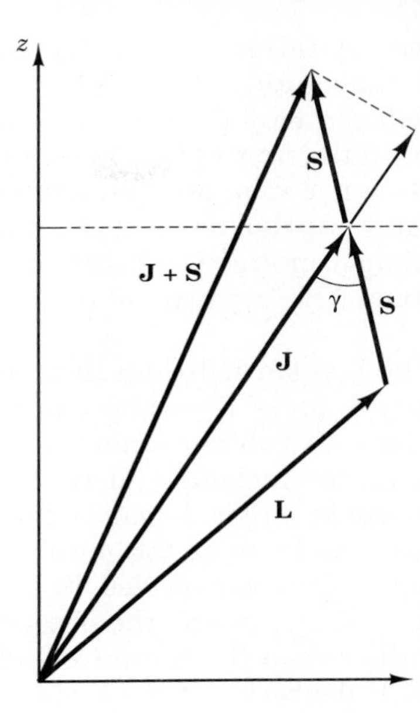

Figure 21•4–3. Geometry of total angular momentum and total magnetic moment.

with the proportionality factor

$$g_J = 1 + \frac{j(j+1) - l(l+1) + s(s+1)}{2j(j+1)}, \tag{21•4–38}$$

called the **Landé *g*-factor** of the state.

Inasmuch as the z-component of $\mathbf{J}$ is quantized, the same is true for the z-components of $\boldsymbol{\mu}_J$:

$$\mu_z = \frac{e}{2m_e} g_J J_z = m_j g_J \mu_B. \tag{21•4–39}$$

Assume now that a magnetic field B_z is applied as a perturbation. In first-order perturbation theory, the energy perturbation is simply the energy expectation value of the perturbation, calculated for the unperturbed state. This leads to

$$\mathcal{E}^{(1)} = -\mu_z B_z = -m_j g_J \mu_B B_z; \qquad m_j = -j, \ldots, +j. \tag{21•4–40}$$

Hence, we expect that, to the first order, an energy level with the three angular momentum quantum numbers j, l, and s will split up into $2j + 1$ equidistant sub-levels, with a separation

$$\Delta\mathcal{E} = g_J \mu_B B_z. \tag{21•4–41}$$

These splittings are confirmed by the full-fledged perturbation treatment—and by experimental observations of the splitting of the lines of atomic spectra in a magnetic field, the so-called *Zeeman effect*.

Although our above treatment was for a single electron, it remains a useful approximation for the multi-electron systems of many atoms, especially the atoms in the upper left corner of the periodic table. The spin-orbit interaction in those atoms can, to the first order, be treated by first combining the spins of all electrons into a net spin, and the orbital angular momenta of all electrons into a net orbital angular momentum, and then treating the spin-orbit interaction as a simple interaction between the net spin and the net orbital angular momentum. This approximation is called **Russell-Saunders coupling**.

The magnitude of the Zeeman splittings then depends on the factor g_J. In the case of zero net spin ($s = 0, j = l$)—which can occur only in systems with an even number of electrons—we obtain again the result $\Delta\mathcal{E} = \mu_B B$, as was to be expected. In all other cases—including all systems with an odd number of electrons—the splittings will be larger. In particular, the splittings for s-states ($l = 0, j = s \neq 0$) are twice as large as the purely orbital splittings and are equal to those for a free spin, again as expected. In all other cases, the splittings will fall in between. Historically, it was the observation of the double-sized splittings for s-states, rather than the Stern-Gerlach experiment, that led to the realization—by G. E. Uhlenbeck and S. Goudsmit, in 1925—that the electron has an *intrinsic* angular momentum, with an intrinsic g-factor of 2.

By carefully measuring the magnetic splittings of the atomic energy levels and comparing them with the theory contained in (21•4–38) and (21•4–39), it is often possible to determine experimentally the quantum numbers j, l, and s associated with a given energy level. For more details, including specifically a discussion of cases where Russell-Saunders coupling is *not* applicable, the reader is referred to texts on atomic physics. More exact full-fledged quantum-mechanical treatments of the Zeeman effect are found in many of the more advanced quantum mechanics texts, particularly the older ones, which tend to stress the application of quantum mechanics to atomic physics more than modern texts do—or than we wish to do.

◆ PROBLEM TO SECTION 21.4

21•4-1: Scattering at a Step

Study, in detail, the spin-orbit coupling effects in the reflection of a spinor plane wave from a one-dimensional potential step

$$V(x) = \begin{cases} 0 & \text{for} \quad x < 0 \\ V_0 & \text{for} \quad x > 0, \end{cases} \tag{21•4–42}$$

under oblique incidence. Specifically, assume that the incident plane wave propagates in the x-y plane, that is,

$$\overline{\psi} = \begin{pmatrix} \psi_1 \\ \psi_2 \end{pmatrix} \propto e^{i(k_x x + k_y y)}, \tag{21•4–43}$$

with

$$k_x = k_0 \cos\alpha, \qquad k_x = k_0 \sin\alpha, \qquad k_z = 0. \tag{21•4–44}$$

What, particularly, is the change in spin direction of the reflected wave? Explain! Express the results in terms of the amplitude reflection coefficient R_0 for scalar waves, the angle of incidence α, and the ratio-of kinetic energy to rest mass energy of the electron,

$$r = \frac{\hbar^2 k_0^2}{2m_e} \Big/ m_e c^2 \ll 1, \tag{21•4–45}$$

Note that we have assumed a non-relativistic electron.

21.5 ODDS AND ENDS

21.5.1 Spin as an Inherently Non-Classical Property

Having shown that the properties of the spin are in many ways analogous to those of a classical charged particle spinning about its axis, it is important also to show the limitations of this analogy: Although there certainly *are* classical particle aspects to the electron spin, in the last analysis, spin is an inherently non-classical phenomenon, in the sense that its properties are *different* from those of a quantized classical charged particle rotating about its own axis.

The properties of the latter would be the same as those of the orbital angular momentum (and magnetic moment) of a particle in a rotationally symmetric potential. That is, its angular momentum would be an integer, rather than a half-integer, multiple of $\hbar$. More important, the angular momentum would not be restricted to just two nonzero values, but would include arbitrary multiples of $\hbar$, including zero! There is probably nothing that indicates the inherently non-classical value of the electron spin more strongly than its two-valuedness!

Even if these formal objections could somehow be overcome, severe quantitative difficulties remain, arising from the sheer magnitude of the electron spin and of the angular momentum. To appreciate the problem, assume that an electron could be treated as a rigidly rotating charge and mass distribution. For simplicity, assume further that the entire charge is concentrated along the equator of a rotating sphere of radius r. This would lead to a magnetic moment

$$\mu = \tfrac{1}{2} evr, \tag{21•5–1}$$

where v is the rotation speed at the equator. Equating this to the actual magnetic moment μ_B leads to a lower limit for the rotation speed,

$$v = \frac{2\mu_B}{er} = \frac{\hbar}{m_e r}. \tag{21•5–2}$$

Now, it is clear that this speed cannot exceed the speed of light. But this puts a lower limit on the radius of the electron in such a model, namely,

$$r > \frac{\hbar}{m_e c} = \alpha a_0, \tag{21•5–3}$$

where α is again the fine-structure constant.

The experimental evidence against an electron of such a large, finite size is overwhelming. An electron of that size would show up as a large deviation from a pure point-charge Coulomb potential in the Hamiltonian for the interaction of an electron with other charged particles, including other electrons. Such a deviation would be readily observable experimentally. For example, it would lead to a removal of the l-degeneracy in the hydrogen atom, which in turn could be measured with an extremely high accuracy. Now, as we stated earlier, the l-degeneracy is indeed split slightly, but the observed splittings agree to an extremely high degree of accuracy with the splitting calculated from various relativistic effects, *without* invoking any deviations from a pure Coulomb potential itself. In fact, invoking such deviations would destroy the agreement. Without going into any detail, it is clear from the hydrogen spectrum alone that there are no deviations from a pure Coulomb potential for the electron on a scale at least down to the proton itself, about 10^{-13} cm.

Furthermore, scattering experiments with high-energy electrons with de Broglie wavelengths much shorter than this value have not exhibited any deviations down to a much smaller scale, and the empirical data do not appear to contradict most theorists' belief that the electron is a true point charge. What all of this means is that the spin of the electron *cannot* be understood by any "shishkebab" model of a classical solid body rotating about some axis. Spin must be viewed as a true quantum phenomenon *without* a true quantitative classical limit: It simply disappears in the limit $\hbar \to 0$.

21.5.2 Other Spin-1/2 Particles

Although we have stated our entire spin theory in terms of electrons, it is natural to ask whether its success extends to other particles. If so, we would expect an intrinsic magnetic moment for the proton, of a magnitude equal to the **nuclear magneton**,[9]

$$\mu_N = \frac{e\hbar}{2m_p} = \frac{m_e}{m_p}\mu_B, \tag{21•5–4}$$

obtained from the Bohr magneton by simply replacing the electron mass with the proton mass m_p. We would also expect no magnetic moment at all for the neutron. Instead, one finds that for the proton,

$$\mu_p \sim 2.79\ \mu_N, \tag{21•5–5a}$$

and for the neutron,

$$\mu_n \sim -1.91\ \mu_N, \tag{21•5–5b}$$

where the minus sign refers to an opposite alignment of the magnetic moment, relative to the angular momentum.

[9] And with a direction parallel rather than antiparallel to that of the angular momentum.

These discrepancies—particularly the large magnetic moment for the electrically neutral neutron—indicate that both the proton and the neutron are not truly elementary, but are composite particles made up of several more elementary particles that are extremely tightly bound together.[10] By assuming, for example, that the neutron is composed of two or more *different* elementary particles whose charges cancel, but whose magnetic moments do not, it is readily possible to reconcile the existence of a magnetic moment with the absence of a *net* charge, and the deviation of the magnitude of the magnetic moment of the proton from the naive value μ_N is explained similarly. In the case of the proton, more direct evidence exists for an internal structure: If electrons of very high energy ($>10^{10}$ eV) are scattered off protons, and the results are analyzed, one is forced to conclude that there are strong deviations from a simple Coulomb law between the two particles for distances less than about 10^{-13} cm. No such deviations have ever been found for the electron-electron Coulomb interaction. This observation, along with the agreement of the properties of the electron with our simple spin theory, presumably indicate that the electron is a particle that is truly without internal structure and, hence, truly elementary.

The electron and its antiparticle, the positron, are not the only particles that are structure-less by this criterion. Two other classes of such particles are the muon (and the antimuon) and the different neutrinos. The muon is an unstable particle that disintegrates, with a mean life of about 2.2×10^{-6} sec, into an electron and two neutrinos, but otherwise it behaves *exactly* like an electron with about 207 times the mass of the electron. In particular, its magnetic moment is exactly what one would expect from our simple theory, given the different mass, even up to the level of the quantum-electrodynamic corrections to the magnetic moment or the g-factor. It is a nuisance particle, in the sense that it doesn't seem to serve any important role in the theory of elementary particles other than causing theorists to wonder why it is even there. Physics as we know it today would be simpler without the muon, but the muon continues to exist just the same.

The other structureless spin-1/2 particles, the neutrinos, are stable particles without charge and mass, moving inherently with the speed of light, like photons, but interacting only extremely weakly with other forms of matter. They are involved in radioactive β-decay, such as the decay of the neutron or of the muon. In many ways, they are the simplest of all particles. Their discussion, too, goes beyond the scope of this text.

Collectively, the electron, the muon, the neutrinos, and their respective antiparticles are referred to as **leptons**, meaning light particles.

---

[10] It is accepted today that protons, neutrons, and mesons are composed of a more elementary class of particles, called *quarks*. The latter are spin-1/2 particles with fractional charges, $-e/3$ and $+2e/3$, with oppositely charged antiquarks. The proton is assumed to be made up of two $+2e/3$ quarks and one $-e/3$ quark, the neutron of one $+2e/3$ quark and two $-e/3$ quarks. Mesons are assumed to be made up of one quark and one antiquark.

Chapter 22

INDISTINGUISHABLE PARTICLES: FERMIONS AND BOSONS

22.1 THE OCCUPATION NUMBER REPRESENTATION

22.1.1 Multi-Particle Basis States

In our initial treatment of multi-particle systems in chapter 11, we pointed out that a problem arises if the particles are indistinguishable: The numbering of particles in the configuration space formalism then no longer corresponds to any physical reality. Many of the formal solutions of the multi-particle Schroedinger wave equation predict different properties for particles with different "serial numbers," when assigning distinct numbers to indistinguishable particles is in fact as meaningless as assigning serial numbers to the dollars in a bank account. Only those solutions can represent physically admissible states for which all observable properties are invariant under particle exchange; the remainder must be discarded.

This non-physical numbering leads to a more complicated formalism for what should be a simpler problem. What is needed is a formalism that never introduces any unphysical numbering in the first place. Such a formalism, called the **occupation number representation**, is developed in the present chapter. The basic idea is simple: Loosely speaking, a given *multi-particle* state is fully specified by stating which *single-particle* states are occupied.

To develop this formalism, consider the energy eigenstates of a system containing an arbitrary number of *non-interacting* indistinguishable particles. In the absence of any interaction, each particle must then itself be in an energy eigenstate of a single-particle Hamiltonian $\hat{H}$, which is the same for all particles. It is a useful common practice to refer to *single*-particle states as **orbitals**, to distinguish them from overall *system* states. The single-particle energy eigenstates serve as the **basis orbitals** for the description of more complicated states.

Suppose next that we arrange the basis orbitals into some convenient order—it need not be the order of increasing energy—and that we label these orbitals by positive integers

$$k = 1, 2, 3, 4, \text{etc.}, \tag{22•1–1}$$

up to infinity. Given this notation, the overall *system* state is then fully specified by listing which orbitals are "occupied" and with how many particles. We can identify such states unambiguously by designating them in the form of an occupation number string

$$|\Psi\rangle = |N_1, N_2, N_3, N_4, \ldots\rangle, \tag{22•1–2}$$

where each of the N_k is an integer that stands for the number of particles occupying an orbital $|k\rangle$ and the dots following N_4 stand for the orbitals beyond $|4\rangle$, occupied or not.

We shall often abbreviate this notation further by defining

$$|\mathbf{N}\rangle \equiv |N_1, N_2, \ldots, N_k, \ldots\rangle, \tag{22•1–3}$$

where the vector symbol $\mathbf{N}$ is shorthand for the set of all occupation numbers. We are designating that set as a vector, because it can be viewed as the set of components of a vector in an abstract infinite-dimensional space similar to the Hilbert space, in which the different "orthogonal" coordinate axes represent the occupation numbers for the different orbitals. This space is called either the **occupation number space** or the **Fock space**, after the Russian physicist V. Fock, who introduced it.

We postulate that the set of states $|\mathbf{N}\rangle$ defined by (22•1–3) forms a complete set of *basis states* for the formal description of *all* physically admissible multi-particle system states, in exactly the same sense in which the set of orbitals $|k\rangle$ forms a complete set of all physically admissible single-particle states. By *complete,* we mean that any physically admissible system state $|\Psi\rangle$ may be written as a linear superposition of the basis states $|\mathbf{N}\rangle$:

$$|\Psi\rangle = \sum_{\mathbf{N}} c(\mathbf{N})|\mathbf{N}\rangle. \tag{22•1–4}$$

In particular, the states of systems of *interacting* particles may in this way be expressed as a superposition of the basis states, similar to the way the states of a single particle in a potential $V(\mathbf{r})$ may be expressed in terms of a superposition of plane-wave basis states.

The physical interpretation of the expansion coefficients is that

$$P(\mathbf{N}) = |c(\mathbf{N})|^2 \tag{22•1–5}$$

is the probability of "finding" the system in the basis state $|\mathbf{N}\rangle$, provided that the sum over all probabilities is normalized,

$$\sum_{\mathbf{N}} |c(\mathbf{N})|^2 = 1. \tag{22•1–6}$$

The total number of particles contained in the system is, of course,

$$\sum_k N_k = N. \tag{22•1–7}$$

The formalism based on (22•1–2) through (22•1–6) is the **occupation number representation**. It may be used regardless of whether there are any restrictions on the occupation numbers for any of the orbitals. We are, of course, familiar with the Pauli exclusion principle for electrons in its most elementary formulation, which specifies that no "state" (meaning: no single-particle state) can be occupied by more than one electron. Hence, for electrons, the occupation numbers in (22•1–2) are restricted to 0 and 1. The same is true for all other elementary objects with half-integer spin; they are collectively called **fermions**, because in statistical thermodynamics they obey Fermi-Dirac statistics. No restriction on the occupation numbers exists for objects with zero or integer spin, collectively called **bosons**, because they obey Bose-Einstein statistics. Recall from chapter 19 that photons have integer spin and hence are bosons. We shall see later how the distinction between fermions and bosons arises naturally within our formalism.

Some Special Basis States

We will often need to refer to single-particle states in which only one orbital $|k\rangle$ is occupied. In Dirac-Fock vector notation, the occupation number space vectors corresponding to the basis orbitals $|k\rangle$ are the "unit basis vectors in the direction k," which may be written symbolically as $\mathbf{1}_k$, with a bold-faced numeral **1**, and the orbitals themselves may be written as

$$|k\rangle = |\mathbf{1}_k\rangle = |0, 0, 0, \ldots, \underset{\substack{\uparrow \\ \text{orbital } k}}{1}, \ldots\rangle. \tag{22•1–8}$$

Note that in this notation the state $|\mathbf{1}_j + \mathbf{1}_k\rangle$ is a two-particle state in which the orbitals $|j\rangle$ and $|k\rangle$ are occupied.

For bosons, we will also encounter the vectors $\mathbf{N}_k$, corresponding to N_k particles in orbital $|\,k\,\rangle$ and none in any other orbital. In terms of the $\mathbf{N}_k$, we may write (22•1–3) as

$$\mathbf{N} = \sum_k \mathbf{N}_k, \quad |\mathbf{N}\rangle = \left|\sum \mathbf{N}_k\right\rangle, \tag{22•1–9a,b}$$

for both bosons and fermions. Note that in (22•1–9b) the sum symbol should not be pulled out of the Dirac brackets: The notation

$$|\,\Psi\rangle = \sum |\,\mathbf{N}_k\rangle \tag{22•1–10}$$

would correspond, *not* to the state $|\,\mathbf{N}\rangle$ with a sharp number of particles in each orbital, but to a non-normalized equal-weight linear superposition of the states $|\,\mathbf{N}_k\rangle$ which, if normalized, would correspond to a state in which none of the occupation numbers had a sharp value.

A notation we will frequently use is $|\,\mathbf{N} \pm \mathbf{1}_k\rangle$, representing a basis state that has one particle more $(+)$ or less $(-)$ in orbital $|k\rangle$ than the state $|\,\mathbf{N}\rangle$ itself—provided that such a state exists, of course.

Finally, we will often need what is commonly called the **vacuum state**,

$$|\,\mathbf{0}\rangle \equiv |\,0, 0, 0, 0, \ldots\rangle, \tag{22•1–11}$$

a state describing a system that *happens* to be empty. It is a perfectly legitimate system state, not to be confused with the system not being present at all. In effect, we encountered the vacuum state earlier, in our discussion of spontaneous emission, where we argued that an electromagnetic field oscillator that happens to be in its ground state (zero photons) is still perfectly capable of interacting with an electron system and hence behaves quite differently from a system in which the oscillator is not present at all.

22.1.2 Examples of Superposition States

Before proceeding, it is instructive to discuss briefly a variety of possible kinds of superposition states of the form (22•1–4).

Single-Particle States

The occupation number formalism contains, within itself, the ordinary single-particle formalism. As an example, consider the single-particle superposition state

$$|\,\psi\rangle = \alpha|\,1\rangle + \beta|2\rangle \quad \text{with} \quad |\,\alpha\,|^2 + |\,\beta\,|^2 = 1, \tag{22•1–12}$$

written here in ordinary single-particle notation. It corresponds to a single particle in a mixed state, with a probability $|\,\alpha\,|^2$ that the particle will materialize in orbital $|\,1\rangle$ during a suitable measurement, a probability $|\,\beta\,|^2 = 1 - |\,\alpha\,|^2$ that it will materialize in orbital $|\,2\rangle$, and zero probability that it will materialize in any other orbital. In Dirac-Fock multi-particle notation, such a state would be written as

$$|\Psi\rangle = \alpha|\mathbf{1}_1\rangle + \beta|\mathbf{1}_2\rangle = \alpha|1, 0, 0, \ldots\rangle + \beta|0, 1, 0, \ldots\rangle. \tag{22•1–13}$$

Some Two-Particle States

The state

$$|\Psi\rangle = \frac{1}{\sqrt{3}}[|1, 1, 0, \ldots\rangle + |0, 1, 1, \ldots\rangle + |1, 0, 1, \ldots\rangle] \tag{22•1–14}$$

is evidently a two-particle state in which the two particles are randomly distributed over orbitals $|1\rangle$ through $|3\rangle$, with no orbital having multiple occupancy. On the other hand,

$$|\Psi\rangle = \frac{1}{\sqrt{3}}[|2, 0, 0, \ldots\rangle + |0, 2, 0, \ldots\rangle + |0, 0, 2, \ldots\rangle] \tag{22•1–15}$$

is a highly correlated two-particle boson state in which both particles are always in the same orbital, $|1\rangle$ through $|3\rangle$, but it is uncertain in *which* orbital, with a probability of 1/3 for each of the participating three orbitals.

States with an Unsharp Number of Particles

The occupation number representation readily permits the description of states for which the number of particles is not sharp. For example,

$$|\Psi\rangle = \frac{1}{2}[\sqrt{2}\cdot|0, 0, 0, \ldots\rangle + |1, 0, 0, \ldots\rangle + |2, 0, 0, \ldots\rangle] \tag{22•1–16}$$

is a state with a 50% probability of containing no particles at all and a 25% probability each of containing either one or two particles in orbital $|1\rangle$. This means that the formalism is also capable of handling processes in which the number of particles is not a fixed sharp number, but where particles can be annihilated and created. Examples would be the annihilation and creation of photons or of electron-positron pairs. In fact, it was the need to describe such processes that first gave rise to the development of the occupation number formalism during the late twenties, in the framework of the development of quantum electrodynamics and of relativistic quantum field theory. However, the advantages of this formalism are not restricted to processes in which the particles can be annihilated or created; it is quite generally the best existing formalism for the description of systems containing indistinguishable particles.

22.1.3 Inner Products and Orthogonality

For the formal manipulations, we need a generalization of the concept of the inner product between two states to multi-particle system states:

1. Let $|\mathbf{M}\rangle$ and $|\mathbf{N}\rangle$ be two—not necessarily different—multi-particle basis states. We then *define* the inner product

$$\langle \mathbf{M} | \mathbf{N} \rangle = \begin{cases} 1 & \text{if} \quad \mathbf{M} = \mathbf{N}, \\ 0 & \text{if} \quad \mathbf{M} \neq \mathbf{N}. \end{cases} \tag{22•1–17}$$

2. For states that are linear superpositions of basis states of the form (22•1–4), we stipulate that (22•1–17) be applied on a term-by-term basis,

$$\langle \mathbf{M} | \Psi \rangle = \sum_{\mathbf{N}} c(\mathbf{N}) \langle \mathbf{M} | \mathbf{N} \rangle = c(\mathbf{M}), \tag{22•1–18a}$$

with the Hermitian conjugate relation

$$\langle \Psi | \mathbf{N} \rangle = \sum_{\mathbf{M}} c^*(\mathbf{M}) \langle \mathbf{M} | \mathbf{N} \rangle = c^*(\mathbf{N}). \tag{22•1–18b}$$

The expansion coefficients are obtained in the usual way by pre-multiplying with $\langle \mathbf{N} |$ and making use of the orthonormality conditions (22•1–17).

The completeness of the set of basis states $|\mathbf{N}\rangle$ can again be expressed, as in chapter 8, in the form of a closure relation:

$$\sum_{\mathbf{N}} | \mathbf{N} \rangle \langle \mathbf{N} | = 1. \tag{22•1–19}$$

22.2 ANNIHILATION AND CREATION OPERATORS FOR BOSONS AND FERMIONS

22.2.1 Annihilation and Creation Operators

Having set up the "static" framework for the description of multi-particle states, we turn to the formalism for the description of *transitions* between those states. Any form of interaction between the particles will, of course, cause such transitions.

The central idea of the occupation number formalism is to treat all such transitions as *sequences* of elementary transitions during which the occupation of only one orbital at a time is changed by ± 1. As an example, consider a transition from the single-particle basis state $|\mathbf{1}_1\rangle = |1, 0, 0, 0, \ldots\rangle$ to the single-particle basis state $|\mathbf{1}_2\rangle = |0, 1, 0, 0, \ldots\rangle$. We may view this transition in (at least) two ways:

1. We might first create a particle in orbital $|2\rangle$ and then annihilate the particle in orbital $|1\rangle$, with a two-particle state as an intermediate state:

$$|1, 0, 0, 0, \ldots\rangle \to |1, 1, 0, 0, \ldots\rangle \to |0, 1, 0, 0, \ldots\rangle. \qquad (22{\cdot}2\text{–}1a)$$

2. Alternatively, we might first annihilate the particle in orbital $|1\rangle$ and then create a particle in orbital $|2\rangle$, with a zero-particle state as an intermediate vacuum state:

$$|1, 0, 0, 0, \ldots\rangle \to |0, 0, 0, 0, \ldots\rangle \to |0, 1, 0, 0, \ldots\rangle. \qquad (22{\cdot}2\text{–}1b)$$

In fact, if we allow the *temporary* occupation of one or more additional orbitals, an infinite number of sequences can be constructed.

Exercise: The transition from the state $|2, 0, \ldots\rangle$ to $|0, 2, \ldots\rangle$ may be viewed as being performed in a number of different ways. One possible sequence involves first annihilating both particles in orbital $|1\rangle$ and then creating two particles in orbital $|2\rangle$:

$$|2, 0, 0, \ldots\rangle \to |1, 0, 0, \ldots\rangle \to |0, 0, 0, \ldots\rangle \to |0, 1, 0, \ldots\rangle \to |0, 2, 0, \ldots\rangle \qquad (22{\cdot}2\text{–}2)$$

List all other possible sequences that do not involve the temporary occupation of any orbital. How many different such sequences are there?

Each of the individual steps in the above procedure involves either the creation or the annihilation of one particle in one of the orbitals. To describe such single-orbital occupation changes, we define, for each orbital, a Hermitian conjugate pair of operators, called **creation** and **annihilation** operators, modeled on the raising and lowering operators of the harmonic oscillator.

We begin our discussion with the **annihilation operators**. We *define* a set of such operators $\hat{a}_k$ by requesting that, for every physically admissible basis state $|\mathbf{N}\rangle$,

$$\boxed{\hat{a}_k|\mathbf{N}\rangle = A_k(\mathbf{N})|\mathbf{N} - \mathbf{1}_k\rangle,} \qquad (22{\cdot}2\text{–}3a)$$

where $A_k(\mathbf{N})$ is a normalization factor the value of which we will specify later. Written out, (22•2–3a) stands, of course, for

$$\hat{a}_k|N_1, \ldots, N_k, \ldots\rangle = A_k(N_1, \ldots, N_k, \ldots)|N_1, \ldots, N_k - 1, \ldots\rangle \qquad (22{\cdot}2\text{–}3b)$$

If $N_k = 0$, any nonzero value of $A_k(\mathbf{N})$ would imply the generation of an impossible state with a negative occupancy of the orbital $|k\rangle$. Hence, we must impose the truncation constraint

$$A_k(\mathbf{N}) = 0 \quad \text{if} \quad N_k = 0, \qquad (22{\cdot}2\text{–}4)$$

similar to the truncation constraint for the harmonic oscillator. Otherwise, we leave the values of the A's open—at least for now.

Following the definition of Hermitian conjugate operator pairs in chapter 9, the Hermitian conjugates of the annihilation operators can be defined by requiring that, for any *two* admissible basis states $|\mathbf{M}\rangle$ and $|\mathbf{N}\rangle$—which need not be different—we have

$$\langle \mathbf{N} | \hat{a}_k^\dagger | \mathbf{M} \rangle = (\langle \mathbf{M} | \hat{a}_k | \mathbf{N} \rangle)^*. \tag{22•2–5}$$

If we insert the definition (22•2–3a) on the right-hand side of this, we obtain

$$\langle \mathbf{N} | \hat{a}_k^\dagger | \mathbf{M} \rangle = A_k^*(\mathbf{N})(\langle \mathbf{M} | \mathbf{N} - \mathbf{1}_k \rangle)^*, \tag{22•2–6}$$

for every admissible $|\mathbf{M}\rangle$. Assume now that $A_k(\mathbf{N}) \neq 0$, in which case $|\mathbf{N} - \mathbf{1}_k\rangle$ must be an admissible state along with $|\mathbf{N}\rangle$. Evidently, (22•2–6) then implies that

$$\langle \mathbf{N} | \hat{a}_k^\dagger | \mathbf{M} \rangle = \begin{cases} A_k^*(\mathbf{N}), & \text{if} \quad \mathbf{M} = \mathbf{N} - \mathbf{1}_k, \\ 0 \text{ otherwise.} \end{cases} \tag{22•2–7}$$

But this implies, in turn, that

$$\boxed{\hat{a}_k^\dagger | \mathbf{N} - \mathbf{1}_k \rangle = A_k^*(\mathbf{N}) | \mathbf{N} \rangle.} \tag{22•2–8a}$$

The operator $\hat{a}_k^\dagger$ evidently acts as a particle **creation operator** for orbital $|k\rangle$.

So far, our formalism does not require a distinction between bosons and fermions. The way this distinction enters is as follows. Like any operator formalism, the creation and annihilation operator formalism requires the specification of the commutation relations involving pairs of operators. As we shall see shortly, there are two, and only two, distinct possibilities:

Either all annihilation operators, as well as all creation operators, commute,

$$\boxed{\hat{a}_j^\dagger \hat{a}_k^\dagger = +\hat{a}_k^\dagger \hat{a}_j^\dagger \quad \text{and} \quad \hat{a}_k \hat{a}_j = +\hat{a}_j \hat{a}_k,} \tag{22•2–9a,b}$$

where the second relation is simply the Hermitian conjugate of the first.

Or all annihilation operators, as well as all creation operators, anti-commute,

$$\boxed{\hat{a}_j^\dagger \hat{a}_k^\dagger = -\hat{a}_k^\dagger \hat{a}_j^\dagger \quad \text{and} \quad \hat{a}_k \hat{a}_j = -\hat{a}_j \hat{a}_k.} \tag{22•2–10a,b}$$

Furthermore, we shall see that these relations remain valid when $j = k$. Consider, then, the case (22•2–10a) of anti-commuting creation operators for $j = k$:

$$(\hat{a}_k^\dagger)^2 = -(\hat{a}_k^\dagger)^2. \tag{22•2–11}$$

But an operator can be equal to its own negative only if the result of the operation is zero:

$$(\hat{a}_k^\dagger)^2 = 0. \tag{22•2–12}$$

This means that with anti-commuting creation operators, a state with two particles in any orbital can never be created! This conclusion is obvious for the simple case of a two-particle state. To see that it holds generally, consider an M-particle basis state $|\mathbf{N}\rangle$ in which the orbital $|k\rangle$ is already occupied once. Any such state may be expressed in terms of a string of M different creation operators operating on the vacuum state and containing the operator $\hat{a}_k^\dagger$ once somewhere inside that string:

$$|\mathbf{N}\rangle \propto \underbrace{\ldots \hat{a}_k^\dagger \ldots |\mathbf{0}\rangle}_{M\text{ terms}}. \tag{22•2–13}$$

Because of the anti-commutation relation (22•2–10a), the operator $\hat{a}_k^\dagger$ may always be "pulled through" to the leftmost position, with at most a change of sign in the overall result. If we then attempt to create a second particle in orbital $|k\rangle$, by applying the operator $\hat{a}_k^\dagger$ a second time, we obtain a vanishing result.

Evidently, this prohibition of multiple occupancy of any orbital is nothing other than the Pauli exclusion principle for fermions, and we hence identify those particles that obey the anti-commutation relations (22•2–10a,b) as fermions. By default, those particles that obey the commutation relations (22•2–9a,b) are identified as bosons. In our framework, this distinction arises naturally as a consequence of the commutation relations for the annihilation and creation operators.

In terms of the coefficients $A_k(\mathbf{N})$, the prohibition of multiple occupancy for fermions implies that, for them, the lower-end truncation constraint (22•2–4) must be sharpened into the requirement that

$$A_k(\mathbf{N}) = 0, \quad \text{if} \quad N_k \neq 1. \tag{22•2–14}$$

No such additional upper-end truncation occurs for bosons.

It is often convenient to re-label the states and replace $\mathbf{N}$ by $\mathbf{N} + \mathbf{1}_k$, which leads to

$$\boxed{\hat{a}_k^\dagger|\mathbf{N}\rangle = A_k^*(\mathbf{N} + \mathbf{1}_k)|\mathbf{N} + \mathbf{1}_k\rangle.} \tag{22•2–8b}$$

If the particles are fermions, and the orbital $|k\rangle$ is already occupied in $|\mathbf{N}\rangle$, then $|\mathbf{N} + \mathbf{1}_k\rangle$ is not an admissible state, and the result of the operation is zero, in accordance with (22•2–14).

As an example of the use of the annihilation and creation operators, consider the simple case of the two transition sequences (22•2–1a) and (22•2–1b). In terms of the annihilation and creation operators, the sequence (22•2–1a) may be written

$$\hat{a}_1\hat{a}_2^\dagger|1, 0, 0, 0, \ldots\rangle = A_1(1, 1, \ldots) \cdot A_2^*(1, 1, \ldots) \cdot |0, 1, 0, 0, \ldots\rangle, \tag{22•2–15a}$$

implying the application of the creation operator $\hat{a}_2$ to the initial state $|1, 0, 0, 0, \ldots\rangle$, followed by the application of the annihilation operator $\hat{a}_1$ to the result of the first operation. For the sequence (22•2–1b), the two operators are reversed:

$$\hat{a}_2^\dagger \hat{a}_1 |1, 0, 0, 0, \ldots\rangle = A_1(1, 0, \ldots) \cdot A_2^*(0, 1, \ldots) \cdot |0, 1, 0, 0, \ldots\rangle. \tag{22•2–15b}$$

Note that the two sequences involve *formally* different normalization factors, a point to which we will address ourselves next.

22.2.2 Commutation Relations

The question as to the differences—if any—between the normalization factors in (22•2–15a,b) is, of course, related to the commutation relations between the various annihilation and creation operators we have introduced, a central aspect of the algebra of these operators.

We consider first the commutation relation for the creation operators alone. As we noticed in (22•2–13), these operators may be used to express the basis states themselves in terms of a suitable sequence of creation operators operating successively on the vacuum state. Let us look at this procedure in somewhat more detail. As a first step, by applying the various creation operators to the vacuum state, we obtain the various basis orbitals:

$$\hat{a}_k^\dagger |\mathbf{0}\rangle = A_k^*(\mathbf{1}_k)|\mathbf{1}_k\rangle \equiv A_k^*(\mathbf{1}_k)|k\rangle. \tag{22•2–16}$$

By next applying the various creation operators to the various single-particle basis states, all two-particle basis states can be generated and can be expressed as the result of two creation operators operating on the vacuum state successively. For example, the two-particle state $|\mathbf{1}_j + \mathbf{1}_k\rangle$ in which both orbitals $|j\rangle$ and $|k\rangle$ are occupied, may be generated as follows:

$$\hat{a}_k^\dagger \hat{a}_j^\dagger |\mathbf{0}\rangle = A_j^*(\mathbf{1}_j) \cdot \hat{a}_k^\dagger |\mathbf{1}_j\rangle = A_j^*(\mathbf{1}_j) A_k^*(\mathbf{1}_j + \mathbf{1}_k)|\mathbf{1}_j + \mathbf{1}_k\rangle. \tag{22•2–17a}$$

But if we apply the two creation operators in inverse order, we obtain

$$\hat{a}_j^\dagger \hat{a}_k^\dagger |\mathbf{0}\rangle = A_k^*(\mathbf{1}_k) \cdot \hat{a}_j^\dagger |\mathbf{1}_k\rangle = A_k^*(\mathbf{1}_j) A_j^*(\mathbf{1}_j + \mathbf{1}_k)|\mathbf{1}_j + \mathbf{1}_k\rangle, \tag{22•2–17b}$$

which may or may not be the same as (22•2–17a). Any difference between the two cases reflects, of course, the commutation relations for the operators.

The extension of this "bootstrapping" procedure to basis states with more particles is obvious, and it is clear that it is in this way possible to express all basis states by the repeated application of creation operators to the vacuum state. Consider, then, a general $(M + 2)$-particle superposition state of the form

$$|\Phi\rangle = \sum_{j,k} c_{j,k}\hat{a}_j^\dagger \hat{a}_k^\dagger |\mathbf{N}\rangle, \tag{22•2–18}$$

where $|\mathbf{N}\rangle$ is any physically admissible M-particle basis state. The sum goes over all values of j and k independently, including $j = k$. The expansion coefficients are arbitrary, except for the requirement that $|\Phi\rangle$ be normalized.

We next define an **exchange operator** $\hat{X}$ by requesting that it interchange the two creation operators preceding $|\mathbf{N}\rangle$ in each term in (22•2–18):

$$\hat{X}\hat{a}_j^\dagger \hat{a}_k^\dagger |\mathbf{N}\rangle \equiv \hat{a}_k^\dagger \hat{a}_j^\dagger |\mathbf{N}\rangle. \tag{22•2–19}$$

Because each basis state $|\mathbf{N}\rangle$ may itself be written in the form (22•2–13), involving a string of M creation operators, the definition (22•2–19) implies that $\hat{X}$ interchanges the two leftmost such operators in any string of at least two creation operators.

If we now apply $\hat{X}$ to the state $|\Phi\rangle$, the result is just another representation of the same *physical* state, which means that $\hat{X}|\Phi\rangle$ can differ from $|\Phi\rangle$ at most by a constant,

$$\hat{X}|\Phi\rangle = \xi|\Phi\rangle. \tag{22•2–20}$$

Put differently, $|\Phi\rangle$ must be an eigenstate of the exchange operator $\hat{X}$.

Because $\hat{X}$ is a cyclic operator of order 2, its eigenvalues must satisfy

$$\xi = \pm 1. \tag{22•2–21}$$

Hence, under exchange of the operator sequence, $|\Phi\rangle$ must go over either into itself or into its negative. But with the different expansion coefficients $c_{j,k}$ in (22•2–18) being arbitrary, except for the normalization constraint, we conclude that each operator pair in the expansion (22•2–18) must itself satisfy (22•2–20), *with the same exchange eigenvalue for all operator pairs*. That is, we must have either

$$\hat{a}_j^\dagger \hat{a}_k^\dagger |\mathbf{N}\rangle = \hat{a}_k^\dagger \hat{a}_j^\dagger |\mathbf{N}\rangle \tag{22•2–22a}$$

for all operator pairs, or

$$\hat{a}_j^\dagger \hat{a}_k^\dagger |\mathbf{N}\rangle = -\hat{a}_k^\dagger \hat{a}_j^\dagger |\mathbf{N}\rangle \tag{22•2–22b}$$

for all pairs. Which of the two cases applies *cannot* depend on the particular pair of operators.

Furthermore, it cannot depend on the state $|\mathbf{N}\rangle$: The entire argument remains applicable if we replace the basis state $|\mathbf{N}\rangle$ everywhere with an arbitrary superposition state $|\Psi\rangle$ of the form (22•1–4), which need not even have a sharp number of particles. The resulting states $|\Phi\rangle$ must again satisfy (22•2–20) with $\xi = \pm 1$. But this will be the case for arbitrary $|\Psi\rangle$ only if we have *always* $\xi = +1$ or *always* $\xi = -1$.

We may therefore drop the state $|\mathbf{N}\rangle$ from (22•2–22a,b) altogether and write these relations as holding for the operators themselves:

$$\boxed{\hat{a}_j^\dagger \hat{a}_k^\dagger = \xi \hat{a}_k^\dagger \hat{a}_j^\dagger.} \tag{22•2–23}$$

Here we have always $\xi = +1$ or always $\xi = -1$. Which of the two cases applies can only depend on the particle species.

Hence, our formalism predicts that there may be two kinds of particles. Both possibilities occur in nature: Those with $\xi = +1$ are called bosons, and those with $\xi = -1$ are called fermions, with the commutation relations (22•2–9a) and (22•2–10a) given earlier.

The commutation relations for the annihilation operators are simply the Hermitian conjugates of those for the creation operators. They, too, were already given earlier, in (22•2–9b) and (22•2–10b). Our argument leaves open the "mixed" commutation relations involving one creation and one annihilation operator. We will turn to those shortly.

22.2.3 Occupation Number Operator

Apart from the truncation requirement (22•2–4) and from whatever constraints are imposed by the commutation relations, we have so far left open the specifications of the coefficients $A_k(\mathbf{N})$ in the definitions of the creation and annihilation operators. We now fill in this gap.

Recall that in the case of the harmonic oscillator, the product $\hat{a}^\dagger\hat{a}$ serves as **occupation number operator**. We request that the same be true for the annihilation and creation operators, in the sense that the operator $\hat{a}_k^\dagger\hat{a}_k$ serve as the occupation number operator $\hat{N}_k$ for orbital $|k\rangle$:

$$\boxed{\hat{N}_k|\mathbf{N}\rangle \equiv \hat{a}_k^\dagger\hat{a}_k|\mathbf{N}\rangle = N_k|\mathbf{N}\rangle.} \tag{22•2–24}$$

If we use (22•2–8a) to operate on (22•2–3a), we obtain

$$\boxed{\hat{a}_k^\dagger\hat{a}_k|\mathbf{N}\rangle = |A_k(\mathbf{N})|^2|\mathbf{N}\rangle.} \tag{22•2–25}$$

Taken together, (22•2–24) and (22•2–14) require the normalization

$$|A_k(\mathbf{N})|^2 = N_k, \tag{22•2–26a}$$

or, assuming real values for the A_k's,

$$A_k(\mathbf{N}) = \pm\sqrt{N_k}. \tag{22•2–26b}$$

Note that this normalization automatically satisfies the downward truncation requirement (22•2–4) and that it is not in conflict with the upward truncation requirement (22•2–14) for fermions, provided $|\mathbf{N}\rangle$ is an admissible state, which requires $N_k \le 1$.

The relations (22•2–26a,b) do not specify the sign. The simplest possibility is to pick the $(+)$-sign for all orbitals,

$$A_k(\mathbf{N}) = +\sqrt{N_k}. \tag{22•2–27}$$

With this choice, the creation operators $\hat{a}_k^\dagger$ evidently commute; hence, this assignment corresponds to bosons, and it is in fact the assignment inevitably made for bosons. Written out, (22•2–3a) and (22•2–8a) become

$$\hat{a}_k|\mathbf{N}\rangle = \sqrt{N_k}\,|\mathbf{N} - \mathbf{1}_k\rangle, \tag{22•2–28a}$$

$$\hat{a}_k^\dagger|\mathbf{N}\rangle = \sqrt{N_k + 1}\,|\mathbf{N} + \mathbf{1}_k\rangle. \tag{22•2–28b}$$

The case of fermions is more complicated. Because the fermions operators must anti-commute, some of the normalization factors must be negative. Furthermore, the sign of $A_k(\mathbf{N})$ in (22•2–26b) cannot be a function of the occupation number N_k alone (this would again lead to commuting operators), but must depend on the occupation numbers of *other* orbitals contained in $|\mathbf{N}\rangle$.

The simplest sign convention leading to the correct commutation relation is the following. Let ν_k be the number of occupied orbitals $|j\rangle$ that precede $|k\rangle$, in whatever ordering scheme has been adopted for the orbitals:

$$\nu_k = \sum_{j<k} N_j. \tag{22•2–29}$$

If ν_k is an even number, we pick the (+)-sign in (22•2–26b), otherwise a (−)-sign. Because the values of N_k are restricted to 0 and 1, we may drop the square root in (22•2–26b) and write simply

$$A_k(\mathbf{N}) = (-1)^{\nu_k} N_k, \tag{22•2–30a}$$

and

$$A_k(\mathbf{N} + \mathbf{1}_k) = (-1)^{\nu_k}(1 - N_k). \tag{22•2–30b}$$

If we insert these relations into (22•2–3a) and (22•2–8b), we obtain the explicit expressions

$$\hat{a}_k|\mathbf{N}\rangle = (-1)^{\nu_k} N_k\,|\mathbf{N} - \mathbf{1}_k\rangle, \tag{22•2–31a}$$

$$\hat{a}_k^\dagger|\mathbf{N}\rangle = (-1)^{\nu_k}(1 - N_k)\,|\mathbf{N} + \mathbf{1}_k\rangle. \tag{22•2–31b}$$

Exercise: Consider two fermion creation operators $\hat{a}_m^\dagger$ and $\hat{a}_n^\dagger$, with $m < n$, operating successively on a basis state $|\mathbf{N}\rangle$ for which

$$\nu_m = \sum_{j<m} N_j \le \nu_n = \sum_{j<n} N_j. \tag{22•2–32}$$

Show that the choice (22•2–30) satisfies the anti-commutation relation (22•2–10a) for fermions.

22.2.4 Mixed Commutation Relations

We finally turn to the commutation relations for products of one annihilation and one creation operator. Our point of departure is our requirement (22•2–24) that the operator $\hat{N}_k = \hat{a}_k^\dagger \hat{a}_k$ be the occupation number operator for orbital $|k\rangle$ and that $\hat{a}_k^\dagger$ and $\hat{a}_k$ serve as upward and downward stepping operators for $\hat{N}$, with steps of ± 1. But in that case, the three operators must satisfy the general commutation relations for unit stepping operators, introduced in chapter 9:

$$[\hat{N}_k, \hat{a}_k^\dagger] = +\hat{a}_k^\dagger, \qquad [\hat{N}_k, \hat{a}_k] = -\hat{a}_k. \tag{22•2–33a,b}$$

The reader may confirm easily that the assignments (22•2–3), (22•2–28a,b), and (22•2–31a,b) satisfy these requirements.

However, if the occupation number operator $\hat{N}_j$ and the annihilation or creation operators $\hat{a}_k$ or $\hat{a}_k$ refer to different orbital $|j\rangle \neq |k\rangle$, then $\hat{N}_j$ commutes with $\hat{a}_k$ or $\hat{a}_k^\dagger$: Because every basis state is an eigenstate of $\hat{N}_j$, we find easily that in this case,

$$[\hat{N}_j, \hat{a}_k^\dagger]|\mathbf{N}\rangle = 0, \qquad [\hat{N}_j, \hat{a}_k]|\mathbf{N}\rangle = 0. \tag{22•2–34a,b}$$

This can be true for *all* basis states only if the operators commute when they refer to *different* orbitals:

$$[\hat{N}_j, \hat{a}_k^\dagger] = 0, \qquad [\hat{N}_j, \hat{a}_k] = 0. \tag{22•2–35a,b}$$

The relations (22•2–33a) and (22•2–35a) may be combined into

$$\hat{N}_j \hat{a}_k^\dagger - \hat{a}_k^\dagger \hat{N}_l = \hat{a}_j^\dagger \delta_{jk}, \tag{22•2–36}$$

with an analogous relation for (22•2–33b) and (22•2–35b).

Inserting (22•2–34) and drawing on the exchange relation (22•2–23) converts (22•2–36) into

$$\hat{a}_j^\dagger \hat{a}_j \hat{a}_k^\dagger - \xi \hat{a}_j^\dagger \hat{a}_k^\dagger \hat{a}_j = \hat{a}_j^\dagger \delta_{jk}, \tag{22•2–37}$$

which may also be written

$$\hat{a}_j^\dagger [\hat{a}_j \hat{a}_k^\dagger - \xi \hat{a}_k^\dagger \hat{a}_j - \delta_{jk}] = 0. \tag{22•2–38}$$

This has a solution only if the square bracket vanishes, which leads finally to

$$\boxed{\hat{a}_j \hat{a}_k^\dagger - \xi \hat{a}_k^\dagger \hat{a}_j = \delta_{jk}.} \tag{22•2–39}$$

The boson commutation relations contained in (22•2–23) and (22•2–39) for $\xi = +1$ can be expressed in terms of the conventional commutator brackets as

$$\boxed{[\hat{a}_j^\dagger, \hat{a}_k^\dagger] = [\hat{a}_k, \hat{a}_j] = 0, \qquad [\hat{a}_j, \hat{a}_k^\dagger] = \delta_{jk}.} \tag{22•2–40a,b}$$

The corresponding relations for fermions ($\xi = -1$) are sometimes written in a similar way by introducing **anti-commutator brackets** $\{\ldots\}$ defined via[1]

$$\{\hat{A}, \hat{B}\} \equiv \hat{A}\hat{B} + \hat{B}\hat{A}. \tag{22•2–41}$$

In terms of these, the fermion anti-commutation relations are

$$\boxed{\{\hat{a}_j^\dagger, \hat{a}_k^\dagger\} = \{\hat{a}_k, \hat{a}_j\} = 0, \qquad \{\hat{a}_j, \hat{a}_k^\dagger\} = \delta_{jk}.} \tag{22•2–42a,b}$$

The commutation and anti-commutation relations are often referred to *collectively* as commutation relations.

22.2.5 The Effect of the Stepping Operators on Mixed States

In (22•2–3) and (22•2–8b), we defined the stepping operators $\hat{a}_k$ and $\hat{a}_k^\dagger$ mathematically in terms of their action on the basis states $|\mathbf{N}\rangle$. In this case, the operators simply annihilate or create a particle in orbital $|k\rangle$, provided the resulting state is physically admissible. Their action can become surprisingly complex when applied to mixed states that are linear superpositions of several basis states. As an example, consider a state in which three bosons are distributed over two orbitals in such a way that all three particles occupy either orbital $|1\rangle$ or orbital $|2\rangle$:

$$|\Psi\rangle = \frac{1}{\sqrt{2}}[|3, 0, 0, \ldots\rangle + |0, 3, 0, \ldots\rangle]. \tag{22•2–43}$$

The average occupancy of each of the two orbitals is

$$\langle N_1\rangle = \langle N_2\rangle = 3/2. \tag{22•2–44}$$

This is clearly a highly correlated state, quite different from, say,

$$\begin{aligned}|\Phi\rangle = \frac{1}{\sqrt{8}}[&|3, 0, \ldots\rangle + \sqrt{3}|2, 1, \ldots\rangle \\ &+ \sqrt{3}|1, 2, \ldots\rangle + |0, 3, \ldots\rangle],\end{aligned} \tag{22•2–45}$$

which corresponds to a statistically independent distribution of the three particles over the two orbitals.

Suppose now that the annihilation operator $\hat{a}_1$ is applied to $|\Psi\rangle$, leading to

$$\hat{a}_1|\Psi\rangle = \sqrt{\frac{3}{2}}|2, 0, 0, \ldots\rangle. \tag{22•2–46}$$

[1] Some authors write $[\ldots]_-$ and $[\ldots]_+$ instead of $[\ldots]$ and $\{\ldots\}$.

Except for an irrelevant pre-factor, this is the pure state $|2, 0, \ldots\rangle$, with two particles in orbital $|1\rangle$ and none in orbital $|2\rangle$: What has happened is that the action of the operator $\hat{a}_1$ on our highly correlated *mixed* state has removed all contributions to that state from basis states that did not contain any particles in orbital $|1\rangle$ at all, with the net result that the *average* occupancy in orbital $|1\rangle$ amongst the remaining basis states has actually increased, from $\langle N_1 \rangle = 3/2$ to $\langle N_1 \rangle = 2$.

Exercise: What is the effect of $\hat{a}_1^\dagger$ on $|\Psi\rangle$?

It is true that, if $|\Psi\rangle$ is a state with a sharp number of particles, the stepping operators always change the number of particles by exactly one (unless the result of the operation is zero altogether). But to identify the change in particle number caused by $\hat{a}_k$ or $\hat{a}_k^\dagger$ purely with the orbital $|k\rangle$ is an oversimplification if the state is a mixed state. There will, in general, be an indirect "spillover" into other orbitals, with a re-distribution of the particles among the orbitals that are present.

In all fairness, it should be admitted that deviations from the naive expectation as extreme as in our example occur only for highly correlated states. Less extreme examples are given in the problem 22•2–1.

◆ PROBLEM TO SECTION 22.2

#22•2-1: Annihilation Operators and Mixed States

Study the effect of the operators $\hat{a}_1$ and $\hat{a}_1^\dagger$ on the states listed below. Calculate the change in the expectation value of the occupancy of each affected orbital. For those states for which no orbital occupancies larger than 1 are given, calculate the result for both bosons and fermions, and compare.

(a) $\frac{1}{\sqrt{2}}[|1, 0, \ldots\rangle + |0, 1, \ldots\rangle]$.

(b) $\frac{1}{2}[|1, 1, 1, 0, \ldots\rangle + |1, 1, 0, 1, \ldots\rangle + |1, 0, 1, 1, \ldots\rangle + |0, 1, 1, 1, \ldots\rangle]$.

(c) $\frac{1}{\sqrt{2}}[|1, 0, 0, \ldots\rangle + |0, 0, 0, \ldots\rangle]$. (Total number of particles is unsharp.)

22.3 BOSONS, FERMIONS, AND SPIN

22.3.1 Bosons and Fermions

Our deliberations show that the common formulation of the Pauli exclusion principle as a prohibition against multiple occupancy of single-particle electron states is only a special case of the more general principle of the antisymmetry of all fermion state functions under particle exchange. This more general form remains readily applicable to interacting particles, for which the system state function can no longer be expected to be a simple product of single-particle state functions and for which it is therefore not meaningful to speak about a prohibition of multiple occupancy of single-particle states.

The differences between the two classes of particles with $\xi = +1$ and $\xi = -1$ are drastic. There is probably nothing that exhibits this difference more strikingly than the role of the Pauli exclusion principle in causing the electronic shell structure of atoms. This shell structure is responsible for the periodic table of elements and, hence, ultimately for the laws of chemistry. If electrons did not obey the Pauli exclusion principle, these laws would be different beyond recognition. Even though the chemical consequences might be the most obvious ones, they are not the only ones. Other consequences occur in the behavior of electrons in metals and in many other areas of physics.

One pronounced difference between two classes of particles—a difference that has given the two classes of particles their names—concerns their properties in statistical thermodynamics. Consider a quantum system in thermal equilibrium with a thermal reservoir of temperature T and capable of exchanging energy with this reservoir. In such a system, the occupancy N_k of each orbital $|k\rangle$ will fluctuate with time. Nevertheless, this occupancy will exhibit some thermal statistical average $\langle N_k \rangle$. The law for this average occupancy is well known from statistical thermodynamics.[2] For particles that do *not* obey the Pauli exclusion principle ($\xi = +1$), we have

$$\langle N_k \rangle = \frac{1}{\exp[(\mathcal{E}_k - \mu)/k_B T] - 1}, \tag{22•3–1a}$$

and for those that *do* ($\xi = -1$),

$$\langle N_k \rangle = \frac{1}{\exp[(\mathcal{E}_k - \mu)/k_B T] + 1}. \tag{22•3–1b}$$

In both cases, $\mathcal{E}_k$ is the energy of the orbital $|k\rangle$, and μ is the chemical potential for the particle species. The statistical law expressed by the distribution function (22•3–1a) is called **Bose-Einstein statistics**, that expressed by (22•3–1b) **Fermi-Dirac statistics**. It is because of their statistical behavior that the particles with $\xi = +1$ are collectively called bosons, while those with $\xi = -1$ are called fermions.

The question as to what determines which type of statistics a particle satisfies is given by the general principle that

> particles whose spin is an integer multiple of $\hbar$ (including zero), are bosons; those with half-integer spin are fermions.

---

[2] See, for example, C. Kittel and H. Kroemer, *Thermal Physics*, 2d ed. (San Francisco: Freeman, 1979), particularly chapter 7.

The proof of this principle[3]—which had been known empirically long before it was finally derived theoretically—involves relativistic quantum field theory and goes beyond the scope of this text.

Those particles that are the most elementary constituents of ordinary matter, namely, electrons, protons, and neutrons, all have spin 1/2 and are therefore fermions. The simplest boson is the photon. It is clear that the photon *must* be a boson rather than a fermion: A single electromagnetic field mode—the equivalent of a single-particle state for a conventional particle—can contain an arbitrary number of photons. What is not so obvious is the magnitude of the spin associated with the photon. As we saw in chapter 20, the photon is a spin-1 particle. The various mesons are either spin-0 or spin-1 bosons; all of them are unstable and disintegrate into other elementary particles.

The interrelation between spin and statistics applies to composite particles, as well as to elementary particles. Consider the deuteron, a composite particle consisting of two fermions—one proton and one neutron. The spins of the two fermions happen to be parallel rather than anti-parallel, leading to the deuteron's being a spin-1 boson.

It is quite clear that any composite particle that consists of an even (odd) number of elementary fermions must be a boson (fermion), even without considering the relationship to the spin: Interchanging two such composite particles amounts to interchanging an even (odd) number of fermions. Each fermion constituent contributes a factor -1 to the state function of the interchanged composite particle, which leads directly to the rule stated. In the preceding chapter, we mentioned briefly that protons and neutrons are not really elementary particles, but are composed of three quarks each. Evidently, quarks must be fermions.

The different isotopes of the same atomic species frequently differ in their statistics, even though their chemical properties are essentially identical. In the case of the lightest atoms, this leads to noticeable differences in their physical properties, particularly at low temperatures. The most pronounced example is provided by the two stable isotopes of helium, ^{3}He and ^{4}He. The more common ^{4}He atom consists of six fermions: two protons, two neutrons, and two electrons. The spins all cancel each other, leading to a spin-0 boson. By contrast, the ^{3}He atom contains one neutron less; it is a spin-1/2 fermion. At low temperatures, the properties of the two substances differ drastically: ^{4}He becomes a superfluid below 2.14 K, while ^{3}He remains an ordinary fluid down to much lower temperatures. In fact, below about 0.87 K liquid ^{3}He and liquid ^{4}He are not even miscible in arbitrary proportions: They behave somewhat like oil and water.[4]

---

[3] W. Pauli, "The connection between spin and statistics," *Phys. Rev.*, Vol. 58, 1940, pp. 716–722.

[4] See again C. Kittel and H. Kroemer, *Thermal Physics, loc. cit.*, chapter 11.

22.3.2 Configuration Space Re-Visited: The Case of Two Electrons

The formalism developed in section 22.2 remains applicable when the particles have spin, in which case the basis orbitals $|k\rangle$ are spinor rather than scalar wave functions. For fermions, this is necessarily so, and we discuss here in some detail the simplest case, that of two spin-1/2 fermions, such as two electrons. This discussion should enable the reader to extend the treatment to more complicated situations when needed.

In practice, one almost always chooses the basis orbitals to be eigenspinors of the spin operator $\hat{\sigma}_z$. If this is done, the entire formalism can be expressed in a pseudo-scalar form, without explicitly introducing the multicomponent nature of spinors. We can then write each spinor orbital as the product of a scalar Schroedinger wave function ψ and one of the unit spinors $\uparrow$ or $\downarrow$ introduced in chapter 21. As an example, we might write, in obvious notation,

$$|j,\uparrow\rangle \triangleq \psi_j(\mathbf{r})\cdot\uparrow, \qquad |k,\downarrow\rangle \triangleq \psi_k(\mathbf{r})\cdot\downarrow, \tag{22•3–2a}$$

where we have split the orbital designator inside the ket brackets into a letter designating the spatial distribution of the orbital and a spin designator. We refer to the spatial part ψ_j of an orbital such as $|j,\uparrow\rangle$ or $|j,\downarrow\rangle$ as a **space orbital**.

The antisymmetric state function for two particles, one in each of the two orbitals in (22•3–2a), is then

$$\overline{\Psi}_{j\uparrow k\downarrow}(\mathbf{r}_1,\mathbf{r}_2) = \frac{1}{\sqrt{2}}[\psi_j(\mathbf{r}_1)\psi_k(\mathbf{r}_2)\cdot\uparrow_1\downarrow_2 - \psi_j(\mathbf{r}_2)\psi_k(\mathbf{r}_1)\cdot\uparrow_2\downarrow_1]. \tag{22•3–3a}$$

The subscripts 1 and 2 attached to the unit spinors $\uparrow$ and $\downarrow$ designate with which particle the spin is associated: The term $\uparrow_1$ implies that the particle with spin up is that particle whose configuration space coordinate is $\mathbf{r}_1$—that is, particle 1—and analogously for $\uparrow_2$, $\downarrow_1$, and $\downarrow_2$.

In (22•3–3a), the spin-up particle is the one whose *spatial* distribution is given by ψ_j, and the spin-down particle corresponds to ψ_k. In other words, in (22•3–3a) a given spin orientation is correlated with a given orbital, not with the artificial particle numbering. If we were interested in the opposite correlation between space orbitals and spins, we would have to consider the orbitals

$$|k,\uparrow\rangle = \psi_k(\mathbf{r})\cdot\uparrow \quad\text{and}\quad |j,\downarrow\rangle = \psi_j(\mathbf{r})\cdot\downarrow. \tag{22•3–2b}$$

The corresponding two-particle state function would be

$$\overline{\Psi}_{k\uparrow j\downarrow}(\mathbf{r}_1,\mathbf{r}_2) = \frac{1}{\sqrt{2}}[\psi_k(\mathbf{r}_1)\psi_j(\mathbf{r}_2)\cdot\uparrow_1\downarrow_2 - \psi_k(\mathbf{r}_2)\psi_j(\mathbf{r}_1)\cdot\uparrow_2\downarrow_1]. \tag{22•3–3b}$$

If both space orbitals are the same—that is, if $j = k$—the state functions in both (22•3–3a) and (22•3–3b) reduce to

$$\overline{\Psi}_{k\uparrow k\downarrow}(\mathbf{r}_1, \mathbf{r}_2) = \frac{1}{\sqrt{2}} \psi_k(\mathbf{r}_1)\psi_k(\mathbf{r}_2)[\uparrow_1\downarrow_2 - \uparrow_2\downarrow_1]. \qquad (22•3–4)$$

Note that this state achieves its anti-symmetry under particle exchange by having an anti-symmetric spin part, while the spatial part is symmetric.

If both space orbitals are different—that is, if $j \neq k$—then (22•3–3a) and (22•3–3b) refer to distinguishable states that differ in the way the two opposite spin orientations are correlated with the two space orbitals ψ_j and ψ_k. However, many of the physical properties of these two states will be essentially the same as if there were no spin-space correlations. For a description of those properties, it is useful to use an alternative pair of states, constructed from the states (22•3–3a) and (22•3–3b) by linear superposition:

$$\overline{\Phi}_{\pm} = \frac{1}{\sqrt{2}}[\overline{\Psi}_{j\uparrow k\downarrow} \pm \overline{\Psi}_{k\uparrow j\downarrow}]$$

$$= \frac{1}{2}[\psi_j(\mathbf{r}_1)\psi_k(\mathbf{r}_2) \pm \psi_j(\mathbf{r}_2)\psi_k(\mathbf{r}_1)] \cdot [\uparrow_1\downarrow_2 \mp \uparrow_2\downarrow_1]. \qquad (22•3–5)$$

Although these new state functions still describe states of two particles with antiparallel spins in the two different[5] space orbitals ψ_j and ψ_k, they no longer associate a specific spin with a specific orbital. Instead, they have the useful property that they are factored into a single product of a pure position function and a pure spin factor. Note that the spatial part of Φ_+ is symmetric under particle exchange, while the spin part is antisymmetric. For Φ_-, the opposite holds.

This case of two particles with opposite spin in two different space orbitals is sufficiently important that it deserves additional comments. The spatial behavior of the two states Φ_+ and Φ_- is quite different. For example, when $\mathbf{r}_1 = \mathbf{r}_2$ $(= \mathbf{r})$, we have

$$\overline{\Phi}_+ (\mathbf{r}, \mathbf{r}) = [\psi_j(\mathbf{r})\psi_k(\mathbf{r})] \cdot [\uparrow_1\downarrow_2 - \uparrow_2\downarrow_1], \qquad (22•3–6)$$

while

$$\overline{\Phi}_-(\mathbf{r}, \mathbf{r}) = 0. \qquad (22•3–7)$$

In the spatially antisymmetric state Φ_-, the probability of "finding" the two particles very close together vanishes. In this state, the motions of the two particles are correlated in such a way that the particles avoid each other, even in the absence of any repulsive potential between them. Less obviously, the particle motions in the spatially symmetric state Φ_+ are also correlated, but in

[5] Note that the requirement $j \neq k$ is essential: For $j = k$, the spatially antisymmetric combination in (22•3–5) vanishes, and the spatially symmetric one would be incorrectly normalized.

the opposite sense. If we calculate, from (22•3–6), the probability density for finding the particles in the same volume element, we obtain

$$\overline{\Phi}_+^\dagger \overline{\Phi}_+ = |\psi_j(\mathbf{r})|^2 |\psi_k(\mathbf{r})|^2 \cdot [\uparrow_1\downarrow_2 - \uparrow_2\downarrow_1]^\dagger[\uparrow_1\downarrow_2 - \uparrow_2\downarrow_1]. \tag{22•3–8}$$

If the spinor product is multiplied out and spinors of the same particle are grouped together, we can write the result, in a self-explanatory abbreviated notation, as

$$\uparrow^\dagger\uparrow_1 \cdot \downarrow^\dagger\downarrow_2 - \uparrow^\dagger\downarrow_1 \cdot \downarrow^\dagger\uparrow_2 - \downarrow^\dagger\uparrow_1 \cdot \uparrow^\dagger\downarrow_2 + \downarrow^\dagger\downarrow_1 \cdot \uparrow^\dagger\uparrow_2. \tag{22•3–9}$$

But for the products referring to the same particle, we have, of course, the orthonormality relations of chapter 21:

$$\uparrow^\dagger\uparrow = \downarrow^\dagger\downarrow = 1, \qquad \uparrow^\dagger\downarrow = \downarrow^\dagger\uparrow = 0. \tag{22•3–10a,b}$$

With these, the two middle terms in (22•3–9) vanish, and the two outer terms add up to 2, leading to

$$\overline{\Phi}_+^\dagger \overline{\Phi}_+ = 2|\psi_j(\mathbf{r})|^2 |\psi_k(\mathbf{r})|^2. \tag{22•3–11}$$

This is twice as large as if the two particles were statistically independent: The particles effectively attract each other, even in the absence of an attractive potential in the Hamiltonian.

The correlations expressed by (22•3–7) and (22•3–11) are called **exchange correlations**. They are a direct consequence of the antisymmetry requirement for fermion state functions. Similar correlations, but in the opposite direction, occur for bosons. We shall see some of the consequences of the exchange correlations in more detail later, especially for fermions.

We close our discussion with a remark about the simple reduction of the product of spinors in (22•3–8) to a pure number in (22•3–11), by means of the orthonormality relations (22•3–10). Our example illustrates an important point about such reductions. The calculation of physically observable properties always involves products of spinors with Hermitian conjugate spinors, and the factors in these products can always be re-arranged in such a way that a spinor for any one particle occurs multiplied with a Hermitian conjugate spinor for the same particle. These pair products can then always be reduced to a pure number through the orthonormality relations (22•3–10).

22.4 NON-INTERACTING PARTICLES: THE HAMILTONIAN AND THE DENSITY OPERATORS

22.4.1 The Hamiltonian for Non-Interacting Particles

We turn now to the question regarding the form the various Hermitian operators take in the occupation number formalism, especially the Hamiltonian of the system. For the present, we continue to assume that the particles are non-interacting. There is then a sharp energy associated with each of the orbitals of the system. If the system is in one of its basis states $|\mathbf{N}\rangle$, the energy of the system has the sharp value

$$\mathcal{E}(\mathbf{N}) = \sum_k \mathcal{E}_k N_k, \tag{22•4–1}$$

where $\mathcal{E}_k$ is the energy of orbital $|k\rangle$.

The evaluation of (22•4–1) requires a knowledge of the occupation numbers N_k contained in the specific state $|\mathbf{N}\rangle$. Hence, (22•4–1) is not yet the energy *operator* for a multi-particle system. To convert it into such an operator, we must replace the occupation numbers N_k by the occupation number operators $\hat{N}_k$ that extract the occupation number N_k from each basis state $|\mathbf{N}\rangle$ according to

$$\boxed{\hat{N}_k|\mathbf{N}\rangle = N_k|\mathbf{N}\rangle = \hat{a}_k^\dagger \hat{a}_k|\mathbf{N}\rangle.} \tag{22•4–2}$$

With this substitution, we obtain

$$\boxed{\hat{\mathcal{H}} = \sum_k \mathcal{E}_k \hat{N}_k = \sum_k \mathcal{E}_k \hat{a}_k^\dagger \hat{a}_k} \tag{22•4–3}$$

as the operator that will extract the energy of every basis state from the state $|\mathbf{N}\rangle$, in the sense that

$$\hat{\mathcal{H}}|\mathbf{N}\rangle = \mathcal{E}(\mathbf{N})|\mathbf{N}\rangle. \tag{22•4–4}$$

But if $\hat{\mathcal{H}}$ is the correct energy operator for every *basis* state, it must also be the correct energy operator for any linear superposition of basis states. This means that the operator $\hat{\mathcal{H}}$ of (22•4–3) is the *system* Hamiltonian in the occupation number formalism. We write it in script to distinguish it from the *single-particle* Hamiltonian $\hat{H}$.

This formalism calls nowhere for an artificial labeling of the particles or for the introduction of a different position coordinate for each particle; the counting by occupied orbitals eliminates such a need. Nor do we even need to specify the number of particles in the system: The sum over all orbitals automatically counts whatever number of particles are present.

22.4.2 Generalized Set of Basis States

In order to be able to write the Hamiltonian for a system of non-interacting particles in the simple form (22•4–3), we had to assume that the basis orbitals $|k\rangle$ are eigenstates of the single-particle Hamiltonian $\hat{H}$ for the system. This is too restrictive: For all but the simplest systems, these eigenstates are not readily obtainable except as approximations themselves. We therefore need to express the Hamiltonian in terms of simpler orbitals.

Because the Hamiltonian (22•4–3) involves products of the operators $\hat{a}_k$ and $\hat{a}_k^\dagger$, our problem is a special case of the general problem of expressing the operators $\hat{a}_k$ and $\hat{a}_k^\dagger$ for one set of basis orbitals in terms of the operators for another set.

We denote the orbitals of an alternative set by attaching primes to the orbital symbols, such as $|k'\rangle$, and for simplicity, we assume that the set of the $|k'\rangle$ is again a discrete set. The two sets can then be expressed in terms of each other via

$$|k\rangle = \sum_{k'} |k'\rangle\langle k'|k\rangle, \qquad |k'\rangle = \sum_{k} |k\rangle\langle k|k'\rangle. \tag{22•4–5a,b}$$

Both transformations involve a unitary transformation matrix whose elements are $\langle k|k'\rangle = (\langle k'|k\rangle)^*$. We next replace the kets $|k\rangle$ in (22•4–5b) by inserting

$$|k\rangle = \hat{a}_k^\dagger|\mathbf{0}\rangle, \tag{22•4–6a}$$

from (22•2–16), with $A_k(\mathbf{1}_k) = 1$:

$$|k'\rangle = \sum_k \hat{a}_k^\dagger|\mathbf{0}\rangle\langle k|k'\rangle = \left[\sum_k \langle k|k'\rangle\hat{a}_k^\dagger\right]|\mathbf{0}\rangle. \tag{22•4–7a}$$

This is again of the form of a creation operator operating on the vacuum state,

$$|k'\rangle = \hat{a}_{k'}^\dagger|\mathbf{0}\rangle, \tag{22•4–6b}$$

analogous to (22•4–6a), but with the transformed creation operator

$$\hat{a}_{k'}^\dagger = \sum_k \langle k|k'\rangle\hat{a}_k^\dagger. \tag{22•4–8a}$$

Evidently, (22•4–8a) defines a set of creation operators $\hat{a}_{k'}^\dagger$ for the new set of orbitals $|k'\rangle$, in terms of the creation operators $\hat{a}_k^\dagger$ for the old set $|k\rangle$.

The transformation (22•4–8a) is easily inverted. By inserting (22•4–6b) into (22•4–5a), we obtain, similarly to (22•4–7a),

$$|k\rangle = \left[\sum_{k'} \langle k'|k\rangle\hat{a}_{k'}^\dagger\right]|\mathbf{0}\rangle, \tag{22•4–7b}$$

which is of the form (22•4–6a), with

$$\hat{a}_k^\dagger = \sum_{k'} \langle k'|k\rangle\hat{a}_{k'}^\dagger. \tag{22•4–8b}$$

The corresponding transformation equations for the annihilation operators are obtained by taking the Hermitian conjugates of (22•4–8a) and (22•4–8b). With the aid of $\langle k'|k\rangle^* = \langle k|k'\rangle$, we obtain

$$\hat{a}_{k'} = \sum_k \langle k'|k\rangle\hat{a}_k, \qquad \hat{a}_k = \sum_{k'} \langle k|k'\rangle\hat{a}_{k'}. \tag{22•4–9a,b}$$

The new operators satisfy the same commutation or anti-commutation relations as the old ones. We show this here only for the mixed commutation relation (22•2–39), leaving the (simpler) remaining cases to the reader. From (22•4–8a), (22•4–9a), and (22•2–39), we obtain

$$\begin{aligned}\hat{a}_{j'}\hat{a}^{\dagger}_{k'} - \xi\hat{a}^{\dagger}_{k'}\hat{a}_{j'} &= \sum_j \sum_k \langle j'|j\rangle\langle k|k'\rangle[\hat{a}_j\hat{a}^{\dagger}_k - \xi\hat{a}^{\dagger}_k\hat{a}_j] \\ &= \sum_j \sum_k \langle j'|j\rangle\langle k|k'\rangle \cdot \delta_{jk} \\ &= \sum_k \langle j'|k\rangle\langle k|k'\rangle = \langle j'|k'\rangle = \delta_{j'k'},\end{aligned} \tag{22•4–10}$$

which is of the same form as (22•2–39).

Exercise: Show that (22•2–23) also remains valid.

From the commutation relations for the operators $\hat{a}^{\dagger}_{k'}$ and $\hat{a}_{k'}$, it is easily shown that the $\hat{N}_{k'}$ satisfy the same commutation relations (22•2–35) and (22•2–33) as the unprimed occupation number operators. But this means that the $\hat{N}_{k'}$ act as the occupation number operators for the new orbitals $|k'\rangle$.

Exercise: Show that (22•2–35) and (22•2–33) remain valid and that this implies that the $\hat{N}_{k'}$ are the occupation number operators for the $|k'\rangle$.

The new (old) occupation number operators are readily expressed in terms of the old (new) stepping operators:

$$\hat{N}_{k'} = \sum_j \sum_k \langle j|k'\rangle\langle k'|k\rangle\hat{a}^{\dagger}_j\hat{a}_k, \tag{22•4–11a}$$

$$\hat{N}_k = \sum_{j'} \sum_{k'} \langle j'|k\rangle\langle k|k'\rangle\hat{a}^{\dagger}_{j'}\hat{a}_{k'}. \tag{22•4–11b}$$

If (22•4–11b) is inserted into the Hamiltonian (22•4–3), we finally obtain the desired Hamiltonian for a system of non-interacting indistinguishable particles, expressed in an arbitrary (discrete) basis:

$$\hat{\mathcal{H}} = \sum_k \sum_{j'} \sum_{k'} \langle j'|k\rangle\mathcal{E}_k\langle k|k'\rangle\hat{a}^{\dagger}_{j'}\hat{a}_{k'} = \sum_{j'} \sum_{k'} H_{j'k'}\hat{a}^{\dagger}_{j'}\hat{a}_{k'}. \tag{22•4–12}$$

Here

$$H_{j'k'} = \sum_k \langle j'|k\rangle\mathcal{E}_k\langle k|k'\rangle = \langle j'|\hat{H}|k'\rangle. \tag{22•4–13}$$

In the last equality, we have drawn on the relation

$$\hat{H} = \sum_k |k\rangle \mathcal{E}_k \langle k|, \tag{22•4–14}$$

expressing the original single-particle Hamiltonian, in terms of outer products of its own set of eigenstates. Inserting (22•4–13) into (22•4–12) and dropping all primes leads to the final form,

$$\boxed{\hat{\mathcal{H}} = \sum_j \sum_k \langle j|\hat{H}|k\rangle \hat{a}_j^\dagger \hat{a}_k.} \tag{22•4–15}$$

Note that in (22•4–15) the orbital labels j and k no longer refer to the original eigenfunctions, which no longer show up of anywhere; $\hat{\mathcal{H}}$ is expressed completely in terms of whatever (discrete) set of basis orbitals we wish to employ.

The result (22•4–15) is readily generalized to operators other than the Hamilton operator. In deriving it, we made use of only one property of the Hamiltonian, namely, that the energy of a system of non-interacting particles is an **additive single-particle property**, in the following sense:

> The energy of the system is the sum of the contributions of the individual particles, and the contribution of each particle depends only on the properties of that particle and does not involve the properties of any other particle.

Note that this would no longer be the case for interacting particles, the discussion of which we will take up in the next section.

The entire formalism remains valid for other additive single-particle properties, such as the potential energy in an *external* potential, the kinetic energy, the angular momentum, etc. If $\hat{A}$ is the single-particle operator corresponding to such a property, and if $\langle j|\hat{A}|k\rangle$ designates the matrix elements of $\hat{A}$ for a single particle in a particular basis of orbitals, then we may write

$$\hat{\mathcal{A}} = \sum_j \sum_k \langle j|\hat{A}|k\rangle \hat{a}_j^\dagger \hat{a}_k, \tag{22•4–16}$$

where $\hat{a}_j^\dagger$ and $\hat{a}_k$ are annihilation and creation operators for the orbitals in this basis. In writing (22•4–16) as a sum rather than an integral, we imply, of course, that the set of basis orbitals is a discrete set. This is the case of interest to us. The reader should have little difficulty transforming the entire formalism to the case where the basis orbitals or the one-particle eigenstates of $\hat{A}$ form a continuum.[6]

[6] For a continuum of states, the Kronecker delta symbols in the commutation relations for the various operators must be replaced by Dirac delta functions.

22.4.3 Particle Density and Density Operators

One of the simplest and most important additive properties of any system is the probability density of particles at a given point, summed over the contributions from all orbitals. In a single-particle system, this density is, of course, simply

$$\rho_1(\mathbf{r}) = \psi^*(\mathbf{r})\psi(\mathbf{r}). \tag{22•4–17}$$

We would like to bring this into the form of an expectation value of a single-particle density *operator* $\hat{\rho}_1(\mathbf{r})$,

$$\rho_1(\mathbf{r}) = \langle\psi|\,\hat{\rho}_1(\mathbf{r})\,|\,\psi\rangle. \tag{22•4–18}$$

By comparing (22•4–17) with (22•4–18), and by recalling that the Dirac brackets in (22•4–18) represent an integration over all space, we see that the desired density operator inside (22•4–18) is simply a three-dimensional Dirac δ-function,

$$\hat{\rho}_1(\mathbf{r}) = \delta(\mathbf{r}' - \mathbf{r}). \tag{22•4–19}$$

Here $\mathbf{r}'$ is the integration variable inside the integral implied by the Dirac brackets in (22•4–18).

If this operator is inserted into (22•4–16) in place of $\hat{A}$, we obtain the multi-particle generalization $\hat{\rho}(\mathbf{r})$ of the density operator:

$$\hat{\rho}(\mathbf{r}) = \sum_j \sum_k \langle j|\,\hat{\rho}_1(\mathbf{r})\,|\,k\rangle \hat{a}_j^\dagger \hat{a}_k. \tag{22•4–20}$$

The matrix element is readily expressed in terms of the Schroedinger wave functions $\psi_j(\mathbf{r})$ and $\psi_k(\mathbf{r})$ of the basis orbitals. If we recall that the Dirac inner product stands for an integration, and use the properties of the Dirac δ-function, we find that

$$\langle j|\,\hat{\rho}_1(\mathbf{r})\,|k\rangle = \psi_j^*(\mathbf{r})\psi_k(\mathbf{r}). \tag{22•4–21}$$

With this,

$$\hat{\rho}(\mathbf{r}) = \sum_j \sum_k \psi_j^*(\mathbf{r})\psi_k(\mathbf{r})\hat{a}_j^\dagger \hat{a}_k. \tag{22•4–22}$$

Although we have written this operator as a function of the position $\mathbf{r}$, it operate, not on wave functions defined in real space, but on state vectors in Fock space. The dependence of $\hat{\rho}$ on $\mathbf{r}$ is a parametric one, in the sense that the result of the action of the operator $\hat{\rho}$ in Fock space depends on the value of the parameter $\mathbf{r}$. Hence, the meaning of the parameter $\mathbf{r}$ in $\hat{\rho}(\mathbf{r})$ is different from that of the independent variable $\mathbf{r}$ in the operators of *single*-particle quantum theory, such as the potential energy "operator" $V(\mathbf{r})$. Instead, the dependence of $\hat{\rho}$ on $\mathbf{r}$ is analogous to the dependence of the occupation number operator $\hat{N}_k$ on the orbital index k. In fact, the density operator $\hat{\rho}(\mathbf{r})$ is nothing but a generalization of the occupation number operator $\hat{N}_k$ to a continuous basis of δ-function orbitals. The only formal difference between $\hat{\rho}(\mathbf{r})$ and the $\hat{N}_k$ is the different normalization that goes hand in hand with the change from a discrete set of orbitals to a continuous one, a point we discussed already in section 13.4. For continuously distributed orbitals, there will be an infinite number of orbitals for a finite range of parameters. This implies that the number of particles in any single orbital is—on the average—infinitesimally small. Hence, the occupation *number* operator must be changed into an occupation *density* operator, measuring the number of particles per unit parameter *interval,* rather than the number having a single sharp value of the parameter.

The density operator becomes particularly simple in a basis of plane waves

$$|k\rangle \triangleq \psi_k(\mathbf{r}) = \frac{1}{\sqrt{\Omega}} \exp(i\mathbf{k} \cdot \mathbf{r}). \tag{22•4–23}$$

Here Ω is a normalization volume established, for example, by imposing periodic boundary conditions, and $\mathbf{k}$ is one of the (discrete) wave vectors allowed by such boundary conditions. For consistency, we continue to write $|k\rangle$ to count the orbitals. The vector $\mathbf{k}$ in the scalar product $\mathbf{k} \cdot \mathbf{r}$ is whatever wave *vector* is associated with the orbital $|k\rangle$; the number k should not be identified with the *magnitude* of the vector $\mathbf{k}$. With (22•4–23), the density operator becomes

$$\hat{\rho}(\mathbf{r}) = \frac{1}{\Omega} \sum_j \sum_k \exp[i(\mathbf{k} - \mathbf{j}) \cdot \mathbf{r})]\hat{a}_j^\dagger \hat{a}_k. \tag{22•4–24}$$

One of the most important uses of the density operator is to re-express various other operators. As an example, consider the operator $\hat{\mathcal{V}}$ for the total potential energy of a system. Setting $\hat{A} = V$ in (22•4–16) yields

$$\hat{\mathcal{V}} = \sum_j \sum_k \langle j|V|k\rangle \hat{a}_j^\dagger \hat{a}_k, \tag{22•4–25}$$

where $V = V(\mathbf{r})$ is the single-particle potential energy. If the single-particle matrix element is expressed as a conventional integral, (22•4–25) can be rewritten as

$$\begin{aligned} \hat{\mathcal{V}} &= \sum_j \sum_k \left[\int \hat{V}(\mathbf{r}) \cdot \psi_j^*(\mathbf{r})\psi_k(\mathbf{r})\, d^3r \right] \hat{a}_j^\dagger \hat{a}_k \\ &= \int \hat{V}(\mathbf{r}) \left[\sum_j \sum_k \psi_j^*(\mathbf{r})\psi_k(\mathbf{r})\hat{a}_j^\dagger \hat{a}_k \right] d^3r = \int \hat{V}(\mathbf{r})\hat{\rho}(\mathbf{r})\, d^3r. \end{aligned} \tag{22•4–26}$$

The last expression has exactly the same form as the classical potential energy of a system with the local particle density $\rho(\mathbf{r})$, except that in (22•4–26) $\hat{\rho}$ is now not the local density itself, but an operator that will extract that density from a Fock space state vector $|\Psi\rangle$. As we shall see, this use of the density operator will be the key to the expression of the mutual energy of interacting particles in the occupation number formalism.

Exercise: Show that the operator $\hat{N}$ for the total number of particles in a system can be written

$$\hat{N} = \sum_k \hat{N}_k = \int \hat{\rho}(\mathbf{r})\, d^3r. \tag{22•4–27}$$

22.4.4 Field Operators

From the definition (22•4–22) of the density operators, one readily sees that the double sum there can be written as the product of two single sums,

$$\boxed{\hat{\rho}(\mathbf{r}) = \hat{\psi}^\dagger(\mathbf{r})\hat{\psi}(\mathbf{r}),} \tag{22•4–28}$$

where the two factor sums

$$\boxed{\hat{\psi}(\mathbf{r}) = \sum_k \hat{\psi}(\mathbf{r})\,\hat{a}_k, \qquad \hat{\psi}^\dagger(\mathbf{r}) = \sum_k \hat{\psi}^\dagger(\mathbf{r})\,\hat{a}_k^\dagger} \tag{22•4–29a,b}$$

are themselves operators. We have written these operators in a notation that resembles that for single-particle wave functions, to bring out the formal similarity between the density operator in (22•4–28) and the single-particle probability density in (22•4–17). The two factor operators form a Hermitian conjugate pair. Furthermore, as linear superpositions of annihilation operators *or* of creation operators, they are themselves operators that annihilate a particle ($\hat{\psi}$) or create one ($\hat{\psi}^\dagger$), or that have a zero result if applied to states in which annihilation or creation of a particle is physically impermissible.

Because the ψ_k's are position-dependent, the operators $\hat{\psi}$ and $\hat{\psi}^\dagger$ are position-dependent, too. That is, the action of these operators on a system state function $|\,\psi\rangle$ leads to a position-dependent result, not because of any position dependence of $|\,\Psi\rangle$—which is impossible[7]—but because of that of $\hat{\psi}$ itself. This is what is meant by the notation $\hat{\psi}(\mathbf{r})$ and $\hat{\psi}^\dagger(\mathbf{r})$. Such operators, whose action is a continuous function of position, are called **field operators**.

To understand the physical meaning of the operator $\hat{\psi}(\mathbf{r})$, and particularly its position dependence, it is useful to re-write (22•4–29) by noticing that in pure Dirac notation the Schroedinger basis orbitals may be written as inner products,

$$\psi_k(\mathbf{r}) = \langle \mathbf{r}\,|\,k\rangle \quad \text{and} \quad \psi_k^*(\mathbf{r}) = \langle k\,|\,\mathbf{r}\rangle, \tag{22•4–30a,b}$$

[7] The $|\,\Psi\rangle$'s are defined in Fock space, not in real space.

where the notation $|\mathbf{r}\rangle$ represents a three-dimensional δ-function orbital corresponding to a particle localized at the position $\mathbf{r}$. If we insert (22•4–30) into (22•4–29), we obtain

$$\hat{\psi}(\mathbf{r}) = \sum_k \langle \mathbf{r}|k\rangle \hat{a}_k, \qquad \hat{\psi}^\dagger(\mathbf{r}) = \sum_k \langle k|\mathbf{r}\rangle \hat{a}_k^\dagger. \tag{22•4–31a,b}$$

But the right-hand sides of (22•4–31a,b) have the same form as those of the basis transformation equations (22•4–9a,b), with $|k'\rangle$ replaced by $|\mathbf{r}\rangle$. This implies that the operators $\hat{\psi}(\mathbf{r})$ and $\hat{\psi}^\dagger(\mathbf{r})$ act as annihilation and creation operators for the δ-function orbitals $|\mathbf{r}\rangle$. We will elaborate on this point shortly.

The field operators satisfy commutation or anti-commutation relations similar to those of the conventional stepping operators. Let $\mathbf{r}$ and $\mathbf{r}'$ be two position vectors, not necessarily different. We then find easily that, for fermions,

$$\{\hat{\psi}(\mathbf{r}'), \hat{\psi}(\mathbf{r})\} = \{\hat{\psi}^\dagger(\mathbf{r}'), \hat{\psi}^\dagger(\mathbf{r})\} = 0, \tag{22•4–32a,b}$$

$$\{\hat{\psi}(\mathbf{r}'), \hat{\psi}^\dagger(\mathbf{r})\} = \delta(\mathbf{r}' - \mathbf{r}), \tag{22•4–32c}$$

where $\delta(\mathbf{r}' - \mathbf{r})$ is the Dirac delta function, not the Kronecker delta as in (22•2–42a,b).

Exercise: Derive (22•4–32a,b,c) from the definitions (22•4–29a,b) and the anti-commutation relations (22•2–42a,b). In the case of the mixed annihilation-creation operator commutation relation (22•4–32c), the derivation also draws on the closure relation (13•3–15). Show also that, for bosons, the following commutation relations hold:

$$[\hat{\psi}(\mathbf{r}'), \hat{\psi}(\mathbf{r})] = [\hat{\psi}^\dagger(\mathbf{r}'), \hat{\psi}^\dagger(\mathbf{r})] = 0, \tag{22•4–33a,b}$$

$$[\hat{\psi}(\mathbf{r}'), \hat{\psi}^\dagger(\mathbf{r})] = \delta(\mathbf{r}' - \mathbf{r}). \tag{22•4–33c}$$

In the case of a plane-wave basis, Eq. (22•4–23), the field operators become simply

$$\hat{\psi}(\mathbf{r}) = \frac{1}{\sqrt{\Omega}} \cdot \sum_k \exp(i\mathbf{k}\cdot\mathbf{r})\hat{a}_k \tag{22•4–34a}$$

and

$$\hat{\psi}^\dagger(\mathbf{r}) = \frac{1}{\sqrt{\Omega}} \cdot \sum_k \exp(-i\mathbf{k}\cdot\mathbf{r})\hat{a}_k^\dagger \tag{22•4–34b}$$

22.4.5 Example: One-Dimensional Fermi Gas

Following (22•4–3a,b), we stated that the operators $\hat{\psi}(\mathbf{r})$ and $\hat{\psi}^\dagger(\mathbf{r})$ act as annihilation and creation operators for the δ-function orbitals $|\mathbf{r}\rangle$. More specifically, this implies that the application of $\hat{\psi}^\dagger(\mathbf{r})$ to the vacuum state $|\mathbf{0}\rangle$ creates a particle at the sharp position **r**. Similarly, the operator $\hat{\psi}(\mathbf{r})$ would annihilate a particle at the sharp position **r**—*if* the state $|\Psi\rangle$ contained such an infinitely sharply localized particle. However, all physically possible states other than the vacuum state are mixed states with respect to the δ-function orbitals for which $\hat{\psi}$ and $\hat{\psi}^\dagger$ act as annihilation and creation operators. As we pointed out already in section 22.2, the action of annihilation and creation operators on such mixed states inevitably spills over onto other basis states.

To illustrate this point for the action of the density and field operators, we consider the ground state of a system of non-interacting fermions in a one-dimensional box of length L, for which the orbitals are simple plane waves

$$\psi_k(x) = \frac{1}{\sqrt{L}} \exp(ikx). \tag{22•4–35}$$

If we assume periodic boundary conditions for the orbitals,

$$\psi(x + L) = \psi(x), \tag{22•4–36}$$

the k's are confined to the values

$$k = 2\pi n/L, \tag{22•4–37a,b}$$

where n is an arbitrary positive or negative integer. For simplicity, we ignore the fact that there are two spin orientations; hence, there is at most one fermion for every allowed value of k. It will be convenient to assume that the number N of fermions (per spin orientation) is odd, i.e., $N = 2M + 1$. Then, in the ground state of the system, all orbitals in the interval

$$-2M\pi/L \le k \le +2M\pi/L \tag{22•4–38}$$

are occupied. This state can be written

$$|\Psi\rangle = \hat{a}^\dagger_{-M} \ldots \hat{a}^\dagger_{+M}|\mathbf{0}\rangle, \tag{22•4–39a}$$

with the Hermitian conjugate

$$\langle\Psi| = \langle\mathbf{0}|\hat{a}_{-M} \ldots \hat{a}_{+M}. \tag{22•4–39b}$$

The state has a uniform particle density $\langle\rho\rangle = \rho_0 = M/L$. To see this, we calculate the expectation value of the density operator:

$$\langle\Psi|\hat{\rho}(x)|\Psi\rangle = \langle\Psi|\hat{\psi}^\dagger(x)\hat{\psi}(x)|\Psi\rangle$$

$$= \frac{1}{L}\sum_j \sum_k \exp[i(k - j)x]\langle\Psi|\hat{a}_j^\dagger \hat{a}_k|\Psi\rangle. \tag{22•4–40a}$$

In obtaining this result, we have used (22•4–34a,b), adapted to the one-dimensional case. The matrix element $\langle\Psi|\hat{a}_j^\dagger\hat{a}_k|\Psi\rangle$ in (22•4–40a) will be nonzero only if $|k\rangle$ is one of the occupied orbitals (22•4–38) and if $|j\rangle = |k\rangle$. In that case, both the exponential factors and the non-vanishing matrix elements become unity, and the double sum becomes N, independent of x, leading to

$$\langle\Psi|\hat{\rho}(x)|\Psi\rangle = N/L. \tag{22•4–40b}$$

"Burning a Hole" into the Uniform Fermi Gas

To illustrate the action of an individual field operator on the state $|\Psi\rangle$, we consider next the state $|\Phi\rangle$ obtained by "burning a hole" into the uniform electron distribution at $x = 0$, by applying the annihilation operator $\hat{\psi}(0)$ to $|\Psi\rangle$:

$$|\Phi\rangle = \frac{1}{\sqrt{\rho_0}}\hat{\psi}(0)|\Psi\rangle, \tag{22•4–41a}$$

$$\langle\Phi| = \frac{1}{\sqrt{\rho_0}}\langle\Psi|\hat{\psi}^\dagger(0). \tag{22•4–41b}$$

The factor $1/\sqrt{\rho_0}$ normalizes this state, as is readily seen by realizing that

$$\langle\Phi|\Phi\rangle = \frac{1}{\rho_0}\langle\Psi|\hat{\psi}^\dagger(0)\hat{\psi}(0)|\Psi\rangle \tag{22•4–42}$$

and by recalling that the matrix element is the same as the expectation value of the density at $x = 0$, which we saw earlier is equal to ρ_0.

We now calculate the (position-dependent) expectation value of the density for this state:

$$\begin{aligned}\langle\Phi|\hat{\rho}(x)|\Phi\rangle &= \frac{1}{\rho_0}\langle\Psi|\hat{\psi}^\dagger(0)\hat{\psi}^\dagger(x)\hat{\psi}(x)\hat{\psi}(0)|\Psi\rangle \\ &= \frac{1}{\rho_0 L^2}\sum_{j'}\sum_{j}\sum_{k}\sum_{k'}\exp[i(k-j)x]\langle\Psi|\hat{a}^\dagger_{j'}\hat{a}^\dagger_j\hat{a}_k\hat{a}_{k'}|\Psi\rangle.\end{aligned} \tag{22•4–43}$$

We see readily that the operation with $\hat{a}_k\hat{a}_{k'}$ on $|\Psi\rangle$ can lead to a non-vanishing result only if $|k\rangle$ and $|k'\rangle$ are two different orbitals, both of which are selected from the occupied orbitals (22•4–38). The subsequent operation with $\hat{a}^\dagger_{k'}\,\hat{a}^\dagger_k$ must restore the original state $|\Psi\rangle$, or else the matrix element will vanish because of orthogonality. This means that, for a given pair $|k\rangle$, $|k'\rangle$, we must have either

$$|j\rangle = |k\rangle \quad\text{and}\quad |j'\rangle = |k'\rangle \tag{22•4–44a}$$

or

$$|j\rangle = |k'\rangle \quad\text{and}\quad |j'\rangle = |k\rangle. \tag{22•4–44b}$$

For those terms that satisfy (22•4–44a), the exponential factor is unity. The operator product inside the matrix element can then be transformed, with the help of the anti-commutation relations:

$$\hat{a}^\dagger_{k'}\hat{a}^\dagger_k\hat{a}_k\hat{a}_{k'} = -\hat{a}^\dagger_{k'}\hat{a}^\dagger_k\hat{a}_{k'}\hat{a}_k = +\hat{a}^\dagger_{k'}\hat{a}_{k'}\hat{a}^\dagger_k\hat{a}_k = \hat{N}_{k'}\hat{N}_k. \tag{22•4–45}$$

Because both $|k'\rangle$ and $|k\rangle$ are singly occupied orbitals, the four-operator matrix element at the end of (22•4–43) simply has the value $+1$.

For those terms in (22•4–43) that satisfy (22•4–44b), the exponential factor is

$$\exp[i(k' - k)x], \tag{22•4–46}$$

and the operator product inside the four-operator matrix element in (22•4–43) can be transformed as follows:

$$\hat{a}^{\dagger}_{k'}\hat{a}^{\dagger}_{k}\hat{a}_{k'}\hat{a}_{k} = -\hat{a}^{\dagger}_{k'}\hat{a}_{k'}\hat{a}^{\dagger}_{k}\hat{a}_{k} = -\hat{N}_{k'}\hat{N}_{k}. \tag{22•4–47}$$

Evidently, that matrix element is now -1. Putting it all together, we obtain

$$\langle\rho(x)\rangle = \frac{1}{\rho_0 L^2}\sum_{k}\sum_{k'}\{1 - \exp[i(k' - k)x]\}. \tag{22•4–48}$$

The double sum can be transformed as follows:

$$\sum_{k}\left\{N - \sum_{k'}\exp[i(k' - k)x]\right\} = N^2 - \left|\sum_{k}\exp[-ikx]\right|^2. \tag{22•4–49}$$

The remaining sum has the value

$$\frac{\sin(N\pi x/L)}{\sin(\pi x/L)} \xrightarrow[x \ll L/\pi]{} N\cdot\frac{\sin(N\pi x/L)}{N\pi x/L} = N\cdot\frac{\sin(\pi x\rho_0)}{\pi x\rho_0}, \tag{22•4–50}$$

where, in the denominator, we have restricted ourselves to distances

$$x \ll L/\pi, \tag{22•4–51}$$

because we are primarily interested in the case of large N, for which the hole burned in the distribution will be small compared to L.

From (22•4–48) through (22•4–50),

$$\langle\rho(x)\rangle = \rho_0 - \frac{1}{\rho_0 L^2}\left[\frac{\sin(\pi x\rho_0)}{\pi x\rho_0}\right]^2. \tag{22•4–52}$$

The density distribution in the vicinity of $x = 0$ is shown in **Fig. 22•4–1**. The annihilation operator $\hat{\psi}(0)$ has "burned a hole" into the fermion distribution in the vicinity of $x = 0$. Directly at $x = 0$, the density is zero.

A good measure for the width of the hole is the location of the first null in the numerator of (22•4–50),

$$\Delta x_0 \approx 1/\rho_0, \tag{22•4–53}$$

which is simply the average distance between particles.

As we shall see, this hole will play an essential role in understanding the interaction between indistinguishable particles, the topic of the next section.

Exercise: Determine the density distribution of the state generated by applying the field *creation* operator $\hat{\psi}^{\dagger}(0)$ to $|\Psi\rangle$.

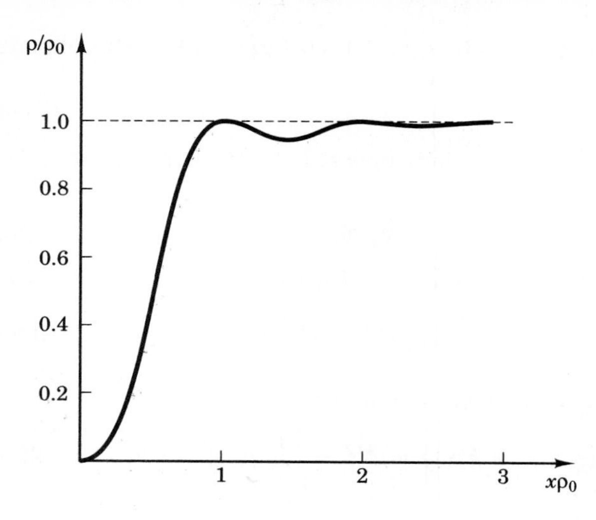

Figure 22•4–1. Density distribution in a one-dimensional Fermi gas after removal of one particle at $x = 0$.

PROBLEM TO SECTION 22.4

22•4-1: Exchange Hole in the Three-Dimensional Fermi Gas

Generalize the discussion of the exchange hole to the three-dimensional Fermi gas.

22.5 INTERACTING PARTICLES

22.5.1 The Interaction Hamiltonian

The introduction of the field and density operators provides us with the formal tools to incorporate the interaction between particles into the overall system Hamiltonian. We will be interested principally in the Coulomb repulsion between electrons, but the basic formalism will cover other forms of interaction, including interactions between bosons.

As our point of departure, we consider briefly how the Coulomb repulsion of a continuous charge distribution is calculated in *classical* electromagnetic field theory. Let $e\rho(\mathbf{r})$ be the classical charge density of a continuous charge distribution,[8] and let us single out two infinitesimal volume elements d^3r and d^3R, located at the positions $\mathbf{r}$ and $\mathbf{R}$. The classical repulsive interaction energy between those volume elements is then

$$d\mathcal{E}_{\text{int}} = \frac{e^2\rho(\mathbf{r})\rho(\mathbf{R})}{4\pi\epsilon_0|\mathbf{r} - \mathbf{R}|}\, d^3r d^3R. \tag{22•5–1}$$

[8] We write $e\rho$ rather than ρ for the space charge density because we would like to identify ρ with the particle density and, later, with the probability density.

The total interaction energy due to the Coulomb repulsion of the charge distribution follows by integration and may be written

$$\mathcal{E}_{\text{int}} = \tfrac{1}{2} \iint V(\mathbf{r} - \mathbf{R})\rho(\mathbf{r})\rho(\mathbf{R})\, d^3r d^3R. \tag{22•5–2}$$

Here

$$V(\mathbf{r} - \mathbf{R}) = \frac{e^2}{4\pi\epsilon_0|\mathbf{r} - \mathbf{R}|}, \tag{22•5–3}$$

and the factor 1/2 preceding the integral compensates for the double counting of the volume element in the double integration. Written in terms of a general pair interaction potential $V(\mathbf{r} - \mathbf{R})$, (22•5–2) applies to arbitrary interaction potentials, not just the Coulomb repulsion.

However, the formulation (22•5–2) fails to account for the finite magnitude of the electron charge and for the fact that an electron does not see its own charge, but only those of the other electrons. Otherwise we would, for example, have to add an energy of the form (22•5–2) even to the energy of a single-electron system, such as the hydrogen atom. This would completely destroy the perfect agreement between theory and experiment for that system and hence cannot be correct. Evidently, in considering the interaction energy of a "test" electron with the rest of a continuous charge distribution, we must first subtract the charge of the test electron from the overall charge distribution.

To allow for the finite magnitude of the electron charge, and for the absence of any self-interaction, we re-formulate the—still classical—problem as follows:

$$\mathcal{E}_{\text{int}} = \tfrac{N}{2} \iint V(\mathbf{r} - \mathbf{R})\rho_1(\mathbf{R})\rho_{N-1}(\mathbf{r}, \mathbf{R})\, d^3r d^3R. \tag{22•5–4}$$

Here $\rho_1(\mathbf{R})$ is the density distribution of a single test particle and $\rho_{N-1}(\mathbf{r}, \mathbf{R})$ the density of the $N - 1$ remaining particles of the system, the latter as a function of position $\mathbf{r}$ and of the position $\mathbf{R}$ of the split-off test particle. The pre-factor $N/2$ accounts for the fact that each of the N particles may serve as test particle and that each particle still gets counted twice, once as a test particle and once as part of the remaining density distribution.

Inasmuch as the particles are indistinguishable, the single-particle density $\rho_1(\mathbf{R})$ is simply $1/N$th of the overall particle density $\rho_N(\mathbf{R})$, and we may simplify (22•5–4) to

$$\mathcal{E}_{\text{int}} = \tfrac{1}{2} \iint V(\mathbf{r} - \mathbf{R})\rho_N(\mathbf{R})\rho_{N-1}(\mathbf{r}, \mathbf{R})\, d^3r d^3R. \tag{22•5–5}$$

The central question is evidently the determination of $\rho_{N-1}(\mathbf{r}, \mathbf{R})$. For a continuous charge density $\rho(\mathbf{r})$, the subtraction of a finite charge e necessarily leads to a reduction in charge density over a nonzero volume in the vicinity of $\mathbf{R}$, and the following question arises: What exactly is the residual charge density distribution attributable to all the other electrons *if* a test electron is located at **R**?

The problem is outside the reach of any classical analogy; it is an inherently quantum-mechanical one, similar to the problem of the exchange hole, discussed in the preceding section. In fact, this similarity is the key to the solution of the problem:

We *postulate* that, in a full quantum-mechanical treatment, the residual charge distribution $\rho_{N-1}(\mathbf{r}, \mathbf{R})$ "seen" by a test particle split off at **R** is the charge density of the state $|\Phi(\mathbf{R})\rangle$ that is obtained by applying the annihilation field operator $\hat{\psi}(\mathbf{r})$, defined in section 22.4, to the original state $|\Psi\rangle$:

$$\rho_{N-1}(\mathbf{r}, \mathbf{R}) = \langle\Phi(\mathbf{R})|\hat{\rho}(\mathbf{r})|\Phi(\mathbf{R})\rangle. \tag{22•5–6}$$

Here

$$|\Phi(\mathbf{R})\rangle = \frac{1}{\sqrt{\rho(\mathbf{R})}}\hat{\psi}(\mathbf{R})|\Psi\rangle. \tag{22•5–7}$$

This is similar to (22•4–41a), except that now the annihilation operator is applied at an arbitrary position **R**, rather than at $\mathbf{R} = 0$. If we insert (22•5–7) for $|\Phi(\mathbf{R})\rangle$ into (22•5–6) and replace the density operator $\hat{\rho}(\mathbf{r})$ by a product of field operators according to (22•4–28), we obtain

$$\rho_{N-1}(\mathbf{r}, R) = \frac{1}{\rho(\mathbf{R})}\langle\Psi|\hat{\psi}^{\dagger}(\mathbf{R})\hat{\psi}^{\dagger}(\mathbf{r})\hat{\psi}(\mathbf{r})\hat{\psi}(\mathbf{R})|\Psi\rangle. \tag{22•5–8}$$

If we next insert this into (22•5–5), the term $\rho_N(\mathbf{R})$ cancels, and we are left with the remarkably simple result

$$\mathcal{E}_{\text{int}} = \tfrac{1}{2}\iint V(\mathbf{r}-\mathbf{R})\cdot\langle\Psi|\hat{\psi}^{\dagger}(\mathbf{R})\hat{\psi}^{\dagger}(\mathbf{r})\hat{\psi}(\mathbf{r})\hat{\psi}(\mathbf{R})|\Psi\rangle\cdot d^3r\cdot d^3R. \tag{22•5–9}$$

We now recall that $|\Psi\rangle$ is a state vector in Fock space, not depending on either **r** or **R**. Hence, we may write (22•5–9) in the form $\langle\Psi|\hat{\mathcal{V}}|\Psi\rangle$, with

$$\hat{\mathcal{V}} = \tfrac{1}{2}\iint V(\mathbf{r}-\mathbf{R})\cdot\hat{\psi}^{\dagger}(\mathbf{R})\hat{\psi}^{\dagger}(\mathbf{r})\hat{\psi}(\mathbf{r})\hat{\psi}(\mathbf{R})\cdot d^3r\cdot d^3R. \tag{22•5–10}$$

This is the desired interaction Hamiltonian. We bring it into a different form by expressing the field operators $\hat{\psi}$ and $\hat{\psi}^{\dagger}$ in terms of the creation and annihilation operators $\hat{a}$ and $\hat{a}^{\dagger}$ for the participating orbitals, via (22•4–29a,b). The result may be written

$$\boxed{\hat{\mathcal{V}} = \tfrac{1}{2}\sum_j\sum_k\sum_m\sum_n\langle j, k|V|m, n\rangle\hat{a}_j^{\dagger}\hat{a}_k^{\dagger}\hat{a}_m\hat{a}_n,} \tag{22•5–11}$$

where

$$\langle j, k|V|m, n\rangle = \iint \psi_j^*(\mathbf{r})\psi_k^*(\mathbf{R}) \cdot V(\mathbf{r} - \mathbf{R}) \cdot \psi_m(\mathbf{R})\psi_n(\mathbf{r}) \cdot d^3r \cdot d^3R. \tag{22•5–12}$$

In defining $\langle j, k|V|m, n\rangle$, we have chosen the order of subscripts in it to be the same as the order of operators in (22•5–11). Note that this requires the order of integration variables to be as given in (22•5–12): One of the two integration variables ($\mathbf{R}$) is shared by the "inner" subscript pair (k, m), the other ($\mathbf{r}$) by the "outer" pair (j, n). Some authors use different conventions, and the reader is urged to ascertain which subscript ordering is used in two-particle matrix elements when encountering such calculations.

22.5.2 The Exchange Energy

Before considering a more general case, we use the formalism to calculate the expectation value of the mutual interaction energy for a general fermion *basis* orbital $|\mathbf{N}\rangle$. From (22•5–11),

$$\langle\mathbf{N}|\hat{\mathcal{V}}|\mathbf{N}\rangle = \tfrac{1}{2}\sum_j\sum_k\sum_m\sum_n\langle j, k|V|m, n\rangle\langle\mathbf{N}|\hat{a}_j^\dagger\hat{a}_k^\dagger\hat{a}_m\hat{a}_n|\mathbf{N}\rangle. \tag{22•5–13}$$

In evaluating this expression, we encounter a situation similar to that in our calculation of the expectation value of the density in (22•4–43). The operation with $\hat{a}_m\hat{a}_n$ on $|\mathbf{N}\rangle$ can lead to a non-vanishing result only if $|m\rangle$ and $|n\rangle$ are two *different* orbitals, both of which are occupied in $|\mathbf{N}\rangle$. The subsequent operation with $\hat{a}_j^\dagger\hat{a}_k^\dagger$ must restore the original state $|\mathbf{N}\rangle$, or else the matrix element will vanish because of orthogonality. This means that for a given pair $|m\rangle$, $|n\rangle$, we must have either

$$|j\rangle = |n\rangle \quad \text{and} \quad |k\rangle = |m\rangle \neq |n\rangle \tag{22•5–14}$$

or

$$|j\rangle = |m\rangle \quad \text{and} \quad |k\rangle = |n\rangle \neq |m\rangle. \tag{22•5–15}$$

The two cases are remarkably different. In (22•5–14), equal orbitals are associated with the same independent variable inside the matrix element (22•5–12), connected via the arrows. We call those terms the **direct interaction** terms. The other terms have one orbital pair exchanged, with each orbital showing up once with one integration variable, once with the other. These terms are called **exchange interaction** terms.

Direct Interaction

For the **direct interaction terms**, which satisfy (22•5–14), the matrix element $\langle j, k|V|m, n\rangle$ in (22•5–12) can be written

$$\begin{aligned}\langle n, m|V|m, n\rangle &= \langle m, n|V|n, m\rangle \\ &= \iint |\psi_n(\mathbf{r})|^2 |\psi_m(\mathbf{R})|^2 \cdot V(\mathbf{r} - \mathbf{R}) \cdot d^3r \cdot d^3R \\ &= \iint \rho_n(\mathbf{r})\rho_m(\mathbf{R}) \cdot V(\mathbf{r} - \mathbf{R}) \cdot d^3r \cdot d^3R. \quad (22{\bullet}5\text{–}16)\end{aligned}$$

But this is nothing other than the classical interaction energy between the two density distributions $\rho_n(\mathbf{r})$ and $\rho_m(\mathbf{R})$. Furthermore, because $m \neq n$, the operator sequence for the direct terms in (22•5–13) becomes

$$\hat{a}_j^\dagger \hat{a}_k^\dagger \hat{a}_m \hat{a}_n = \hat{a}_n^\dagger \hat{a}_m^\dagger \hat{a}_m \hat{a}_n = +\hat{a}_m^\dagger \hat{a}_m \hat{a}_n^\dagger \hat{a}_n = \hat{N}_m \hat{N}_n, \quad (22{\bullet}5\text{–}17)$$

and the direct interaction part of (22•5–13) may be written

$$\langle \mathbf{N}|\hat{\mathcal{V}}|\mathbf{N}\rangle_{\text{dir}} = \sum_m \sum_n \mathcal{E}_{m,n}^{\text{dir}} \langle \mathbf{N}|\hat{N}_m \hat{N}_n|\mathbf{N}\rangle. \quad (22{\bullet}5\text{–}18)$$

Exchange Interaction

For the **exchange interaction** terms, which satisfy (22•5–15), we obtain a very different result,

$$\begin{aligned}\langle m, n|V|m, n\rangle &= \langle n, m|V|n, m\rangle \\ &= \iint \psi_m^*(\mathbf{r})\psi_m(\mathbf{R}) \cdot \psi_n^*(\mathbf{R})\psi_n(\mathbf{r}) \cdot V(\mathbf{r} - \mathbf{R}) \cdot d^3r \cdot d^3R.\end{aligned} \quad (22{\bullet}5\text{–}19)$$

This is not of the form of a quasi-classical interaction between two densities. Although each orbital again shows up twice, the two factors have different independent variables, and their product is no longer the probability density of that orbital as a function of a common position variable. The term (22•5–19) is an inherently non-classical term, called an **exchange energy**.

Furthermore, instead of (22•5–17), we now have

$$\hat{a}_j^\dagger \hat{a}_k^\dagger \hat{a}_m \hat{a}_n = \hat{a}_m^\dagger \hat{a}_n^\dagger \hat{a}_m \hat{a}_n = -\hat{a}_m^\dagger \hat{a}_m \hat{a}_n^\dagger \hat{a}_n = -\hat{N}_m \hat{N}_n, \quad (22{\bullet}5\text{–}20)$$

where we have drawn on the anti-commutation relations for fermions belonging to different orbitals. Combining (22•5–19) and (22•5–4) leads to the exchange contribution to the total interaction energy

$$\langle \mathbf{N}|\hat{\mathcal{V}}|\mathbf{N}\rangle_{\text{exc}} = -\sum_m \sum_n \langle m, n|V|m, n\rangle \langle \mathbf{N}|\hat{N}_m \hat{N}_n|\mathbf{N}\rangle. \quad (22{\bullet}5\text{–}21)$$

Note that the minus sign in front of the exchange energy sum is a consequence of our dealing with fermions; in the case of bosons, the exchange energy would have the opposite sign. The juxtaposition is imperfect, however, because in the case of bosons we could have multiple occupancy of orbitals, and terms with $m = n$ in (22•5–13) could not have been ruled out.

Spin

Our treatment of the exchange interaction ignored the role of spin. When spin is included, we find that the two-particle matrix elements $\langle j, k|V|m, n\rangle$ of (22•5–12) will vanish if the two orbitals at position $\mathbf{r}$ or the two orbitals at position $\mathbf{R}$ have opposite spin. This is of no consequence for the direct interaction, but the exchange interaction terms will vanish if the two particles involved in a term have opposite spin. The sum (22•5–21) will therefore split into two sums, one involving only spin-up orbitals, the other only spin-down orbitals.

22.5.3 Example: The Energy of a Uniform Fermi Gas ("Jellium")

To demonstrate our formalism, we apply it here to one of the simplest cases of a system of interacting particles, the ground state of a gas of interacting electrons. We assume that there are N electrons contained in a box of volume $\Omega = L^3$. To make the system electrically neutral, we also assume that a positive background charge $+Ne$ is present, with a *uniform density* exactly equal to the *average* electronic charge density. Such a system, usually referred to as a **jellium**, is a first approximation to the electrons in a metal, and it gives a good insight into many of the properties of a system of interacting fermions, with a minimum of mathematical complications that are irrelevant to the problems of the electron-electron interaction per se.

We describe the system in terms of plane-wave orbitals of the form

$$|m\rangle = \frac{1}{\sqrt{\Omega}} \cdot \exp(i\mathbf{k}_m \cdot \mathbf{r}), \tag{22•5–22}$$

and we consider *that* particular multi-particle basis state $|\mathbf{N}\rangle$ in which all orbitals up to a certain energy $\mathcal{E}_F$—the **Fermi energy**—are occupied and all orbitals above are empty. In the absence of any interactions, this would be the true ground state of the system.

The energies of the basis orbitals are, of course,

$$\mathcal{E}_m = \frac{\hbar^2 k_m^2}{2m_e}. \tag{22•5–23a}$$

In the ground state, the occupied orbitals form a sphere in **k**-space, the **Fermi sphere**. Its radius k_F is called the **Fermi wave number**; it is related to the Fermi energy via

$$\mathcal{E}_F = \frac{\hbar^2 k_F^2}{2m_e}. \tag{22•5–23b}$$

By imposing periodic boundary conditions on the orbitals (22•5–22), we find that the allowed **k**-vectors form a lattice in **k**-space with spacing $\Delta k = 2\pi/L$.

Hence, the number of orbitals contained within the Fermi sphere, including spin multiplicity, is

$$N = \frac{2}{(\Delta k)^3} \cdot \frac{4\pi}{3} k_F^3 = \frac{\Omega}{3\pi^2} k_F^3. \tag{22•5–24}$$

Insertion into (22•5–23c) gives the Fermi energy as a function of the electron density $n = N/\Omega$:

$$\mathcal{E}_F = \frac{\hbar^2}{2m_e} \cdot (3\pi^2 n)^{2/3}. \tag{22•5–25}$$

This is a central relation in the theory of the degenerate Fermi gas.

We calculate here the expectation value of the shift in energy of the system as a result of the Coulomb repulsion between the electrons and the Coulomb attraction between the electrons and the positive background. As we shall see, the two Coulomb interactions do not cancel out, and exchange effects lead to a substantial lowering of the overall energy.

For plane-wave basis orbitals, the probability density of all orbitals is an orbital-independent constant, and the direct-interaction part of the electron-electron interaction is simply that of the classical self-energy of a uniform-density charge distribution. To this we must add the equally large classical self-energy of the positive background. The sum of the two repulsive self-energies is equal in magnitude, but opposite in sign, to the interaction of the *uniform* negative electron gas with the uniform positive background. Evidently, these three terms cancel, and the expectation value of the total potential energy is given by the exchange interaction alone.

Our task of calculating the total exchange energy consists of two parts: (*i*) calculation of the exchange matrix element (22•5–19) and (*ii*) summing the result.

The Exchange Matrix Element

If the plane-wave orbitals (22•5–23) are inserted into (22•5–19), along with the Coulomb interaction (22•5–3), we obtain, for two orbitals with parallel spin,

$$\langle m, n | V | m, n \rangle = \langle n, m | V | n, m \rangle$$
$$= \frac{1}{\Omega^2} \frac{e}{4\pi\epsilon_0} \iint \frac{1}{|\mathbf{r} - \mathbf{R}|} \exp[i(\mathbf{k}_n - \mathbf{k}_m) \cdot (\mathbf{r} - \mathbf{R})] d^3r d^3R. \tag{22•5–26}$$

The double integral is readily executed; it has the value

$$\iint \cdots = \frac{4\pi\Omega}{(\mathbf{k}_n - \mathbf{k}_m)^2}, \tag{22•5–27}$$

leading to

$$\langle m, n | V | m, n \rangle = \langle n, m | V | n, m \rangle = \frac{1}{\Omega} \cdot \frac{e^2}{\epsilon_0} \cdot \frac{1}{(\mathbf{k}_n - \mathbf{k}_m)^2}. \qquad (22\text{•}5\text{–}28)$$

Summation over All Electron Pairs

We turn to the evaluation of the double sum in (22•5–21). For the system ground state, the matrix elements $\langle \mathbf{N} | \hat{N}_m \hat{N}_n | \mathbf{N} \rangle$ will have the value unity if both orbitals are inside the Fermi sphere and zero otherwise. Hence, if we restrict the sum to orbitals inside the Fermi sphere, we may drop this matrix element.

For a sufficiently large volume, we may replace each single-spin sum by an integral over the Fermi sphere, according to

$$\sum_{\text{one spin}} \cdots \to \frac{1}{(\Delta k)^3} \cdot \int_{k<k_F} \cdots d^3k = \frac{\Omega}{(2\pi)^3} \cdot \int_{k<k_F} \cdots d^3k. \qquad (22\text{•}5\text{–}30)$$

The double sum in (22•5–21) involves two identical double integrals, one for spin up and one for spin down, leading to the overall result

$$\langle \mathbf{N} | \hat{\mathcal{V}} | \mathbf{N} \rangle_{\text{exc}} = -\frac{e^2}{\epsilon_0} \cdot \frac{\Omega}{(2\pi)^6} \iint \frac{d^3k_m d^3k_n}{(\mathbf{k}_n - \mathbf{k}_m)^2}, \qquad (22\text{•}5\text{–}31)$$

where we have inserted (22•5–28) for the two-particle exchange matrix element.

The double integral can again be evaluated in closed form. One finds, after some tedious manipulation, that

$$\iint \frac{d^3k_m d^3k_n}{(\mathbf{k}_n - \mathbf{k}_m)^2} = 4\pi^2 k_F^4, \qquad (22\text{•}5\text{–}32)$$

leading to the final form

$$\langle \mathbf{N} | \hat{\mathcal{V}} | \mathbf{N} \rangle_{\text{exc}} = -\frac{e^2}{\epsilon_0} \cdot \frac{\Omega}{(2\pi)^4} \cdot k_F^4 \cdot \qquad (22\text{•}5\text{–}33)$$

This can be re-formulated in a number of more informative ways; for our purposes, the most useful is the exchange energy *per electron* as a function of the electron density:

$$\mathcal{E}_{\text{exc}} \equiv \frac{1}{N} \langle \mathbf{N} | \hat{\mathcal{V}} | \mathbf{N} \rangle_{\text{exc}} = -\frac{e^2}{\epsilon_0} \cdot \left(\frac{3}{8\pi} \right)^{4/3} \cdot n^{1/3}. \qquad (22\text{•}5\text{–}34)$$

This result is often expressed by expressing the ratio e^2/ϵ_0 in terms of the Rydberg energy and the Bohr radius, via (3•2–9),

$$e^2/\epsilon_0 = 8\pi R_\infty a_0. \qquad (22\text{•}5\text{–}35)$$

If this is inserted into (22•5–34), the result may be written

$$\mathcal{E}_{\text{exc}} = -\left(\frac{3^5}{2^5\,\pi^2}\right)^{1/3} R_\infty \cdot \left(\frac{4\pi}{3} a_0^3 n\right)^{1/3} = -\frac{0.916 R_\infty}{r_s}, \tag{22•5–36}$$

where, in the last equality, we have defined the **radius parameter**

$$r_s = \frac{1}{a_0}\left(\frac{4\pi}{3} n\right)^{-1/3} \tag{22•5–37}$$

This is the radius of a sphere containing one electron, measured in units of the Bohr radius; it is widely used in the description of the energetics of the degenerate electron gas. Evidently, as the density of an electron gas approaches the electron density in the hydrogen atom, the exchange energy gets to be of the order of the Rydberg energy.

Total Energy

The exchange energy is only the net *potential* energy of the degenerate Fermi gas. To obtain the total energy, we must add the kinetic energy of the electrons. It is not difficult to show that the average kinetic energy per electron is $3\mathcal{E}_F/5$. Expressed in terms of the quantity r_s,

$$\begin{aligned}\frac{3}{5}\mathcal{E}_F &= \frac{3}{5} \cdot \frac{\hbar^2}{2m_e} \cdot (3\pi^2 n)^{2/3} = \left(\frac{3^7\,\pi^2}{2^4 5^3}\right)^{1/3} R_\infty \cdot \left(\frac{4\pi}{3} a_0^3 n\right)^{2/3} \\ &= +\frac{2.210 R_\infty}{r_s^2}.\end{aligned} \tag{22•5–38}$$

Combining (22•5–37) and (22•5–38) leads to a total-energy expression:

$$\frac{\mathcal{E}_{\text{total}}}{R_\infty} = \frac{2.210}{r_s^2} - \frac{0.916}{r_s}. \tag{22•5–39}$$

This dependence of the total energy on the dimensionless radius parameter is shown in **Fig. 22•5–1**. For large values of r_s, the negative exchange energy term dominates, corresponding to a binding effect of the exchange energy. For small values, the repulsive effect of the kinetic energy dominates. The total energy has a minimum, $\mathcal{E}_{\min} = -0.095 R_\infty = -1.29$ eV, at a value $r_{\min} = 4.825$, corresponding to a diameter of the one-electron-sphere radius of 0.51 nm, slightly larger than the interatomic distance of typical metals, but of the same order.

The theory we have presented here for the total energy of the jellium system is basically a first-order theory of what holds a metal together. The neglect of the atomic structure of the positive background is, of course, a gross over-simplification. The metals that come closest to this model are the alkali metals, lithium through cesium, and their properties actually straddle the predicted values. Cohesive energies per atom—essentially our $\mathcal{E}_{\min}$—range from -1.63 eV for Li to -0.804 eV for Cs, with Na (-1.113 eV) coming closest to the predicted value of -1.29 eV. Similarly, the radius parameters range from

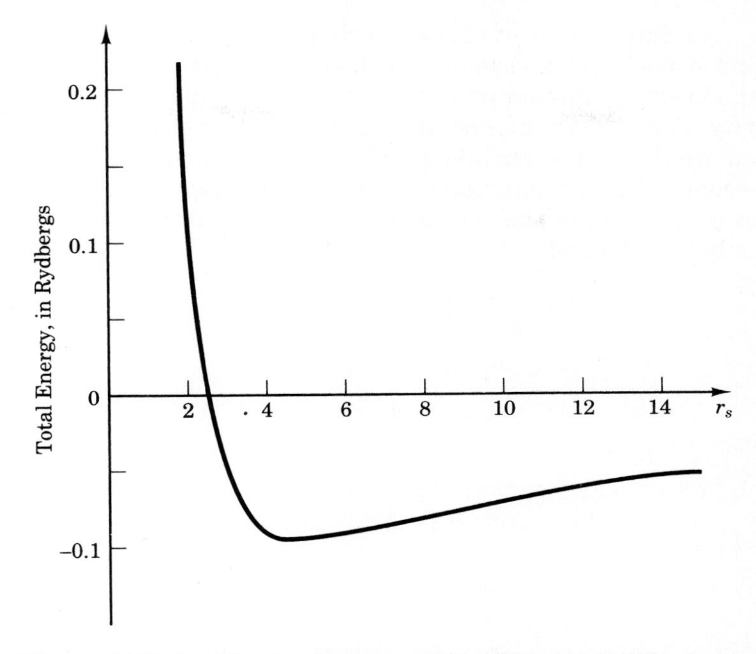

Figure 22•5–1. The total energy of the degenerate Fermi gas, in Rydberg energies, as a function of the radius parameter r_s.

3.25 for Li to 5.63 for Cs; the value for K (4.86) comes within 1% of the predicted value of 4.825.[9]

Note that the minimal energy we have calculated here is not yet the true jellium ground-state energy: Inasmuch as we have used plane-wave orbitals for our estimate, appropriate for non-interacting electrons, our result is simply a variational approximation to the true ground state, which can only be lower. To go beyond our estimate is a much more difficult problem, far beyond the scope of this text, and we must refer the interested reader to the more advanced quantum-mechanics literature.

22.5.4 Epilogue

Our discussion of the exchange interaction in the degenerate Fermi gas brings us to a natural end of any *elementary* treatment of the energetics of systems of interacting indistinguishable particles—and, therefore, of this book. The reader who has followed us up to this point should have little difficulty making the transition to more advanced topics in this huge field, topics that properly belong in more advanced, but also more specialized, texts.

[9] The numerical values listed are taken from C. Kittel, *Introduction to Solid-State Physics*, 6th ed. (New York: Wiley, 1986).

Perhaps the most obvious such application is the physics of the chemical bond, especially the covalent bond, as found, in its simplest form, in the hydrogen molecule. Another important example would be the theory of superconductivity. For readers interested in semiconductor physics, an important application would be the shrinkage of the energy gap of a semiconductor in the presence of high doping concentrations or the reduction in photon energy below the energy gap of the semiconductor in semiconductor lasers. The list could easily be extended.

Appendix 

DIRAC δ-FUNCTION

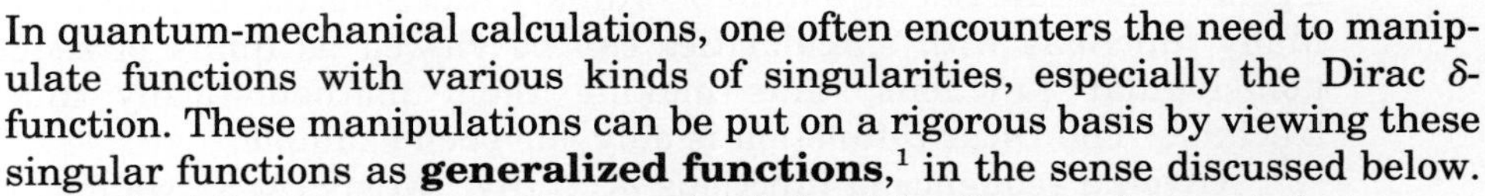

In quantum-mechanical calculations, one often encounters the need to manipulate functions with various kinds of singularities, especially the Dirac δ-function. These manipulations can be put on a rigorous basis by viewing these singular functions as **generalized functions**,[1] in the sense discussed below.

The Dirac δ-function $\delta(x)$ may be defined by the requirement that

$$\int_a^b f(x)\delta(x - x')\,dx = f(x'), \tag{A–1}$$

for any interval $a < x' < b$, no matter how narrow, and for any function $f(x)$ that is continuous and finite in this interval. This definition implies that

$$\delta(x) = 0 \quad \text{for} \quad x \neq 0, \tag{A–2}$$

but no function value can be assigned to $\delta(x)$ for $x = 0$. Simply setting $\delta(0) = \infty$ would still say nothing about the value of the integral in (A–1). Hence, $\delta(x)$ is a function that is not defined in the sense of an ordinary function, by assigning a function value to each argument value. Rather, it is defined in terms of the integral (A–1).

Although no single function in the ordinary sense can be given that satisfies (A–1), it is possible to specify *sequences* of ordinary functions whose limit satisfies (A–1);

$$\lim_{N\to\infty} \int_a^b f(x)\delta_N(x - x')\,dx = f(x'), \tag{A–3}$$

[1] Our treatment is based on the superb little book by M. J. Lighthill, *Introduction to Fourier Analysis and Generalised Functions* (Cambridge: Cambridge Univ. Press, 1958 and later printings).

for any interval $a < x' < b$ and for any function $f(x)$ that is continuous and finite in that interval. The requirement (A–3) does not specify the sequence $\delta_N(x)$; we work here with the Gaussian representation of the δ-function,

$$\delta_N(x) = \sqrt{\frac{N}{2\pi}} \exp(-\tfrac{1}{2}Nx^2). \tag{A–4}$$

Note that $\delta_N(x)$ goes rapidly to zero for $x \neq 0$ as $N \to \infty$ and that the integral in (A–3) goes over into the integral from $-\infty$ to $+\infty$,

$$\int_{-\infty}^{+\infty} \exp(-\tfrac{1}{2}Nx^2)\, dx = \sqrt{\frac{2}{N}} \int_{-\infty}^{+\infty} \exp(-s^2)\, ds = \sqrt{\frac{2\pi}{N}}, \tag{A–5}$$

which cancels the leading factor in (A–4). Note also that all $\delta_N(x)$ are analytical functions in the entire complex x-plane.

Such non-ordinary functions that can be specified as limits of sequences of ordinary functions are referred to as **generalized functions**, or, somewhat non-obviously, as **distributions**. We use the former term here. This concept of a generalized function as a limit of ordinary functions has proven very fruitful. Many functions with singularities can be viewed as limits of sequences of non-singular functions, thus making them mathematically much more tractable. The Dirac δ-function is only one such example.

Given a sequence of functions $\delta_N(x)$ that satisfy (A–3), we may then say that the Dirac δ-function $\delta(x)$ is given by the limit

$$\delta(x) = \operatorname*{ilim}_{N\to\infty} \delta_N(x), \tag{A–6}$$

where the notation "ilim" rather than "lim" denotes that for $x = 0$, $\delta(x)$ has a meaning only under an integral sign, and that the limit is to be performed *after* integration over the argument x in $\delta(x)$. Similar definitions may be given for other generalized functions.

Our results are readily generalized to three dimensions. The three-dimensional Dirac δ-function $\delta(\mathbf{r})$ can be defined analogously to (A–6) as

$$\delta(\mathbf{r}) = \operatorname*{ilim}_{N\to\infty} \delta_N(\mathbf{r}), \tag{A–7}$$

where the δ_N satisfy

$$\lim_{N\to\infty} \int_{\Omega} f(\mathbf{r})\delta_N(\mathbf{r} - \mathbf{r}')\, d^3r = f(\mathbf{r}'), \tag{A–8}$$

with the integration going over any volume Ω that contains $\mathbf{r} = \mathbf{r}'$ and with $f(\mathbf{r})$ being any function that is continuous and finite in that volume. The analog of (A–4) in three dimensions is

$$\delta_N(\mathbf{r}) = \delta_N(x)\delta_N(y)\delta_N(z) = \left(\frac{N}{2\pi}\right)^{3/2} \exp(-\tfrac{1}{2}Nr^2). \tag{A–9}$$

Appendix **B**

POISSON-DISTRIBUTED EVENTS

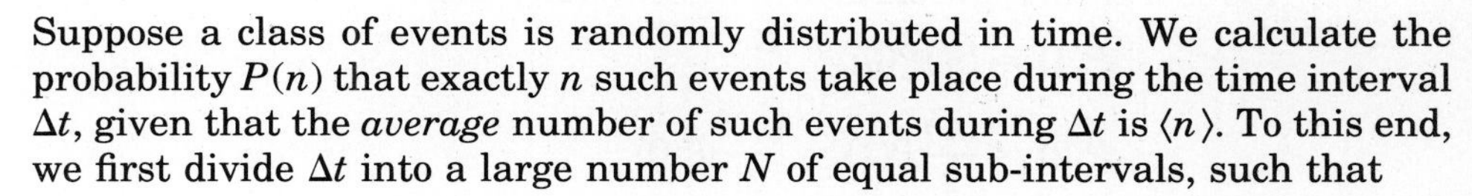

Suppose a class of events is randomly distributed in time. We calculate the probability $P(n)$ that exactly n such events take place during the time interval Δt, given that the *average* number of such events during Δt is $\langle n \rangle$. To this end, we first divide Δt into a large number N of equal sub-intervals, such that

$$N \gg \langle n \rangle. \tag{B–1}$$

The average number of events per sub-interval is then

$$p_0 = \langle n \rangle / N \ll 1. \tag{B–2}$$

If N is chosen large enough, the sub-interval becomes so small that no two events will ever occur in the same sub-interval. In this case, the *probability* that a *single* event takes place inside any specified sub-interval is also given by p_0, and

$$q_0 = 1 - p_0 = 1 - \langle n \rangle / N \tag{B–3}$$

is the probability that *no* event takes place during that sub-interval.

In order for n events to take place during Δt, is is necessary that n sub-intervals contain one event each, while $N - n$ sub-intervals contain none. If the events are statistically independent of each other, the probability that any *specified* combination of n sub-intervals contains exactly n events is then simply

$$p = p_0^n q_0^{N-n}. \tag{B–4}$$

But there are

$$g = \frac{N!}{n!(N-n)!} \tag{B–5}$$

distinguishable ways to select n sub-intervals out of N, and the final probability $P(n)$ must be given by the product $p \cdot g$,

$$P(n) = p_0^n q_0^{N-n} \cdot \frac{N!}{n!(N-n)!}, \tag{B–6}$$

taken in the limit $N \rightarrow \infty$. Note that (B–6) may also be written in the form

$$P(n) = \frac{p_0}{q_0} \frac{N-n+1}{n} P(n-1) = \frac{\langle n \rangle}{N - \langle n \rangle} \frac{N-n+1}{n} P(n-1). \tag{B–7}$$

In the limit $N \rightarrow \infty$, the two terms containing N cancel, and (B–7) becomes

$$P(n) = \frac{\langle n \rangle}{n} P(n-1). \tag{B–8}$$

This is a recursion relation. We see readily that

$$P(1) = \langle n \rangle P(0), \tag{B–9a}$$

$$P(2) = \frac{\langle n \rangle}{2} P(1) = \frac{\langle n \rangle^2}{2!} P(0), \tag{B–9b}$$

$$\vdots$$

$$P(n) = \frac{\langle n \rangle^n}{n!} P(0). \tag{B–10}$$

The sum over all probabilities must add up to unity:

$$\sum_{n=0}^{\infty} P(n) = \left[\sum_{n=0}^{\infty} \frac{\langle n \rangle^n}{n!} \right] \cdot P(0) = e^{\langle n \rangle} \cdot P(0) = 1. \tag{B–11}$$

Hence, we must have $P(0) = \exp(-\langle n \rangle)$, and

$$\boxed{P(n) = \frac{\langle n \rangle^n}{n!} \cdot e^{-\langle n \rangle}.} \tag{B–12}$$

This distribution is called the **Poisson distribution**.

We confirm that the average number of events is indeed $\langle n \rangle$:

$$\sum_{n=0}^{\infty} nP(n) = \sum_{n=1}^{\infty} nP(n) = \langle n \rangle \sum_{n=1}^{\infty} P(n-1) = \langle n \rangle \sum_{m=1}^{\infty} P(m) = \langle n \rangle. \tag{B–13}$$

Here we have first dropped the vanishing term with $n = 0$ from the sum, then used the recursion relation (B–8), and finally substituted $n - 1 = m$ and used (B–11).

The mean of n^2 is obtained similarly:

$$\langle n^2 \rangle = \sum_{n=1}^{\infty} n^2 P(n) = \langle n \rangle \sum_{n=1}^{\infty} nP(n - 1) = \langle n \rangle [\langle n \rangle + 1]. \tag{B–14}$$

The **variance** of the distribution is defined as

$$(\Delta n)^2 = \langle (n - \langle n \rangle)^2 = \langle n^2 \rangle - \langle n \rangle^2. \tag{B–15}$$

By inserting (B–14), we obtain

$$(\Delta n)^2 = \langle n \rangle. \tag{B–16}$$

Taking the square root leads to the **standard deviation**

$$\Delta n = \sqrt{\langle n \rangle}. \tag{B–17}$$

The **relative standard deviation** is given by

$$\frac{\Delta n}{\langle n \rangle} = \frac{1}{\sqrt{\langle n \rangle}}. \tag{B–18}$$

Appendix

SPHERICAL HARMONICS

TABLE OF THE LOWEST-ORDER SPHERICAL HARMONICS

$\boldsymbol{l} = 0$:

$$Y_0^0 = \frac{1}{2}\sqrt{\frac{1}{\pi}} \tag{C–1}$$

$\boldsymbol{l} = 1$:

$$Y_1^0 = +\frac{1}{2}\sqrt{\frac{3}{\pi}} \cos \theta \tag{C–2}$$

$$Y_1^{\pm 1} = \mp\frac{1}{2}\sqrt{\frac{3}{2\pi}} e^{\pm i\phi} \sin \theta \tag{C–3}$$

$\boldsymbol{l} = 2$:

$$Y_2^0 = \frac{1}{4}\sqrt{\frac{5}{\pi}} (3 \cos^2 \theta - 1) \tag{C–4}$$

$$Y_2^{\pm 1} = \mp\frac{1}{2}\sqrt{\frac{15}{2\pi}} e^{\pm i\phi} \sin \theta \cos \theta \tag{C–5}$$

$$Y_2^{\pm 2} = \frac{1}{4}\sqrt{\frac{15}{2\pi}} e^{\pm 2i\phi} \sin^2 \theta \tag{C–6}$$

$\boldsymbol{l} = 3$:

$$Y_3^0 = \frac{1}{4}\sqrt{\frac{7}{\pi}} (5 \cos^3 \theta - 3 \cos \theta) \tag{C–7}$$

$$Y_3^{\pm 1} = \mp\frac{1}{8}\sqrt{\frac{21}{\pi}}e^{\pm i\phi} \sin^2 \theta \, (5 \cos^2 \theta - 1) \tag{C–8}$$

$$Y_3^{\pm 2} = \frac{1}{4}\sqrt{\frac{105}{2\pi}}e^{\pm 2i\phi} \sin^2 \theta \cos \theta \tag{C–9}$$

$$Y_3^{\pm 3} = \mp\frac{1}{8}\sqrt{\frac{35}{\pi}}e^{\pm 3i\phi} \sin^3 \theta \tag{C–10}$$

$l = 4$:

$$Y_4^0 = \frac{3}{16}\sqrt{\frac{1}{\pi}} \, (35 \cos^4 \theta - 30 \cos^2 \theta + 3) \tag{C–11}$$

$$Y_4^{\pm 1} = \mp\frac{3}{8}\sqrt{\frac{5}{\pi}}e^{\pm i\phi} \sin \theta \, (7 \cos^3 \theta - 3 \cos \theta) \tag{C–12}$$

$$Y_4^{\pm 2} = \frac{3}{8}\sqrt{\frac{5}{2\pi}}e^{\pm 2i\phi} \sin^2 \theta \, (7 \cos^2 \theta - 1) \tag{C–13}$$

$$Y_4^{\pm 3} = \mp\frac{3}{8}\sqrt{\frac{35}{\pi}}e^{\pm 3i\phi} \sin^3 \theta \cos \theta \tag{C–14}$$

$$Y_4^{\pm 4} = \mp\frac{3}{16}\sqrt{\frac{35}{2\pi}}e^{\pm 4i\phi} \sin^4 \theta \tag{C–15}$$

Appendix **D**

HYDROGEN RADIAL EIGENFUNCTIONS

TABLE OF THE LOWEST-ORDER HYDROGEN RADIAL EIGENFUNCTIONS

Note: For compactness, the functions $\chi_{nl}(r)$ given below have been expressed in terms of s rather than r. However, their normalization is r-normalization, not s-normalization.

$n = 1$ $(s = 2r/1a_0)$:

$$\chi_{10}(r) = Cs\, L_0^1(s)e^{-s/2} = \frac{1}{\sqrt{a_0}}se^{-s/2} \quad (L_0^1 = 1) \tag{D–1}$$

$n = 2$ $(s = 2r/2a_0)$;

$$\chi_{20}(r) = Cs\, L_1^1\,(s)e^{-s/2} = \frac{1}{2\sqrt{2a_0}}s(2 - s)e^{-s/2} \tag{D–2}$$

$$\chi_{21}(r) = Cs^2 L_0^3\,(s)e^{-s/2} = \frac{1}{2\sqrt{6a_0}}s^2e^{-s/2} \quad (L_0^3 = 1) \tag{D–3}$$

$n = 3$ $(s = 2r/3a_0)$:

$$\chi_{30}(r) = Cs\, L_2^1\,(s)e^{-s/2} = \frac{1}{6\sqrt{3a_0}}s(6 - 6s + s^2)e^{-s/2} \tag{D–4}$$

$$\chi_{31}(r) = Cs^2 L_1^3\,(s)e^{-s/2} = \frac{1}{6\sqrt{6a_0}}s^2(4 - s)e^{-s/2} \tag{D–5}$$

$$\chi_{32}(r) = Cs^3 L_0^5\,(s)e^{-s/2} = \frac{1}{6\sqrt{30a_0}} s^3 e^{-s/2} \quad (L_0^5 = 1) \tag{D–6}$$

$n = 4$ $(s = 2r/4a_0)$:

$$\chi_{40}(r) = Cs\,L_3^1(s)e^{-s/2} = \frac{1}{48\sqrt{a_0}} s(24 - 36s + 12s^2 - s^3)e^{-s/2} \tag{D–7}$$

$$\chi_{41}(r) = Cs^2 L_2^3(s)e^{-s/2} = \frac{1}{16\sqrt{15a_0}} s^2(20 - 10s + s^2)e^{-s/2} \tag{D–8}$$

$$\chi_{42}(r) = Cs^3 L_1^5(s)e^{-s/2} = \frac{1}{48\sqrt{5a_0}} s^3(6 - s)e^{-s/2} \tag{D–9}$$

$$\chi_{43}(r) = Cs^4 L_0^7\,(s)e^{-s/2} = \frac{1}{48\sqrt{35a_0}} s^4 e^{-s/2} \quad (L_0^7 = 1) \tag{D–10}$$

Appendix E

FOURIER INTEGRAL

The Fourier integral is one of the basic mathematical tools of quantum mechanics. The derivations of the Fourier integral given in textbooks often make excessively restrictive assumptions about the functions involved, leaving it unclear to what extent the theorem is applicable to functions containing various kinds of singularites, which often occur in quantum mechanics. It is therefore desirable to put the application of the theorem to such functions on a rigorous basis, drawing on the concept of **generalized functions** introduced in appendix A in the context of the Dirac δ-function.*

Let $f(x)$ be a function that is differentiable at least twice and that, together with its first two derivatives, vanishes sufficiently rapidly at infinity that the integrals

$$\int_{-\infty}^{+\infty} |f|\, dx, \qquad \int_{-\infty}^{+\infty} |f'|\, dx, \qquad \text{and} \qquad \int_{-\infty}^{+\infty} |f''|\, dx$$

are finite. Such a function is clearly a case of a function that can be expressed in terms of itself by means of the Dirac δ-function, as discussed in appendix A:

$$f(x) = \int_{-\infty}^{+\infty} f(x')\delta(x'-x)dx' = \lim_{N\to\infty} \int_{-\infty}^{+\infty} f(x')\delta_N(x' - x)\, dx'. \qquad \text{(E–1)}$$

The Fourier integral theorem is most easily obtained, with a minimum of restrictions, by inserting into (E–1) the following representation of the Dirac δ-function:

$$\delta_N(x) = \frac{1}{2\pi} \int_{-\infty}^{+\infty} \exp\left(-\frac{1}{2N}k^2 - ikx\right) dk. \qquad \text{(E–2)}$$

---

* Our treatment is again based on the book by Lighthill cited in appendix A.

This is nothing other than the representation (A–4) of the Dirac δ-function,

$$\delta_N(x) = \sqrt{\frac{N}{2\pi}} \exp\left(-\frac{1}{2}Nx^2\right), \tag{E–3}$$

but expressed in terms of its own Fourier transform. To show that (E–2) is the same as (E–3), we complete the square in the exponent in (E–2), which readily leads to

$$\begin{aligned} \delta_N(x) &= \frac{1}{2\pi} \exp\left(-\frac{1}{2}Nx^2\right) \int_{-\infty}^{+\infty} \exp\left[-\frac{1}{2N}(k + iNx)^2\right] dk \\ &= \frac{\sqrt{2N}}{2\pi} \exp\left(-\frac{1}{2}Nx^2\right) \int_{-\infty}^{+\infty} \exp[-z^2]\, dz, \end{aligned} \tag{E–4}$$

where, in the last integral, z has been substituted for $(k + iNx)/\sqrt{2N}$, implying an integration path in the complex z-plane along a line parallel to the real axis. Because $\exp(-z^2)$ is analytic for all finite values of $|z|$ and falls off rapidly for $\text{Re}(z) \to \infty$, the integral has the same value as along the real axis, $\sqrt{\pi}$, leading to the form (E–3).

If (E–2) is inserted into (E–1), we obtain

$$f(x) = \lim_{N\to\infty} \frac{1}{2\pi} \int_{-\infty}^{+\infty} f(x') \left[\int_{-\infty}^{+\infty} \exp\left(-\frac{k^2}{2N} - ikx\right) dk\right] dx'. \tag{E–5}$$

Under our assumptions, the integral

$$\boxed{F(k) = \frac{1}{\sqrt{2\pi}} \int_{-\infty}^{+\infty} f(x) e^{-ikx}\, dx} \tag{E–6a}$$

is finite for all values of k, and we can interchange the two integrations in (E–5) and write

$$\begin{aligned} f(x) &= \lim_{N\to\infty} \frac{1}{\sqrt{2\pi}} \int_{-\infty}^{+\infty} \exp\left(-\frac{k^2}{2N}\right) \left[\frac{1}{\sqrt{2\pi}} \int_{-\infty}^{+\infty} f(x') e^{-ikx'}\, dx'\right] e^{+ikx}\, dk \\ &= \lim_{N\to\infty} \frac{1}{\sqrt{2\pi}} \int_{-\infty}^{+\infty} \exp\left(-\frac{k^2}{2N}\right) F(k) e^{+ikx}\, dk. \end{aligned} \tag{E–7}$$

Furthermore, the existence of an absolutely integrable second derivative of f assures that $F(k)$ vanishes at infinity at least as rapidly as $1/k^2$. (This follows from (E–6a) by integrating by parts twice.) Thus, the integral

$$\int_{-\infty}^{+\infty} f(k) e^{+ikx}\, dk \tag{E–8}$$

is also finite for all values of x. We can then interchange the limit as $N \to \infty$ and the integration in (E–7) and obtain

$$f(x) = \frac{1}{\sqrt{2\pi}} \int_{-\infty}^{+\infty} F(k) e^{+ikx}\, dk \tag{E–6b}$$

Equations (E–6a,b) together constitute the **Fourier integral theorem.** The function $F(k)$ is called the (symmetrized) **Fourier transform** of $f(x)$, and conversely, $f(x)$ is the (symmetrized) **inverse Fourier transform** of $F(k)$. Note the almost complete symmetry between the two transforms, differing only by an exchange of k and x and a change in the sign in the exponent. In the literature, one frequently finds the unsymmetric pair

$$F(k) = \frac{1}{2\pi} \int_{-\infty}^{+\infty} f(x) e^{-ikx}\, dx, \tag{E–9a}$$

$$f(x) = \int_{-\infty}^{+\infty} F(k) e^{+ikx}\, dk. \tag{E–9b}$$

For quantum-mechanical purposes, the symmetrized form (E–6) is more useful because it leads to the same normalization for f and F.

To show the latter, we first multiply (E–6a) with $F^*(k)$ and integrate over all k:

$$\int_{-\infty}^{+\infty} |F(k)|^2\, dk = \frac{1}{\sqrt{2\pi}} \int_{-\infty}^{+\infty} \int_{-\infty}^{+\infty} F^*(k) f(x) e^{-ikx}\, dx\, dk. \tag{E–10a}$$

Similarly, multiplying (E–6b) with $f^*(x)$ and integrating over all x yields

$$\int_{-\infty}^{+\infty} |f(x)|^2\, dx = \frac{1}{\sqrt{2\pi}} \int_{-\infty}^{+\infty} \int_{-\infty}^{+\infty} f^*(x) F(k) e^{+ikx}\, dk\, dx. \tag{E–10b}$$

Under our assumptions, all of the integrals in (E–10a) and (E–10b) exist. The right-hand sides of (E–10a) and (E–10b) are simply the complex conjugates of each other, and because both left-hand sides are real, they must in fact be equal to each other:

$$\int_{-\infty}^{+\infty} |F(k)|^2\, dk = \int_{-\infty}^{+\infty} |f(x)|^2\, dx \tag{E–11}$$

Our results are readily generalized to three dimensions. By applying the three-dimensional Dirac δ-function of (A–7) through (A–9) to the function $f(\mathbf{r})$, one finds, by analogous procedures, that

$$F(\mathbf{k}) = \frac{1}{(2\pi)^{3/2}} \int_{-\infty}^{+\infty} f(\mathbf{r}) e^{-i\mathbf{k}\cdot\mathbf{r}}\, d^3\mathbf{r}, \tag{E–12a}$$

$$f(\mathbf{r}) = \frac{1}{(2\pi)^{3/2}} \int_{-\infty}^{+\infty} F(\mathbf{k}) e^{+i\mathbf{k}\cdot\mathbf{r}}\, d^3\mathbf{k}, \tag{E–12b}$$

$$\int_{-\infty}^{+\infty} |F(\mathbf{k})|^2\, d^3\mathbf{k} = \int_{-\infty}^{+\infty} |f(\mathbf{r})|^2\, d^3\mathbf{r}. \tag{E–13}$$

EXTENSION TO GENERALIZED FUNCTIONS

For our derivation of the Fourier integral theorem, we made fairly restrictive assumptions about the function $f(x)$. It is readily possible to relax these restrictions drastically and to generalize the theorem to all functions $f(x)$ that, although they themselves do not satisfy our assumptions, can be written as limits of sequences of functions $f_N(x)$ that do. Accordingly, assume that there exists a sequence $f_N(x)$ of functions such that

$$f(x) = \lim_{N\to\infty} f_N(x) \tag{E–14a}$$

and that each $f_N(x)$ satisfies the assumptions made about $f(x)$ itself at the beginning of this appendix. It is then possible to apply the Fourier integral theorem to each of the $f_N(x)$, leading to a sequence of Fourier transform pairs

$$F_N(k) = \frac{1}{2\pi}\int_{-\infty}^{+\infty} f_N(x)e^{-ikx}\, dx \tag{E–15a}$$

and

$$f_N(x) = \int_{-\infty}^{+\infty} F_N(k)e^{+ikx}\, dk. \tag{E–15b}$$

We can then *define* the Fourier transform of $f(x)$ by the limit

$$F(k) = \lim_{N\to\infty} F_N(k). \tag{E–14b}$$

Our result shows that the Fourier integral theorem holds for all functions $f(x)$ that can be written as generalized functions for which the sequence of defining functions $f_N(x)$ satisfies the conditions stated earlier. The corresponding Fourier transform is itself a generalized function. This range of generalized functions is vastly more general than the range of functions mentioned on page 616 and includes functions with all sorts of singularities, such as discontinuities in the function itself or its derivative and even local infinities.

One warning is in order, though: The normalization integrals in (E–11) need not remain finite in the limit as $N \to \infty$. For example, for the simple Dirac δ-function, the normalization integral becomes infinite.

The extension of this generalization to three dimensions should be obvious.

Appendix F

CONSTRUCTION OF TWO GROUP CHARACTER TABLES

The character table for a given group can often be obtained with surprising ease by drawing on the column and row orthogonality relations for the character vectors given in the relations (16•2–14) and (16•2–15). We demonstrate here the procedure for the group of a square and the group of a cube without inversion symmetry.

GROUP OF A SQUARE

Our point of departure is the knowledge that we have a group of eight elements distributed over five classes, which implies the existence of five irredudible representations. As pointed out in (16•2–13), the dimensionality theorem demands that four of the representations be one-dimensional and the fifth be two-dimensional. We know that the identity representation is always present, for which all characters are $+1$; this specifies the top row of the 5×5 character table. We also know that the character associated with the identity element $\hat{E}$ in each representation is simply the dimensionality of the representation; this specifies the leftmost column. Hence, we have the following incomplete "starter table:"

	$\hat{E}$	$\hat{C}_2$	$2\hat{C}_4$	$2\hat{\sigma}_x$	$2\hat{\sigma}_d$
A_1	+1	+1	+1	+1	+1
A_2	+1				
B_1	+1				
B_2	+1				
E	+2				

Consider next the three remaining one-dimensional representations. Inasmuch as the characters must be symmetry eigenvalues, all characters must be ± 1. The possibility $\pm i$ for the class containing $\hat{C}_4$ can be ruled out, because with that choice, row orthogonality with the top row cannot be obtained. Given that all characters must be ± 1, and given the number of elements in the various classes, row orthogonality requires that the character of $\hat{C}_2$ must remain $+1$, but two of the three remaining two-element classes must reverse the sign of their characters. There are three ways to pick two classes out of three, leading to the characters shown in Table 16•2–2 for the three remaining one-dimensional representations. The remaining characters in the last row of Table 16•2–2 follow most easily from column orthogonality with the leftmost column. It is left to the reader to show that the table also meets all other orthogonality relations not explicitly invoked in its construction.

GROUP OF A CUBE WITHOUT INVERSION SYMMETRY

As a second example, consider the character table of the cube rotation group without inversion, Table 16•2–4. Again we have five classes, but now the number of group elements is 24, which demands two one-dimensional representations, one two-dimensional one, and two three-dimensional ones, leading to the following "starter table:"

	$\hat{E}$	$8\hat{C}_3$	$3\hat{C}_{2x}$	$6\hat{C}_{2e}$	$6\hat{C}_{4x}$
A_1	+1	+1	+1	+1	+1
A_2	+1				
E	+2				
T_1	+3				
T_2	+3				

As in the case of the group of a square, the characters in the second row must be ± 1; thus, complex values for the classes containing $\hat{C}_3$ and $\hat{C}_{4x}$ can again be ruled out. Row orthogonality of the representation A_2 demands the values $+1$, $+1$, $+1$, -1, -1 in the A_2 row. It is left to the reader to derive the rest of the table.

Appendix G

SELECTED GENERAL REFERENCES

The following is a list of selected references for supplemental reading on a variety of topics. The list has been kept short on purpose; it reflects my own limited knowledge and prejudices, but it contains those references that I myself have found most useful and that I invariably recommend first before turning anywhere else.

1 HISTORICAL AND CONCEPTUAL DEVELOPMENT OF QUANTUM MECHANICS

1.1 General Reviews

M. Jammer. *The Conceptual Development of Quantum Mechanics.* New York: McGraw-Hill, 1966.

M. Jammer. *The Philosophy of Quantum Mechanics.* New York: Wiley-Interscience, 1976.

The first of these is *the* standard work on the subject of its title. The second singles out, for a more complete scholarly discussion, the conceptual and philosophical problems and difficulties arising out of the non-intuitive nature of quantum mechanics and its interpretation.

1.2 Reprint Collections

The following three paperbacks contain reprints (all in English) of some of the most important original papers, supplemented by excellent discussions:

D. ter Haar. *The Old Quantum Theory.* Oxford: Pergamon Press, 1967.

G. Ludwig. *Wave Mechanics.* Oxford: Pergamon Press, 1968.

B. L van der Waerden. *Sources of Quantum Mechanics.* New York: Dover, 1968.

The first two volumes belong together. Ter Haar covers Planck's and Einstein's work on black-body radiation and on light quantization, as well as the Rutherford-Bohr atomic model with its consequences. Ludwig continues the coverage with excerpts from de Broglie's 1924 thesis, Schroedinger's four key 1926 papers, and three papers by Heisenberg, Born, and Jordan.

Van der Waerden's work covers the development of the algebraic form of quantum mechanics. It contains an excellent analysis by its author, with numerous otherwise unpublished quotes from letters, etc.

1.3 Dirac's and Heisenberg's Monographs

P. A. M. Dirac. *The Principles of Quantum Mechanics,* 4th ed. Oxford: Oxford University Press, 1967.

A classic (the first edition dates back to 1930), this is *the* standard axiomatic treatment of the principles of quantum mechanics. For the beginner, it is hard to read: Dirac simply postulates the principles of quantum mechanics axiomatically and proceeds from there deductively, without explaining where those principles come from and without even attempting to resolve the (apparent) contradictions between quantum mechanics and common sense. Perhaps one might say that one of the objectives of a first course on quantum mechanics should be to learn how to read Dirac. This is a book that summarizes knowledge, not a text to be used to acquire it.

W. Heisenberg. *The Physical Principles of the Quantum Theory.* Dover Reprints, 1949.

Written in 1930 by one of the originators of quantum mechanics, this paperback classic was the first book devoted specifically to a discussion of the conceptual difficulties of the subject. It is still worthwhile reading and still important because it is frequently quoted in the quantum mechanics literature.

2 TEXTBOOKS

2.1 Strictly Undergraduate-Level; for Background

A. P. French and E. F. Taylor. *An Introduction to Quantum Physics.* New York: W. W. Norton, 1978.

One of the best modern textbooks for an undergraduate-level introduction to quantum mechanics. The first two chapters contain a superb survey of the experimental basis of quantum mechanics.

R. P. Feynman, R. B. Leighton, and M. Sands. *The Feynman Lectures on Physics, Vol. 3: Quantum Mechanics.* Reading, MA: Addison-Wesley, 1965.

A "must-have" classic. The discussion of wave-particle duality is the best found anywhere. Unsurpassed in its discussion of the physical ideas underlying quantum mechanics.

2.2 Graduate Level; to Supplement the Present Text

E. Merzbacher. *Quantum Mechanics,* 2d ed. New York: Wiley, 1970.

This is my personal preference amongst the existing general graduate-level texts. Excellent coverage, very "physical," with numerous good exercises and problems.

L. K. Schiff. *Quantum Mechanics,* 3d ed. New York: McGraw-Hill, 1968.

For many years, this text and its earlier editions used to be *the* general graduate-level text, and to a few instructors, it still is. Not the most readable book, it is still an excellent reference work containing a vast amount of material, particularly on problem-solving techniques.

2.3 A Problem Collection

S. Flügge. *Practical Quantum Mechanics,* Springer Study Edition. New York: Springer, 1974.

A collection of 219 problems, worked out in detail. It is one of the most useful books to supplement *any* text.

2.4 Advanced General Text

J. J. Sakurai. *Advanced Quantum Mechanics.* New York: Addison-Wesley, 1967.

To me, this is the most readable of the more advanced general-purpose texts. Designed as a follow-up to such standard texts as Merzbacher and Schiff, it goes right to the forefront of current research.

Appendix H

FUNDAMENTAL CONSTANTS

Name	Symbol	Value	Units
Speed of light	c	2.99792×10^{8}	m/sec
Vacuum permittivity	ϵ_0	8.85419×10^{-12}	coul/volt·m
Planck's constant	h	6.62608×10^{-34}	joule·sec
		$=4.13567 \times 10^{-15}$	eV·sec
	$\hbar = h/2\pi$	1.05457×10^{-34}	joule·sec
		$=6.58212 \times 10^{-16}$	eV·sec
Elementary charge	e	1.60218×10^{-19}	coul
Electron mass	m_e	9.10939×10^{-31}	kg
Avogadro's number	N_A	6.02214×10^{23}	mol^{-1}
Boltzmann's constant	k	1.38066×10^{-23}	joule/Kelvin
		$=8.61739 \times 10^{-5}$	eV/Kelvin
Atomic mass unit	u	1.66054×10^{-27}	kg

INDEX

A

B

C

D

I

P

Q

R

S

T

U

V

W

Z